1987

W9-ABM-276
3 0301 00054194 2

THE ANTHROPIC COSMOLOGICAL PRINCIPLE

THE ANTHROPIC COSMOLOGICAL PRINCIPLE

JOHN D. BARROW
Lecturer, Astronomy Centre, University of Sussex

and

FRANK J. TIPLER
Professor of Mathematics and Physics, Tulane University, New Orleans

With a foreword by John A. Wheeler

CLARENDON PRESS · *Oxford*
OXFORD UNIVERSITY PRESS · *New York*
1986

LIBRARY
College of St. Francis
JOLIET, ILL.

Oxford University Press, Walton Street, Oxford OX2 6DP
Oxford New York Toronto
Delhi Bombay Calcutta Madras Karachi
Kuala Lumpur Singapore Hong Kong Tokyo
Nairobi Dar es Salaam Cape Town
Melbourne Auckland
and associated companies in
Beirut Berlin Ibadan Nicosia

Oxford is a trade mark of Oxford University Press

Published in the United States
by Oxford University Press, New York

© *John D. Barrow and Frank J. Tipler, 1986*

All rights reserved. No part of this publication may be reproduced, stored in a retrieval system, or transmitted, in any form or by any means, electronic, mechanical, photocopying, recording, or otherwise, without the prior permission of Oxford University Press

British Library Cataloguing in Publication Data
Barrow, John D.
The anthropic cosmological principle.
1. Man
I. Title II. Tipler, Frank J.
128 BD450
ISBN 0-19-851949-4

Library of Congress Cataloging in Publication Data
Barrow, John D., 1952—
The anthropic cosmological principle.
Bibliography: p.
Includes index.
1. Cosmology. 2. Man. 3. Teleology. 4. Intellect.
5. Life on other planets. 6. Science—Philosophy.
I. Tipler, Frank J. II. Title.
BD511.B34 1985 113 85–4824
ISBN 0-19-851949–4

Printing (last digit): 9 8 7 6 5 4 3 2 1

Printed in the United States of America

113
B272

To Elizabeth and Jolanta

125,243

Foreword

John A. Wheeler, Center for Theoretical Physics,
University of Texas at Austin

'Conceive of a universe forever empty of life?' 'Of course not', a philosopher of old might have said, contemptuously dismissing the question, and adding, over his shoulder, as he walked away, 'It has no sense to talk about a universe unless there is somebody there to talk about it'.

That quick dismissal of the idea of a universe without life was not so easy after Copernicus. He dethroned man from a central place in the scheme of things. His model of the motions of the planets and the Earth taught us to look at the world as machine. Out of that beginning has grown a science which at first sight seems to have no special platform for man, mind, or meaning. Man? Pure biochemistry! Mind? Memory modelable by electronic circuitry! Meaning? Why ask after that puzzling and intangible commodity? 'Sire', some today might rephrase Laplace's famous reply to Napoleon, 'I have no need of that concept'.

What is man that the universe should be mindful of him? Telescopes bring light from distant quasi-stellar sources that lived billions of years before life on Earth, before there even was an Earth. Creation's still warm ashes we call 'natural radioactivity'. A thermometer and the relative abundance of the lighter elements today tell us the correlation between temperature and density in the first three minutes of the universe. Conditions still earlier and still more extreme we read out of particle physics. In the perspective of these violences of matter and field, of these ranges of heat and pressure, of these reaches of space and time, is not man an unimportant bit of dust on an unimportant planet in an unimportant galaxy in an unimportant region somewhere in the vastness of space?

No! The philosopher of old was right! Meaning is important, is even central. It is not only that man is adapted to the universe. The universe is adapted to man. Imagine a universe in which one or another of the fundamental dimensionless constants of physics is altered by a few percent one way or the other? Man could never come into being in such a universe. That is the central point of the anthropic principle. According to this principle, a life-giving factor lies at the centre of the whole machinery and design of the world.

What is the status of the anthropic principle? Is it a theorem? No. Is it a mere tautology, equivalent to the trivial statement, 'The universe has to be such as to admit life, somewhere, at some point in its history, because *we* are here'? No. Is it a proposition testable by its predictions? Perhaps. Then what is the status of the anthropic principle? That is the issue on which every reader of this fascinating book will want to make his own judgement.

Nowhere better than in the present account can the reader see new thinking, new ideas, new concepts in the making. The struggles of old to sort sense from nonsense in the domain of heat, phlogiston, and energy by now have almost passed into the limbo of the unappreciated. The belief of many in the early part of this century that 'Chemical forces are chemical forces, and electrical forces are electrical forces, and never the twain shall meet' has long ago been shattered. Our own time has made enormous headway in sniffing out the sophisticated relations between entropy, information, randomness, and computability. But on a proper assessment of the anthropic principle we are still in the dark. Here above all we see how out of date that old view is, 'First define your terms, then proceed with your reasoning'. Instead, we know, theory, concepts, and methods of measurement are born into the world, by a single creative act, in inseparable union.

In advancing a new domain of investigation to the point where it can become an established part of science, it is often more difficult to ask the right questions than to find the right answers, and nowhere more so than in dealing with the anthropic principle. Good judgement, above all, is required, judgement in the sense of George Graves, 'an awareness of all the factors in the situation, and an appreciation of their relative importance'.

To the task of history, exposition, and judgement of the anthropic principle the authors of this book bring a unique combination of skills. John Barrow has to his credit a long list of distinguished contributions in the field of astrophysics generally and cosmology in particular. Frank Tipler is widely known for important concepts and theorems in general relativity and gravitation physics.

It would be difficult to discover a single aspect of the anthropic principle to which the authors do not bring a combination of the best thinking of past and present and new analysis of their own.

Philosophical considerations connected with the anthropic principle? Of the considerations on this topic contained in Chapters 2 and 3 perhaps half are new contributions of the authors.

Why, except in the physics of elementary particles at the very smallest scale of lengths, does nature limit itself to three dimensions of space and one of time? Considerations out of past times and present physics on this topic give Chapter 4 a special flavour. In Chapter 6 the authors provide one of the best short reviews of cosmology ever published. In Chapter 8 Barrow and Tipler not only recall the arguments of L. J. Henderson's famous 1913 book, *The fitness of the environment.* They also spell out George Wald's more recent emphasis on the unique properties of water, carbon dioxide, and nitrogen. They add new arguments to Wald's rating of chlorophyll, an unparalleled agent, as the most effective photosynthetic molecule that anyone could invent. Taking account of biological considerations and modern statistical methods, Barrow and Tipler derive with new clarity Brandon Carter's striking anthropic-principle inequality. It states that the length of time from now, on into the future, for which the earth will continue to be an inhabitable planet will be only a fraction of the time, 4.6 billion years, that it has required for evolution on earth to produce man. The Carter inequality, as thus derived, is still more quantitative, still more limiting, still more striking. It states that the fraction of these 4.6 billion years *yet to come* is smaller than 1/8th, 1/9th, 1/10th, ... or less, according as the number of critical or improbable or gateway steps in the *past* evolution of man was 7, 8, 9, ... or more. This amazing prediction looks like being some day testable and therefore would seem to count as 'falsifiable' in the sense of Karl Popper.

Chapter 9, outlining a space-travel argument against the existence of extraterrestrial intelligent life, is almost entirely new. So is the final Chapter 10. It rivals in thought-provoking power any of the others. It discusses the idea that intelligent life will some day spread itself so thoroughly throughout all space that it will 'begin to transform and continue to transform the universe on a cosmological scale', thus making it possible to transmit 'the values of humankind... to an arbitrarily distant futurity... an Omega Point... [at which] life will have gained control of all matter and forces...'.

In the mind of every thinking person there is set aside a special room, a museum of wonders. Every time we enter that museum we find our attention gripped by marvel number one, this strange universe, in which we live and move and have our being. Like a strange botanic specimen newly arrived from a far corner of the earth, it appears at first sight so carefully cleaned of clues that we do not know which are the branches and which are the roots. Which end is up and which is down? Which part is nutrient-giving and which is nutrient-receiving? Man? Or machinery?

Everyone who finds himself pondering this question from time to time will want to have Barrow and Tipler with him on his voyages of thought. They bring along with them, now and then to speak to us in their own words, a delightful company of rapscallions and wise men, of wits and discoverers. Travelling with the authors and their friends of past and present we find ourselves coming again and again upon issues that are live, current, important.

Preface

This book was begun long ago. Over many years there had grown up a collection of largely unpublished results revealing a series of mysterious coincidences between the numerical values of the fundamental constants of Nature. The possibility of our own existence seems to hinge precariously upon these coincidences. These relationships and many other peculiar aspects of the Universe's make-up appear to be necessary to allow the evolution of carbon-based organisms like ourselves. Furthermore, the twentieth-century dogma that human observers occupy a position in the Universe that must not be privileged in any way is strongly challenged by such a line of thinking. Observers will reside only in places where conditions are conducive to their evolution and existence: such sites may well turn out to be special. Our picture of the Universe and its laws are influenced by an unavoidable selection effect—that of our own existence.

It is this spectrum of ideas, its historical background and wider scientific ramifications that we set out to explore.

The authors must confess to a curious spectrum of academic interests which have been indulged to the full in this study. In seemed to us that cosmologists and lay persons were often struck by the seeming novelty of this collection of ideas called the Anthropic Principle. For this reason it is important to display the Anthropic Principle in a historical perspective as a modern manifestation of a certain tradition in the history of ideas that has a long and fascinating history involving, at one time or another, many of the great figures of human thought and speculation.

For these reasons we have attempted not only to describe the collection of results that modern cosmologists would call the 'Anthropic Principle', but to trace the history of the underlying world-view in which it has germinated, together with the diverse range of subjects where it has interesting but unnoticed ramifications. Our discussion is of necessity therefore a medley of technical and non-technical studies but we hope it has been organized in a manner that allows those with only particular interests and uninterests to indulge them without too much distraction from the parts of the other sort. Roughly speaking, the degree of difficulty increases as the book goes on: whereas the early chapters study the historical antecedents of the Anthropic Principle, the later ones investigate modern developments which involve mathematical ideas in cosmology, astrophysics, and quantum theory.

There are many people who have played some part in getting this project started and bringing it to some sort of conclusion. In particular, we are grateful to Dennis Sciama without whose encouragement it would not have begun, and to John Wheeler without whose prodding it would never have been completed. We are also indebted to a large number of individuals for discussions and suggestions, for providing diagrams or reading drafts of particular chapters; for their help in this way we would like particularly to thank R. Alpher, M. Begelman, M. Berry, F. Birtel, S. Brenner, R. Breuer, P. Brosche, S. G. Brush, B. J. Carr, B. Carter, P. C. W. Davies, W. Dean, J. Demaret, D. Deutsch, B. DeWitt, P. Dirac, F. Drake, F. Dyson, G. F. R. Ellis, R. Fenn, A. Flew, S. Fox, M. Gardner, J. Goldstein, S. J. Gould, A. Guth, C. Hartshorne, S. W. Hawking, F. A. Hayek, J. Hedley-Brooke, P. Hefner, F. Hoyle, S. Jaki, M. Jammer, R. Jastrow, R. Juszkiewicz, J. Leslie, W. H. McCrea, C. Macleod, J. E. Marsden, E. Mascall, R. Matzner, J. Maynard Smith, E. Mayr, L. Mestel, D. Mohr, P. Morrison, J. V. Narlikar, D. M. Page, A. R. Peacocke, R. Penrose, J. Perdew, F. Quigley, M. J. Rees, H. Reeves, M. Ruderman, W. Saslaw, C. Sagan, D. W. Sciama, I. Segal, J. Silk, G. G. Simpson, S. Tangherlini, R. J. Tayler, G. Wald, J. A. Wheeler, G. Whitrow, S.-T. Yau, W. H. Zurek, and the staff of Oxford University Press.

On the vital practical side we are grateful to the secretarial staff of the Astronomy Centre at Sussex and the Departments of Mathematics and Physics at Tulane University, especially Suzi Lam, for their expert typing and management of the text. We also thank Salvador Dali for allowing us to reproduce the example of his work which graces the front cover, and finally we are indebted to a succession of editors at Oxford University Press who handled a continually evolving manuscript and its authors with great skill and patience. Perhaps in despair at the authors' modification of the manuscript they had cause to recall Dorothy Sayers' vivid description of what Harriet Vane discovered when she happened upon a former tutor in the throes of preparing a book for publication by the Press... 'The English tutor's room was festooned with proofs of her forthcoming work on the prosodic elements in English verse from Beowulf to Bridges. Since Miss Lydgate had perfected, or was in process of perfecting (since no work of scholarship ever attains a static perfection) an entirely new prosodic theory, demanding a novel and complicated system of notation which involved the use of twelve different varieties of type; and since Miss Lydgate's handwriting was difficult to read and her experience in dealing with printers limited, there existed at that moment five successive revises in galley form, at different stages of completion, together with two sheets in page-proof, and an appendix in typescript, while the important Introduction which afforded the key to the whole argument still remained to be written. It was only when a section had advanced to page-proof condition that Miss Lydgate became fully convinced of the necessity of transferring large paragraphs of argument from one chapter to another, each change of this kind naturally demanding expensive over-running on the page-proof, and the elimination of the corresponding portions in the five sets of revises...'

Brighton
July, 1985

J. D. B.
F. J. T.

Acknowledgements

The authors gratefully acknowledge the following sources of illustrations and tables reproduced in this book, and thank authors and publishers who have granted their permission.

Figures: 5.1 adapted from B. Carr and M. Rees, *Nature, Lond.* 278, 605 (1979); **5.2** based on A. Holden, *Bonds between atoms*, p. 15, Oxford University Press (1977); **5.3**. V. S. Weisskopf, 'Of atoms, mountains and stars: a study in qualitative physics', *Science* 187, 602–12, Diagram 21 (February 1975); **5.5** R. D. Evans *The atomic nucleus*, p. 382, Fig. 3.5, McGraw Hill, New York (1955); **5.6** and **5.7** P. C. Davies, *J. Physics* A, 5, 1296 (1972); **5.9** redrawn from D. Clayton, *Principles of stellar evolution and nucleosynthesis*, p. 302, Fig. 4–6, University of Chicago Press (1968 and 1983); **5.11** adapted from M. Harwit, *Astrophysical concepts*, p. 17, Wiley, New York; **5.12** adapted from B. Carr and M. Rees, *Nature, Lond.* 278, 605 (1979); **5.13** reproduced, with permission, from the *Annual Reviews Nuclear and Particle Science* **25** © 1975 by Annual Reviews Inc; **5.14**. M. Begelman, R. Blandford, and M. Rees, *Rev. mod. Phys.* **56,** 294, Fig. 15, with permission of the authors and the American Physical Society; **6.4** redrawn from D. Woody and P. Richards, *Phys. Rev. Lett.* **42,** 925 (1979); **6.5** and **6.6** C. Frenk, M. Davis, G. Efstathiou, and S. White; **6.7** adapted from B. Carr and M. Rees, *Nature, Lond.* 278, 605 (1979); **6.10** based on M. Rees, Les Houches Lectures; **6.12** based on H. Kodama 'Comments on the chaotic inflation', *KEK Report* 84–12, ed. K. Odaka and A. Sugamoto (1984); **7.1** B. De Witt, *Physics Today* **23,** 31 (1970); **8.2** J. D. Watson, *Molecular biology of the gene*, W. A. Benjamin Inc., 2nd edn, copyright 1970 by J. D. Watson; **8.3** M. Arbib, in *Interstellar communication: scientific prospects*, ed. C. Ponnamperuma and A. Cameron, Houghton Mifflin, Boston (1974); **8.4, 8.5, 8.6, 8.7, 8.8** adapted from Linus Pauling in *General chemistry*, W. H. Freeman, New York (1956); **8.9** adapted from Linus Pauling and R. Hayward in *General chemistry*, W. H. Freeman, New York (1956), **8.10** J. Edsall and J. Wyman, *Biophysical chemistry*, Vol. 1, p. 178, Academic Press (1958); **8.11, 8.12, 8.13** adapted from Linus Pauling in *General chemistry*, W. H. Freeman, New York (1956); **8.14** adapted from J. Edsall and J. Wyman, *Biophysical chemistry*, Vol. 1, p. 3, Academic Press (1958); **8.15** reprinted from Linus Pauling *The nature of the chemical bond*, third edition, copyright © 1960 by Cornell University, used by permission of the publisher, Cornell University Press; **8.16** F. H. Stillinger 'Water revisited', *Science* **209,** 451–7 (1980), © 1980 by the American Association for the Advancement of Science; **8.17** Albert L. Lehninger in *Biochemistry*, Worth Publishers Inc., New York (1975); **8.18** adapted from G. Wald, *Origins of life* **5,** 11 (1974) and in *Conditions for life* ed, A. Gabor, Freeman, New York (1976); **8.20** J. Lovelock, *J. E. Gaia: a new look at life on earth*, Oxford University Press (1979).

Tables: 8.1–8.7 A. Needham, *The uniqueness of biological materials*, Pergamon Press, Oxford (1965); **8.8** J. Lovelock, *Gaia: a new look at life on earth, Oxford University Press* (1979); **8.9** J. Edsall and J. Wyman, *Physical chemistry*, Vol. 1, p. 24, Academic Press (1958); **8.10** Albert L. Lehninger, *Biochemistry*, Worth Publishers Inc., New York (1975).

Preparation for publication of the Foreword was assisted by the Center for Theoretical Physics, University of Texas at Austin and by NSF Grants PHY 8205717 and PHY 503890.

Contents

1 INTRODUCTION
1.1 Prologue 1
1.2 Anthropic Definitions 15

2 DESIGN ARGUMENTS
2.1 Historical Prologue 27
2.2 The Ancients 31
2.3 The Medieval Labryrinth 46
2.4 The Age of Discovery 49
2.5 Mechanical Worlds 55
2.6 Critical Developments 68
2.7 The Devolution of Design 83
2.8 Design in Non-Western Religion and Philosophy 92
2.9 Relationship Between The Design Argument and the Cosmological Argument 103

3 MODERN TELEOLOGY AND THE ANTHROPIC PRINCIPLES
3.1 Overview: Teleology in the Twentieth Century 123
3.2 The Status of Teleology in Modern Biology 127
3.3 Henderson and the Fitness of the Environment 143
3.4 Teleological Ideas and Action Principles 148
3.5 Teleological Ideas in Absolute Idealism 153
3.6 Biological Constraints on the Age of the Earth: The First Successful Use of an Anthropic Timescale Argument 159
3.7 Dysteleology: Entropy and the Heat Death 166
3.8 The Anthropic Principle and the Direction of Time 173
3.9 Teleology and the Modern 'Empirical' Theologians 180
3.10 Teleological Evolution: Bergson, Alexander, Whitehead, and the Philosophers of Progress 185
3.11 Teilhard de Chardin: Mystic, Paleontologist and Teleologist 195

4 THE REDISCOVERY OF THE ANTHROPIC PRINCIPLE
4.1 The Lore of Large Numbers 219
4.2 From Coincidence to Consequence 220
4.3 'Fundamentalism' 224
4.4 Dirac's Hypothesis 231
4.5 Varying Constants 238
4.6 A New Perspective 243
4.7 Are There Any Laws of Physics? 255
4.8 Dimensionality 258

5 THE WEAK ANTHROPIC PRINCIPLE IN PHYSICS AND ASTROPHYSICS
5.1 Prologue 288
5.2 Atoms and Molecules 295
5.3 Planets and Asteroids 305
5.4 Planetary Life 310
5.5 Nuclear Forces 318
5.6 The Stars 327
5.7 Star Formation 339
5.8 White Dwarfs and Neutron Stars 340
5.9 Black Holes 347
5.10 Grand Unified Gauge Theories 354

6 THE ANTHROPIC PRINCIPLES IN CLASSICAL COSMOLOGY
6.1 Introduction 367
6.2 The Hot Big Bang Cosmology 372
6.3 The Size of the Universe 384
6.4 Key Cosmic Times 385
6.5 Galaxies 387
6.6 The Origin of the Lightest Elements 398
6.7 The Value of *S* 401
6.8 Initial Conditions 408
6.9 The Cosmological Constant 412
6.10 Inhomogeneity 414
6.11 Isotropy 419
6.12 Inflation 430
6.13 Inflation and the Anthropic Principle 434
6.14 Creation *ex nihilo* 440
6.15 Boundary Conditions 444

7 QUANTUM MECHANICS AND THE ANTHROPIC PRINCIPLE
7.1 The Interpretations of Quantum Mechanics 458
7.2 The Many-Worlds Interpretation 472
7.3 The Friedman Universe from the Many-Worlds Point of View 490
7.4 Weak Anthropic Boundary Conditions in Quantum Cosmology 497
7.5 Strong Anthropic Boundary Conditions in Quantum Cosmology 503

8 THE ANTHROPIC PRINCIPLE AND BIOCHEMISTRY
8.1 Introduction 510
8.2 The Definitions of Life and Intelligent Life 511
8.3 The Anthropic Significance of Water 524
8.4 The Unique Properties of Hydrogen and Oxygen 541
8.5 The Anthropic Significance of Carbon, Carbon Dioxide and Carbonic Acid 545
8.6 Nitrogen, Its Compounds, and other Elements Essential for Life 548
8.7 Weak Anthropic Principle Constraints on the Future of the Earth 556

9 THE SPACE-TRAVEL ARGUMENT AGAINST THE EXISTENCE OF EXTRATERRESTRIAL INTELLIGENT LIFE
9.1 The Basic Idea of the Argument 576
9.2 General Theory of Space Exploration and Colonization 578
9.3 Upper Bounds on the Number of Intelligent Species in the Galaxy 586
9.4 Motivations for Interstellar Communication and Exploration 590
9.5 Anthropic Principle Arguments Against Steady-State Cosmologies 601

10 THE FUTURE OF THE UNIVERSE
10.1 Man's Place in an Evolving Cosmos 613
10.2 Early Views of the Universe's Future 615
10.3 Global Constraints on the Future of the Universe 621

10.4 The Future Evolution of Matter: Classical Timescales 641
10.5 The Future Evolution of Matter: Quantum Timescales 647
10.6 Life and the Final State of the Universe 658

INDEX 683

THE ANTHROPIC COSMOLOGICAL PRINCIPLE

Ah Mr. Gibbon, another damned, fat, square book.
Always scribble, scribble, scribble, eh?

THE DUKE OF GLOUCESTER

[on being presented with volume 2
of *The Decline and Fall of the
Roman Empire*]

1 Introduction

The Cosmos is about the smallest hole that a man can hide his head in

G. K. Chesterton

1.1 Prologue

What is Man, that Thou art mindful of him?

Psalm 8:4

The central problem of science and epistemology is deciding which postulates to take as fundamental. The perennial solution of the great idealistic philosophers has been to regard Mind as logically prior, and even materialistic philosophers consider the innate properties of matter to be such as to allow—or even require—the existence of intelligence to contemplate it; that is, these properties are necessary or sufficient for life. Thus the existence of Mind is taken as one of the basic postulates of a philosophical system. Physicists, on the other hand, are loath to admit any consideration of Mind into their theories. Even quantum mechanics, which supposedly brought the observer into physics, makes no use of intellectual properties; a photographic plate would serve equally well as an 'observer'. But, during the past fifteen years there has grown up amongst cosmologists an interest in a collection of ideas, known as the Anthropic Cosmological Principle, which offer a means of relating Mind and observership directly to the phenomena traditionally within the encompass of physical science.

The expulsion of Man from his self-assumed position at the centre of Nature owes much to the Copernican principle that we do not occupy a privileged position in the Universe. This Copernican assumption would be regarded as axiomatic at the outset of most scientific investigations. However, like most generalizations it must be used with care. Although we do not regard our position in the Universe to be central or special in every way, this does not mean that it cannot be special in *any* way. This possibility led Brandon Carter[1] to limit the Copernican dogma by an 'Anthropic Principle' to the effect that 'our location in the Universe is necessarily privileged to the extent of being compatible with our existence as observers'. The basic features of the Universe, including such properties as its shape, size, age and laws of change, must be *observed* to be of a type that allows the evolution of observers, for if intelligent life did not evolve in an otherwise possible universe, it is obvious that no one would be asking the reason for the observed shape, size, age and so forth of the Universe. At first sight such an observation might appear true but trivial. However, it has far-reaching implications for physics. It is a restatement, of the fact that any observed properties of the Universe that may initially appear astonishingly improbable, can only be seen in their true perspective after we have accounted for the fact that certain properties of the Universe are necessary prerequisites for the evolution and existence of any observers at all. The measured values of many cosmological and physical quantities that define our Universe are circumscribed by the necessity that we observe from a site where conditions are appropriate for the occurrence of biological evolution and at a cosmic epoch exceeding the astrophysical and biological timescales required for the development of life-supporting environments and biochemistry.

What we have been describing is just a grandiose example of a type of intrinsic bias that scientists term a 'selection effect'. For example, astronomers might be interested in determining the fraction of all galaxies that lie in particular ranges of brightness.[2] But if you simply observe as many galaxies as you can find and list the numbers found according to their brightness you will not get a reliable picture of the true brightness distribution of galaxies. Not all galaxies are bright enough to be seen or big enough to be distinguished from stars, and those that are brighter are more easily seen than those that are fainter, so our observations are biased towards finding a disproportionately large fraction of very bright galaxies compared to the true state of affairs. Again, at a more mundane level, if a ratcatcher tells you that all rats are more than six inches long because he has never caught any that are shorter, you should check the size of his traps before drawing any far-reaching conclusions about the length of rats. Even though you are most likely to see an elephant in a zoo that does not mean that all elephants are in zoos, or even that most elephants are in zoos. In section 1.2 we shall restate these ideas in a more precise and quantitative form, but to get the flavour of how this form of the Anthropic Principle can be used we shall consider the question of the size of the Universe to illustrate how our own existence acts as a selection effect when assessing observed properties of the Universe.

The fact that modern astronomical observations reveal the visible Universe to be close to fifteen billion light years in extent[3] has provoked many vague generalizations about its structure, significance and ultimate purpose. Many a philosopher has argued[4] against the ultimate importance of life in the Universe by pointing out how little life there appears to be compared with the enormity of space and the multitude of distant galaxies. But the Big Bang cosmological picture shows this up as too simplistic a judgement. Hubble's classic discovery[5] that the Universe is in a dynamic state of expansion reveals that its size is inextricably bound up with its age.[6] The Universe is fifteen billion light years in size because it is fifteen billion years old. Although a universe the size of a single galaxy would contain enough matter to make more than one hundred billion stars the size of our Sun, it would have been expanding for less than a single year.

We have learned that the complex phenomenon we call 'life' is built upon chemical elements more complex than hydrogen and helium gases. Most biochemists believe that carbon, on which our own organic chemistry is founded, is the only possible basis for the *spontaneous* generation of life. In order to create the building blocks of life—carbon, nitrogen, oxygen and phosphorus—the simple elements of hydrogen and helium which were synthesized in the primordial inferno of the Big Bang must be cooked at a more moderate temperature and for a much longer time than is available in the early universe.[7] The furnaces that are available are the interiors of stars. There, hydrogen and helium are burnt into the heavier life-supporting elements by exothermic nuclear reactions. When stars die, the resulting explosions which we see as supernovae, can disperse these elements through space and they become incorporated into planets and, ultimately, into ourselves. This stellar alchemy takes over ten billion years to complete. Hence, for there to be enough time to construct the constituents of living beings, the Universe must be at least ten billion years old and therefore, as a consequence of its expansion, at least ten billion light years in extent. We should not be surprised to observe that the Universe is so large. No astronomer could exist in one that was significantly smaller. The Universe needs to be as big as it is in order to evolve just a single carbon-based life-form.

We should emphasize that this selection of a particular size for the universe actually does *not* depend on accepting most biochemists' belief that only carbon can form the basis of spontaneously generated life. Even if their belief is false, the fact remains that *we are a carbon-based intelligent life-form which spontaneously evolved on an earthlike planet around a star of G2 spectral type, and any observation we make is necessarily self-selected by this absolutely fundamental fact.* In particular, a life-form which evolved spontaneously in such an environment must necessarily see the Universe to be at least several billion years old and hence see it to be at least several billion light years across. This remains true even if non-carbon life-forms abound in the cosmos. Non-carbon life-forms are not necessarily restricted to seeing a minimum size to the universe, but *we* are. Human bodies are measuring instruments whose self-selection properties *must* be taken into account, just as astronomers *must* take into account the self-selection properties of optical telescopes. Such telescopes tell us about radiation in the visible band of the electromagnetic spectrum, but it would be completely illegitimate to conclude from purely optical observations that all of the electromagnetic energy in the Universe is in the visible band. Only when one is aware of the self-selection of optical telescopes is it possible to consider the possibility that non-visible radiation exists. Similarly, it is essential to be aware of the self-selection which results from our being *Homo sapiens* when trying to draw conclusions about the nature of the Universe. This self-selection principle is the most basic version of the Anthropic Principle and it is usually called the *Weak Anthropic Principle.* In a sense, the Weak Anthropic Principle may be regarded as the culmination of the Copernican Principle, because the former shows how to separate those features of the Universe whose appearance depends on anthropocentric selection, from those features which are genuinely determined by the action of physical laws.

In fact, the Copernican Revolution was initiated by the application of the Weak Anthropic Principle. The outstanding problem of ancient astronomy was explaining the motion of the planets, particularly their retrograde motion. Ptolemy and his followers explained the retrograde motion by invoking an epicycle, the ancient astronomical version of a new physical law. Copernicus showed that the epicycle was unnecessary; the retrograde motion was due to an anthropocentric selection effect: we were observing the planetary motions from the vantage point of the moving Earth.

At this level the Anthropic Principle deepens our scientific understanding of the link between the inorganic and organic worlds and reveals an intimate connection between the large and small-scale structure of the Universe. It enables us to elucidate the interconnections that exist between the laws and structures of Nature to gain new insight into the chain of universal properties required to permit life. The realization that the possibility of biological evolution is strongly dependent upon the global structure of the Universe is truly surprising and perhaps provokes us to consider that the existence of life may be no more, but no less, remarkable than the existence of the Universe itself.

The Anthropic Principle, in all of its manifestations but particularly in its Weak form, is closely analogous to the self-reference arguments of mathematics and computer science.[54] These self-reference arguments lead us to understand the limitations of logical knowledge: Gödel's Incompleteness Theorem demonstrates that any mathematical system sufficiently complex to contain arithmetic must contain true statements which cannot be proven true, while Turing's Halting Theorem shows that a computer cannot fully understand itself. Similarly, the Anthropic Principle shows that the observed structure of the Universe is restricted by the fact that we are observing this structure; by the fact that, so to speak, the Universe is observing itself.

The size of the observable Universe is a property that is changing with time because of the overall expansion of the system of galaxies and clusters. A selection effect enters because we are constrained by the timescales of biological evolution to observe the Universe only after billions of years of expansion have already elapsed. However, we can take this consideration a little further. One of the most important results of twentieth-century physics has been the gradual realization that there exist invariant properties of the natural world and its elementary components which render the gross size and structure of virtually all its constituents quite inevitable.[8] The sizes of stars and planets, and even people, are neither random nor the result of any Darwinian selection process from a myriad of possibilities. These, and other gross features of the Universe are the consequences of necessity; they are manifestations of the possible equilibrium states between competing forces of attraction and repulsion. The intrinsic strengths of these controlling forces of Nature are determined by a mysterious collection of pure numbers that we call the *constants of Nature.*[9]

The Holy Grail of modern physics is to explain why these numerical constants—quantities like the ratio of the proton and electron masses for example—have the particular numerical values they do. Although there has been significant progress towards this goal during the last few years[10] we still have far to go in this quest. Nevertheless, there is one interesting approach that we can take which employs an Anthropic Principle in a more adventurous and speculative manner than the examples of self-selection we have already given.

It is possible to express some of the necessary or sufficient conditions for the evolution of observers as conditions on the relative sizes of different collections of constants of Nature. Then we can determine to what extent our observation of the peculiar values these constants are found to take is necessary for the existence of observers. For example, if the relative strengths of the nuclear and electromagnetic forces were to be slightly different then carbon atoms could not exist in Nature[11] and human physicists would not have evolved. Likewise, many of the global properties of the Universe, for instance the ratio of the number of photons to protons,[12] must be found to lie within a very narrow range if cosmic conditions are to allow carbon-based life to arise.

The early investigations of the constraints imposed upon the constants of Nature by the requirement that our form of life exist produced some surprising results. It was found that there exist a number of unlikely coincidences between numbers of enormous magnitude that are, superficially, completely independent; moreover, these coincidences appear essential to the existence of carbon-based observers in the Universe.[13] So numerous and unlikely did these coincidences seem that Carter proposed[1] a stronger version of the Anthropic Principle than the Weak form of self-selection principle introduced earlier: that the Universe *must* be such 'as to admit the creation of observers within it at some stage.' This is clearly a more metaphysical and less defensible notion, for it implies that the Universe could not have been structured differently—that perhaps the constants of Nature could not have had numerical values other than what we observe. Now, we create a considerable problem. For we are tempted to make statements of comparative reference regarding the properties of our observable Universe with respect to the alternative universes we can imagine possessing different values of their fundamental constants. But there is only one Universe; where do we find the other possible universes against which to compare our own in order to decide how fortunate it is that all these remarkable coincidences that are necessary for our own evolution actually exist?

There has long been an interest in the idea that our Universe is but one of many possible worlds. Traditionally, this interest has been coupled with the naive human tendency to regard our Universe as optimal, in some sense, because it appears superfically to be tailor-made for the presence of living creatures like ourselves. We recall Leibniz' claim that ours is the 'best of all possible worlds'; a view that led him to be mercilessly caricatured by Voltaire as Pangloss, a professor of 'metaphysico-theologo-cosmolo-nigology'. Yet, Leibniz' claims also led Maupertuis to formulate the first Action Principles of physics[14] which created new formulations of Newtonian mechanics and provided a basis for the modern approach to formulating and determining new laws of Nature. Maupertuis claimed that the dynamical paths through space possessing non-minimal values of a mathematical quantity he called the Action would be observed if we had less perfect laws of motion than exist in our World. They were identified with the other 'possible worlds'. The fact that Newton's laws of motion were equivalent to bodies taking the path through space that minimizes the Action was cited by Maupertuis as proof that our World, with all its laws, was 'best' in a precise and rigorous mathematical sense.

Maupertuis' ensemble of worlds is not the only one that physicists are familiar with. There have been many suggestions as to how an ensemble of different hypothetical, or actual' universes can arise.[15,16] Far from being examples of idle scholastic speculation many of these schemes are part and parcel of new developments in theoretical physics and cosmology. In general, there are three types of ensemble that one can appeal to in connection with various forms of the Anthropic Principle and they have rather different degrees of certitude.

First, we can consider collections of different possible universes which are parametrized by different values of quantities that do not have the status of invariant constants of Nature. That is, quantities that can, in principle, vary even in our observed Universe. For example, we might consider various cosmological models possessing different initial conditions but with the same laws and constants of Nature that we actually observe. Typical quantities of this sort that we might allow to change are the expansion rate or the levels of isotropy and spatial uniformity in the material content of the Universe. Mathematically, this amounts to choosing different sets of initial boundary conditions for Einstein's gravitational field equations of general relativity (solutions of these equations generate cosmological models). In general, arbitrarily chosen initial conditions at the Big Bang do not necessarily evolve to produce a universe looking like the one we observe after more than fifteen billion years of expansion.[17] We would like to know if the subset of initial conditions that does produce universes like our own has a significant intersection with the subset that allows the eventual evolution of life.

Another way of generating variations in quantities that are not constants of Nature is possible if the Universe is infinite, as current astronomical data suggest. If cosmological initial conditions are exhaustively random and infinite then anything that can occur with non-vanishing probability will occur somewhere; in fact, it will occur infinitely often.[18] Since our Universe has been expanding for a finite time of only about fifteen billion years, only regions that are no farther away than fifteen billion light years can currently be seen by us. Any region farther away than this cannot causally influence us because there has been insufficient time for light to reach us from regions beyond fifteen billion light years. This extent defines what we call the '*observable*, (or *visible*), *Universe*'. But if the Universe is randomly infinite it will contain an infinite number of causally disjoint regions. Conditions within these regions may be different from those within our observable part of the Universe; in some places they will be conducive to the evolution of observers but in others they may not. According to this type of picture, if we could show that conditions very close to those we observe today are absolutely necessary for life, then appeal could be made to an extended form of natural selection to claim that life will only evolve in regions possessing benign properties; hence our observation of such a set of properties in the finite portion of the entire infinite Universe that is observable by ourselves is not surprising. Furthermore, if one could show that the type of Universe we observe out to fifteen billion light years is necessary for observers to evolve then, because in any randomly infinite set[19] of cosmological initial conditions there must exist an infinite number of subsets that will evolve into regions resembling the type of observable Universe we see, it could be argued that the properties of our visible portion of the infinite Universe neither have nor require any further explanation. This is an idea that it is possible to falsify by detecting a density of cosmic material sufficient to render the Universe finite. Interestingly, some of the currently popular 'inflationary' theories of how the cosmic medium behaves very close to the Big Bang not only predict that if our Universe is infinite then it should be extremely non-uniform beyond our visible horizon, but these theories also exploit probabilistic properties of infinite initial data sets.

A third class of universe ensembles that has been contemplated involves the speculative idea of introducing a change in the values of the constants of Nature, or other features of the Universe that strongly constrain the outcome of the laws of Nature—for example, the charge on the electron or the dimensionality of space.[23] Besides simply imagining what would happen if our Universe were to possess constants with different numerical values, one can explore the consequences of allowing fundamental constants of Nature, like Newton's gravitation 'constant', to vary in space or time. Accurate experimental measurements are also available to constrain the allowed magnitude of any such variations.[24] It has also been suggested[48] that if the Universe is cyclic and oscillatory then it might be that the values of the fundamental constants are changed on each occasion the Universe collapses into the 'Big Crunch' before emerging into a new expanding phase.

A probability distribution can also be associated with the observed values of the constants of Nature arising in our own Universe in some new particle physics theories that aim to show that a sufficiently old and cool universe must inevitably display apparent symmetries and particular laws of Nature even if none really existed in the initial high temperature environment near the Big Bang. These 'chaotic gauge theories', as they are called,[25] allow, in principle, a calculation of the probability that after about fifteen billion years we see a particular symmetry or law of Nature in the elementary particle world.

Finally, there is the fourth and last class of world ensemble. A much-discussed and considerably more subtle ensemble of possible worlds is one which has been introduced to provide a satisfactory resolution of paradoxes arising in the interpretation of quantum mechanics.[26] Such an ensemble may be the only way to make sense of a quantum cosmological theory. This 'Many Worlds' interpretation of the quantum theory introduced by Everett and Wheeler requires the simultaneous existence of an infinite number of equally real worlds, all of which are more-or-less causally disjoint, in order to interpret consistently the relationship between observed phenomena and observers.

As the Anthropic Principle has impressed many with its apparent novelty and has been the subject of many popular books and articles,[27] it is important to present it in its true historical perspective in relation to the plethora of Design Arguments beloved of philosophers, scientists and theologians in past centuries[28] and which still permeate the popular mind today. When identified in this way, the idea of the Anthropic Principle in many of its forms can be traced from the pre-Socratics to the founding of modern evolutionary biology. In Chapter 2 we provide a detailed historical survey of this development. As is well known, Aristotle used the notion of 'final causes' in Nature in opposition to the more materialistic alternatives promoted by his contemporaries. His ideas became extremely influential centuries later following their adaption and adoption by Thomas Aquinas to form his grand synthesis of Greek and Judaeo-Christian thought. Aquinas used these teleological ideas regarding the ordering of Nature to produce a Design Argument for the existence of God. Subsequently, the subject developed into a focal point for both expert and inept comment. The most significant impact upon teleological explanations for the structure of Nature arose not from the work of philosophers but rather from Darwin's *Origin of Species*, first published in 1859. Those arguments that had been used so successfully in the past to argue for the anthropocentric purpose of the natural world were suddenly turned upon their heads to demonstrate the contrary: the inevitable conditioning of organic structures by the local environment via natural selection. Undaunted, some leading scientists sought to retain purpose in Nature by subsuming evolutionary theory within a universal teleology.

We study the role played by teleological reasoning in twentieth-century science and philosophy in Chapter 3. There we show also how more primitive versions of the Anthropic Principles have led in the past to new developments in the physical sciences. In this chapter we also describe in some detail the position of teleology and teleonomy in evolutionary biology and introduce the intimate connection between life and computers. This allows us to develop the striking resemblance between some ideas of modern computer theorists, in which the entire Universe is envisaged as a program being run on an abstract computer rather than a real one, and the ontology of the absolute idealists. The traditional picture of the 'Heat Death of the Universe', together with the pictures of teleological evolution to be found in the works of Bergson, Alexander, Whitehead and the other philosophers of progress, leads us into studies of some types of melioristic world-view that have been suggested by philosophers and theologians.

We should warn the professional historian that our presentation of the history of teleology and anthropic arguments will appear Whiggish. To the uninitiated, the term refers to the interpretation of history favoured by the great Whig (liberal) historians of the nineteenth century. As we shall discuss in Chapter 3, these scholars believed that the history of mankind was teleological: a record of slow but continual progress toward the political system dear to the hearts of Whigs, liberal democracy. The Whig historians thus analysed the events and ideas of the past from the point of view of the present rather than trying to understand the people of the past on their own terms.

Modern historians generally differ from the Whig historians in two ways: first, modern historians by and large discern no over-all purpose in history (and we agree with this assessment). Second, modern historians try to approach history from the point of view of the actors rather than judging the validity of archaic world-views from our own Olympian heights. In the opinion of many professional historians, it is not the job of historians to pass moral judgments on the actions of those who lived in the past. A charge of Whiggery—analysing and judging the past from our point of view—has become one of the worse charges that one historian can level at another; a Whiggish approach to history is regarded as the shameful mark of an amateur.[49]

Nevertheless, it is quite impossible for any historian, amateur or professional, to avoid being Whiggish to some extent. As pointed out by the philosopher Morton White,[51] in the very act of criticizing the long-dead Whig historians for judging the people of the past, the modern historians are themselves judging the work of some of their intellectual forebears, namely the Whig historians. Furthermore, every historian must always select a finite part of the infinitely-detailed past to write about. This selection is necessarily determined by the interests of people in the present, the modern historian if no one else. As even the arch critic of Whiggery, Herbert Butterfield, put it in his *The Whig Interpretation of History*:

> The historian is something more than the mere external spectator. Something more is necessary if only to enable him to seize the significant detail and discern the sympathies between events and find the facts that hang together. By imaginative sympathy he makes the past intelligible to the present. He translates its conditioning circumstances into terms which we today can understand. It is in this sense that history must always be written from the point of view of the present. It is in this sense that every age will have to write its history over again.[50]

This is one of the senses in which we shall be Whiggish: we shall try to interpret the ideas of the past in terms a modern scientist can understand.[55] For example, we shall express the concepts of absolute idealism in computer language, and describe the cosmologies of the past in terms of the language used by modern cosmologists.

But our primary purpose in this book is not to write history. It is to describe the modern Anthropic Principle. This will necessarily involve the use of some fairly sophisticated mathematics and require some familiarity with the concepts of modern physics. Not all readers who are interested in reading about the Anthropic Principle will possess all the requisite scientific background. Many of these readers—for instance, theologians and philosophers—will actually be more familiar with the philosophical ideas of the past than with more recent scientific developments. The history sections have been written so that such readers can get a rough idea of the modern concepts by seeing the parallels with the old ideas. Such an approach will give a Whiggish flavour to our treatment of the history of teleology.

There is a third reason for the Whiggish flavour of our history: we *do* want to pass judgments on the work of the scientists and philosophers of the past. Our purpose in doing so is not to demonstrate our superiority over our predecessors, but to learn from their mistakes and successes. It is essential to take this approach in a book on a teleological idea like the Anthropic Principle. There is a general belief that teleology is scientifically bankrupt, and that history shows it always has been. We shall show that on the contrary, teleology has on occasion led to significant scientific advances. It has admittedly also led scientists astray; we want to study the past in order learn under what conditions we might reasonably expect teleology to be reliable guide.

The fourth and final reason for the appearance of Whiggery in our history of teleology is that there *are* re-occurring themes present in the history of teleology; we are only reporting them. We refuse to distort history to fit the current fad of historiography.

We are not the only contemporary students of history to discern such patterns in intellectual history. Such patterns are particularly noticeable in the history of science: the distinguished historian of science Gerald Holton[52] has termed such re-occurring patterns *themata*. To cite just one example of a re-occurring thema from the history of teleology, the cosmologies of the eighteenth-century German idealist Schelling, the twentieth-century British philosopher Alexander, and Teilhard de Chardin are quite similar, simply because all of these men believed in an evolving, melioristic universe; and, broadly speaking, there is really only one way to constuct such a cosmology. We shall discuss this form of teleology in more detail in Chapters 2 and 3.

In Chapter 4 we shall describe in detail how the modern form of the Anthropic self-selection principle arose out of the study of the famous Large Number Coincidences[29] of cosmology. Here the Anthropic Principle was first employed in its modern form to demonstrate that the observed Large Number Coincidences are necessary properties of an observable Universe. This was an important observation because the desire for an explanation of these coincidences had led Dirac[30] to conclude that Newton's gravitation constant must decrease with cosmic time. His suggestion was to start an entirely new sub-culture in gravitation research. We examine then in more detail the idea that there may exist ensembles of different universes in which various coincidences between the values of fundamental constants deviate from their observed values. One of the earliest uses of the Anthropic self-selection idea was that of Whitrow[31] who invoked it as a means of explaining why space is found to possess three dimensions, and we develop this idea in the light of modern ideas in theoretical physics. One of the themes of this chapter is that the recognition of unusual and suggestive coincidences between the numerical values of combinations of physical constants can play an important role in framing detailed theoretical descriptions of the Universe's structure.

Chapter 5 shows how one can determine the gross structure of all the principal constituents of the physical world as equilibrium states between competing fundamental forces. We can then express these characteristics solely in terms of dimensionless constants of Nature aside from inessential geometrical factors like 2π. Having achieved such a description one is in a position to determine the sensitivity of structures essential to the existence of observers with respect to small changes in the values of fundamental constants of Nature. The principal achievement of this type of approach to structures in the Universe is that it enables one to identify which fortuitous properties of the Universe are real coincidences and distinguish them from those which are inevitable consequences of the particular values that the fundamental constants take. The fact that the mass of a human is the geometric mean of a planetary and an atomic mass while the mass of a planet is the geometric mean of an atomic mass and the mass of the observable Universe are two striking examples.[32] These apparent 'coincidences' are actually *consequences* of the particular numerical values of the fundamental constants defining the gravitational and electromagnetic interactions of physics. By contrast the fact that the disks of the Sun and Moon have virtually the same angular size (about half a degree) when viewed from Earth is a pure coincidence and it does not appear to be one that is necessary for the existence of observers. The ratio of the Earth's radius and distance from the Sun is another pure coincidence, in that it is not determined by fundamental constants of Nature alone, but were this ratio slightly different from what it is observed to be, observers could not have evolved on Earth.[33]

The arguments of Chapter 5 can be used to elucidate the inevitable sizes and masses of objects spanning the range from atomic nuclei to stars. If we want to proceed further up the size-spectrum things become more complicated. It is still not known to what extent properties of the whole Universe, determined perhaps by initial conditions or events close the Big Bang, play a role in fixing the sizes of galaxies and galaxy clusters. In Chapter 6 we show how the arguments of Chapter 5 can be extended into the cosmological realm where we find the constants of Nature joined by several dimensionless cosmological parameters to complete the description of the Universe's coarse-grained structure. We give a detailed overview of modern cosmology together with the latest consequences of unified gauge theories for our picture of the very early Universe. This picture enables us to interrelate many aspects of the Universe once regarded as independent coincidences. It also enables us to highlight a number of extraordinarily finely tuned coincidences upon which the possible evolution of observers appears to hinge. We are also able to show well-known Anthropic arguments regarding the observation that the Universe is isotropic to within one part in ten thousand are not actually correct.[17]

In order to trace the origin of the Universe's most unusual large scale properties, we are driven closer and closer to events neighbouring the initial singularity, if such there was. Eventually, classical theories of gravitation become inadequate and a study of the first instants of the Universal expansion requires a quantum cosmological model. The development of such a quantum gravitational theory is the greatest unsolved problem in physics at present but fruitful approaches towards effecting a marriage between quantum field theory and general relativity are beginning to be found. There have even been claims that a quantum wave function for the Universe can be written down.[34]

Quantum mechanics involves observers in a subtle and controversial manner. There are several schools of thought regarding the interpretation of quantum theory. These are described in detail in Chapter 7. After describing the 'Copenhagen' and 'Many Worlds' interpretations we show that the latter picture appears to be necessary to give meaning to any wave function of the entire Universe and we develop a simple quantum cosmological model in detail. This description allows the Anthropic Principle to make specific predictions.

The Anthropic Principles seek to link aspects of the global and local structure of the Universe to those conditions necessary for the existence of living observers. It is therefore of crucial importance to be clear about what we mean by 'life'. In Chapter 8 we give a new definition of life and discuss various alternatives that have been suggested in the past. We then consider those aspects of chemical and biochemical structures that appear necessary for life based upon atomic structures. Here we are, in effect, extending the methodology of Chapter 5 from astrophysics to biochemistry with the aim of determining how the crucial properties of molecular structures are related to the invariant aspects of Nature in the form of fundamental constants and bonding angles. To complete this chapter we extend some recent ideas of Carter[35] regarding the evolution of intelligent life on Earth. This leads to an Anthropic Principle prediction which relates the likely time of survival of terrestrial life in the future the number of improbable steps in the evolution of intelligent life on Earth via a simple mathematical inequality.

In Chapter 9 we discuss the controversial subject of extraterrestrial life and provide arguments that there probably exists no other intelligent species with the capability of interstellar communication within our own Milky Way Galaxy. We place more emphasis upon the ideas of biologists regarding the likelihood of intelligent life-forms evolving than is usually done by astronomers interested in the possibility of extraterrestrial intelligence. As a postscript we show how the logic used to project the capabilities of technologically advanced life-forms can be used to frame an Anthropic Principle argument against the possibility that we live in a Steady-State Universe. This shows that Anthropic Principle arguments can be used to winnow-out cosmological theories. Conversely, if the theories which contradict the Anthropic Principle are found to be correct, the Anthropic Principle is refuted; this gives another test of the Anthropic Principle.

Finally, in Chapter 10, we attempt to predict the possible future histories of the Universe in the light of known physics and cosmology. We describe in detail the expected evolution of both open and closed cosmological models in the far future and also stress a number of global constraints that exist upon the structure of a universe consistent with our own observations today. In our final speculative sections we investigate the possibility of life surviving into the indefinite future of both open and closed universes. We define life using the latest ideas in information and computer theory and determine what the Universe must be like in order that information-processing continue indefinitely; in effect, we investigate the implications for physics of the requirement that 'life' never becomes extinct. Paradoxically, this appears to be possible only in a *closed* universe with a very special global causal structure, and thus the requirement that life never dies out—which we define precisely by a new '*Final Anthropic Principle*'—leads to definite testable predictions about the global structure of the Universe. Since indefinite survival in a closed universe means survival in a high-energy environment near the final singularity, the Final Anthropic Principle also leads to some predictions in high-energy particle physics.

Before abandoning the reader to the rest of the book we should make a few comments about its contents. Our study involves detailed mathematical investigations of physics and cosmology, studies of chemistry and evolutionary biology as well as a considerable amount of historical description and analysis. We hope we have something new to say in all these areas. However, not every reader will be interested in all of this material. Our chapters have, in the main, been constructed in such a way that they can be read independently, and the notes and references are collected together accordingly. Scientists with no interest in the history of ideas can just skip the chapters in which they are discussed. Likewise, non-scientists can avoid mathematics altogether they wish. One last word: the authors are cosmologists, not philosophers. This has one very important consequence which the average reader should bear in mind. Whereas philosophers and theologians appear to possess an emotional attachment to their theories and ideas which requires them to believe them, scientists tend to regard their ideas differently. They are interested in formulating many logically consistent possibilities, leaving any judgement regarding their truth to observation. Scientists feel no qualms about suggesting different but mutually exclusive explanations for the same phenomenon. The authors are no exception to this rule and it would be unwise of the reader to draw any wider conclusions about the authors' views from what they may read here.

1.2 Anthropic Definitions

> Definitions are like belts. The shorter
> they are, the more elastic they need to be.
> S. Toulmin

Although the Anthropic Principle is widely cited and has often been discussed in the astronomical literature, (as can be seen from the bibliography to this chapter alone), there exist few attempts to frame a precise statement of the Principle; rather, astronomers seem to like to leave a little flexibility in its formulation perhaps in the hope that its significance may thereby more readily emerge in the future. The first published discussion by Carter[1] saw the introduction of a distinction between what he termed 'Weak' and 'Strong' Anthropic statements. Here, we would like to define precise versions of these two Anthropic Principles and then introduce Wheeler's Participatory Anthropic Principle[6] together with a new Final Anthropic Principle which we shall investigate in Chapter 10.

The Weak Anthropic Principle (WAP) tries to tie a precise statement to the notion that any cosmological observations made by astronomers are biased by an all-embracing selection effect: our own existence. Features of the Universe which appear to us astonishingly improbable, *a priori*, can only be judged in their correct perspective when due allowance has been made for the fact that certain properties of the Universe are necessary if it is to contain carbonaceous astronomers like ourselves.

This approach to evaluating unusual features of our Universe first re-emerges in modern times in a paper of Whitrow[31] who, in 1955, sought an answer to the question '*why does space have three dimensions*?'. Although unable to explain why space actually has, (or perhaps even why it must have), three dimensions, Whitrow argued that this feature of the World is not unrelated to our own existence as observers of it. When formulated in three dimensions, mathematical physics possesses many unique properties that are necessary prerequisites for the existence of rational information-processing and 'observers' similar to ourselves. Whitrow concluded that only in three-dimensional spaces can the dimensionality of space *be* questioned. At about the same time Whitrow also pointed out that the expansion of the Universe forges an unbreakable link between its overall size and age and the ambient density of material within it.[36] This connection reveals that only a very 'large' universe is a possible habitat for life. More detailed ideas of this sort had also been published in Russian by the Soviet astronomer Idlis.[37] He argued that a variety of special astronomical conditions must be met if a universe is to be habitable. He also entertained the possibility that we were observers merely of a tiny fraction of a diverse and infinite universe whose unobserved regions may not meet the minimum requirements for observers that there exist hospitable temperatures and stable sources of stellar energy.

Our definition of the WAP is motivated in part by these insights together with later, rather similar ideas of Dicke[13] who, in 1957, pointed out that the number of particles in the observable extent of the Universe, and the existence of Dirac's famous Large Number Coincidences '*were not random but conditioned by biological factors*'. This motivates the following definition:

Weak Anthropic Principle (WAP): The observed values of all physical and cosmological quantities are not equally probable but they take on values restricted by the requirement that there exist sites where carbon-based life can evolve and by the requirement that the Universe be old enough for it to have already done so.

Again we should stress that this statement is in no way either speculative or controversial. It expresses only the fact that those properties of the Universe we are able to discern are self-selected by the fact that they must be consistent with our own evolution and present existence. WAP would not necessarily restrict the observations of non-carbon-based life but *our* observations are restricted by our very special nature.

As a corollary, the WAP also challenges us to isolate that subset of the Universe's properties which are *necessary* for the evolution and continued existence of our form of life. The entire collection of the Universe's laws and properties that we now observe need be neither necessary nor sufficient for the existence of life. Some properties, for instance the large size and great age of the Universe, do appear to be necessary conditions; others, like the precise variation in the distribution of matter in the Universe from place to place, may not be necessary for the development of observers at some site. The non-teleological character of evolution by natural selection ensures that *none* of the observed properties of the Universe are sufficient conditions for the evolution and existence of life.

Carter,[35] and others, have pointed out that as a self-selection principle the WAP is a statement of Bayes' theorem. The Bayesian approach[38] to inference attributes *a priori* and *a posteriori* probabilities to any hypothesis before and after some piece of relevant evidence, E, is taken into account. In such a situation we call the before and after probabilities p_B and p_A, respectively. The fact that for any particular outcome O, the probability of observing O before the evidence E is known equals the probability of observing O given the evidence E, after E was accounted for, is expressed by the equation,

$$p_B(O) = p_A(O/E) \tag{1.1}$$

where / denotes a conditional probability. Bayes' formula[38] then gives the relative plausibililty of any two theories α and β in the face of a piece of evidence E as

$$\frac{p_E(\alpha)}{p_E(\beta)} = \frac{p_A(E/\alpha)p_A(\alpha)}{p_A(E/\beta)p_A(\beta)} \tag{1.2}$$

Thus the relative probabilities of the truth of α or β are modified by the conditional probabilities $p_A(E/\alpha)$ and $p_A(E/\beta)$ which account for any bias of the experiment (or experimenter) towards gathering evidence that favours α rather than β (or *vice versa*). The WAP as we have stated it is just an application of Bayes' theorem.

The WAP is certainly not a powerless tautalogical statement because cosmological models have been defended in which the gross structure of the Universe is predicted to be the same on the average whenever it is observed. The, now defunct, continuous creation theory proposed by Bondi, Gold and Hoyle is a good example. The WAP could have been used to make this steady-state cosmology appear extremely improbable even before it came into irredeemable conflict with direct observations. As Rees points out,[12]

> the fact that there is an epoch when [the Hubble time, t_H, which is essentially equal to the age of the Universe] is of order the age of a typical star..... is not surprising in any 'big bang' cosmology. Nor is it surprising that we should ourselves be observing the universe at this particular epoch. In a steady-state cosmology, however, there would seem no *a priori* reason why the timescale for stellar evolution should not be *either* [much less than] t_H (in which case nearly all the matter would be in dead stars or 'burnt-out' galaxies) or [much greater than] t_H (in which case only a very exceptionally old galaxy would look like our own). Such considerations could have provided suggestive arguments in favour of 'big bang' cosmologies...

We can also give some examples of how the WAP leads to synthesizing insights that deepen our appreciation of the unity of Nature. Observed facts, often suspected at first sight to be unrelated, can be connected by examining their relation to the conditions necessary for our own existence and their explicit dependence on the constants of physics. Let us reconsider, from the Bayesian point of view, the classic example mentioned in section 1.1, relating the size of the Universe to the period of time necessary to generate observers. The requirement that enough time pass for cosmic expansion to cool off sufficiently after the Big Bang to allow the existence of carbon ensures that the observable Universe must be relatively old and so, because the boundary of the observable Universe expands at the speed of light, very large. The nuclei of carbon, nitrogen, oxygen and phosphorus of which we are made, are cooked from the light primordial nuclei of hydrogen and helium by nuclear reactions in stellar interiors. When a star nears the end of its life, it disperses these biological precursors throughout space. The time required for stars to produce carbon and other bioactive elements in this way is roughly the lifetime of a star on the 'main-sequence' of its evolution, given by

$$t_\star \sim \left(\frac{Gm_N^2}{hc}\right)^{-1} \frac{h}{m_N c} \sim 10^{10} \text{ yrs} \tag{1.3}$$

where G is Newton's gravitation constant, c is the velocity of light, h is Planck's constant and m_N is the proton mass. Thus, in order that the Universe contain the building-blocks of life, it must be at least as old as $t_\star$ and hence, by virtue of its expansion, at least $ct_\star$ (roughly ten billion light years) in extent. No one should be surprised to find the Universe to be as large as it is. We could not exist in one that was significantly smaller. Moreover, the argument that the Universe should be teeming with civilizations on account of its vastness loses much of its persuasiveness: the Universe has to be as big as it is in order to support just one lonely outpost of life. Here, we can see the deployment of (1.2) explicitly if we let the hypothesis that *the large size of the Universe is superfluous for life on planet Earth* be α and let hypothesis β be that *life on Earth is connected with the size of the Universe.* If the evidence E is that the Universe is observed to be greater than ten billion light years in extent then, although $p_B(E/\beta) \ll 1$, the hypothesis is not necessarily then improbable because we have argued that $p_A(E/\beta) \simeq 1$.

We also observe the expansion of the Universe to be occurring at a rate which is irresolvably close to the special value which allows it the smallest deceleration compatible with indefinite future expansion. This feature of the Universe is also dependent on the epoch of observation. And again, if galaxies and clusters of galaxies grow in extent by mergers and hierarchical clustering,[2] then the characteristic scale of galaxy clustering that we infer will be determined by the cosmic epoch at which it is observed.

Ellis[39] has stressed the existence of a spatial restriction which further circumscribes the range of observed astronomical phenomena. What amounts to a universal application of the principle of natural selection would tell us that observers may only exist in particular regions of a *spatially inhomogeneous* universe. Since realistic mathematical models of inhomogeneous universes are extremely difficult to construct, various unverifiable cosmological 'Principles' are often used by theoretical cosmologists to allow simple cosmological models to be extracted from Einstein's general theory of relativity. These Principles invariably make statements about regions of the Universe which are unobservable not only in practice but also in principle (because of the finite speed of light). Principles of this sort need to be used with care. For example, Principles of Mediocrity like the Copernican Principle or the Principle of Plenitude (see Chapter 3) would imply that if the Universe did possess a preferred place, or centre, then we should not expect to find ourselves positioned there. However, general relativity allows possible cosmological models to be constructed which not only possess a centre, but which also have conditions conducive to the existence of observers only near that centre. The WAP would offer a good explanation for our central position in such circumstances, whilst the Principles of Mediocrity would force us to conclude that we do not exist at all!

According to WAP, it is possible to contemplate the existence of many possible universes, each possessing different defining parameters and properties. Observers like ourselves obviously can exist only in that subset containing universes consistent with the evolution of carbon-based life.

This approach introduces necessarily the idea of an ensemble of possible universes and was suggested independently by the Cambridge biologist Charles Pantin in 1965. Pantin had recognized that a vague principle of amazement at the fortuitous properties of natural substances like carbon or water could not yield any *testable* predictions about the World, but the amazement might disappear if[40]

> we could know that our Universe was only one of an indefinite number with varying properties, [so] we could perhaps invoke a solution analogous to the principle of Natural Selection; that only in certain universes which happen to include ours, are the conditions suitable for the existence of life, and unless that condition is fulfilled there will be no observers to note the fact

However, as Pantin also realized, it still remains an open question as to why *any* permutation of the fundamental constants of Nature allows the existence of life, albeit a question we would not be worrying about were such a fortuitous permutation not to exist.

If one subscribes to this 'ensemble interpretation' of the WAP one must decide how large an ensemble of alternative worlds is to be admitted. Many ensembles can be imagined according to our willingness to speculate—different sets of cosmological initial data, different numerical values of fundamental constants, different space-time dimensions, different laws of physics—some of these possibilities we shall discuss in later chapters.

The theoretical investigations initiated by Carter[1] reveal that in some sense the subset of the ensemble containing worlds able to evolve observers is very 'small'. Most perturbations of the fundamental constants of Nature away from their actual numerical values lead to model worlds that are still-born, unable to generate observers and become cognizable. Usually, they allow neither nuclei, atoms nor stars to exist.

Whatever the size and variety of permutations allowed within a hypothetical ensemble of 'many worlds', one might introduce here an analogue of the Drake equation[41] often employed to guess the number of extraterrestrial civilizations in our Galaxy. Instead of expressing the probability of life existing *elsewhere* as a product of independent probabilities for the occurrence of processes like planetary formation, protocellular evolution and so forth, one could express the probability of life existing *anywhere* as a product of probabilities that encode the fact that life is only possible if parameters like the fine structure constant or the strong coupling constant lie in a particular numerical range.[42,43]

The existence of the fundamental cosmic timescale like (1.3), fixed only by invariant constants of Nature, c, h, G, and m_N, was exploited by Dicke[13] to produce a powerful WAP argument against Dirac's conclusion[30] that the Newtonian gravitation constant, G, is decreasing with time. Dirac had noticed that the dimensionless measure of the strength of gravity

$$\alpha_G \equiv \frac{Gm_N^2}{hc} \sim 10^{-39} \tag{1.4}$$

is roughly of order the inverse square root of the number of nucleons in the observable Universe, $N(t)$, at the present time $t_0 \sim 10^{10}$ yrs. At any time, t, the quantity $N(t)$ is simply

$$N(t) \equiv \frac{M_U}{m_N} = \frac{4\pi\rho_U(ct)^3}{3m_N} \sim \frac{c^3 t}{Gm_N} \sim 10^{78}\left(\frac{t}{10^{10}\text{ yrs}}\right) \tag{1.5}$$

if we use the cosmological relation that the density of the Universe, ρ_U, is related to its age by $\rho_U \sim (Gt^2)^{-1}$. (The present age of roughly 10^{10} yrs is displayed in the last step.) Dirac argued that it is very unlikely that these two quantities should possess simply related dimensionless magnitudes which are both so vastly different from unity and yet be independent. Rather, there must exist an approximate *equality* between them of the form

$$N(t) \sim \alpha_G^{-2} \tag{1.6}$$

However, whereas α_G is a *time-independent* combination of constants, $N(t)$ increases linearly with the time of observation, t, which for us is the present age of the Universe. The relation (1.6) can only hold for all times if one component of α_G is time-varying and so Dirac suggested that we must have $G \propto t^{-1}$ so that $N(t) \propto \alpha_G^{-2} \propto t^2$. The quantities $N(t)$ and α_G^{-2} are now observed to be of the same magnitude because (as a result of some unfound law of Nature) they are actually *equal*, and furthermore, they are of such an enormous magnitude because they both increase linearly in time and the Universe is very old—although this 'oldness' can presumably only be explained by the WAP even in this scheme of 'varying' constants for the reasons discussed above in connection with the size of the Universe.

However, the WAP shows Dirac's radical conclusion of a time-varying Newtonian gravitation constant to be quite unnecessary. The coincidence that today we observe $N \sim \alpha_G^{-2}$ is necessary for our existence. Since we would not expect to observe the Universe either before stars form or after they have burnt out, human astronomers will most probably observe the Universe close to the epoch $t_\star$ given by (1.3). Hence, we will observe the time-dependent quantity $N(t)$ to take on a value of order $N(t_\star)$ and, by (1.3) and (1.4), this value is necessarily just

$$N(t_\star) \sim \frac{t_\star}{Gm_N} \sim \alpha_G^{-2} \tag{1.7}$$

where the second relation is a *consequence* of the value of $t_\star$ in (1.3). If we let δ be Dirac's hypothesis of time-varying G, while γ is the hypothesis that G is constant while the 'evidence', E, is the coincidence (1.6); then, although the *a priori* probability that we live at the time when the numbers $N(t)$ and α_G^{-2} are equal is very low, ($p_B(E/\gamma) \ll 1$), this does not render hypothesis γ (the constancy of G) implausible because there is an anthropic selection effect which ensures $p_A(E/\gamma) \simeq 1$. This selection effect is the one pointed out by Dicke. We should notice that this argument alone explains why we must observe $N(t)$ and α_G^{-2} to be of equal magnitude, but not why that magnitude has the extraordinarily large value $\sim 10^{79}$. (We shall have a lot more to say about this problem in Chapters 4, 5 and 6).

As mentioned in section 1.1, Carter[1] introduced the more speculative Strong Anthropic Principle (SAP) to provide a 'reason' for our observation of large dimensionless ratios like 10^{79}; we state his SAP as follows:

Strong Anthropic Principle (SAP): The Universe must have those properties which allow life to develop within it at some stage in its history.

An implication of the SAP is that the constants and laws of Nature must be such that life can exist. This speculative statement leads to a number of quite distinct interpretations of a radical nature: firstly, the most obvious is to continue in the tradition of the classical Design Arguments and claim that:

(A) *There exists one possible Universe 'designed' with the goal of generating and sustaining 'observers'.*

This view would have been supported by the natural theologians of past centuries, whose views we shall examine in Chapter 2. More recently it has been taken seriously by scientists who include the Harvard chemist Lawrence Henderson[44] and the British astrophysicist Fred Hoyle, so impressed were they by the string of 'coincidences' that exist between particular numerical values of dimensionless constants of Nature without which life of any sort would be excluded. Hoyle[45] points out how natural it might be to draw a teleological conclusion from the fortuitous positioning of nuclear resonance levels in carbon and oxygen:

> I do not believe that any scientist who examined the evidence would fail to draw the inference that the laws of nuclear physics have been deliberately designed with regard to the consequences they produce inside the stars. If this is so, then my apparently random quirks have become part of a deep-laid scheme. If not then we are back again at a monstrous sequence of accidents.

The interpretation (A) above does not appear to be open either to proof or to disproof and is religious in nature. Indeed it is a view either implicit or explicit in most theologies.

This is all we need say about the 'teleological' version of the SAP at this stage. However, the inclusion of quantum physics into the SAP produces quite different interpretations. Wheeler[6] has coined the title '*Participatory Anthropic Principle*' (PAP) for a second possible interpretation of the SAP:

(B) *Observers are necessary to bring the Universe into being.*

This statement is somewhat reminiscent of the outlook of Bishop Berkeley and we shall see that it has physical content when considered in the light of attempts to arrive at a satisfactory interpretation of quantum mechanics.[46] It is closely related to another possibility:

(C) *An ensemble of other different universes is necessary for the existence of our Universe.*

This statement receives support from the 'Many-Worlds' interpretation of quantum mechanics and a sum-over-histories approach to quantum gravitation because they must unavoidably recognize the existence of a whole class of *real* 'other worlds' from which ours is selected by an optimizing principle.[47] We shall express this version of the SAP mathematically in Chapter 7, and we shall see that this version of the SAP has consequences which are potentially testable.

Suppose that for some unknown reason the SAP is true and that intelligent life must come into existence at some stage in the Universe's history. But if it dies out at our stage of development, long before it has had any measurable non-quantum influence on the Universe in the large, it is hard to see why it *must* have come into existence in the first place. This motivates the following generalization of the SAP:

Final Anthropic Principle (FAP): Intelligent information-processing must come into existence in the Universe, and, once it comes into existence, it will never die out.

We shall examine the consequences of the FAP in our final chapter by using the ideas of information theory and computer science. The FAP will be made precise in this chapter. As we shall see, FAP will turn out to require the Universe and elementary particle states to possess a number of definite properties. These properties provide observational tests for this statement of the FAP.

Although the FAP is a statement of physics and hence *ipso facto*[53] has no ethical or moral content, it nevertheless is closely connected with moral values, for the validity of the FAP is the physical precondition for moral values to arise and to continue to exist in the Universe: no moral values of any sort can exist in a lifeless cosmology. Furthermore, the FAP seems to imply a melioristic cosmos.

We should warn the reader once again that both the FAP and the SAP are quite speculative; unquestionably, neither should be regarded as well-established principles of physics. In contrast, the WAP is just a restatement, albeit a subtle restatement, of one of the most important and well-established principles of science: that it is essential to take into account the limitations of one's measuring apparatus when interpreting one's observations.

References

1. B. Carter, in *Confrontation of cosmological theories with observation*, ed. M. S. Longair (Reidel, Dordrecht, 1974), p. 291.
2. See for example P. J. Peebles, *The large scale structure of the universe*, (Princeton University Press, Princeton, 1980).
3. A. Sandage and E. Hardy, *Astrophys. J.* **183,** 743 (1973). S. Weinberg, *Gravitation and cosmology* (Wiley, NY, 1972).
4. We examine some of these claims in Chapters 2 and 3.
5. E. Hubble, *Proc. natn. Acad. Sci., U.S.A.* **15,** 169 (1929).
6. J. A. Wheeler, in *Foundational problems in the special sciences*, ed. R. E.

Butts and J. Hintikka (Reidel, Dordrecht, 1977), p. 3; in *The nature of scientific discovery*, ed. O. Gingerich (Smithsonian Press, Washington, 1975), pp. 261–96 and pp. 575–87.

7. D. Clayton, *Principles of stellar evolution and nucleosynthesis* (McGraw-Hill, NY, 1968). R. J. Tayler, *The stars: their evolution and structure* (Wykeham, London, 1970).
8. For reviews of a number of examples see, in particular, B. J. Carr and M. J. Rees, *Nature* **278,** 605 (1979), V. Weisskopf, *Science* **187,** 605 (1975); we shall investigate this in detail in Chapters 5 and 6.
9. For an interesting overview of constants see *The constants of nature*, ed. W. H. McCrea and M. J. Rees (Royal Society of London, London, 1983). This book was originally published as the contents of *Phil. Trans. R. Soc.* Vol. A **310** in 1983 See also J. M. Levy-Leblond, *Riv. nuovo Cim.* **7,** 187 (1977).
10. P. Candelas and S. Weinberg, *Nucl. Phys.* B **237,** 397 (1984).
11. P. C. W. Davies, *J. Phys.* A **5,** 1296 (1972).
12. M. J. Rees, *Comm. Astrophys. Space Phys.* **4,** 182 (1972).
13. R. H. Dicke, *Rev. Mod. Phys.* **29,** 355 and 363 (1977); *Nature* **192,** 440 (1961).
14. P. L. M. Maupertuis, *Essai de cosmologie* (1751), in *Oeuvres*, Vol. 4, p. 3 (Lyon, 1768). We discuss these developments in detail in sections 2.5 and 3.4.
15. J. D. Barrow, 'Cosmology, the existence of observers and ensembles of possible universes', in *Les voies de la connaissance* (Tsukuba Conference Proceedings, Radio-France Culture, Paris 1985).
16. J. D. Barrow, *Quart. J. R. astron. Soc.* **23,** 146 (1983).
17. C. B. Collins and S. W. Hawking, *Astrophys. J.* **180,** 317 (1973); J. D. Barrow, *Quart. J. R. astrom. Soc.* **23,** 344 (1982). For a popular discussion see Chapter 5 of J. D. Barrow and J. Silk, *The left hand of creation* (Basic Books, NY, 1983 and Heinemann, London, 1984).
18. This is an old argument applied to cosmology by G. F. R. Ellis and G. B. Brundrit, *Q. J. R. astron. Soc.* **20,** 37 (1979). See F. J. Tipler, *Quart. J. R. astron. Soc.* **22,** 133 (1981) for a discussion of the history of this argument.
19. Notice the infinity alone is not a sufficient condition for this to occur; it must be an exhaustively random infinity in order to include all possibilities.
20. If the visible part of the Universe is accurately described by Friedman's equation without cosmological constant (as seems to be the case, see ref. 3) then a density exceeding about $2 \,.\, 10^{-29}$ gm cm^{-3} is required.
21. A. Guth, *Phys. Rev.* D **23,** 347 (1981); K. Sato, *Mon. Not. R. astron. Soc.* **195,** 467 (1981); A. Linde, *Phys. Lett.* B **108,** 389 (1982). For an overview see G. Gibbons, S. W. Hawking, and S. T. C. Siklos, *The very early universe* (Cambridge University Press, Cambridge, 1983).
22. A. Linde, *Nuovo Cim. Lett.* **39,** 401 (1984).
23. J. D. Barrow, *Phil. Trans. R. Soc.* A **310,** 337 (1983); E. Witten, *Nucl. Phys.* B **186,** 412 (1981).
24. R. D. Reasonberg, *Phil. Trans. R. Soc.* A **310,** 227 (1983).
25. H. B. Nielsen, *Phil. Trans. R. Soc.* A **310,** 261 (1983); J. Iliopoulos, D. V. Nanopoulos, and T. N. Tamvaros, *Phys. Lett.* B **94,** 141 (1980); J. D. Barrow and A. C. Ottewill, *J. Phys.* A **16,** 2757 (1983).
26. J. A. Wheeler and W. H. Zurek, *Quantum theory and measurement*, (Prince-

ton University Press, Princeton, 1982); H. Everett, *Rev. Mod. Phys.* **29,** 454 (1957); F. J. Tipler, 'Interpreting the wave function of the universe', *Phys. Rep.* (In press.) B. Espagnet, *Scient. Am. Nov* (1979), p. 128.

27. V. Trimble, *Am. Scient.* **65,** 76 (1977); F. Dyson, *Scient. Am.* **224,** No. 3, pp. 50–9 (Sept 1971); J. Leslie, *Am. Phil. Quart.* **19,** 141 (1982), *Am. Phil. Quart.* **7,** 286 (1970), *Mind* **92,** 573 (1983); in *Scientific explanation and understanding: essays on reasoning and rationality in science,* ed. N. Rescher (University Press of America, Lanham, 1983), pp. 53–83; in *Evolution and creation,* ed. E. McMullin (University of Notre Dame Press, Notre Dame 1984); P. J. Hall, *Quart. J. R. astron. Soc.* **24,** 443 (1983); J. Demaret and C. Barbier, *Revue des Questions Scientifique* **152,** 181, 461 (1981); E. J. Squires, *Eur. J. Phys.* **2,** 55 (1981); P. C. W. Davies, *Accidental universe* (Cambridge University Press, Cambridge, 1982); R. Breuer, *Das anthropische Prinzip* (Meyster, München, 1981); J. D. Barrow and J. Silk, *Scient. Am.*, April (1980), p. 98; A. Finkbeiner, *Sky & Telescope,* Aug. (1984), p. 107; J. D. Barrow and F. J. Tipler, *L'homme et le cosmos* (Imago-Radio France, Paris, 1984); J. Eccles, *The human mystery* (Springer, NY, 1979); B. Lovell, *In the centre of immensities,* (Harper & Row, NY, 1983); J. A. Wheeler, *Am. Scient.* **62,** 683 (1974); G. Gale, *Scient. Am.* **245** (No. 6, Dec.), 154 (1981); M. T. Simmons, Mosaic (March-April 1982) p. 16; G. Wald, *Origins of Life* **5,** 7 (1974); S. W. Hawking, *CERN Courier* **21** (1), 3 (1981); G. F. R. Ellis, *S. Afr. J. Sci.*, **75,** 529 (1979); S. J. Gould, *Natural History* **92,** 34 (1983); J. Maddox, *Nature* **307,** 409 (1984); P. C. W. Davies, *Prog. Part. Nucl. Phys.* **10,** 1 (1983); F. J. Dyson, *Disturbing the universe* (Harper & Row, NY, 1979).
28. For a representative general bibliography of Design Arguments see: H. Baker, *The image of man* (Harper, NY, 1947); P. Bertocci, *An introduction to the philosophy of religion,* (Prentice Hall, NY, 1951); *The cosmological arguments,* ed. D. R. Burnill (Doubleday, NY, 1967); E. A. Burtt, *The metaphysical foundations of modern physical science* (Harcourt Brace, NY, 1927); C. Hartshorne, *A natural theology for our time* (Open Court, NY, 1967); L. E. Hicks, *A critique of Design Arguments* (Scribners, NY, 1883); R. H. Hurlbutt III, *Hume, Newton and the Design Argument,* (University Nebraska Press, Lincoln, Nebraska, 1965); P. Janet, *Final causes* (Clark, Edinburgh, 1878); D. L. LeMahieu, *The mind of William Paley* (University Nebraska Press, Lincoln, Nebraska, 1976); A. O. Lovejoy, *The Great Chain of Being: a study in the history of an idea,* (Harvard University Press, Cambridge, 1936); J. D. McFarland, *Kant's concept of teleology,* (University Edinburgh Press, Edinburgh, 1970); T. McPherson, *The argument from design* (Macmillan, Edinburgh, 1972); L. Stephen, *English thought in the eighteenth century,* Vol. 1 (Harcourt Brace, NY, 1962); R. G. Swinburne, *Philosophy* **43,** 164 (1968); F. R. Tennant, *Philosophical theology,* 2 vols (Cambridge University Press, Cambridge, 1930); A. Woodfield, *Teleology* (Cambridge University Press, Cambridge, 1976); L. Wright, *Teleological explanations* (University of California Press, Berkeley, 1976).
29. J. D. Barrow, *Quart. J. R. astron. Soc.* **22,** 388 (1981).
30. P. A. M. Dirac, *Nature* **139,** 323 (1937).
31. G. Whitrow, *Br. J. Phil. Sci.* **6,** 13 (1955).
32. B. J. Carr and M. J. Rees, *Nature* **278,** 605 (1979).
33. M. H. Hart, *Icarus* **33,** 23 (1978).
34. J. Hartle and S. W. Hawking, *Phys. Rev.* D **28,** 2960 (1983); F. J. Tipler, preprint (1984).

125, 243

College of St. Francis Library
Joliet, Illinois

35. B. Carter, *Phil. Trans. R. Soc.* A **310,** 347 (1983).
36. Whitrow, cited in E. Mascall, *Christian theology and natural science: 1956 Bampton lectures* (Longmans Green, London, 1956).
37. G. Idlis, *Izv. Astrophys. Inst. Kazakh. SSR* **7,** 39 (1958), in Russian.
38. See, for example, P. L. Meyer, *Introductory probability and statistical applications* (Addison-Wesley, NY, 1971).
39. G. F. R. Ellis, *Gen. Rel. Gravn.* **11,** 281 (1979); G. F. R. Ellis, R. Maartens, and S. D. Nel, *Mon. Not. R. Soc.* **184,** 439 (1978).
40. C. F. A. Pantin, in *Biology and personality*, ed. I. T. Ramsey (Blackwell, Oxford, 1965), pp. 103–4.
41. I. S. Shklovskii and C. Sagan, *Intelligent life in the universe* (Dell, NY, 1966).
42. J. D. Barrow, in ref. 23.
43. T. L. Wilson, *Quart. J. R. astron. Soc.* **25,** 435 (1984).
44. L. J. Henderson, *The fitness of the environment* (Smith, Gloucester, Mass., 1913; reprinted Harvard University Press, Cambridge, Mass., 1970) and *The order of Nature* (Harvard University Press, Cambridge, Mass., 1917).
45. F. Hoyle, in *Religion and the scientists* (SCM, London, 1959).
46. M. Jammer, *The philosophy of quantum mechanics* (Wiley, NY, 1974).
47. R. P. Feynman and A. R. Hibbs, *Quantum mechanics and path integrals* (McGraw-Hill, New York, 1965); L. S. Schulman, *Techniques and applications of path integration* (Wiley, NY, 1981).
48. C. W. Misner, K. S. Thorne, and J. A. Wheeler, *Gravitation* (Freeman, San Francisco, 1973), Chapter 44.
49. For a lucid recent expression of this view from a professional historian see 'Whigs and professionals', by C. Russell in *Nature* **308,** 777 (1984).
50. H. Butterfield, *The Whig interpretation of history* (G. Bell, London, 1951), p. 92. This book was first published in 1931. Butterfield later toned down his opposition to Whiggery. See, for example, his book *The Englishman and his history* (Cambridge University Press, Cambridge, 1944).
51. M. White, unpublished lecture at Harvard University (1957); quoted by W. W. Bartley III, in *The retreat to commitment* (Knopf, NY, 1962), pp. 98–100.
52. G. Holton, *Thematic origins of scientific though* (Harvard University Press, NY, 1973).
53. Physical assumptions and moral assumptions belong to different logical categories. Physical assumptions, like all scientific statements, are statements of matters of fact: syntactically, they are declarative sentences. Moral assumptions, on the other hand, are statements concerning moral obligation: syntactically, they are imperative sentences, which contain the word 'ought' or its equivalent. For further discussion of this point see any modern textbook on moral philosophy, for instance H. Reichenbach, *The rise of scientific philosophy* (University of California Press, Berkeley, 1968).
54. For further discussion of self-reference arguments in mathematics, see D. Hofstadter, *Gödel, Escher, Bach: an eternal golden braid* (Basic Books, NY, 1979).
55. For a more recent defence of the idea that the past has to be discussed in terms the present can understand, see D. Hull, *History & Theory* **18,** 1 (1979). We are grateful to Professor S. G. Brush for this reference.

2 Design Arguments

> What had that flower to do with being white,
> The wayside blue and innocent heal-all?
> What brought the kindred spider to that height,
> Then steered the white moth thither in the night?
> What but design of darkness to appall?—
> If design govern in a thing so small.
>
> Robert Frost

2.1 Historical Prologue

> Original ideas are exceedingly rare and the most that philosophers have done in the course of time is to erect a new combination of them.
>
> G. Sarton

The Anthropic Principle is a consequence of our own existence. Since the dawn of recorded history humankind has used the local and global environment to good advantage; the soil and its fruits for food, the heavenly bodies for navigation, and the winds and waves for power. Such beneficiaries might naturally be led to conclude that the world in all its richness and subtlety was contrived for their benefit alone; uniquely designed for them rather than merely fortuitously used by them. From such inclinations and the natural attraction they appear to hold for those seeking meaning and significance in life, simple design arguments grew in a variety of cultures, each fashioned by the knowledge and sophistication of the society around it and nurtured by the religious and scientific beliefs of the day. In the Hebrew writings that form our Old Testament, we see the idea of providential design as a key feature of the Creation narratives and the epic poetry of the Wisdom and prophetic writings. The idea of a partially anthropocentric universe with teleological aspects is the warp and woof of the Judaeo-Christian world-view that underlies the growth of Western civilization. Another important aspect of our heritage is the growth of science and logic in early Greece, where the early Greeks also generated a detailed teleological view of the world which was, in time, wedded by the Scholastics to the poetic view of the Judaeo-Christian tradition.

Astronomers and physicists who first encounter the collection of results and observations that exist under the collective label of the Anthropic Principle are usually surprised by the novelty of such an anthropocentric approach to Nature. Yet, the Anthropic Principle is just the latest manifestation of a style of argument that can be traced back to ancient times when philosophy and science were conjoined and 'metaphysics' was concerned with the method as well as the meaning of science. In this chapter we shall follow these arguments from ancient to modern times and attempt to display the recurrent polarization of opinion regarding the meaning of the order perceived in the overall constitution of the world and the apparent teleological relationship between living creatures and their habitats. We shall see many foreshadowings of modern 'Anthropic' arguments.

The Strong Anthropic Principle of Carter has strong teleological overtones. It suggests that 'observers' must play a key role in (if not be the goal of) the evolution of the Universe. This type of notion was extensively discussed in past centuries and was bound up with the question of evidence for a Deity. The search for supporting circumstantial evidence focussed primarily upon the biological realm. Indeed, to such an extent did organic analogies permeate the ideas of most Greeks that the entire universe was viewed as an organism wherein the constituent parts were constantly adjusting for the benefit of the whole and in which the lesser members were meaningful only through their function as part of the whole. The most notable supporter of such a view, whose ideas were to dominate Western thought for nearly two thousand years, was Aristotle. He was aware that any phenomenon could be associated with various types of cause, among them an 'efficient' cause (which is what modern physicists would call a 'cause'). But Aristotle did not believe one could claim a true understanding of any natural object or artefact unless one knew also its 'final cause'—the end for which it exists. This he believed to be the pre-eminent quality of things. Rival philosophers denied the relevance of such a notion and even Aristotle's pupils occasionally urged moderation in the deployment of final causes as a mode of explanation. It was, unfortunately, apt to produce 'laws' of Nature that tell us things are as they are because it is their natural place to be so!

Aristotle's ideas emerge in Western culture through the channel of medieval scholasticism. Scholars like Aquinas realized the power of teleological reasoning as support for an *a posteriori* 'Design Argument' for the existence of a Deity to whom the 'guidedness of things' might be attributed.

Broadly speaking, the Greeks viewed the world as an organism, a view based in part upon the analogy between the natural world and human society. The renaissance view which superseded the Greek view was no less analogical but the paradigm had changed from the organic to the mechanical. The new picture of the clockwork 'watch-world' displayed both the religious conviction in a created order for the world and the desire to find a Creator playing the role of the watch-maker. Wheras the teleological view accompanying the organic world-picture supported a general 'guidedness of things', the element of design in the mechanical picture was evidenced by the God-given intrinsic properties of things and the regularity of the laws of Nature. This development leads us to draw a distinction between *teleological* arguments—which argue that because of the laws of causality order must have a consequent purpose, and *eutaxiological* arguments—which argue that order must have a cause, which is planned. Whereas teleological arguments were based upon the notion that things were constructed for either our immediate benefit or some ultimate end, the eutaxiological arguments point just to their co-present, harmonious composition. There is a clear distinction: the intricate construction of a watch can be appreciated without knowing anything of the 'end' for which it has been made. This important distinction, and the terminology, was introduced by Hicks[79] in 1883.

The growth of design arguments was, of course, accompanied by the efforts of persuasive and eloquent dissenters to discredit the notion of premeditated design in every or any facet of the natural world. Many of these expressions of scepticism have proven to be overwhelmingly compelling in the biological realm where environmental adaption is now seen to play a key role through the mechanism of natural selection. However, when originally proposed they fell largely upon deaf ears in the face of an impressive array of observational data marshalled in support of 'design'. Scientists rarely take philosophers seriously and they did not often do so in these matters either. One of the strengths of the teleological argument for the layperson is its compelling simplicity; for as one nineteenth-century reviewer remarked, 'Imagine two men debating in public, one affirming and the other denying that eyes were intended to see with'. Commonsense superficially appears to affirm the teleological view very convincingly. Closer examination reveals that the argument contains all manner of hidden assumptions and associations, not least of which is a confusion between the ideas of purpose and function. The eutaxiological argument so popular with Newton and his disciples, on the other hand, *is* logically simpler than the teleological one and hides no linguistic subtleties; but to appreciate the existence of the mathematical beauty and harmony it exhibits and verify the examples cited in support of its claims requires considerable scientific knowledge. For this reason the logically simpler, but conceptually more difficult and more interesting, eutaxiological arguments appealed less to the popular mind. The eutaxiological Design Argument is most similar to the Weak Anthropic Principle. Teleological Design Arguments are analogous to the Final Anthropic Principle, and the Strong Anthropic Principle has something in common with both forms of Design Argument. As a rule, teleological arguments go hand in hand with a holistic, synthetic and global world view whilst the eutaxiological approach is wedded to the local and analytic perspective that typifies modern physics. To those brought up with the modern scientific method and its emphasis upon concepts like verification, experiment, falsification and so forth, it is surprising that science made as much progress as it did when inbred by teleological ideas. Yet it is clear that even the naivest Design Arguments, unlike the philosophical objections to them, were steeped in observations of the natural world. Indeed, Darwin attributes much of his initial interest in the problem of natural adaption to William Paley's meticulous recording of design in the plant and animal kingdoms. There are other striking examples of teleological reasoning producing significant advances in experimental and theoretical science; for example, Harvey's discovery of the human circulatory system, Maupertuis' discovery of the Principle of Least Action and von Baer's discovery of the mammalian ovum.

We shall see that the simpler teleological arguments concerning biological systems were supplanted by Darwin's work, but the system of eutaxiological arguments regarding 'coincidences' in the astronomical make-up of the Universe and in the fortuitous form of the laws of Nature were left unscathed by these developments and it is these arguments that have evolved into the modern Anthropic Principles. But careful thinkers would not jump now so readily to the conclusions of the early seekers after Design. The modern view of Nature stresses its unfinished and changing character. This is the real sense in which our world differs from a watch. An unfinished watch does not work and the discovery of time's role in Nature led to an abandonment of Design arguments based upon omnipresent harmony and perfection in favour of those that concentrated upon current co-present coincidences. The other modern view that we must appreciate is that we have come to realize the difference between the world as it really is ('reality') and our scientific theories about it and models of it. In every aspect our physical theories are approximations to reality, they claim merely to be 'realistic' and so we hesitate to draw far-reaching conclusions about the ultimate nature of reality from models which must be, at some level, inaccurate descriptions of reality. Scientists have not always recognized this, and some do not even today. We see good examples of the consequences of this weakness when we look back at the religious fervour with which Newton's equations of motion and gravitation were regarded by those eighteenth-century scientists intent upon demonstrating that God, like Newton, was also a mathematician. Whilst this group were claiming that the constancy and reliability of the laws of Nature witnessed a Creator, another was citing the breakdown of their constancy, or miracles, as the prime evidence for a Deity.

Our treatment of these questions regarding 'design' will be largely chronological and our aim is to chart the history of ideas concerning design and teleology and to bring into focus the similarity between these ancient ideas and the way modern 'Anthropic' arguments are framed. The Anthropic Principle, we shall argue, is a consequence of a certain symmetry in the history of ideas. We shall also see that many other contemporary issues that today are tangent to the Anthropic Principles were also associated with Design Arguments of the past. For example, the question of the plurality of worlds and the construction of proofs of the existence of God (or gods), the uniqueness of man in anthropocentric Christian teleology and the logical status of our perceptions of the natural world were all of continual fascination. There is also a detectable and recurrent trend revealed by our study: students of Nature build a model to describe its workings based on observations; if this description is successful the model becomes an article of faith, some aspect of absolute truth comes to be taken as embodied within it. The descriptive model then becomes almost an idol of worship and a proliferation of Design Arguments arise as expressions of a faith that would claim no comparable or superior descriptions could exist (the fate, perhaps, of a 'paradigm' in ancient times). Thus the modern anthropic principles can be seen partly as natural consequences of the fact that current physical theories are extremely successful. This success is itself still a mystery; after all there is no obvious reason why we should find ourselves able to understand the fundamental structure of Nature. It is also, in part, a consequence of the fact that we have found Nature to be constructed upon certain immutable foundation stones, which we call fundamental constants of Nature. As yet, we have no explanation for the precise numerical values taken by these unchanging dimensionless numbers. They are not subject to evolution or selection by any known natural or unnatural mechanism. The fortuitous nature of many of their numerical values is a mystery that cries out for a solution. The Anthropic Principle is but one direction of inquiry, albeit, as we shall now see, a surprisingly traditional one.

2.2 The Ancients

> You all know the argument from design: everything in the world is made just so that we can manage to live in the world, and if the world was ever so little different, we could not manage to live in it. This is the argument from design.
>
> B. Russell

Our inquiry into the Western predecessors of the modern Anthropic Principle begins on the Mediterranean island of Ionia during the sixth century BC within a culture that valued both curiosity and abstraction for their own sakes. Here, a tiny society nurtured some of the first natural philosophers to pose abstract problems completely divorced from any technological, nautical, agricultural or authoritarian stimuli. Their primary goal was to elucidate the primary forms and functions at the root of all natural phenomena. To realize that ambition they had to understand both the nature of man and the structure of his environment.

Anaxagoras of Clazomenae[1] (500–428 BC) is a pivotal figure, a mediator between the ancient Ionian philosophical tradition and the emergence of the Greek tradition. In 480 BC he migrated to Athens, probably as a member of Xerxes' militia, and there remained for thirty years as the first teacher of philosophy among the Athenians. Eventually, like Socrates, his career there was to end with charges of heresy; but unlike his famous successor he chose to leave, and fleeing to Ionia, worked there for a further twenty-five years.

Unfortunately we possess only fragments of Anaxagoras' writings in their original form[2] and these seem to be of an introductory and general nature, but later writers provide sufficient commentary for a fragmentary 'identikit' portrait of his ideas to be composed. Both Plato and Aristotle regard him as the first to attribute the evident structural harmony and order in Nature to some form of intelligent design plan rather than the chance concourse of atoms. Since Anaxagoras appears to be first of the known pre-Socratics to dwell upon the presence of order in Nature, it is perhaps no surprise that he was among the first to attempt to explain this observation by some primary cause. Anaxagoras sought some all-embracing dynamical influence which would provide him with an explanation for the mysterious harmony he saw about him. He believed the Universe and all matter to have always existed, at first a mindless confusion of infinitesimal particles, but destined to become ordered by the influence of a cosmic 'Mind'. This 'Mind' (*νους*) intervened to eradicate the state of primeval chaos by the induction of a vortical motion in space[3], which first led to a harmonious segregation of natural things and then slowly abated leaving quiescence, harmony and order. The rotation of the heavenly bodies in the solar system remain as the last vestige of the action of cosmic 'Mind'. Anaxagoras aims to explain the orderly motion and arrangement of matter by some subtle and fluid entity which exercises a guiding influence upon the Universe like a man's mind controls his body. These ideas are relevant because they signal the first introduction of 'Mind' in conjunction and in competition with 'Matter' for the explanation of phenomena; a problem to be much discussed by subsequent generations of philosophers and scientists. Our interest is attracted by this simple feature of his thinking because it forges a link with later Platonic and Aristotelian ideas.

Unfortunately, if the extant writings provide a fair sample, *νους* appears to have been a rather vaguely defined entity. It is employed to order all things initially, but thereafter plays no direct role in the temporal development of things nor is it ever used to explain the specific order and design displayed by an individual object or organism. Anaxagoras' description places its influence at the boundary of the Universe, its role cosmological and metaphysical,

And what was to be, and what was and is not now, and what is now and what will be—all these mind ordered.[4]

This initial and purposeful cause contrasts sharply with the metaphysical edifices that were constructed later by Plato and Aristotle. The latter postulated an 'end' (τελος), neither personal nor purposefully goal-directed, to which phenomena were magnetically directed. Anaxagoras' lack of a teleological emphasis provokes criticism from Aristotle who highlights what appears to moderns the plain common sense of the Anaxagorean view. The disagreement between Anaxagoras and Aristotle is interesting because it will appear again and again through the centuries, albeit suitably camouflaged by the though-forms and categories of contemporary thinking,

Now Anaxagoras says that it is due to his possessing hands that man is of all things the most intelligent. But it may be argued that he comes into possession of hands because of his outstanding intelligence. For hands are a tool, and Nature always allots each tool, just as any sensible man would do, to whosoever is able to make use of it[5]

The root of Aristotle's discontent with Anaxagoras is a suspicion that his predecessor was merely advocating a pre-Socratic version of the 'God-of-the-Gaps' methodology in his approach to the natural world. 'Mind' appears only as a form of metaphysical mortar to fill the gaps and cracks of ignorance in his otherwise entirely deterministic world model. For, Aristotle claims

Anaxagoras uses mind as a theatrical device for his cosmogony; and whenever he is puzzled over the explanation of why something is from necessity, he wheels it in; but in the case of other happenings he makes anything the explanation rather than mind.[6]

This criticism had in fact been voiced in a disconsolate commentary a little earlier by Socrates, who describes how objections slowly dawned upon him as he read one of Anaxagoras' books in search of ideas on design in the Universe. He recalls the moment of anticlimax vividly,

Well, I heard someone reading once out of a book, by Anaxagoras he said, how mind is really the arranger and cause of all things; I was delighted with this cause, and it seemed to me in a certain way to be correct that mind is the cause of all, and I thought if this is true, mind arranging all things in places as is best. If, therefore, one wishes to find out the cause of anything, how it is generated or perishes or exists, what one ought to find out is how it is best for it to exist or to do or feel everything. . . . I was glad to think I had found a teacher of the cause of things after my own mind in Anaxagoras. . . . For I did not believe that when he said all this was ordered by mind, he would bring in any other cause for them that it was best that they should be as they are. . . . I got his books eagerly. . . . How high I soared, how low I fell! When as I went on reading I saw the man using mind not at all; and stating no valid cause of the arrangement of all things, but giving airs and ethers and waters no causes, and many other strange things.[7]

Whilst these earliest notions concerning order and motion were being incubated, a Sicilian contemporary, Empedocles of Argigentum (492–435 BC), was developing some radically different ideas about the origin of ordered organic structures and their survival over the course of time. Unlike many of his contemporaries, Empedocles was a keen and careful observer of Nature and despite sporadic delusions of divinity combined this with the general study of magic, poetry and medicine. His key insight was to intertwine the notions of change and temporal evolution with physical processes rather than conceive of them possessing some time-invariant meaning. These evolutionary processes he imagined to be somehow connected with the presence of order and design in Nature. In modern biological parlance we would say that he proposed the mechanism of 'normalizing selection'. Initially, creatures of all possible forms and genetic permutations were imagined to exist but over the passage of time only some were able to reproduce and multiply. Gradually the centaurs and half-human monsters eliminate themselves through sterility. He imagines that eventually only the ordered, and therefore 'normal,' beings survive. This type of selection only maintains an invariant species against mutant invasion and is really quite distinct from Darwin's idea of natural selection wherein no species is immune to change. Again we learn more of these ideas through Aristotle's condemnation of them; he quotes Empedocles' summary

On [the earth] many heads sprung up without necks and arms wandered bare and bereft of shoulders. Eyes strayed up and down want of foreheads. . . . Shambling creatures with countless hands. . . . While others, again arose as offspring of men with the heads of oxen, and creatures in whom the nature of women and men was mingled, furnished with sterile parts.[8]

Parmenides (*c*.480 BC) the founder of the school of Elea in Southern Italy was one of the earliest logicians. Although he seems to have written in verse, it is of a sufficiently prosaic nature to allow his principal theses to be extracted. He hoped to explain what is 'intelligible' and wanted to show it was impossible to make a negative existential judgement. Parmenides claimed that a 'many-worlds' interpretation of nature is necessary because of the non-uniqueness of the subjective element in our perception and understanding of the world. As a corollary to this he maintained that what is inconceivable must actually be impossible—empty space cannot exist! Over two thousand years later these ideas will appear in a new guise in debates concerning the role of the observer in quantum theory and the theory of measurement.[9] The more immediate, but no less important consequence of these ideas was the early atomists' abandonment of trust in the senses as a certain and invariant gauge of world structure. In order to avoid this awkward perceptive subjectivity they sought objective reality in imperceptible 'atomic' microphenomena that they believed to be independent of the observer and absolute in character.

Socrates (470–399 BC) and his student Plato (427–347 BC) later reacted against this trend towards purely materialistic explanations of natural phenomena and attempted to show that material order not only sprang from 'Mind' but was actively sustained by it. Plato argued that because matter cannot induce motion itself, the observed presence of motion is evidence of a mental presence and Cause underpinning the whole natural world. He also conceived of a particular hierarchical cosmological model exhibiting this doctrine. In the beginning the outer sphere of his hierarchical universe was perturbed into motion by an obliging deity and thereafter remained in ordered motion and displayed a completely invariant structure. In the '*Laws*' this regular structure is cited as evidence of the gods. For, when asked how one might prove the existence of the gods, Cleinas replies with one of the most explicit early design arguments:

> How? In the first place, the earth and the sun, and the stars and the Universe, and the fair order of the seasons, and division of them into years and months, furnishes proofs of their existence.[10]

However, this appeal to astronomical phenomena has a slightly hollow ring to it in the light of Socrates' attitude towards all experimental philosophy and astronomy. We see that he was aware of the ability of 'physical philosophers' to provide many different but equally plausible explanations of a single observation but has no notion that perhaps further observations might narrow down the number of 'conflicting opinions':

> With regard to astronomy Socrates considered a knowledge of it desirable to the extent of determining the day of the year or of the month and the hour of the night; but as for learning the course of the stars, [he regards] occupying oneself with the planets or inquiring about their distance from the earth or about their orbits or the causes as a waste of time. He dwelt on the contradictions and conflicting opinions of the physical philosophers ... and, in fine, he held that speculators on the Universe and on the laws of the heavenly bodies were hardly better than madmen.[11]

Plato opposed contemporary ideas that attempted to explain the observed structures and contrivances in Nature as a result of either chance or mechanism, and this opposition was grounded on the evidence for design in the natural world. He preferred a picture of the Universe as an organic and teleologically ordered structure.

Socrates gives the first clear statement of an anthropocentric design argument with a distinctly eighteenth-century flavour to it when he is reported by Xenophon extolling the human eye as a proof of the wisdom of the gods:

> But which seems to you most worthy of admiration Astrodemus? The artist who forms images devoid of motion and intelligence, or who had skill to produce animals that are endued, not only with activity, but understanding? . . . But it is evidently apparent that he who at the beginning made man endued him with senses because they were good for him . . . Is not that providence, Aristodemus, in a most eminent manner conspicuous, which because the eye of man is delicate in its contexture, hath therefore prepared eyelids like doors, whereby to screen it, which extend themselves whenever it is needful, and again close when sleep approaches? . . . Is it not to be admired . . . that the mouth through which the food is conveyed should be placed so near the nose and eyes as to prevent the passage unnoticed of whatever is unfit for nourishment? And cans't thou still doubt Aristodemus, whether a disposition of parts like this should be the work of chance, or of wisdom and contrivance.[12]

Another very early commentator on the beneficial and superficially purposeful contrivance of natural things toward our perennial well-being was the Cretan philosopher, Diogenes (400–325 BC). Working about a century after Anaxagoras, he appears to be one of the earliest thinkers who appealed to a teleological principle behind natural phenomena on the basis of their optimal arrangements. In particular, he was impressed by the regular cycle of the seasons,

> Such a distribution would not have been possible without Intelligence, that all things should have their measure: winter and summer and night and day and rain and winds and periods of fine weather; other things also, if one will study them closely, will be found to have the best possible arrangement.[13]

He claims that 'air' must be this ordering 'Intelligence' because 'man and the other animals that breathe live by air . . .'.[14]

The earliest opponents of these teleological notions were Democritus (450–? BC) and Leucippus of Elea (440–? BC). Leucippus appears as a rather obscure fifth-century figure reputed to have founded the school at Abdera in Thrace where Democritus was born. Again our knowledge of their work derives principally from secondary sources—through Aristotle, Epicurus, and others. Leucippus proposed the early 'atomic' theory which was then developed more 'scientifically' by Democritus before being tenuously extrapolated into the realm of ethics and philosophy by

Epicurus. Their development of the mechanism of causation and an atomic view of the world was entirely ateleological; the only causes admitted were atomic collisions (although later Epicurus and Lucretius were to appeal to a mysterious intrinsic atomic property, 'swerve', which enabled atoms to collide). As with Empedocles we see inklings of some parallels with modern evolutionary biology and the 'many worlds' interpretation of quantum theory in their writings. Democritus understands the link between life and its local environment and has the notion of an ensemble of planetary systems:

> There are worlds infinite in number and different in size. In some there is neither sun nor moon, in others there are more than one sun and moon. The distance between the worlds are unequal, in some directions there are more of them ... Their destruction comes about through collision with one another. Some worlds are destitute of animal and plant life and of all moisture.[15]

The pre-eminent proponent of a teleological world view amongst the ancients was Aristotle (384–322 BC) and his commentary on the ideas of others provides a valuable source of information. The Stagirite's teleological view was to become tremendously influential, some would claim out of all proportion to its profundity, because it became amalgamated with the Judaeo-Christian revelation in the Scholastic synthesis. By this indirect route his ideas were able to shape the thought of Western Europe for nearly two thousand years. Unlike Socrates and Plato, Aristotle was not an Athenian. His father was a physician at the court of the Macedonian royal family and his keen observation of and life-long interest in flora and fauna may have derived from early paternal influence. Whilst still a teenager he went to Athens to study as a student of Plato at the Academy. There he worked for twenty years, principally on ethics, mathematics, politics and philosophy, but then left for the coastal region of Asia Minor where he rekindled his interest in observation through studies in zoology and biology. So much did he learn during that period that on his return to Athens he was able to establish a thriving school of botanical and biological investigation which laid the foundations of modern study in these disciplines.

Aristotelian science was based upon presupposition of an 'intelligent natural world that functions according to some deliberate design'. Its supporters were therefore very critical of all those pre-Socratic thinkers who regarded the world structure as simply the inevitable residue of chance or necessity. Aristotle's own detailed observational studies in botany, biology and zoology led him to take up the organic analogy as the most fruitful description of the world and he regarded it as superior to the mechanistic paradigm.

In his *Metaphysics*, Aristotle works through the ideas of earlier philosophers and rejects them one by one. He strongly opposes a recurrent idea, held for example by the Atomists, that a thing is explained when one knows what it is made of. For, he argues, its material composition provides us with its 'Material Cause', but to explain it completely we require an understanding of three further 'Causes'. A 'Formal Cause' must be identified. This relates to the form or pattern intrinsic to the object which prevents it from behaving like another; for example, it distinguishes sculptures from lumps of unformed metal (or at least it did!). Next, the 'Efficient Cause' should be recognized as the agent which produces the object, transferring the mental notion of a statue from the sculptor's mind into solid material bronze; the 'Efficient Cause' is what moderns mean when they use the word 'cause'. Finally, there exists that 'Cause' which Aristotle regarded as the most important: the 'Final Cause'—the purpose for which the object exists. Even at this stage it is evident that this multiplicity of causes leads very quickly to metaphysical ideas of supreme initial causes or ultimate final ends.

The common preoccupation with the presence of order in the Universe meant there were many similarities between the cosmologies of Aristotle and Plato. Where Aristotle differed was in his attitude towards initial conditions. He argued that knowledge of the 'beginning' is not relevant to our understanding of the present configuration—that initial conditions did not matter—and furthermore, there were reasons for supposing there never was an origin in time—the natural order should be eternal and unchanging. Aristotle's cosmology was the first 'steady-state' Universe.

There, the similarity with any modern cosmological model very abruptly ends. Aristotle imagined the Universe to possess a spherical boundary with the earth resting at its centre. Surrounding the earth were a whole series of concentric shells; the three closest to the centre contained water, air and fire respectively. Now, the idea behind this hierarchical structure was to explain why, for example, flames 'naturally' rose whilst other objects, like stones, always fell to the earth. The outer shell of fire was encompassed by a succession of seven solid and crystalline spheres; they carried the Moon, Mercury, Venus, the Sun, Mars, Jupiter, Saturn and finally the fixed stellar background. This outer stellar sphere was endowed with a dynamical rotation which it is communicated to the inner spheres and thereby to the planets themselves.[16]

Aristotle's guiding principle was that the ultimate meaning of things was to be divined from their 'end' (τελος) rather than their present configuration—that is, by learning of their final rather than their material causes. This 'end' was the most perfect and fitting purpose,

> ... it belongs to physical science to consider the purpose or end for which a thing subsists. The poet was led to say 'An end it has for which it was produced'. This is absurd, for not that which is last deserves the name of end, but that which is most perfect.[17]

Although, as we saw above, Aristotle credits Anaxagoras for germinating this view, he upbraids him strongly for employing it in so limited and sterile a fashion. In contrast, he energetically develops his own scheme of final causes in combination with the Platonic teleology and uses it to interpret his own detailed observations of Nature. Although he is not often credited for it, he carried through this programme with something of the modern scientific philosophy:

The actual facts are not yet sufficiently made out. Should further research ever discover them, we must yield to their guidance rather than to that of theory; for theories must be abandoned, unless their teachings tally with the indisputable results of observation.[18]

He is clearly anxious to derive support for his teleological ideas from observational facts and wants to avoid the approach of those of his predecessors who have adopted the methodology of armchair natural philosophers.

From the idea of a 'Final Cause' there emerged the Aristotelian idea of an internal perfecting principle or 'entelechy' which directs things toward some terminal point characterized by its unique harmony. In any individual object all its sub-components are united for its greatest benefit and are coherently organized with this 'perfect' end in view. The evidence for such an opinion, he argues, is much more readily obtained from astronomical observations than from biological ones. For, in the former system, the time-scale over which significant changes occur is so much longer:

For order and definiteness are much more plainly manifest in the celestial bodies than in our own frame; while change and chance are characteristic of the perishable things of earth. Yet there are some who, while they allow that every animal exists and was generated by nature, nevertheless hold that the heaven was constructed to be what it is by chance and spontaneity; the heaven, in which not the faintest sign of haphazard or of disorder is discernible! Again whenever there is plainly some final end to which a motion tends, should nothing stand in the way, we always say that such final end is the aim or purpose of the motion and from this it is evident that there must be a something or other really existing, corresponding to what we call by the name of Nature.[19]

Aristotle also displays an objectivity and breadth of view in his discussion of the limitations and conceivable objections to his teleology that was to prove all too rare in the later work of his many followers. He realizes, for example, that development could play an important role in generating organic structures:

In plants, also there is purpose, but it is less distinct; and this shows that plants were produced in the same manner as animals, not by chance, as by the union of olives upon grape-vines. Similarly, it may be argued, that there should be an accidental generation [or production] or the germs of things; but he who asserts this subverts Nature herself, for Nature produces those things which, being continually moved by a certain principle contained in themselves, arrive at a certain end.[20]

and that necessity must be considered as an influence upon their development

We have ... to inquire whether necessity may not also have a share in the matters and it must be admitted that these mutual relations could not from the very beginning have possibly been other than they are.[21]

On another occasion he recapitulates the antiteleological position of the atomists in a convincing fashion:

But here a doubt is raised. Why, it is said, may not nature act without having an end, and without seeking the best of things? Jupiter, for instance, does not send rain to develop and nourish the grain, but it rains by a necessary law; for in rising, the vapour must grow cool, and the cooled vapour becoming water must necessarily fall. But if, this phenomenon taking place, the wheat profits by it to germinate and grow, it is a simple accident. And so again, if the grain which someone has put into the barn is destroyed in the consequence of rain, it does not rain apparently in order to rot the grain, and it is a simple accident if it be lost. What hinders us from saying as well, that in nature the bodily organs themselves are subject to the same law and that the teeth, for instance, necessarily grow ... What hinders us from making the same remark for all the organs where there seems to be an end and a special destination.[22]

Whereas Plato had been interested in order and structural design within the Universe principally as manifestations of its static, permanent and unchangeable nature, Aristotle's view was clearly more dynamic. The Aristotelian world was endowed with a process of temporal evolution acting solely for the sake of the entities finally evolved.

Following the death of Aristotle, peripatetic thinking was dominated for a period of thirty-five years by Tyrtamus of Eresos (372–287 BC). Now regarded as one of the founders of systematic botanical study, Tyrtamus is better known to us by his nickname 'Theophrastus' which he received from Aristotle because of his stimulating conversation. Like others before him, Theophrastus was struck by a dichotomy in his experience. On the one hand he was conscious of the orderliness of his mental processes whilst on the other he perceived a natural world of enormous complexity. He felt that if some link could be forged between these disjoint areas of experience then light might be shed upon them both.

Despites his long association with Aristotle, first as a fellow student of Plato at the Academy and then as a co-worker at the Lyceum, he was critical of his master's teleological mode of thinking and recognized the strongly subjective elements that were incorporated in its application:

As regards the view that everything has a purpose and nothing is in vain, first of all the definition of purpose is not so easy, as is often said; for where should we begin and where decide to stop? Moreover, it does not seem to be true of various things, some of which are due to chance and others to a certain necessity, as we see in the heavens and in the many phenomena on earth.[23]

He then goes on the give many examples of natural phenomena, like drought, flood, and famine, which yield no discernible end, interpreting them as casting doubt upon Aristotle's perfecting principle as a useful practical guide into the nature of things. He concludes that natural science will only make sure and sound progress if it moderates[24] its appeal to final causes, for

We must try to set a limit to the assigning of final causes. This is the prerequisite for all scientific enquiry into the universe, that is into the conditions of existence of real things, and their relations with one another.[23]

The contemporary counter to the peripatetic school's teleology was the radical alternative of Epicurus of Samos (341–270 BC) and his followers. Following in the footsteps of Democritus and Leucippus, these later atomists emphasized the importance of assuming a complete state of statistical disorder at the moment of the World's creation. They claimed this chaotic initial state subsequently evolved by natural forces into an ordered system characterized by regular and steady rotations. They argued that the infinite time allowed for creation makes it inevitable that it should eventually develop into a stable configuration capable of remaining in a constantly ordered state. The Epicureans were, of course, anxious to scotch any notions of supernatural causation or the appeal to any entity who controls or ordains events. Interestingly, no useful scientific structure was erected upon this materialistic foundation because Epicurus had a very low view of mundane scientific investigation. Indeed, he excluded many of its basic tools—logic, mathematics, grammar and history—from his school's curriculum. He was particularly hostile to the study of astronomy because celestial phenomena seemed to him to admit of so many equally consistent and indistinguishable explanations:

First of all then we must not suppose that any other object is to be gained from the knowledge of the phenomena of the sky, whether they are dealt with in connection with other doctrines or independently, than peace of mind and a sure confidence, just as in all other branches of study.[25]

The most remarkable spokesman for the Epicurean position was the Roman poet Titus Lucretius Carus (99–55 BC). His great poem *De Rerum Natura*[26] aimed to bury all superstitious speculation and philosophical dogma by outlining the vast scope of a purely materialistic doctrine. It reveals an uncanny intuition regarding the future conceptual development of physics and displays such a good knowledge of flora and fauna that one is led to wonder whether Lucretius wrote other prosaic and systematic studies of these subjects which are now lost to us.

Lucretius believed life to have originated at some definite moment in the past by natural processes but that the created beings included 'a host of monsters, grotesque in build and aspect' who were subsequently eliminated by their sterility:

In those days, again, many species must have died out altogether and failed to reproduce their kind. Every species that you now see drawing the breath of the world survived either by cunning or by prowess or by speed. In addition, there are many that survive under human protection because their usefulness has commended them to our care.[27]

As his poem unfolds the entire materialistic methodology is eloquently restated and the logical difficulty inherent in a teleological approach is forcefully presented to his patron, Memmius: to put it bluntly, he claims that teleologists like Aristotle have simply been putting the cart before the horse:

There is one illusion that you must do your level best to escape—an error to guard against with all your foresight. You must not imagine that the bright orbs of our eyes were created purposely, so that we might be able to look before us . . . and helpful hands attached at either side, in order that we might do what is needful to sustain life. To interpret these or any other phenomena on these lines is perversely to turn the truth upside down. In fact, nothing in our bodies was born in order that we might be able to use it, but the thing born creates the use . . . The ears were created long before a sound was heard . . . They cannot, therefore, have grown for the sake of being used.[28]

Yet this critical approach ground to a temporary halt with Lucretius whilst the teleological aspect of Aristotle's philosophy he criticized so strongly, being more adaptable to the theistic Islamic and Christian cultures, was to grow in influence and extent.

Another group who inherited some of Aristotle's teleological ideas were the Stoics; a school which was founded by Zeno of Citium (334–262 BC) during the fourth century BC and which took its name from a painted corridor on the north side of the market place in Athens where it was the custom of the school to meet for discussion. Teleological ideas appear in Stoic physics under the guise of 'Providence'. For the Stoics this concept embodied the notion that all was the best; the idea was carefully gauged to temper the harsher Stoic dictum of 'fate' within which was enshrined the absolute rule of causality. They replaced Aristotle's infinitely old, 'steady-state' Universe with one possessing a cyclic recurrence.[29] Their conviction regarding the innate order and rationality of Nature, which became the basis of their ethics, made the Stoics fervent supporters of the cosmological Design Argument in all its forms. Although they rejected the mechanical world-view in favour of a more Aristotelian organic analogy, they nevertheless developed their Design Arguments via the analogy between the workings of the world and familiar mechanical models. The Roman lawyer, orator and popularizer of Greek philosophy, Marcus Cicero, records that[30]

> The Stoics, however, most assuredly did consider man to be at the very apex of the hierarchy of beings and felt that the rest of the Universe was geared to his benefit.

Cicero (106–43 BC) himself devotes much of his famous work *De Natura Deorum* to arguments for the existence of the gods drawn from the beneficial contrivance of the world. He also signals the start of a tendency for teleological design arguments to be employed to establish not only the existence but also the character traits of a deity or deities. *De Natura Deorum* describes the conversations between two disciples of Plato, namely Cotta and Cicero; a Stoic, Balbus; and an Epicurean atomist, Velleius. As might be anticipated from our discussion so far, Balbus provides various teleological arguments for the gods' existence and is backed up by the Platonists in the face of Velleius' continuous opposition. For example, Balbus criticizes the Epicurean view that things could have fallen out so nicely just by chance and reveals a new type of numerical perspective on the likelihood of ordered configurations arising spontaneously:

> Can I but wonder here that anyone can persuade himself that certain solid and individual bodies should move by their natural forces and gravitation in such a manner that a world so beautiful adorned should be made by their fortuitous concourse. He who believes this possible may as well believe, that if a great quantity of the one and twenty letters, composed either of gold or any other matter, were thrown upon the ground, they would fall into such order as legibly to form the '*Annals of Ennius*'. I doubt whether fortune could make a single verse of them ... Thus if we every way examine the Universe, it is apparent from the greatest reason that the whole is admirably governed by a divine providence for the safety and preservation of all beings.[31]

These arguments were inspired by a lost work of Aristotle (*De Philosophia*) in which he reportedly argued that our familiarity with the remarkable aspects of Nature has removed our sense of wonder at them. If we had spent our lives underground and then suddenly came to the surface we would be so struck by the structure of the heavens and the beauty of the Earth that we would be inevitably and 'immediately convinced of the existence of the gods and that all these wonders were their handiwork'.[32]

Cicero couples a purely mechanical view of the world with a good anatomical knowledge and even gives the now classic design argument based upon the watch analogy that was used so persistently by Boyle, Niewentyt, Paley and others over fifteen hundred years later

When we see some example of a mechanism, such as a globe or clock or some such device, do we doubt that it is the creation of a conscious intelligence? So when we see the movement of the heavenly bodies, ... how can we doubt that these too are not only the works of reason but of a reason which is perfect and divine?[33]

These and many other examples adorn an argument for the 'gods' that is eutaxiological rather than teleological in character; that is, it is based upon the presence of discernible order and mutual harmony in Nature rather than the recognition of any conscious or unconscious anthropocentric purposes. It is a type of argument that was to be repeated regularly in future centuries.

Another, whose ideas were later to form the basis of many eighteenth- and nineteenth-century treatises on the 'Wisdom of God' as evidenced by anthropocentric teleology, was the Greek physician Galen (131–201). Although Galen was eclectic in his philosophical outlook he clearly favoured the Aristotelian picture as the most natural backdrop for his monotheistic views. He developed the doctrine of Final Causes in a more specific and teleological manner than Cicero, arguing that the purpose of the deity could be ascertained by detailed inspection of his assumed works in Nature. Specifically, his study of the specialized design of the human hand was a classic piece of anatomical analysis that became the basis of Bell's Bridgewater Treatise on the teleological aspects of this organ over sixteen hundred years later, so little were later workers able to add to his insights. Of the human body he writes:

Let us, then, scrutinize this member of our body, and inquire, not simply whether it be in itself useful for all the purposes of life and adapted to an animal endued with the highest intelligence, but whether its entire structure be not such that it could not be improved upon by any conceivable alteration.[34]

His approach was wholly teleological and maintained that all the bodily processes were divinely and optimally planned in every respect. This anthropocentric tenor also runs through the encyclopaedic natural history of the Roman, Pliny (23–79), who also usually described nature by drawing on its relation to man:

Nature and earth fill us with admiration ... as we contemplate the great variety of plants and find that they are created for the wants or enjoyments of mankind.[35]

Despite their great administrative, legal and military skills the Romans produced little in the way of lasting abstract ideas. The most relevant character to our study is perhaps Boethius (470–525) who mediates the transition from Roman to Scholastic thinking. For many years a prominent Roman statesman and philosopher he was to write his influential manual[36] *The Consolation of Philosophy* whilst incarcerated in Pavia gaol awaiting execution. This work is one of the few threads of contact between classical learning and the Dark Ages and is written in an unusual medley of poetry and prose (the author speaks in prose whilst philosophy replies in verse). Boethius' support of the teleological doctrine of Final Causes is clear from the outset of his work where he hails Socrates, Plato and Aristotle as the only true philosophers and sets them in opposition to the spurious Stoic and Epicurean thinkers:

Thinkest thou that this world is governed by haphazard and chance? Or rather doest thou believe that it is ruled by reason?[37]

His answer ensured that the teleological argument was handed on safely to the emerging civilizations of Northern Europe, for Boethius' book was probably the most widely read scholarly work of the medieval period. It played a major role in shaping the philosophical vocabulary and perspective of those times—it is even fabled that Alfred the Great (849–901) had it translated into Anglo-Saxon. Although the world-*view* it presents is teleological and anthropocentric through and through, the world *model* it presumes most definitely is not. Boethius saw and stated that despite the implication of final causes, the astronomical position of man was both infinitesimal and insignificant; a view that would have become familiar to his later pre-Copernican readership:

Thou hast learnt from astronomical proofs that the whole earth compared with the Universe is no greater than a point; that is, compared with the sphere of the heavens, it may be thought of as having no size at all. Then, of this tiny corner, it is only one-quarter that, according to Ptolemy, is habitable to living things. Take away from this quarter the seas, marshes, and other desert places, and the space left for man hardly even deserves the name of infinitesimal.[38]

This completes the sketch of Greek and Roman origins, showing how the Design and anti-Design arguments began there. (The dates of the principal protagonists are shown in Figure 2.1.) But, these seeds would have fallen on stony ground had it not been for their adoption by the inheritors of an entirely different tradition.

During the next seven hundred years Greek learning was first perpetuated by the Arabic schools who translated many of the early texts. This Eastern influence reached its zenith during the tenth century and through it Aristotelian ideas slowly diffused into the European culture to be moulded into a Christian form by Aquinas as easily as it was fitted into the Muslim perspective of the early Arabic philosophers.

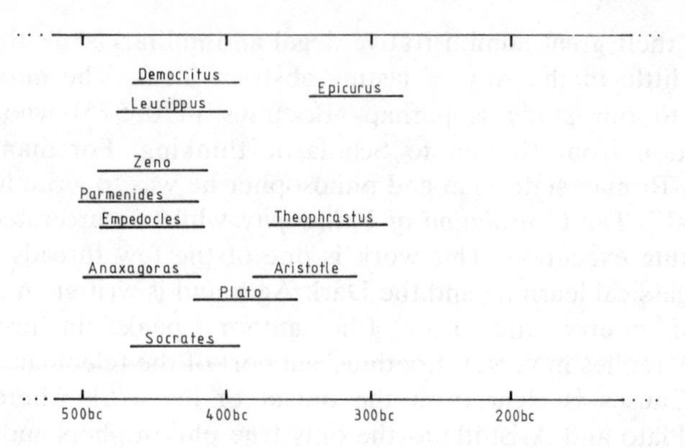

Figure 2.1. The chronology of some of the early contributors to the question of design in nature. Where precise dates of birth and death are unknown estimates have been used.

2.3 The Medieval Labryrinth

> The human imagination has seldom had before it an object so sublimely ordered as the medieval cosmos ... it is perhaps ... a shade too ordered. Is there nowhere any vagueness? No underdiscovered byways?
>
> C. S. Lewis

What characterizes the Medieval mind most uniquely for the modern spectator is its absolute respect for written authorities. All writers tried to base their works on ancient authority—most notably that of Aristotle. Also, in C. S. Lewis' words,[39] 'Medieval man was not a dreamer nor a wanderer. He was an organizer, a codifier, a builder of systems. He wanted "a place for everything and everything in the right place." Distinction, definition, tabulation were his delight.' These two powerful traits proved perfect, not only for the preservation of the ancient Design arguments, but for their subsequent elevation to the status of ecclesiastical dogma. The nearest one gets to a parallel of the atomists versus the teleologists is, at first, the division of opinion concerning whether science, religion and metaphysics should be conjoined with the blessing of the Design Argument.

Averroes of Cordova (1126–1198) was a Mohammedan member of the early Hispan-Arabic school of philosophy and medicine who opposed such a scholastic synthesis. He wanted to separate the basis of religion from experimental science and logic because of the pseudo-conflicts he saw inherent in such a union. He still maintains a teleological view but it is only partially anthropocentric, for he feels it is unreasonable to say that all Nature exists solely for the luxury of humankind:

Why did God create more than one sort of vegetable and animal soul? The reason is that the existence of most of these species rests upon the principle of perfection (completeness). Some animals and plants can be seen to exist only for the sake of man, or of one another; but of others this cannot be granted, for example, of the wild animals which are harmful to men.[40]

Looking to another culture one finds the Jewish rabbi Maimonides (1135–1204), an astronomer, philosopher and physician who, like the Arabs, sought to reconcile Aristotelian philosophy with his own religious heritage.[41] This led to his construction of a Jewish Scholastic system that developed the 'proof' of God from contingent being following analogous earlier work by Avicenna (980–1037) and others. Maimonides wrote an apologetic work as a spiritual guide for atheistic philosophers entitled *Guide for the Perplexed* wherein he states an objection to anthropocentric teleology which is based on the enormous size of the Universe:

Consider then how immense is the size of these bodies, and how numerous they are. And if the earth is thus no bigger than a point relative to the sphere of the fixed stars, what must be the ratio of the human species to the created Universe as a whole? And how then can any of us think that these things exist for his sake, and that they are meant to serve his uses?[44]

By the middle of the thirteenth century the Dominican scholars, Albert the Great and Thomas Aquinas (1225–74) had completed Aristotle's conversion to Christianity. Aquinas, the 'angelic doctor', was born after the major rediscovery and translation of many of Aristotle's works into Latin and his own unique contribution was a vast unification of Aristotle's philosophy with the Judaeo-Christian doctrine of the Catholic church. The Scholastic ideal held that the nature of ultimate things was accessible to reason alone without revelation from a divine source. Therefore Scholasticism preserved a strong belief in the intrinsic intelligibility of Nature and in the presence of an underlying rationality in an age full of astrological and magical notions. Ironically, this rationality would in the future backfire against some of the more negative aspects of Scholastic dogma.

Specifically, Aquinas uses a teleological design argument for the existence of a unique God as the basis of his famous 'Fifth Way' to prove the existence of God and attributes the idea to St. John of Damascene:

The fifth way begins from the guidedness of things. For we observed that some things which lack knowledge, such as natural bodies, work towards an end. This is apparent from the fact they always or most usually work in the same way and move towards what is best. From which it is clear that they reach their end not by chance but by intention. For these things which do not have knowledge do not tend to an end, except under the direction of someone who knows and understands: the arrow, for example, is shot by the archer. There is therefore, an intelligent personal being by whom everything in nature is ordered to its end.[45]

His argument does not appeal to any specific pieces of empirical evidence or detailed examples of adaption but to a single aspect of world order—the general trend of natural behaviour.

Alongside Thomist philosophy there began to develop, through a number of eminent Franciscan friars, an approach to science that has a more modern flavour. Roger Bacon (1214–94) was the most far-sighted—and the most persecuted—of the advocates for this new emphasis. His foresight influenced many fields of learning that are today quite distinct. He argued, for example, that the use of *original* texts in historical and linguistic study was essential for scholarship whilst in the sciences he saw that useful progress could only be made through a combination of mathematical reasoning and *experimental* investigation. Yet, alongside this new and modernistic philosophy of the scientific method Bacon held what was, for his time, a typical view of final causation and mankind's pre-eminent position within the natural world:

Man, if we look to final causes, may be regarded as the centre of the world; in so much that if man were taken away from the world, the rest would seem to be all astray, without aim or purpose . . . and leading to nothing. For the whole world works together in the services of man; and there is nothing from which he does not derive use and fruit . . . in so much that all things seem to be going about man's business and not their own.[46]

The strength of his position was that he did not allow such finalistic inclinations to usurp the place of direct observations in the practice of physical science. Final causes were relegated entirely to the metaphysical domain.

Conscious of the ease with which we adopt preconceived and fallacious modes of reasoning, Bacon ear-marked four explicit sources of erroneous deduction; undue regard for established doctrines and authorities, habit, prejudice and the 'false conceit of knowledge'. Uncritical adoption of Aristotelian metaphysics in the area of physical science was clearly the paradigm for the first of these pitfalls.

The Scholastics, in addition to introducing the term 'final cause' (*causa finalis*) into philosophy, were also the first to use the appellation 'natural theology' (*theologia naturalis*) which was to prove so popular during the seventeenth and eighteenth centuries. It originates in the work of Raymonde of Sebonde (*c.* 1400), an obscure scholar who was persuaded to remain in Toulouse as the university professor of medicine, philosophy and theology whilst passing through on a journey to Paris from his home in Barcelona. His book[47] *Theologia Naturalis sive Liber Creaturarum* was clearly not wholly orthodox because it was placed on the Index in 1595, but the reasons for this are still not altogether clear. It later became influential following its translation by Montaigne in 1569 and was reprinted thereafter in France on several occasions. The author's guiding theme is the kinship of mankind with the natural world and is slightly reminiscent of St. Francis. This unity between man and his environment speaks to him of both design and a unique Designer:

> There could not be so great an agreement and likeness between man and the trees, plants and animals, if there were two designers, rulers or artificers in nature; nor would the operations of plants and trees be carried on so regularly after the manner of human operations, nor would they all be so much in man's likeness, except that He which guided and directed the operations of these trees and plants were the same Being that gave man understanding and that ordered the operations of trees which are after the manner of works done by understanding, since in trees and plants there is no reason nor understanding. And of far more strength is the oneness of matter and sameness of life in man, animals, trees and plants an evidence of the oneness of their Maker.[47]

2.4 The Age of Discovery

> Inquiry into final causes is sterile,
> and, like a virgin consecrated to God,
> produces nothing.
>
> F. Bacon

The developments heralding the birth of what has become known as the Renaissance view of the world have been exhaustively discussed by scholars. With hindsight, Nicholas Copernicus (1473–1543) appears to us a pivotal figure, the last of the Aristotelians and the harbinger of a fully mechanical model of the Universe. What is now equally clear is that his classic,[48] *De revolutionibus orbium celestium*, had negligible influence until the seventeenth century. Few copies of it were sold and even fewer read in the early years after Copernicus' death; other great events, like the Portuguese voyages of discovery, completely overshadowed it. Although Copernicus' world model was new and heliocentric, his world-view was extremely anthropocentric and he appears a little reticent about relinquishing even the physical centrality of Man, but assures us that Man's displacement is really only very slight, given the immense size of the cosmos:

> So it is also as to the place of the earth; although it is not at the centre of the world, nevertheless the distance [to that centre] is as nothing in particular when compared to that to the fixed stars.[49]

It is also interesting that Copernicus uses various tenets of Aristotelian teleology concerning the necessary harmony and order of the Universe to guide him in the construction of a purely mechanical model.

Spherical configurations were appropriate for the celestial motions because 'this figure is the most perfect of all' and the coalescence of falling bodies inevitable because 'nothing is more repugnant to the order of the whole and to the form of the world than for anything to be outside of its place'.

Following the heliocentric insights of Copernicus, a route was opened for philosophers to develop the notion of a 'plurality of worlds'. The Aristotelian cosmology could not have countenanced such an asymmetry and periodicity because of its hierarchical and geocentric structure. To the early Greeks the notion of 'many worlds' carried with it, not the more modern picture of additional solar systems and habitable planets, but rather reproductions of the entire Universe. This latter view was characteristic of the early Epicureans but the possibility of its extension into the Aristotelian cosmology was vigorously opposed by Aquinas on logical and aesthetic grounds. For, he claimed, if all worlds were similar then all bar one were superfluous, whilst if they were dissimilar then a semantic and logical contradiction has arisen because a world does not then contain all that is possible. The notion of 'multiverses' in both of the above-mentioned senses was to be an enduring consideration, generating new arguments both for and against the naive anthropocentric teleologies.

Copernicus' famous scientific successors, Galileo (1564–1642) and Kepler (1571–1630), held strong but diametrically opposed views on the subject of anthropocentric design. Whereas Galileo felt such ideas were simply unthinking manifestations of human presumption:

> We arrogate too much to ourselves if we suppose that the care of us is the adequate work of God, an end beyond which the divine wisdom and power does not extend,[50]

his contemporary, Kepler, was thoroughgoing teleologist in outlook, holding that 'all things have been made for man'. Furthermore, Kepler appealed to the obvious presence of order in the Universe to substantiate such a belief. Paul Janet,[51] a nineteenth-century French philosopher, records this amusing domestic exchange between Kepler and his wife which was recounted in Bertrand's *Les Foundateurs de l'Astronomie Moderne*:[52]

> Dost think, that if from the creation plates of tin, leaves of lettuce, grains of salt, drops of oil and vinegar, and fragments of hard-boiled eggs were floating in all directions and without order, chance could assemble them today to form a salad?' 'Certainly not so good a one' replied my fair spouse, 'nor so well seasoned as this'.

Kepler was convinced that God had created the Universe in accord with some perfect numerological or geometrical principle. In his astronomical work Kepler strove to use this Platonic conviction to search for the ultimate causes of the planetary motions.[53]

Not surprisingly, many other sixteenth-century scholars had little sympathy for this classical Design Argument drawn from the superficial order of the World. Indeed Kepler's contemporaries contrived some of the most cogent objections to teleology since those of the ancients. The French essayist Montaigne (1533–92) argued that most teleological arguments were too anthropocentric to be taken seriously and amusingly, he parodied Man's grand self-image with an ornithocentric teleology, arguing that we simply do not know for whom or what purpose natural contrivances are geared,

> Why should not a gosling say thus: All the parts of the Universe regard me; the earth serves me for walking, the sun to give me light, the stars to inspire one with their influences. I have this use of the winds, that of the waters; there is nothing which this vault so favourably regards as me; I am the darling of nature. Does not man look after, lodge, and serve me? It is for me he sows and grinds: if he eat me, so does he his fellow-man as well; and so do I the worms that kill and eat him . . . .[54]

And he uses an objection to teleology that we remember was also cited by Velleius in Cicero's *De Natura Deorum*,

> Who has persuaded himself that this motion of the celestial vault, the eternal light of these lamps revolving so proudly above his head, the awful movements of this infinite sea, were established and are maintained so many ages for his convenience and service?.[55]

More vehement was the criticism of Francis Bacon (1561–1626), one of the patrons of the modern inductive method and a pioneer in the logical systematization of scientific procedure. He felt most strongly that philosophy and theology should remain completely disjoint rather than fall confused and conjoined within some elaborate Scholastic synthesis. This made him extremely hostile to all aspects of Aristotelian science and a strong supporter of the early atomists.[56] Although Bacon certainly did not wish to deny that Nature may both possess and display some divine purpose, he objected to the use of this belief in generating teleological 'explanations' which then became intermingled with the empirical investigations of the physical sciences. His attitude towards the fruitlessness of teleological and finalistic explanations in natural science is summarized by his famous jibe,[57] which is the epigram for this section.

For Bacon, final causes have a role to play only in metaphysics. In physics, experience guides us to exclude them. With Bacon's ideas we see the beginning of a trend that has continued to the present day with most scientists *qua* scientists ignoring 'ultimate' questions; and instead, concentrating on more limited local problems and the interconnection between material and efficient causes; Bacon claims this is advantageous because,

the handling of final causes mixed with the rest in physical inquiries, hath intercepted the severe and diligent inquiry of all real and physical causes, and given men the occasion to stay upon these satisfactory and specious causes, to the great arrest and prejudice of further discovery. For this I find done not only in Plato, who ever anchoreth upon that shore, but by Aristotle, Galen and others. For to say that . . . the clouds are for watering of the earth; or that the solidness of the earth is for the station and mansion of living creatures, and the like, is well enquired and collected in Metaphysic; but in Physic they are impertinent . . . the search of the Physical Cause hath been neglected and passed in silence . . . Not because those final causes are not true, and worthy to be enquired, being kept within their own province; but because their excursions into the limits of physical causes hath bred a vastness and solitude in that track.[58]

In the course of his work Bacon isolated a number of 'idols' of natural or man-made origin which could cause us to stumble from the path to sure knowledge. Two are strikingly reminiscent of the snares pointed out by his medieval namesake: *Idola Tribus*—fallacies generically inherent in human thought, notably the proneness to perceive in Nature a greater degree of order than is actually present, and *Idola Theatri*—idols constructed around received and venerated systems of thought. The classical design argument has points of contact with each and Bacon's demarcation helps us to trace some of the psychological origins of this argument.

Yet despite the good sense of Bacon's advice, there was amongst his contemporaries a notable Aristotelian; and one whose contribution to science will be remembered after Bacon is long forgotten. William Harvey (1578–1657) made his monumental discovery of the human circulatory system by employing the very style of reasoning derided by Bacon. Harvey was not an atomist and he regarded the facts uncovered by his studies of embryology as a refutation of any scientific philosophy devoid of purpose. In his final publication[59] he claims that 'The authority of Aristotle has always had such weight with me that I never think of differing from him inconsiderably'. The way in which this respect for Aristotle was realized in Harvey's work seems to have been in the search for discernible purpose in the workings of living organisms—indeed, the expectation of purposeful activity—rather than any association in his mind with a vast labyrinth of metaphysical ideas about the structure of the World and the living organisms within it. Harvey's discovery of the human circulatory system actually arose as a consequence of his Aristotelian approach: on the one hand he wondered if the motion of human blood might be circular—with all the significance such a geometry would have for Aristotelians—whilst on the other he tried to conceive of how a purposeful designer would have constructed a system of motion. Robert Boyle records[60] a conversation in which he asked Harvey how he had hit upon such an idea as circulation. Harvey replied that when he had noticed how carefully positioned were the valves within the veins so as to allow blood to pass towards the heart but not away from it, he was

> ... invited to imagine, that so Provident a cause as Nature had not so placed so many values without Design: and no Design seem'd more possible than that, since the Blood could not well, because of the interposing valves, be sent, by the veins to the limbs; it should be sent through the Arteries and return through the veins.

Elsewhere in Harvey's writings,[61] we find even a desire to interpret the internal structure of the body as a form of mini solar system with the heart at the centre along the lines of an Aristotelian cosmology.

These motivations were clearly not the sole reason for Harvey's success. He was also among the first of a new generation of physicians[62] who did not look simply to Galen for their instruction but dissected, examined and recorded, and carried out their own experimental investigations. By his successful synthesis of teleology and experiment Harvey appears as the forerunner of a new type of teleologist, those with a special interest in the observation of the minute intricacy of Nature.

Another illustrious contemporary of Bacon who was deeply concerned with the unverifiable and imprecise nature of the foundations of all types of philosophy was the founder of modern critical philosophy, René Descartes (1596–1650). Like Galileo and many other renaissance scientists he was convinced that the primary qualities of the Universe were mathematical in nature. This led him firmly to reject final causation as a useful scientific concept because it was associated with an anthropocentric and subjective view of the world, reflecting little more than our presumption in supposing we could unravel the purposes of God. Things have many ends, Descartes says, but most of these have no interaction with Man at all:

> It is not at all probable that all things have been created for us in such a manner that God had no other end in creating them ... Such a supposition would, I think, be very inept in reasoning about physical questions; for we cannot doubt that an infinitude of things exist, or did exist, though they have now ceased to do so, which have never been beheld or comprehended by man, and have never been of any use to him.[63]

This view was reinforced by his belief that the Universe was infinite.

Descartes's approach to natural philosophy was an attempt to deduce the essence of the world structure from self-evident primary principles solely by the methods of mathematical reasoning. The Cartesian world-view was 'Deistic'; that is, it maintained that order was inherent in the

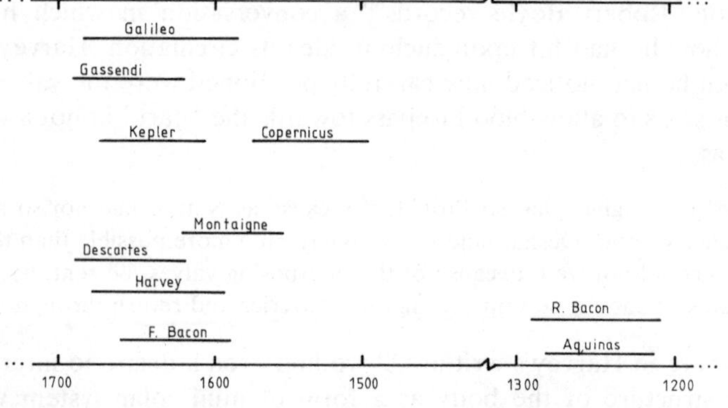

Figure 2.2. The chronology of the principal contributors to our discussion of the Design Argument during the thirteenth to the sixteenth centuries.

properties of inorganic material and endowed at the moment of creation; thereafter all operates by mechanical causes alone:

God has so wondrously established these laws that even if we suppose that he creates nothing more than I have said [matter and motion], and even if he puts into this no order nor proportion, but makes of it a chaos as confused and perplexed as the poets could describe, they are sufficient to cause the parts of this chaos to unravel themselves, and arrange themselves in so good an order that they shall have the form of a very perfect world.[64]

Whereas Bacon had banished final causes to the metaphysical world, Descartes wished to exorcise them from this realm as well. Following Francis Bacon's example, he made no attempt to deny that Nature may possess some ultimate end of premeditated design, but claimed that it is simply beyond our ken to identify it; for,

the capacity of our mind is very mediocre, and not to presume too much on ourselves, as it seems we would do were we to persuade ourselves that it is only for our use that God has created all things, or even, indeed, if we pretended to be able to know by the force of our mind what are the ends for which he has created them.[65]

The reason why the concept of teleology has arisen in our minds, Descartes claimed, is due to muddled thinking about the relationship between causes and effects rather than the reality of different types of cause as Aristotle would have it. By contrast the Cartesian approach would[65] 'explain effects by causes, and not causes by effects'. Yet Descartes did seem to allow just one final cause; for he believed God has provided Man with a closely correlated body and mind to evade danger—mankind's end was survival.

2.5 Mechanical Worlds

> But of this frame, the bearing and the ties,
> The strong connections, nice dependencies,
> Gradations just, has thy pervading soul
> Look'd thro? Or can a part contain the whole?
>
> A. Pope

The seventeenth century saw a gradual change from an organic to a mechanical world picture; the opinion that an entity which generates life must therefore itself be alive steadily receded in the wake of the manifest success that flowed from the mechanistic paradigm. This appears as an important metamorphosis and one which we are apt to skip over, so familiar are we with the comings and goings of the theoretical models in modern physical science. In modern science, models and descriptions of natural phenomena are taken up and discarded solely according to their transient usefulness, whereas for early scientists they represented not just a model but the very essence of the Universe, the 'thing in itself'. Because of this attitude the new mechanical perspective brought with it a more interesting and enthusiastic form of eutaxiological argument which found support principally amongst British physicists. Although their arguments were strongly motivated by their theistic outlook, their arguments also grew out of careful observations and an experimental interrogation of the new clockwork world.

It was Robert Boyle (1627–91) who became the most eloquent expositor and spirited supporter[67] of the 'new' design argument. Boyle laid emphasis upon specific examples and coincidences in Nature, claiming them as 'curious and excellent tokens and effects of divine artifice'. His cosmological view required the Deity to initiate the primordial motion of atoms and thereafter remain in lawful and beneficent control to 'contrive them into the world he designed they should compose'; this establishes why the laws of nature bear the hallmark of design. Yet Boyle's approach was consistently mechanical throughout and, like Descartes, he rejected the Aristotelian world-view, based as it was upon an organic model of the Universe, along with the concepts of the Schoolmen which he saw, were an obstacle to the progress of science because they[68] 'do neither oblige nor conduct a man to deeper searches into the structures of things.'

Despite his admiration for many aspects of Descartes's work, Boyle disagreed strongly with him regarding his blanket exclusion of final causes, for to do thus would:

> throw away an argument, which the experience of all ages shews to have been the most successful [and in some cases the only prevalent one] to establish, among philosophers, the belief and veneration of God.[69]

Whilst he agreed with Descartes that one could not hope to ascertain all the underlying purposes in Nature, he did not see why some, at least, could not be fathomed. But, unlike Descartes, Boyle felt that a major reason for the existence of the world was its service to man, though he certainly granted it could have other ends as well, for he writes,

And here it may not be amiss to take notice, in relation to the opinion, that the whole material world was made for man, that though the arguments we have used may be more probable than others hitherto proposed, against the Vulgar Opinion, especially as it relates to the celestial region of the world, yet amongst the ends designed in several of his works, especially plants, animals and metals, the usefulness of them were designed chiefly for men, yet God may design several ends in several creatures, which may find other, and more noble uses for several creatures than have yet been discovered.[70]

Opponents of the Design Argument, like Montaigne, had highlighted the presumption attached to any affirmation of anthropocentric design in Nature; but as a corollary Boyle claimed that, given our fragmentary understanding, it was equally presumptuous of them to deny it.

Another original aspect of Boyle's approach to final causes was his claim that the discovery of features pointing to design in Nature is promoted principally by experimental science and provides a strong motivation for these empirical investigations. It is because of lack of good experimental evidence that Boyle shows so little enthusiasm for arguing for manifest design in the astronomical world. He has serious reservations here, for

I am apt to fear that men are wont, with greater confidence than evidence, to assign the systematical ends and uses of the celestial bodies, and to conclude them to be made and moved only for the service of the earth and its inhabitants.[71]

Instead, he preferred to find indications of design from the minutiae of flora and fauna, because of their more allegorical nature and the stronger possibility of deciding the purpose of their composite structures.

For there seems more admirable contrivance in the muscles of a man's body, than the celestial orbs; and the eye of a fly seems a more curious piece of work than the body of the sun[72]

Such deductions were less obvious in the extraterrestrial realm:

I think that, from the ends and uses of the parts of living bodies, the naturalist may draw arguments, provided he do it with due cautions of which I shall speak. That the inanimate bodies here below that proceed not from seminal principles have a more parable texture . . . and will not easily warrant ratiocinations drawn from their supposed ends.[73]

Like Aristotle before him, Boyle searched for particular examples of micro-engineering in the structure of animals and insects; such examples had, at that time, received a lot of publicity following the publication of Hooke's *Micrographia* in 1665. The invention of the microscope had, for the first time, allowed people to see the intricacy of the smallest organisms. In no small way this advance gave added momentum to the Design Argument. Boyle's discussions of these matters appeared in 1688 in a work bearing a rather intimidating title:[73] *Disquisition about the Final Causes of Natural Things: wherein is inquired whether and (if at all) with what caution a naturalist should admit them.* There he attempted to classify the various ends one could discern in Nature into four categories: the 'universal' (divine), the 'cosmical' (which govern the celestial motions), the 'animal' ('which are those that the peculiar parts of animals are destinated to, and for the welfare of the animal itself') and 'human' (mental and corporeal). Each category provoked Design Arguments but they differed in character and force according to the quality of the evidence available and the impact they made on the imagination.

Following Cicero's employment of the horological analogy of design, Boyle replied to Descartes's claim that final causes are irresolvable, dissipated in a sea of vague possibilities:

> Suppose that a peasant entering in broad daylight the gardens of a famous mathematician, finds there one of those curious gnomonic instruments which indicate the position of the sun in the zodiac, its declination from the equator, the day of the month, the length of the day and so on; it would, no doubt, be a great presumption on his part, ignorant alike of mathematical science and of the intentions of the artist, to believe himself capable of discovering all the ends in view of which this machine, so curiously wrought, has been constructed; but when he remarks that it is furnished with an index, with lines and horary numbers, in short, with all that constitutes a sun-dial, and sees successively the shadow of the index mark in succession the hour of the day, there would be in his part as little presumption as error in concluding that this instrument, whatever may be its other uses, is certainly a dial made to tell the time.[74]

Boyle argues that in many circumstances no ambiguity arises about the object and purpose of natural contrivances. The world is like a mechanism, and like all known mechanisms, is built for a specific purpose that can almost always be elucidated by a thoughtful inspection of its inner workings.

In this contention he was supported by his continental contemporary Gassendi (1592–1655) who also disagreed with Descartes,

> You say that it does not seem to you that you could investigate and undertake to discover, without rashness, the ends of God. But although that may be true, if you mean to speak of ends that God has willed to be hidden, still it cannot be the case with those which he has, as it were, exposed to the view of all the world, and which are discovered without much labour.[75]

The specific influence of the new mechanical world model can be seen in an interesting way: Boyle is so impressed by the correspondence between the internal workings of the world and a timepiece, that he believes behind the world lurks a designer of mechanisms with a measure of human intelligence:

Thus, he who would thoroughly understand the nature of a watch, and not rest satisfy'd with knowing, in general, that a man made it for such uses, but he must, particularly, know of what materials the spring, the wheels, the chain, and the balance are made, he must know the number of the wheels, their magnitude, shape, situation and connexion in the engine, and after what manner or part moves another ... In short, the neglect of efficient causes would render philosophy useless; but the studious search after them will not prejudice the contemplation of final causes.[76]

The end of his statement reveals his stance: although immediate efficient causes of phenomena were entirely mechanical in Boyle's physics, their ultimate and final causes were seen as entirely supernatural. He hoped that his crusade for such a complementarity in the scientific view of the world would not die with him. To support and perpetuate teleological studies he bequeathed a sum of fifty pounds 'forever, or at least for a considerable number of years' to support a series of public lectures on Natural Theology.

At this time those Protestant scientists who, like Boyle, supported the experimental approach advocated by Bacon were rapidly becoming impatient with the methodological dogmas of the Schoolmen. The lead given by Descartes and Boyle was enthusiastically followed by others who were more colourful in their condemnations as this extract from John Webster's view of Scholastic reasoning rather vividly indicates!

What is it else, but a confused chaos of needless, frivolous, fruitless, trivial, vain, curious, impertinent, knotty, ungodly, irreligious, thorny and hell-hatch'd disputes, altercations, doubts, questions and endless janglings, multiplied and spawned forth even to monstrosity and nauseousness.[77]

The development of the new mechanized physics was to carry with it a design argument based upon the observation of meticulous contrivances in Nature and the conviction of an underlying order of its universal laws. But in biology the organic approach still held sway. An exceptional scientist who remained unconvinced of the mechanical analogy in all its facets was John Ray (1628–1704), the greatest of seventeenth-century English naturalists. In his famous teleological study, *The Wisdom of God manifested in the works of Creation*,[78] he amassed a wealth of observational data to argue that animals were pre-adapted to survive in special environments. His comprehensive work also reviewed both the astronomical and terrestrial sciences and stressed the manner in which Man's welfare is ensured by the special properties of water, fire, air and wind. It was Ray's meticulous botanical and biological observations that led him to reject the mechanical analogy as too simplistic a view of Nature because it gave no insight into the reasons for the enormous differences in scale between intricately constructed organisms and the Universe as a whole. He challenged Boyle's contention that Nature originally possessed all the intrinsic properties necessary for its multivarious outworkings; rather he appealed to a vitalist force to provide for its constant orchestration: He concludes that

I therefore incline to Dr Cudworth's opinion, that God uses for these effects the subordinate ministry of some inferior plastic nature . . .[79]

The novelty of 'Dr. Cudworth's opinion' was the concept he termed 'Plastic Nature'[80] which possessed a measure of irrational motion independent of the immediate direction of the Deity. This property enabled it to be employed as an explanation for the aberrations as well as the successes of Nature. Even the lack of design could now be attributed to design.

A strong continental opponent of these attempts to introduce some finalistic design principle into physics was Benedict de Spinoza (1632–77). His antagonism toward any deployment of final causes or inferences from supposed design in the world is spelt out in an appendix to his *Ethics*[81] published in the year of his death. Such notions, he claims, have only arisen because of our ignorance of mechanical laws of Nature and our gullibility regarding the prejudices of anthropocentric philosophy. Far from being in a position to determine the causes and effects of most things we tend to react in amazement, thinking that however these things have come out, they cannot but be for our benefit. This is why, he says, everyone who 'strives to comprehend natural things as a philosophere, in place of admiring them as a stupid man, is at once regarded as impious'.

Those who employ finalistic reasoning simply confuse causes with effects because,

It remains to be shown that nature does not propose to itself any end in its operations, and that all final causes are nothing but pure fictions of human imagination. I shall have little trouble to demonstrate this; for it has already been firmly established . . . I will, however, add a few words in order to accomplish the total ruin of final causes. The first fallacy is that of regarding as a cause that which is by nature anterior, it makes posterior . . .[82]

Also, if the doctrine of final causes is correct he argues, then those most perfect things we are seeking as irrefutable evidences of the 'perfect principle' must, by definition, lie in the unobservable future, for

If the things which God immediately produces were made in order to attain an end, it would follow that those which God produces last would be the most perfect of all, the others having been made in order to these.[83]

Spinoza claims that our deductions of final causes are probably nothing more than mere wish-fulfillment; expressing, not the nature of the real world, but the nature we hope it has:

When we say that the final cause of a house is to provide a dwelling, we mean thereby nothing more than this, that man, having represented to himself the advantages of the domestic life, has had the desire to build a house. Thus, then this final cause is nothing more than the particular desire just mentioned ...[84]

Such metaphysical and logical objections seemed to carry very little weight on the other side of the English Channel where the greatest scientific genius of his age, Isaac Newton (1642–1727), was giving his support to anthropocentric teleology:

Can it be an accident that all birds, beasts and men have their right side and left side alike-shaped (except in their bowels) and just two eyes and no more, on either side of the face; and just two ears on either side of the head ...? Whence arises this uniformity in all their outward shapes but from the counsel and contrivance of an Author? .. Did blind chance know that there was light, and what was its refraction, and fit the eyes of all creatures after the most curious manner to make use of it?[85]

Underlying all Newton's thinking was his deeply-held belief that order was 'created by God at first and conserved by him to this Day in the same state and condition'. Our observation of the planetary orbits, he argued, should convince us that their arrangement did not simply 'arise out of chaos by the mere laws of Nature, though being once formed it may continue by those laws for many ages'.

Whereas Robert Boyle had been a critic of Cartesian metaphysics, Newton also opposed Cartesian physics and in particular Descartes's vortex theory of celestial motions, which he showed, by employing angular momentum conservation, to be in conflict with Kepler's observed laws of planetary motion. In his last works Newton voices his exasperation at the omission of final causes in the Cartesian explanations, which he clearly felt to be incomplete because they provided no explanation for the economy and special constitution of Nature:

Whence is it that Nature does nothing in vain; and whence arises all that Order and Beauty which we see in the world? To what end are comets ... How come the bodies of animals to be contrived ... For what ends are their several parts? ... Was the eye contrived without skill in optics? ...[86]

The Newtonian theory of the world, so carefully and impressively argued in his *Principia*, became the foundation for a steady stream of design arguments based upon optical and gravitational phenomena. Indeed Newton remarked that in writing the treatise he had an 'eye upon arguments' for belief in a deity and in the introduction to his *Opticks* he claims that the main business of natural philosophy is to deduce causes from effects until we arrive at the 'First Cause'. However, one man became inextricably linked with Newton in the propagation of these teleological interpretations of Newtonian physics; that man's name was Richard Bentley.

Richard Bentley (1662–1742) was a Yorkshireman from humble beginnings who later, principally because of his successful Christian apologetics and classical scholarship, became the Master of Trinity College, Cambridge. Bentley came first into the public eye in 1691, when, while still chaplain to Edward Stillingfleet, the Bishop of Worcester, he was invited to give the inaugural Boyle Lectures on Natural Theology. They were entitled the[87] *Confutation of Atheism from the Origin and Frame of the World* and in giving them he displayed an excellent knowledge and understanding of Newton's mathematical physics, a familiarity known to have been fostered by his close correspondence[89] and dialogue on such matters with Newton himself. Bentley was to argue that design is most clearly witnessed by elegant mathematical laws of a general and invariant character rather than by the specific, but relative, adaptations we see in the animal world. He attempted to construct a eutaxiological design argument based upon our knowledge rather than, as often had been the case, a teleological argument founded upon our ignorance. The cornerstone of his argument, Newton's gravitational theory, derived, for the first time, what we still consider to be one of the fundamental constants of nature: the gravitational constant. It was this underlying universal constant that was responsible for the apparently universal nature of Newton's deductions and explanations in gravitation physics and it led to the belief that there was something absolute about the entire model of the world it gave rise to—a model that was mechanical, like the workings of a watch.

In retrospect it is perhaps predictable that outstanding success in scientific model-building and explanation should lead to an accompanying proliferation of teleological and eutaxiological design arguments. One sees it in the Aristotelian period and in the twentieth-century study of cosmology and elementary particles. Whenever absolute deductions are possible from a theoretical model, and successfully explain what is seen, then some form of absolute credence tends to be attributed to the mathematical model responsible.

Newton's authority was also extensively employed by other apologists, notably Hales, Clarke, Whiston and MacClaurin, all with Newton's blessing according to David Gregory's report

> In Mr. Newton's opinion a good design of a publick speech . . . may be to show that the most simple laws of nature are observed in the structure of a great part of the Universe, that the philosophy ought there to begin, and that Cosmic Qualities are as much easier as they are more Universal than particular ones, and the general contrivance simpler than that of animals, plants . . .[89]

The result of this enthusiasm and its widespread influence was to make Newton and his followers the principal target of Hume's attack in the[113] *Dialogues concerning Natural Religion.* In his *History of England* Hume describes Newton and his achievement in two-edged terms 'the greatest and rarest genius that ever rose for the ornament and instruction of the species', but yet 'while Newton seemed to draw off the veil from some of the mysteries of nature, he shewed at the same time the imperfections of the mechanical philosophy; and thereby restored her ultimate secrets to that obscurity in which they ever did and ever will remain'. The statement of the Design Argument used by Hume in his work is in fact that given by Colin MacClaurin (1698–1746) in his book *An Account of Sir Isaac Newton's Philosophical Discoveries* wherein he remarks

the plain argument for the existence of the Deity, obvious to all and carrying irresistable conviction with it, is from the evident contrivance and fitness of things for one another, which we find throughout the Universe.[90]

At this point, it is worth pausing to mention a gradual transition that has occurred in the nature of design arguments from the Scholastics to Newton. For the Schoolmen the *causa finalis* of Nature was God himself; the unmoved mover was Omega as well as Alpha. The future succession of effects must come to an end just as surely as the past procession of causes must have had a beginning and Man, they argued, should use this insight to know God. For Newton and his colleagues the ordered laws of motion themselves appear to be the end of Nature. God exists to uphold and perpetuate them, defending the world system from falling into chaos and irrationality.

The second Boyle lecturer was another Newtonian, Samuel Clarke, but it is not for this that Clarke is chiefly remembered. Rather, it is for his dialogue with another scientist who was not so readily seduced by the Newtonian design arguments. Clarke's formidable opponent was Gottfried Leibniz (1646–1716) and throughout their famous correspondence Clarke was undoubtedly being coached by his compatriot, Newton. Leibniz believed that mechanistic science alone left no room for theocentric purpose. Such a purpose could only be evident through the recognition and incorporation of perfect geometrical principles into physics. In principle, Leibniz argued, there were many possible worlds that were logically self-consistent but the reason for the selection of the existing cosmos was its maximal degree of perfection; it was 'the best of all possible worlds'.[91] He argued that the use of this principle of perfection was quite essential in physical modelling and[92] 'So far from excluding final causes and the consideration of a Being acting with wisdom, it is from this that everything must be deduced in physics'. In conjunction with mechanical explanation the use of final causation and teleology provides a parallel line of analysis and it is to everyone's benefit that they be conjoined.

In order to make use of his 'perfecting principle' Leibniz gave examples of laws in Nature that he believed were not metaphysically necessary. For example, the principle of continuity in the motion of physical systems which appears to be generic when one might have anticipated discontinuities ('leaps') to be prevalent *a priori*:

The hypothesis of leaps cannot be refuted except by the principle of order, by the help of supreme reason, which does everything in the most perfect way.[93]

So, in the beginning God established all things harmoniously and thereafter they maintained their harmony and mutual consistencies even though they were causally disjoint. The maintenance of order in this fashion was proposed by Leibniz as 'a new proof of the existence of God, which is one of surprising clearness'; it was an *a posteriori* argument from an initially established ordering. He is convinced of it because there seems to exist coordination between things that have never been in causal contact with one another (a dilemma known[94] to modern cosmologists as the 'horizon problem')

This perfect harmony of so many substances which have no communication with each other can only come from a common cause.[95]

Leibniz' perfect harmony does not necessarily have any anthropocentric bias and because of that it is not surprising that 'we find in the world things that are not pleasing to us', we would expect it because 'we know that it was not made for us alone'. In this contention Leibniz would have been supported by some Newtonians but the area where disagreement with Clarke, and thereby Newton himself, rested was in the manner of the *maintenance* of the world order. Clarke was an 'occasionalist' believing that God constantly intervenes to correct aberrations in the order of Nature just as the watchmaker occasionally finds it necessary to regulate or repair his watch. Leibniz held that such a view implied either that the laws of Nature and creation were in some way imperfect or the Deity was lacking in foresight; he could not believe the world needed repair 'otherwise we must say that God bethinks himself again'.[96] Clarke retorted that Leibniz had turned the Deity into an absentee landlord and relegated the sphere of divine action to that of a limited initial cause but received the reply that, to the contrary, His dynamic role was the constant maintenance of the world order.

Besides the two scientific giants of the age, there were several other more off-beat contributors to the Design Argument debate; not least the botanist Nehemiah Grew (1641–1712). In his study *Cosmologia Sacra* he gave not only many ingenious examples of design in crystallography but also an argument from the large scale regularity of Nature to the existence of extraterrestrial planetary systems:

> there can be no manner of symmetry in finishing so small a part of the Universal expansion with so noble an apparatus as aforesaid, and letting innumerable and far greater intervals lie waste and void. If then there are many thousands of visible and invisible fixed stars, or suns, there are also as many planetary systems belonging to them, and many more planetary systems belonging to them, and many more planetary worlds.[97]

An unusual continental commentary is provided in the famous drama[98] *Le Festin de Pierre* by Molière (1622–73). There, the Design Argument found itself on the lips of a pious valet who says to his unbelieving master:

> This world that we see is not a mushroom that has come of itself in a night ... Can you see the inventions of which the human machine is composed, without admiring the way in which it is arranged, one part within another? ... My reasoning is that there is something wonderful in man, whatever you may say, and which all the savants cannot explain.[98]

Another famous French author with interesting opinions on final causes, who was also a vehement opponent of Leibniz' entire world-view was François-Marie Arouet (1694–1778), better known by his *nom-de-plume*, Voltaire. Voltaire is perhaps most succinctly categorized as an anti-Epicurean, anti-Christian, Newtonian Deist. His opinion of the order of Nature was that 'a watch proves a watch-maker, and that a Universe proves a God'. It was unthinkable to him that one could attribute the existence of the human mind to blind chance:

> We are intelligent beings, and intelligent beings could not have been formed by a blind, brute, insensible thing ...[99]

Furthermore, he maintained, the evident presence of intelligence in Nature made it necessary to consider final causes in Nature.

Although he believed that normalizing selection could explain the adaption that animals displayed with respect to their environments it could account neither for their mental faculties nor the intricacy of the design actually engineered within them, and, as for chance as a feasible mechanism he claimed

> The disposition of a fly's wings or of the feelers of a snail is sufficient to confound you.[99]

Yet Voltaire was a scathing opponent of *anthropocentric* design arguments because he felt that our scanty knowledge made the objects and beneficiaries of design indeterminate and inevitably, the subject provided excellent material for his *Dictionary* article on 'Ignorance'. In the same volume he argues against the synthesis of Final Causes with these anthropocentric delusions on the grounds that things could not have been set up long ago with our present specific and unpredictable day-to-day needs in view,

In order to become certain of the true end for which a cause acts, that effect must be at all times and in all places. There have not been vessels at all times and on all seas: thus it cannot be said that the ocean has been made for vessels. One feels how ridiculous it would be to allege that nature had wrought from the earliest times to adjust itself to our arbitrary inventions, which have all appeared so late; but it is very evident that if noses have not been made for spectacles, they have been made for smelling, and that there have been noses ever since there have been men.[100]

We also recall the caricature of Leibniz and his 'best of all possible worlds' philosophy through Dr. Pangloss, the professor of 'metaphysico-theologo-cosmolonigology' in *Candide.*[101]

One of Voltaire's co-editors of the *Encyclopédie* and the author of its mathematical content was D'Alembert (1717–83). He was, like Voltaire, sceptical of the numerous metaphysical bases to mathematical physics. Also interesting is his distinction between the intrinsic laws of nature and the mathematical models we use to represent them. This distinction he develops when discussing the form of the laws of motion,

It seems to me that these thoughts can serve to make us evaluate the demonstrations given by various philosophers of the laws of motion as being in accord with the principle of final causes, that is to say with the designs of the Author of Nature in establishing these laws. Such proofs can be convincing only insofar as they are preceded and supported by direct demonstrations and have been derived from principles which are within our reach; otherwise they could often lead us into error. It is for having followed that path, for having believed that it was the Creator's wisdom to conserve always the same quantity of motion in the Universe, that Descartes was mistaken about the laws of collision. Those who imitate him run the risk of either being deceived like him, or taking for a general principle something that takes place only in special cases, or finally of regarding a purely mathematical consequence of some formula as a fundamental law of nature.[102]

For modern mathematicians D'Alembert's name is linked with that of his younger contemporary Moreau de Maupertuis (1698–1759) through their important contributions to the variational principles of mechanics. Such variational principles are remarkable quantitative examples of teleological reasoning being directly and predictively employed in mathematical physics and we shall discuss them in a little more detail in Chapter 3.4. Here, we just mention how they enabled Maupertuis to arrive at a quantification of the notion of 'the best of all possible worlds': the optimal configuration or state within an ensemble of logically consistent possibilities.

In general, a variational principle indicates how the actual motion or state of a system differs from all of the kinematically possible motions permitted by its constraints. This principle may be *differential*, giving the difference between the actual and the optimal systems at each instant of time; or, less generally, it may be *integral*. Integral variational principles establish the difference between the actual motion of a system and all of its kinematically possible motions during a finite time interval. Maupertuis' name is associated with the famous integral principle of variation—the Least Action Principle. Maupertuis used this idea to argue for a system of God-inspired final causes in Nature and claimed that it was a mathematically precise version of Leibniz' doctrine of 'the best of all possible worlds'. Formerly, Design Arguments had been implicitly making statements of comparative reference without any other "worlds" being available; the novelty of Maupertuis' Design Argument is that the other worlds do exist. They are the paths with non-stationary action.

Yet, Maupertuis was well aware that the growth of accurate mathematical models of nature had spawned many over-zealous metaphysical extrapolations:

> For all ages proofs of the wisdom and power of Him who governs the Universe have been formed by those who applied themselves to the study of it. The greater the progress in physics, the more numerous have these proofs become. Some struck with amazement at the divine tokens which we behold every moment in nature, other through a zeal misnamed religious, have given certain proofs greater weight than they ought to have, and sometimes taken for proof that which was not conclusive.[103]

and he believed Newton to be the originator of this uncritical approach because,

> That great man believed that the movements of the celestial bodies sufficiently demonstrate the existence of Him who governs them; such uniformity must result from the Will of a Supreme Being.[104]

and other less distinguished authors, Derham, Fabricus and Lesser were chastised for their unimaginative repetition of earlier platitudes,

> Almost all the modern authors in physics and natural history have done little else than expand the proofs drawn from the organization of animals and plants, and push them in the details of nature ... A crowd of physicists since Newton have found God in stars, in insects, in plants, and in water; not to mention those who find him in the wrinkles of the rhinoceros' hide ... leave such bagatelles to those who do not perceive their folly.[105]

The only people with whom he appears to have less sympathy are those who would outlaw Final Causes at the behest of chance and mechanism.

His own approach was grounded in a search for general regulatory principles and for physical laws generated by the precise formulation of a Least Action Principle. He argues that the only objective approach to evaluating the tendencies of nature is to dwell on the form of its laws—not its artefacts and organisms,

I review the proofs drawn from the contemplation of nature, and I add a reflection: it is, that those which have the greatest strength have not been sufficiently examined as regards their validity and extent. That the cosmos presents trains of agencies convergent to an end on a thousand occasions, is no proof of intelligence and design ... skill in the extension is not sufficient .. the purpose must be rational ... The organization of animals, the multiplicity and minuteness of the parts of insects, the immensity of celestial bodies, their distances and revolutions are better suited to astonish the mind than to enlighten it ... Let us search for him in the fundamental laws of the cosmos, in those universal principles of order which underlie the whole, rather than in the complicated results of those laws.[106]

A number of Maupertuis' criticisms were directed specifically at William Derham (1657–1735) the Boyle lecturer for 1711–12, a minor scientist and an enthusiast for the Newtonian world-view. His Boyle Lectures consisted of sixteen sermons delivered at St. Mary-le-Bow Church which appeared in book form a year later under the title *Physico-Theology*.[107] He considered all the usual good fortunes of the world, the suitability of the terrestrial environment, the diurnal and seasonal variations and so on, all from an anthropocentric perspective. Extraordinarily, he pauses to wonder if the eye might have been more efficiently situated on the hand, but upon reflection, considers it safer from injury on the head! Another unusual trend in his argument is an attempt to persuade the reader that many minor disasters, which one might at first sight have found difficult to reconcile with providential design, were actually beneficial in staving off even graver catastrophes! For instance,

To instance the very worst of all things named, viz., the volcanoes ignivomous mountains: although they are some of the most terrible shocks of the globe and dreadful scourges of the sinful inhabitants thereof .. Nay, if the hypothesis of a central fire and waters be true, these outlets seem to be of the greatest use to the peace and quiet of the terraqueous globe in venting the subterraneous heat and vapours, which, if pent up, would make dreadful and dangerous commotions of the earth and waters.[108]

Later he was to abandon this anthropocentric bias, referring to it as the 'old vulgar opinion that all things were made for man'. His more sophisticated teleological outlook was written-up in a later work *Astro-Theology*. There, in contrast to his earlier work, he realizes the need to consider the role of the heavenly bodies whose motions appear to be of no possible relevance to ourselves. He uses their existence to support a eutaxiological argument by appeal to the manifest design of their orderly motions:

For where we have such manifest strokes of wide order and management, of the observance of mathematical proportions, can we conclude there was anything less than reason, judgement and mathematical skill in the case? Or that this could be effected by any other power but that of an intelligent Being.[108]

Eighteenth-century biologists were beginning to think more carefully about the progressive development of forms but came to widely differing conclusions. The Swiss naturalist Bonnet (1720–93) introduced the term *evolution* to describe the ontogenetic development of an individual from fetus to adult and argued that the entire inorganic world was similarly preprogrammed. Further this complete determinism was sufficient to explain the match of living things to their local environment. Yet, his French contemporary, the zoologist Buffon (1707–88), believed that no useful information about animal function could be gleaned from the doctrine of Final Causes so commonly employed by the physicists:

Those who believe they can answer these questions by final causes do not perceive that they take the effect for the cause.[109]

2.6 Critical Developments

> The believers in Cosmic Purpose make much of our supposed intelligence but their writings make one doubt it. If I were granted omnipotence, and millions of years to experiment in, I should not think Man much to boast of as the final result of all my efforts.
>
> Bertrand Russell

Besides Maupertuis, the most original approach to the metaphysical problems at the core of the mechanical world-view issued from the pen of Giovanni Vico (1688–1744), a Neapolitan professor of Jurisprudence.[110] In his own time his work was not widely discussed, but retrospectively he is seen, by philosophers of science, as a forerunner of Kant. Vico was interested in refuting the Cartesian dogma that all science required in order to unravel the working of the World was an axiomatic basis for reasoning and a sound mathematical methodology. His approach was to establish a clear distinction between the world as it really is and the world which we create and cognize through the use of mathematical models and physical experiments. He realized that the understanding one has of something created by oneself is of a different nature to that understanding gleaned from simple observation. This distinction means we can never be free from subjectivism. Vico saw that mathematical models appear intelligible and coherent to our minds because our minds alone have made them. All our enquiry is necessarily anthropocentric because we employ man-made tools and human reason in its pursuit. Vico believed the 'real' world of nature, which obeyed knowable but inaccessible rules, differed in kind from our do-it-yourself model of intelligible but man-made laws;

> Create the truth which you wish to cognize, and I, in cognizing the truth that you have proposed to one, will 'make' it in such a way that there will be no possibility of my doubting it, since I am the very one who has produced it.[111]

Vico recognized four distinct types of knowledge and warned against abstracting conclusions drawn from information within one category of enquiry into others. One of his categories is *Scienze*: *a priori* knowledge of the real nature of things, which one can only possess of artefacts or models we have made. God alone possesses this type of knowledge of everything. Vico himself was a Christian teleologist who believed that we could only know the ultimate ends of Nature by revelation, (which would endow us with the third of his four types of knowledge).[111,112] Yet, his ideas provide a natural prologue to the more critical analyses of the Design Argument and the theory of knowledge which were to be developed by David Hume and Immanuel Kant.

In his posthumous publication, the *Dialogues Concerning Natural Religion*,[113] David Hume (1711–76) mounted a sceptical attack on the logical structure of many naive design arguments and indeed also upon the rational basis of any form of scientific enquiry. In the *Dialogues*, and in other works, Hume calls the Design Argument 'the religious hypothesis' and proceeds to attack its foundation from a variety of directions. Hume's approach was entirely negative; whereas most of his contemporaries accepted the rationality and ordered structure of the world without question, Hume did not. A common-sense view of the world, along with the metaphysical trimmings that had been added to the Newtonian world model, Hume rejected. His *Dialogues* are analogous to Cicero's *De Natura Deorum*; the *Dialogues* describe a debate in which the sceptical Philo umpires and examines the argument between two supporters of different types of 'religious hypothesis'. On the one hand there is Demea, representing the school of *a priori* truth and revelation and on the other Cleanthes, who reasons in *a posteriori* manner, employing the fashionable synthesis between Final Causes and the mechanical world-view. The views of Newton's supporters[114] are voiced through Cleanthes who actually adopts MacClaurin's statement of the Newtonian Design Argument when summarizing his position:

> I shall briefly explain how I conceive this matter. Look round this world: Contemplate the whole and every part of it. You will find it to be nothing but one

great machine, subdivided into an infinite member of lesser machines. . . All these various machines and even their most minute parts, are adjusted to each other with an accuracy, which ravishes into admiration all men who have ever contemplated them. The curious adapting of means to ends, throughout all nature, resembles exactly, though it much exceeds, the productions of human contrivance; of human design, thought, wisdom and intelligence. . .[115]

The principal objections which Hume allows to surface during the course of the discussion are threefold: Firstly, the Design Argument is unscientific; there can be no causal explanation for the order of Nature because the uniqueness of the world removes all grounds for comparative reference. Secondly; analogical reasoning is so weak and subjective that it could not even provide us with a reasonable conjecture, never mind a definite proof. And finally: all negative evidence has been conveniently neglected. Hume maintains that a dispassionate approach could argue as well for a disorderly cause if it were to concentrate upon the disorderly aspects of the world's structure. His aim is not so much to refute the Design Argument as to show it only raises questions that are undecidable from the evidence available.

Hume's spokesmen question the anthropocentric bias of the Design Argument

. . . we are guilty of the grossest, and most narrow partiality, and make ourselves the model of the Universe . . . What peculiar privilege has this little agitation of brain which we call thought, that we must thus make it the model of the whole Universe.[116]

Hume also draws attention to the tautological nature of the deductions from animal structure. For if the harmonious interrelation of organs is a necessary condition for life how could we fail to inhabit a world of harmonious appearances

It is vain . . . to insist upon the uses of the parts in animals or vegetables and their curious adjustments to each other. I would fain know how an animal could subsist, unless its parts were so adjusted?.[117]

An alternative explanation of order is suggested: perhaps the development of the world is random but has had an infinite amount of time available to it so all possible configurations arise until eventually a stable self-perpetuating form is found:

. . . let us suppose it [matter] finite. A finite number of particles is only susceptible to finite transpositions. And it must happen in an eternal duration, that every possible order or position must be tried an infinite number of times . . . a chaos ensues; till finite though innumerable revolutions produce at last some forms, whose parts and organs are so adjusted as to support the forms amidst a continued succession of matter.[118]

Despite these counter-arguments Cleanthes' support for the Design Argument was so carefully built up that there has even been scholarly debate as to where Hume's own sympathies really lay.[113] Elsewhere Hume[119] appears to display a vitalist view, believing matter to possess some intrinsic self-ordering property:

... that order, arrangement, or the adjustment of final causes is not, of itself, any proof of design; but only in so far as it has been experienced to proceed from that principle. For aught we can know *a priori* matter may contain the source or spring of order originally, within itself, as mind does ... It is only to say, that such is the nature of material objects and that they are originally possessed by a faculty of order and proportion.

Hume's most telling remarks in the *Dialogues* seek to convince the reader that problems of design simply cannot be meaningfully posed. Our position in the Universe introduces natural limitations upon our powers of generalization:[120]

A very small part of this great system, during a very short time is very imperfectly discovered to us: And do we thence pronounce decisively concerning the origin of the whole? ... Let us remember the story of the Indian philosopher and his elephant. It was never more applicable than to the present subject. If the material world rests upon a similar ideal world this ideal world must rest upon some other; and so on, without end. It were better, therefore, never to look beyond the present material world.[121]

At the conclusion of the dialogue the sceptical Philo admits to 'a deeper sense of religion impressed on the mind', for even though the arguments he has heard in support of design are logically unsound they still have considerable psychological impact upon him; they strike him, he says, 'with irresistible force'.

History shows that the Humean tirade against the simple design arguments of the English physicists fell, for the time being, upon deaf ears. There were probably a number of reasons for this. Many English intellectuals, for instance Samuel Johnson[122] and Joseph Priestly, felt that Hume was being merely mischievous or downright frivolous in an attempt to ensure literary fame and he was an isolated and ignored figure in literary circles even during his own lifetime.[121] His *Dialogues* were published posthumously.

More significant hurdles to Hume's acceptance by the scientific community were his eccentric scientific ideas. His unusual theory of causality and the serious suggestion that the Universe may be organic rather than mechanical in nature must have seemed rather naïve when held up against the staggering quantitative achievements of the Newtonian system. Those, like Maupertuis, who subscribed to more sophisticated systems of final causation would not have regarded his objections as relevant and some of his arguments could be falsified by detailed scientific examples.[123] However, although his objections to the Design Argument were to lie temporarily dormant, they were to prove extremely significant for the future spirit of critical inquiry.

At least one zoologist, Erasmus Darwin (1731–1802), who was Charles Darwin's grandfather, enthusiastically took up Hume's intimations concerning the organic nature of the World. Erasmus Darwin was starting to take the early steps towards an evolutionary theory of animal biology, maintaining that the components of an animal or plant were not designed for the use to which they are currently applied, but rather, have grown to fit that use by a process of gradual improvement. However, in order to maintain his belief in theistic design Darwin had to subsume this evolutionary development within some deeper all-embracing plan—a Universal Teleology, an idea common amongst romantic philosophers of this period:

> The late Mr. David Hume ... concludes that the world itself might have been generated, rather than created; that is, it might have been gradually produced from very small beginnings increasing by the activity of its inherent principles, rather than by a sudden evolution of the whole by the Almighty fiat—What a magnificent idea of the infinite power to cause the causes of effects, than to cause the effects themselves.[124]

Of the few other thinkers who saw deeper possibilities and challenges to the Design Argument growing from David Hume's work the most famous is Immanuel Kant (1724–1804). He read Hume's *Dialogues* in a translated manuscript form in 1780 and subsequently acknowledged his debt to him for awaking him 'from his dogmatic slumbers'. Kant's early work had attempted to reconcile the mechanical and teleological views of the world contained in the works of Leibniz and Newton. There he displayed a cautious respect for the Design Argument and the way in which it had been deployed to deduce the existence of a Supreme Being as, for example, in Aquinas' Fifth Way.

> In our humble opinion this cosmological proof is as old as the reason of man.... In this respect the endeavours of Derham, Nieuwentyt, and many others, though they sometimes betray much vanity in giving all sorts of physical insights or even chimeras a venerable semblance by the signal of religions, do human reason honour.[125]

Kant's later critical works take up the claims of Hume concerning the impossibility of deriving sure and necessary principles of a universal nature from empirical data. Independently of Vico he recognizes the irreducible subjectivity of our observations and interpretations. In the *Critique of Pure Reason* Kant summarizes the Design Argument in detail and calls it the 'Physico-Theological Argument':

(1) In the world we everywhere find clear signs of an order in accordance with a determinate purpose, carried out with great wisdom; and this in a Universe which is indescribably varied in content and unlimited in extent.
(2) This purposive order is quite alien to the things of the world and only belongs to them contingently; that is to say, the diverse things could not of themselves have co-operated, by so great a combination of diverse means, to the fulfilment of determinate final purposes, had they not been chosen and designated . . .
(3) There exists, therefore, a sublime and wise cause . . . .
(4) The unity of this cause may be inferred . . . with certainly in so far as our observation suffices for its verification, and beyond these limits with probability in accordance with the principle of analogy.[126]

He admits great respect for this argument because of its stimulus to scientific enquiry: he realizes that many biological investigations have been motivated by the expectation of purpose in organic structures,

It enlivens the study of nature . . . It suggests ends and purposes, where our observation would not have detected them by itself, and extends our knowledge of nature by means of the guiding concept of a special unity, the principle of which is outside Nature . . .[137]

However, Kant then goes on to undermine the logical foundation of any contention that design exists in nature, arguing that we can neither prove nor disprove statements about the real world by pure reason alone. For, in reaching our conclusions we inevitably introduce facts and observations and employ our, possibly erroneous, 'practical reason'. It is only with respect to the 'practical reason' that the Design Argument can maintain its cogency:

It would therefore be not only extremely sad, but utterly vain to diminish the authority of that proof . . . we have nothing to say against the reasonableness and utility of this line of argument, but wish on the contrary to commend and to encourage it, yet we cannot approve of the claims which this proof advances of apodictic certainty.[128]

Then he explains how this lack of 'certainty' arises by pointing out that all our empirical enquiries into the structure of Nature regard it as an entity which incorporates within itself a system of empirical laws. These laws are unified and naturally adapted to the faculties of our own cognition. The design we perceive must be necessarily mind-imposed and subjective to our innate categories of thought. Although the 'things in themselves' are mind-independent, our act of understanding completely creates the categories in terms of which we order them. Inevitably we view the world through rose-coloured spectacles. These self-created categories cannot themselves be ascertained by observation; they are *a priori*, conditions of the experience we have, like the perception of the space-time continuum. We could not through our experience hope to ascertain the conditions of such experience. Our observation of order and structure in the Universe, he argues, arises inevitably because we have introduced such concepts into our analysis of experience. We must not then proceed to rederive them from it. We can say nothing stronger than that the world is such as to make its perception by our minds in any form but ordered, impossible.

Kant claimed morality as the final end of nature, for when we consider moral beings, he writes,

> we have a reason for being warranted to regard the world as a system of final causes.[129]

He thought that only through this ethico-teleology could the final cause of the world be discerned; but its nature is disjoint from the arena of 'physico-theological' design arguments because the latter do not concentrate on the character of final ends, only the transient ends that benefit ourselves here and now:

> Now I say that no matter how far physico-theology may be pushed it can never disclose to us anything about a final end of creation; for it never even begins to look for a final end.[130]

Kant's notion of teleology[129,188,189] had an enormous influence on the work of German biologists in the first half of the nineteenth century. Like Kant, for the most part these biologists did not regard teleology and mechanism as polar opposites, but rather as explanatory modes complementary to each other. Mechanism was expected to provide a completely accurate picture of life at the chemical level, without the need to invoke 'vital forces'. Indeed, Kant and many of the German biologists were strongly committed to the idea that all objects in Nature, be they organic or inorganic, are completely controlled by mechanical physical laws. These scientists had no objection to the idea that living beings are brought into existence by the mechanical action of physical laws. What they objected to was the possibility of constructing a scientific theory, based on mechanism alone, which described that coming into being, and that could completely describe the organization of life. The impossibility of such a scientific theory was not due to non-mechanical processes in Nature, but rather it lay in the inherent limitations of the human mind. In Kant's view, a mechanical explanation, which was equivalent to a causal explanation in Kant's philosophy, could be given only when there is a clear separation between cause and effect. In living beings, causes and effects are inextricably mixed. An effect in a living being cannot be *completely* understood without describing every reaction in the being: ultimate biological explanations require a special non-mechanical notion of causality—teleology—in which each part is simultaneously cause and effect. Parts related to the whole in this way transcend mechanical causality. The order and arrangement of the organism is, according to Kant, a fundamental explanatory mode in biological science.

The limitation of explanation in terms of mechanical causality can perhaps be best understood by comparing a living being to a computer. As Michael Polanyi has pointed out[190,191] the internal workings of the computer can of course be completely understood in terms of physical laws. What cannot be so explained is the computer's program. To explain the program requires reference to the *purpose* of the program, that is, to teleology.

Even the evolution of a deterministic Universe cannot be completely understood in terms of the differential equations which govern the evolution. The boundary conditions of the differential equations must also be specified. These boundary conditions are not determined by the laws of physics which are the differential equations. The universal boundary conditions are as fundamental as the physical laws themselves; they must be included in any explanation on a par with the physical laws.

In a biological organism, the analogues of the computer program are the processing and organizational plans coded in the organism's DNA. The German biologists who followed Kant's program—the historian Lenoir has named them the *teleomechanists*—sought to discover the plan in the over-all organization of the organism. As the physiologist Hermann Lotze put it,

> Thus all parts of the animal body in addition to the properties which they possess by virtue of their material composition also have *vital* properties; that is, mechanical properties which are attributable to them only so long as they are in combination with the other parts . . . Life belongs to the whole but it is in the strictest sense a combination of inorganic processes[192] . . . Biological organization is, therefore, nothing other than a particular direction and combination of pure mechanical processes corresponding to a natural purpose. The study of organization can only consist therefore in the investigation of the particular ways in which nature combines those processes and how in contrast to artificial devices she unites a multiplicity of divergent series of phenomena into complex atomic events.[193]

The study of biological organization by the teleomechanists led to a number of important discoveries, particularly in embryology, which they studied because the action of an organism's organizational plan is most manifest when the creature is being formed. For example, such studies led to the discovery of the mammalian ovum by the teleomechanist von Baer.[188]

In spite of such scientific feats, by the latter part of the nineteenth century the teleomechanists had been eclipsed by the reductionists, led by

Hermann Helmholtz.[188] The great weakness of the teleomechanists was their tendency to think of teleology not only as a plan of organization but also as an actual life force, a tendency which Kant warned against. This led them to believe that it was impossible for organisms to change their fundamental plan of organization, that is, to evolve, under the action of inorganic forces. As a consequence, they later opposed Darwin's theory of evolution by natural selection, and as the evidence for such evolution became overwhelming, they ceased to exert an influence on the development of biology.

Kant's important ideas in critical philosophy and the theory of knowledge which grew out of his work were to have little or no effect upon the growing momentum of the Design Argument in England. The first books describing Kant's work began to appear in English from about 1796 onwards but the logical difficulties they highlighted were not taken seriously by allies of William Paley (1743–1805), author of the famous *Natural Theology*, a work that was to become something of a minor classic in its own time and synonymous with the gospel according to anthropocentric design.

Paley had a distinguished early career at Cambridge; the Senior Wrangler in 1763, he was later greatly admired by his students for a lucid and memorable lecturing style but his progressive social views prevented him rising to high office in the Church of England. On reading his work one is struck by the clarity of his explanation, the skill with which he marshalls his material and the naïviety with which he uses his biological examples. This last trait actually led some European supporters of the Design Argument to disown him in embarrassment.[132] However, because of its lucidity and the widespread support for its conclusions, *Natural Theology* was for many years a set text at Cambridge and a special edition was even produced with essay questions bound into it for undergraduate study. Charles Darwin was to recall how he 'was charmed and convinced by the long line of argumentation' on reading it during his undergraduate years. Where Kant was a model of obscurity Paley is a paragon of literary clarity.

Paley bases his case for design entirely upon the constitution rather than the development of natural things and interprets this constitution in a completely anthropocentric fashion: everywhere in Nature, he claims, we see elements of design and purpose. Design implies a Designer. Therefore Nature is the result of a Designer who is, by implication, God. Paley claims that, wielded in this manner, teleology 'has proved a powerful and perhaps indispensible organ of physical discovery' but he expresses a dislike for the notion of 'Final Causes' largely because of its Scholastic undertones:

... it were to be wished that the scholastic phrase 'final cause' could, without affectation, be dropped from our philosophical vocabulary and some more unexceptional mode of speaking be substituted instead of it.[133]

His central argument appears dramatically in the opening lines of his book.

In crossing a heath, ... suppose I had found a watch upon the ground, and it should be inquired how the watch happened to be in that place ... For this reason, and for no other, viz. that, when we come to inspect the watch, we perceive ... that its several parts are framed and put together for a purpose.[134]

The analogy of the watch-world had been the watchword of many earlier workers. The advantage of the analogy, Paley claims, is that it makes his point regardless of whether one knows the origin of the watch or understands every facet of its machinery. Furthermore, he believed it evaded other well-known objections: even though the world (watch) occasionally malfunctions it would be peculiar not to attribute its mechanism to contrivance. It would be senseless, he says, to claim it was merely 'one of possible combinations of material forms, a result of the laws of metallic nature' or the inevitable consequence of there having 'existed in things a principle of order.' Nothing, he argues, is to be gained

by supposing the watch before us to have been produced from another watch, that from a former, and so on indefinitely ... A designing mind is neither supplied by this supposition, nor dispensed with.[135]

The idea that postulating 'laws' of Nature gave explanations of design he thought to be a form of mysticism, 'a mere substitution of words for reasons, names for causes.' The so-called 'laws' of Nature may be, even now, nothing more than a way of codifying observations that have been made. They do not guarantee anything will take place in the future. They do not provide an explanation of the sort Paley required.

Paley continues to consistently and obliviously mix analogies from the organic and mechanical realms; for example, when discussing explanations of order via evolutionary development and summarizing the general nature of his methodology he admits:

The generations of the animal no more account for the contrivance of the eye or ear, than, upon the supposition stated ..., the production of a watch by the motion and mechanism of a former watch, would account for the skill and intention evidenced in the watch so produced ... Every observation which was made ... concerning the watch, may be repeated with strict propriety concerning the eye; concerning animals; concerning plants; concerning, indeed all the organized parts of the works of Nature.[136]

This complete faith in the mechanistic analogy, even in the organic realm, convinces Paley that we can infer ultimate causes from local effects because of the string of causal and mechanical connections that will exist between them. He brushes aside the critique of Hume, Spinoza and Descartes regarding the transposition of causes for effects:

'Of a thousand other things,' say the French academicians, 'we perceive not the contrivance, because we understand them only by the effects, of which we know not the causes': but we here treat of a machine, all the parts whereof are visible; and which need only be looked upon to discover the reasons of its motion and action . . .[137]

Like Galen, Boyle, Newton and many others before him Paley concentrates upon the internal structure of the human eye as the example of design *par excellence*; so enamoured is he by the eye's remarkable structure that he exclaims,

Were there no example in the world of contrivance, except that of the eye, it would be alone sufficient to support the conclusion which we draw from it.[138]

There is much that is humorous in his examples of design: he dwells upon the foresight displayed by the provision of the epiglottis in the human throat; the following passage has been dubbed the 'devotional hymn to the epiglottis'![139]

Reflect how frequently we swallow, how constantly we breathe. In a city feast, for example, what deglutition, what anhelation! Yet does this little cartilage, the epiglottis, so effectually interpose its office, so securely guard the entrance of the wind-pipe that . . . Not two guests are choked in a century.[140]

More noteworthy are the passing parries he aims at two alternative explanations of order. In accordance with his whole approach, firmly grounded in observation, (and we note in passing that Paley was keen amateur naturalist[141]) he excludes them on the basis of current observations. Concerning the argument that orderly forms were the inevitable result of normalizing selection from an array of randomly constituted organisms, he takes a blinkered approach to fossilized remains and remarks that:

[chance] . . . would persuade me to believe . . . every organized body which we see, are only so many out of the possible varieties and combinations of being, which the lapse of infinite ages has brought into existence; that the present world is the relic of that variety; millions of other bodily forms and other species having perished, being by the defect of their constitution incapable of preservation, of continuance by generation. Now there is no foundation whatever for this conjecture in anything which we observed in the works of nature; no such experiments are going on at present; no such energy operates . . . A countless variety of animals might have existed, which do not exist.[142]

Paley felt that chance was not a mechanism, as many regarded it at that time, but merely a label for 'the ignorance of the observer.' He also claimed that appeal to some inherent and universal ordering principle in Nature was in conflict with observation:

. . . a principle of order, acting blindly and without choice, is negatived by the observation, that order is not universal; which it would be, if it issued from a constant and necessary principle . . . where order is wanted there we find it; where order is not wanted, i.e. where, if it prevailed, it would be useless, there we do not find it . . . No useful purpose would have arisen from moulding rocks and mountains into regular solids, bounding the channel of the ocean by geometrical curves; or form a map of the ocean resembling a table of diagrams in Euclid's Elements, or Simpson's Conic Sections.[143]

The second half of Paley's *Natural Theology* is much more interesting to post-Darwinians than the first. Here he moves from the world of zoology and anatomy to consider the laws of motion and gravitation and their role in astronomy. The first interesting remarks concern the velocity of light: because of its enormous value he infers that the mass of the photon needs to be extremely small to be compatible with our existence:

Light travels from the sun at the rate of twelve millions of miles in a minute . . . It might seem to be a force sufficient to shatter to atoms the hardest bodies. How then is this effect, the consequence of such prodigious velocity, guarded against? By a proportionable minuteness of the particles of which light is composed.[144]

He continues with a discussion of astronomical phenomena, gratefully acknowledging his debt to Rev. J. Brinkley, Professor of Astronomy at Dublin[145] for assistance with many details. He confesses that he feels there to be severe disadvantages as well as advantages in this new line of reasoning:

My opinion of astronomy has always been, that it is *not* the best medium through which to prove the agency of an intelligent Creator; but that, this being proved, it shows, beyond all other sciences, the magnificence of his operations . . . but it is not so well adapted as some other subjects are to the purpose of argument. We are destitute of the means of examining the constitution of the heavenly bodies. The very simplicity of their appearance is against them.[146]

In this area Paley feels adrift from the practice of direct observation he so values and is also relieved of his principal dialectical device because he feels[147] 'we are cut off from one principal gound of argumentation—analogy'. Undoubtedly, he also feels a little less confident of his assertions in an area where he must seek considerable guidance from others.

Now separated from his false analogical guide he proceeds with Brinkley's help to make a number of insightful observations concerning the stability of the solar system and the form of the law of gravitation. Many of these have been subsequently re-derived in connection with the question of whether we could, from the fact of our own existence alone, actually deduce that the world possesses precisely three spatial dimensions, (see section 4.8).

Paley also points out that the evolution of the Sun rules out the possibility of an infinite steady-state history without evolutionary change:

it follows, that the sun also himself must be in his progress towards growing cold; which puts an end to the possibility of his having existed, as he is from eternity.[148]

He goes on to describe the manner in which the terrestrial oblateness and ocean content sensitively determine the local environment and shows how the present topographical circumstances are necessary for our own existence.

The next observations he makes are the most intriguing from a modern perspective: he points out the unique features that are intrinsic to Newton's inverse square law of gravitation. The basis for his comparative study is an imaginary ensemble containing all possible power laws of variation for the gravitational force. The size of the subset of this collection which are consistent with our existence can then be examined in Anthropic fashion:

whilst the possible laws of variation were infinite, the admissible laws, or the laws compatible with the preservation of the system, lie within narrow limits. If the attracting force had varied according to any direct law of the distance, let it have been what it would, great destruction and confusion would have taken place. The direct simple proportion of the distance would, it is true, have produced an ellipse; but the perturbing forces would have acted with so much advantage, as to be continually changing the dimensions of the ellipse, in a manner inconsistent with our terrestrial creation.[149]

This enables Paley to quantify that formerly rather vague, qualitative notion of the mechanical optimality in the World's structure and laws. Next he considers the fitness of the various possible force laws in connection with the stability of the elliptical planetary orbits which he assumes are a necessary condition of our existence:

Of the inverse laws, if the centripedal force had changed as the cube of the distance, or in any higher proportion ... the consequence would have been, that the planets, if they once began to approach the sun, would have fallen into its body; if they once, though by every so little, increased their distance from the centre, would forever have receded from it ... All direct ratios of the distance are excluded, on account of the danger from perturbing forces; all reciprocal ratios, except what lie beneath the cube of the distance, ... would have been fatal to the repose and order of the system ... the permanency of our ellipse is a question of life and death to our whole sensitive world.[149]

Having thus narrowed down the form of the force law to an *inverse* power law he claims that the inverse square is uniquely selected because it allows extended bodies to behave gravitationally as point particles with an equal mass concentrated at the centre of mass of the body, (see section 4.8),

> whilst this law prevails between each particle of matter, the united attraction of a sphere, composed of that matter obeys the same law . . . it is a property which belongs to no other law of attraction that is admissible . . . expected attraction varying directly as the distance.[150]

The possibility of precisely circular orbits are also excluded on the grounds of stability and Paley argues that the selection of a force law which optimally serves 'to guard against [perturbations] running to destructive lengths, is perhaps the strongest evidence of care and foresight that can be given.' His case for anthropocentric design rests upon the concurrence in our solar system of the four circumstances required for the stability of the planetary orbits against perturbations of a 'periodical or vibrating' nature:

> . . . viz. that the force shall be inversely as the square of the distance; the masses of the revolving bodies small, compared with that of the body at the centre; the orbits not much inclined to one another; and their eccentricity little.[151]

To complete this intriguing collection of mathematical arguments for anthropocentric design Paley makes some remarks similar to those of Newton in his correspondence with Bentley concerning the gravitational stability of the Universe. This provides him with a simple argument for the finite age of the Universe:

> If the attraction acts at all distances, there can be only one quiescent centre of gravity in the universe: and all bodies whatever must be approaching this centre, or revolving around it . . . if the duration of the world had been long enough to allow of it, all its parts, all the great bodies of which it is composed, must have been gathered together in a heap round this point.[152]

Despite the naïvety of its earlier treatment of some of the human sciences, Paley's widely read work was to play an important role in summarizing and clearly placing before scientists' eyes the simple facts of adaption in the natural world. In order to supersede his teleological thesis another theory would be required to give a convincing explanation for the vast array of detailed examples he catalogues. The lack of a viable and positive alternative may possibly explain the negligible impact that the afore-mentioned metaphysical objections to the Design Argument actually had. Hume offered no such explanations or deductions with clear observational consequences whereas the hypothesis which was to displace the Paleyean branch of teleology—natural selection—did provide a plausible alternate explanation for the very facts upon which the anthropocentric design argument was based. The relevance of Paley's organic examples of 'design' was later recognized by Huxley who went so far as to remark that Paley 'proleptically accepted the modern doctrine of evolution'. It is also worth noting that Paley's astronomical examples—which are so similar to modern Anthropic arguments—are clearly of a different and inorganic nature and lie entirely outside the jurisdiction of Darwinian natural selection. Strangely, they have been ignored in subsequent evaluations of his work.

Paley's work opened the floodgates for apologetic treatises on every conceivable aspect of 'design', although few of these had anything new to say. The most encyclopaedic and systematic arose at the bequest of the Rev. Francis Egerton, the Eighth Earl of Bridgewater who died in 1829. Egerton charged his executors with the duty of selecting eight eminent scientific authors to demonstrate:

> The Power, Wisdom and Goodness of God, as manifested in the Creation; illustrating such work by all reasonable arguments, as for instance, the variety and formation of God's creatures in the animal, vegetable, and mineral Kingdoms; the effect of digestion, and thereby of conversion; the construction of the hand of man, and an infinite variety of other arguments.[153]

The scholars chosen to carry out this task were Charles Bell, William Buckland, Thomas Chalmers, John Kidd, William Kirby, William Prout, Peter Roget and William Whewell[154] with a later independent contribution by the mathematician Charles Babbage. They were all eminent scholars of their day; several held university lectureships in the sciences and some like the chemist William Prout are now famous for their scientific work—and almost everyone has Roget's *Thesaurus* on their bookshelves. Despite their varying subject matter the Bridgewater Treatises have two things in common: they were all published in London and all sold out almost at once, subsequently going through many editions. With the exception of Babbage's numerical study, the style of the contributions is reminiscent of earlier eighteenth-century works and marked by a dogmatically anthropocentric bias that may be ascertained from their fairly explicit titles. It has been suggested[155] that the whole collection is well summed-up by a sentence in Prout's contribution, 'The argument of design is necessarily cumulative; that is to say, is made up of many similar arguments!'

Whereas in England this teleological spirit appears to have been firmly entrenched in the minds of many scientists, evolutionary ideas were beginning to germinate elsewhere. The biologist von Baer (1792–1876) remarked in his 1834 lectures that 'only in a very childish view of nature could organic species be regarded as permanent and unchangeable types'. Another articulate critic of teleology who was considering the consequences of an evolutionary perspective was Goethe (1749–1832). A widely gifted man who was responsible for important contributions in anatomy, botany, poetry and philosophy, Goethe tried to introduce an evolutionary perspective into every one of these disciplines. As a student he studied in Leipzig and Strasbourg where his thinking was strongly influenced by the works of Bacon, Spinoza, and Kant. Like Francis Bacon, Goethe detects and rejects that systematic bias in Man's self-image which tempts him to elevate himself relative to the world at large:

Man is naturally disposed to consider himself as the centre and end of creation, and to regard all the beings that surround him as bound to subserve his personal profit . . . He cannot imagine that the least blade of grass is not there for him.[156]

2.7 The Devolution of Design

The apparent uniqueness of the Universe primarily depends upon the fact that we can conceive of so many alternatives to it.
C. Pantin

The seventy-fifth section of Kant's *Critique of Judgement* bears the title 'The conception of an objective finality of nature is a critical principle of reason for the use of the reflective judgement', and in it Kant made a confident claim:

It is . . . quite certain that we can never get a sufficient knowledge of organized beings and their inner possibility, much less get an explanation of them, by looking merely to mechanical principles of nature . . . we may confidently assert that it is absurd for me even to entertain any thought of so doing or to hope that maybe another Newton may some day arise, to make intelligible to us even the genesis of but a blade of grass from natural laws that no design has ordered. Such insight we must absolutely deny to mankind.[129]

When the young Charles Darwin (1809–82) began his theological studies at Christ's College Cambridge, where Paley had been both a student and a fellow, he did not study Kant; but for Darwin the study of Paley's various works was compulsory.[157] Many years later Darwin was to recall in his autobiography these early studies:

In order to pass the B.A. examination, it was also necessary to get up Paley's . . . *Evidences*. The logic of this book and, as I may add, of his *Natural Theology* gave me as much delight as did Euclid. The careful study of these works, without attempting to learn any part by rote, was the only part of the academical course which, as I then felt and as I still believe, was of the least use to me in the education of my mind. I did not at that time trouble myself about Paley's premises; and taking these on trust, I was charmed and convinced by the long line of argumentation.[158]

Following his monumental development of the theory of natural selection in parallel with Wallace, Darwin remarked on its interaction with the traditional design arguments:

The old argument from design in nature, as given by Paley, which formerly seemed to me so conclusive, fails, now that the law of Natural Selection has been discovered.[159]

As he grew older Darwin became more agnostic, especially with regard to the awkward problem of the evolution of intelligence. He considered:

the impossibility of conceiving this immense and wonderful universe, including man . . . as the result of blind chance or necessity. When thus reflecting I feel compelled to look to a First Cause having an intelligent mind in some degree analogous to that of man and I deserve to be call a Theist. But then arises the doubt, can the mind of man, which has, as I fully believe, been developed from a mind as low as that possessed by the lowest animal, be trusted when it draws such grand conclusions.[160]

Many have looked to the relegation of Man's special status in relation to the animal world as the principal cause of hostility between Darwinians and those of an orthodox religious persuasion.[161] But it appears that the possible demolition of the Design Argument may have been an equally strong motivation for opposition. Charles Hodge made this explicit at the time in his book *What is Darwinism*:

It is, however, neither evolution nor natural selection which gives Darwinism its peculiar character and importance. It is that Darwin rejects all teleology, or the doctrine of final causes.[162]

The nineteenth-century philosopher Winston Graham also pointed out that primarily Darwin had launched a successful assault on the Design Argument of the natural theologians:

Now it appears that Darwin has at last enabled the extreme materialist to attempt and carry the design argument, the last and hitherto impregnable fortress behind which natural theology has entrenched herself.[163]

Ideas of a general evolutionary development had of course been in the wind and were suggested by many previous workers, but it was only Darwin's introduction of the concept of natural selection along with a vast collection of observational evidence that finally displaced the anthropocentric design arguments drawn from biology. The stress laid upon the many precise adaptions visible in Nature by writers like Paley and the Bridgewater authors can be seen to have played an interesting role in this development. Their claims for design were usually based upon a systematic study of biological and botanical observations and, whether or not the Design Argument was regarded as true, they served to focus the attention of naturalists upon a set of remarkable adapted features.[164]

The new evolutionary world-view led predictably to a re-evaluation of the teleological interpretation and the conception of a universal teleology that used the process of natural selection to direct events towards a final cause. Most notable amongst the supporters of this view was the American botanist and Calvinist, Asa Gray (1810–88). Gray had been appointed professor of natural science at Harvard in 1842 and through his exchange of ideas[165] with Darwin before the publication of the *Origin of Species* in 1859 had confirmed its thesis by his own independent botanical studies. His approach to teleology was to use the Darwinian hypothesis as a panacea to solve many of the problems which had formerly been brushed under the carpet by supporters of the Design Argument, for:

> Darwinian teleology has the special advantage of accounting for the imperfections and failures as well as for successes. It not only accounts for them, but turns them to practical account... So the most puzzling things of all to the old-school teleologists are the principles of the Darwinian,... it would appear that in Darwinian evolution we may have a theory that accords with, if it does not explain, the principal facts, and a teleology that is free from the common objection ... if [a theist] cannot recognize design in Nature because of evolution, he may be ranked with those of whom it was said 'Except ye see signs and wonders ye will not believe'.[166]

In a letter to de Candolle in 1863 Gray offered his

> ... hearty congratulations of Darwin for his striking contributions to teleology... knowing well that he rejects the idea of design, while all the while he is bringing out the neatest illustrations of it.[167]

Darwin liked Gray's interpretation of his work, but perhaps only because it helped soothe the public antagonism to his ideas; he remarked in a private letter to Gray that 'what you say about Teleology pleases me especially and I do not think anyone else has ever noticed the point'. In the later editions of the *Origin* he even acknowledges Gray as 'a celebrated author and divine' who had:

> gradually learnt to see that it is just as noble a conception of the Deity to believe that he created a few original forms capable of self-development into other and needful forms...

Another American who recognized the impact of evolution on Design was the philosopher and science writer John Fiske (1842–1901), who gave a series of thirty-five lectures on Darwinian evolution at Harvard in 1871; they subsequently appeared in revised and expanded book-form as the *Outlines of Cosmic Philosophy*. Fiske was another to realize that it was the overthrow of the anthropocentric design arguments by the mechanism of natural selection that made Darwin's work so unpopular:

> From the dawn of philosophic discussion, Pagan and Christian, Trinitarian and Deist, have appealed with equal confidence to the harmony pervading nature as the surest foundation of their faith in an intelligent and beneficient Ruler of the

universe. We meet the argument in the familiar writing of Xenophon and Cicero, and it is forcibly and eloquently maintained by Voltaire as well as by Paley, and, with various modifications by Agassiz as well as by the authors of the Bridgewater Treatises. One and all they challenge us to explain, on any other hypothesis than that of creative design, these manifold harmonies, these exquisite adaptions of means to ends, whereof the world is admitted to be full, and which are, especially conspicuous among the phenomena of life . . . , in natural selection there has been assigned and adequate cause for the marvellous phenomena of adaption, which has formerly been regarded as clear proofs of beneficent creative contrivance.[168]

Like Gray, Fiske believed that natural selection did not necessitate the rejection of a teleology that was conceived on a large enough scale. Fiske's development of these ideas was looked upon with approval by his friend Thomas Huxley, to whose memory his subsequent work[169] *Through Nature to God* was dedicated. Huxley (1825–95) had set himself up as the principal public defender of the evolutionary 'faith' in England on Darwin's behalf, but was himself surprisingly sympathetic to the teleological interpretation of evolutionary theory.

Huxley foresaw the demise of natural theology but was at first taken aback by the manner in which the evolutionary hypothesis had received a teleological interpretation from some of his colleagues:

It is singular how one and the same book will impress different minds. That which struck the present writer most forcibly on his first perusal of the *Origin of Species* was the conviction that teleology, as commonly understood, had received its death-blow at Mr. Darwin's hands.[170]

Huxley was the first to draw attention to the contribution which the earlier teleological ideas had made in focusing attention upon a number of remarkable organic adaptions. This common interest of teleology and evolution, he said, meant that Darwin

. . . has rendered a most remarkable service to philosophic thought by enabling the student of nature to recognize, to the fullest extent, those adaptions to purpose which are so striking in the organic world, and which teleology has done good service in keeping before our minds . . . The apparently diverging teachings of the teleologist and of the morphologist are reconciled by the Darwinian hypothesis.[171]

More interesting still is Huxley's recognition of an awkward problem for the idea of natural selection—determinism. He saw that because the mechanistic view of the world must regard the later products of natural selection as a completely determined function of the initial molecular configurations, it reduces to a specification of the initial conditions. Natural selection appeared to offer an 'explanation' that things are as they are only because they were as they were:

...there is a wider teleology which is not touched by the doctrine of evolution. This proposition is that the whole world, living and not living, is the result of the mutual interaction, according to definite laws, of the forces possessed by the molecules of which the primitive nebulosity of the universe was composed...The teleological and mechanical views of nature are not, necessarily, mutually exclusive. On the contrary, the more purely a mechanist the speculator is, the more firmly does he assume a primordial molecular arrangement of which all the phenomena of the universe are the consequences and the more completely is he thereby at the mercy of the teleologist, who can always defy him to disprove that this primordial molecular arrangement was not intended to evolve the actual phenomena of the universe...Evolution has no more to do with theism than the first book of Euclid has.[172]

Huxley also speculated that the evolutionary approach to Nature might have a far wider applicability. For, suppose the laws of motion and energy conservation were also just the results of natural selection acting upon a collection of possibilities:

Of simplest matter and definitely operating energy...it is possible to raise the question whether it may not be the product of evolution from a universe of such matter, in which the manifestations of energy were not definite—in which for example laws of motion held good for some units and not for others, or for some units at one time and not another.[173]

However, neither Huxley nor any of his colleagues addressed the astronomical design arguments based upon the co-presence of a number of coincidental features in the solar system dynamics and upon which the stability our environment so delicately hinges. The only debate that took place with physicists concentrated upon other more fundamental problems like reconciling evolutionary development with contemporary views on the age and origin of the earth. In that conflict the most critical opponent of Darwin's theory amongst the ranks of the physicists was Lord Kelvin who argued that the geophysical evidence pointed towards a terrestrial age too brief for natural selection to evolve the observed spectrum of living creatures. This objection against evolution, which at the time Darwin called 'the gravest yet advanced' generated an extremely significant debate which we shall present in extended form in Chapter 3 since it led to the first modern prediction derived from an Anthropic Principle. Kelvin's deepest sympathies were with design couched in a suitable form because of the difficulties inherent in making any observational test of the Darwinian evolutionary hypothesis:

The essence of science consists in inferring antecedent conditions and anticipating future evolutions from phenomena which have actually come under observation.

In biology the difficulties of successfully acting up to this ideal are prodigious ... I have always felt that the hypothesis of 'the origin of species through natural selection' does not contain the true theory of evolution ... I feel convinced that the argument of design has been greatly too much lost sight of in recent zoological speculations.[174]

As we shall see, Kelvin's opposition was extremely influential because of his pre-eminent position amongst British scientists of his day and the greater respect most scientists had for arguments based upon mathematical physics rather than the purely qualitative hypothesis of natural selection.

Another outstanding physicist who contributed to the argument concerning the place of final causes in the evolutionary view was James Clerk Maxwell. Maxwell focused his attention upon molecules, which were then regarded as invariant and fundamental structures. He argued that their time invariance and identical structure proves they could not have developed from some natural process in a statistical fashion. These invariance properties gave them 'the stamp of the manufactured article' and signalled a cut-off in the applicability of a principle of natural selection. His address to the British Association in 1873 contains a statement of these ideas:

No theory of evolution can be formed to account for the similarity of molecules, for evolution necessarily implies continuous change, and the molecule is incapable of growth, or decay, of generation or destruction. None of the processes of Nature, since the time when Nature began, have produced the slightest difference in the properties in the operation of any of the causes which we call natural ... the molecules out of which these systems are built—the foundation stones of the material universe—remain unbroken and unknown.[175]

These are the first glimmerings of a more sophisticated twentieth-century approach to the invariant properties of crucial molecular structures and their relevance to the existence of a life-supporting environment. This approach was later to be developed in a remarkable way by the American biochemist Lawrence Henderson whose work we shall discuss at length in Chapter 3.[176]

One of Henderson's forerunners both in advocating such a view and as Professor of Chemistry at Harvard was Josiah Cooke. Cooke[177] appealed strongly to the form of laws of Nature and the special properties of particular chemical compounds (for example, water) as evidences for order in Nature. However, he keeps these eutaxiological arguments distinct from those which appeal to purposeful design:

We can see that each property of water has been designed for some purpose ... [But] the strength of our argument lies ... in the harmonious working of all the separate details.

To me the laws of nature afford the strongest evidences ... I do not, therefore, regard the constitution of water as some-thing apart from law ... nor do I believe that this argument from general plan could supply the place of the great argument from design. The last lies at the basis of natural theology . . .[178]

We have seen that from the very earliest times there have been strong criticisms of attempts to 'explain' the structure of inorganic and organic phenomena on the basis of teleological or eutaxiological design arguments. Most antagonistic objectors attempted to show that the principal arguments for design were confused or vacuous whilst sceptical or agnostic commentators held that all such issues were undecidable. Very few of the treatises on natural theology or teleological science ever attempted to deal with these criticisms in a convincing or systematic fashion. One interesting exception, whose work signals the end of the pre-modern approach to the question of final causes, was the French philosopher Paul Janet. His *Causes Finales* was translated into English[51] in 1878, several years after its publication in France and it provides a careful and moderately critical summary of ideas up to and including the Darwinian 'revolution'. Janet's work is characterized by a broad and undogmatic discussion of possible objections to a rightly conceived system of final causation which he defines at the outset in three points:

(I) There is no *a priori* principle of final causes. The final cause is an induction, a hypothesis, whose probability depends on the number and character of observed phenomena.
(II) The final cause is proved by the existence in fact of certain combinations, such that the accord of these combinations with a final phenomenon independent of them would be a mere chance, and that nature altogether must be explained by an accident.
(III) The relation of finality being once admitted as a law of the universe, the only hypothesis appropriate to our understanding that can account for this law, is that it is derived from an intelligent cause.[179]

In his second point we see that Janet seeks to exclude any arguments based upon development and concentrates instead upon the simultaneous realization of inorganic configurations. The system is not intended to possess the anthropocentric orientation of Paley of whom he does not approve because,

This anthropocentric doctrine as it has been called, appears to be connected with the geocentric doctrine, that made the earth the centre of the world, and ought to disappear with it.[180]

Janet then attempts to counter a number of criticisms, both ancient and modern, against the accusation that finalists have consistently confused causes for effects. He cites an example of the 'chicken and egg problem' in which the effect of reproduction is then the cause of further reproduction and acts

To perpetuate and to immortalize the species. Here, the order of causes is manifestly reversed, and whatever Lucretius and Spinoza may say, it is the causes that are the effects.

Janet then proceeds to argue against the claim, which he attributes to Maupertuis, (although Maupertuis merely cites Lucretius), that normalizing selection could have ensured the inevitable survival of ordered beings from random permutations. Like Paley, Janet asserts that there is no observational evidence for such a claim, but he glosses over the significance of the recent fossil finds. The theory of progressive evolutionary development, on the other hand, he cites approvingly as an excellent manifestation of final causes:

The progressive development of forms, far from being opposed to the theory of finality, is eminently favourable to it. What more simple and more rational law could have presided over creation than that of a progressive evolution, in virtue of which the world must have seen forms, more and more finished, successively appear?[181]

Janet hopes to follow Boyle and Leibniz in propounding a doctrine of complementarity where both mechanism and finalism provide different, but equally valid complementary descriptions of the same phenomena, each complete within its own sphere of reference. Janet then continues his discussion with an evaluation of what he terms certain 'contrary facts'; these include the presence of apparently useless or vestigial organs in animals. Interestingly, he discusses them in relation to the Least Action Principle, suggesting that they may be byproducts of the quest for the most economical path of development. He believes that the variational principles have some application in deciding the pathway of evolutionary development:

For that certain pieces of the organism have ceased to serve is no reason why they should entirely disappear. The law of economy is only a particular application of the metaphysical principle of the simplicity of ways, appealed to by Malebranche, or of the mathematical principle of least action, defended by Euler and Maupertuis.[182]

Janet then turns to discuss the status of final causation in a completely deterministic mechanical system, using Laplace's Nebular Hypothesis as the mechanical paradigm. He points out the logical equivalence of setting initial or final data for the evolution of a completely determined physical system. He also questions the notion of 'chaos' in completely determined systems[183] because however a random a system might *appear*, it should still have evolved deterministically from definite initial conditions and will likewise evolve towards a definite final state:

The primitive nebula was, then already the actual world potentially . . . But let it be observed, the nebula is not a chaos; it is a definite form, whence there is to issue later, in virtue of the laws of motion, an ordered world . . . If you do not admit anything that guides and directs phenomena, you at the same time admit that they are absolutely undetermined, that is to say, disordered: now how are you to pass from this absolute disorder to any order whatever?[184]

Janet has turned the argument against the evolutionist and the mechanist. In effect he is saying that determinism means we must suppose the Universe to have possessed very special initial conditions if human life is to result.

There follows a discussion of the pros and cons of evolutionary theories of organic development and the principle of natural selection. Janet argues against the sufficiency of the latter hypothesis on two grounds: first, he claims that although such ideas work in the context of forced breeding experiments—unnatural selection—the probability of a sufficient number of advantageous selections occurring naturally in the real world is extremely small. Secondly, he argues that adaptations tend not to be propagated, but rather are diluted in their offspring and this tends to keep a species invariant.

Janet's final discussions centre around the consequences of various theories of knowledge for his doctrine of final causes. Of particular interest is his discussion of Kant's claim that our knowledge of the world is a property of the observer not the observed. In the course of a lengthy discussion he cites a number of contemporary objections to Kant's thesis from the works of Trendelenburg and Herbart. If ordering is an inevitable selection effect created by our act of perception, why, he asks, do we find some things unintelligible and why do we not see *everything* as a teleological structure?

How is it . . . that the convenience of the arrangement of nature is only made evident in certain cases; that very often this convenience appears doubtful to us; in fine, that nature offers us a certain mechanical regularity, or even simple facts, of which it is impossible for us to give an account?[185]

Janet closes his work with a discussion of the final end of Nature. He has already rejected the anthropocentric notion that this end is Man, and now he also rules out the possibility that the Deity might have created all for himself for this would suggest his privation—a contradiction. Janet then meanders through various lesser possibilities in a style reminiscent of the Scholastics, before concurring with Kant that ethical goals provide the only ultimate meaning for Nature:

. . . if there are no ends in the universe, there are none for man any more than for nature; that there is no reason why the series of causes should be mechanical up to the appearance of man, and become teleological from man upwards. If

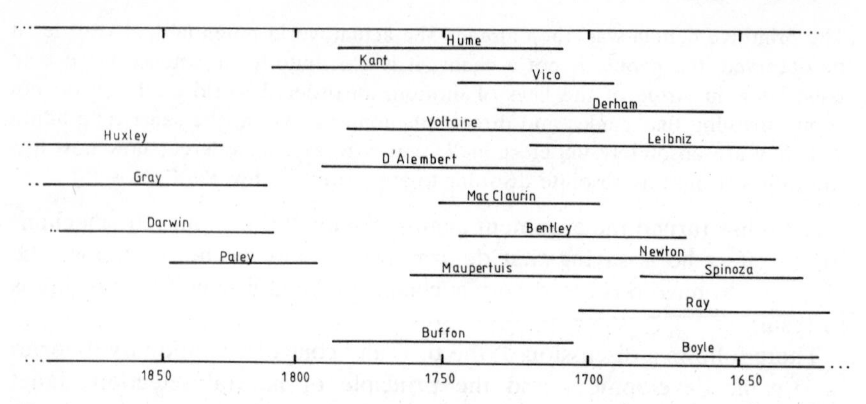

Figure 2.3. The chronology of the principal contributors to discussions of the Design Argument from the sixteenth until the end of the nineteenth centuries.

mechanism reigns in nature, it reigns everywhere, and in ethics as well as in physics ... Morality is, therefore, at once the accomplishment and the ultimate proof of the law of finality.[186]

Finally, we cannot resist citing our favourite Design Argument which is due to Bernadin de Saint-Pierre[187] which is of a type that distresses Janet very greatly. Indeed, Janet feels that it is a member of a class of examples which 'one could believe ... invented to ridicule the theory itself'. Bernadin claims that 'dogs are usually of two opposite colours, the one light and the other dark, in order that, wherever they may be in the house, they may be distinguished from the furniture, with the colour of which they might be confounded'.

2.8 Design in Non-Western Religion and Philosophy

> There was no confidence that the code
> of Nature's laws could ever be unveiled
> and read, because there was no
> assurance that a divine being, even
> more rational than ourselves, had ever
> formulated such a code capable of being read.
>
> J. Needham

Recently, the paleontologist Stephen Jay Gould characterized the Anthropic Principle as the latest manifestation of '... that age-old pitfall of Western intellectual life—the representation of raw hope gussied up as rationalized reality'.[194] He further warned: 'Always be suspicious of conclusions that ... follow comforting traditions of Western thought'.[194]

Actually, the idea that humanity is important to the cosmos and indeed the idea that the material world was created for man both seem to be present in many cultural traditions; they may even be universal. Although no study of non-Western teleology has ever been done, a cursory search of the anthropological literature shows teleological notions defended in Mayan,[195] Zuñi (New Mexico Indian),[196] the 'Thompson' Indian of the North Pacific coast,[197] Iroquois,[198] Sumerian,[199] Bantu,[205] ancient Egyptian,[200] Islamic-Persian,[201] and Chinese[204] traditions.

In the *Popal Vuh*, the most important surviving work of Mayan literature, it is recorded that the dry Earth and all life thereon was created by the gods for the benefit of mankind:

Let it be thus done. Let the waters retire and cease to obstruct, to the end that earth exist here, that it harden itself and show its surface, to the end that it be sown, and that the light of day shine in the heavens and upon the earth; for we shall receive neither glory nor honour from all that we have created and formed until human beings exist, endowed with sentience.[195]

In the Zuñi Indian creation myth, much of the material world, including the moon, planets, rain, and vegetation, was formed for the benefit of both Mankind and animals, who were viewed as the children of the Creator gods:

Thus, as a man and woman, spake [the Earth-mother and Sky-father], one to the other. 'Behold!' said the Earth-mother as a great terraced bowl appeared at hand and within it water, 'this is as upon me the homes of my tiny children shall be. On the rim of each world-country they wander in, terraced mountains shall stand, making in one region many, whereby country shall be known from country, and within each, place from place. Behold again! said she as she spat on the water and rapidly smote and stirred it with her fingers. Foam formed, gathering about the terraced rim, mounting higher and higher. 'Yea', said she, 'and from my bosom they shall draw nourishment, for in such as this shall they find the substance of life whence we were ourselves sustained, for see!' Then with her warm breath she blew across the terraces; white flecks of the foam broke away, and, floating over the water, were shattered by the cold breath of the Sky-father attending, and forthwith shed downward abundantly fine mist and spray! 'Even so, shall white clouds float up from the great waters at the borders of the world, and clustering about the mountain terraces of the horizons be borne aloft and abroad by the breaths of the surpassing of soulbeings, and of the children, and shall hardened and broken be by the cold, shedding downward, in rain spray, the water of life, even into the hollow places of my lap! For therein chiefly shall nestle our children mankind and creature-kind, for warmth in thy coldness'.

'Even so!' said the Sky-father; 'Yet not alone shalt thou helpful be unto our children, for behold!' and he spread his hand abroad with the palm downward and into all the wrinkles and crevices thereof he set the semblance of shining yellow corn grains; in the dark of the early world-dawn they gleamed like sparks of fire, and moved as his hand was moved over the bowl, shining up from and also moving in the depths of the water therein. 'See!' said he, pointing to the seven grains clasped by his thumb and four fingers', by such shall our children be guided; for behold, when the Sun-father is not nigh, and thy terraces are as the dark itself (being all hidden therein), then shall our children be guided by lights . . . Yea! and even as these grains gleam up from the water, so shall seed-grains like to them, yet numberless, spring up from thy bosom when touched by my waters, to nourish our children'. Thus and in other ways many devised they for their offspring.[196]

The 'Thompson' Indians of the North Pacific coast believed that the parts of the world were formed from five hairs which the Creator pulled from his head. The first two hairs chose to become women, the third the Earth, and

The fourth chose to be Fire in grass, trees, and all wood, for the good of man. The fifth became Water, to 'cleanse and make wise' the people. 'I will assist all things on earth to maintain life'.[197]

In the Iroquois origin myth the Earth was created primarily for the benefit of mankind by the people of the Sky World. The sky god Sapling created the first man out of red clay, and then made a compact between the Earth people and the people of Sky World:

I have made you master over the Earth and over all that it contains. It will continue to give comfort to my mind. I have planted human beings on the Earth for the purpose that they shall continue my work of creation by beautifying the Earth, by cultivating it and making it more pleasing for the habitation of man.[198]

Thus the Iroquois believed they had a mandate to change the Earth, in order to make it 'more pleasing for the habitation of man'.[198] A similar motif appears in some of the Sumerian origin legends. Human beings were created to serve the gods primarily by offering sacrifices and homage, but also by imitating the gods in creating and preserving the cosmic order.[199]

According to the Boshongo, a Bantu Tribe in central Africa, the Universal Creator Bumba walked among mankind, saying unto them 'Behold [the] wonders [of the Earth]. They belong to you.'[205]

The ancient Egyptian text *The Instruction of King Meri-ka-Re* (written *c.* 2000 BC) records

Men, the cattle of God, have been well provided for. He [the sun god] made the sky and the earth for their benefit . . . He made the air to vivify their nostrils, for they are his images, issued from his flesh. He shines in the sky, he makes plants and animals for them, birds and fish to feed them.[200]

This passage appears to represent the typical Egyptian tradition concerning the origin and purpose of mankind.

Islam is closely related to Christianity, for both are rooted in Judaism and both were influenced by Greek philosophy. Thus it is not surprising to find in Islam certain teleological ideas similar to those in Judaism and Christianity. Teleological concepts are prominent in the works of one of the most outstanding Muslim scientists, the Persian al-Bīrūnī. (*c*. 1000 AD). This scholar held that Man's intellect made him God's vice-regent (Khalīfat Allāh) on earth. Because Man is God's vice-regent, the world is ordered for his benefit, and he is granted power over God's creation.[201] The more abstract teleological ideas are also present in al-Bīrūnī's works. In his view, everything in Nature was ordered according to God's plan. As al-Bīrūnī put it: 'Praise therefore be unto Him who has arranged creation and created everything for the best.[202] ... there is no waste or deficiency in His Work'.[202] The idea in these passages are strikingly similar to the view of the Christian philosopher Leibniz, who contended that God has created the best of all possible worlds. The same notion of a perfectly ordered cosmos is found in both Christianity and Islam, for both religions have an omnipotent, omniscient, and perfect god who would naturally create a perfect world, a world in which no event or thing would be outside the Divine plan.

More subtle notions of teleology were evolved in Chinese civilization, a civilization which never possessed the concept of a Supreme Deity.[204] Like other peoples, the Chinese developed the idea that the Earth was made for Man, but early in their civilized history they were aware of the arguments against this rather naive form of teleology. The following story, taken from the book *Lich Tzu* attributed to the semi-legendary Taoist philosopher Lieh Yü-Khou (much of the book probably comes from the third century BC[197]) illustrates both:

> Mr. Thien, of the State of Chhi, was holding an ancestral banquet in his hall, to which a thousand guests had been invited. As he sat in their midst, many came up to him with presents of fish and game. Eyeing them approvingly, he exclaimed with unction; 'How generous is Heaven to man! Heaven makes the five kinds of grain to grow, and brings forth the finny and the feathered tribes, especially for our benefit'. All Mr. Thien's guests applauded this sentiment to the echo, except the twelve-year-old son of a Mr. Pao, who, regardless of seniority, came forward and said; 'It is not as my Lord says. The ten thousand creatures [in the universe] and we ourselves belong to the same category, that of living things, and in this category there is nothing noble and nothing mean. It is only by reason of size, strength or cunning, that one particular species gains mastery over another, or that one feeds upon another. None of them are produced in order to subserve the uses of others. Man catches and eats those that are fit for food, but how [could it be maintained that] Heaven produced them just for him? Mosquitoes and gnats suck [blood through] his skin; tigers and wolves devour his flesh—but we do not therefore assert that Heaven produced man for the benefit of mosquitoes and gnats, or to provide food for tigers and wolves'.[206]

Needham[206] cites this passage as an indication of the denial of general teleology by the Taoists, but we think it indicates an acceptance of naive teleology by most Chinese. Note that *all* except the boy agree with the teleological sentiments expressed by Mr. Thien. The criticism of teleology is probably placed in the mouth of a boy by the Taoist author in order to emphasize that the argument against naive teleology should be obvious even to a child.

China had two major indigenous philosophical systems: Taoism and Confucianism. The former was concerned primarily with the order of Nature while the latter concerned itself primarily the proper ordering of human society. These two branches of Chinese philosophy and Buddhism were partially merged by the Neo-Confucian philosophers in the Sung dynasty (eleventh and twelfth centuries AD). Among these scholars, the most important was Chu Hsi (1131–1200). In Neo-Confucian philosophy, social order was placed in Nature, but Nature took on certain aspects of social order. In the view of Chu Hsi, the vast 'pattern' of Nature was normal because it was inevitable that moral values and moral behaviour would appear when the Universe had developed sufficiently far.[208] Nevertheless, this Natural spontaneous moral order was not the result of conscious design:

> Someone also asked, 'When Heaven brings into being saints and sages, is it only the effect of chance, and not a matter of design?' The philosopher replied, 'How could Heaven and Earth say: 'We will now proceed to produce saints and sages? It simply comes about that the required quantities [of matter-energy] meet together in perfect mutual concordance, and thus a saint or a sage is born. And when this happens it looks as if Heaven had done it by design'.[209]

Chu Hsi's spontaneous ordering principle seems strikingly similar to Leibniz' pre-established harmony. Needham himself considers the emergent moral order in Chu Hsi's work as closely analogous to the Western idea of emergent evolution, defended in particular by Herbert Spencer, Henri Bergson, and Alfred North Whitehead (whose work we shall discuss at length in Chapter 3), wherein the moral order appears at later stages in the Universe's history. We see that a spontaneous ordering principle may or may not be teleological. It can properly be regarded as teleological only if the spontaneous order is generated as a consequence of the purposeful interaction of goal-directed organisms, or if the final state of the ordering process is emphasized over the initial and intermediate states. Otherwise, the spontaneous ordering principle is more properly regarded as eutaxiological.

The concept of spontaneous order has been central in Chinese philosophy from the dawn of Chinese civilization to the twentieth century. The idea probably arose as a result of the close observation of the growth of plants and the non-coercive social organization which develops spontaneously among human beings in primitive farming communities.[204] The following passage, by Liu Tsung-Yuan (773–819) a T'ang dynasty naturalist, illustrates both:

One day a customer asked ['Camel-Back' Kuo, a famous market-gardener, how he was so successful in growing plants], to which he replied: 'Old Camel-Back cannot make trees live or thrive. He can only let them follow their natural tendencies. In planting trees, be careful to set the root straight, to smooth the earth around them, to use good mould and ram it down well. Then, don't touch them, don't think about them, don't go and look at them, but leave them alone to take care of themselves, and Nature will do the rest. I only avoid trying to make my trees grow. I have no special method of cultivation, no special means for securing luxuriance of growth. I just don't spoil the fruit. I have no way of getting it either early or in abundance. Other gardeners set with bent root, and neglect the mould, heaping up either too much earth or too little. Or else they like their trees too much and become anxious about them, and are for ever running back and forth to see how they are growing; sometimes scratching them to make sure they are still alive, or shaking them to see if they are sufficiently firm in the ground; thus constantly interfering with the natural bias of the tree, and turning their care and affection into a bane and a curse. I just don't do those things. That's all'.

'Can these principles of yours be applied to government?' asked his listener. 'Ah', replied Camel-Back, 'I only understand market-gardening; government is not my trade. Still, in the village where I live, the officials are constantly issuing all kinds of orders, apparently out of compassion for the people, but really to their injury. Morning and night the underlings come round and say, 'His Honour bids us urge on your ploughing, hasten your planting, supervise your harvest. Do not delay with spinning and weaving. Take care of your children. Rear poultry and pigs. Come together when the drum beats. Be ready when the rattle goes'. Thus we poor people are badgered from morning till night. We haven't a moment to ourselves. How could anyone develop naturally under such conditions? It was this that brought about my deformity. And so it is with those who carry on the gardening business'.

'Thank you', said the listener. 'I simply asked about the management of trees, but I have learnt about the management of men. I will make this known, as a warning to government officials.'[210]

We have quoted at length a minor T'ang writer, but the same notion of spontaneous order appears over and over again in most Chinese philosophical writing, including the most influential works. For example, the *Tao Tê Ching* of Lao Tzu (fourth century BC), the most important of the Taoist books, considers the Tao to be simply a spontaneous ordering principle:

The supreme Tao, how it floods in every direction! This way and that, there is no place where it does not go. All things look to it for life, and it refuses none of them; Yet when its work is accomplished it possesses nothing. Clothing and nourishing all things, it does not lord it over them. Since it asks for nothing from them It may be classed among things of low estate; But since all things obey it without coercion It may be named supreme. It does not arrogate greatness to itself And so it fulfils its Greatness.[211]

In this passage, the action of the spontaneous ordering principle of Nature, the Tao, is contrasted with the order brought about by the conscious design of a ruler. The superiority of the order brought about spontaneously by human interaction over the order imposed from above by force is also a central motif in Confucian works. In fact, the early Confucians felt that the Tao of mankind was to be good, or rather to order naturally their relations with each other in mutually beneficial ways. They believed the ideal ruler would govern his people most effectively by his upright example rather than by force, as the following passage from the *Analects of Confucius* illustrates:

Chi Khang Tzu asked Confucius about the art of ruling. Confucius said, 'Ruling is straightening. If you lead along a straight way, who will dare go by a crooked one?'.[212]

Chi Khang Tzu asked Confucius about government, saying, 'Supposing we liquidated all those people who have not the Tao in order to help those who do have the Tao, what would you think about it?' Confucius replied, 'You are there to rule, not to kill. If you desire what is good, the people will be good. The natural ruler [chüntzu] has the virtue of wind, the people the virtue of grass. The grass must needs bend when the wind blows over it'.[213]

Similar remarks can be found throughout the works of the Confucians, at least through the tenth century AD; (see Chapter 9 of ref. 204 for representative examples). Politically, the Confucians can be regarded as China's native liberals. They were able to prevent the continuation, though not the formation, of a totalitarian state in China: the Chin Empire (second century BC). The advocates of such a state, the Legalists, were in the end defeated with the overthrow of the Chin and its replacement by the Han dynasty. The Legalists argued that the people should be governed according to positive law, *fa*, which were written rules expressing the arbitrary will of the supreme autocrat, while the Confucians, true to their tradition, countered that society should be ordered spontaneously according to evolved good customs, called *li*. It has been *li* rather than *fa* that has been the most significant force governing the day-to-day actions of the Chinese people from the formation of the Han Empire to the founding of the Republic in 1912. Needham argues[214] that such an emphasis on *li*, as opposed to *fa*, made it impossible for the Chinese to develop the concept of natural laws, which in the West, he believes, were originally pictured as decrees from the Supreme Ruler of the Universe, God. However this may be, it would be difficult for the notion of teleology, to be developed in Chinese philosophy and applied to the cosmos since cosmic teleology involves planning in some sense by a thinking being.

Nevertheless, there is a deep connection between teleology and spontaneous social order, a connection which has been pointed out by philosophers of the classical liberal tradition,[215] of whom Fredrick Hayek is the most distinguished representative in our own time. Hayek received his original university training in Vienna in Law, but spent the first thirty or so years of his career in economics (he was awarded the Nobel prize in economics in 1974) at the University of London. He has concentrated his attention on questions of social organization during the last thirty years. Like the Confucians, Hayek is primarily concerned with spontaneous order in human society. Human language is the most obvious example of such an order. It was not formed by the conscious design of any individual or group of individuals. Rather, it just grew. It is growing and changing now by the daily interactions of countless numbers of human beings. Hayek argues in scores of articles and many books (e.g. refs. 216–22) that the free market is a similar sort of order, an order which is created by the decentralized action of many minds, using far more information than is available or could be available to any one mind, thus generating an order much more complex than any one mind could even imagine. The market order cannot be said to have an overall purpose in the naive sense of the word. As Hayek puts it:

> Most important . . . is the relation of a spontaneous order to the conception of purpose. Since such an order has not been created by an outside agency, the order as such also can have no purpose, although its existence may be very serviceable to the individuals which move within such order. But in a different sense it may well be said that the order rests on purposive action of its elements, when 'purpose' would, of course, mean nothing more than their actions tend to secure the preservation or restoration of that order.[217]

In effect, the different and often conflicting purposes of the many human beings interacting via the market are woven together into an orderly whole; the entire system evolves in a direction none can foresee, because the knowledge dispersed throughout the system, and sustaining the order, is much greater than any individual can comprehend:

> Certainly nobody has yet succeeded in deliberately arranging all the activities that go on in a complex society. If anyone did ever succeed in fully organizing such a society, it would no longer make use of many minds, but would be altogether dependent on one mind; it would certainly not be very complex but extremely primitive—and so would soon be the mind whose knowledge and will determined everything. The facts which could enter into the design of such an order could be only those which were known and digested by this mind; and as only he could decide on action and thus gain experience, there would be none of that interplay of many minds in which alone mind can grow.[218]

Teleology is definitely present, for the human actors all have their own purposes, but it is teleology in the small, not a global teleology. The market system harmonizes these individual purposes, but it has none of its own. The image of the market and its spontaneous order developed by Hayek appears strikingly similar to the picture of spontaneously-ordering human society given by the Chinese sages in the quotes above. Hayek himself points out that his notion of spontaneous social order is closely analogous to the Greek *Kosmos*, which originally meant 'a right order for a community'.[219]

The precise and subtle relationship between a spontaneously ordered social system and the teleology of the beings who comprise it has recently been worked out by the political scientist Robert Axelrod. He has shown[220] on the basis of game theory that the spontaneous formation of a cooperative social order actually requires a very strong teleology to be acting at the individual level. That is, such an order can form spontaneously only if the future expectations of the individuals in the society are dominant over their immediate expectations in determining their present actions.

The barrier to the spontaneous formation of cooperation in a population of individuals without teleology is illustrated by the famous Prisoner's Dilemma. Two prisoners are in separate gaol cells and not permitted to communicate. Their gaoler urges each to confess, telling each that if he confesses to the crime and his partner does not, then the party that confesses will go free, while the other will get the maximum punishment of five years. If both confess, the confession of each will be worth less, so they both will get three years. If neither confess, then both will be convicted of only a minor charge, and each will get only one year. What action should the prisoners take? Consider the strategy of prisoner A. If the other prisoner B confesses, then A has no choice but to confess also since otherwise he would get five years rather than three. On the other hand, if B does not confess, then it is in the interest of A to confess since then he would go free. Thus, whatever B does, it is in A's interest to confess. Since the same analysis applies to B, we conclude that the best strategy for each to adopt is to confess. But the joint confession results in both getting three years rather than the one year they both would have received if they had cooperated. Nevertheless, it would be against the self-interest of each not to confess, even though both would be better off if neither confesses. The Prisoner's Dilemma is faced by every individual in many, if not every, interaction with other individuals; for it is always in the self-interest of an individual to get something for nothing; it is always in the self-interest of an individual to cheat another in any given interaction even though both might be better off if neither cheated! How then is it possible for cooperation to arise spontaneously in a group of individuals each pursuing his own interests?

Cooperation can arise because in general individuals will interact with a given individual not just once but many times. In the language of game theory, the Prisoner's Dilemma two-person positive-sum game must be replaced with a sequence of such games; the resulting game is termed an iterated Prisoner's Dilemma game. The pay-off matrix for the Prisoner's Dilemma game is

		Player (prisoner) A: Cooperate	Player (prisoner) A: Don't cooperate
Player B	Cooperate	Both players get R	A gets S and B gets T
	Don't cooperate	A gets T and B gets S	Both players get P

To fix ideas, let us choose $R=3$, $P=1$, $S=0$, and $T=5$. Then as in the example above, it is in the rational interest of both players not to cooperate, even though this means that they receive the pay-off P rather than the pay-off R which they would have both received if they had cooperated. In general, the Prisoner's Dilemma arises when $T>R>P>S$, with $R>(T+S)/2$, and both players must choose their strategy before they know what strategy the other chooses.

In the iterated Prisoner's Dilemma, the above game is played many times and the total pay-off is achieved over many games. However, the present value of a future pay-off is not as great as the present value of a present pay-off because a future good is not as valuable as a present good (if one is to receive a thousand pounds, it is better to receive it now rather than ten years from now), and also because there is some chance that the game will halt after a finite number of steps (in real life, interactions eventually will cease because one of the players dies, moves away, or becomes bankrupt). Therefore, the pay-off of each game is discounted relative to the previous game by a discount parameter w, where $0<w<1$. The expected cumulative pay-off of an infinite number of games is obtained by adding all the expected pay-offs from each game, where the expected pay-off of each game is obtained by multiplying the pay-off of the immediately preceding game by w. For example, the expected cumulative pay-off accruing to both players if they cooperate in all games would be given by $R+Rw+Rw^2+Rw^3+\ldots=R/(1-w)$ when the sum is an infinite geometrical progression. Cooperation becomes a possible rational strategy, because although a given player does not know the other's choice on the present game, he does know what the other chose on previous games. He can choose his strategy for the nth game according to what the other player has chosen on the preceding $(n-1)$ games with him.

The discount parameter w measures the importance of the future. One can prove[220,221] that only if w is sufficiently close to 1—i.e., only if the present value of future pay-offs is sufficiently high—is it possible for 'nice' strategies (those which have the player cooperate until the other player doesn't cooperate) to be a collectively stable strategy. Axelrod calls a strategy 'collectively stable' if, when everyone is using such a strategy, then no one can do better by adopting a different strategy.

In order for a strategy to persist in Nature, it must be collectively stable, for there will always arise by mutation individuals who try different strategies. In an evolutionary context, the collective stability of some cooperative strategies shows that, if a population of individuals using such strategies ever forms, it can persist and grow. Collectively stable strategies are essentially the same as what the evolutionist John Maynard Smith[223,224] has called 'evolutionarily stable strategies'. Furthermore, Axelrod shows that a population of non-cooperators can be successfully invaded by clusters of cooperators if w is high enough and if the relative frequency with which the cooperators interact with each other rather than with the non-cooperators is sufficiently high. For example, if we have $T=5$, $R=3$, $P=1$, $S=0$, and $w=0.9$ then a cluster of individuals using the 'nice' strategy of 'cooperate until the other does not, then don't cooperate for one game, then cooperate until the other does not cooperate again' can successfully invade a population of non-cooperators if only 5 per cent of their interactions are with other members of the cluster. (Individual cooperators cannot invade a population of non-cooperators because the strategy of total non-cooperation is also collectively stable. But a cluster of non-cooperators cannot invade a population of cooperators using a collectively stable strategy.)

These ideas have been applied to the evolution of cooperative behaviour by Axelrod and Hamilton,[220,222] and by Maynard Smith.[223,224,225] More speculatively, these results showing the importance of teleology for the formation of order suggest that if one wishes to model the physical cosmos after a biological or social cosmos—this is the idea underlying Wheeler's Participatory Anthropic Principle—then the state of the universe at any given time must be determined not only by the state of the universe an instant before, which is the usual physical assumption, but rather it must be a function of all the preceding states and all the future states.

2.9 Relationship Between The Design Argument and The Cosmological Argument

> To someone who could grasp the Universe from a unified standpoint the entire creation would appear as a unique truth and necessity.
>
> J. D'Alembert

> The unrest which keeps the never stopping clock of metaphysics going is the thought that the non-existence of the world is just as possible as its existence.
>
> W. James

The cosmological argument is today probably the most often used theological existence proof. It was the main argument used by Father Copleston against Bertrand Russell in their famous BBC debate on the existence of God,[226] and there have been several books written recently defending this argument; see refs 227–9. The argument in its most common version is based on two assumptions: (1) something exists, and (2) there must be a sufficient reason for everything that exists. The argument begins with the claim that the existence of everything in the Universe is merely contingent; that is, it could be otherwise than it is. To use an example of Matson,[230] this book could have been as it is except for one extra typo. The book as it is is contingent, since we would think that the number of typos in the book is not a logically necessary feature of the Universe; we would not expect the Universe to be logically inconsistent with the extra typo. Thus by the principle of sufficient reason, there must be a reason why that typo is not there, namely our sharp eyes. But we are also apparently contingent, which means there has to be a reason (or cause) for our own existence. And so it goes for everything in the Universe upon which we are dependent. It is now contended that these other objects—which at this stage in the argument include everything in the Universe—must have a reason (cause) for existence. Since it has been argued that everything in the Universe must be explained in terms of something else, this other reason must be outside the Universe. Furthermore, this transcendental reason must be the final reason, for otherwise the hierarchy of causes would continue without end, and this hierarchy would itself want a reason. In order to avoid the charge that the Final Cause itself needs a cause, the defenders claim it is its own cause. We should emphasize that the hierarchy of 'causes' which is referred to in the cosmological argument does not refer to a hierarchy of causes in the sense of a series of causes preceding effects in time. It refers rather to hierarchy of underlying 'reasons' for events which are perceived by the human mind. Another example of such a hierarchy is as follows: (first level) our writing of these words has as its physical cause the muscle fibres in our arms, the cells of which (second level) obey the laws of chemistry, the laws of chemistry being derived from (third level) atomic physics, and finally the laws of atomic physics are determined by (fourth level) quantum physics. At present, we are forced to accept quantum physics as a brute fact, but physicists feel in their heart of hearts that there is some reason why the Universe runs according to quantum physics rather than, say, Newtonian physics. (There is a temporal version of the cosmological argument, with 'cause' being followed by 'effect' which is another cause, but as this version is easily demolished, no major theologian in the past thousand years has defended it. For instance, Aquinas did not accept the temporal version: he did not believe it was possible to show by reason alone whether or not the Universe had a beginning.)

There are many problems with the non-temporal version of the cosmological argument, which is often called the argument from contingency. We shall focus only on those difficulties which are relevant to the Anthropic Principle; the reader can consult references 231 and 232 for a more complete discussion. One obvious problem with the argument is, why should we accept its minor premise? Why should the principle of sufficient reason be true? The defenders of the cosmological argument feel that the Universe must at bottom be rational, but again, why should it? Antony Flew, who is the most profound of the contemporary critics of theism, points out[231] that not only is the principle of sufficient reason unjustified, but it is actually demonstrably false! *Any* logical argument must start with some assumptions, and these assumptions must themselves be unjustified. We might of course be able to justify these particular assumptions in the context of another demonstration from which the particular assumptions are deduced, but this just pushes the problem to a higher level; the basic underlying assumptions in the higher level argument are themselves unjustified. At some point we have to just accept some postulates for which we can give no reason why they should be true. Thus the nature of logic itself requires the principle of sufficient reason to be false.[233]

Nevertheless, by insisting that the Universe is rational—which really means that the Universe has a causal structure which can be ordered by the human mind, and further that the ultimate reason for the existence of the Universe can be understood by human beings—the defenders of the cosmological argument are taking an Anthropic Principle position. In its insistence that there is an actual hierarchy of causes in the Universe which is isomorphic to the pyramid of causes constructed by human beings, the cosmological argument is analogous to the teleological argument, for the latter argument asserts that the order observed in the Universe is isomorphic to the order produced by human beings when they construct artifacts. In both arguments the mental activities of human beings are used as a model for the Universe as a whole.

The major premise in the cosmological argument is that things exist, and further, that contingent things exist. But Hume in his *Dialogues* pointed out contingency could be just an illusion of our ignorance:

> To a superficial observer so wonderful a regularity [as a complex property of the integers] may be admired as the effect either of chance or design; but a skilled algebraist immediately concludes it to be the work of necessity, and demonstrates that it must forever result from the nature of these numbers. Is it not probable, I [Philo] ask, that the whole economy of the Universe is conducted by a like necessity, though no human algebra can furnish a key which solves the difficulty?[234]

Most defenders of the cosmological argument have dismissed Hume's objection, but many modern cosmologists are coming to the conclusion that there is only one logically possible universe. These modern cosmologists have hubris that Hume's alter-ego, Philo, would have blanched at: some of them believe they have found the key (or rather, keys) which will permit a mathematical description of this single logically possible universe! For example, Hartle and Hawking[235] and Hawking[236] have obtained an expression for the 'wave function of the Universe', using path integral techniques. The wave function of the Universe, regarded as a function of three spatial variables and a time variable, is essentially a list of all possible histories, classical and otherwise, through which the universe could have evolved to its present quantum state, which itself includes all logically possible particles and arrangements of particles that could exist in the Universe at the present time. If we accept the Many-Worlds interpretation of quantum mechanics—as Hartle and Hawking do—then all these possibilities *actually* exist. In other words, the Universe, which is defined as everything that actually exists, is equal to all logically consistent possiblities. What more could there possibly be? Furthermore, there are strong indications that the mathematical structure of quantum mechanics requires that all observationally distinguishable possibilities are actually realized. More precisely, the Universal wave function can be shown to have only isolated zeros, if it is an eigenstate of the energy—the Hartle–Hawking Universal wave function is such an eigenstate—and if the Universal Hamiltonian is a self-adjoint operator.[237,238] This means that such a wave function is non-zero almost everywhere on the domain of possibilities. It is impossible to distinguish observationally between a function which is non-zero everywhere and one which is non-zero almost everywhere. If it could be proved that the mathematical structure assumed for quantum mechanics were logically necessary, then we would have a proof that only one unique Universe—the one we live in—is logically possible. The above discussion sounds a bit woolly, but it is possible to make predictions by restricting attention to a few parameters of the domain of possibilities. See Chapter 7 for a detailed analysis, from the point of view of the Many-Worlds interpretation, of a Universal wave function in which the only possibility considered is the radius of the sidereal Universe. As we shall point out in Chapter 3, the mathematician-philosopher A. N. Whitehead was the first to suggest that the problem of contingency might be solved if the actual Universe realized all possibilities: if this were the case, there would be no contingency in the large.

The remainder of the cosmological argument's major premise is the assertion 'something exists'. This is a rather unobjectionable postulate, as it wins the assent of realists, idealists, and even solipsists. Nevertheless, the nerve of the cosmological argument lies in creating the suspicion that the entire Universe, even if it is necessarily unique, may want a reason for existence: that is, it suggests we should ask the question, 'why is there something rather than nothing?'

This question has an answer only if there is something whose existence is logically necessary, which is to say, that the denial of its existence would be a logical contradiction. This brings us to the ontological argument for the existence of God, a proof claiming to deduce the existence of God from His definition.

The ontological argument has had a rather mixed reception by theologians and philosophers since its introduction by St. Anselm in 1078.[239] Aquinas did not accept it as valid, nor have the vast majority of modern philosophers, but both Descartes and Leibniz did believe it to be valid. As Kant and, more recently, Antony Flew[231] have pointed out, the explicit rejection of the ontological argument puts those theologians who accept the cosmological argument in a difficult position, because the cosmological argument assumes that there is a Final Cause who is Its own reason for existence, and only if the existence of this Final Cause is logically necessary will it be superfluous to find a reason for its existence. At bottom, the cosmological argument presupposes the validity of the ontological argument.

The reason Kant gave for the invalidity of the argument is basically the one which modern philosophers find convincing: existence is not a property, but rather it is a precondition for something to have properties. An example will make this clearer. It certainly makes sense to say 'some black lions exist', but the statement 'some black lions exist, and some don't' is conceptually meaningless. It is meaningless because although 'black' can be a property of lions, 'existence' cannot. Modern logicians have constructed a notation that makes it impossible to formulate in the notation existential sentences like 'some black lions exist, and some don't', which are grammatically correct in English but conceptually meaningless. In this notation, 'X exists' means 'X has an instance'. Another criticism that can be levelled against the ontological argument is that the concept of logical necessity applies to propositions, not to questions of existence.

In our opinion, these criticisms of the ontological argument are correct; it is not possible to deduce the existence of any single being from its definition. But a caveat must be made. If the Universe is by definition the totality of everything that exists, it is a logical impossibility for the entity 'God,' whatever He is, to be outside the Universe if in fact He exists. By definition, nothing which exists can be outside the Universe. This is a viewpoint which more and more twentieth-century theologians are coming to hold: they are beginning to adopt a notion of deity which insofar as questions of existence are concerned, is indistinguishable from pantheism.[245] As Paul Tillich succinctly put it, 'God is being-itself, not *a* being'.[240] We do not concern ourselves with whether it is appropriate for a theologian defending such a position to call himself a theist, as most of them do. (The atheist George Smith has subjected such theologians to a very witty and scathing criticism on this point in ref. 232.) Rather, we are interested in the truly important implication of this notion of deity, which is that in the context of such a notion, the purpose of the ontological argument is to establish the existence of the Universe, or equivalently, the existence of something, as logically necessary. This is the caveat to the above-mentioned refution of ontological argument which we wish to consider: granted that the existence of no single being is logically necessary, could it nevertheless be true that it would be a logical contradiction for the entire Universe, which is not *a* being, but all being considered as a whole, not to exist? If the Universe must necessarily exist, then modern logical notation cannot be applied to the single unique case of the ontology of the Universe, but it would be valid in every other situation.

Even philosophical atheists differ as to the validity of the cosmological/ontological argument interpreted in such a way. David Hume, in the *persona* of Cleanthes, admitted that *if* the logic of the cosmological argument were sound, then

> ... why may not the material universe be the necessarily existent Being, according to this pretended explication of necessity? We dare not affirm that we know all the qualities of matter; and, for aught we can determine, it may contain some qualities which, were they known, would make its non-existence appear as great a contradiction as that twice two is five.[234]

Bertrand Russell, on the other hand, thought we had to accept the existence of the Universe as an irreducible brute fact. As he expressed it in his BBC debate with Copleston:

COPLESTON: Then you would agree with Sartre that the universe is what he calls 'gratuitous'? RUSSELL: Well, the word 'gratuitous' suggests it might be something else; I should say that the universe is just there, and that's all.[241]

The reason 'something exists' may be necessarily true arises from a close analysis of what the word 'existence' means. An entity X is said to exist if it is possible, at least in principle, to observe it, or to infer it from observation of other entities. If it is claimed that X had, has, and will have no influence whatsoever on *anything* we can possibly observe, then by definition X does not exist.[242] But again by definition, there must exist this ill-defined entity we have termed 'the observer' to act as an arbiter for the existence of everything else, which implies that something—the observer, at least—actually exists. This argument will be immediately recognized as a self-reference argument, a category of arguments to which all the Anthropic Principle arguments belong. To put the argument another way, the phrase 'nothing exists' is logically contradictory because the phrase 'X has an instance' (equivalent to 'X exists' in modern logic) means that there is an 'observer' who can, at least in principle, observe X. Thus 'nothing has an instance' would mean that an observer has not observed anything. But the observer has observed himself, or at least *he* himself exists, which means it is not true that 'nothing has an instance.' *Cognito ergo sum*. Assuming the truth of 'nothing has an instance' implies its falsity, which means that it is contradictory and hence false.

We do not defend this self-reference argument: we merely note it because of its Anthropic Principle flavour. The philosopher Charles Hartshorne, who is generally recognized as the most influential defender of the ontological argument in the twentieth century,[239] is a pantheist in the sense described above in his ontology, and he believes that 'something exists' is a logically necessary truth.[243] For Hartshorne, the phrase, 'God exists necessarily' means that the non-existence of the Universe is a logical contradiction. (His critics, e.g. Hick,[239] seem unaware of this, and base their refutation of his arguments on another, more traditional concept of deity.) If one does not accept the non-existence of the Universe as logically contradictory, then one is forced into Bertrand Russell's position of regarding the Universe's existence as a brute fact.[244] But if the speculations of some modern cosmologists are correct, there may be only one unique Universe which is logically possible, and the assumption of the Universe's existence is the only assumption we have to make.

In this chapter we have attempted to outline the history of Design Arguments and the philosophical debates surrounding them. In this way we have been able to introduce some of the questions touched upon by the modern Anthropic Principles. At the very least we aim to have shown that the Anthropic Principle is not the new and revolutionary idea that many scientists see it to be. We have argued that the Anthropic Principles are but a modern manifestation of the traditional tendency to frame design arguments around successful mathematical models of Nature. Investigation reveals there to have existed quite distinct teleological and eutaxiological Design Arguments whose divide mirrors the divide between different varieties of Anthropic Principle. We found that both Western and Eastern cultures acquired an interest in the question of design. What characterizes the European interest especially is the use of these arguments to prove the existence of a deity from the apparent purpose or harmonious workings of the machinery of Nature. The surprising persuasiveness of such arguments can be traced to the dramatic success of the Newtonian approach to science to which they were wedded. This led us to consider the famous Cosmological Argument for the existence of God in some detail and discuss its connections with the Anthropic Principles. The blow dealt by Darwin to the traditional design arguments founded upon the existence of environmental adaption revealed two interesting features. On the one hand the early Design Arguments played a key role in leading Darwin to develop his theory of natural selection but on the other we must recognize that this advance still left untouched most of the design arguments of the day that were framed around non-biological phenomena. It is this class of eutaxiological design argument that has evolved into the more precise examples which motivated the modern Anthropic arguments. One of the most interesting features of the world to emerge from study of biological populations has been the possibility that order can develop spontaneously. Modern ideas concerning the spontaneous generation of order in social systems were discussed together with the relevance of this for the teleological behaviour of their members. This departure prepares the ground for a more detailed investigation of the use of teleological arguments in science and philosophy in the next chapter.

References

1. D. E. Gershenson and D. A. Greenberg, *Anaxagoras and the birth of physics* (Blaisdell, NY, 1964).
2. These fragments are listed in ref. 1 along with subsequent citations. They are also listed in G. S. Kirk and J. E. Raven, *The presocratic philosophers: a critical history with a selection of texts* (Cambridge University Press, 1957). Hereafter we shall reference fragments according to their catalogue number in Kirk and Raven, for example as KR999.
3. See Simplicius, *Physics* **164,** 24; KR504 'mind controlled also the whole rotation, so that it began to rotate in the beginning. And it began to rotate first from a small area ... Mind arranged ... this rotation in which we are now rotating, the stars, the sun and the moon'.
4. Anaxagoras, KR504.

5. Aristotle, *Parts of animals*, 687a7.
6. Aristotle, *Metaphysics*, 985a18.
7. Plato, *Phaedo*, 98B7 (KR522). For some further criticism, see Simplicius, *Physics* **327,** 26.
8. Aristotle, *De Caelo* **2,** 300b20. See also KR444151617.
9. J. A. Wheeler, 'Genesis and observership', in *Foundational problems in the special sciences*, ed. R. Butts and J. Hintikka (Reidel, Dordrecht, 1977).
10. Plato, *The Republic*, Bk7.
11. Xenophon, *Memorabilia*, I(1), 10.
12. Xenophon, *Memorabilia*, IV(3), 5.
13. Diogenes, KR604.
14. KR505.
15. KR564.
16. For a discussion of Aristotle's cosmology see *Theories of the universe*, ed. M. K. Munitz (I.U. Free Press, Glencoe, 1957).
17. Aristotle, *Physics* Bk2, §2.
18. Aristotle, cited in *On man in the universe* (W. Black, NY, 1943) p. xiii.
19. Aristotle, *Parts of animals* Bk1, 1. There exists at least one example of how these unusual metaphysical ideas, so different from those of modern scientists, could be effective in making predictions. The Alexandrian, Hero (*c.* 50) was able to deduce the law of optical reflection from the Aristotelian 'axiom' that everything strives towards an optimal, perfect end. He interpreted this to mean that light rays should always traverse the shortest path available to them—this was the first use of a variational principle. See § 3.4 for further discussions of Hero's work.
20. Aristotle, *Physics*, cited by H. Osborne in *From the Greeks to Darwin*, 2nd edn (Scribners, NY, 1929), p. 85.
21. Aristotle, *Parts of animals*, II, 1.
22. Aristotle, *Physics*, II, 8.
23. Theophrastus, *Metaphysics*, transl. W. D. Ross and F. H. Forbes (Clarendon Press, Oxford, 1929).
24. Note that he does not wish to outlaw the study of teleology completely because he feels the study of phenomena apparently exhibiting design or remarkable contrivance to be legitimate. He merely wishes to encourage a little more scepticism in the deployment of explanations based on final causation. This is rather similar to modern approaches to 'design'. The Anthropic Principle picks on a large number of remarkable coincidences of Nature for examination but does not aim to use teleological reasoning to explain specific local phenomena.
25. Epicurus, *Letter to Pythocles; Epicurus: the extant remains*, transl. and notes by C. Bailey (G. Olms, Hildesheim, 1970).
26. Lucretius, *On the nature of the universe*, transl. R. E. Latham (Penguin, London, 1951).'
27. ibid., p. 196.
28. ibid., p. 156.
29. S. Jaki, *Science and creation* (Scottish Academic Press, Edinburgh, 1974); F.

J. Tipler in *Essays in general relativity*, ed. F. J. Tipler (Academic Press, San Francisco, 1980), p. 21.

30. Cicero, *The nature of the gods*, transl. H. C. P. McGregor (Penguin, London, 1972), p. 89.
31. ibid., p. 161.
32. ibid., 162.
33. ibid., p. 163.
34. Galen, *On the usefulness of the parts of the body*, transl. M. T. May (Cornell University Press, NY, 1968).
35. Pliny, cited in C. Singer, *A short history of scientific ideas to 1900* (Oxford University Press, Oxford, 1959), p. 106.
36. A. M. S. Boethius, *The consolation of philosophy*, transl. J. F. Steward and E. K. Rand (Wm. Heinemann, NY, 1918).
37. ibid., BkI, vi.
38. ibid., BkII, prose vii.
39. C. S. Lewis, *The discarded image: an introduction to medieval and renaissance literature*, (Cambridge University Press, Cambridge, 1964), p. 10.
40. C.U.M. Smith, *The problem of life: an essay in the origins of biological thought*, (Macmillan, London, 1976).
41. Averroes, *Metaphysics*, transl. M. Horten (M. Niemeyer, Halle, 1972), p. 200.
42. The Old Testament is, of course, full of particular Design Arguments. For the Jews, as God's chosen race, the idea of teleology would have been completely accepted and such notions as the 'Day of the Lord' evidence of ultimate Final Causation.
43. Maimonides, *Guide for the perplexed*, 2nd edn, transl. M. Friedländer (Routledge, London, 1956).
44. ibid., BkIII, Chapter 14.
45. Aquinas, *Summa theologica*, Q.2, Art 3; see also F. C. Copleston, *Aquinas*, (Penguin, London, 1955).
46. R. Bacon, *De sapientia veterum*, in *Works*, ed. R. Ellis, J. Spedding, and D. Heath (Longmans, London, 1875–9), VI, p. 747.
47. Raymonde, '*Theologia naturalis sive liber creaturarum*', see C. C. J. Webb, *Studies in the history of natural theology* (Cambridge University Press, Cambridge, 1915).
48. N. Copernicus, *On the revolution of the heavenly spheres*, transl. C. G. Wallis, ed. R. M. Hutchins (Encyclopaedia Britannica, London, 1952). See also *The nature of scientific discovery: Copernicus symposium*, ed. O. Gingerich (Smithsonian, Washington, 1975).
49. ibid., Bk1, Chapter 6.
50. G. Galileo, *Dialogues concerning two new sciences* (Dover, NY, 1953), III, p. 400.
51. P. Janet, *Final causes*, transl. W. Affleck (Clark, Edinburgh, 1878).
52. The quote is taken from p. 154.
53. For an interesting biography of Kepler by a strong admirer of his work see, A. Koestler, *The sleepwalkers* (Grosset & Dunlap, NY, 1970). The section

on Kepler in this work was previously published as *The watershed: a biography of Johannes Kepler* (Doubleday, NY, 1960).

54. M. Montaigne, *Essays*, ii, xii, transl. E. J. Trechmann (Oxford University Press, London, 1927).
55. ibid., ii, xii.
56. The earliest revival of 'atomism' may be N. Hill, *Philosophia epicurea* (Paris, 1601), who tried to make it theologically respectable by maintaining that the atoms and their motions were initiated by God, see G. McColly, *Ann. Sci.* **4,** 390 (1949). The more familiar atomic revivalist is P. Gassendi, *Observations on the tenth book of Diogenes Laertius* (Lyon, 1649) and it appears he also took the view that a Deity must make the atoms.
57. F. Bacon, *De augmentis scientiarum* BkIII, Chapter 5 (1623), and *The philosophical works of Francis Bacon*, ed. J. M. Robertson (Routledge, London, 1905).
58. ibid., p. 96–7.
59. W. Harvey, *Anatomical exercises in the generation of animals* (London, 1653), exercise 11.
60. R. Boyle, *A disquisition about the final causes of natural things* (London, 1688), p. 157.
61. W. Harvey, The motion of the heart and blood. Chapter 8, in *Great books of the Western World*', ed. R. M. Hutchins (Encyclopaedia Britannica, London, 1952), Vol. 28.
62. The foundation of modern anatomical study began with Vesalius', *De humani corporis fabrica* (1543, Basilaae; reprint by Culture et Civilization, Brussels, 1964) which contained the results of a large number of dissections and which corrected many accepted dogmas of Galen. It was also the first printed book to contain diagrams. Formerly these were absent from medical treatises, and illustrations were simply reproductions of classical anatomical configurations.
63. R. Descartes, *Principles of philosophy*, III, 3; *The philosophical works of Descartes*, ed. and transl. E. S. Haldane and G. R. T. Ross, 2 vols (Cambridge University Press, Cambridge, 1911–12).
64. R. Descartes, *Le monde*, ed. V. Cousin (F. Levrault, Paris, 1824–6), Chapter 6, p. 249.
65. R. Descartes, *Principles of philosophy* III, 2, op. cit.
66. ibid., III, 4.
67. The design argument from a mechanical world model was seen also in Cicero and in many early Stoic writings.
68. R. Boyle, from *The Christian virtuoso*, cited in *Anglicanism: the thought and practice of the Church of England, illustrated from the religious literature of the seventeenth century*, ed. P. E. More and F. L. Cross (SPCK, London, 1935), p. 235.
69. R. Boyle, *A disquisition about the final causes of natural things* (London, 1688), p. 522.
70. R. Boulton (ed.) *The theological works of the Honourable Robert Boyle* (London, 1715), II, p. 235.
71. ibid., II, pp. 211–12.
72. ibid., II, pp. 221, 251.

73. R. Boyle, ref. 69, p. 528.
74. R. Boyle, *Letter on final causes*, a reply to Descartes cited in Janet, ref. 51, p. 481.
75. P. Gassendi, *Objections to the 4th meditation*, Vol. II (Amsterdam, 1642), p. 179.
76. R. Boyle, ref. 69, p. 520.
77. J. Webster, *Academiarum examen* (London, 1654), p. 15.
78. J. Ray, *The widsom of God manifested in the works of creation* (London, 1691). This work went through twenty editions between 1691 and 1846.
79. J. Ray, cited in L. E. Hicks, *A critique of Design Arguments* (Scribner, NY, 1883).
80. R. Cudworth, *The intellectual system of the universe*, with notes and dissertations of J. L. Moshiem (Thomas Tegg, London, 1845). For further discussion of 'Plastic Nature' and Ray's interpretation of it see C. E. Raven, *John Ray: naturalist: his life and works* (Cambridge University Press, Cambridge, 1950) and C. E. Raven, *Organic design: a study of scientific thought from Ray to Paley*, 7th Dr. Williams' Library Lecture (Oxford University Press, Oxford, 1953).
81. B. Spinoza, *Ethics* (see *The chief works of Benedict de Spinoza*, transl. R. H. Elves, 2 vols, Bohn's Philosophical Library, repr. Dover, NY, 1951), appendix to Part I. The argument of Spinoza against the Design Argument is exactly the same as Cicero's *for* it—that men from underground would with 'stupid amazement' ascribe the regularities of Nature to a Deity.
82. B. Spinoza, ibid.
83. B. Spinoza, ibid.
84. B. Spinoza, ibid.
85. I. Newton, *The reasonableness and certainty of the Christian religion* (London, 1700) BkII, 18. Elsewhere it appears that, although extremely pious, Newton preferred to allow others to make use of his ideas in support of the Design Argument rather than defend it himself directly. His views here were those of the Protestant orthodoxy of his day. The philosophical influence of his discovery of the *universal* law of gravity and the first constant of Nature can be seen in William Whiston's work *Astronomical principles of religion* (London, 1717) which was dedicated to Newton. On p. 131 he writes, 'The Universe appears thereby to be evidently One Universe; governed by One Law of Gravity through the whole; and observing the same laws of motion everywhere so that this Unity of God is now for ever established by that more certain knowledge we have of the Universe'.
86. I. Newton, *Opticks*, 4th edn (London, 1730), query 28.
87. R. Bentley, *A confutation of atheism from the Origin and frame of the world* (London, 1693).
88. The Newton–Bentley letters are reprinted in I. B. Cohen, *Isaac Newton's papers and letters on natural philosophy and related documents* (Harvard University Press, Cambridge, Mass., 1958).
89. D. Gregory, cited in H. Guerlac and M. C. Jacob, *J. Hist. Ideas* **30,** 307 (1969).
90. C. MacClaurin, *An account of Sir Isaac Newton's philosophical discoveries* (London, 1748), p. 405, see also p. 400.
91. G. Leibniz, in *The philosophic works of Leibniz*, ed. G. M. Duncan (London,

1890). On p. 101 he writes, 'the present state exists because it follows from the nature of God that he should prefer the most perfect'.

92. ibid., p. 36; this was a letter to Boyle.
93. G. Leibniz, in letter to de Volder; C. I. Gerhardt, *Die philosophischen Schriften von G. W. Leibniz* (Georg. Olms, Berlin, 1960), Vol. 2, p. 193.
94. Parts of the present Universe which have apparently never been in causal contact during its entire past history exhibit the same large-scale density and temperature to within one part in ten thousand, see J. D. Barrow and J. Silk, *Scient. Am.* (April 1979). The reason for this synchronization is a key problem of modern cosmology, see J. D. Barrow and M. S. Turner, *Nature* **292,** 35 (1981).
95. G. Leibniz, *Refutations of Spinoza* in ref. 91, p. 176.
96. Second letter to Clarke, ref. 91, pp. 241–2.
97. N. Grew, *Cosmologia sacra* (London, 1701).
98. J. B. Molière, *Dom Juan, ou, Le festin de pierre,* Act 3, scene 1 (1665), cited in P. Janet, *Final causes,* see ref. 51, p. 291; a modern edition is ed. W. Howarth (Blackwell, Oxford, 1958).
99. Voltaire, 'Atheist atheism', *Philosophical dictionary* (1769), transl. and ed. P. Gay, 2 vols (Basic Books, NY, 1955).
100. Ariticle on 'Causes finales' in ref. 99, **1,** 271.
101. Voltaire, *Candide* (Washington Square Press, NY, 1962).
102. J. D'Alembert, *Traite de dynamique*: *discours preliminaire* (1742), transl. Y. Elkane. The particular errors of Descartes he is referring to were his rules of collision, that (1) if two bodies have equal mass and velocity before collision then after any collision they will have the same speeds as before, and (2) if two bodies have different mass then the lighter body is reflected and its velocity becomes equal to that of the larger one. Leibniz showed that these rules contradict the requirement of continuity on approach to the situation where the two masses are equal. This inconsistency was first pointed out by Leibniz in 1692 although it was not published until 1844.
103. P. L. M. Maupertuis, 'Essai de cosmologie' (1751) in *Oeuvres,* Vol. 4, p. 3 (1768), (Lyon).
104. ibid.
105. ibid.
106. ibid.
107. W. Derham, *Physico-Theology* (London, 1714).
108. W. Derham, *Astro-theology* (London, 1795). The titles of both Derham's books have a familiar ring to them. During the eighteenth century countless such works of natural theology were written. The naturalist Lesser wrote a sequence of books entitled *Helio-theologie* (1744), *Testaceo-theologie* (1757), '*Insecto-theologie*' (1757), whilst Fabricus authored a *Theologie de l'eau* (1741).
109. G. de Buffon, 'History of animals', (in *Natural history,* transl. H. D. Symonds, London, 1797), Chapter 1. Buffon also doubted that it could be argued that the entire collection of celestial bodies and motions could be conceived as contrived for the service of humans.

110. For a discussion of his work see I. Berlin, *Vico and Herder: two studies in the history of ideas* (Hogarth Press, London, 1976).
111. G. Vico, cited in *On the study methods of our time* (The Library of Liberal Arts, Bobbs-Merrill, NY, 1965).
112. The other categories were '*Conscienze*'—the behavioural and everyday information common to all men—then intuition regarding ultimate things, and finally, human psychology.
113. D. Hume, *Dialogues concerning natural religion*, ed. N. Kemp Smith (Bobbs Merrill, Indiana, 1977). First published in 1779, probably in Edinburgh, but this is not certain. See J. V. Price, *Papers Bibliog. Soc. Am.* **68,** 119 (1974).
114. For some discussion of the relation between Newton's influence and Hume's writings see R. H. Hurlbrutt III, *Hume, Newton and the Design Argument* (University of Nebraska Press, Lincoln, 1965).
115. Hume, op. cit., Part II.
116. ibid., Part II.
117. ibid., Part VIII.
118. ibid., Part VIII.
119. D. Hume, *An enquiry concerning human understanding* (London, 1748), § 11. Hume would probably have appeared a 'crank' to the Newtonians because of these outmoded vitalist views.
120. He also draws attention to the fallacy of composition which is latent in the classical argument that if everything in the Universe has a cause then so must the Universe. For example, although every member of a club has a mother it certainly does not follow that every club has a mother! The original argument is based upon a simple confusion between different logical classes. See section 2.9 for further discussion.
121. Ref. 113.
122. Boswell recalls that Johnson once left the room when Hume entered, see J. Boswell, *The life of Samuel Johnson* (Oxford University Press, London 1953). Johnson was, of course, an ardent admirer of Newton and his work.
123. For example, as any applied mathematician knows, an infinitesimal perturbation (cause) can often have an arbitrarily large effect when a system is unstable. For a discussion of the logical status of the Design Argument following Humes' work see R. G. Swinburne, *Philosophy* **43,** 164 (1968).
124. E. Darwin, *Zoonomia, or the laws of organic life*, 2 vols (London, 1794). A discussion of his evolutionary ideas can be found in E. Krause, *Erasmus Darwin* (transl. W. S. Dallas, NY 1880).
125. I. Kant, *Der einzig mogliche Beweisgrund au eine Demonstration des Daseyns Gottes* (Königsberg, J. J. Kanter, 1763) in *Werke* (Suhrkamp edn, Frankfurt, 1960), Vol. 1, p. 734. No full translation appears to exist. The translation we give is due to Hicks (ref. 79, p. 210) but it is wrongly attributed and referenced by him.
126. I. Kant, *Kritik der reinen Vernunft* (Riga, 2nd edn, 1787), *Critique of pure reason*, ed. N. Kemp Smith (Macmillan, London, 1968), p. 521.
127. ibid.

128. ibid.
129. I. Kant, *Kritik der Urteilskraft* (Berlin and Leibau, 1790), transl. J. C. Meredith, *The critique of judgement*, ed. R. M. Hutchins (Encyclopaedia Britannica Inc., Chicago, London, Toronto, 1952), § 85.
130. ibid.
131. W. Paley, *Natural theology* (1802) in *The works of William Paley*, 7 vols, ed. R. Lynam (London, 1825). *Natural theology* was so successful that an expanded edition containing notes and further illustrations was published in 1836; *Paley's natural theology with illustrative notes*, 2 vols, ed. H. Brougham and C. Bell (London, 1836). Paley began work on *Natural theology* in the 1770's when he delivered a series of sermons entitled *The being of God demonstrated in the works of creation*, *Works*, Vol. 7, pp. 405–44.
132. For example, Paul Janet in ref. 51.
133. Cited in Hicks, ref. 79, p. 232.
134. There has been considerable debate as to the source of Paley's watch story. It is probable that its source was in B. Nieuwentyt's, *Religious philosopher*, (Vols 1–3), transl. J. Chamberlayne (London, 1719), where it appears in Vol. 1, p. xlvi; however, S. Leslie (*History of English thought in the eighteenth century* (ed. Harbinger, 1962), first publ. 1876, p. 347) thinks he abstracted it from Abraham Tucker (apparently Paley's favourite author), *Light of Nature pursued* (London, 1768–1778), 7 vols; i, 523, ii, 83. To confuse things further, Henry Brougham, *Discourse of Nature theology*, (London, 1835) says Paley's work is chiefly taken from the writings of Derham! Remarkably, *Encyclopaedia Britannica* articles on 'Nieuwentyt' in the nineteenth century claim that Paley 'appropriated' Nieuwentyt's ideas and arrangement 'without anything like honourable acknowledgement'.
135. *Natural theology*, p. 8–9.
136. ibid., p. 40–5.
137. ibid., p. 30.
138. ibid., p. 60.
139. J. Clive, *Scotch reviewers: The Edinburgh review* (1802–15) (Faber & Faber, London, 1957), p. 149; cited in D. L. LeMahieu, *The mind of William Paley: a philosopher and his age* (University of Nebraska Press, Lincoln and London, 1976), p. 74. This book contains further interesting biographical information, as does M. L. Clarke, *Paley: evidences for the man* (SPCK, London, 1974).
140. Paley, op. cit., p. 152.
141. Ref. 131, *Works*, Vol. 1, p. 320.
142. *Natural theology*, p. 52.
143. ibid., p. 58.
144. ibid., p. 317. Today we would say that the fine structure constant, which gives the strength of interactions between matter and light, must be small.
145. Brinkley was probably responsible for almost all the material in Chapter 22 of *Natural theology*. Like Paley he had been a Senior Wrangler at Cambridge. He became professor at Dublin in 1792 and met Paley shortly afterwards through their mutual friend John Law.
146. *Natural theology*, p. 318–9.

147. ibid., p. 319.
148. ibid., p. 323. This idea seems to have originated with Buffon and Newton who estimated the age of the Earth using a law of cooling for a metal ball initially at red heat. Paley quotes Buffon's work on p. 339.
149. ibid., p. 332.
150. ibid., p. 333. The other case he mentions, with attraction varying as distance, corresponds to allowing a cosmological constant term in Newtonian gravity.
151. ibid., p. 334–5.
152. ibid., p. 341.
153. see C. C. Gillespie, *Genesis and geology* (Harper, NY, 1951).
154. The original titles allocated and written under were: T. Chalmers, *On the power, wisdom and goodness of God as manifested in the adaption of external nature to the moral and intellectual constitution of Man*, 2 vols (London, 1833), eight editions by 1884; J. Kidd, *On the adaption of external Nature to the physical condition of Man* (London, 1833), seven editions by 1887; W. Whewell, *Astronomy and general physics, considered with reference to natural theology* (London, 1833), nine editions by 1864; C. Bell, *The hand, its mechanism and vital endowments, as evincing design* (London, 1833), seven editions by 1865; P. M. Roget, *Animal and vegetable physiology, considered with reference to natural theology*, 2 vols (London, 1834), five editions by 1870; W. E. Buckland, *Geology and mineralogy, considered with reference to natural theology*, 2 vols (London, 1836), nine editions by 1860; W. Kirby, *On the power, wisdom and goodness of God as manifested in the creation of animals, and in their history, habits and instincts* (London, 1835), six editions by 1853; W. Prout, *Chemistry, meteorology and the function of digestion* (London, 1834), four editions by 1855. The unusual independent addition was that of C. Babbage, *The 9th Bridgewater treatise; a fragment*, 2nd edn (London, 1838).
155. C. C. Gillespie, ref. 153.
156. J. W. Eckermann, *Conversations of Goethe with Eckermann and Soret*, transl. K. Oxenford (Bell, London, 1892), Vol. 2, p. 282. See also C. S. Sherrington, *Goethe on Nature and on science* (Cambridge University Press, Cambridge, 1949).
157. Darwin met Whewell, Sedgewick, and Babbage at Christ's College, Cambridge. All were strong supporters of the Design Argument.
158. C. Darwin, *The autobiography of Charles Darwin*, ed. N. Barlow (Dover, NY, 1958 [first published 1898]), p. 19.
159. ibid.
160. F. Darwin, *The life and letters of Charles Darwin*, 3 vols (Appleton, NY, 1897) Vol. 1, p. 282.
161. J. C. Greene, *The death of Adam* (Iowa State University Press, NY, 1961); J. R. Moore, *The post-Darwinian controversies* (Cambridge University Press, Cambridge, 1979).
162. C. Hodge, *What is Darwinism* (Princeton, NY, 1874), p. 52.
163. W. Graham, *The creed of science* (London, 1881), p. 319.
164. It would be nice to attribute the discovery of natural selection to this impetus from the Design Argument, but it would not be correct; Darwin

attributed his inspiration to reading T. Malthus, *An essay on the principle of population* (London, 1798), although his writings show the idea of natural selection was clear to him before he read Malthus; see S. Smith, *Adv. Sci.* **16,** 391 (1960), and R. M. Young, *Past & Present* **43,** 109 (1969).

165. C. R. Darwin, *The origin of species* (Murray, London, 1859).
166. A. Gray, *Darwiniana* (Appleton, NY, 1876).
167. C. Darwin, *Autobiography and selected letters*, ed. F. Darwin (Dover, NY, 1958), p. 308.
168. J. Fiske, *Outlines of cosmic philosophy* (Mifflin, Boston, 1874).
169. J. Fiske, *Through Nature to God* (Mifflin, Boston, 1899).
170. T. H. Huxley, *Lay sermons, addresses and reviews* (Appleton, NY, 1871), p. 301. Huxley was commenting on Kolliker's claim that 'Darwin is, in the fullest sense of the word, a teleologist'.
171. T. H. Huxley, *Critiques and addresses* (Macmillan, London, 1873), p. 305.
172. T. H. Huxley, 'On the reception of the Origin of Species' in *Life and letters of Charles Darwin*, ref. 160, Vol. 2, p. 179.
173. T. H. Huxley, 'The progress of science', in *Method and results* (Macmillan, London, 1894), p. 103–4.
174. Lord Kelvin, cited by A. Ellegard in *Darwin and the general reader* (Göteborg, 1958), p. 562.
175. C. Maxwell, Address of the British Association (1873) in *Scientific papers* (Cambridge University Press, Cambridge, 1890), Vol. 2, p. 376.
176. L. J. Henderson, *The fitness of the environment* (Harvard University Press, Mass., 1913) discusses the fitness of various chemical elements for incorporation of living systems. In a later work, *The order of Nature* (Harvard University Press, Mass., 1917), he discusses the philosophical background to the apparent teleology of chemistry, but study of his unpublished papers shows that his interpretation of chemical 'fitness' changed in later life; see Chap. 3, ref. 339.
177. J. P. Cooke, *Religion and chemistry* (Scriven, NY, 1880), rev. edn. We were unable to locate a copy of the first edition anywhere in the United Kingdom.
178. ibid., p. 161. Cooke's analysis of water provided the basis of several other American publications supporting the Design Argument; P. A. Chadbourne, *Lectures on natural theology* (Putnam, NY, 1870) has detailed discussion of the advantageous properties of carbon, nitrogen, oxygen and carbonic acid, whilst the later M. Valentine, *Natural theology; or rational theism* (Griggs, NY, 1885) summarizes Cooke's ideas about chemistry and also has extensive discussion of Darwinian evolution as a system of final causes.
179. Ref. 51, p. 3.
180. ibid., p. 415.
181. ibid., p. 59.
182. ibid., p. 148–9.
183. Only recently has the connection between chaos and determinism been carefully considered by physicists. Behaviour that appears random to us—for example, fluid turbulence—is described by mathematical models that exhibit a very sensitive dependence on initial conditions. These mathematical models are deterministic in principle but not in practice: in order to

know the state of the system precisely at any future time one must know its initial state *exactly*. In practice, there always exists some minute error in our knowledge of the initial state, and this error is amplified exponentially in the evolution time of the system, so that very soon we have no idea where the state of the system resides. Laplacian determinism is impossible; see Chapter 3 for further discussion of the meaning of determinism. It is interesting to note that, as early as 1873, Maxwell urged natural philosophers to study 'the singularities and instabilities, rather than the continuities of things ... [which] ... may tend to remove that prejudice in favour of determinism which seems to arise from assuming that the physical science of the future is a mere magnified image of the past.' [We are grateful to M. Berry for drawing our attention to this passage]. Maxwell's remarks are unusual, given the prevailing fascination of Victorians for the clockwork predictability of the world legislated by the Newtonian mechanical description.

184. ibid., p. 202.
185. ibid., p. 323–4.
186. ibid., p. 424.
187. ibid., p. 495, from Bernadin de St. Pierre, *Etudes de la Nature*—Bernadin also makes a memorable statement of what we might call 'natural anti-selection'—'wherever fleas are, they jump on white colours. This instinct has been given them, that we may more easily catch them'.
188. T. Lenoir, *The strategy of life: teleology and mechanics in nineteenth century biology* (Reidel, Dordrecht, 1982).
189. J. D. McFarland, *Kant's concept of teleology* (University of Edinburgh Press, Edinburgh, 1970), pp. 69–139.
190. M. Polanyi, *Science* **113,** 1308 (1968).
191. M. Polanyi, *The tacit dimension* (Doubleday, NY, 1966).
192. H. Lotze, 'Lebenskraft', in *Handwörterbuch der Physiologie*, Vol. 1, ed. Rudolph Wagner (Göttingen, 1842).
193. This English translation of Lotze's remarks taken from ref. 188, 170–1.
194. S. J. Gould, *Natural History* **92** (1983), 34–8.
195. M. Eliade, *From primitives to Zen: a thematic source book of the history of religions* (Harper & Row, NY, 1967), p. 94.
196. ibid., pp. 131–2.
197. ibid., p. 135.
198. D. Blanchard, unpublished notes, reported by Sol Tax, in *Free Inquiry* **2** (1982), No. 3, 45.
199. M. Eliade, *A history of religious ideas*, Vol. 1 (University of Chicago Press, Chicago, 1978), pp. 59–60.
200. M. Eliade, ref. 199, p. 90.
201. S.-H. Nasr, *An introduction to Islamic cosmological doctrines* (Shambhala, Boulder, 1978), p. 150.
202. Al-Bīrūnī, *Alberni India*, quoted in ref. 201, p. 123.
203. Al-Bīrūnī, *Kitāb, al-jamāhir*, quoted in ref. 201, p. 123.
204. J. Needham, *Sciences and civilization in China*, Vol. 2 (Cambridge University Press, Cambridge, 1956).

205. M. Eliade, *Gods, goddesses, and myths of creation* (Harper & Row, NY, 1974), p. 92.
206. Ref. 204, pp. 55–6.
207. Ref. 204, p. 36.
208. Ref. 204, p. 453.
209. Chu Hsi, *Chu Tzu Chhüan Shu* (*Collected works of Chu Hsi*), Chapter 43, quoted in ref. 204, p. 489.
210. Liu Tsung-Yuan, quoted in ref. 204, p. 577.
211. Lao Tzu, *Tao Tê Ching*, Chapter 34, quoted in ref. 204, p. 37.
212. Confucius, *Lun Yü* (*Analects*), XII, 17, quoted in ref. 204, p. 10.
213. Confucius, *Lun Yü* (*Analects*), XII, 19, quoted in ref. 204, p. 10.
214. Ref. 204, Chapter 18.
215. N. Barry, 'The tradition of spontaneous order', *Literature of Liberty* **5,** (1982) No. 2, pp. 7–58.
216. F. A. Hayek, *The constitution of liberty* (University of Chicago Press, Chicago, 1960); *Individualism and economic order* (University of Chicago Press, Chicago, 1948); *Studies in philosophy, politics, and economics* (Routledge & Kegan Paul, London, 1967); *New studies in philosophy, economics, and the history of ideas* (Routledge & Kegan Paul, London, 1978); *The counter-revolution of science* (Liberty Press, Indianapolis, 1979); *The road to serfdom* (University of Chicago Press, Chicago, 1944).
217. F. A. Hayek, *Law, legislation, and liberty*, Vol. 1: *Rules and order* (University of Chicago Press, Chicago, 1973), p. 37.
218. Ref. 217, p. 49.
219. Ref. 217, p. 37.
220. R. Axelrod, *The evolution of cooperation* (Basic Books, NY, 1984).
221. R. Axelrod, *Am. Political Sci. Rev.* **75,** 306 (1981).
222. R. Axelrod and W. Hamilton, *Science* **211,** 1390 (1981).
223. J. Maynard Smith, *J. Theor. Biol.* **47,** 209 (1974).
224. J. Maynard Smith, *Evolution and the theory of games* (Cambridge University Press, Cambridge, 1982).
225. R. Dawkins, *The selfish gene* (Oxford University Press, Oxford, 1976).
226. F. C. Copleston and B. Russell, 'Debate on the existence of God', repr. in *The existence of God*, ed. J. Hick (Macmillan, NY, 1964).
227. W. L. Rowe, *The cosmological argument* (Princeton University Press, Princeton, 1975).
228. W. L. Craig, *The cosmological argument from Plato to Leibniz* (Macmillan, NY, 1980).
229. R. Swinburne, *The existence of God* (Oxford University Press, Oxford, 1979).
230. W. I. Matson, *The existence of God* (Cornell University Press, Ithaca, 1965).
231. A. Flew, *God and philosophy* (Harcourt Brace & World, NY, 1966).
232. G. H. Smith, *Atheism: the case against God* (Prometheus, Buffalo, 1979).
233. Not every theologian agrees with Flew; see, for example, the discussion in ref. 227.
234. D. Hume, ref. 113, part IX.

235. J. Hartle and S. W. Hawking, *Phys. Rev.* D **28,** 2960 (1983).
236. S. W. Hawking, in *Les Houches lectures* 1983, ed. C. DeWitt (Addison-Wesley, NY, 1984).
237. L. Landau and E. Lifshitz, *Quantum mechanics: non-relativistic theory*, 2nd edn (Pergamon Press, London, 1965), p. 60.
238. R. Courant and D. Hilbert, *Methods of mathematical physics*, Vol. 1 (Interscience, NY, 1953), pp. 451–64.
239. J. Hick, *Arguments for the existence of God* (Macmillan, London, 1970).
240. P. Tillich, *Systematic theology*, Vol. 1 (University of Chicago Press, Chicago, 1967), p. 236; quoted in ref. 323, p. 33.
241. Ref. 226, p. 175.
242. Antony Flew used this definition of 'existence' to great effect against theism in his classic paper 'Theology and falsification', repr. in John Hick's *The existence of God*, ref. 226. It has also been reprinted in A. Flew and A. MacIntyre, *New essays in philosophical theology* (Macmillan, NY, 1964), together with commentary by theologians.
243. C. Hartshorne, *A natural theology for our time* (Open Court, La Salle, 1967), pp. 50, 83. Professor Hartshorne disagrees with our interpretation of his work (private communication to FJT). However, his objections seem to be due to the meaning he gives to the word 'Universe', which differs from ours. By 'Universe' he means 'a particular collection of laws and particles', whereas we mean 'all collections of laws and particles which ever did, does, or ever will exist'; see also ref. 245 below.
244. As the epigrams to this section suggest, the literature on the logical necessary (or lack of it) of the Universe is immense. Any philosopher worthy of the name discusses it. For recent guides to the literature on the question, the interested reader might consult M. K. Munitz, *The mystery of existence: an essay in philosophical cosmology* (Appleton-Century-Crofts, NY, 1965); Anna-Teresa Tymieniecka, *Why is there something rather than nothing?* (Van Gorcum, Amsterdam, 1966); and M. Gardner, *Scient. Am.* **232** (No. 2, Feb.), 98 (1975). The interested reader is advised to avoid R. Nozick, *Philosophical investigations* (Harvard University Press, Harvard, 1981).

   Some philosophers are fond of arguing that nothing naturally engenders something. For a method of generating the entire real number system from the empty set, see J. H. Conway, *On numbers and games* (Academic Press, NY, 1976).
245. If the Universe is defined to be the totality of everything that exists, then every believer in God—however one defines God—is a pantheist by definition. The traditional distinctions between theism, deism, pantheism, etc. can be made only if it is possible to distinguish between God and the physical Universe. Most modern, philosophically minded theologians would contend that such a distinction can be made, but that the physical universe is actually a proper subset of God: the physical universe is *in* God, but God is more than the physical universe. This position was termed *panentheism* (literally, all-in-God) by the eighteenth-century German philosopher K. C. F. Krause.

   The philosopher R. C. Whittemore has pointed out (in an unpublished manuscript, 'The universal as spirit') that most of the philosophers traditionally regarded as pantheists (that is, it has been thought that these men identified God with the physical universe) were actually panentheists. For example, Spinoza asserted in a letter that 'I assert that all things live and move in God . . . However, those who think that the *Tractatus Theologico-Politicus* rests on this, namely, that God and Nature (by which they mean a certain mass, or corporeal matter) are one and the same, are entirely mistaken'. (p. 343 of *The correspondence of Spinoza* (Letter LXIII), transl. and ed. A. Wolf (Dial Press, NY, 1955). Panentheism is distinguished from theism by saying that the latter contends that God is wholly other than the world. A problem with this distinction is that it is difficult to find a 'theistic' philosopher or theologian who really believes in theism in this sense of the word. For example, mystical Judaism is perhaps best described as panentheistic (G. G. Scholem, *Major trends in Jewish mysticism* (Thames & Hudson, London, 1955); H. Weiner, *Nine and one-half mystics: the Kabbala today* (Macmillan, NY, 1969)). Traditional Christianity also claims that not only is God everywhere in the physical universe, but—following St. Paul—that everything is also 'in' Him.

The philosophical difficulty which panentheism must overcome is showing that it makes sense to talk about something which is outside (or 'transcendent to') the physical universe. One approach is to say that the organization (= information content) of the universe is distinguishable from and transcendent to the substance of which it is composed. This approach has been defended by Charles Hartshorne, whose work we discuss in Chapter 3. See also his book *Omnipotence and other theological mistakes* (State University of New York Press, Albany, 1984). According to J. Barr, *Biblical words for time* (SCM Press, London, 1962), pp. 80 and 145, the philology of the Biblical texts does not allow any distinction to be drawn between the ideas of God being external in time or eternal in the sense that he is beyond time.

3 Modern Teleology and the Anthropic Principles

Once he has grasped this, he will no longer have to look at teleology as a lady without whom he cannot live but with whom he would not appear in public.

E. von Brücke

3.1 Overview: Teleology in the Twentieth Century

Science cannot solve the ultimate mystery of Nature. And it is because in the last analysis we ourselves are part of the mystery we are trying to solve.

M. Planck

Teleological modes of explanation, which for some two thousand years after Aristotle were regarded as vastly preferable to efficient causes as modes of explanation, have been severely denigrated by the great majority of twentieth-century scientists. So far has the prestige of teleology fallen that the French molecular biologist Jacques Monod claimed that the 'cornerstone of biology', which he termed 'the Postulate of Objectivity', is 'the systematic or axiomatic denial that scientific knowledge can be obtained on the basis of theories that involve, explicitly or not, a teleological principle'.[1] The rather violent hostility with which most scientists regard teleology is partly due to the failure of teleological arguments to account for adaptation in living things—evolution by natural selection is a much better explanation—but it is also due to the perceived paucity of significant scientific advances derived from teleological arguments. Most scientists would in fact claim that the attempt to introduce teleology into science has been positively harmful: not only has it led to no results, but it has seduced an enormous number of otherwise competent workers, who might have made important additions to *true* science, into wasting their lives exploring *cul-de-sacs*.

We shall show in this chapter that although there is much truth in the above criticism of the use of teleology, it is not the whole truth: teleological ideas *did* on occasion lead to correct predictions, and in some cases these predictions were contrary to the ones obtained from Monod's 'Postulate of Objectivity'. In other cases, teleological arguments were able to obtain results—correct results—which the non-teleological methods of the time were too poorly developed to obtain. Even more significant were the broad philosophical questions which teleology led people to ask early in this century, questions which were not followed up at the time perhaps because of the disrepute of teleology, but which bear a striking resemblance to some of the questions now being attacked on the frontiers of modern cosmology and high energy particle physics. It will be the purpose of this chapter to discuss many of the teleological predictions made and philosophical insights which derived from the teleological approach.

We shall open our discussion of modern teleology with a summary of the use of this concept in contemporary biology. Monod notwithstanding, living creatures *do* exhibit purpose in their behaviour, and it is also obvious that bodily organs are most easily described in terms of the bodily purposes (functions) they serve. It is simply not possible to avoid using teleological concepts in biology, and in section 3.2 we shall describe the attempts of a number of biologists to prune teleology of the dubious features to which Monod objects.

One feature of traditional teleology that modern biologists find particularly unscientific is its claim that mankind is the inevitable and foreordained outcome of the evolutionary process. One most often meets this claim in connection with the question of whether intelligent life exists on other planets. On the contrary, the consensus of modern evolutionists is that the evolution of intelligent life on Earth was not only not foreordained, it is so improbable that it is most unlikely to occur elsewhere in our Galaxy. We can understand its presence on Earth only by using the WAP: only on that unique planet on which it occurs is it possible to wonder about the likelihood of intelligent life. In section 3.2 we shall discuss briefly the reasons evolutionists have for believing intelligence to be an incredibly improbable accident. Additional arguments against the existence of extraterrestrial intelligent life will be found in section 8.7 and Chapter 9.

Intelligent life can appear only where more primitive life has evolved first and, as we shall see in Chapter 8, it is likely that primitive life of the type which can later evolve to intelligence can arise spontaneously only if it is based on certain very special properties of a few elements. This fact was first pointed out by the Harvard University chemist Lawrence Henderson early in this century, and we shall discuss his work in section 3.3, and in Chapter 8.

Monod's most serious charge against teleology is that it does not yield testable predictions, and is thus sterile and *ipso facto* unscientific. We shall begin a rebuttal of this claim in section 3.4, where we shall discuss *action principles*, a teleological formulation of physics. It is often claimed that action principles are fully equivalent to the standard non-teleological formulation of physical laws, but we shall demonstrate this is not entirely true. Action principles have occasionally led to predictions which the standard non-teleological formulations of the day had been unable to make. Fermat was able to predict the law of light propagation through a material media correctly using an action principle argument, while Newton's non-teleological calculation led to an incorrect prediction. We ourselves shall point out that the very existence of a globally defined action for the universe requires it to be closed, a prediction which, as is well-known, cannot be obtained by non-teleological arguments.

The Anthropic Principle, particularly in the form of SAP and FAP, suggests that mind is in some way essential to the cosmos. If this is so, it is natural to ask if mind is in fact *everything*. We have seen in the past how this question was posed and answered by Berkeley. Berkeley's empiricism had a strong influence on Kant, whose most significant German followers—Fichte, Schelling and Hegel—were led to a position vaguely analogous to Berkeley's which they called *absolute idealism*. We present an analysis of absolute idealism in section 3.5, using the concepts of computer theory to give a meaning to the basic undefined terms—such as 'thought' and 'mind'—of absolute idealism. We point out that there is a striking resemblance between certain speculations of modern computer theorists, in which the entire Universe is envisaged as a program being run on an abstract computer rather than a real one, and the ontology of the absolute idealists. As we shall show, the most important contribution made by Schelling was his introduction of a temporal notion of teleology into Western philosophy.

Modern Anthropic Principle arguments, particularly those which lead to testable predictions, use evolutionary timescales as a crucial step. We have briefly mentioned in Chapter 1 Wheeler's argument that the Universe must be at least as large as it is in order for it to exist long enough for life to evolve (see also § 6.3). An analogous argument led Dicke to invent the WAP. In Chapter 7 timescale arguments will be important in obtaining SAP constraints on the wave function of the universe, and an evolutionary timescale will actually be derived as a testable WAP prediction in section 8.7. However, it is not often realized that an evolutionary timescale anthropic argument was used in the nineteenth century by the famous University of Chicago geologist Thomas Chamberlain to predict that the power source of the Sun was a force inside atoms. This prediction, which was ignored at the time, we count as the first successful Anthropic Principle prediction, and we discuss its genesis in detail in section 3.6.

Chamberlain based his prediction on the Second Law of thermodynamics: in the absence of an atomic power source, the Sun could radiate for too short a period to be consistent with the evolutionary timescale. Another implication of the Second Law was the inevitable extinction of life on Earth. Such an implication clearly conflicted with teleological contentions that life was important, indeed essential, to the cosmos; it indicated the cosmos was not only non-teleological, it was dysteleological! The Second Law extinction was called the 'Heat Death'; we discuss and compare the opinions of the nineteenth and twentieth century philosophers and scientists on the teleological implications of the Heat Death in section 3.7. One suggestion, made by the Austrian physicist Boltzmann, that the Heat Death is only a local phenomenon and that it is connected with a WAP selection of a local time direction, is sufficiently important to warrant discussion in a separate section 3.8. Also in this section, we discuss two failures of Anthropic arguments to yield correct predictions, and we show one failure was due to an incorrect use of the physics known at the time, and the other was based on incorrect observational data.

The dysteleology of the Heat Death and the collapse of Paleyian teleology under the impact of the Darwinian revolution has forced theologians to modify drastically the traditional religious teleology. We discuss some of these new theological views of teleology in section 3.9. In general, the new views are much more abstract and less connected with the science of the day than were the older views. The primary exception was E. Barnes, an Anglican bishop and mathematician, who predicted in the early twientieth century that, on teleological grounds, the then currently accepted theory for the formation of the solar system had to be wrong, and he was correct.

The most interesting defences of teleology in Nature were made in the post-Darwinian period by speculative philosphers rather than by theologians. We discuss the work of a number of these men—Marx, Spencer, Bergson, Alexander, Whitehead and Hartshorne—in section 3.10. Like Schelling, these philosophers in their different ways believed in a progressive Cosmos, evolving towards a higher state. To Bergson and Hartshorne belongs the credit for using a temporal version of teleology to infer that there had to exist a uniquely defined global time-ordering, that the lack of such a unique global temporal ordering meant that special relativity could not apply globally, though it might apply locally. This is now known to be correct; general relativity applied to cosmology allows the existence of such a unique universal time, although such a time is not permitted in special relativity.

When asked by the American philosopher Dudley Shapere[1] for examples of teleology in biology which could be ruled out by his 'Postulate of Objectivity', Monod gave the Marxian and Spencerian theories of progress, but he singled out the teleological cosmological theory of Teilhard de Chardin as being particularly untestable and hence unscientific. We shall discuss the Teilhardian theory at some length in section 3.11. We point out that far from being unstable, it actually makes a prediction about the nature of thought, and this prediction has been falsified! Nevertheless, the structure of Teilhard's teleological cosmos has certain features which must appear in *any* theory of a melioristic cosmos that is consistent with modern science. He was really the first philosopher of optimism who faced the problem of the dysteleological Heat Death head-on. Although his specific cosmological model failed to correspond to reality, it is by no means impossible to construct a testable theory of a progressive cosmos which is roughly analogous to the Teilhardian theory. For illustrative purposes, we shall construct such a theory in Chapter 10.

In general, it can be said that teleology failed, and gave either incorrect predictions or untestable nonsense, when it was applied in the small, to the details of the evolutionary history of the single species *Homo sapiens*, or to questions of the physical structure of living things, which is to say, when it degenerated into vitalism. This was the erroneous use of teleology which Kant warned against in the eighteenth century. When teleology was restricted to global arguments—its true domain, according to Kant and according to T. H. Huxley, as we saw in Chapter 2—its predictions have, as we described briefly above and as we shall see in detail in this chapter, been by and large correct.

3.2 The Status of Teleology in Modern Biology

> We are the products of editing, rather than authorship.
>
> G. Wald

In the time of Paley and the *Bridgewater Treatises*, teleology was the explanation for most facts in the biological world. The marvellous adaptation of living creatures to their environment was attributed to the providential care and design of a Creator who constructed them to fit into their environment, just as a human watchmaker purposefully manufactures the components of a timepiece. The purpose of this intelligent Creator in constructing such creatures was also thought to be understood: the Universe and the creatures in it were created for both the enjoyment of the creatures and for the glory of the Creator.

The Darwinian revolution changed all this. Recalling the words of T. H. Huxley, whom we quoted in Chapter 2: 'That which struck the present writer most forcibly on his first perusal of the *Origin of Species* was the conviction that Teleology, as commonly understood, had received its deathblow at Mr. Darwin's hands'.[2] Adaptation of living beings was now seen to be due to natural selection acting over billions of years on modifications of organic structures created by random mutation. Some biologists, notably Asa Gray,[3] attempted to retain the purpose of God in

Nature by giving Him the credit for causing—and directing—the mutations, but this view died out in the face of enormous evidence that the variations of genotype were truly random: a chance collision of a cosmic ray with a DNA molecule could in principle give rise to a wholly new biological structure.

The nineteenth-century biologists also saw teleology at work not only in the adaptation of living things, but also in the over-all relationship of living beings to each other. As the historian A. O. Lovejoy[4] has pointed out, pre-evolutionary biology regarded the living world as organized into a 'Great Chain of Being' with single-celled organisms at the bottom of the Chain, mankind somewhere in the middle, the Angels above him, and God at the top. This picture of living creatures was static; the species were created to fit into this ordering at the beginning of time and were ordained to remain so ordered for all time. God's purpose never changed since He was unchanging. A species could never become extinct.

The non-extinction of species was justified by an assumption which Lovejoy termed the *Principle of Plenitude*: '. . . that no genuine potentiality of being can remain unfulfilled, [and] that the extent and abundance of creation must be as great as the possibility of existence, and commensurate with the productive capacity of a 'perfect' and inexhaustible Source'.[4] The extinction of a species would mean that a gap in the Great Chain of Being would appear, and a possible species would not exist. The Principle of Plenitude was almost universally accepted by philosophers until well into the nineteenth century.

The Darwinian revolution broke the Great Chain of Being and shattered the teleology-in-the-large of the Principle of Plenitude. Species arose in time and died out, to be replaced by other species. Over the past hundred years a number of biologists have attempted to retain teleology-in-the-large by changing the Great Chain of Being from a static relation in space to a dynamic relation in time. The picture developed by these men—primarily vitalists such as Driesch[5] and J. S. Haldane[6] in the early part of this century and du Noüy,[7] Sinnott,[8] Wright[9] and Teilhard de Chardin[10] in the post World War II period—is of an inevitable development commencing three billion years ago from the simplest single-celled organisms then living to produce the incredible complexity of a human being today.

These views, which the evolutionist George Gaylord Simpson has termed 'the new mysticism'[11] have a certain beauty and emotional attraction, but are contradicted by a detailed examination of the evolutionary record. As Simpson[12,13] and Ayala[14,15] have discussed at length, there is no generally purposeful pattern evident in the collection of all lineages. Most lineages have died out, a few have regressed in the sense of becoming less complex, while some—including the branch of the evolutionary tree which has led to Man—have increased the complexity of their nervous systems dramatically. One can adopt any of a number of criteria of progress—complexity of general structure, complexity of the nervous system, number of species in existence at a given time, number of ecological niches occupied—by any of these criteria, the collection of all lineages has at times advanced, but at some times retrogressed. As pointed out by Dobzhansky *et al.*,[16] the biosphere of the Earth is probably more advanced now than in Cambrian times in terms of the latter two categories, and some species—especially Man—are more advanced at present than any species in Cambrian times in terms of the former two categories.[17] Other criteria of progress could be advanced (see refs. 12 and 13 for an extensive list) and by almost all of these criteria, the biosphere sometimes progresses and sometimes retrogresses. The major problem with most of these criteria is that they involve a value element. What is progression from the point of view of one species would be retrogression from the view of another. Human beings tend to take an anthropocentric position, and regard any development which leads to human characteristics as progressive, and any other line of development as either retrogressive or neutral. Given the WAP observation that Man exists, it follows that there must exist a lineage which is progressive by one of the anthropocentric definitions of progression, but there is no guarantee that a planet which contains living things must inevitably evolve an intelligent species, and so there is no guarantee that a biosphere anywhere would be progressive in this sense, and no guarantee that an intelligent species would continue to develop in intelligence.

It is often claimed, particularly by believers in the existence of intelligent life on other planets, that because intelligence is advantageous in the struggle for life, natural selection will act to force an increase in the complexity of the nervous system at least in some lineages, and that as a consequence the intelligence of the most intelligent creature on Earth in a given epoch will increase with time. However, this is not necessarily true, because it is not intelligence alone which generates selective advantage; a sophisticated nervous system requires a huge number of support systems—such as eyes, manipulative organs, organs for transport, and so on—if it is to be effective. It is quite possible that no lineage on an earthlike planet will evolve the necessary support systems for a human-level intellect, and possible that even if they do, the genetic coding of the support systems will be such that an increase in the complexity of the nervous system will necessarily be offset by degeneration of some essential support organs in all possible lineages on the earthlike planet.

That such an outcome is quite possible can be seen by reference to several lineages on Earth. No lineage in the entire plant kingdom has shown a significant increase in its ability to process information since the metazoan ancestors of the plants first appeared some 500–1000 million years ago. Such increase as has occurred—the ability to orient towards the light, the ability of certain plants such as the Venus fly-trap to react to tactile sensations, for instance—have developed so slowly that were the increase to be projected into the future at the rate inferred in the past, it would require many trillions of years for the information-processing ability to reach the human level. Compare this with about 10 billion years, which is the total time the Sun will remain on the Main Sequence and radiate energy at an approximately constant rate. And it is most unlikely the rate of increase of information-processing ability could increase at the present rate; for plant metabolism simply cannot supply sufficient energy to supply a large nervous system.

Even in *Homo sapiens*, the brain is difficult to supply; it requires about 20% of the energy consumed by the body when resting.[18] This fraction is comparable to the over-all metabolic requirements of active reptiles of comparable body size, and for this reason the paleontologist D. A. Russell has concluded that 'a large brain is incompatible with a reptilian metabolism'.[19] On the Earth, out of many millions of lineages, only birds and mammals have a sufficiently high metabolic rate to support a large brain. For reptiles, the advantages of intelligence are irrelevant; they are unable to evolve human-level intelligence no matter how advantageous it is, unless they first evolve a non-reptilian metabolism.

It is nevertheless true that on the Earth, there has been an increase of encephalization, which is the ratio of brain weight to body weight, in some lineages since the evolution of metazoans.[20] Encephalization is thought to be a better measure of intelligence, or information-processing ability, than brain weight, because much of the brain is used to control body functions, and the larger the animal the larger the brain it must have in order to control these functions.[21] The increase in the encephalization in the human lineage is in accord with an evolutionary trend established 200 million years ago.

However, the encephalization rate was altered dramatically some 230 million years ago at approximately the same time as the massive extinction which defines the Permo-Triassic boundary:[19] the rate of encephalization was much faster prior to this extinction than it was afterwards. Had the older rates persisted, a human level of encephalization would have been reached 60 million years ago, while the more recent rates of encephalization would have required 20 billion years to attain a human level from the level characteristic of primitive metazoans.[19] The higher rate of encephalization characteristic of the pre-Triassic period was essential for the evolution of humanoid intelligence on Earth; the later rate would have been quite inadequate. Much of the earlier encephalization occurred in the oceans, and it is not at all clear it could have continued to the human level. Technology requires a terrestrial environment.

There is some evidence that encephalization goes just so far in marine lineages, and then stops increasing. For example, the encephalization of the cetacean (dolphin) lineage, whose encephalization is comparable to that of humans, reached its present level some 20–30 million years ago, but has undergone no significant change since.[20] The dolphins are believed to have intelligence comparable to that of dogs by most biologists.[22]

Very little is known about the evolution of the cephalopods, such as squid and octopi, which are often cited as highly intelligent creatures with large brains, for such soft-bodied animals leave little trace in the fossil record. However, the encephalization of the cephalopods has certainly not increased as rapidly as the vertebrates over the past 500 million years.[19] What is known of their evolution is consistent with a rapid early encephalization, followed by essentially no increase in encephalization, as happened with the dolphins.

Even if it evolves, high encephalization by no means guarantees the survival of a species or evolution to a higher grade. The Proboscidea (elephants) have an encephalization markedly higher than most other mammals,[20] and yet they have been in decline since the Miocene, being represented by only two living species. They are survived by equally large but less-encephalized animals in similar ecological zones.[23] Survival requires a good many animal body systems—and a benign environment—in addition to intelligence.

In fact, as the evolutionist C. O. Lovejoy points out,[23] an increased information-processing capacity in the nervous system is actually a reproductive liability both pre-natally (since a complex nervous system requires a long gestation period) and post-natally (since it takes longer to raise and teach the young). Intelligence has no *a priori* advantage, but it is a clear and unmistakable reproductive hazard. Thus for this reason alone we would expect such capacity to be selected for 'only in rare instances'. Primates are such an instance, but in this order of mammals encephalization is to a great extent directly related to highly unusual feeding strategies and locomotion.[23] Furthermore, primate encephalization cannot be regarded as a typical trend of the mammals, because the primates are unusually primitive in the majority of mammalian traits. Even amongst the primates a well-defined limit on the degree of encephalization was reached in the Miocene in all primate lineages except that leading to *Homo sapiens*, and the other homonid primates were replaced by less encephalized, more reproductively successful cercopithecoids.[24]

In short, the evolution of 'cognition', or intelligence and self-awareness of the human type, is most unlikely even in the primate lineage. As C. O.

Lovejoy puts it:

. . . man is not only a unique animal, but the end product of a completely unique evolutionary pathway, the elements of which are traceable at least to the beginnings of the Cenozoic. We find, then, that the evolution of cognition is the product of a variety of influences and preadaptive capacities, the absence of any one of which would have completely negated the process, and most of which are unique attributes of primates and/or homonids. Specific dietary shifts, bipedal locomotion, manual dexterity, control of differentiated muscles of facial expression, vocalization, intense social and parenting behaviour (of specific kinds), keen stereoscopic vision, and even specialized forms of sexual behaviour, all qualify as irreplaceable elements. It is evident that the evolution of cognition is neither the result of an evolutionary trend nor an event of even the lowest calculable probability, but rather the result of a series of highly specific evolutionary events whose ultimate cause is traceable to selection for unrelated factors such as locomotion and diet.[25]

The believers in the existence of beings on other planets with human-level intelligence often cite the convergent evolution, (which means the independent evolutionary invention of a trait in two unrelated lineages), of eyes in vertebrates and cephalopods as indicating that the convergent evolution of intelligent life on different planets is not too improbable. The response to this argument by the great evolutionist Ernst Mayr is worth quoting in full:

. . . the case of the evolution of eyes is [indeed] of decisive importance in the argument about the evolution of intelligence. The crucial point is that the evolution of eyes is not at all that improbable. In fact whenever eyes were of any selective advantage in the animal kingdom, they evolved. Salvini-Plawen and myself have shown[26] that eyes have evolved no less than at least 40 times independently in the animal kingdom. Hence a highly complicated organ can evolve independently, if such evolution is at all probable.

Let us apply this case to the evolution of intelligence. We know that the particular kind of life (system of macromolecules) that exists on Earth can produce intelligence. We have no way of determining whether there are any other macromolecular systems elsewhere in the universe that would have the capacity to develop intelligence. We know however, as I have said, that we do have such a system on Earth and we can now ask what was the probability of this system producing intelligence (remembering that the same system was able to produce eyes no less than 40 times). We have two large super-kingdoms of life on Earth, the prokaryote evolutionary lines each of which could lead theoretically to intelligence. In actual fact none of the thousands of lines among the prokaryotes came anywhere near it. There are 4 kingdoms among the eukaryotes, each again with thousands or ten thousands of evolutionary lineages. But in three of these kingdoms, the protists, fungi, and plants, no trace of intelligence evolved. This leaves the kingdom of Animalia to which we belong. It consists of about 25 major branches, the so-called phyla, indeed if we include extinct phyla, more than 30 of them. Again, only one of them developed real intelligence, the chordates. There are numerous Classes in the chordates, I would guess more than 50 of them, but only one of them (the mammals) developed real intelligence, as in Man. The mammals consist of 20-odd orders, only one of them, the primates, acquiring intelligence, and among the well over 100 species of primates only one, Man, has the kind of intelligence than would permit [the development of advanced technology]. Hence, in contrast to eyes, an evolution of intelligence is not probable.[27]

For the above reasons, and many others which we omit for reasons of space, there has developed a general consensus among evolutionists that the evolution of intelligent life, comparable in information-processing ability to that of *Homo sapiens*, is so improbable that it is unlikely to have occurred on any other planet in the entire visible universe.[28] The consensus view has been defended by many of the leading evolutionists such as Dobzhansky,[29] Simpson,[30] Francois,[31] Ayala *et al.*[16] and Mayr.[32] The only evolutionist of any standing who has disagreed with the consensus is Stephen Jay Gould,[33] and even Gould claims conscious intelligence is sufficiently unlikely to evolve that, should Mankind blow itself to bits, 'Conscious intelligence ... has no real prospect for repetition [on the Earth]'.[33] (We might also mention that Mayr called Gould's arguments in favour of his anti-consensus position—and in reality, it does not differ that much from the consensus—a 'sleight of hand',[27] and we agree with Mayr's assessment.)

In short, there is no indication in the geological record that the evolution of intelligence is at all inevitable; in fact, quite the reverse.

It is true that, in the words of Simpson 'there is in evolution a tendency for life to expand, to fill in all available spaces in environments, including those created by the expansion of life itself'. But in so far as this occurs—and 'it does seem certain that life has, on the average, expanded throughout most of the evolutionary process',[34]—this is due to the capacity of life to expand exponentially, combined with the fact that as more species come into existence, more ecological niches are formed so more species can come into being and so forth. There is absolutely no evidence to show it is due to some obvious over-riding Plan which is guiding the entire development. Furthermore, there is a definite limit to the expansion of life on Earth. The biomass is ultimately restricted by the efficiency of the basic metabolic processes which govern all living things, the mass of the Earth, and the amount of sunlight which strikes the Earth.

Thus the evidence is against some of the traditional conclusions of teleological explanation in biology, and this has led a number of well-known biologists, such as Mayr,[35,36] to try to eliminate the concept of teleology from biology altogether. However, this is difficult to do. Animals, especially man, *do* show purposeful behaviour. In fact, as Monod[37] has argued, 'purposeful behaviour is essential to the very definition of living things'. (This does not contradict his anti-teleological views quoted in the introduction to this chapter, for Monod is only opposed to teleology in the large, to the idea that evolution has a plan.) Mayr and the other anti-teleological biologists are of course aware of this, and Mayr[35] proposes to use the word 'teleonomic' to describe purposeful action in living creatures. This word allows Mayr to discuss purpose in biology without implying Design in the living world, as the use of the word 'teleological' in this context would tend to do. Dobzhansky *et al.*[16] and Ayala,[14] on the other hand, feel that such a terminological innovation would introduce more confusion than clarity into the analysis of the purposeful behaviour of living beings. In their opinion, we may as well admit that individual living things do exhibit teleology. Ayala[14] has distinguished two valid uses of the teleological concept in biology. The first, which he calls *artificial teleology* (external teleology in the nomenclature of Dobzhansky[38]), is purposeful behaviour, or the teleology exhibited by objects constructed for a definite purpose. The watch—the favourite example in the Design Argument—fits into this category. The nests of birds, the hives of bees, and the burrows of certain rodents are examples of objects constructed for a definite purpose by non-human living creatures; they are also said to exhibit artificial teleology. The purposeful actions of living beings—a man making a watch or a mountain lion hunting a deer—are also examples of artificial teleology. In all cases of artificial teleology, it is possible to discover the action of some nervous system which either directs the behaviour toward some discernible end, or controls the construction of an object which will be used for some discernible purpose.

The hand of a man and the wing of a bird also serve definite purposes: the former is used for manipulation and the latter for flying. However, they were not constructed under the guidance of a complex nervous system with a view to serving these purposes. They were created by natural selection acting upon the phenotypic results of random mutations in the genotype. Nevertheless they do serve a discernible purpose, and so are said by Ayala[14] to exhibit *natural teleology* (internal teleology in the nomenclature of Dobzhansky[38]).

One can subdivide natural teleology into two types: *determinate natural teleology* and *indeterminate natural teleology*.[14,16] Determinate teleology occurs when the end purpose is achieved independently of small environmental fluctuations. Examples are the development of an egg into a chick, and a human zygote into a baby. In the terminology of Ayala,[14] indeterminate natural teleology occurs when the final state is not uniquely determined from the initial state and indeed the final state of the system is just one of several possible final states which could have arisen from the system's initial state. We use the term 'indeterminate natural teleology' in those cases where we are trying to discuss the evolution of a system in terms of its final stage, but where this final state is not the goal of a directing nervous system, nor the result of a deterministic developmental process. The evolution of a primate lineage into *Homo sapiens* is an example of indeterminate natural teleology. We are extremely interested in knowing just how the final state—mankind—came about, but this final state was *not* an inevitable evolutionary outcome of any of the primate species which existed ten million years ago. Had the environmental pressures or the sequence of mutations been slightly different at any point during this period, the human species never would have arisen. Nevertheless, from our WAP viewpoint we want to know the steps in the evolutionary process leading to Man, so an explanation of this process is crucially dependent upon the final stages. We sift through the complex interaction of the closely-related hominid lineages to find the unique class that leads to *Homo sapiens*: the development of the others are of much less interest. This explanation is thus a teleological one; in fact, one of indeterminate teleology since the specific environmental pressures and mutations which arose along the way could not be predicted (by biological means) from the initial biological state, and also one of natural teleology since no nervous system was guiding the evolution of the primates toward the goal of mankind.

One can draw an analogy with the study of human history. In the nineteenth century a major school of British historians (and the philosopher Spencer, as we shall see in section 3.10) regarded liberal democracy as the apex of human development and viewed political history as progress toward this state. These scholars picked out those earlier events which led to liberal democracy, and de-emphasized or ignored those occurrences which did not seem to contribute to this development, even though some of those excluded events were regarded as most important at the time. The historian Herbert Butterfield[39] felt this 'natural teleological' interpretation of history—he called it the 'Whig interpretation of history'—was a serious distortion of cultural development. We agree; and it is a distortion which arises from not taking WAP into account. Only if liberal democracy (the Whig utopia) arises is it possible to believe that it will inevitably develop from earlier political systems. Judging from the historical record, it is more reasonable to say that from the information available to observers at a given epoch, the structure of the succeeding political system is unpredictable. Nevertheless, the people in the succeeding civilization are interested in the events that led to them, even if that history was most improbable, just as we are extremely interested in knowing the steps that led to the evolution of *Homo sapiens*, even though those steps were exceedingly improbable. Political history, like biological history, can be regarded from a teleological point of view if it is remembered that the teleology in question is indeterminate. The philosopher and biologist Grene[40] has called the natural selection process which produces teleological structure in living things '*historical teleology*'. She calls the teleology of organs which act in a useful way, like the wing of a bird '*instrumental teleology*', and she calls determinate natural teleological processes, like the development of an egg into a chick, '*development teleology*.' The evolutionary biologists seem to agree on the natural divisions of the teleological concept in biology, even though the terms used for these various divisions differ from one biologist to another.

The question of whether teleological explanations in biology can be translated into causal explanations is a subtle one. In company with other natural scientists, evolutionary biologists have generally assumed that such a translation could in fact be made, though perhaps only with great difficulty. The natural teleological development of the egg into a chick could in principle be explained in terms of a series of complex biochemical interactions among the molecules comprising the egg. A similar description could be made of the working of the human hand or a bird's wing. It might even be possible to explain the purposeful behaviour of human beings in terms of physical interactions, with the brain regarded as merely an extremely complex chemical computer.

But it seems likely that such a purely causal, non-teleological and complete explanation of purposeful biological behaviour would be so complex that no such explanation will ever be achieved. The justification for this assertion is a simple numerical estimate of the complexity of living beings. The amount of information that can be stored in a human brain is estimated to be between 10^{10} and 10^{15} bits, with the lower number assuming there is one bit stored on the average for each of the brain's 10^{10} cells. Now about 1% to 10% of the brain's cells are firing at any one time, at a rate of about 100 hertz. This gives a computation rate of 10 to 1000 gigaflops (a gigaflop is 10^9 (*flo*ating *p*oint computatio*ns* per second). The lower bound of 10 gigaflops is about the rate at which the eye processes information before it is sent to the brain.[41] For comparison, the fastest computer in existence today, the Cray-2, has a speed of 1 gigaflop and storage capacity for 2×10^{10} bits (in 64 bit words).[42] (The IBM-AT personal computer can have up to 10^7 bits of RAM. Currently available 32-bit personal computer central processors can address about 10^{10} bits of RAM. However, currently available RAM chips can store only about 10^6 bits, but 10^9 bit RAM chips should be available by the year 2000.[42]) So the most powerful computer has a storage capacity and information processing rate between 10 and 1000 times less than that of a human being.

But only the information which a human being can process *consciously*, or hold in the forefront of the mind, can be used in forming a humanly acceptable explanation. We don't know exactly how much this would be, but it is comparable in order of magnitude to the information coded in a single book, which is typically 1 to 10 million bits. No explanation humans have ever dealt with has been as complex as this. The content of most science books has been concerned with justifying the explanation rather than presenting it. Furthermore, there is an enormous redundancy in books. 10^6 bits is at least 4 orders of magnitude below the amount of information required for a numerical simulation of a human brain, assuming it could be done. The amount of information required for a numerical simulation of a higher mammal is within two orders of magnitude of that of a human being.

This argument assumes of course that we require at least 10^{10} bits—the lower bound to the brain capacity of the human mind—in order to carry out a numerical simulation of a human being. If anything, this is a wild underestimate, because it ignores round-off errors. Even more important, in fact the essential point in estimating the difficulty of carrying out a numerical simulation of a living creature, is that the actions of living creatures are unstable from the causal (numerical simulation) point of view: a tiny change in the initial input or stored information can lead to a drastic change in the macroscopic behavior. For this reason it is not possible to reduce substantially the amount of data required in a simulation much below 10^{10} bits. We can drastically reduce the amount of data we require to understand our fellows because we know that they will typically react in certain ways to certain stimuli. *But this drastic reduction in the data set is precisely what is accomplished by teleological explanation*! Using teleology, we learn that certain data, processed via teleological concepts, are sufficient for us to understand human beings and animals. In contrast, a purely causal explanation *cannot* make use of the same simplifications in the data, for by assumption such an explanation is not allowed to organize the data teleologically.

It will be possible, we believe, to construct a computer that can process information at the human level; that is to say, be as intelligent as a human being. In fact, our arguments in Chapters 9 and 10 will assume such a computer to be possible. But we will *never* be able to completely understand such a machine at the causal level; it will sometimes act unpredictably, and we will find teleological explanations of its actions more useful than causal ones, at least in understanding its most complex behaviour. This is not to advocate vitalism in computers; we assume *of course* that computer elements obey the laws of physics, and that there are no 'vital' forces acting anywhere in Nature. A similar view of teleology in computers can be found in a paper by the mathematician Norbert Wiener.[43]

A position of our sort was perhaps best defined by Ayala[44] who distinguishes between three types of reductionism: ontological, methodological, and epistemological.

Ontological reductionism claims that the 'stuff' comprising the world can be reduced ultimately to the particles and forces studied by physics; the vast majority of biologists (and we ourselves) are ontological reductionists.

Methodological reductionism holds that in the study of living phenomena, researchers should always look for explanations at the lowest level of complexity, ultimately at the level of atoms and molecules (or even the elementary particles that compose them.) We partially support this form of reductionism, noting however, that such methods have definite limits, and other methods will often yield better results. In fact, many advances are due to letting different levels of explanation interact, and this will be our strategy in this book.

Epistemological reductionism holds that theories and experimental laws formulated in one field of science can *always* be shown to be special cases of laws formulated in other areas of science. It is this form of reductionism which we deny. We do not think teleological laws either in biology or physics can be fully reduced to non-telelogical laws, for the reasons given above. We note that even in physics the Second Law of Thermodynamics cannot be derived from molecular mechanics without anthropic assumptions, (see section 3.8). The most indefatigable modern critic of methodological and epistemological reductionism has been Michael Polanyi,[45,46] whose work we discussed briefly in Chapter 2. He always emphasized that he was an ontological reductionist.

The distinction which Ayala draws between various forms of reductionism suggests the following distinctions between various forms of determinism:

Ontological determinism claims that the evolution equations which govern the time development of the ultimate constituents of the world are deterministic; that is, the state of these constituents at a given time in the future is determined uniquely by the state of these constituents now. All theories of physics which have ever been proposed as fundamental—Newtonian particle physics, the electromagnetic field equations of Maxwell, Einstein's general relativity theory for gravity, and even quantum mechanics—all of these are ontologically deterministic theories. They differ only in the nature of the entities which are claimed as fundamental. For Newtonian physics, particles were fundamental; for Maxwell and Einstein, physical fields were fundamental; and for quantum mechanics, the wave function is fundamental. Although the fundamental constituents of the world have changed with each successive scientific revolution, the fundamental evolution equations for these entities have always been deterministic. Thus there is no evidence whatsoever that the fundamental equations are not deterministic; in fact, to the extent that we believe the fundamental equations to be true, we are forced by the evidence to be ontological determinists.

Methodological determinism holds that in the study of complex phenomena, such as living beings, we should always look for deterministic laws governing the phenomena. In our opinion, this form of determinism is much too strong. It is often the case that complex phenomena are better described by statistical laws in which chance is fundamental. In fact, the laws of classical thermodynamics are statistical laws which are often more useful in describing heat engines and living things than the deterministic laws from which they are often 'derived'.

Epistemological determinism holds that it is possible, using the deterministic fundamental evolution equations (which are assumed to exist), to compute and hence predict the future behaviour of complex systems, in particular the future behaviour of living organisms. This form of determinism we also deny. The theory of quantum mechanics itself tells us that it is impossible to get the necessary information to predict the future wave function, even though the future wave function is in fact determined. We have argued at length above that the behaviour of living organisms like ourselves is too complex to be predictable by beings of similar complexity.

There is considerable evidence that the behaviour of living beings cannot be predicted for any significant length of time by any intelligent being, no matter how intelligent. Computer scientists term[120,121] a computation problem *intractable* if the number of computations needed to solve the problem grows exponentially with the length of time over which the prediction is to be made. Intractable problems are effectively unsolvable by computer no matter how powerful. Wolfram[341] has recently shown that intractable problems are quite common in simple physical models; tractable problems may be the exception rather than the rule. In fact, the instability of living systems, which we noted above, probably makes the calculation of their future behaviour an intractable problem.

The difficulty of translating the teleology of living systems into the usual causal language of physical science has led the economic philosopher Ludwig von Mises to draw a fundamental distinction between these sciences and the 'science' of human action, which is basically teleological. His view of human history is similar to the biologists' view of evolution: it is an example of indeterminate natural teleology. There are no 'historical forces' in the sense of Marx. There are only the plans of individual people who only frame their purposes in the short term. These plans and their resulting actions interact to produce a development which has no regularity after the manner of physical laws, and which is unpredictable in the long run. He asserts that 'it is ideas that make history, and not history that makes ideas',[47] and ideas which originate amongst a small number of intellectuals can be transmitted very rapidly and begin to strongly influence the actions of senior government officials and other members of the society. The result of this amplification of ideas is on occasion to change drastically the course of social evolution. Von Mises' student, Friedrich A. Hayek (whose work on spontaneous order we discussed in section 2.8), attributed the cause of this indeterminate teleology of a human social system to the inherent organizational complexity of the system:

> Organized complexity here means that the character of the structures showing it depends not only on the properties of the individual elements of which they are composed, but also on the manner in which the individual elements are connected with each other. In the explanation of the working of such structures we can for this reason not replace the information about the individual elements by statistical information, but require full information about each element if from our theory we are to derive specific predictions about individual events. Without such specific information about the individual elements we shall be confined to what on another occasion I have called mere pattern predictions—predictions of some of the general attributes of the structures that will form themselves, but not containing specific statements about the individual elements of which the structures will be made up.
>
> This is particularly true of our theories accounting for the determination of the systems of relative prices and wages that will form themselves on a well-functioning market. Into the determination of these prices and wages there will enter the effects of particular information possessed by every one of the participants in the market process—a sum of facts which in their totality cannot be known to the scientific observer, or to any other single brain. It is indeed the source of the superiority of the market order, and the reason why, when it is not suppressed by the powers of government, it regularly displaces other types of order, that in the resulting allocation of resources more of the knowledge of particular facts will be utilized which exists only dispersed among uncounted persons, than any one person can possess.[48]

Hayek is concerned with describing the behaviour of a free-market economic system, but it is clear that the teleological behaviour of this system is exactly the same as the teleological behaviour of the entire living world: the teleology is there, but it occurs only on the level of the individual, who has purposes planned only for the short-term future. The entire system has a teleological structure only in so far as these individual teleologies interact to govern the dynamical behaviour of the entire system. The long-term evolution of a biological or economic system is unpredictable and any trends which may be visible at a given time could be reversed in the future. This makes it impossible for evolutionists to make long-term predictions about the future of the human race.[11,49,50]

The local, short-sighted teleology of biological and economic systems does tend to increase the complexity of the systems, however. In part this occurs as a consequence of the increased stability of complex systems.[17] As Paul Ehrlich, one of the leaders of the ecological movement, puts it:

> . . . we have both observational and theoretical reasons to believe that the general principle holds: complexity is an important factor in producing stability. Complex communities, such as the deciduous forests that cover much of the eastern United States, persist year after year if man does not interfere with them . . . A cornfield, which is a man-made stand of a single kind of grass, has little natural stability and is subject to almost instant ruin if it is not constantly managed by man.[51]

In the same work Ehrlich points out that attempts by Man to artificially stabilize such a simplified ecosystem often increases its instability. Of course, Hayek[52] and Milton Friedman[53] make the same point in regard to government attempts to stabilize the economy and the money supply: the effect of attempting to stabilize a complex system artificially often *increases* the instability rather than decreases it. (The Ehrlich statement is very similar to the Hayek statement above if the words 'Man' and 'ecology' in the former are replaced by 'government' and 'economic system' respectively). In both ecology and economics the maximum use of information—and the maximum stability—occurs when no attempt is made to simplify the system by imposing a single or small number of goals upon it. Maximum stability and maximum teleological development of the entire system occur when the teleology inherent in the system—the different interacting goals of *all* living things in an ecology or *all* humans in an economy—is maximized. Yet ecologists like Ehrlich seem unable to extend their correct observations and correct biological theories into political economy, even though their descriptions of ecological system and their moral arguments in favour of natural systems are exactly the same as the descriptions of economic systems and the arguments in favour of free markets by Mises, Hayek, and Friedman. A similar criticism was made against ecologists like Ehrlich by William Havender, a biologist at the University of California at Berkeley.[54] Ehrlich wants government to impose a single goal upon the whole of mankind:

> Perhaps the major necessary ingredient that has been missing from a solution to the problems of both the United States and the rest of the world is a goal, a vision of the kind of Spaceship Earth that ought to be and the kind of crew that should man her.[55]

The general complexity theory of the ecologists themselves shows that this attempt to impose a goal would have the same effect on the political-economic system as Man's interference has on the ecology. A complex system like an ecology or a market economy cannot have a goal in the sense that a single individual can, and any attempt to impose one leads to disaster. Since complex systems tend to be more stable than simple ones—this improves their selective advantage amongst systems and makes a given complex system difficult to replace except by one of increased complexity—there does seem to be a long-term trend of increasing complexity in the evolutionary record, according to Stebbins.[17] However, this trend can be reversed—indeed, it occasionally has been[13]—and cannot be regarded as a uni-directional teleological trend.

In recent years a number of philosophers of science have attempted to describe the 'progressive' teleological development of science in terms of Darwinian evolutionary concepts. However, as Stephen Toulmin has emphasized,[56] most of these philosophers have depicted the teleology as acting in the large to cause an inevitable development of science towards ultimate truth. Both Toulmin[57] and Thomas Kuhn[58] have attempted to argue that the teleology is local just as in evolutionary biology; theories compete in the sense that scientists decide between them on the basis of such things as explanatory and predictive power amongst the theories which are known to the scientists at the time the decision is made, but there is no evidence that the historical sequence of physical theories is approaching some limit which could be termed 'Ultimate Truth'. As Kuhn puts it:[59] 'Comparison of historical theories give no sense that their ontologies are approaching a limit: in some fundamental ways Einstein's general relativity resembles Aristotle's physics more than Newton's'. This vision of the scientific enterprise was best summed-up by Kuhn in the concluding pages of his famous work *The Structure of Scientific Revolutions*:[58]

> The developmental process described in this essay has been a process of evolution *from* primitive beginnings—a process whose successive stages are characterized by an increasingly detailed and refined understanding of nature. But nothing that has been or will be said makes it a process of evolution *toward* anything ... need there be any such goal? Can we not account for both science's existence and its success in terms of evolution from the community's stage of knowledge at any given time? Does it really help to imagine that there is some one full, objective, true account of nature and that the proper measure of scientific achievement is the extent to which it brings us closer to that ultimate goal? ... the entire [scientific development] process may have occurred, as we now suppose biological evolution did, without benefit of a set goal, a permanent fixed scientific truth, of which each stage in the development of scientific knowledge is a better exemplar.
>
> Anyone who has followed the argument this far will nevertheless feel the need to ask why the evolutionary process should work. What must nature, including man, be like in order that science be possible at all? .... It is not only the scientific community that must be special. The world of which that community is a part also possess quite special characteristics, and we are no closer than we were at the start to knowing what these must be. That problem—What must the world

be like in order that man may know it?—was not, however, created by this essay. On the contrary, it is as old as science itself, and remains unanswered.

It is the goal of the Anthropic Principle to answer it, at least in part.

3.3 Henderson and The Fitness of The Environment

What is matter?—Never mind
What is mind?—It doesn't matter.
Anon

Lawrence J. Henderson was a professor of biological chemistry at Harvard at the turn of the century, and he published his two seminal books on teleology, *The Fitness of the Environment*,[60] and *The Order of Nature*[61] in 1913 and 1917, respectively, before quantum mechanics was available to provide the basis for the understanding of the physical underpinnings of chemistry. Nevertheless, his discussion of what we might term 'physical teleology' was grounded on physical principles sufficiently general that the core of his argument has withstood the buffetings of several scientific revolutions which have occurred between his time and ours. His work, as updated by several modern biochemists, notably George Wald,[62] still comprises the foundation of the Anthropic Principle as applied to biochemical systems. We shall discuss more modern work in Chapter 8.

Henderson was led to reflect on teleology in the biochemical world through his work on the regulation of acidity and alkalinity in living organisms.[63] He noticed that of all known substances, phosphoric acid and carbonic acid (CO_2 dissolved in water) possessed the greatest power of automatic regulation of neutrality. Had these substances not existed, such regulation in living things would be much more difficult. Henderson searched the chemical literature and uncovered a large number of substances whose peculiar properties were essential to life. Water, for example, is absolutely unique in its ability to dissolve other substances, in its anomalous expansion when cooled near the freezing point, in its thermal conductivity among ordinary liquids, in its surface tension, and numerous other properties. Henderson showed that these strange qualities of water made it necessary for any sort of life. Furthermore, the properties of hydrogen, oxygen, and carbon had a number of quirks amongst all the other elements that made these elements and their properties essential for living organisms. These quirks were discussed in detail in his book *Fitness of the Environment.* These properties were so outstanding in the role they played in living things that '. . . we were obliged to regard this collocation of properties as in some intelligible sense a preparation for the process of planetary evolution . . . . Therefore the properties of the elements must for the present be regarded as possessing a teleological character.'[64] Henderson never actually asserted that no life would be possible in the absence of the elements hydrogen, oxygen, and carbon, just that ... 'No other element or group of elements possesses properties which on any account can be compared with these. All such are deficient at many points, both qualitatively or quantitatively.[65] ... The unique properties of water, carbonic acid, and the three elements constitute, among the properties of matter, the fittest ensemble of characteristics for durable mechanism.'[66]

In earlier days such observations would be cited as evidence of a Designer—indeed, Henderson himself quotes the Bridgewater Treatise of William Whewell as pointing out many of the unique properties of water—but Henderson takes a distinctly modern approach. He discusses the theories of vitalism and mechanism at length in both of his books, and strongly criticizes the former, arguing that a scientist must always assume that living things operate according to physical laws, that there are no laws like the vital forces of Bergson and Driesch which operate in life only. In short, in the living world evolution is controlled by efficient causes and by efficient causes only. He, in contrast to the directed-evolution philosophers discussed earlier, bases his analysis on the assumption—which all moderns accept—that the development of life was at all times the result of natural selection acting on changes in the hereditary structure. Thus ultimately there is no teleology acting in a living organism; the planning which a living creature undertakes to guide his future actions can ultimately be reduced to mechanism, to the interaction of the elements in accordance with ascertainable physical laws. Furthermore, concerning the existence of a Designer, Henderson remained an agnostic.[67,339] However, from the apparent 'preparation' of the elements and their properties for the eventual evolution of life he could not escape:

> [we want a term] ... from which all implication of design or purpose is completely eliminated. By common consent that term has come to be recognized as *teleology*. Thus we say that adaptation is teleological, but do not say that it is the result of design or purpose. I shall therefore ... assert that the connection between the properties of the three elements and the evolutionary process is teleological and non-mechanical.[68]

(Henderson was unaware of the distinction, which we introduced in Chapter 2, between teleology and eutaxiology, although this distinction had been introduced in 1883 by the American philosopher L. E. Hicks. Clearly, Henderson was impressed by eutaxiological, not teleological, order.)

But how can this connection be non-mechanical if all interactions in the Universe, both living and non-living, are mechanical? The answer is simple: this teleological order of the three elements which is a preparation for life, was imposed in the beginning: 'For no *mechanical* cause whatever is conceivable of those original conditions, whatever they may be, which unequivocally determine the changeless properties of the elements and the general characteristics of systems alike'.[69] One might think that this state of affairs would make the scientific study of the teleological order impossible, for it would seem that the work of science consists of finding efficient causes. The conditions at the beginning are, as Henderson said, presently beyond investigation. Nevertheless, Henderson argued that one can study the teleological order by the probabilistic analysis standard in other areas of physical science, and that therefore conclusions reached through this analysis have a similar force:

The chance that this unique ensemble of properties should occur by 'accident' is almost infinitely small (i.e., less than any probability which can be practically considered). The chance that each of the unit properties [heat capacity, surface tension, number of possible molecules, etc.] of the ensemble, by itself and in cooperation with the others, should 'accidentally' contribute a maximum increment is also almost infinitely small. Therefore there is a relevant causal connection between the properties of the elements and the 'freedom' of evolution.[70]

The 'freedom of evolution' was '... freedom of development. This freedom is, figuratively speaking, merely the freedom of trial and error. It makes possible the occurrence of a great variety of trials and a large proportion of successes'.[71] That is, the peculiar properties of the three elements hydrogen, oxygen, and carbon permitted a large number of molecules to be formed, and this enormous number of molecules allowed a large number of possible organisms to be based on these molecules. If the properties of the elements were slightly different, if there were no carbon atoms in the world, if for instance living things attempted to substitute silicon instead, then vastly fewer molecules would be possible, and evolution by natural selection on different genotypes would be impossible. Probably no organisms as complex as a single cell would arise, and certainly no creatures as complex as human beings would evolve. 'Hence the operations of a final cause, if such there be, can only occur through the evolution of systems. Therefore the greatest possible freedom for the evolution of systems involves the greatest possible freedom for the operation of a final cause'.[72] Thus the theory of evolution by natural selection—i.e., evolution by trial and error and not goal-directed evolution—was essential to Henderson's argument. Note that Henderson's concept of final cause, since it operates by allowing many possible developments rather than making one particular development inevitable, allows evolution to Man, but does not require it. It thus subsumes the notion of 'indeterminate natural teleology' of the modern biologists.

Although the properties of the elements allowed the maximum possible freedom of evolutionary development, the properties themselves were not free to interact with living things and so evolve themselves. In Henderson's opinion, this precluded a mechanical explanation of the elemental properties, and required a teleological explanation:

It cannot be that the nature of this relationship [between the elements which allows life to evolve] is, like organic adaptation, mechanically conditioned. For relationships are mechanically conditioned in a significant manner only when there is opportunity for modification through interaction. But the things related are supposed to be changeless in time, or, in short, absolute properties of the Universe.[73]

This argument for the non-mechanical determination of the elemental properties assumes these properties to be unchanging:

Nothing is more certain than that the properties of hydrogen, carbon, and oxygen are changeless through-out time and space. It is conceivable that the atoms may be formed and that they may decay. But while they exist they are uniform, or at least they possess perfect statistical uniformity which leads to absolute constancy of all their sensible characteristics, that is to say of all the properties with which we are concerned....[74] Accordingly, the properties of the elements to be regarded are fully determined from the earliest conceivable epoch and perfectly changeless in time. This we may take as a postulate.[75]

Although we now believe that the elements have evolved in the sense of changing their numbers relative to hydrogen, we still believe their properties to be fixed by natural laws just as Henderson did. Thus the elements cannot 'evolve' in the sense of having the freedom to take different evolutionary pathways like living creatures can. This portion of Henderson's argument must still be regarded as sound.

The part of his argument which is more questionable is his contention that the various unique properties of matter which make the Earth's environment the fittest for the evolution of life, are statistically independent. This difficulty is the bugbear of all Anthropic Principle arguments. One can never be sure that future developments in physics will not show that the supposedly independent properties of matter are in fact subtly related, and that only one very particular collection of material properties are logically possible. To his great credit, Henderson was aware of this difficulty, and he attempted to meet it in two ways. First, he contended[76] that there was a fundamental distinction between the laws of nature properly speaking, which might be deduced *a priori* from the laws of thought and the properties of matter, which are not laws of thought. Henderson said the Second Law of thermodynamics might be an example of a law of thought: 'Possibly the second law of thermodynamics ... might have been worked out by a mathematician in perfect ignorance of how energy should be conceived.' Although the laws of thermodynamics may be laws of thought, the types of interactions and types of matter to which they apply are not; forces other than those which actually exist are possible. In particular, '... the prediction of electrical phenomena by one ignorant of all such phenomena seems to be quite impossible;[76] see also ref. 77. Second, Henderson argued that an application of Gibbs' Phase Rule, a general theorem of thermodynamics which is so fundamental that it is unlikely ever to be overthrown, indicated that the elemental properties were indeed independent:

...since the whole analysis is founded upon the characteristics of systems and therefore upon concepts which according to Gibbs are independent of and specify nothing about the properties of the elements, it is unnecessary to examine the possibility of the existence of other groups of properties which may otherwise be unique.[78]

In short, Henderson presented what must still be regarded as a powerful argument that the properties of matter are, in a fundamentally teleological sense, a preparation for life:

The properties of matter and the course of cosmic evolution are now seen to be intimately related to the structure of the living being and to its activities; they become, therefore, far more important in biology than has previously been suspected. For the whole evolutionary process, both cosmic and organic, is one, and the biologist may now rightly regard the Universe in its very essence as biocentric.[79]

Henderson's work on the Fitness of the Environment had very little impact on his scientific contemporaries. *The Fitness of the Environment* was reviewed in *Nature*,[80] but without critical comment; only a paragraph appeared summarizing the argument. The physiologist J. S. Haldane reviewed *The Order of Nature* for *Nature*.[81] Haldane, who was at heart a vitalist and believed in goal-directed evolution, gave the book fulsome praise, but did not appreciate Henderson's arguments. The main effect of Haldane's reading was in influencing his son, J. B. S. Haldane, who in a number of letters to *Nature* used Henderson's ideas to explain why the laws of Nature are seen to have the properties they do. The greatest early impact of Henderson's ideas on his contemporaries was not in science but in theology, as we shall see in section 3.9. By and large, Henderson's work did not lead to any new work on the question of the fitness of the environment by scientists, although a few biologists, for instance Joseph Needham[82] and George Wald[62] occasionally mention his work with approval. Most biologists, however, either ignored his work or took the attitude of the zoologist Homer Smith:

One should not be surprised that there is a remarkable 'fitness' between life and the world it lives in, for the fitness of the living organism to its environment and the fitness of the environment to the living organism are as the fit between a die and its mould, between the whirlpool and the river bed.[83]

3.4 Teleological Ideas and Action Principles

> The great end of life is not Knowledge
> but Action.
>
> T. Huxley

Teleological ideas have played a role in mathematical physics mainly in the form of 'minimal' principles. In 'minimal' (or more precisely, extremum) principles, one deduces the behaviour of a physical system between times t_1 and t_2 by requiring that the evolution of the system be such as to *minimize* a certain quantity. For example, in the first use of a minimum principle in physics, Hero of Alexandria[84] showed in the first century AD that if a light ray goes from an object to a mirror, and from the mirror to an observer's eye, the path taken by the ray is *shorter* than any other path from the object via the mirror to the eye. Putting the observed behaviour into teleological language, we would say the light ray seems to know that its goal is the observer's eye, and it picks out among all paths from the object to the mirror to the eye the shortest one—its behaviour is teleological in other words, since its behaviour is determined by its final destination. Hero did not discover anything new about the behaviour of light rays through use of the minimal principle, for the path taken by light during reflection from a mirror was already known. He did, however, regard his teleological principle as an *explanation* for the behaviour of light. Hero's explanation fitted in well with Aristotle's dictum that final causes were to be regarded as the primary causes. Furthermore, Aristotle himself had argued[85] that planets moved in circular orbits because, of all closed curves bounding a given area, the circle is the *shortest.* Both Aristotle and Hero connected these shortest paths with the maximum speed of motion; that is, the motion also attempts to minimize the time spent in motion.

The principle of least time was the basis of the next use of a minimal principle, by the seventeenth-century French mathematician and lawyer Fermat. He argued[86] that the behaviour of a ray of light in both reflections *and* refractions could be understood by assuming that it always travels from one point to another so as to make the time of travel a minimum. For reflection, Fermat's *Principle of Least Time* reduces to Hero's law of reflection; but for refraction, Fermat was able to show that his Principle implied both Snell's law (which was known at the time), and the fact that light travels more slowly in a medium with a higher refractive index (which was not shown experimentally until two centuries after Fermat). This is the first known case in mathematical physics where thinking about physics teleologically led to an experimentally verifiable (and correct) prediction.

Fermat's work led the German philosopher Leibniz to argue in a letter[87] written in 1687 that in as much as the concept of purpose was basic to true science, the laws of physics should and could be expressed in terms of minimum principles. It is not known whether he followed-up this suggestion with an explicit reformulation of the laws of mechanics in terms of a minimum principle, but if he did, it was never published.

The first such formulation was given by the French scientist Maupertuis who in 1744 presented a paper[88] to the French Academy of Sciences showing that the behaviour of bodies in an *impact* could be predicted by assuming the product mvs, where m is the mass, v the velocity, and s the distance, to be a minimum. He contended[89] that his formulation indicated the operation of final causes in Nature, and that final causes imply the existence of a Supreme Being. Maupertuis, following Leibniz and Wolff, called the quantity mvs, which has dimensions of energy times time, the *action.*

Maupertuis' Principle of Least Action was immediately generalized[90] by his friend, the brilliant mathematician Leonhard Euler, into an integral theorem, valid for the continuous motion of a single particle acted on by an arbitrary conservative force. Euler showed that if the mass of the particle was assumed constant, then the integral $\int v\,ds$ taken along the path of a particle between its *initial* and *final* positions, was an extremum along the actual path of the particle. That is, if the value of $\int v\,ds$ along the actual path were subtracted from the value of $\int v\,ds$ along paths infinitesimally close to the actual path (the particle energy having the same value on all paths), this difference would be an infinitesimal quantity of second order. This vanishing of the difference to first order is a necessary (but not a sufficient) condition for the integral $\int v\,ds$ to be an actual minimum on the real path, and in fact there are cases in which the action is not a minimum for the actual path, but a *maximum.* The action principle of Euler was later extended to the case of motion of a system of interacting particles by Lagrange[91] in 1760 and given a particularly useful formulation by Hamilton[92] in 1835. In both the Lagrangian and Hamiltonian formulations the behaviour of a general physical system is determined by the requirement that the time integral of a function of the system be an extremum.

Thus Maupertuis was incorrect in calling his discovery a principle of *least* action, though he was quite right in interpreting the principle as a teleological formulation of physics, since the motion of a physical system is determined in the action principle formulation by both the initial and the final states of the system. This aspect was also emphasized by Euler.[93]

Physicists have disagreed on the significance of the fact that mechanics can be formulated in telelogical language. Like Maupertuis, Euler was attracted to the Action Principle formulation of mechanical laws because of its teleological aspects. Euler also believed that the Action Principle formulation could solve problems which were intractable in the usual approach to mechanics. As Euler put it:

> All the greatest mathematicians have long since recognized that the [least action] method . . . is not only extremely useful in analysis, but that it also contributes greatly to the solution of physical problems. For since the fabric of the universe is most perfect, and the work of a most wise Creator, nothing whatsoever takes place in the universe in which some relation of maximum and minimum does not appear . . . there is absolutely no doubt that every effect in the universe can be explained as satisfactorily from final causes, by the aid of the method of maxima and minima, as it can from the effective causes themselves . . . since, therefore, two methods of studying effects in Nature lie open to us, one by means of effective causes, which is commonly called the direct method, the other by means of final causes, the mathematician uses each with equal success. Of course, when the effective causes are too obscure, but the final causes are more readily ascertained, the problem is commonly solved by the indirect method; on the contrary, however, the direct method is employed whenever it is possible to determine the effect from the effective causes. But one ought to make a special effort to see that both ways of approach to the solution of the problem be laid open; for thus not only is one solution greatly strengthened by the other, but, more than that, from the agreement between the two solutions we secure the very highest satisfaction.[90]

On the other hand, Poisson,[94] Hertz,[95] and Mach[96] felt that such a formulation was merely a mathematical curiosity, rather than something fundamental about the world. In particular, these men emphasized that the usual approach to mechanics and the action principle approach are really mathematically equivalent, but the usual approach—which calculates the future state from initial data—is much easier to handle in practical problems. Even those who believe in the fundamental nature of action principles rarely do calculations by computing the minimum of the action integral. Instead, they use the action principle only to infer the *differential* equations of motion which allow one to calculate a future state from the present state. Once they have obtained the equations by motion, they proceed in the usual way. In addition, the opponents of the action principle have expressed a hostility toward introducing the concept of teleology into physics, for this notion has usually served as a wedge to infiltrate religious and metaphysical ideas into what should be a purely physical discussion. D'Abro,[97] and, as we shall see in more detail in section 3.10, Henri Bergson have pointed out that in a deterministic system, there is no real difference between a teleological description and a 'mechanistic' description—a description which deduces the future states from initial state information via the equations of motions. If the system is deterministic, then one could calculate the initial state from the data available in the final state; thus in this sense the initial state is determined by the final state. Finally, Yourgrau and Mandelstam[98] have argued that for *any* set of evolution equations, an 'action' can be defined which is an extremum for the actual path. This would mean that action principles in general have no physical content.

Nevertheless, many physicists have contended that the action principle formulation of mechanics is more fundamental than the mechanistic formulation. In the latter part of the nineteenth century Helmholtz argued[99] that an action principle could act '... as a heuristic and guiding principle in our endeavour to formulate the laws governing new classes of phenomena.' Max Planck[100,101] also felt the action formulation was a more fundamental view of natural phenomena than the mechanistic approach, primarily because he was partial to teleological explanations for religious reasons, but also because action principles expressed the laws of physics in a relativistic manner—the action was a scalar, and so its value did not depend on the choice of the coordinate system,—and because action appeared to play a fundamental role in quantum mechanics. Planck's constant has the dimensions of action.

Helmholtz' assertion that action principles can suggest new physical laws has been confirmed in the twentieth century. The German mathematician Hilbert discovered the final form of the Einstein field equations independently of Einstein by combining hints coming from earlier attempts by Einstein to construct gravitational field equations with the requirement that the equations be derived from a 'simple' action integral.[102] In this case adopting the attitude that the action—and hence by implication, a teleological process—is basic to nature led to a major discovery. Nevertheless, the teleological aspects of the action were really not paramount in Hilbert's thinking.

The explicitly teleological aspect of the action was, however, basic to the early work of Richard Feynman. While still a graduate student at Princeton, Feynman developed with his teacher John Wheeler a theory of classical electrodynamics in which the radiation reaction of an electrically charged particle is explained in terms of an interaction of the particle with other particles in the past *and* in the future.[103,104] Thus the motion of a particle today depends on what the other particles in the Universe will be doing in the far future. The action principle formulation of Wheeler–Feynman electrodynamics is conceptually simpler than the usual field-and-particles formulation, in that it does not need to introduce the notion of an electromagnetic field—electrodynamics is due to the direct action of the particles on themselves. In the Wheeler–Feynman picture, the electromagnetic field is not a real physical entity, but just a book-keeping device constructed to avoid having to talk about the particles teleologically. In the conventional particles-and-fields electrodynamics, the future behaviour of the particles and fields is determined by information given at one instant of time. In contrast, it is not possible to determine the future behaviour of the particles alone solely by giving the initial position and velocities of the particles. One must also specify some information about their future and past behaviour; that is, one must discuss the particles teleologically. To date, the Wheeler–Feynman formulation of electrodynamics has not led directly to any important new discoveries (see, however, refs. 105 and 106). However, this teleological way of thinking about the motion of charged particles led Feynman to develop his sum-over-histories formulation of quantum mechanics,[107] which is a method of expressing quantum mechanics in terms of an action principle. In this formulation the wave function $\psi(x_1, t_1)$ of a particle at the present time t_1 is determined from the wave function $\psi(x_0, t_0)$ at an earlier time t_0 by summing a function of the classical action of the particle over all possible paths the particle could take in going from x_0 to x_1 in time $t_1 - t_0$. Using this formulation of quantum mechanics, Feynman was able to derive the so-called Feynman Rules for the scattering of elementary particles. As happened in previous centuries, many physicists (for instance, S. Weinberg) felt that teleological formulations of physical theories such as the sum-over-histories method were unphysical, and these physicists soon developed alternative ways of deriving the Feynman rules. The value of the sum-over-histories method over the alternative methods was demonstrated, however, by the proof of 't Hooft and others, using the sum-over-histories method, that exact spontaneously broken gauge symmetry theories would be renormalizable.[108] This proof encouraged experimenters to test the gauge theories, particularly the gauge theory for the electro-weak interaction of Weinberg and Salam, with the result that the Weinberg–Salam theory has now been confirmed. Weinberg now asserts[109] that the sum-over-histories method is the best way to prove the renormalizability of the gauge theories, and he no longer feels that the sum-over-histories method is unphysical. Since the whole of contemporary particle physics is now formulated in terms of gauge theories,[109] and since these theories must be analysed in some respects in terms of the sum-over-histories action principle method, it would seem that teleological thinking has become essential to modern mathematical physics. The sum-over-histories technique can be formulated in a non-teleological language[110] but the other formulations lack the great heuristic power of the sum-over-histories approach, as the inventors of the alternative formulations admit.[110]

We shall use the Feynman sum-over-histories method in Chapter 7 to obtain an expression for the wave function of the Universe. We approach the problem of finding the Universal wave function via the action principle because the action principle can enormously simplify the problem of the boundary conditions: as we shall point out in section 7.3, the action principle formulation strongly suggests the Universe is closed, since only closed universes have finite action and no difficulties with the boundary conditions at infinity. We could have another teleological prediction: the Universe must be closed. This prediction depends crucially on taking the action formulation as fundamental, and it cannot be obtained from a non-teleological approach employing differential equations. More discussion on the closed universe prediction can be found in ref. 111.

3.5. Teleological Ideas in Absolute Idealism

> It is no use arguing with a prophet;
> you can only disbelieve him.
>
> Winston Churchill

German absolute idealism arose at the end of the eighteenth century in part as a reaction to Kant's notion of 'thing-in-itself'. Kant had argued that we could not know an object as it actually is, but rather our minds act to force our sensory experience of the object into certain patterns which may or may not resemble the actual object being experienced. There was, nevertheless, a real object underlying our experience of the object. This 'real object' was the 'thing-in-itself'.

The difficulty with the notion of a thing-in-itself is of course the fact that by definition, it is absolutely unknowable. No possible experiment is capable of giving us any information at all about the thing-in-itself. As the first absolute idealist, J. G. Fichte, put it in 1797:

> A finite rational being has nothing beyond experience; it is this that comprises the entire staple of his thought. The philosopher is necessarily in the same position; it seems, therefore, incomprehensible how he could raise himself above experience[112] ... the thing-in-itself is a pure invention and has no reality whatever. It does not occur in experience for the system of experience is nothing other than thinking...[113]

Fichte and the other absolute idealists proposed to eliminate the concept of the thing-in-itself altogether; thought comprises all of reality:

> [an object]...is nothing else but the *totality of* [*all*] *relations* [*of the object*] *unified by the imagination*, [Fichte's emphasis] and that all these relations constitute the thing; the object is surely the original synthesis of all these concepts. Form and matter are not separate items; the totality of form is the matter ...[114]

Fichte's notion that a real object consists of all possible experiences it can generate in the mind of a potential observer is in all essentials the same as Niels Bohr's view of what is meant by a 'objectively real' property of a quantum mechanical object.[115] It is also similar to the economist F. Hayek's idea that the total capital in an economic system can be adequately described by listing all the possible products it could generate.[116]

But if everything is thought, what is thought? Fichte pointed out that one must be careful not to think of thought as a sort of substance, for this would get us nowhere:

> [the intellect] has no *being* proper, no subsistence, for this is the result of an interaction and there is nothing ... with which the intellect could be set to interact. The intellect, for idealism, is an *act*, and absolutely nothing more; we should not even call it an *active* something, for this expression refers to something subsistent in which activity inheres.[117]

Since the absolute idealists claim everything is thought, we shall attempt to make sense of this and other passages by translating the statements of this philosophical school into a rigorous modern language: abstract computer theory.

The central concept of computer theory is the idea of a program, or procedure. A program can be regarded abstractly as a map $f: N \rightarrow N$ from the set of natural numbers, N, into itself. That is, an input data set will be specified by an integer, and the program will generate from this number an output which is another number. The whole of computer theory can be said to be concerned with deciding what constitutes an *effective* procedure, and with describing the attributes of an effective procedure. By the Turing Test,[118,119] which we shall discuss at length in section 8.2, a human intellect can be equated with a particular type of program. But it is often pointed out (e.g. ref. 119) that we can go further. We can in fact simulate—in the computer language sense of representing the evolution of—the entire Universe with a program, for the Universe evolves deterministically from an initial state (input data set of the program) into a final state (output data set) and the Universal states are operationally denumerable. (We should mention that even a quantum mechanical Universe is deterministic; see Chapter 7. The evolution equation (7.37) for the Universe is a deterministic equation, since a state at any time is determined uniquely from its initial state.)

The absolute idealists want to make the step which many computer scientists have taken (see ref. 119 for examples) and equate the Universe with its simulation. This is not as unreasonable as it sounds at first hearing. If a simulation is perfect, then those subprograms which are isomorphic to human beings in the general Universal Program act the same in the simulation as do human beings in the actual Universe. They laugh, they cry, they live, and they die. By the underlying logic of the Turing Test, they have to be regarded as persons.

Now the Universal simulation need not be run on an actual computer; it can be regarded as an abstract sequence of mappings from one input set to another. The actual Universe is a representation of the abstract Universal Program in the same sense that the written Roman numerical III is a representation of the abstract Idea of three, or as an actual physical computer is a representation of an abstract program.[120,121] A rational subprogram inside the Universal Program cannot by any logically possible operation distinguish between the abstract running of the Universal Program, and a physically real, evolving Universe. Such a physically real Universe would be equivalent to the Kantian thing-in-itself. As empiricists, we are forced to dispense with such an inherently unknowable object: the Universe must *be* an abstract program, or Absolute Idea, which is of the same nature as the human intellect, or program. Fichte's *act*, the undefined basic property of the intellect, can be thus equated with the basic map (= procedure = program) that takes one state into another, or more precisely, equated with the class of basic operations of an abstract universal machine.[120,121]

The human mind is a very complex yet very special type of program. It is capable, in particular, of forming a model of itself as a subprogram, and studying this subprogram. This model-building and analysing process is called consciousness. The model is only a rough model, for Gödel's Theorem shows an exact model to be impossible even for infinite machines such as the universal Turing machine. The problem the absolute idealists had to deal with was explaining why the Universal Program is as complex as it is observed to be, involving many subprograms, including those which can be called rational. This difficulty can be attacked in one of two ways. The first approach, which could be termed *subjective idealism*, would take the finite rational subprogram as the basic entity, and try to construct the Universal Program out of the inherent logical nature of the rational subprogram. This was Fichte's approach. The obvious problem with this approach is that it is difficult to avoid solipsism. The second approach, *objective idealism*, which was the one preferred by Fichte's successors, Schelling and Hegel, is to take the Universal Program as basic, and to argue that rational subprograms are produced by the very nature of the Universal Program. As Schelling put it:

> Fichte could side with idealism from the point of view of reflection. I, on the other hand, took the viewpoint of production with the principle of idealism. To express this contrast most distinctly, idealism in the subjective sense had to assert, the ego is everything, while conversely idealism in the objective sense had to assert: Everything = ego and nothing exists but what = ego. These are certainly different views, although it will not be denied that both are idealism.[122,123]

None of the absolute idealists would however, accept Berkeley's solution that a God external to the universe held the non-rational order in existence when no rational ego was observing. Rather, they all in their different ways attempted to construct reality from itself.

Fichte argued that a rational self-conscious mind must of logical necessity have experiences which it would interpret as a world external to itself: the subjective posits the objective. But to posit an object means that the self limits itself. Furthermore, other finite rational minds must exist in order for the freedom of a rational program to be fully realized.[124,125] Translating this into computer language, we would say that a program sufficiently complex to count as rational would have to act as if embedded in a larger program which would contain other rational subprograms and submappings which would be interpreted by the rational subprograms as an external world. But we know at least one rational program exists because each individual knows himself to be one by self-reflection. According to Fichte, each rational being by its innate nature must have goals, which is to say it must be teleological. The ends of all rational beings in the Universal Program must impart a limited teleology to Nature itself, which as Mind must also have a Purpose Itself, but Fichte did not investigate this Purpose.

Schelling, who regarded the Absolute Ego, or Universal Program as fundamental, *was* concerned with its Ultimate Purpose:

Has creation a final purpose at all, and if so why is it not attained immediately, why does perfection not exist from the very beginning? There is no answer to this except the one already given: because God is a *life*, not a mere being. All life has a destiny and is subject to suffering and development. God freely submitted himself to this too, in the very beginning ... in order to become personal ... for being is only aware of itself in becoming.... All history remains incomprehensible without the concept of a humanly suffering God. Scripture, too, ... puts that time into a distant future when God will be all in all, that is, when He will be completely realized. For this is the final purpose of creation, that which could not be in itself, shall be in itself....[126] Succession itself is gradual. I.e., it cannot in any single moment be given in itself entirely. But the farther succession proceeds, the more fully the universe is unfolded. Consequently, the organic world also, in proportion as succession advances, will attain to a fuller extension and represent a greater part of the universe....[127]

In the opinion of the great historian of ideas Arthur O. Lovejoy,[128] it is this first introduction into philosophy of an evolutionary metaphysics, or more particularly, the notion of an evolving God, who at the final state of the Cosmos will be both fully realized and one with the Cosmos, that is the chief contribution of Schelling to human thought. In his celebrated debate with the philosopher Jacobi, who was defending the traditional conception of a perfect, unchanging Deity, Schelling put it thus:

I posit God, as the first and the last, as the Alpha and the Omega; but as Alpha he is not what he is as Omega, and in so far as he is only the one—God 'in an eminent sense'—he cannot be the other God, in the same sense, or in strictness, be called God. For in that case, let it be expressly said, the unevolved [*unentfaltete*] God, Deus implicitus, would already be what, as Omega, the *Deus explicitus* is.[129]

In Schelling's view the Universal Program would give rise to self-conscious subprograms which would, in the fullness of time, merge together into one self-knowing Mind. Nature is teleological for two reasons: the rational subprograms are presently an image of the Universal Program, and further—as a consequence of being an image of an intrinsically teleological entity—the Universal Program has the goal of universal self-consciousness.

Hegel agreed with Schelling that the Absolute Idea (= Universal Program) is fundamentally teleological: the Universe, or totality, is ultimately self-thinking thought; or to put it another way, the process of Nature is the teleological movement toward the Universe becoming aware of itself. The human species is the means whereby the Universe becomes aware of itself. In fact Hegel contended the struggle of the Universe to become aware of itself was the purpose of human history:

... the final cause of the World at large, we allege to be the consciousness of its own freedom on the part of Mind [*Geist*], and ipso facto, the reality of that freedom.[130] ... substance is essentially subject ... the Absolute is Mind ... Mind alone is reality.[131]

In contrast to Schelling, Hegel did not believe in a perpetually evolutionary cosmos.[132] In the words of the English idealist John McTaggart, 'while [Hegel] did not explicitly place any limits to the development of the universe in time, he seems to have regarded its significance ... as pretty well exhausted when it had produced the Europe of 1820',[133] which is to say, with the development of Hegelian philosophy.

Absolute idealism went into a decline with the deaths of Schelling in 1854 and Hegel in 1831, but it flourished anew at the end of the nineteenth century in both the United States and Great Britain. McTaggart was one of the leaders of the British idealist school, which also included F. H. Bradley and B. Bosanquet. These men were influenced mainly by Hegel rather than by Fichte or Schelling; they were regarded as neo-Hegelians by contemporary British realists such as Russell. Nevertheless, toward the end of his career McTaggart had moved from the static absolute idealism of Hegel to the cosmic evolutionary idealism of Schelling, of which he was apparently unaware. McTaggart argued that *value*, or the good in the universe is increasing with time, and that it must become infinite in finite time. This infinite good is the ultimate goal of a teleological universe. Most of McTaggart's idealist contemporaries, like

Bosanquet,[134] retained Hegel's static comology in which Man was the ultimate knowing subject; the Universe was teleological only through Man, and because teleological Man was the image of the Universe. McTaggart felt that

> ... those Idealists ... seem generally unwilling to adopt a view which makes the selves that we know numerically insignificant in the universe ... the conclusion that the time to be passed through before the goodness of the final state is reached may have any finite length, cannot be altogether attractive to those who feel how far our present life is from that great good.... Hegel is perhaps the strongest example of this unwillingness to accept the largeness of the universe.... But the universe *is* large, whether we like that largeness or not.[133]

The American idealist school included the Harvard philosopher Josiah Royce, and to a certain extent Charles Sanders Peirce, considered by many to be the greatest American philosopher. Peirce held a view which he termed 'tychistic idealism', in which life, regarded as being a sort of intrinsic chance or spontaneity, is a fundamental aspect of everything.[136] In some of his writings, Peirce argued that the Universe was too vast to have any character, teleological or otherwise.[136] In other writings, Peirce defended a 'Cosmogonic Philosophy', in which the very development of life would cause it to gradually lose its spontaneous character, and thus life would eventually totally order an initial universal chaos:

> [Cosmogonic Philosophy] would suppose that in the beginning—infinitely remote—there was a chaos of unpersonalized feeling, which being without connection or regularity would properly be without existence. This feeling, sporting here and there in pure arbitrariness, would have started the germ of a generalizing tendency. Its other sportings would be evanescent, but this would have a growing virtue. Thus the tendency to habit would be started; and from this, with the other principles of evolution, all the regularities of the universe would be evolved. At any time, however, an element of pure chance survives and will remain until the world becomes an absolutely perfect, rational, and symmetrical system, in which mind is at last crystallized in the infinitely distant future.[137]

Royce, on the other hand, always defended a cosmic teleology; for example, he did so in his Gifford Lectures.[135] In Royce's view, Nature arises from a sort of mutual interaction between the knower and the known:

> Reality is not the world apart from the activity of knowing beings, it is the world of the fact and the knowledge in one organic whole.[135]

Royce's most significant contribution to teleology, however, was not contained in his published work, but rather lay in his discussions with Lawrence Henderson on the subject. Royce had organized a private evening seminar, which included Henderson and a number of other scholars at Harvard, including for a time even T. S. Eliot. H. T. Costello took minutes of these meetings throughout the period 1913–1914, and they reveal that Henderson's *Fitness of the environment* and the possible interpretations of the chemical concept of 'fitness' which Henderson proposed was the topic of presentations and debate at the seminar for over three months.[339] Although Henderson did not obtain his idea of the fitness of the environment from Royce or others at the seminar, Henderson acknowledged that his insight was sharpened by the debate.[340]

3.6 Biological Constraints on The Age of The Earth: The First Successful Use of An Anthropic Timescale Argument

> *Anthropic*, (or *Anthropical*): of, or relating to mankind or the period of man's existence on earth.
>
> Webster's Dictionary, 1975

The Anthropic Principle imposes constraints on the types of physical processes allowed in the Universe by requiring that these processes must be of such an age that slow evolutionary processes will have had time to produce intelligent beings from non-living matter. Thus one sort of physical prediction which can be made using the Anthropic Principle would be a prediction of the types of energies and materials which can be present in the Earth and Sun, with the prediction being based on purely biological arguments of the minimum time needed for the evolution of intelligence. This is in fact the approach we shall use in later chapters to study constraints on the physical constants. We shall take from biology the estimate that a lower bound of a billion years is required for the evolution of intelligence, which implies that stars must be stable for at least that long, and so on. However, the first Anthropic prediction of this sort was actually made in the latter part of the nineteenth century in the course of a debate on the age of the Earth between biologists and physicists. This debate was initiated by Lord Kelvin, one of the most influential physicists of the nineteenth century.

The first scientific attempt to measure the age of the Earth was made in the late eighteenth century by the great French scientist Buffon. Buffon adopted the point of view that the Sun's heat was insufficient to warm the Earth; heat from the Earth's interior was essential to provide enough heat for organic life. He also assumed that the Earth's internal heat was not being continuously generated, but was residual—the Earth had been initially very hot, but has been cooling down ever since its formation. Earlier Newton had pointed out in the *Principia* that a globe of red hot iron the size of the Earth would need at least 50,000 years to cool. Buffon confirmed Newton's estimate by measuring the time required for balls made of various substances to cool from red heat to the absence of glow and then to room temperature. Extrapolating to a globe the size of the Earth, Buffon estimated that an initially molten Earth would need about 36,000 years before it would be cool enough for organic life to begin, and that about 39,000 years had passed from this beginning of organic life to the present day.[138,139] This attempt by Buffon to calculate the age of the Earth attracted a great deal of attention, and a desire to put Buffon's cooling calculations on a more rigorous basis and thus to put an estimate of the age of the Earth on a more secure foundation was what led Fourier to develop his theory of heat conduction.[140,141]

Fourier's work was the basis of Lord Kelvin's well-known estimate of the age of the Earth and Sun. In his 1863 paper, '*On the Secular Cooling of the Earth*',[142] Kelvin assumed that the cooling of the Earth could be modelled by that of an infinite homogeneous solid. That is, Fourier's heat conduction equation was solved by assuming that the temperature varied in one direction only, the x-direction say. For a constant value of x the temperature was the same for any values of y and z, the two orthogonal directions. Kelvin also assumed that initially the Earth was a solid sphere of uniform temperature throughout. He justified this assumption on the basis that he felt solid rock would be denser than molten rock, and so rock cooling near the Earth's surface would sink before solidifying, thereby creating convection currents which would maintain a constant temperature throughout the entire Earth until its interior was solid throughout. The initial constant temperature would thus be the melting temperature of rock, which Kelvin estimated to be between 7000 and 10,000 degrees Fahrenheit. The centre of the Earth would still be at this temperature (i.e., $T = 10{,}000°\mathrm{F}$ at $x = 4000$ miles). A final assumption made by Kelvin was that the thermal conductivity of the Earth was constant throughout, and equal to a suitable average of the conductivities of various surface rocks. These assumptions allowed Kelvin to calculate the thermal gradient on the surface of the Earth as a function of time. It was generally accepted that a thermal gradient of one degree Fahrenheit per 50 feet of depth was a probable mean over the present surface of the Earth, so Kelvin's formula yielded the estimate of 98 million years since the solidification of the Earth. Because of the uncertainties, Kelvin extended the limits of this period to between 20 million and 400 million years. Fourier had actually given the same formula for the age of the Earth and suggested roughly the same data in 1820, but had not written down the resulting age of the Earth. In the opinion of the historian Stephen Brush,[140] Fourier apparently felt that 100 million years was such an incredibly large number it was not even worth writing down!

In a paper published a year earlier,[143] Kelvin had also obtained an estimate of the Sun's age. By assuming that the source of the Sun's heat was gravitational potential energy—the Sun was envisaged to have been formed from meteors initially very far apart and with zero kinetic energy—Kelvin was able to place a lower limit to the original supply of solar energy at 10^7 times the present annual heat loss. Because of the uncertainties involved—the Sun's density, its specific heat, and the amount of its present-day contraction were not known—the upper limit could be up to 10 times higher.

Kelvin summarized his result as follows:

It seems, therefore, on the whole most probable that the sun has not illuminated the earth for 100,000,000 years, and almost certain that he has not done so for 500,000,000 years. As for the future, we may say, with equal certainty, that inhabitants of the earth cannot continue to enjoy the light and heat essential to their life, for many million years longer, unless sources now unknown to us are prepared in the great storehouse of creation.[143]

Kelvin concluded in his later paper:

... most probably the sun was sensibly hotter a million years ago than he is now. Hence geological speculation assuming somewhat greater extremes of heat, more violent storms ... are more probable than those of the extreme quietest, or 'uniformitarian' school ... it is impossible that hypotheses assuming an equality of sun and storms for a million years can be wholly true.[142]

These papers by Kelvin appeared some three years after the first edition of Darwin's *Origin of Species*, and although Kelvin pointed out in his papers the basic incompatibility of his chronology and Darwin's theory (a desire to refute Darwin was his motivation for writing the papers), biologists did not immediately respond to Kelvin's challenge. The first important reference to this incompatibility was Fleming Jenkin's review of the *Origin* in 1867. The Scot Jenkin was himself a physicist, and a close personal friend of Kelvin.[144] Jenkin pointed out that

... Darwin's theory requires countless ages, during which the earth shall have been habitable, and he claims geological evidence as showing an inconceivably great lapse of time, and as not being in contradiction with inconceivably greater periods that are even geologically indicated—periods of rest between formation, and periods anterior to our so-called first formations, during which the rudimentary organs of the early fossils became degraded from their primeval uses.[145]

As to a numerical estimate for the timescale, Jenkin claimed

... we doubt whether a thousand times more change than we have any reason to believe has taken place in wild animals in historic times, would produce a cat from a dog, or either from a common ancestor. If this be so, how preposterously inadequate are a few hundred times this unit for the action of the Darwinian theory.[146]

Jenkin emphasized the inconsistency between this Darwinian timescale, and Kelvin's chronology: 'From the earth we have no very safe calculation of past time, but the sun gives five hundred million years as the time separating us from a condition inconsistent with life.'[147]

The arguments of Kelvin and his minions gradually began to tell on the biologists; by 1871 both Wallace, the co-discoverer of natural selection, and Huxley, the chief fighter for evolution in the public arena, had yielded to Kelvin's arguments to the extent of admitting that evolutionary change may have occurred much more rapidly in the past than now, with the result that the entire evolution of living things occurred within Kelvin's timescale of 100 million years. In the sixth and last edition of the *Origin*, Darwin made a similar concession:

It is, however, probable, as Sir William Thomson [*sic*; Thomson was ennobled as Lord Kelvin] insists, that the world at a very early period was subjected to more rapid and violent changes in its physical conditions than those now occurring; and such changes would have tended to induce changes at a corresponding rate in the organisms which then existed.[148]

Later in the book, however, Darwin included a hedge:

With respect to the lapse of time not having been sufficient since our planet was consolidated for the assumed amount of organic change . . . , I can only say firstly that we do not know at what rate species change as measured in years, and secondly that many philosophers are not as yet willing to admit that we know enough of the constitution of the universe and of the interior of our globe to speculate with safety on its past duration.[149]

Although the biologists and geologists were willing to accept Kelvin's limit of 100 to 400 million years for the age of the Earth and Sun, several physicists and astronomers began to argue in the 1870's that Kelvin had been far too generous in assigning his upper limit, and that in fact it was much lower. Kelvin's friend and fellow Scot, the physicist Tait, contended in a series of public lectures delivered in 1874 that further calculations of the Earth's cooling indicated that the time since the Earth's solidification could be 10 to 15 million years at most, that evidence from tidal friction implies *less* than 10 million years, and that the Sun had heated the Earth for no more than 15 to 20 million years.[150] In 1878 the American astronomer Simon Newcomb reviewed Kelvin's arguments on the Sun's heat, and came to a conclusion similar to Tait's, that the Sun could not have supported life for more than 10 million years.[151]

The American Clarence King, an unconventional field geologist who served as the first director of the United States Geological Survey, obtained[152] a figure similar to Tait's for the age of the Earth—22 to 24 million years. Kelvin himself agreed that the age of the Earth and Sun should be reduced below his original estimate, but not quite to the drastic reduction of Tait. This reduction to 20 million years was more than the geologists and biologists could accept as consistent with the observations in their own fields, and the new number provoked cries of outrage. Darwin, in particular, referred to Tait's new number as 'monstrous'.[153] The Scottish geophysicist James Croll argued in response to Tait's number that the evidence of geology alone show 'without absolute certainty that [the Earth's age] must be far greater than 20 million years'.[154] He went on to say

> ... it does not follow as a necessary consequence, as is generally supposed, that [the Sun's initial] store of energy must have been limited to the amount obtained from gravity in the condensation of the Sun's mass. The utmost that any physicist is warranted in affirming is simply that it is impossible for him to *conceive* of any other source. His *inability*, however, to conceive of another source cannot be accepted as a proof that there is no other source. But the physical argument that the age of our earth must be limited by the amount of heat which could have been received from gravity is in reality based upon this assumption—that, because no other source can be conceived, there is no other source.
>
> It is perfectly obvious, then, that this mere negative evidence against the possibility of the age of our habitable globe being more than 20 to 30 million years is of no weight whatever when pitted against the positive evidence [from geology] that its age must be far greater.[154]

Even Archibald Geikie, the director of the Geological Survey of Scotland and a friend of Kelvin, was moved to reply to these later estimates, though he had originally accepted Kelvin's earlier estimate of 100 million years. Indeed, Geikie was a major cause of the widespread acceptance of the earlier estimate among geologists. In his 1892 Presidential Address to the British Association for the Advancement of Science, Geikie asserted:

> After careful reflection on the subject, I affirm that the geological record furnishes a mass of evidence which no arguments drawn from other departments of nature can explain away, and which it seems to me, cannot be satisfactorily interpreted save with an allowance of time much beyond the narrow limits which recent physical speculation would concede.[155] ... that there must be some flaw in the physical argument I can, for my own part, hardly doubt, though I do not pretend to be able to say where it is to be found. Some assumption, it seems to me, has been made, or some consideration has been left out of sight, which will eventually be seen to vitiate the conclusions, and which when duly taken into account will allow time enough for any reasonable interpretation of the geological record.[155]

As Geikie pointed out, the arguments of the geologists and paleontologists for a vaster timescale were based on observations of the present rate of geological and biological change, and the total absence of any evidence that these rates had changed during the history of the geological record. On the contrary, there was positive evidence that these rates had *not* changed over time. Edward B. Poulton, professor of zoology at Oxford, listed[156] some of this positive evidence in his address as president of the zoological section of the British Association in 1896. For example, many insects in the Carboniferous period had large wings, but insects in stormy areas today are wingless. Thus storms could not have been more violent then, as Kelvin's argument would require. Poulton asserted that if the rate of deposition of sediment were constant, then 400 million years must have passed since the Cambrian period, and this number must be further increased, 'perhaps doubled', to account for evolution prior to the Cambrian. Poulton contended that natural selection takes much longer to alter simple organisms than more complex ones, and that since except for the vertebrates the origin of no phylum can be found in the fossil record (since the beginning of the Cambrian), it follows that a very long period must have preceded the Cambrian—Darwin had taken a similar position in the first edition of the *Origin*. John G. Goodchild, curator of the Geological Survey Collections at the Edinburgh Museum of Science and Art, also made a calculation[157] of the age based on geological and biological evidence, and concluded that 700 million years had passed since the beginning of the Cambrian, with at least an equal period for the pre-Cambrian.

Even if Kelvin's arguments were wrong—and by the end of the nineteenth century most biologists and geologists who thought about the matter were convinced that he was, when Kelvin began to defend a very low age of the Earth—there remained the problem of where the error lay. Many writers took the point of view that there must be some source of energy which Kelvin had overlooked, as Darwin implied in the last edition of the *Origin*. (For references to these writers see ref. 144.) However, it was also possible that some of Kelvin's approximations were in error. This possibility was first discussed in detail in 1895 by John Perry, a former assistant of Kelvin. Perry pointed out[158,159] that Kelvin's age of the Earth was sensitive to his assumption that the thermal conductivity was a constant throughout the Earth, and that if, instead, it increased by a factor of ten from the Earth's surface to the centre, then Kelvin's time limit had to be increased by a factor of fifty-six. Perry also contended that some degree of fluidity must exist in the Earth's interior, and so heat conduction would be augmented by convection, which would have the effect of increasing the heat flow and hence the effective conductivity. Furthermore, he gave a mechanism for increasing the amount of energy available to the Sun.

In his reply[160] Kelvin expressed doubts that the internal conductivity of the Earth could be as high as Perry's argument would require, but he admitted that on the basis of the Earth's heat alone, an upper limit to the Earth's age could be set at 4000 million years. However, he still insisted that the Sun's heat limited the Earth's age to a few score million years.

As this limit was based on the available amount of gravitational energy (and Kelvin had long before pointed out that all other known forms of energy were even more inadequate) and the assumption that the solar output had to be essentially the same as it was today—and this was supported by the biological evidence itself—then this limit had to be accepted, if one granted that all sources of energy were known.

The most emphatic denial that all sources were known was made in 1899 by Thomas C. Chamberlain, professor of geology at the University of Chicago, who is best known today for his development of a solar system formation theory somewhat similar to Laplace's nebular theory. Chamberlain's argument amounted to an Anthropic Principle prediction:

Is present knowledge relative to the behavior of matter under such extraordinary conditions as obtained in the interior of the sun sufficiently exhaustive to warrant the assertion that no unrecognized sources of heat reside there? What the internal constitution of the atoms may be is yet open to question. It is not improbable that they are complex organizations and seats of enormous energies. Certainly no careful chemist would affirm either that the atoms are really elementary or that there may not be locked up in them energies of the first order of magnitude. No cautious chemist would probably venture to assert that the component atomecules, to use a convenient phrase, may not have energies of rotation, revolution, position, and be otherwise comparable in kind and proportion to those of the planetary system. Nor would they probably be prepared to affirm or deny that the extraordinary conditions which reside at the center of the sun may not set free a portion of this energy.[161]

As is well-known, the extreme conditions at the Sun's centre cause thermonuclear fusion of hydrogen into helium, and this in effect sets free a portion of the energy locked up in these atoms (in the form of mass). This process was first discussed some thirty years later, long after it was realized that radioactive decay in the Earth's interior also invalidated Kelvin's argument on the cooling of the Earth. In principle, Chamberlain's arguments could have led to experiments on the behaviour of matter at very high energies which could have led to the discovery of nuclear fusion reactions much earlier. Thus Anthropic constraints—evolutionary time scales—on the behaviour of matter in effect predicted nuclear sources of energy. We shall use analogous evolutionary timescale arguments to make some other predictions in later chapters, in particular Chapters 7 and 8.

3.7 Dysteleology: Entropy and the Heat Death

A man said to the Universe:
'Sir, I exist.'
'However', replied the Universe,
'The fact has not instilled in me a
sense of obligation'
Stephen Crane

Modern science presents a critical problem for teleological arguments. The very notion of teleology, that there is some goal to which the Universe is heading, strongly suggests a steady improvement as this goal is approached. Although progress was not strictly allowed by the Newtonian physics of the day,[162,163] the defenders of the teleological argument before the nineteenth century generally held this optimistic view. Meliorism even survived Darwin's destruction of traditional teleology. Darwin himself felt that his theory of evolution justified such an optimistic view. As he wrote in the closing pages of the first edition of *On the Origin of Species*:

> As all the living forms of life are the lineal descendants of those which lived long before the Silurian epoch, we may feel certain that the ordinary succession by generation has never once been broken, and that no cataclysm has desolated the whole world. Hence we may look with some confidence to a secure future of equally inappreciable length. And as natural selection works solely by and for the good of each being, all corporeal and mental endowments will tend to progress towards perfection.[164]

Darwin wrote these words in 1859, just slightly after the formulation of the Second Law of Thermodynamics, but before its dysteleological implications became generally known. The great German physicist Hermann von Helmholtz was the first to point out, in an article published in 1854,[165] that the Second Law suggested the Universe was using up all its available energy, and thus within a finite time all future changes must cease; the Universe and all living things therein must die when the Universe reaches this final state of maximum entropy. This is the famous 'Heat Death' of the Universe. It strongly denies the Universe is progressing toward some goal; but rather is using up the store of available energy which existed in the beginning. The Universe is actually moving from a higher state to a lower state. The Universe, in other words, is not teleological, but *dysteleological*!

As the historian of science Stephen Brush has pointed out,[140] this Heat Death concept had a profoundly negative effect on the optimism of the late nineteenth and early twentieth centuries. The popular books on cosmology written in the 1930's by the British astronomers Jeans[167] and Eddington[168] were particularly important in making the general public aware of the Heat Death. The new attitude this produced concerning the relationship between Man and the Cosmos was epitomized in 1903 in a famous passage of Bertrand Russell's:

... the world which science presents for our belief is even more purposeless, more void of meaning, [than a world in which God is malevolent]. Amid such a world, if anywhere, our ideas henceforward must find a home. That man is the product of causes which had no prevision of the end they were achieving; that his origin, his growth, his hopes and fears, his loves and his beliefs, are but the outcome of accidental collocations of atoms; that no fire, no heroism, no intensity of thought and feeling, can preserve an individual life beyond the grave; that all the labours of the ages, all the devotion, all the inspiration, all the noonday brightness of human genius, are destined to extinction in the vast death of the solar system, and the whole temple of Man's achievement must inevitably be buried beneath the debris of a universe in ruins—all these things, if not quite beyond dispute, are yet so nearly certain that no philosophy which rejects them can hope to stand. Only within the scaffolding of these truths, only on the firm foundation of unyielding despair, can the soul's habitation henceforth be safely built.[169]

The dysteleology of the long-term evolution of the Universe did not worry Russell. He suggested it meant we should take a short-term view of life:

I am told that that sort of view is depressing, and people will sometimes tell you that if they believed that, they would not be able to go on living. Do not believe it; it is all nonsense. Nobody really worries much about what is going to happen millions of years hence. ... Therefore, although it is of course a gloomy view to suppose that life will die out—at least I suppose we may say so, although sometimes when I contemplate the things that people do with their lives I think it is almost a consolation—it is not such as to render life miserable. It merely makes you turn your attention to other things.[170]

But some people were unable to take a short-term view. For example, by the end of his life, Charles Darwin's own optimism had been severely shaken by the prospect of the Heat Death, which he learned about in the course of the late nineteenth-century debates on the age of the Earth. As Darwin recorded in his *Autobiography*:

[consider] ... the view now held by most physicists, namely that the sun with all the planets will in time grow too cold for life, unless indeed some great body dashes into the sun and thus gives it fresh life—there is a clash between 'life' and 'Believing'. Believing as I do that man in the distant future will be a far more perfect creature than he now is, it is an intolerable thought that he and all other sentient beings are doomed to complete annihilation after such long-continued slow progress.[171]

Most philosophers, especially those who defended teleology in Nature, were, like Darwin, unable to take Russell's indifferent attitude. For instance the mathematician and controversial Anglican bishop E. W.

Barnes, whose work we discuss in section 3.9, was much troubled by the Heat Death.[172] The dilemma it creates for a value system based on science was clearly expressed by the paleontologist and mystical theologian Teilhard de Chardin:

> ... what disconcerts the modern world at its very roots is not being sure, and not seeing how it ever could be sure, that there is an outcome—*a suitable outcome*—to ... evolution. ... And without the assurance that this tomorrow exists, can we really go on living, we to whom has been given—perhaps for the first time in the whole story of the universe—the terrible gift of foresight?[173] Either nature is closed to our demands for futurity, in which case thought, the fruit of millions of years of effort, is stifled or else an opening exists—that of the super-soul above our souls...[174]

As we shall see in section 3.11, Teilhard accepted the notion of an evolving God. William R. Inge, the Dean of St. Paul's, (and known as the 'gloomy Dean'!) preferred the other horn of the dilemma: he rejected the possibility of an ethics based on the scientific world-view.

Inge wrote in the 1930's an entire book, *God and the Astronomers*, to discuss the Heat Death theory presented by Jeans and Eddington. He called the Heat Death 'the new Götterdämmerung' in reference to the Norse myth which held that the world would end with the destruction of everything, including the gods.[175] Inge was not bothered by the Heat Death; indeed, he welcomed it:

> The idea of the end of the world is intolerable only to modernist philosophy, which finds in the idea of unending temporal progress a pitiful substitute for the blessed hope of everlasting life, and in an evolving God a shadowy ghost of the unchanging Creator and Sustainer of the Universe. It is this philosophy which makes Time itself an absolute value, and progress a cosmic principle. Against this philosophy my book is a sustained polemic. Modernist philosophy is, as I maintain, wrecked on the Second Law of Thermodynamics; it is no wonder that it finds the situation intolerable, and wriggles piteously to escape from its toils.[176]

In other words, theologians should welcome the Heat Death, for such a future for the Universe precludes the possibility of the Universe being an emotionally acceptable home for Man. People will be forced to return to the traditional Christian static God, who is wholly outside the Universe, and hence not subject to the Heat Death.[177] The opposing views of Teilhard and Inge are but an echo of the debate between Schelling and Jacobi in the previous century (see section 3.5).

The views of Inge himself were echoed by the British mathematical physicist E. T. Whittaker, best known for his monumental history of electromagnetism, in his 1942 Riddell Lectures, which he entitled *The Beginning and the End of the World*:

> The knowledge that the world has been created in time, and will ultimately die, is of primary importance for metaphysics and theology: for it implies that God is not

Nature, and Nature is not God; and thus we reject every form of pantheism, the philosophy which identifies the Creator with creation, and pictures him as coming into being in the self-unfolding or evolution of the material universe. For if God were bound up with the world, it would be necessary for God to be born and to perish.... The certainty that the human race, and all life on this planet, must ultimately be extinguished is fatal to many widely held conceptions of the meaning and purpose of the universe, particularly whose central idea is progress, and which place their hope in an ascent of man.[343]

Whittaker nevertheless believed that there was a purpose in the Universe, and he felt the Heat Death itself indicated what that purpose was. Although Man and all his works would eventually vanish, the universe began with a sufficient amount of free-energy to permit his emergence, and thus

The goal of the entire process of evolution, the justification of creation, is the existence of human personality: of all that is in the universe, this alone is final and has abiding significance, and we believe that this has been granted, in the eternal purpose of God, in order that the individual man, born into the new creation of the Church, shall know, serve, and love Him forever.[343]

However, an evolving cosmos, particularly a cosmos evolving toward a bad end like the Heat Death, poses the following problem for teleology pointed out by Bertrand Russell:

... why should the best things in the history of the world [such as mankind] come late rather than early? Would not the reverse order have done just as well?[178] ... Before the Copernican revolution, it was natural to suppose that God's purposes were specially concerned with the Earth, but now this has become an unplausible hypothesis. If the purpose of the Cosmos is to evolve mind, we must regard it as rather incompetent in having produced so little in such a long time. It is, of course *possible* that there will be more mind later on somewhere else, but of this we have no jot of scientific evidence.[179]

This criticism will be recognized as the standard, centuries-old argument against an evolving, melioristic cosmology. We have previously seen it directed against Schelling's cosmos. It has been recently repeated by Roger Penrose[180] as a criticism of the Anthropic Principle. The only possible answer to the criticism, as pointed out by Schelling, is that the evolutionary process is logically necessary; the most advanced forms of life could *not* appear in the very beginning.

The Heat Death is most often discussed today in terms of the ecology of the planet Earth. The leaders of the 'ecology movement', for instance the Stanford biologist Paul Ehrlich (whose work we mentioned in section 3.2), have argued that Second Law limitations on terrestrial energy-flow require humanity to switch from a steadily growing economy to a steady-state one in which the energy use is constant and comparable in order of magnitude to the current total human energy use, about 3×10^{20} joules per year.[181] For comparison, the net amount of energy stored by all the photosynthetic plants on the Earth is about 3×10^{21} joules per year, an order of magnitude higher,[182] and a single human being requires about 4×10^{9} joules per year (2500 calories per day) in food energy.[181] Ehrlich points out, quite correctly, that the exponential growth in energy use and population size which has been typical of recent times cannot continue indefinitely; in fact, an exponential growth rate of one per cent per year in either population, or energy use, or anything else would exhaust all conceivable resources in the entire solar system in the order of a thousand years.

Unfortunately, Ehrlich's proposed steady-state economy will also eventually run out of resources: a civilization restricted to the Earth will in the end succumb to the Heat Death. It is a simple matter to derive some upper bounds to the length of time an Earth-restricted civilization, or indeed Earth-restricted and carbon-based life, could survive. The total energy available to such a civilization is equal to the energy-equivalent of the mass of the Earth, 5.4×10^{41} joules. A single human being, with the above-mentioned food energy requirement could survive at most 2×10^{32} years;[183] a billion people at most 2×10^{23} years. The human species could continue to use energy at the rate which is is currently doing for *all* purposes for at most 2×10^{21} years. A *single* cell, with an energy requirement roughly 10^{-10} that of a single human being, could survive at most 2×10^{42} years. The entire biosphere, with energy-use approximately that of the total net energy stored by photosynthesis, could survive at most 2×10^{20} years. If we imagine future human civilization limited to the entire solar system, then the above upper bounds to survival times are increased by a factor of about 3×10^{5}, which is the ratio of the mass of the Sun to the mass of the Earth. These upper-bounds to our survival time are summarized in Table 3.1.

These survival upper-bounds are of course extremely large in comparison to the timescales with which human beings normally concern themselves: even economists who try to project economic trends into the 'extreme far future' generally limit themselves to the next 100–500 years.[184] They are nevertheless the order of, or in many cases substantially less than, many timescales of physical processes which physicists are now measuring: for example, the expected proton lifetime on the basis of the SU(5) grand unified gauge theory is about 10^{31} years (see Chapters 5 and 6 for a detailed discussion).

But the essential point is the fact that the survival times are *finite*. No matter what we, or any other form of life based on DNA do, we (or rather our descendants) are doomed if we restrict our operations to a single planet, or even a single solar system. We shall discuss the question of unlimited survival of life in more detail in Chapter 10; it is sufficient

TABLE 3.1
Table of upper bounds to survival times for carbon-based life forms in our solar system

Type of life and its energy usage	Survival time: upper bound using mass-energy of the Earth	Survival time: upper bound using mass-energy of the solar system
1 living cell, using just food energy	2×10^{42} years	5×10^{47} years
1 person, using just food energy	2×10^{32} years	5×10^{37} years
10^9 people, using just food energy	2×10^{23} years	5×10^{28} years
Human civilization, using energy at the rate the whole of mankind used energy in 1973	2×10^{21} years	6×10^{26} years
Entire biosphere, using energy at rate provided by net photosynthesis on Earth today	2×10^{20} years	6×10^{25} years

Other significant timescales

Estimated proton lifetime, predicted by minimal SU(5) gauge theories	10^{31} years
Length of time Earth-based civilization can use energy at current rate and at current price, using uranium in Earth's crust as energy source[192]	1×10^{10} years
Period the Sun will remain on main sequence	5×10^{9} years
Upper bound to future life of biosphere (see Chapter 8)	5×10^{8} years
Average survival time of mammalian species	1×10^{6} years
Length of time modern man (*Homo sapiens sapiens*) has existed.	4×10^{4} years

for now to note that such survival requires expansion beyond our solar system, and that carbon-based life is doomed in any case.

As Ehrlich himself admits,[185] 'almost all economists' disagree with him and most other ecologists on the necessity for a steady-state economy. The economists' argument has perhaps been best presented by Julian Simon in his book *The Ultimate Resource*.[186] The basic difference between the ecologists and the economists is the fact that the former view the ecological and economic system in terms of a flow of energy and material resources, while the latter view it in terms of a flow of information. According to the economists, the economic system is concerned with producing not specific goods, but services: as consumers we are interested in the *services* we can get from energy and material resources rather than in the resources themselves. To use an example of Simon's, the copper in a cooking-pot can be replaced by other materials as technology develops substitutes, for we desire a cooking service rather than a pot made of a certain metal. Thus the important cost is the cost of providing the cooking service rather than the cost of copper. As human knowledge grows, the number of materials we can use to perform a given service and our ability to obtain any given material grows also, with the result that 'The cost trends of almost every natural resource—whether measured in labour time required to produce the energy, in production costs, in the proportion of our incomes spent for energy, or even in the price relative to other consumer goods—have been downward over the course of recorded history'.[187] In fact, as Simon documents,[186] the price of raw materials and energy have, on the long term average, been decreasing exponentially over the past two centuries (the period for which we have good data) with a decay-constant of about 50 years. This means that a project whose cost is dominated by raw material costs will be much cheaper to carry out in the future than it is now, if past experience is any guide. The implications of these price trends will be important when we consider the likelihood of interstellar travel in Chapter 9. The modern economists' view of the economic system as being concerned with the production and transfer of services (utilities), goes back at least to the foremost English economist of the nineteenth century, John Stuart Mill.[188]

Both Simon's analysis (and similar analyses by almost all economists who have considered trends in raw material costs) indicate that the costs of all services are controlled almost in their entirety by information located in human minds and elsewhere in human civilization. The *prices* of the services *are* controlled in their entirety by their subjective valuation in human minds, as shown by marginal utility theory.[189] Indeed, it was thought for thousands of years that the price of a product was an objective feature like its weight, but modern economic theory demonstrates it is a purely subjective quantity generated by the collective interaction of the human race via the product's marginal utility. The price of an object is an example of an apparently objective feature of the world which actually exists only in human minds. We can regard the price structure of an economic system as a Participatory Anthropic Principle in operation.

In general, we may say that *all* services—the entire output of the economic system—may be each equated with a form of 'information' in the sense this word is used in information theory. We can make this clearer by returning to Simon's cooking-pot example. Ultimately, we do not buy the pot to obtain even a cooking service, but rather to obtain a release from hunger and to obtain the sensation of having eaten a delicious meal. It is possible in principle to obtain the same service by direct transfer of material directly to the body cells while causing nerve pulses to be sent to the brain which fools the mind into believing it has enjoyed a real meal. One could go even further, along the lines discussed in the section on absolute idealism, and imagine the program which corresponds to a human mind being run on a Universal Turing machine[118,119] with input to the mind-program being chosen so that the input gives rise in the program to the complete sensation of eating a delicious meal. Both of these possibilities are of course far, far beyond current technology. But the Turing machine example demonstrates that, ultimately, services are a form of information input for a very complex program called a human mind.

Thus the ultimate limits of economic systems and civilizations are exactly the same as the ultimate limits of minds: they are all ultimately limited by the amount of information that can be read, processed, and stored. As yet we are ignorant of how many bits of information a given economic service and a given amount of human knowledge correspond to, but for our limited purposes it is sufficient to know just that both are forms of information. We shall make use of this fact in Chapter 10 to calculate some very interesting constraints on the behaviour of life in the far future. Civilization can continue to grow in the far future only if it eventually leaves the solar system, as the economists also grant.[186,187] The ecologists seem unwilling to admit this;[190] see however ref. 191.

The bare fact that the economic system is wholly concerned with generating and transferring information has an interesting ethical implication. If we assume (as intellectuals generally do) that the government should not interfere with the generation and transfer of information, then does it not follow that the government should not interfere with the operation of the economic system? Furthermore, if it is argued (as scientists often do) that the growth of knowledge is maximized by information generation and flow being unimpeded by government intervention, does it not follow that the growth of economic services would be maximized if unimpeded by government intervention? Conversely, if social utility may sometimes require governmental restrictions on the evolution of the economic system, may it not likewise require governmental restrictions on academic freedom and the growth of scientific knowledge? Both the unlimited growth of scientific knowledge and unlimited economic growth may be regarded as undesirable but an argument for restricting one is automatically an argument for restricting the other and an argument for not restricting one is automatically an argument for not restricting the other.

3.8 The Anthropic Principle and The Direction of Time

> Time is defined so that motion looks simple.
>
> J. A. Wheeler

The Weak Anthropic Principle was used by the Austrian physicist Ludwig Boltzmann to explain the direction of time. By the middle of the nineteenth century, it had been realized that there was only one physical law which defined a time direction, and that was the Second Law of thermodynamics. In the latter part of the nineteenth century, Boltzmann began a research programme to deduce the Second Law of thermodynamics from classical mechanics.[193] By applying the statistical techniques of Maxwell to atomic collisions, Boltzmann 'deduced' his so-called H-Theorem. The H-Theorem asserted that a quantity denoted by H, which was a function of the positions and velocities of the atoms of the system, must always decrease with time or remain constant. Identifying the function H with the negative of the entropy, Boltzmann claimed to have a proof of the Second Law.[194,195]

However, many physicists[193] were a bit dubious about a 'proof' which deduced irreversibility—the fact that H never increased—from reversible classical mechanics. In particular, Loschmidt, Boltzmann's colleague at the University of Vienna pointed out that for every evolution of a system of atoms in which H decreased, one could obtain an evolution in which it increased by reversing the velocities of all the particles. Therefore, it would seem impossible to prove that H *never* increased whatever the initial conditions. Boltzmann admitted[196] that one could not show that H never increased whatever the initial conditions, but he contended that almost all initial states which were far from a Maxwellian equilibrium state (the Heat Death state of maximum entropy) would approach this equilibrium state in which H would be at a minimum, and that this is sufficient to account for the Second Law of thermodynamics. This law, he asserted, has only statistical validity; furthermore, its validity is due to the fact that atoms are so small relative to human beings. A being the size of the molecules would not see a continuous increase of entropy. Maxwell later gave a striking example of this.[197] An intelligent being—a demon—the size of a molecule could violate the Second Law by using the fact that, even in equilibrium, a gas of atoms would contain atoms with a range of velocities. This demon could station itself beside a door between two containers initially at the same temperature. The demon would allow only fast-moving atoms to pass in the other direction. After a while one container would contain atoms with a higher average velocity than the other, and so it would attain a higher temperature, since temperature is a measure of average atomic kinetic energy. This demon would thus create a temperature difference without doing work, which would violate the Second Law. Hence, as Lord Kelvin[166] first pointed out in 1874, this example suggested that the Second Law was not an absolute law of nature, but a human artefact resulting from the relative size of Man to atom and of the Law of Large Numbers.[198]

Planck's student Zermelo pointed out[199,200] that Poincaré had proven a theorem showing that almost any mechanical system with finite potential energy, finite kinetic energy, and bounded in space must necessarily return to any previous initial state. Thus, whatever the state of the Universe now, the entropy as defined by Boltzmann would almost certainly have to decrease in the future back to its present value. Thus the observed entropy increase which the Universe is presently undergoing could occur only if it is assumed that for some mysterious reason the Universe just happens to be in one of the extremely rare low entropy initial states. Zermelo went on to say: 'But as long as one cannot make comprehensible the *physical origin* [Zermelo's emphasis] of the initial state, one must merely assume what one wants to prove; instead of an explanation one has a renunciation of any explanation'.[201]

In his reply[202], Boltzmann acknowledged that one could prove an H-Theorem only for those initial states which are far from equilibrium. However, one is not necessarily forced thereby to assume a special Universal initial state:

> One has the choice of two kinds of pictures. One can assume that the entire universe finds itself at present in a very improbable state. However, one may suppose that the eons during which this improbable state lasts, and the distance from here to Sirius, are minute compared to the age and size of the universe. There must then be in the universe, which is in thermal equilibrium as a whole and therefore dead, here and there relatively small regions of the size of our galaxy (which we call worlds), which during the relatively short time of eons deviate significantly from thermal equilibrium. Among these worlds the state probability [the H-function] increases as often as it decreases. For the universe as a whole the two directions of time are indistinguishable, just as in space there is no up or down. However, just as at a certain place on earth's surface we can call 'down' the direction toward the centre of the earth, so a living being that finds itself in such a world at a certain period of time can define the time direction as going from less probable to more probable states (the former will be the 'past' and the latter the 'future') and by virtue of this definition he will find that this small region, isolated from the rest of the universe, is 'initially' always in an improbable state. This viewpoint seems to me to be the only way in which one can understand the validity of the Second Law and the heat death of each individual world without invoking a unidirectional change of the entire universe from a definite initial state to a final state. The objection that it is uneconomical and hence senseless to imagine such a large part of the universe as being dead in order to explain why a small part is living—this objection I consider invalid. I remember only too well a person who absolutely refused to believe that the sun could be 20 million miles from Earth, on the grounds that it is inconceivable that there could be so much space filled only with aether and so little with life.[202,203]

Boltzmann wrote the above words in 1897, and as is well-known, within 20 years the statistical interpretation of the Second Law became universally accepted. Thus physicists were implicitly forced to choose between Boltzmann's 'two pictures' for the origin of the observed present-day improbable universal state. (We say 'implicitly' because most physicists simply ignored the problem.) One could either adopt the 'creation' interpretation of the Second Law, which held that the Universe at some initial time was simply 'given' in an improbable initial state, or one could adopt the 'anthropic-fluctuation' interpretation, which claimed the Second Law is observed to hold because intelligent life can exist only in regions where the initial conditions allow the Second Law to hold.

Both pictures have had their advocates. Boltzmann himself adroitly avoided committing himself definitely to either picture; he even gave credit to his old assistant Dr. Schuetz for the anthropic-fluctuation interpretation.[204] The French physicist Poincaré was mildly attracted to the anthropic-fluctuation interpretation because of the promise it held for avoiding the Heat Death of the Universe, and he pointed out[205,206] that intelligent life would probably be impossible in a world in which the entropy decreased with time. In such a world prediction would be impossible. For instance, friction would be a destabilizing force rather than a damping force. Two bodies initially at the same temperature would later acquire different temperatures, and it would be essentially impossible to predict in advance which one would become the warmer. Thus intelligent action would be impossible. The American mathematician Norbert Wiener also emphasized that communication between worlds with different directions of entropy increase would be impossible.[207]

The creation interpretation became dominant after the discovery that the universe is expanding, since the expansion defined a natural time—the beginning of the expansion—at which to impose initial conditions.[208] Zermelo's problem of the origin of these initial conditions would then be solved (or avoided) by noting that the laws of motion and the universal initial conditions came into being at the same instant, and so the origin of the latter would be no more mysterious than the former. The expansion of the Universe would cause matter to become spread out over an ever-increasing volume, and thus Poincaré recurrence would not be inevitable.[209] Even closed universes, which do not expand forever, will avoid Poincaré recurrence because the momentum space is unbounded in this type of universe.[209] Thus Poincaré recurrence does not hold in any cosmology governed by general relativity. The statistical interpretation of the Second Law could be combined with the idea of an irreversible Heat Death.

The wide diffusion of these ideas stimulated a few non-physicists to revive and defend the anthropic-fluctuation interpretation. The British biologist J. B. S. Haldane calculated,[210] from Jeans' estimate of the size of the Universe, that the time needed for a run-down universe (one at maximum entropy) to return to an atomic distribution as improbable as the one observed at present is $10^{10^{100}}$ years. 'During all but a fraction of eternity of this order of magnitude, nothing definite happens. But on a

Materialistic view there is no one to be bored by it'. Haldane went on to say:

If this view is correct, we are here as the result of an inconceivably improbable event, and we have no right to postulate it if any less improbable hypothesis will explain our presence. If there are other stars on which intelligent beings are wondering about their origin and destiny, a far smaller and therefore vastly more probable fluctuation would be enough to account for the existence of the human race.[211]

Haldane argued on the basis of solar system formation theories current in the 1920's that planets with life are very rare, and hence '... it becomes fairly likely that our planet is the only abode of intelligent life in space'. He concluded that

... if this is correct, the [anthropic-fluctuation interpretation] becomes plausible. We have not assumed a more improbable fluctuation than is necessary to account for our being there to marvel at its improbability. If the future progress of astronomy substantiates the uniqueness of our earth, the [anthropic-fluctuation interpretation] of course will gain likelihood.[211]

Haldane's argument will be recognized as a Weak Anthropic Principle argument. It is a variant of Wheeler's argument that the Universe must be at least as big as it is in order to contain intelligent life, and it is an argument we shall be using on many occasions in this book. Actually, the fluctuations could not occur because of gravitational instabilities, as we shall discuss in Chapter 10. Thus if Boltzmann and Haldane had used the correct physics which was known in their day, they would not have reached an incorrect conclusion using the Weak Anthropic Principle. Furthermore, the physicist Richard Feynman, in a 1965 lecture, levelled an objection to the fluctuation theory which is sufficiently general to apply against *any* Anthropic size argument. He called the anthropic-fluctuation theory 'ridiculous'[212] on the grounds that a fluctuation much smaller than the entire visible universe would account for the existence of an inhabited planet, and thus it is most unlikely that the entire visible universe would be in an improbable state, as it is observed to be. Only if intelligent life ultimately requires a space much larger than a single planet can the Anthropic size argument be defended against Feynman's objection. We shall show why a much larger space is needed in Chapter 6.

In the past 40 years the anthropic-fluctuation interpretation has been defended mainly by philosophers,[213,214] while physicists and astronomers have generally developed versions of the creation interpretation shaped to fit the observed fact of universal expansion. The only major exceptions to this rule were those astrophysicists who supported the steady-state theory. Since this theory explicitly and intentionally violated both the First and Second laws of thermodynamics, these men were not forced to choose between the two pictures. (They simply assumed that matter was created in a low entropy state.) The discovery of the microwave background radiation in 1965 ruled out the steady state theory, and in recent years debate had centred on what sort of initial conditions were imposed in the beginning on the initial singularity. There are two schools of thought. The 'orderly singularity' school,[215] represented by the British mathematician Roger Penrose,[216] contend that the initial singularity had a very regular structure, with just enough irregularity to give rise to the stars and galaxies. The other opinion has been dubbed 'chaotic cosmology' by its chief proponent, the American cosmologist Charles Misner.[217] In this view the Universe would have its approximately regular aspect now no matter what the initial condition of the singularity because dissipative processes in the early universe would have smoothed out major irregularities by the present epoch (when intelligent life has arisen). Since irregular initial states are much more numerous than regular initial states one would expect the initial singularity to be very chaotic in structure. The attractiveness of the chaotic cosmology idea lies in the fact that it obviates the necessity of explaining the initial conditions, while the orderly singularity school is faced with explaining them and with Zermelo's problem. We shall give an anthropic explanation for initial conditions which give a globally defined direction of time, in Chapters 7 and 10, thus combining the two possible pictures of Boltzmann into one. The chaotic initial condition model will be discussed in more detail in Chapter 6. The universal direction of time, which is determined by the conditions imposed on the initial singularity, is ultimately explained anthropically.

It is possible in principle to test whether or not the Universe has an overall time direction, in which entropy always increases no matter how far into the future we go. We shall assume in our arguments elsewhere in this book that entropy always increases; but, suppose on the contrary that entropy were to rise to a maximum at the point of maximum expansion of a closed universe (see Chapter 6 for a discussion of the behaviour of the various cosmological models) and thereafter begin to decrease, with a return at the final singularity to the conditions which prevailed at the initial singularity. In order for such a return to occur, the disintegration of radioactive materials (for example) must be counterbalanced by a spontaneous regeneration even today, and this could be searched for. John A. Wheeler[218] and W. J. Cocke[219] have considered the experimental implications of this regeneration in some detail. Since such a reversal of entropy would make the continued increase of knowledge by intelligent life impossible, it would contradict FAP, we predict that any experiment which looked for a spontaneous regeneration will have negative results.

There remains the question as to whether Maxwell's Demon could create, in the small, a direction of time in reverse to the large-scale universal direction of time by violating the Second Law. He *cannot*, for it was shown in the 1930's and 1940's that Maxwell's Demon cannot operate.

The concept of Maxwell's Demon assumes that it is possible for an intelligent being to operate—to gather information and act on this information—on a scale much smaller than the atoms from which the everyday world is constructed. The Demon might be subjected to a set of thermodynamic laws appropriate to his own scale, but would be oblivious to those of our scale. This was not nonsense when the idea of Maxwell's Demon was first developed in the course of an exchange of letters between Maxwell and Tait[220,221,222] in the 1860's. At that time there were actually some observations, namely the absorption of starlight in interstellar space, which were interpreted by Kelvin's friend, the physicist Tait[223] as evidence of a leakage of energy from our everyday world into another 'world' with its own laws of thermodynamics. Tait, in fact, later tried[223,224,225] to use this concept of a hierarchy of 'worlds' to prove the existence of angels, not demons! Tait's cosmology was developed in order to allow intelligent life to escape the Heat Death by moving from one 'world' to another one of an infinite set. The egregious failure of Tait's idea is a warning example to those who would construct a cosmology wherein life can escape the Heat Death, as many have tried to do after him, from the semi-mystical approach of Teilhard de Chardin (section 3.11) and to the more scientific approach of ourselves (Chapter 10). The failure of Tait's theory is a failure of the Anthropic Principle applied in the large, where we have argued at length the AP should be valid. It counts as evidence *against* the AP as a methodological principle. Nevertheless, it can be said in defence of the AP that Tait's theory was based on false observations. *No* scientific principle can yield correct theories if false information is used.

Once it is accepted that such a hierarchical structure does not exist, and that any intelligent being would have to use the materials and physical laws of a single unique scale in his operations, it can be shown that Maxwell's Demon cannot exist. Szilard,[226] and later Brillouin,[227,228] pointed out that in order to separate the fast-moving molecules from the slow-moving ones, the Demon would first have to measure the speeds of the molecules moving toward his door. This measurement necessarily increases the entropy of the system more than the separation of the molecules would decrease it, and so the Second Law would not be violated[229]. The arguments for the non-existence of Maxwell's Demon have suggested to Bohr's student Leon Rosenberg[230] that the observer whose existence gives rise to the complimentarity principle in quantum mechanics, must have a size and complexity comparable to human beings. In the course of his proof that Maxwell's Demon cannot operate, Brillouin obtained a formula for the minimum amount of energy that must be expended to obtain a bit of information. This formula will be crucial in obtaining ultimate limits to the activities and indeed the existence of intelligent life, which we will do in Chapter 10.

Since the statistical interpretation of the Second Law is often cited as an example of the reduction of one theory to a more fundamental theory, we might mention that in reality no one has ever been able to deduce the Second Law rigorously from either classical or quantum mechanics without using anthropic arguments or unphysical assumptions[193,231]. The key problem is that in both quantum and classical mechanics, the phase space occupied by the system, measured by Boltzmann's factor H, is an *exact* constant of motion: it does not change with time. Only by assuming that the *observer* only roughly measures the factor H can it be shown that H increases with time. The *exact* H cannot change with time, and this is true whatever the initial conditions.

3.9 Teleology and the Modern 'Empirical' Theologians

> But then arises the doubt, can the mind of man, which has, as I fully believe, been developed from a mind as low as that possessed by the lowest animal, be trusted when it draws such grand conclusions?
>
> Charles Darwin

The 'Empirical' theologians, those theologians who address the question of the purpose of the physical Universe—if any—and the place of Man in it, are a vanishing breed in the twentieth century. Having been burned by the Darwinian refutation of the Paleyian design teleology, most modern theologians try to avoid altogether discussion of this question, and the few that do consider the question, generally answer it by making sweeping assertions with very few actual examples from the physical world either to back-up or illustrate those assertions. For instance, Andrew Pringle-Pattison, a Scottish theologian who is considered[232] to have been a major figure in natural theology at the turn of the century, claimed: '... my contention is ... that man is organic to the world, or ... the world is not complete without him. The intelligent being is, as it were, the organ which the Universe beholds and enjoys itself'.[233] He argued that philosophy would be defective if it did not indicate a purpose in the Universe, and that 'philosophy must be unflinchingly *humanistic*, anthropocentric'[234]. The purpose which Pringle-Pattison found in Nature is akin to Henderson's, although independently conceived. Pringle-Pattison admitted that the crude teleology of Paley was finished, and he held that this was a good thing: *cosmic* teleology is the only teleology which can now be defended:

A teleological view of the Universe means the belief that reality is a significant whole. When teleology in this sense is opposed to a purely mechanical theory, it means intelligible whole as against the idea of reality as a mere aggregate or collocation of independent facts.[235]

This notion of 'cosmic teleology' was developed in far more detail by a British theologian, F. R. Tennant, in an influential book *Philosophical Theology*, first published in 1930, and still in print today. His basic argument for teleology is now familiar:

The forcibleness of Nature's suggestion that she is the outcome of intelligent design lies not in particular cases of adaptedness in the world, nor even in the multiplicity of them ... [it] consists rather in the conspiration of innumerable causes to produce, by either united and reciprocal action, and to maintain, a general order of Nature. Narrower kinds of teleological arguments, based on surveys of restricted spheres of fact, are much more precarious than that for which the name of 'the wider teleology' may be appropriated in that the comprehensive design-argument is the outcome of synopsis or conspection of the knowable world.[236]

According to Tennant, there were three types of natural evidence in favour of teleology acting on a cosmic scale: (*1*), the fact that the world can be analysed in a rational manner; (*2*), 'the fitness of the inorganic to minister to life';[237] and (*3*), 'progressiveness in the evolutionary process culminating in the emergence of man with his rational and moral status'.[238] Both type (*1*) and type (*2*) are essentially Anthropic Principle arguments. In defence of the first type, Tennant points out[239] that it is logically possible to imagine a world which is nothing but a chaos in which similar events never occurred, in which there were no laws. Since the events of the world can be ordered into what Tennant calls[240] 'anthropic categories'—this appears to be the first use of the word 'anthropic' in this context—it follows that the world is selected out of all possible universes to allow the existence of a reasoning creature; 'anthropocentrism, in some sense, is involved in cosmic teleology'.[241] In short, there is a relation between '... the intelligibility of the world to the specifically anthropic intelligence possessed by us, and ... the connection between the conditioning of that intelligibility, on the one hand, and the constitution and process of Nature, on the other hand'.[241] Note that it is the entire orderliness of Nature that shows teleology in Tennant's view. The Universe is a Cosmos in the Greek sense of the word.[242] Tennant emphasizes that 'anthropic' in his use of the term did not necessarily mean that Man as a species was the ultimate purpose of creation. He meant that the Universe was anthropocentric in the sense of being consistent with rational being: 'it is of course, a matter of indifference to teleology and anthropocentrism whether the material heavens contain a plurality of worlds'.[243]

In defence of the evidence of type (*2*), Tennant cited Henderson's work with approval, and essentially repeated Henderson's arguments. The result of this approach is a teleological picture that '... no longer plants its God in the gaps between the explanatory achievements of natural science, which are apt to get scientifically closed up'.[242] The disadvantage of this is a more abstract notion of teleology which is apt to lose all connection with Nature.

Type (*3*) evidence sounds a bit like the directed evolution of the philosophers discussed earlier, but is really a concept intermediate between this and the no-global-teleology of modern biologists. Tennant asserted that 'the forthcoming alternative views, between which facts scarcely enable us to decide, may be briefly mentioned':

> The divine purposing may be conceived as pre-ordination, in which every detail is foreseen. An analogy is presented in Mozart's (alleged) method of composition, who is said to have imagined a movement—its themes, development, embroidery, counterpoint, and orchestration—in all its detail and as a simultaneous whole, before he wrote it. If God's composition of the cosmos be regarded as similar to this, all its purposiveness will be expressed in the initial collocation, and evolution will be preformation. On the other hand, God's activity might be conceived as fluent, or even as 'increasing', rather than as wholly static, purpose. It might then be compared, in relevant respects, with the work of a dramatist or a novelist such, perhaps, as Thackeray, who seems to have moulded his characters and plot, to some extent, as he wrote.[244]

Although Tennant granted that the question of 'what the ultimate purpose or goal of the world-process is ... may admit of no complete answer by man,[238] nevertheless, in Tennant's view we can say that '... man is the culmination, up to the present stage of the knowable history of Nature, of a gradual ascent'.[245]

The type (*2*) evidence for teleology as given above by Tennant has, in the intervening half-century, been echoed by a number of distinguished theologians: Laird,[246] Gibson,[247] Bertocci,[248] and Raven.[249] These authors always cite Henderson's work as evidence for such cosmic teleology, but it is clear that they learned of Henderson from Tennant, and they add nothing very original to the argument.

The type (*1*) argument did not originate with Tennant, however. For example, the psychologist James Ward, asserted in his Gifford Lectures delivered in 1896–1898 and published in two volumes with the title *Naturalism and Agnosticism*, that:

> ... we are now ... entitled to say that this unity and regularity of Nature proves that Nature itself is teleological, and that in two respects: (1) it is conformable to

human intelligence and (2), in consequence, it is amenable to human ends. Such is the new step in our [teleological] argument, and it contains all that is essential to complete it.[250]

W. R. Matthews, the Dean of St. Paul's, put this somewhat differently in 1935:

The facts from which the [general teleological] arguments start are general characters of the universe as experienced by us. There is first the impression of an order which is both rational and sublime; there is secondly, the fact that the universe, when interrogated by reason, seems to be a coherent system . . . .[251]

Matthews gave an interesting defence of this perceived order to Hume's criticism, which we discussed in Chapter 2, that it could all be due to some anthropic selection out of chaos—it is well-known that a finite number of elements would in infinite time go through all possible combinations, some of which would have order.[252] Matthews responds to Hume in two ways: first, the assumption that the number of things in the Universe is finite is in itself an assumption of order. If the number of things is indefinite, then there need not be a repetition of all events. Second, Matthews points out that according to modern science, the Universe has only existed for a finite time, and in this finite time only a finite number of events could have occurred, all of which seem to be orderly. (Henderson makes a similar appeal to observation in the visible Universe; as far as we can tell, the entire universe is orderly, which is contrary to what we would expect if we merely lived in an island of order.)

This direct appeal to experimental fact in support of cosmic teleology is unfortunately rare among modern natural theologians. Both Peacocke,[253,254] who is not only a theologian but also a physical biochemist, and Mascall[255]—to list only two of the more well-known of the recent writers on the relation between religious topics and science—defend cosmic teleology by arguing that the continuing operation of physical laws needs some teleological justification. If, as we have reason to believe, chaos is much more probable than any form of order, why does the Universe not lapse into chaos the next instant? Why are our expectations of seeing the familiar types of order tomorrow always fulfilled. This sort of argument is so general that it would be consistent with *any* scientific result, and so, although interesting, it is completely useless. Indeed, Mascall goes out of his way to argue that both a steady-state Universe and the Big-Bang cosmology would be consistent with it.[255] Both Peacocke[254] and Mascall[255] mention the Anthropic Principle—Peacocke in its modern Dicke-Carter form, and Mascall in a primitive version due to Whitrow (which we shall discuss in section 4.8)—but only in passing.

However, there was actually one heroic attempt in the 1930's to combine the anthropic viewpoint that intelligence is important or even essential to the Universe with the science of the day, and actually make a testable prediction. Ernest W. Barnes, the Bishop of Birmingham, was both a theologian and a mathematician with a Sc.D. from Cambridge. His Gifford lectures, delivered in the years 1927–1929, are probably unique amongst modern lectures on theology: about half of the book form of the lectures—which take up about 650 pages—consists of tensor equations together with an exposition of the quantum theory. The book could be used as a textbook in mathematical cosmology *circa* 1930. At the time Barnes was giving the lectures, the generally accepted theory of planetary formation was binary collisions between stars; of this idea he writes,

> But, if planetary systems originate in actual collisions, there may be merely a few hundred of such systems in our Universe . . . this number seems utterly disproportionate to the size of the galactic universe, *if we regard that universe as having been created with a view to the evolution of intelligent beings.* [Barnes' emphasis] . . . and the suggestion forces itself upon us insistently that the cosmos was made for some end other than the evolution of life. Certainly, however, no such end is apparent to us.
>
> My own feeling that the cosmos was created as a basis for the higher forms of consciousness leads me to speculate that our theory of the formation of the solar system is incorrect.[256]

It is well-known that Barnes was correct; the theory of solar system formation held in the 1930's was incorrect. This is a correct prediction obtained by Anthropic Principle reasoning. (We shall argue in Chapters 8 and 9 that intelligent life is most unlikely to evolve on any other earthlike planet. But we shall also claim in Chapter 10 that the other solar systems could serve a purpose for intelligent life. Furthermore, if the evolution of intelligence is improbable, many solar systems must exist if there is to be a reasonable chance of intelligent life arising at least once.)

Although he believed that the purpose of the Universe was to be a home for intelligent beings, Barnes did not regard mankind as the apex of intelligent life:

> But as the millions of years go by, so to, if we may judge the future by the past, will humanity as we know it ultimately yield place to some other animal form? What form? Whence evolved? We cannot say. But some Cosmic Intellect, watching the mature capacities of this unknown form, will almost certainly judge it to be more highly evolved, of greater value in the scheme of things, than ourselves. On Earth man has no permanent home; and if, as I believe, absolute values are never destroyed, those which humanity carries must be preserved elsewhere than on this globe.[257]

3.10 Teleological Evolution: Bergson, Alexander, Whitehead and the Philosophers of Progress

> Why is it that you physicists always require so much expensive equipment? Now the Department of Mathematics requires nothing but money for paper, pencils and waste paper baskets and the Department of Philosophy is better still. It doesn't even ask for waste paper baskets.
>
> Anonymous University President

By the start of the nineteenth century, evolutionary concepts had begun to seep into philosophical systems, and in some cases, like that of Schelling, they formed the basis of the system. The idea of an evolutionary cosmos came initially not from the observation of Nature, but rather from a new view of human history. The scholastics of the Middle Ages considered themselves inferior to, or at best equal to, the ancient Greek philosophers. In their opinion, there had been no significant change in basic knowledge or any other fundamental aspect of human society over the whole of human history, which in duration had been about the same as the length of time the Universe had been in existence. Thus there was no reason to believe in an evolutionary Universe.

In contrast, the philosophers of the Enlightenment believed themselves vastly more knowledgeable than the Greeks, as shown by the very name of this period. Their scientific knowledge, particularly Newtonian physics and astronomy, was clearly superior to anything the ancients had developed. This indicated an evolutionary change in human knowledge, and it was a change for the better. Progress had obviously occurred in human history, and it required but a short leap of the imagination to go from a progressive humanity to a progressive Universe. By the late nineteenth and early twentieth centuries, it had become generally accepted that any realistic picture of the Universe had to be evolutionary. The evolutionary world view is not totally dominant even in this century, and the idea of a static cosmos has held a strong attraction: Einstein's first cosmological model, proposed in 1917, was globally static.[258] More recently, the steady-state model both in its original 1950's form[258] and in its contemporary inflationary universe guise[259]—which we discuss at length in Chapter 6—are both attempts to retain a cosmos which on a sufficiently large scale does not change[252]. In general, however, philosophical systems and cosmological models of the present day are fundamentally evolutionary.

But an evolving cosmos can be either teleological or non-teleological. There are three ways in which a philosophical system could be teleological. First, some event which the philosopher regards as supremely good—such as the eventual evolution of the human species and its progress to higher and higher levels of civilization—could be considered an inevitable eventual outcome of the evolutionary process. Second, it could be held that the entire Universe is evolving toward some goal. Third, the Universe could be pictured as an organism which by its very nature is teleological. In this section we shall discuss three of the four most influential such teleological systems to be formulated in the twentieth century: the systems of Bergson, Alexander, and Whitehead. The fourth system, the one developed by Teilhard de Chardin, is sufficiently unusual to warrant a separate treatment in section 3.11.

These philosophers did not work in a vacuum; they had an enormous number of nineteenth-century predecessors who we might term the 'philosophers of progress'. As the distinguished historians Bury[260] and Nisbet[261] have demonstrated, the belief that human history is progressive reached its height in the nineteenth century, although such a view was not unknown even in classical antiquity.[262] The two best known of the philosophers of progress were Karl Marx and Herbert Spencer. The Marxist theory of human development, in which the human social system evolves from the capitalist society of the nineteenth century into first a socialist and finally a communist society, egalitarian and anarchist, is sufficiently familiar to twentieth-century readers to make a detailed discussion unnecessary, but Spencer and his philosophy are almost unknown.

The reverse would have been true in the late nineteenth century; Spencer was the most celebrated philosopher of his day: a rationalist, an anti-imperialist, and the last of the great *laissez-faire* liberals.[263] Spencer's theory of an evolving cosmos is probably derived ultimately from his political philosophy, although he claimed the deduction proceeded in the opposite direction. His first publication, *The Proper Sphere of Government*,[264] which he published in 1843 at the age of twenty-three, was a defence of the individual against the power of the state: in this short book he opposed not only state interference with religion and the press, but also government schools and government support of the poor. He reluctantly granted the state the right to make war, but he wished to impose more restrictions on this power than any government has ever accepted. For Spencer, progress occurred through the voluntary cooperation of individuals. The level of advancement of human society could be measured by the amount of restriction it imposed on voluntary cooperation. According to the classical economic theory of Adam Smith in which Spencer believed, a voluntary or free market society would inevitably develop an increasing amount of human heterogeneity due to the increasing division of labour. A more heterogeneous society would contain more net knowledge than a homogeneous society, for each individual could concentrate on being expert in one area, rather than having to know everything. If everybody possesses essentially the same information that is possessed by everyone else, then the amount of information in the entire system is no greater than the information a single individual has. Only the division of labour could permit the growth of civilization: the more heterogeneous a society is, the more advanced it is, if the differentiation arises by voluntary cooperation.

In contrast to Spencer, Marx regarded the division of labour not as an essential feature of an advanced civilization, but merely as a mark of class exploitation. Marx, and his followers to the present day, believed that in a communist industrial society, every individual could do *all* jobs:

> ... where nobody has one exclusive sphere of activity but each can become accomplished in any branch he wishes ... [it would be] possible for me to do one thing today and another tomorrow, to hunt in the morning, fish in the afternoon, rear cattle in the evening, criticize after dinner, just as I have a mind, without having ever becoming hunter, fisher, shepherd, or critic[265] ... the enslaving subjugation of individuals to the division of labour, and thereby the antithesis between intellectual and physical labour have disappeared ... when the all-around development of individuals has also increased their productive powers.[266]

Similar views of advanced societies can be found today as a general rule only among those socialists ignorant of economics. Those socialists who are knowledgeable of economics (e.g., refs 267, 268) recognize the necessity of division of labour for an advanced society, as do the vast majority of economists of all political beliefs.[269] It thus seems reasonable to assume that Spencer was correct at least in this respect about the social organization of all possible advanced societies. This will be important in the analysis of the likely behaviour of advanced extraterrestrial civilizations, which will be covered in Chapter 9.

Spencer divided social systems into two types: *military* and *industrial*. The former are characterized by rigid hierarchical social classes like an army. Cooperation and the resulting division of labour is restricted in such societies by force, for cooperation would interfere with the privileges of the ruling classes. The industrial society is the form of free market society ushered in by the industrial revolution. Since the industrial society is both more knowledgeable and based on cooperation rather than violence, it is morally superior to the military society. Being able to use more knowledge, industrial society is competitively superior to the more primitive military society, and consequently it should eventually replace the military society. Thus human social evolution is clearly progressive; evolution has a goal, and this goal is freedom for the individual.[261] Spencer's cosmology is teleological in the first sense defined above.

Spencer argued[270,271] that the driving force behind progressive human social evolution—increasing differentiation—was also operative in non-human biological and inorganic realms of the Universe. The Spencerian cosmos began with a homogeneous cloud of matter, which the force of gravity differentiated into stars and planets.[271,272] Inorganic matter differentiated under the action of electrical forces into first, complex forms of non-living compounds, and later, life. In Spencer's opinion, the increasing complexity of living creatures seen in the fossil record is best understood by comparing it to the increasing complexity which occurs in a developing embryo: it begins as a single cell, which divides and differentiates into the various cell types required by the cell division of labour in the metazoan. The cosmic differentiation process has now progressed to the human level, and it should continue to improve the human type. Spencer never considered the possibility that the differentiation process might eventually generate a species superior to *Homo sapiens*. He did, however, worry about the ultimate fate of his cosmology, for there might be a limit to the heterogeneity of matter, and he was aware also of the Heat Death problem. He concluded that the Universe is fundamentally cyclic, and that eventually the Universe would re-homogenize.[140,252,271]

As mentioned above, Spencer's ideas had an immense influence on intellectuals the world over at the beginning of the twentieth century. Amongst them was the American palaeobotanist and sociologist Lester Ward, who argued that the next stage of evolution, which Western Man was just entering, was characterized by 'telic evolution', or 'social telesis', in which government would provide more precise guidance to progress.[261,273] Ward's ideas were echoed in Britain by L. T. Hobhouse.[261,274] Ward, Hobhouse, and later John Dewey, were the main philosophers of progress who changed liberalism from its classical or *laissez-faire* form, in which progress would result from the unregulated free market, into its modern form, in which the goal is best obtained by government oversight. In either form, liberalism claims human social development is inevitably melioristic, and hence liberalism is teleological in the first sense defined above.

None of 'inevitable' social developments predicted by any of the above mentioned philosophers of progress actually happened. Spencer would be shocked by the increase of government control of the economy in this century, a development he would have regarded as reactionary, while Marx would be shocked by the continued existence and expansion of the free market, a development which *he* would have regarded as reactionary. Social philosophers such as Karl Popper[275] and Friedrich Hayek[276,277] have argued that the future evolutionary history of a complex social system is inherently unpredictable in the long run because a prediction would have to be based on an accurate model of society, and a sufficiently accurate model would be too complex to be coded in any mind or computer in the society. The memory of a finite state machine is inadequate to describe everything including itself. One of Hayek's arguments[276] was actually a formal mathematical proof that a finite state machine could not predict its future evolution. A similar proof for an infinite state machine was first obtained by the famous computer scientist Alan Turing some years after Hayek. Popper[278] has developed this argument that unpredictability in social evolution is due to the impossibility of complete self-reference.

Henri Bergson (1859–1941) is generally regarded as the foremost French philosopher of the twentieth century. His philosophy is based on 'Becoming', or the temporal aspects of reality, as the fundamental metaphysical concept. 'Being', or existence, is the basic metaphysical entity in the Cartesian philosophical tradition which was the dominant influence in French philosophy before Bergson. In philosophies of Being, time or more generally evolution, is regarded as illusory or of no fundamental importance. Teleology, which is basically temporal, is also not regarded as primary. The most significant contribution of Bergson was to make French philosophy take evolution seriously, an effect Schelling had earlier had on German philosophy. In fact, the historian Lovejoy, whose classic work *The Great Chain of Being* is largely an analysis of the tension between the ideas of Being and Becoming in Western philosophy and theology, considers[4] Bergson's philosophy to be largely a reworking of Schelling's. Bergson's influence on such French evolutionary philosophers as Teilhard de Chardin was immense.

Bergson carefully distinguished his version of teleology, or finalism, from the versions which were at bottom really equivalent to mechanism:

> ... Radical mechanism implies a metaphysic in which the totality of the real is postulated complete in eternity, and in which the apparent duration of things expresses merely the infirmity of a mind that cannot know everything at once... we reject radical mechanism.
>
> But radical finalism is quite as unacceptable.... The doctrine of teleology, in its extreme form, as we find it in Leibniz, for example, implies that things and beings merely realize a programme previously arranged. But there is nothing unforeseen, no invention or creation in the universe; time is useless again. As in the mechanistic hypothesis, here again it is supposed *all is given.* Finalism thus understood is only inverted mechanism.[279]
>
> Yet finalism is not, like mechanism, a doctrine with fixed rigid outlines ... It is so extensible, and thereby so comprehensive, that one accepts something of it as soon as one rejects pure mechanism. The theory we shall put forward ... will therefore necessarily partake of finalism to a certain extent... [the doctrine of finality] realizes that if the universe as a whole is the carrying out of a plan, this cannot be demonstrated empirically...[280]

Bergson's version of teleology was what he termed 'external finality', by which he meant that all living beings were ordered for each other:

In [our theory], finality is external, or it is nothing at all, ,[281] If there is finality in the world of life, it includes the whole of life in a single indivisible embrace.[282]

Evolution in Bergson's opinion was fundamentally creative in the sense that it always engendered something wholly new, something whose nature and whose coming-into-being could not have been foreseen by knowledge of what had come before. Only if evolution worked in this way could Becoming and not Being be regarded as metaphysically primary. Nature was an organic whole, ultimately teleological because it is driven by a non-physical Life Force, but whose future and goals are ultimately unknowable:

Never could the finalistic interpretation, as we ... propose it, be taken for an anticipation of the future.[283] ... the universe is not made, but is being made continually. It is growing, perhaps indefinitely, by the addition of new worlds.[284]

Bergson was aware of the difficulty which the Heat Death posed for his philosophy through the books of the French physicist Meyerson, but he tried to play down the problem. He could only suggest[284] that life may be able to take a form in which the ultimate Heat Death, the final use of *all* free energy, was delayed indefinitely. He also suggested that 'considerations drawn from our solar system' might not apply to the Universe as a whole. These were good guesses, as we shall see in Chapter 10.

Samuel Alexander (1859–1938) was a metaphysician who was born in Australia, but who spent his adult life in England. His most noteworthy contribution was an attempt to infer on philosophical grounds the future evolutionary history of the most advanced branch of life. In 1930 he was made a member of Order of Merit (an honour which is more highly regarded by British academics than winning a Nobel prize) for his work. He presented a fully developed version of his theory as a series of Gifford Lectures[285] at the University of Glasgow in 1916–1918. His system had a great influence on speculative British philosophy in the early part of this century.[286,287] Whitehead's metaphysical system can be regarded as an elaboration and extension of Alexander's from a somewhat different perspective.

For Alexander, the fundamental entity was Space-Time, which engenders first matter, then life, and finally mind. But there is a stage beyond mind, termed 'deity' by Alexander, which is as superior to mind as mind is to life without mind. Just as a mind exists in a living being, but most living beings (all non-human living beings, in fact) do not have mind, so deity is a property which will exist in mind, but most minds will not possess deity.[288] The purpose of the universe is to bring deity into being:

> Deity is thus the next higher empirical quality to mind, which the universe is engaged in bringing to birth. That the universe is pregnant with such a quality we are speculatively assured . . . .[289] There is a *nisus* in Space-Time which, as it has borne its creatures forward through matter and life to mind, will bear them forward to some higher level of existence.
>
> . . . our supposed angels are finite beings with the quality of deity, that quality which contemplates mind as mind contemplates life and matter . . . beings with finite deity are finite Gods.[290]

With Alexander, the notion of an evolving God, who does not always exist but rather comes into existence, first appears in English philosophy. In the distant past there was no deity, just as there once was no mind, and even further back in time there was no life. For Alexander,

> God is the whole universe as possessing the quality of deity. Of such a being the whole world is the 'body' and deity is the 'mind' . . . .[291] God includes the whole universe, but his deity, though infinite, belongs to, or is lodged in, only a portion of the universe.[292]

Alexander's concept of an evolving God will be recognized as similar to that of Schelling and Teilhard de Chardin. However, Alexander did not leave behind him a school which developed his particular brand of evolutionary and teleological metaphysics.

Alfred North Whitehead (1861–1947) was trained as a mathematical physicist, and his metaphysics reflects his training, in the sense that it was far more consistent with the physical science of his day—relativity and quantum mechanics—than were the systems of Alexander and Bergson. Whitehead's cosmology received its most comprehensive expression in his Gifford Lectures delivered at the University of Edinburgh during 1927–1928, which were published under the title *Process and Reality*: *An Essay in Cosmology*.[293] Throughout this work Whitehead constantly asserts the natural world to be an organism, by which he meant that it resembles an organism in that the essence of each object lies not in its intrinsic nature, but rather in its relation to the whole: his view was quite similar to that of the nineteenth-century German biologists and the ancient Chinese Taoists and Confucians whose philosophies we discussed in Chapter 2. Like the Chinese, Whitehead applied his philosophy of organism not only to living things, but also to the inorganic physical universe. Whitehead used the very suggestive word 'society' to refer to the order which results:

> The members of the society are alike because, by reason of their common character, they impose on other members of the society the conditions which lead to that likeness.[294]

The entities of physical science formed such a society:

> Maxwell's equations of the electromagnetic field hold sway by reason of the throngs of electrons and of protons. Also each electron is a society of electronic occasions, and each proton is a society of protonic occasions. These occasions are the reasons for the electromagnetic laws; but their capacity for reproduction, whereby each electron and each proton has a long life, and whereby each electron and each proton come into being, is itself due to these same laws.... Thus in a society, the members can only exist by reason of the laws which dominate the society, and the laws only come in to being by reason of the analogous characters of the members of the society.[295]

In other words, the laws of physics and the elementary particles come into existence spontaneously by a sort of mutual self-consistency requirement. But a self-ordered society is not forever:

> But there is not any perfect attainment of an ideal order whereby the indefinite endurance of a society is secured. A society arises from disorder, where 'disorder' is defined by reference to the ideal for that society; the favourable background of a larger environment either itself decays, or ceases to favour the persistence of the society after some stage of growth: the society ceases to reproduce its members, and finally after a stage of decay passes out of existence. Thus a system of 'laws' determining reproduction in some portion of the universe gradually rises into dominance; it has its stage of endurance, and passes out of existence....[295]

Thus the laws governing the elementary particles which exist today together with the elementary particles themselves will gradually pass out of existence, and they will be replaced by other types of elementary particles governed by different laws. Whitehead explicitly lists as something bound to pass away not only the laws of electromagnetism, but also the four-dimensional nature of the space-time continuum, the axioms of geometry, and even the dimensional character of the continuum.[295] A period of universal history in which a definite self-consistent set of physical laws holds sway was termed a *cosmic epoch* by Whitehead. In the fullness of time all logically possible universes will exist; our own Universe—our own cosmic epoch—is just one of many which will eventually pass away.[296] Whitehead rejected Leibniz' theory of the best of all possible worlds (which is Leibniz' explanation of why just one of all logically possible worlds exists) as 'an audacious fudge'.[297]

Broadly speaking, Whitehead's cosmology is the same as the globally static cosmologies generating so much interest today. His picture of cosmic epochs is similar to the 'bubble universe' model developed by Fred Hoyle and by Richard Gott,[259] in which the visible portion of the Universe is just one of an infinite number of bubbles in an over-all chaotic universe, or the bubble universes in the inflationary universe models[298,299] (see Chapter 6 for a detailed discussion). In the inflationary universe, our bubble universe has certain physical properties because of the particular way in which it condensed out of a chaotic medium. As in Gott's theory, the bubble may disappear, with the material in the bubble returning once again to chaos. On a sufficiently large scale, the universe is pictured as chaotic;[300] for, assuming global chaos obviates the problem of assuming certain initial conditions for the field equations. This is the modern analogue of Whitehead's solution of why just one of all logically possible worlds exists. There is no problem if they all exist, as we mentioned in section 2.9. We shall discuss another version of this solution in Chapter 7, when we describe the Many-Worlds interpretation of quantum mechanics. In contrast to the bubble universe model, the Many-Worlds model allows evolution to occur on a global scale while simultaneously allowing all logically possible universes to exist. The difference is due to the fact that the different universes exist in the bubble universe model in physical space, whereas in the Many-Worlds model the different universes exist in a Hilbert space of realized possibilities. Whitehead can be regarded as the first philosopher who appreciated the advantages of the Many-Worlds ontology. The ontology of the Many-Worlds cosmology and Whitehead's cosmology is that which was implied by what A. O. Lovejoy called the 'Principle of Plenitude', (see section 3.2). We can regard these cosmologies as a modern expression of this principle.

Whitehead's cosmology is also remarkably similar to Wheeler's earlier proposal[252,301] (which he no longer believes) that a closed universe may go through an infinite number of cycles, with the physical laws being different in each cycle. Wheeler's proposal is often mentioned as a possible model for WAP: our particular type of intelligent life is consistent only with a very special set of physical laws, so we naturally exist in that cycle in which such laws hold sway.

The basic mechanism for change in Whitehead's cosmology is teleological. When an object A changes into an object B, the change is not pictured as random, but rather A is to be thought of as orienting its changes *toward* B.[302] Furthermore, processes occur in a cosmic epoch because 'eternal objects'—which are somewhat analogous to the Platonic forms that exist in the realm of ideas—act as a 'lure' (Whitehead's term) for the process. This teleological process at the most fundamental level gives rise to the efficient causes which scientists investigate. Whitehead regarded efficient and final causes as complimentary modes of explanation:

A satisfactory cosmology must explain the interweaving of efficient and final causation. Such a cosmology will obviously remain an explanatory arbitrariness if our doctrine of the two modes of causation takes the form of a mere limitation of the scope of one mode by the intervention of the other mode. What we seek is such an explanation of the metaphysical nature of things that everything determinable by efficient causation is thereby determined, and that everything determinable by final causation is thereby determined. The two spheres of operation should be interwoven and required, each by the other. But neither sphere should arbitrarily limit the scope of the alternative mode.[342]

Whitehead's main physical evidence for the existence of final causation was the very existence of a 'bubble' Universe:

Our scientific formulation of physics displays a limited universe in the process of dissipation. We require a counter-agency to explain the existence of a Universe in dissipation within a finite time.[342]

Nevertheless, in a more fundamental sense, Whitehead's cosmology is not teleological in the large, as the cosmology of Alexander is, for the Universe in the sense of the totality of everything that exists is not evolving toward some goal.

An anthropic prediction was made in the 1960's by Whitehead's follower, the philosopher Charles Hartshorne, whose work on the ontological argument we discussed in section 2.9. Hartshorne did not accept the overall lack of progress in Whitehead's cosmology, and he modified the Whiteheadian Universe so that there was net progress from one time to the next.[303] This requires that it be possible to define globally a time coordinate, for otherwise it would not possible in the large to define the temporal sequence of events. In special relativity it is not possible to define a unique global time-coordinate because of the global properties of the Poincaré group. As a mathematical physicist, Whitehead was aware of this, and we submit that this awareness prevented him from endowing his cosmology with progress in the large.[304] Hartshorne, as a philosopher, did not feel the constraints of physics as strongly as Whitehead. Nevertheless, he was aware that the popular view of relativity posed problems for his cosmology:

Relativity physics is a puzzling case for my thesis, the most puzzling indeed of all. If reality is ultimately a self-surpassing process, embraced in a self-surpasssing divine life, there must be something like a divine past and future. According to relativity physics, there is indeed, for our localized experience, a definite cosmic past and a definite cosmic future, but not a definite cosmic present. We may have two contemporaries out in space, one of which is years in the past of the other. And there seems no way to divide the cosmic process as a whole into past and future. Yet if neoclassical theism is right, it seems there must, for God at least, be a way. What is God's 'frame of reference', if there is no objectively right frame of reference for the cut between past and future? I can only suppose that we have in this apparent conflict a subtler form of the illicit extrapolation to the absolute from observational facts. Somehow relativity as an observational truth must be compatible with divine unsurpassability.[303]

As we mentioned in section 2.9, in so far as his ontology is concerned,

Hartshorne is basically a pantheist. As he put it, 'Pantheism in this sense is simply theism aware of its implications ... [on ontological questions] there is indeed no real issue between theism and pantheism'.[305] Thus his deity must be subject to the rules of temporal succession implied by physics. Since special relativity does not permit the required temporal succession, Hartshorne insists that the temporal succession rules of special relativity do not apply globally to the Universe.

Hartshorne is quite correct; it doesn't. The frame of reference in which the cosmological background radiation has the same temperature in all directions defines a unique global time coordinate in which the notions of past, present and future of a given event can be defined. Furthermore, the global time defined by this frame is essentially the same as the global time defined by the constant mean curvature foliation in a Universe which is approximately homogeneous and isotropic.[306] This point will be discussed in more detail in Chapter 10. In special relativity no unique global time can be defined, but general relativity is Lorentz invariant only locally, not globally, and thus a global time can be defined. The existence of a global time in cosmology was actually pointed out by Sir Arthur Eddington[307] in 1920.

Hartshorne was unaware of this, so his argument counts as a correct prediction about the global temporal structure based on the Anthropic Principle, for in his view men were to be thought of as 'nerve cells' of God.[308] Hartshorne even tried to justify the speed of light limitation anthropically:

> There is a conceivable teleological justification for relativity. What good would it do us to be able to transmit messages with infinite velocity? It is bad enough being able to learn about troubles around the world in seconds, but to get bad news quickly from remote planets, and have to reply almost at once—that would be too much. Thank God we are isolated by the cosmically slow speed of light—we have enough complexity on our hands with this planet. Thus, once more, the heavens declare the glory of God ....[309]

3.11 Teilhard de Chardin: Mystic, Paleontologist and Teleologist

> Schopenhauer was a degenerate, unthinking, unknowing, nonsense scribbling philosopher, whose understanding consisted solely of empty, verbal trash.
>
> Ludwig Boltzmann

As stressed by Joseph Needham,[310] one of the reasons for the widespread and continuing popular interest in the philosophical work of Teilhard de Chardin is the man himself. Many theologians and philosophers before him had attempted to make their religious beliefs or philosophical systems consistent with or even based on the fact of an evolving cosmos. Many devout scientists before him had tried to show their evolutionary science was consistent with their religion. Previously we have given examples of both. But Teilhard combined in one person the scientist and theologian: he had acquired a world-wide reputation as a paleontologist specializing in the evolution of Man; he was also a Jesuit priest. When it came to reconciling science and religion, a scientist and theologian could speak perhaps with double authority.

Our society tends simultaneously to respect authority and to distrust it. An authority who is silenced by authority is thus especially interesting. Teilhard had begun in the 1920's to lecture about his speculations on combining Catholicism with evolution. The leaders of the Jesuit order exiled him to China to prevent further discussion of these views in his native France. He was forbidden to publish any of his philosophical works in his lifetime. When a chair in paleontology became vacant at the College de France, he was not permitted to apply for the position. He moved to New York City, where he died in 1955. He is exiled even in death: he is buried in the cemetery of a small monastery some 50 miles from New York, far from his beloved France.[311]

When Teilhard's ideas on evolutionary Christianity were published in the year of his death, his friends (of which there were many, for by all accounts he was an extraordinarily likeable man) spread far and wide the *pathos* of his life-story. Undoubtedly, this resulted in his ideas being given a vastly more sympathetic hearing than they might otherwise have received (and than they probably deserve!).

Nevertheless, it would be a mistake to think that the enormous initial and continuing interest in the work of Teilhard is due entirely or even primarily to mere psychological and social factors. His evolutionary theological cosmology has certain key features which distinguish it from the somewhat similar systems of Schelling, Alexander, and Bergson. Many of the theologians in the English speaking world, notably Philip Hefner,[312,313] Arthur R. Peacocke,[254] and Charles Raven[314] have been very sympathetic to Teilhard's work.

Teilhard opens what is generally regarded as his most significant philosphical work, *The Phenomenon of Man*, with the statement: "if this book is to be properly understood, it must be read not as a work on metaphysics, still less as a sort of theological essay, but purely and simply as a scientific treatise. The title itself indicates that'.[10] His critics—for instance the evolutionary biologist G. G. Simpson,[11] and the zoologist Sir Peter Medawar[315] have taken him to task for this assertion, but we believe a close reading of Teilhard's central work will justify his claim. The work was admittedly not written in standard scientific style; Teilhard used a more mystical language, which certainly annoyed many scientists.

Medawar, for example, was so put off by the language that he charged the book '... cannot be read without a feeling of suffocation, a gasping and flailing around for sense ... the greater part of it is nonsense, tricked out by a variety of tedious metaphysical conceits, and its author can be excused of dishonesty only on the grounds that before deceiving others he has taken great pains to deceive himself'.[315]

Most of *The Phenomenon of Man* is devoted to a poetic description of an evolving Earth, beginning with the formation of the planet, and then moving on to the development of life from its most primitive manifestation to the emergence of Man. On a phenomenological level Teilhard's picture is the standard scientific one of the late 1930's when the book was written. Some phyla of single-celled organisms eventually develop into metazoans, some phyla of which in turn develop organisms with highly developed nervous systems, and one lineage of these creatures finally acquires intelligence: the 'hominisation'—Teilhard's word—of the world has at last occurred. If the picture is standard, the physical mechanism behind the ascent of life is not.

Teilhard argued that energy existed in two basic modes, 'tangential' and 'radial'. The former is essentially the energy measured by the instruments of the physicist, while the latter can be regarded as a sort of psychic or spiritual energy. Teilhard's motivation for introducing the latter variety is two-fold: First, his cosmological system evolves higher and higher order in its biota as time proceeds, and this seemed to him to be forbidden by the Second Law of Thermodynamics which he admits governs the evolution of the usual variety of energy.[316] Furthermore, the eventual Heat Death predicted by the thermodynamicists would undermine any hope of having Ultimate Intelligence permanently immanent in the Cosmos. He is well-aware that if intelligence is at bottom completely dependent on tangential energy, it must be doomed to extinction in the end, however powerful it becomes, if in fact the Heat Death occurs. Therefore, his radial energy is subject to a universal law contrary to the Second Law of tangential energy: radial energy becomes more concentrated, more available with time, and it is this concentration that drives the evolution of life to Man, and beyond.[317] Radial energy—psychic energy—is as ubiquitous as tangential energy. It is present in all forms of matter at least to a rudimentary extent, and so all forms of matter have a low-level sort of life.[318] To modern scientists this vitalism seems archaic, even occult, but such a concept was held by a number of distinguished thinkers at the time Teilhard was writing.

In the opinion of Teilhard, '... the idea of the *direct* [his emphasis] transformation of one of these two energies into the other ... has to be abandoned. As soon as we try to couple them together, their mutual independence becomes as clear as their interrelation'.[319] His reasons for this view are as follows:

> . . .'To think, we must eat'. But what a variety of thoughts we get out of one slice of bread! Like the letters of the alphabet, which can equally well be assembled into nonsense as into the most beautiful poem, the same calories seem as indifferent as they are necessary to the spiritual values they nourish.
>
> The two energies—of mind and matter—spread respectively through the two layers of the world (the *within* and the *without*) have, taken as a whole, much the same demeanour. They are constantly associated and in some way pass into each other. But it seems impossible to establish a simple correspondence between their curves. On the one hand, only a minute fraction of 'physical' energy is used up in the higher exercise of spiritual energy; on the other, this minute fraction, once absorbed, results on the internal scale in the most extraordinary oscillations.
>
> A quantitative disproportion of this kind is enough to make us reject the naive notion of 'change of form' (or direct transformation)—and hence all hope of discovering a 'mechanical equivalent' for will or thought. Between the *within* and the *without* of things, the interdependence of energy is incontestable. But it can in all probability only be expressed by a complex symbolism in which terms of a different order are employed.[320]

Since this passage was written, we *have* discovered in effect the 'mechanical equivalent' for will or thought. These manifestations of mind are just two types of information, and the minimum amount of energy that must be dissipated in order to generate a given number of thoughts (or bits of information) can be calculated rather simply. The detailed theory will be presented in Chapter 10, but here we can remark that, to take Teilhard's example, a piece of bread can generate at most about 10^{25} bytes of thought. Information theory thus removes a cornerstone of Teilhard's theory, and *qua* scientific theory it crashes to the ground. Medawar[315] mentioned that 'Teilhard's radial, spiritual or psychic energy may be equated to "information" or "information content" in the sense that has been made reasonably precise in the sense of communications engineers',[321] and he realized that information did not avoid the restrictions of the Second Law. However, the fact that it is possible to demolish Teilhard's theory by reference to physics shows it was in fact a scientific theory as Teilhard claimed, for a general conceptual scheme which is in principle falsifiable *is* a scientific theory. Many modern philosophers of science would not agree with Karl Popper[322] that falsifiability is a necessary condition for a theory to count as scientific, but most, if not all, would agree that it is a sufficient condition. Although the specific theory advanced by Teilhard has been refuted, his basic meta-theoretical notion of a melioristic cosmos, a universe which evolves God, has not been refuted and indeed cannot be. No mere experiment can destroy a general conceptual scheme. Furthermore, any evolving universe theory which rejects dysteleology must be broadly similar to Teilhard's (or Schelling's or Alexander's). In Chapter 10 we shall present a mathematical model of a different sort of melioristic cosmos.

According to his biographer Claude Cuénot, Teilhard became fascinated in the 1950's with computers and the relation between information and entropy.[323] Cuénot also claims[324] that Teilhard's new knowledge of information theory led him to go beyond the distinction which he had drawn earlier in *The Phenomenon of Man* between radial and tangential energy. (We believe Teilhard would have been unable to replace radial energy with information, for reasons we discuss below.) In a letter of May 15, 1953, Teilhard wrote:

What really interests me in cybernetics is the transformation of materialism it suggests to us. A machine is not (or is no longer) an affair primarily of energy set in motion, but of information put together and transmitted.[324]

Teilhard felt that the information-processing of a computer was analogous to human thought.[324,325] He did not believe that computers would replace Man 'for a variety of biological reasons'.[325] Rather, he envisaged Man and the computer in a partnership which would enormously expand human mental powers.[323,325]

In Teilhard's theory the radial energy generated single-celled organisms on the newly condensed Earth, then drove these organisms to cover the Earth and combine to form the metazoans. More than half of *The Phenomenon of Man* is devoted to describing the expansion and combination process in terms, which in rough outline, do not differ significantly from standard evolutionary textbooks. The great evolutionists Simpson[11] and Dobzhansky[50] differ about whether Teilhard believed evolution to be orthogenetic, (orthogenesis means the development of life throughout the entire past history of the Earth is nothing but a predetermined unfolding of characteristics already present in the beginning of organized life), or whether he believed, with the vast majority of contemporary evolutionists, that the evolutionary process is opportunistic, with no foresight. Teilhard certainly *described* evolution as 'orthogenetic'. From this and the *apparent* fact that in Teilhard's picture the development of Man is inevitable, Simpson dismisses Teilhard's work as 'evolutionary mysticism'. Medawar[315] and many others are also particularly hard on Teilhard's theory because it apparently requires orthogenesis. However, we must agree with Dobzhansky that '... in spite of himself, Teilhard was not an exponent of orthogenesis'.[326] Teilhard himself said he used orthogenesis '... for singling out and affirming the manifest property of living matter to form a system in which 'terms *succeed each other* experimentally, following constantly increasing degrees of centro-complexity'. [Teilhard's stress and quotes] ... Without orthogenesis life would only have spread; with it there is an ascent of life that is invincible'.[327]

The key word in this passage is 'experimentally'. Teilhard, as Dobzhansky emphasizes, is quite aware that the success of new forms of life is not guaranteed. New species are experiments of radial energy—the life force, as it 'gropes'—Teilhard's word—its way to higher and higher complexity. Only increased organization of life is inevitable, because of the inherent centralizing properties of the radial energy. As a devout Catholic priest convinced of Man's free will, Teilhard could not possibly be advocating anything that would resemble determinism. Orthogenesis would entail determinism if Teilhard used the word in its standard sense. Most orthogenetic theories imply the inevitable evolution of intelligent life. But Teilhard believed it would be unlikely 'that if the human branch disappeared, another thinking branch would soon take its place'.[328] He also thought the evolution of extraterrestrial intelligent life to have a '... probability too remote to be worth dwelling on'.[329]

However, there is a weak determinism acting: although any individual or species may fail, the most complex organism cannot do so until it has engendered its even more complex successor:

> ... we must not forget that since the birth of thought man has been the leading shoot of the tree of life. That being so, the hopes for the future ... (of biogenesis, which in the end is the same as cosmogenesis) is concentrated exclusively upon him as such. How then could he come to an end before his time, or stop, or deteriorate, unless the universe committed abortion upon itself, which we have already decided to be absurd? ... *Man is irreplaceable.* Therefore, however improbable it might seem, *he must reach the goal*, not necessarily, doubtless, but infallibly [Teilhard's emphasis].[330]

We could claim that Teilhard's distinction between 'necessary success' and 'infallible success' is his way of distinguishing between strong and weak determinism in the manner we discussed above. This distinction is a traditional one in Catholic theology (it has also been drawn by Leibniz): it is essentially Aquinas' distinction between absolute and hypothetical necessity.[331] Such a distinction is mandatory if a metaphysics is to contain both free will and an omniscient and omnipotent Deity, as Catholicism does.

What is the goal of mankind, according to Teilhard? Just as non-sapient life covered the Earth to form the *biosphere*, so mankind—thinking life—has covered the Earth to form what Teilhard terms the *noosphere*, or cogitative layer. At present the noosphere is only roughly organized, but its coherence will grow as human science and civilization develop, as 'planetization'—Teilhard's word—proceeds. Finally, in the far future, the radial energy will at last become totally dominant over, or rather independent of, tangential energy, and the noosphere will coalesce into a super-sapient being, the *Omega Point*. This is the ultimate goal of the tree of life and of its current 'leading shoot', *Homo sapiens*. As Teilhard poetically puts it in *The Phenomenon of Man*:

This will be the end and the fulfilment of the spirit of the Earth.

The end of the world: the wholesale internal introversion upon itself of the noosphere, which has simultaneously reached the uttermost limit of its complexity and centrality.

The end of the world: the overthrow of equilibrium [read, 'Heat Death'], detaching the mind, fulfilled at last, from its material matrix, so that it will henceforth rest with all its weight on God-Omega.[332]

So speaks Teilhard the Catholic mystic, who has identified the Omega Point with the Christian God (or rather with Christ, who in the Catholic doctrine of the Trinity is regarded as the manifestation of God in the physical Universe). But Teilhard claims to be writing *qua* scientist in this book, and in fact some phenomenological properties of the Omega Point can be gleamed from some of the book's passages.

One key property of the Omega Point is that It, in contrast to the dysteleology of the Second Law of Thermodynamics as understood by the physicists of the early twentieth century, *must* allow mankind to finally escape the Heat Death, the inevitable end of the forces of tangential energy:

The radical defect in all forms of belief in progress, as they are expressed in positivist credos, is that they do not definitely eliminate death. What is the use of detecting a focus of any sort in the van of evolution if that focus can and must one day disintegrate? To satisfy the ultimate requirements of our action, Omega must be independent of the collapse of the forces with which evolution is woven ... Thus something in the cosmos escapes from entropy....[333]

The Omega Point must in some sense be in the future, at the end or boundary of time, after the end of matter;

... Omega itself is ... at the end of the whole process, in as much as in it the movement of synthesis culminates. Yet we must be careful to note that under this evolutive facet Omega still only reveals *half of itself*. While being the last term of its series, it is also *outside all series*. Not only does it crown, but it closes .... If by its very nature it did not escape from time and space which it gathers together, it would not be Omega.[333]

The details of the transition from the disorganized noosphere to the unity of the Omega Point are (not surprisingly!) few. Teilhard speaks, however, of the transition from animal existence to reflecting, thinking life in terms which make us suspect he was envisaging an analogous process for the origination of the Omega Point:

... taking a series of sections from the base towards the summit of a cone, their area decreases constantly; then suddenly, with another infinitesimal displacement, the surface vanishes leaving us with a *point* [Teilhard's emphasis]. . . . what was previously only a centered surface became a center . . . Thus by these remote comparisons we are able to imagine the mechanism involved in the critical threshold of reflection.[334]

In other words, the Omega Point could be compared to a conical *singularity*. Coincidentally, this is essentially the view of the end of time one finds in modern cosmology for closed universes, and indeed in another Omega Point theory developed in the final chapter of this book; there the Omega Point will actually be identified with a point on the c-boundary of space-time. It is essential that the Universe be closed—that is, be finite in spatial extent—if the future c-boundary is to have a point-like structure. Interestingly, Teilhard's Omega Point theory also seems to require a boundedness of the spatial structure—the Earth in his theory—in order for the Omega point to be generated out of the coalescence of mankind:

. . . there intervenes a fact, commonplace at first sight, but through which in reality there transpires one of the most fundamental characteristics of the cosmic structure—the roundness of the Earth. The geometrical limitation of a star closed, like a gigantic molecule, upon itself . . . What would have become of humanity, if, by some remote chance, it had been free to spread indefinitely on an unlimited surface, that is to say left only to the devices of its internal affinities? Something unimaginable, certainly something altogether different from the modern world. Perhaps even nothing at all, when we think of the extreme importance of the role played in its development by the forces of compression.[335]

The 'forces of compression' about which Teilhard speaks are the social forces which arise from Man communicating with his fellows. It is the requirement of ceaseless communication in the future Universe that implies a point c-boundary structure for the future end of the Universe, as shown in Chapter 10. In the theory developed there as well as in Teilhard's theory, an Omega Point can evolve only in a bounded world.

Teilhard's bounded world was the finite Earth. He did not believe that space travel would ever be an important phenomenon in the future evolution of mankind.[336] Indeed as the immediately preceding passage makes clear, a mankind freed from the confines of the Earth would probably never combine into the Omega Point. Teilhard made this point explicitly in a private conversation in 1951. As recorded by J. Hyppolite, a professor of philosophy at the Sorbonne, Teilhard said:

Following in the steps of [J.B.S.] Haldane, the neo-Marxist tends to escape into the perspectives of a vital *expansion*, in other words, into a *vitalization* of the *Totality* of stellar *Space*. Let me stress this second point a little. From his own viewpoint, the Marxist will approach willingly and with an open mind the idea of an eschatology for a classless society in which the Omega Point is conceived as the point of natural convergence for humanity. But suppose we remind him that our Earth, because of the implacable laws of entropy, is destined to die; suppose we ask him what will be the outcome allowed humanity in such a world. Then he replies—in terms that H. G. Wells has already used—by offering perspectives of interplanetary and intergalactic colonization. This is one way to dodge the mystical notion of a Parousia, and the gradual movement of humanity towards an ecstatic union with God.[337]

The necessity of restricting mankind to the Earth in Teilhard's Omega Point theory is one major difference between his theory and the one developed in Chapter 10, which is closer to the 'neo-Marxist' theory. We believe the entropy problem and the finiteness of the Earth would have made it impossible for Teilhard to give up radial energy for information, as Cuénot suggested he might have.

Teilhard did not consider the non-terrestrial part of the Universe to be very important. What was truly significant was life, and this was apparently restricted to the Earth:

...what matters the giddy plurality of the stars and their fantastic spread, if that immensity (symmetrical with the infinitesimal) has no other function but to equilibrate the intermediary layer where, and where only in the medium range of size, life can build itself up chemically?[343]

This is strikingly similar to Wheeler's idea that the Universe must be at least as large as it is in order for any intelligent life at all to exist in it. In a sense, the large amount of matter in the Universe 'equilibrates'—permits the existence over long periods of time—the planetary envirionment upon which life must arise.

Teilhard continually uses spatial images to describe the Omega Point:

...[the noosphere] must somewhere ahead [in time] become involuted to a point which we might call *Omega* [Teilhard's emphasis], which fuses and consumes [it] integrally in itself. However, immense the sphere of the world may be, it only exists and is finally perceptible in the directions in which its radii meet—even if this were beyond space and time.[338]

In a closed universe, the radii of the Universe meet beyond space and time in the final singularity—the mathematical Omega Point defined rigorously in Chapter 10 of this book.

Teilhard made only one drawing of the Omega Point (Diagram 4 in the *Phenomenon of Man*), and amusingly, it is quite similar to the Penrose diagram for a closed universe whose future c-boundary is a single point (see Figure 10.5)! In the Penrose diagram, the convergence of the lines into a point is a mathematical expression of unlimited communication between spatially separated regions. By the convergence of the lines in his figure, Teilhard intended to signify the integration by communication of the entire noosphere.

Teilhard's original theory was conceived before the advent of information theory (which made the idea of radial energy at least a possibility at the time), and of modern cosmology. His original theory has been refuted, or perhaps we should say it has become obsolete. However, the basic framework of his theory is really the only framework wherein the evolving Cosmos of modern science can be combined with an ultimate meaningfulness to reality. As the dysteleologists have argued at length, if in the end all life becomes extinct, meaning must also disappear. In the final chapter we construct a mathematical Omega Point theory and by so doing we suggest that value may be able to avoid extinction.

In this chapter we have investigated what we consider to be the most influential uses of teleological reasoning in science, philosophy and theology. The way in which local teleological ideas are used in modern biology and physics was carefully distinguished from their indiscriminate global deployment in past centuries. The developments in physics during early years of this century saw examples where essentially Anthropic arguments led to successful physical predictions. However, since that time, the study of teleology was dominated by an interesting collection of philosophers and theologians whose work we have tried to unravel and present in a logical progression. Interesting connections with the ideas of some modern economists can also be traced. A time-chart displaying the lifespans of the principal individuals whose ideas have been discussed in this chapter is given in Figure 3.1.

This completes our non-mathematical survey of teleological ideas in science and philosophy and provides a back-drop against which to view the modern form of the Anthropic Principle enunciated by cosmologists

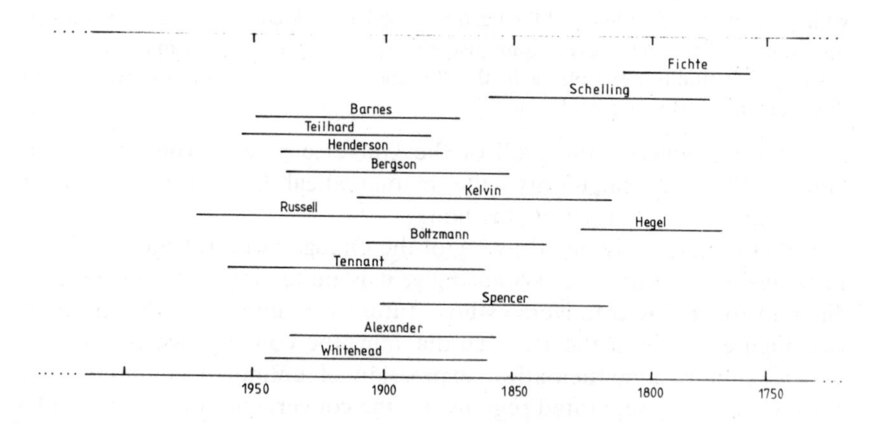

Figure 3.1. The chronology of some of the principal scientists and philosophers whose work is discussed in this chapter.

interested in the existence of a collection of surprising numerical coincidences in the make-up of the physical world.

References

1. J. Monod, in *Studies in the philosophy of biology*, ed. F. J. Ayala and T. Dobzhansky (University of California Press, Berkeley, 1974).
2. T. Huxley, *Lectures and essays* (Macmillan, NY, 1904), pp. 178–9.
3. J. R. Moore, *The post-Darwinian controversies* (Cambridge University Press, Cambridge, 1979).
4. A. O. Lovejoy, *The Great Chain of Being* (Harvard University Press, Cambridge, Mass., 1936).
5. H. A. E. Driesch, *The history and theory of vitalism* (Macmillan, London, 1914); *Man and the universe* (Allen & Unwin, London, 1927); *Mechanism, life and personality* (J. Murray, London, 1914).
6. J. S. Haldane, *The philosophy of a biologist* (Clarendon Press, London, 1935); *The philosophical basis of biology* (Doubleday, Garden City, 1931).
7. P. Lecomte du Noüy, *Human destiny* (Longmans, NY, 1955).
8. E. W. Sinnott, *The biology of the spirit* (Viking Press, NY, 1955).
9. S. Wright, *Monist* **48,** 265 (1964).
10. P. Teilhard de Chardin, *The phenomenon of Man*, rev. English transl. (Harper & Row, Colophon edn, NY, 1975), p. 29.
11. G. G. Simpson, 'Evolutionary theology: theology: the new mysticism', in *This view of life: the world of an evolutionist* (Harcourt Brace & World, NY, 1964), p. 213.
12. G. G. Simpson, ref. 11, section 3.
13. G. G. Simpson, *The meaning of evolution* (Yale University Press, New Haven, 1967).
14. F. J. Ayala, *Phil. Sci.* **37,** 1 (1970).
15. F. J. Ayala, 'The concept of biological progress', in *Studies in the philosophy of biology*, see ref. 1.
16. T. Dobzhansky, F. J. Ayala, G. L. Stebbins, and J. W. Valentine, *Evolution* (Freeman, San Francisco, 1977).
17. G. L. Stebbins, *The basis of progressive evolution* (University of North Carolina Press, Chapel Hill, 1969).
18. L. Sokoloff, in *Basic neurochemistry*, 2nd edn, ed. G. J. Siegel, R. W. Alberts, R. Katzman, and B. W. Arganoff (Little Brown, Boston, 1976).
19. D. A. Russell, in *Life in the universe*, ed. J. Billingham (MIT Press, Cambridge, Mass. 1981).
20. H. J. Jerison, *The evolution of the brain and intelligence* (Academic Press, NY, 1973).
21. H. J. Jerison, *Current Anthropol.* **16,** 403 (1975).
22. E. O. Wilson, *Sociobiology* (Harvard University Press, Cambridge, 1974).
23. C. O. Lovejoy, in *Life in the universe*, see ref. 19.
24. C. O. Lovejoy, *Science* **211,** 341 (1981).

25. Ref. 23, p. 326.
26. L. v. Salvini-Plawen and E. Mayr, in *Evolutionary biology*, ed. M. K. Hecht, W. C. Steere, and B. Wallace (Plenum, NY, 1977).
27. Letter from Ernst Mayr to FJT dated December 23, 1982.
28. The existence of such a consensus is attested to in ref. 27; see also ref. 23.
29. T. Dobzhansky, in *Perspectives in biology and medicine*, Vol. 15, p. 157 (1972); T. Dobzhansky, *Genetic diversity and human equality* (Basic Books, NY, 1973), pp. 99–101.
30. G. G. Simpson, *This view of life* (Harcourt Brace & World, NY, 1964), Chapters 12 and 13.
31. J. Francois, *Science* **196,** 1161 (1977); see also W. D. Mathew, *Science* **54,** 239 (1921).
32. E. Mayr, *Scient. Am.* **239** (Sept), 46 (1978).
33. S. J. Gould, *Discovery* **4** (No. 3, March), 62 (1983). Gould's argument was part of a three-way debate on extraterrestrial intelligence between himself, Carl Sagan, and FJT.
34. Ref. 13, p. 512.
35. E. Mayr, 'Teleological and teleonomic, a new analysis', in *Boston studies in the philosophy of science*, Vol. 14 ed. R. S. Cohen Wartofsky (Reidel, Dordrecht, 1974), p. 91.
36. E. Mayr, 'Cause and effect in biology', in *Cause and effect*, ed. D. Lerner (Free Press, NY, 1965).
37. J. Monod, *Chance and necessity* (Knopf, NY, 1971).
38. T. Dobzhansky, *The genetics of the evolutionary process* (Columbia University Press, NY, 1970), p. 4.
39. H. Butterfield, *The Whig interpretation of history* (Bell, London, 1931).
40. M. Grene, *The understanding of Nature: essays in the philosophy of biology* (Reidel, Dordrecht, 1974).
41. M. M. Waldrop, *Science* **224,** 1225 (1984).
42. I. Goodwin, *Physics Today* **37** (No. 5, May), 63 (1984). The estimate that 10^9 bit RAM chips will be available by the year 2000 was made by S. Chou, director of Intel's Portland Technology Development; see Computers and Electronics **22** (No. 8, Aug.), 16 (1984). Computer speeds can also be measured in MIPS (million instructions per second). The relationship between 'flops' and MIPS is difficult to define exactly, for they depend in a complicated way upon machine architecture. Roughly speaking, a MIPS is equal to a megaflop to within an order of magnitude, and the two measures of speed become closer the the faster the machine described. A typical fast mainframe computer like the IBM 3081 has a speed of about 10 MIPS.
43. A. Rosenblueth, N. Wiener, and J. Bigelow, in *Purpose in nature*, ed. J. Canfield (Prentice-Hall, Englewood Cliffs, NJ, 1966).
44. F. J. Ayala, 'Introduction', in *Studies in the philosophy of biology*, see ref. 1.
45. M. Polanyi, *Personal knowledge* (University of Chicago Press, Chicago, 1958); see especially pp. 140, 158, 394, and 396.
46. M. Polanyi, *The study of man* (University of Chicago Press, Chicago, 1959), pp. 48–51.
47. L. von Mises, *Human action: a treatise on economics*, 3rd edn (Henry Regnery, Chicago, 1966), p. 83.

48. F. A. Hayek, 'The pretence of knowledge' (1974 Nobel Lecture), reprinted in *New studies in philosophy, politics, economics, and the history of ideas* (Routledge & Kegan Paul, London, 1978), pp. 26–27.
49. T. Dobzhansky, *Mankind evolving* (Yale University Press, New Haven, 1962).
50. T. Dobzhansky, *The biology of ultimate concern* (New American Library, NY, 1967).
51. P. R. Ehrlich and A. H. Ehrlich, *Population, resources, and environment* (Freeman, San Francisco, 1970), p. 159.
52. F. A. Hayek, *Unemployment and monetary policy* (Cato Institute, San Francisco, 1979).
53. M. Friedman and R. Friedman, *Free to choose* (Harcourt Brace Jovanovich, NY, 1980).
54. W. Havender, *Reason* **14** (December), 52 (1982). We are grateful to Mr. H. Palka for this reference.
55. Ref. 51, p. 324.
56. S. Toulmin, *Human understanding, Vol. I: The collective use and evolution of concepts* (Princeton University Press, Princeton, 1972), Chapter 5.
57. S. Toulmin, ref. 56, pp. 324–40.
58. T. Kuhn, *The structure of scientific revolutions*, 2nd edn (University of Chicago Press, Chicago, 1970), pp. 170–3.
59. T. Kuhn, 'Reflections of my critics', in *Criticism and the growth of knowledge*, ed. I. Lakatos and A. Musgrave (Cambridge University Press, Cambridge, 1972), p. 265.
60. L. J. Henderson, *The fitness of the environment*, reprint with an introduction by George Wald (Peter Smith, Gloucester, Mass., 1970). The original edition was published by Harvard in 1913.
61. L. J. Henderson, *The order of Nature* (Harvard University Press, Cambridge, Mass., 1917).
62. G. Wald, *Origins of Life* **5**, 7 (1974).
63. Ref. 60, p. vi.
64. Ref. 61, p. 192.
65. Ref. 61, p. 184.
66. Ref. 61, p. 185.
67. Ref. 61, p. 208.
68. Ref. 61, p. 204.
69. Ref. 61, p. 203.
70. Ref. 61, p. 191.
71. Ref. 61, p. 183.
72. Ref. 61, p. 200.
73. Ref. 61, p. 211.
74. Ref. 61, p. 198.
75. Ref. 61, p. 201.
76. Ref. 61, p. 146.
77. Ref. 61, p. 181.
78. Ref. 61, p. 184.
79. Ref. 60, p. 312.

80. Anon., *Nature* **91,** 292 (1913).
81. J. S. Haldane, *Nature* **100,** 262 (1917).
82. J. Needham, *Order and life* (Yale University Press, New Haven, 1936), p. 15.
83. H. W. Smith, *Kamongo* (Viking Press, NY, 1932) p. 153. We are grateful to Professor J. A. Wheeler for this reference.
84. Hero of Alexandria, Catoptics 1–5, in I. B. Cohen and I. E. Drabkin, *A source book in Greek science* (McGraw-Hill, NY, 1948).
85. Aristotle, *De caelo* II. 4, 287a.
86. P. de Fermat, *Oeuvres* (1679).
87. G. Leibniz, in *Leibniz selections*, ed. P. P. Wiener (Scribner's, NY, 1951), p. 70.
88. P. L. M. de Maupertuis, *Accord de différentes lois de la Nature*, in, Vol. 4, p. 3 (1768).
89. P. L. M. de Maupertuis, *Essai de cosmologie*, in *Oeuvres*, Vol. 1, p. 5 (1768).
90. L. Euler, *Methodus inveniendi lineas curvas maximi minimive proprietate gaudentes, additamentum*, in *Collected works*, Vol. 24 (ed. C. Caratheodory, 1952). English translation by W. A. Oldfather, C. A. Ellis, and D. M. Brown, *Isis* **20,** 72 (1933). We are grateful to Professor S. G. Brush for this reference.
91. J. L. Lagrange, *Mécanique analytique*, *Oeuvres*, Vol. 11 (1867).
92. W. R. Hamilton, 'Second essay on a general method in dynamics', *Phil. Trans. R. Soc.* **1,** 95 (1835).
93. L. Euler, in Maupertius, *Oeuvres*, Vol. 4 (1768).
94. S. D. Poisson, quoted in ref. 97.
95. H. Hertz, quoted in ref. 97.
96. E. Mach, *The science of mechanics* (Dover, NY, 1953).
97. A. d'Abro, *The rise of the new physics*, Vol. I (Dover, NY, 1951).
98. W. Yourgrau, and S. Mandelstam, *Variational principles in dynamics and quantum theory* (Pitman, London, 1960).
99. H. von Helmholtz, quoted in ref. 98.
100. M. Planck, *A survey of physical theory* (Dover, NY, 1960), pp. 69–81.
101. M. Planck, *Scientific autobiography and other papers* (Greenwood, Westport, 1971), pp. 176–87.
102. J. Mehra, *Einstein, Hilbert, and the theory of gravitation* (Reidel, Boston, 1974).
103. R. P. Feynman, *Science* **153** 699 (1966).
104. J. A. Wheeler, and R. P. Feynman, Rev. Mod. Phys. **17** 157 (1945).
105. F. J. Tipler, *Nuovo Cim.* **28** 446 (1975).
106. R. B. Partridge, *Nature* **244** 263 (1973).
107. R. P. Feynman, *Rev. Mod. Phys.* **20** 267 (1948).
108. E. S. Abers and B. W. Lee, *Phys. Rep.* C **9** 1 (1973).
109. S. Weinberg, quoted in *Physics Today* **32** (No. 12) 18 (1979).
110. L. S. Schulman, *Techniques and applications of path integration* (Wiley, NY, 1981), pp. 7, 12.

111. J. D. Barrow and F. J. Tipler, 'Action in Nature', preprint, 1985.
112. J. G. Fichte, *Science of knowledge*, transl. of *Wissenschaftslehre* by P. Heath and J. Lachs (Cambridge University Press, Cambridge, 1982), p. 8.
113. Ref. 112, p. 10.
114. Ref. 112, p. 23.
115. N. Bohr, in *Albert Einstein: philosopher-scientist*, Vol. 1 (Harper & Row, NY, 1959), p. 234.
116. F. A. Hayek, *The pure theory of capital* (University of Chicago Press, Chicago, 1934).
117. Ref. 112, p. 21.
118. A. M. Turing, *Mind* **59,** 433 (1950); reprinted in ref. 119.
119. D. R. Hofstadter and D. C. Dennett, *The mind's I* (Basic Books, NY, 1981).
120. F. S. Beckman, *Mathematical foundations of programming* (Addison-Wesley, London, 1980).
121. M. Machtey and P. Young, *An introduction to the general theory of algorithms* (Elsevier North-Holland, Amsterdam, 1978).
122. F. W. J. von Schelling, *Darstellung meines Systems der Philosophie*, English transl. in ref. 123, p. 15.
123. F. W. J. von Schelling, *The ages of the World*, transl. with notes by F. Bolton (Columbia University Press, NY, 1942).
124. T. P. Hohler, *Imagination and reflection: intersubjectivity—Fichte's 'Grundlage' of 1794* (Martinus Nijhoff, The Hague, 1982). This reference interprets Fichte as taking the finite ego as fundamental. Fichte has also been interpreted by other philosophers, particularly those who have approached absolute idealism through Hegel, as taking the Infinite Will, or Universal Program, as fundamental (see ref. 125 for such an interpretation). We follow our own reading, and what seems to be a general consensus of Fichte scholars, in the 'finite ego as fundamental' interpretation.
125. F. Copleston, *A history of philosophy*, Vol. 7: *Fichte to Hegel* (Doubleday, NY, 1965).
126. F. W. J. von Schelling, *Of human freedom*, transl. of *Über das Wesen meschlichen Freiheit*, by J. Gutmann (Open Court, Chicago, 1936).
127. F. W. J. von Schelling, *System of transcendental idealism*, English transl. in ref. 126, p. 318.
128. Ref. 4, p. 323.
129. F. W. J. von Schelling, *Denkmal der Schrift von den gottlichen Dingen* (1812); English transl. of quoted passage in ref. 4, p. 323.
130. G. W. F. Hegel, *The philosophy of history* (Colonial Press, NY, 1899), p. 19.
131. G. W. F. Hegel, *The phenomenology of mind*, 2nd edn, English transl. of *Die Phaenomenologie des Geistes*, by J. B. Baillie (Allen & Unwin, London, 1931), p. 85.
132. G. W. F. Hegel, *Philosophy of nature* (Oxford University Press, Oxford, 1970), pp. 20 and 284.
133. J. M. E. McTaggart, *The nature of existence*, Vol. II, ed. C. D. Broad (Cambridge University Press, Cambridge, 1927), p. 478–9.
134. B. Bosanquet, *Proc. Br. Acad.* **2,** 235 (1905–6).

135. J. Royce, *The world and the individual* (Macmillan, NY, 1908). We are grateful to Prof. R. Whittemore for this extract.
136. C. S. Peirce, *Collected papers, Vol. VI: Scientific metaphysics*, ed. C. Hartshorne and P. Weiss (Harvard University Press, Cambridge, 1935), pp. 174 and 299.
137. Ref. 136, p. 26.
138. G. L. Buffon, *Oeuvres complétes de Buffon*, Vol. 9 (Paris, 1854).
139. F. C. Haber, *The age of the world: Moses to Darwin* (Johns Hopkins Press, Baltimore, 1959).
140. S. G. Brush, *The temperature of history* (Franklin, NY, 1978).
141. J. B. J. Fourier, *Théorie analytique de la chaleur* (Paris, 1822).
142. Lord Kelvin, (W. Thomson), *Phil. Mag.* (ser. 4) **25** (1863), 1; also in Kelvin's *Mathematical papers*, Vol. 3, p. 295.
143. Lord Kelvin, (W. Thomson), *Macmillan's Mag.* (5 March, 1862), 288.
144. J. O. Burchfield, *Lord Kelvin and the age of the Earth* (Macmillan, London, 1975), p. 73. This is the most important book on the subject, a gold-mine of information.
145. F. Jenkin, *North British Review* (June, 1867), 277; quote from p. 304.
146. Ref. 145, p. 301.
147. Ref. 145, p. 305.
148. Quoted in ref. 144, p. 77.
149. Quoted in ref. 144, p. 79.
150. P. G. Tait, *Lectures on some recent advances in physical science* (Macmillan, London, 1976).
151. S. Newcomb, *Popular astronomy* (Harper, NY, 1878), pp. 505–11.
152. C. King, *Am. J. Sci.* **145** (1893), 1.
153. C. Darwin, Letter to Reade, 9 Feb. 1877, in *More letters of Charles Darwin*, ed. F. Darwin and A. C. Seward (Appleton, NY, 1903), vol. 2, pp. 211–12; see also ref. 144, p. 110.
154. J. Croll, *Quart. J. Sci.* **7** (1877), 307. Quotes are on pp. 307 and 317–18.
155. A. Geikie, *Landscape in history and other essays* (Macmillan, London, 1905). Quotes are on pp. 172 and p. 186 respectively.
156. E. B. Poulton, *Essays in evolution* (Oxford University Press, Oxford, 1908), pp. 1–45.
157. J. G. Goodchild, *Proc. R. Phys. Soc. Edin.* **13** 259 (1896).
158. J. Perry, *Nature* **51,** 224 (1895).
159. J. Perry, *Nature* **51,** 582 (1895).
160. Kelvin, Lord (W. Thomson), *Nature* **51,** 438 (1895).
161. T. C. Chamberlin, *Science* **10,** 11 (1899). See S. G. Brush, *J. Hist. Astron.* **9,** 13 (1978) for a discussion of Chamberlin's views on the thermal history of the Earth. Chamberlin's views were based in part on a belief in global teleology; see H. C. Winnik, *J. Hist. Ideas* **31,** 441 (1970) for some discussion of Chamberlin's teleological views.
162. D. Kubrin, *J. Hist. Ideas* **28,** 325 (1967).
163. R. H. Hurlbutt III, *Hume, Newton and the Design Argument* (University of Nebraska Press, Lincoln, 1965).

164. C. Darwin, *On the origin of species by means of natural selection*, 2nd edn (John Murray, London, 1860), p. 486.
165. H. von Helmholtz, 'On the interaction of the natural forces', rep. in *Popular scientific lectures*, ed. M. Kline (Dover, NY, 1961).
166. W. Thomson, *Proc. R. Soc. Edin.* **8,** 325 (1874); repr. in ref. 195, p. 176.
167. J. Jeans, *The universe around us* (Cambridge University Press, Cambridge, 1929).
168. A. S. Eddington, *The nature of the physical world, Gifford lectures 1927* (Cambridge University Press, Cambridge, 1928).
169. B. Russell, *Why I am not a Christian* (George Allen & Unwin, NY, 1957), p. 107.
170. Ref. 169, p. 11.
171. N. Barlow (ed.), *The autobiography of Charles Darwin* (Harcourt Brace, NY, 1959), p. 92.
172. E. W. Barnes, *Scientific theory and religion, Gifford lectures 1927–1929* (Cambridge University Press, Cambridge, 1933).
173. Ref. 10, p. 229.
174. Ref. 10, p. 233.
175. W. R. Inge, *God and the astronomers, Warburton lectures 1931–1933* (Longmans Green, London, 1934), p. 24.
176. Ref. 175, p. 28. We suspect the last word of this quote was a misprint of 'coils' in the original text.
177. E. Hiebert, 'Thermodynamics and religion: a historical appraisal' in *Science and contemporary society*, ed. F. J. Crosson (University of Notre Dame Press, London, 1967).
178. B. Russell, *Religion and science* (Oxford University Press, NY, 1968), p. 210.
179. Ref. 178, p. 216.
180. R. Penrose, in *General relativity: an Einstein centenary survey*, ed. S. W. Hawking and W. Israel (Cambridge University Press, Cambridge, 1979).
181. P. R. Ehrlich, A. H. Ehrlich, and J. P. Holdren, *Ecoscience: population, resources, environment* (Freeman, San Francisco, 1977), p. 393.
182. Ref. 181. p. 74.
183. Obviously a single human is not immortal. The calculation envisages only that a single human being is alive at any given time. This could be the case only if, for example, a baby were born the instant the previous solitary inhabitant of the Earth died.
184. B. L. Cohen, *Before it's too late: a scientist's case for nuclear energy* (Plenum, London, 1981), p. 181.
185. P. and A. Ehrlich, *Extinction: the causes and consequences of the disappearance of species* (Random House, NY, 1981), p. xii.
186. J. L. Simon, *The ultimate resource* (Princeton University Press, Princeton, 1981).
187. J. L. Simon, *Science* **208,** 1431 (1980). Anne and Paul Ehrlich do not regard Simon's work highly. They assert that this paper in particular '...would have been a fine centerpiece for an April Fools' issue of any scientific journal'. (ref. 185, p. 291, note 14).

188. J. S. Mill, *Principle of political economy*, Vol. 1, 5th edn (Appleton, NY, 1895), p. 71.
189. Marginal utility is discussed at length in any modern economics textbook; e.g. *Economics*, 10th edn, by P. Samuelson (McGraw-Hill, NY, 1975).
190. P. R. Ehrlich, *Coevolution Quart.*, Spring, 1976.
191. Ref. 181, p. 823.
192. B. L. Cohen, *Am. J. Phys.* **51,** 75 (1983).
193. S. G. Brush, *The kind of motion we call heat*, 2 vols (North-Holland, Amsterdam, 1976).
194. L. Boltzmann, *Sber. Akad. Wiss. Wien*, part II, **66** (1872), 275, English translation in ref. 195, p. 88.
195. S. G. Brush, *Kinetic theory: Vol. 2—Irreversible processes* (Pergamon, Oxford, 1966).
196. L. Boltzmann, *Sber. Akad. Wiss. Wien*, part II, **75** (1877), 67; English translation in ref. 195, p. 188.
197. J. C. Maxwell, *Theory of heat*, 3rd edn (Longmans, London, 1872), p. 208.
198. K. Pearson, *The grammar of science* (Dent, London, 1937).
199. E. Zermelo, *Ann. Physik* **57,** 485 (1896); English translation in ref. 195, p. 208.
200. E. Zermelo, *Ann. Physik* **59,** 793 (1896); English translation in ref. 195, p. 229.
201. Ref. 195, p. 235.
202. L. Boltzmann, *Ann. Physik* **60,** (1897); English translation in ref. 195, p. 238.
203. Ref. 195, p. 242.
204. L. Boltzmann, *Nature* **51,** 413 (1895).
205. H. Poincaré, *Rev. de Metaphysique et de Morale* **1** (1893), 534; English transl. in ref. 195, p. 203.
206. H. Poincaré, *The foundations of science* (Science Press, Lancaster, 1946).
207. N. Wiener, *Cybernetics*, 2nd edn (MIT Press, Cambridge, Mass., 1961), pp. 34–5.
208. E. A. Milne, *Modern cosmology and the Christian idea of God* (Oxford University Press, Oxford, 1952).
209. F. J. Tipler, *Nature* **280,** 203 (1979) see also ref. 252.
210. J. B. S. Haldane, *Nature* **122,** 808 (1928).
211. J. B. S. Haldane, *The inequality of man* (Chatto & Windus, London, 1932), p. 169.
212. R. P. Feynman, *The character of physical law* (MIT Press, Cambridge, Mass., 1965).
213. A. Grünbaum, *Philosophical problems of space and time* (Knopf, NY, 1963), p. 227.
214. H. Reichenbach, *The direction of time* (University of California Press, Berkeley, 1971).
215. P. J. Peebles, *Comm. Astrophys. Space Phys.* **4** (1972), 53; however, see J. D. Barrow and R. A. Matzner, *Mon. Not. R. astron. Soc.* **181,** 719 (1977)

for a discussion of this paper.

216. R. Penrose, in *General relativity: an Einstein centenary survey*, ed. S. W. Hawking and W. Israel (Cambridge University Press, Cambridge, 1979).
217. C. W. Misner, *Astrophys. J.* **151** (1968), 431.
218. J. A. Wheeler, *Frontiers of time* (North-Holland, Amsterdam, 1978) pp. 54–73.
219. W. J. Cocke, *Phys. Rev.* **160,** 1165 (1967).
220. Ref. (193), p. 589.
221. M. J. Klein, *Am. Scient.* **58,** 84 (1970).
222. E. E. Daub, *Stud. Hist. Phil.* **1,** 213 (1970).
223. B. Stewart, and P. G. Tait, *The unseen universe; or physical speculations on a future state* (Macmillan, London, 1875).
224. W. K. Clifford, *Fortnightly Rev.* **23,** 776 (1875).
225. P. M. Heimann, *Br. J. Hist. Sci.* **6,** 73 (1972).
226. L. Szilard, *Z. Physik* **53,** 840 (1929). English translation in *Behavioral Science* **9,** 301 (1964).
227. L. Brillouin, *Science and information theory* (Academic Press, NY, 1962).
228. L. Brillouin, *Appl. Phys.* **22,** 334 (1951).
229. R. Carnap, and A. Shimony, *Two essays on entropy* (University of California Press, Berkeley, 1977).
230. L. Rosenberg, *Nature* **190,** 384 (1961).
231. I. Prigogine, *Science* **201,** 772 (1978).
232. P. A. Bertocci, *The empirical argument for God in late British thought* (Harvard University Press Cambridge, Mass. 1938).
233. A. S. Pringle-Pattison, *The idea of God in the light of recent philosophy* (Oxford University Press, Oxford, 1917), p. 111.
234. A. S. Pringle-Pattison, *Man's place in the cosmos*, 2nd edn (William Blackwood, London, 1902), p. 42.
235. Ref. 233, p. 330.
236. F. R. Tennant, *Philosophical theology*, Vol. II (Cambridge University Press, Cambridge, 1930), p. 79.
237. Ref. 236, p. 80.
238. Ref. 236, p. 81.
239. Ref. 236, p. 82.
240. Ref. 236, p. 83.
241. Ref. 236, p. 113.
242. Ref. 236, p. 104.
243. Ref. 236, p. 114.
244. Ref. 236, p. 117.
245. Ref. 236, p. 101.
246. J. Laird, *Theism and cosmology, Gifford lectures 1939* (Allen & Unwin, London, 1940).
247. A. B. Gibson, *Theism and empiricism* (Schocken Books, NY, 1970).
248. P. A. Bertocci, *Introduction to the philosophy of religion* (Prentice-Hall, NY, 1951).

249. C. E. Raven, *Natural and Christan theology, Gifford lectures 1952* (Cambridge University Press, Cambridge, 1953).
250. J. Ward, *Naturalism and agnosticism, Gifford lectures 1896–1898* (Adam & Charles Black, London, 1906), 3rd edn, Vol. 2, p. 254.
251. W. R. Matthews, *The purpose of God, Robertson lectures 1935* (Nisbet, London, 1935), p. 64.
252. F. J. Tipler, in *Essays in general relativity*, ed. F. J. Tipler (Academic Press, NY, 1980).
253. A. R. Peacocke, *Science and the Christian experiment* (Oxford University Press, Oxford, 1971).
254. A. R. Peacocke, *Creation and the world of science, Bampton lectures 1978* (Oxford University Press, Oxford, 1979).
255. E. L. Mascall, *Christian theology and natural science, Bampton lectures 1956* (Longmans Green, London, 1956).
256. Ref. 172, p. 402.
257. Ref. 172, p. 503.
258. F. J. Tipler, C. J. S. Clarke, and G. F. R. Ellis, in *General relativity and gravitation*, Vol. II, ed. A. Held (Plenum, NY, 1980).
259. R. Gott, *Nature* **295,** 304 (1982); F. Hoyle and J. V. Narlikar, *Proc. R. Soc.* A **290,** 162, 177 (1966).
260. J. B. Bury, *Idea of progress: an inquiry into its origins and growth* (Macmillan, London, 1921).
261. R. Nisbet, *History of the idea of progress* (Basic Books, NY, 1980). Strangely, this book has no bibliography. The bibliography was published as part of a separate article in *Literature of Liberty* **2,** 7 (1979).
262. L. Edelstein, *The idea of progress in classical antiquity* (Johns Hopkins University Press, Baltimore, 1967).
263. H. Spencer, *The Man versus the State*, ed. E. Mack (Liberty Press, Indianapolis, 1981).
264. H. Spencer, *The proper sphere of government*, repr. in ref. 263; original publication in 1843.
265. K. Marx and F. Engels, *The German ideology*, ed. C. J. Arthur (International Publishers, NY, 1970).
266. K. Marx, *The Gotha program* (New York Labor News Company, NY, 1935).
267. H. D. Dickinson, *Economics of socialism* (Oxford University Press, Oxford, 1939).
268. O. Lange and F. M. Taylor, *On the economic theory of Socialism*, ed. B. E. Lippincott (University of Minnesota Press, Minneapolis, 1948).
269. F. A. Hayek, 'Socialist calculation', in *Individualism and economic order* (University of Chicago Press, Chicago, 1948).
270. H. Spencer, 'Progress: its law and cause', in *Essays*, Vol. 1 (Appleton, NY, 1901).
271. H. Spencer, *First principles* (Appleton, NY, 1901). On p. 473 he shows that he appreciates the effect of gravitational instability in bringing about structure in the Universe, for 'any finite homogeneous aggregate must inevitably lose its homogeneity, through the unequal exposure of its parts to incident [gravitational] forces'.

272. H. Spencer, 'The nebular hypothesis', in *Essays*, Vol. 1 (Appleton, NY, 1901).
273. L. Ward, *Applied sociology: a treatise on the conscious improvement of society by society* (Ginn, NY, 1906).
274. L. T. Hobhouse, *Development and purpose: an essay towards a philosophy of evolution* (Macmillan, London, 1913).
275. K. Popper, *The open society and its enemies*, rev. edn (Princeton University Press, Princeton, 1950).
276. F. A. Hayek, *The sensory order* (University of Chicago Press, Chicago, 1952). Although this book was first published in 1952, it was written in the 1920's, before Turing's work on the Halting Problem. We are grateful to Professor Hayek for pointing out this reference to us.
277. F. A. Hayek, 'The use of knowledge in society', rep. in ref. 269.
278. K. Popper, *Br. J. Phil. Sci.* **1,** 117 (1950); and **1,** 173 (1950). This argument has various interesting consequences for the question of whether human action can be predicted and also known to be predicted, see D. M. Mackay, *Freedom of action in a mechanistic universe* (Cambridge University Press, Cambridge, 1967).
279. H. Bergson, *Creative evolution*, transl. A. Mitchell (Macmillan, London, 1964), p. 41.
280. Ref. 279, p. 42.
281. Ref. 279, p. 43.
282. Ref. 279, p. 46.
283. Ref. 279, p. 54.
284. Ref. 279, p. 255.
285. S. Alexander, *Space, time, and Deity: Gifford lectures at Glasgow, 1916–1918*, Vol. II (Macmillan, London, 1966).
286. J. Macquarrie, *Twentieth century religious thought* (Harper & Row, NY, 1963).
287. R. G. Collingwood, *The idea of Nature* (Oxford University Press, Oxford, 1945).
288. Ref. 285, p. 355.
289. Ref. 285, p. 347.
290. Ref. 285, p. 346.
291. Ref. 285, p. 353.
292. Ref. 285, p. 357.
293. A. N. Whitehead, *Process and reality: an essay in cosmology*, corrected edition, ed. D. R. Griffin and D. W. Sherburne (Free Press, NY, 1978).
294. Ref. 293, p. 89; see also p. 34.
295. Ref. 293, p. 91.
296. Ref. 293, p. 168.
297. Ref. 293, p. 47.
298. A. H. Guth, *Phys. Rev.* D **23,** 347 (1981).
299. J. D. Barrow and M. Turner, *Nature* **298,** 801 (1982).
300. A. Guth, private communication to FJT.
301. C. W. Misner, K. S. Thorne, and J. A. Wheeler, *Gravitation* (Freeman, San Francisco, 1973).

302. Ref. 287, p. 167.

303. C. Hartshorne, *A natural theology for our time* (Open Court, La Salle, 1967), 93.

304. The philosopher Milic Capek has pointed out in his book *Bergson and modern physics* (Reidel, Dordrecht, 1971), pp. 252–3, that Whitehead was inconsistent in his various writings on the question of whether a global time exists. In *Science and the modern world*, (p. 172), Whitehead argues that in special relativity there is no 'unique present instant'. But in his book *The concept of nature*, Whitehead distinguishes between what he terms 'the creative advance of nature', which seems to be something like a global time sequence, and the 'discordant time-systems' in special relativity. At bottom, Whitehead knew, like his follower Hartshorne, that a progressive cosmos required a globally defined time.

Bergson also realized this. He was very much worried about the lack of a global time in special relativity, as demonstrated by the Twin Paradox. Bergson wrote an entire book *Duration and simultaneity*, transl. L. Jacobson (Bobbs-Merill, Indianapolis, 1965), in which he tried to argue that *physically* a global time existed even in special relativity, and that in particular, the Twin Paradox could not occur. He was wrong. The Twin Paradox is a valid (and experimentally confirmed) prediction of both general and special relativity. We should mention that the analysis of the Twin Paradox by Capek is incorrect, due to a misunderstanding of what is meant by the terms 'special' and 'general' relativity. *Special* relativity is an analysis of the space-time (η, R^4), where η is the Minkowski metric and R^4 is the Euclidean four-manifold. *General* relativity is an analysis of the *general* space-time (g, M), where g is a general non-degenerate metric with signature -2, and M is a general 4-manifold. Clearly, general relativity reduces to special relativity when $\eta = g$ and $M = R^4$. But equally clearly, it is possible to analyse accelerated motion in special relativity, and to talk about accelerated reference frames in special relativity, just as it is possible to use non-linear coordinate systems in Euclidean space. Properly speaking, the realm of general relativity is those space-times in which the Riemann curvature tensor is not identically zero. The Twin Paradox can be completely analysed in a region of space where the curvature is essentially zero, and so it is a purely special relativity effect. See E. F. Taylor and J. A. Wheeler, *Space-time physics* (Freeman, San Francisco, 1966) for a very nice, clear discussion of the Twin Paradox.

In general relativistic cosmologies a global time exists, and all the times of *all* observers advance according to this global time. But the *rates* of advance depend on the individual observer, and this is the import of the Twin Paradox.

305. C. Hartshorne, *Anselm's discovery: a re-examination of the ontological proof for God's existence* (Open court, La Salle, 1965), p. 109.

306. J. E. Marsden and F. J. Tipler, *Phys. Rep.* **66**, 109 (1980).

307. A. S. Eddington, *Space, time, and gravitation* (Cambridge University Press, Cambridge, 1920), p. 163.

308. Ref. 303, p. 98.

309. Ref. 303, p. 96.

310. J. Needham, 'Cosmologist of the future: a review of *The phenomenon of man* by Teilhard de Chardin', *New Statesman* **88** (1959), 632.

311. M. Lukas and E. Lukas, *Teilhard* (Doubleday, NY, 1977).
312. P. Hefner, *The promise of Teilhard* (Lippincott, Philadelphia, 1970).
313. Professor Hefner recently (February, 1984) remarked to FJT that 'Teilhard was wrong in many respects, but his heart was in the right place'.
314. C. Raven, *Teilhard de Chardin: scientist and seer* (Harper & Row, NY, 1962).
315. P. B. Medawar, 'Critical review of *The phenomenon of man*', in *Mind* **70** (1961), 99–106.
316. See in particular the passages on pp. 43, 52, and 66 of ref. 10.
317. See, for example, his remarks to this effect on p. 149 of ref. 10.
318. This opinion is expounded on p. 57 (see especially the footnote) and on pp. 71 and 301 of ref. 10.
319. Ref. 10, pp. 63–4.
320. Ref. 10, p. 64.
321. Ref. 315, p. 103.
322. K. Popper, *The logic of scientific discovery* (Harper & Row, NY, 1959); rev. edn (Hutchinson, London, 1968).
323. C. Cuénot, *Teilhard de Chardin* (Helicon, Baltimore, 1965), p. 290.
324. Ref. 323, p. 352.
325. Teilhard de Chardin, *Études* **264** (1950), 403–4. This note has no by-line, but Cúenot (ref. 323, p. 453), lists Teilhard as the author.
326. Ref. 50, p. 120. See also Dobzhansky's articles on Teilhard in *Zygon* **3,** 242 (1968), and in *Beyond chance and necessity*, ed. J. Lewis (Humanities Press, Atlantic Highlands, NJ, 1974).
327. Ref. 10, pp. 108–9.
328. Ref. 10, p. 275.
329. Ref. 10, p. 286. There is some indication that Teilhard changed his view on extraterrestrial life in the 1950's. His biographer Claude Cuénot records him as saying: 'The more we expand the world and the potentialities of the biosphere, the more out of character and even unworthy of God it seems that all the energy of matter and its combinations should be dispersed over an immense universe for just one single living human kind'. (Ref. 323, p. 365.) However, as we discuss in the text, this quote is inconsistent with Teilhard's Ω-point theory as developed in the *Phenomenon of man*. Even more inconsistent is an essay on the subject of extraterrestrial life which Teilhard wrote in 1953 (first published in the collection of essays entitled *Christianity and evolution* (Harcourt Brace & Jovanovich, NY, 1971), p. 229). In this essay Teilhard argues that *orthogenesis* makes the evolution of intelligent life inevitable on numerous planets throughout the cosmos! Furthermore, in this essay Teilhard asserts that the noosphere on Earth is just one of many noospheres scattered throughout the universe; presumably this implies that the Ω-point on the Earth is not the ultimate goal of life, but rather a goal which will be achieved on many planets. Thus, in this essay Teilhard gives up the idea of universal evolution (because he believes the noospheres on the various planets cannot communicate, and hence cannot combine, and in any case Teilhard has identified the Ω-point achieved on Earth with Christ, who has no higher stage)! As we point out in the text, Teilhard's Ω-point theory cannot be extended consistently beyond the Earth.

330. Ref. 10, p. 276.
331. Ref. 4, p. 74.
332. Ref. 10, pp. 287–8.
333. Ref. 10, pp. 270–1.
334. Ref. 10, pp. 168–9.
335. Ref. 10, pp. 239–40.
336. Ref. 10, pp. 286–7, 307.
337. J. Hyppolite, letter of June 24, 1957 to Claude Cuénot. Quoted in ref. 323, pp. 254–5.
338. Ref. 10, p. 259.
339. G. Smith, ed. *Josiah Royce's seminar, 1913–1914: as recorded in the notebooks of Harry T. Costello* (Rutgers University Press, New Brunswick, 1963).
340. J. Parascandola, *J. Hist. Biol.* **4,** 64 (1971).
341. S. Wolfram, *Phys. Rev. Lett.* **54,** 735 (1975).
342. A. N. Whitehead, *The function of reason: Louis Clark Vanuxem lectures 1929* (Princeton University Press, Princeton 1929), pp. 22–3.
343. E. T. Whittaker, *The beginning and the end of the World* (Oxford University Press, Oxford, 1942), pp. 40–2.

4 The Rediscovery of the Anthropic Principle

I believe there are 15, 747, 724, 136, 275, 002, 577, 605, 653, 961, 181, 555, 468, 044, 717, 914, 527, 116, 709, 366, 231, 425, 076, 185, 631, 031, 296 protons in the Universe and the same number of electrons.

A. S. Eddington

4.1 The Lore of Large Numbers

Then feed on thoughts that voluntary move Harmonious numbers.

John Milton

The modern form of the Weak Anthropic Principle[1] arose from attempts to relate the existence of invariant aspects of the Universe's structure to those conditions necessary to generate 'observers'. Our existence imposes a stringent selection effect upon the type of Universe we could ever expect to observe and document. Many observations of the natural world, although remarkable *a priori*, can be seen in this light as inevitable consequences of our own existence.

Cosmological interest in such a perspective arose from attempts to explain the ubiquitous presence of large dimensionless ratios in combinations of micro and macrophysical parameters. Whereas most local dimensionless physical constants lie within an order of magnitude or so of unity, there exist a number of notorious and flagrant exceptions: the ratio of the electric and gravitational forces between a proton and electron is approximately 10^{40} whatever their separation; the number of nucleons in the Universe is $\sim 10^{80}$; the ratio of the action of the Universe to the quantum of action is $\sim 10^{120}$; and so forth. In this chapter we shall describe some of the background to these and other cosmological 'coincidences' and show how, in the period 1957–1961, they led to Dicke's proposal[1] of an anthropomorphic mode of explanation.

En route to this goal we shall describe a variety of numerical coincidences which have attracted the attention of physicists. We shall also give some historical examples to show how purely numerological relations, although originally viewed as coincidental, have occasionally stimulated the development of precise casual explanations for the interrelations they display. The above-mentioned 'large numbers' will be a recurrent theme in our discussion, and it is amusing to recall that such huge magnitudes first found their way into the pages of scientific papers as early as about 216 BC.

Archimedes wrote two papers[2] on the problems of arithmetic enumeration. No copies of the first survive but this work, entitled *Principles*, (ἀρχαι), was addressed to his colleague Zeuxippus and appears to have[3] proposed a system of symbolic representation for integers of arbitrarily large magnitude. The famous follow-up to this work was addressed to Gelon, then King of Syracuse. It bears the title *The Sand Reckoner* (Ψαμμιτης) and besides meeting some objections brought against the scheme outlined in his earlier paper, Archimedes devoted it to enumerating the number of sand grains in the Universe as a worked example to display the economy of his new notation. He argues that previous mystical claims to the effect that the number of grains of sand on the Sicilian sea-shore are beyond the power of man to number are completely groundless. Moreover, his system of accounting could not only perform this enumeration quite compactly but was capable of enumerating the number of sand grains in the entire Universe!

Archimedes' Universe consisted of a sphere enclosing the Sun and the fixed stars with its centre at the Earth. Using a series of geometrical arguments he is able to calculate the diameter of this celestial sphere in terms of the distance from the Earth to the Sun and the terrestrial and solar diameters. The latter was estimated experimentally by the parallax method of Aristarchus. Following these steps Archimedes was led to conclude that the Universe is a sphere of diameter 10^{14} stadia ($\sim 10^{18}$ cm) and contains 10^{63} sand grains; the average sand grain he assumes to extend about $\sim 2.5 \times 10^{-6}$ of a finger's breath. Assuming Archimedes' finger is about one centimetre wide his calculation implies[4] that the Universe contains $\sim 10^{80}$ nucleons!

If we were to make Archimedes' (false!) assumption that the average density of the solar system is that of a sand grain ~ 1 gm cm^{-3} then the number of nucleons in a sphere of radius $\sim 10^{14}$ cm enclosing the outer planetary orbits and centred on the Sun would be $\sim 10^{66}$, quite close to his estimate of 10^{63}.

4.2 From Coincidence to Consequence

> . . . And thus they spend
> The little wick of life's poor shallow lamp
> In playing tricks with nature giving laws
> To distant worlds, and trifling in their
> own.
>
> W. Cowper

Numerological and mystic speculation was especially rife amongst German Romantics and *Naturphilosphen* during the nineteenth century[5] and grew out of ancient teleological speculations concerning the harmonious distribution of the heavenly bodies.[6] Such speculation was by no means confined to the celestial motions; in 1818 the Kantian mineralogist Christian Weiss[7] even argued for a link between aspects of rhombic-dodecahedral crystal structure and the musical scale of tones! However, such flights of imaginative fancy generally had little impact upon the work of serious scientists; with one notable exception.

In 1766 Johann Daniel Titius von Wittenberg was preparing a German translation[8] of Charles Bonnet's *Contemplation de la Nature.*[9] To the section on planetary motions he added a now famous footnote pointing out that the radii of all the planetary orbits can be generated by the following simple algorithm, (where r is measured in astronomical units ($1\text{ AU} = 1.496 \times 10^{13}$ cm):

$$r_n = 0.4 + 0.3 \times 2^n\,;\; n = 0,1,2,\dots\,. \tag{4.1}$$

This formula provided a striking approximation for the distance from the Sun to the six then-known planets: Mercury, Venus, Earth, Mars, Jupiter and Saturn. Their distances from the Sun at the time of the Law's inception are indicated together with Titius' predictions as follows:

Planet	Measured r_n(AU)	n	'Predicted' r_n
Mercury	0.39	—	0.4
Venus	0.72	0	0.7
Earth	1.00	1	1.0
Mars	1.52	2	1.6
Jupiter	5.20	4	5.2
Saturn	9.55	5	10.0

In 1772 Johann Bode[12] came across Titius' footnote and inserted it into the new edition of his own astronomy book, but without a reference to Titius and this led to Bode becoming erroneously associated with its discovery.

Titius' purely numerical relation initially had two great successes. First, it successfully predicted the discovery of the next planetary body, Uranus, at a distance $r_6 \sim 19.2$ AU from the Sun. This planet was in fact named by Bode following its discovery in 1781 by Herschel.[13] Later, an extensive search revealed that the 'gap' in the Titius sequence at $r_3 \sim 2.8$ AU was filled by the asteroid belt and since it was conceivable that the bodies filling this band arose from a past planetary disintegration this was counted as another significant success for the formula.

However, if we calculate $r_7 \sim 38.8$ AU and $r_8 \sim 77.2$ AU there is a dramatic disagreement with the observed orbits of Neptune (30.1 AU) and Pluto (39.5 AU). In the final analysis if we account for the original input in (4.1) of three parameters (0.4, 0.3, 2), which ensures at least three predictions must accord with observation, we are left with five successes and two outright failures. Whether this means that the Titius law is physically significant or compatible with any reasonably spaced sequence of purely random numbers remains a matter of some debate amongst planetary astronomers to this day.[14]

An example of numerology with a more fruitful outcome is provided by the Balmer formula for the spectral frequency of hydrogen. By the 1880's the hydrogen spectrum was seen to possess an obvious pattern and this tempted various physicists to suggest an empirical formula which would summarize its structural features. In 1885 the Swiss Johann Balmer suggested the following numerical law[15]

$$\lambda = \lambda_0 \frac{m^2}{m^2 - 4}, \qquad m = 3, 4, 5, 6, \ldots, \qquad \lambda_0 = \text{constant} \tag{4.2}$$

for the wavelengths, λ, of H_α (6563 Å), H_β (4861 Å), H_γ (4340 Å) and H_δ (4102 Å) spectra. This was generalized for all alkali spectra by Rydberg[16] in 1890 and various similar algorithms were found to fit spectral series in other wavebands by Lyman, Paschen, Brackett and Pfund. These purely numerical formulae were later found to have a beautiful and precise explanation in Bohr's quantum theory of hydrogen atom. According to Rosenfeld[17], Bohr was significantly guided by these empirical formulae. He records Bohr remarking to him about the problem of atomic structure that 'as soon as I saw Balmer's formula, the whole thing was immediately clear to me' and recalls how in 1911–12, according to Bohr's recollection, he was asked by the young Danish physicist Hans Marius Hansen how atomic theory could explain the spectra. In Bohr's view, the experimental spectra were too complicated for a simple explanation to exist but Hansen disputed this and simply pointed to Balmer's formula.

Another closely related numerological debate began at the turn of the century when, in May 1899, Planck first stated a value for the fundamental constant that now bears his name ($h = 6.62 \times 10^{-27}$ erg sec). Six years later he wrote in a letter to Paul Ehrenfest claiming that[18]

> it seems to me not completely impossible . . . h has the same order of magnitude as e^2/c.'

and regarded it as plausible that there might exist some link between electrical processes and the new quantum of action. In 1909 Einstein took this suggestion a little further; he realized that e^2/c possessed the dimension of an action and was, to within a reasonable numerical factor, of order Planck's new constant h. He remarked that[19]

It seems to me that we can conclude from $h = e^2/c$ that the same modification of theory that contains the elementary quantum e as a consequence, will also contain as a consequence the quantum structure of radiation.

Soon afterwards these words were read by Haas who was motivated to equate quantities with the dimensions of potential and kinetic energy in Thomson's model of the atom, obtaining $e^2/a \sim h\nu$, where a is the atomic radius and ν some characteristic frequency. With one more dimensional estimate he gave Planck's constant in terms of the electron mass, m_e, the atomic radius a and electric charge e, as[20]

$$h = 2\pi e(am_e)^{1/2} \tag{4.3}$$

This, in February 1910, is actually Bohr's formula for the ground-state radius of the hydrogen atom. Few took his result seriously although Lorentz[17] did refer to it as a 'daring hypothesis'. Bohr emphasized on various occasions that he had no knowledge of Haas' early work but he was clearly influenced indirectly by Sommerfeld's knowledge of it. Sommerfeld[21] was the first to spell-out clearly the physical significance of the dimensionless parameter e^2/hc. Again, we see an interesting chain of events sparked by purely dimensional and numerological speculation but culminating in rigorous quantitative developments.

In 1856 Weber and Kohlrausch[22] made the first experimental determination of the ratio between the units of electric and magnetic charge. They obtained the value 3.107×10^{10} cm s^{-1} and the proximity of this number to the measured value for the velocity of light was noticed by Kirchhoff in 1857. Maxwell and Riemann were also singularly impressed by this numerical 'coincidence' and the following year Riemann presented a paper to the Göttingen Academy in which he formally deduced their equality and so began the development of a unified theory of electricity and magnetism.

As a final, and more recent, example of such numerological serendipity it is interesting to recall the development of black hole thermodynamics. It had been known for some time prior to 1974[23] that the theoretical relations governing mechanical interactions between black holes bore an uncanny formal resemblance to the laws of thermodynamics. In fact, if one associated an entropy with the area of the black hole event horizon and a temperature with its surface gravity then the zeroth, first and second laws of thermodynamics were simply known properties of black hole mechanics in disguise. For some while these analogies were treated as curiosities devoid of any real physical content because no particles could emerge from a classical black hole to endow it with the thermal properties of an object at non-zero temperature. Eventually, intrigued by these analogical concidences, Hawking[24] made a monumental discovery, namely, that black holes are black bodies. They radiate particles with thermal characteristics. Their surface area and gravity *do* precisely determine the entropy and temperature of the radiated particles and they obey the laws of equilibrium thermodynamics. This realization has prompted a tremendous concentration of effort by theoretical physicists to investigate the unsuspected interconnections between quantum mechanics, general relativity and thermodynamics. It could be said that this fruit has grown principally from the roots of coincidence.

4.3 'Fundamentalism'

> He thought he saw electrons swift
> Their charge and mass combine.
> He looked again and saw it was
> The cosmic sounding line.
> The population then said he,
> must be 10^{79}.
>
> H. Dingle

In modern times the first scientist to notice the presence of large dimensionless numbers in Nature appears to have been the mathematical physicist Hermann Weyl. As an aside to his early discussion of general relativity, published in 1919, he remarks on the huge difference between the electric and gravitational radii of the electron:[25]

> It is a fact that pure numbers appear with the electron, the magnitude of which is totally different from 1; so for example, the ratio of the electron radius to the gravitational radius of its mass, which is of order 10^{40}; the ratio of the electron radius to the world radius may be of similar proportions.[25]

The idea of explaining such occurrences, and indeed exploiting them to pursue a programme which had as its goal a calculation of all the fundamental physical constants of Nature, was suggested by Arthur Eddington in 1923.[26] The quest for his '*Fundamental Theory*' of the physical world in which the basic interaction strengths and elementary particle masses would be predicted entirely combinatorically by simple counting processes was vigorously pursued until his death in 1944. Although still fragmentary even then, to our modern eyes this work appears mysterious, if not slightly eccentric. Yet despite its peculiar nature it had some interesting consequences and served to isolate many problems which still cry out for an explanation.

Whittaker has described the guiding principle of Eddington's approach to the fundamental constants of Nature in the following words[27]:

> All the quantitative propositions of physics, that is, the exact values of the pure numbers that are constants of science, may be deduced by logical reasoning from

qualitative assertions without making any use of quantitative data derived from observation.

This is truly a 'philosopher's dream' and Eddington, in the 1923 edition of his book *The Mathematical Theory of Relativity*,[26] began to ponder the disconcerting presence of large dimensionless numbers in the local and global model of the universe he had done so much to construct:

among the constants of Nature there is one which is a very large pure number; this is typified by the ratio of the radius of an electron to its gravitational mass $= 3.10^{42}$. It is difficult to account for the occurrence of a pure number (of order greatly different from unity) in the scheme of things; but this difficulty would be removed if we could connect it with the number of particles in the world—a number presumably decided by pure accident'.

Through this speculation, and the ways in which it was developed in his later work, Eddington was the first to suggest that the total number of particles in the Universe, N, might play a part in determining other fundamental constants of Nature. He evaluated this number to high precision and it is now often termed the 'Eddington number'[28]

$$N \equiv 2.136 \times 2^{256} \sim 10^{79} \tag{4.4}$$

One of the attractions of this quantity for Eddington was the necessity that its value be *integral*. This meant that it could, in principle, be calculated exactly.

In these early days when the weak and strong interactions were still unknown Eddington set about constructing a model of the Universe from the following collection of dimensional physical constants: G, c, m_e, m_N, e, h which denote the gravitation constant, the velocity of light, the electron and proton masses, the electron charge and Planck's constant respectively. From them he derived three independent dimensionless ratios

$$m_N/m_e \sim 1840; \qquad \hbar c/e^2 \sim 137; \qquad e^2/Gm_Nm_e \sim 10^{39} \tag{4.5}$$

To these he added two cosmological parameters: the Eddington number, $N \sim 10^{79}$, and Einstein's cosmological constant, Λ. From the latter he constructed a further dimensionless ratio

$$\frac{c}{H_0}\left(\frac{m_Nm_e}{\Lambda}\right)^{1/2} \sim 10^{39} \tag{4.6}$$

(where the numerical value is that used by Eddington, who believed the Hubble constant H_0 to be $\sim 500 \text{ kms}^{-1} \text{ Mpc}^{-1}$). The last expression gives the ratio of the radius of curvature of the de Sitter space-time to the geometric mean of the electron and proton Compton wavelengths. Through the introduction of these two cosmological parameters he could begin to develop a set of Machian interconnections between the micro and macro-physical worlds by exploiting the dual numerical coincidences between (4.5), (4.6) and $N^{1/2}$.

In isolating these dimensionless ratios Eddington highlighted the fact that their values are not uniformly distributed over the entire range of real numbers but reside, within a factor of a hundred or so, around 1, 10^{40} and 10^{80}. His subsequent work sought to ascertain whether or not these quantities were reducible to simpler forms or calculable from first principles. If these numbers are necessarily fixed by the internal consistency of Nature they could, in principle, be determined by theory. However, if they are completely arbitrary then only experiment can reveal their values to us.

A typical example of Eddington's methodology, which displays the manner in which he sought to employ the number of particles in the Universe as a mediator between gravitational and atomic phenomena, is given by his attempt to calculate a fundamental mass. His argument went like this:

Since most of the particles in the Universe interact very infrequently they may be represented by plane waves with a uniform probability distribution. If their positions are random, each with positional uncertainty R then, by the law of large numbers, the centroid of this distribution also possesses a positional uncertainty, Δx, where

$$\Delta x \sim R/N^{1/2}.$$

If we employ the Uncertainty Principle of Heisenberg, a mass scale m_0 can be associated with this uncertainty, $m_0 \sim hN^{1/2}/Rc$. Eddington claimed that this mass uncertainty arises entirely as a consequence of the finite space in which the N particles reside.

Now if R is the gravitational mass of the Universe so that we have

$$R \sim GM/c^2 \tag{4.7}$$

where $M \sim Nm_N$ is the mass of the Universe, and if the limit of precision measurement of each particle is taken to be the classical electron radius, r_e, where

$$r_e \sim \frac{e^2}{m_e c^2}$$

then we have the prediction that,

$$\frac{e^2}{Gm_e m_N} \sim N^{1/2} \tag{4.8}$$

and, (keeping account of all the numerical factors), Eddington calculated the associated 'fundamental' mass m_0 to lie close to the proton mass

$$m_0 \simeq \frac{\hbar N^{1/2}}{Rc} \sim 3 \times 10^{-25} \text{ gm} \tag{4.9}$$

The conclusion drawn from relations like (4.8) and (4.9) was that the 'large' numbers $\sim 10^{40}$, and powers thereof, are of this huge order of magnitude because they are determined by N. Dimensionless quantities with values neighbouring unity are simply those whose values are not explicitly conditioned by N.

Exact versions of the formula (4.9) initiated a later numerological excursion culminating in a 'determination' of the electron and proton masses. These were determined as the roots of a certain quadratic equation[28]

$$10m^2 - 136mm_0 + \left(\frac{137}{136}\right)^{5/6} m_0^2 = 0. \tag{4.10}$$

This gave the two solutions for m as:

$$\begin{aligned} m_e &= 9.10924 \times 10^{-28} \text{ gm} \\ m_N &= 1.67227 \times 10^{-24} \text{ gm} \end{aligned} \tag{4.11}$$

Another version of this calculation employed the roots of

$$10m^2 - 136m + 1 = 0 \tag{4.12}$$

which lie in the ratio 1847.6.

Other arguments of this ilk were arranged to display the fine structure constant as the reciprocal of the number of terms in a symmetric 16-dimensional tensor[30]

$$\alpha^{-1} = \frac{16^2 - 16}{2} + 16 = 136 \tag{4.13}$$

Later, unity was added to this value to align it better with the experimental value 137.036. Such *post facto* changes in some of his combinatorical predictions damaged the credibility of much of this work. Despite a sceptical reaction from other scientists Eddington worked very seriously throughout a long period of his life on arguments of this nature and generated a vast array of results that still lack a coherent basis.[27]

A fair idea of how some notable physicists viewed this work at the time can be obtained from two 'spoofs' which were specifically designed to parody the Eddington methodology. The following article entitled 'Concerning the quantum theory of absolute zero' was written by Beck, Bethe and Riezler[31] and appeared in the 9 January issue of *Naturwis-*

senschaften in 1931:

Let us consider a hexagonal crystal lattice. The absolute zero of this lattice is characterized by the fact that all degrees of freedom of the system are frozen out, i.e., all inner movements of the lattice have ceased, with the exception, of course, of the motion of an electron in its Bohr orbit. According to Eddington every electron has $1/\alpha$ degrees of freedom where α is the fine structure constant of Sommerfeld. Besides electrons our crystal contains only protons and for these the number of degrees of freedom is obviously the same since, according to Dirac, a proton is considered to be a hole in a gas of electrons. Therefore to get to the absolute zero we have to remove from the substance per neutron (= 1 electron plus 1 proton; our crystal is to carry no net charge) $2/\alpha - 1$ degrees of freedom since one degree of freedom has to remain for the orbital motion. We thus obtain for the zero point temperature $T_0 = -(2/\alpha - 1)$ degrees. Putting $T_0 = -273°$, we obtain for $1/\alpha$ the value 137, in perfect agreement within the limits of accuracy with the value obtained by totally independent methods. It can be seen very easily that our result is independent of the particular crystal lattice chosen.

In his 1944 lectures on '*Experiment and Theory in Physics*' Max Born writes[32] of Eddington's numerology,

Eddington connects the dimensionless physical constants with the number n of the dimensions of his E-spaces, and his theory leads to the function $f(n) = n^2(n^2+1)/2$ which, for consecutive even numbers $n = 2, 4, 6, \ldots$ assumes the values 10,136,666 .... Apocalyptic numbers, indeed. It has been proposed that certain well-known lines of St. John's Revelation ought to be written in this way: 'And I saw a beast coming up out of the sea having $f(2)$ horns ... and his number is $f(6)$ ...' but whether the figure x in
'... and there was given to him authority to coninue x months ...' is to be interpreted as $1 \times f(3) - 3 \times f(1)$ or as $\frac{1}{3}[f(4) - f(2)]$ can be disputed ...'

Although Eddington's '*Fundamental Theory*' is very easy to criticize, it is still interesting for the vision of an underlying unity in Nature which it displays. A vision that has since materialized in an entirely different form. Through his work in this area Eddington directed the attention of many other workers to the ubiquity of large dimensionless numbers. This, in turn, stimulated other approaches to cosmological theory that have borne more fruit than their progenitor.

Of the other early contributors to this style of working the most prolific appears to have been Haas who, during the period 1932–8 devoted a whole series of short papers and a large portion of a book to these matters.[33] For example, in 1935 he derived a value for the gravitational mass of the Universe and then, by a similar argument to that of Eddington given above, gives the uncertainty in the Universe's centre of mass as $R/N^{1/2}$. This yields a relation between N and the gravitational coupling similar to (4.8)

$$N^{1/2} = \frac{hc}{Gm_N^2} \tag{4.14}$$

Another early example of a now familiar type of cosmological coincidence was given by Stewart[34] in 1931. Out of the constants e, h, c, G, m_e and m_N he formed the three dimensionless quantities hc/e^2, e^2/Gm_e^2 and m_N/m_e. By trial and error he found a combination roughly equal to the present Hubble radius cH_0^{-1},

$$cH_0^{-1} \sim \left(\frac{e^2}{m_e c^2}\right)\left(\frac{e^2}{Gm_e^2}\right)\left(\frac{e^2}{\hbar c}\right) \sim \frac{e^6}{G\hbar m_e^3 c^3} \tag{4.15}$$

Stewart suggests that this 'formula is simpler than would be expected if it is assumed to represent a relationship due merely to chance'.

More recently Weinberg[35] has pointed out that one can construct a mass close to the pion mass ($m_\pi \sim 140\ \text{MeV}/c^2$) out of $\hbar$, c, G and H_0

$$m_\pi \sim \left(\frac{h^2 H_0}{Gc}\right)^{1/3} \tag{4.16}$$

If we rewrite it in the form

$$cH_0^{-1} \sim \left(\frac{e^2}{m_\pi c^2}\right)\left(\frac{e^2}{Gm_\pi^2}\right)\left(\frac{hc}{e^2}\right)^2 \tag{4.17}$$

we can see its resemblance to the Stewart coincidence and this arises because of the additional numerical coincidence, $e^2/\hbar c \sim (m_e/m_\pi)$. Clearly, one can systematize such coincidences through dimensional analysis: any mass formed from the parameters G, h, c, and H_0 depends on just one free index, λ, as follows:

$$m(\lambda) \propto \left(\frac{Gh}{c^5}\right)^\lambda \frac{hH_0^{1+2\lambda}}{c^2} \tag{4.18}$$

and Weinberg's coincidence is $m(-\frac{1}{3})$. These coincidences have reappeared in some later work that has many similarities with Eddington's use of (4.9). A long series of papers[36] by the Japanese physicists Hayakawa, Tanaka, and Hokkyo have attempted to explain a relation equivalent to (4.6) which has the form

$$h \sim mc^2 t_0 N^{-1/2} \tag{4.19}$$

where $t_0 \sim H_0^{-1}$ is the age of the Universe.

If there exists a dispersion Δm in the mass of elementary particles in the Universe then for a Gaussian distribution they expect its scatter to be of the form

$$\Delta m/m \sim N^{-1/2} \tag{4.20}$$

If the Uncertainty Principle is the origin of this dispersion then

$$\Delta mc^2 \sim h/t_0 \tag{4.21}$$

and (4.20) and (4.21) yield (4.19).

If one combines (4.20) and (4.21) with the relativistic relations $R \sim ct_0 \sim GM/c^2 \sim GmN/c^2$ we obtain the Weinberg coincidence with $m \sim m_\pi$ and

$$m \sim \left(\frac{hc}{G}\right)^{1/2} N^{-1/4} \tag{4.22}$$

Edward Teller[37] appears to have been the first to speculate that there may exist a logarithmic relation between the fine structure constant and the parameter $Gm_N^2/hc \sim 10^{-39}$ of the form

$$\alpha \sim \ln\left(\frac{Gm_N^2}{hc}\right) \tag{4.23}$$

(in fact $\alpha^{-1} = \ln(3.17 \times 10^{60})$ and the formula is too insensitive to be of very much use in predicting exact relations).

Various authors have attempted to place such a relationship on a more formal footing. Salam *et al.*[38] tried to remove the ultraviolet divergence in the electron self-energy by the inclusion of a gravitational self-energy term E_s. This yields

$$E_s \sim m_e c^2 \left(\frac{e^2}{hc}\right) \ln N \tag{4.24}$$

Peebles and Dicke[40] and Landau[39] have derived relations of the form (4.23) by attempting to take into account renormalization terms in the calculation of α.

There exists another whole class of purely numerical concidences whose significance is even harder to assess than those sketched above. Some of the most striking such coincidences[41] are the proximity of m_N/m_e ($= 1836.1515$) to $6\pi^5$ ($= 1836.118$); the ratios of the proton, Λ, Σ and Ξ masses[42] to a regular progression,

$$m_N : m_\Lambda : m_\Sigma : m_\Xi = 1 : 2^{1/4} : 2^{1/3} : 2^{1/2}, \tag{4.25}$$

the mass-splitting coincidence[43] involving the neutron mass, m_n,

$$\frac{m_n - m_N}{m_\Lambda - m_N} \simeq \alpha \tag{4.26}$$

and the ratio of the new J/ψ and ψ' particle masses: $J/\Psi'(3684)/m_J/\Psi(3098) = 1.1891542$, which is roughly $2^{1/4} = 1.1892071$. MacGregor's correlation between powers of α and the life-times of metastable states is another curious trend:[44] many other 'coincidences' of dubious significance undoubtedly exist.[45]

Peres[104] has suggested an instructive mathematical approach to evaluating the real significance of many of these numerical formulae. For example, if we take the numerical coincidence 'calculated' by Wyler[105] for the fine structure constant

$$\alpha^{-1} = 2^{19/4}3^{-7/4}5^{1/4}\pi^{11/4} = 137.036082 \tag{4.27}$$

one might ask a more general question. Given the numbers 2, 3, 5 and π how well can we approximate α by juggling with powers of these four numbers?

Quantitatively we look for integers a, b, c and d so that the relation

$$(1-\varepsilon)\alpha^{-1} < (2^a 3^b 5^c \pi^d)^{1/4} < (1+\varepsilon)\alpha^{-1} \tag{4.28}$$

can be satisfied for very small ε, (e.g., pick $\varepsilon = 1.5 \times 10^{-6}$). Then one is confronted with examining a three-dimensional surface $a \log 2 + b \log 3 + c \log 5 + d \log \pi$ in the four-dimensional lattice space spanned by the integers a, b, c and d. The distance between the two limiting surfaces is calculated to be

$$8\varepsilon[(\log 2)^2 + (\log 3)^2 + (\log 5)^2 + (\log \pi)^2]^{-1/2} = 5.4 \times 10^{-6} \tag{4.29}$$

So, on average, within any three-dimensional area of size 1.85×10^5 one should find one lattice point in the slab (4.29). This corresponds to searching the interior of a sphere of radius 35 and Peres claims that (at the given level of 'surprise' of $\varepsilon = 1.5 \times 10^{-6}$) one would only be surprised to find (4.28) satisfied if the solution set $\{a, b, c, d\}$ had a distance from the origin much smaller than 35. In Wyler's example it is only 23. Such a sphere is large enough to contain a lattice point (solution to (4.28)) with good probability and so (4.27) is likely a real 'numerical' coincidence.

Most of the early work of Eddington and others on the large number coincidences has been largely forgotten. It has little point of contact with ideas in modern physics and is now regarded as a mere curiosity in the history of ideas. Yet in 1937 Paul Dirac suggested an entirely different resolution of the large numbers dilemma which, because of its novelty and far-reaching experimental consequences, has remained an idea of recurrent fascination and fundamental significance.

4.4 Dirac's Hypothesis

> You and I are exceptions to be laws of
> Nature; you have risen by your gravity,
> and I have sunk by my levity.
>
> Sydney Smith

Dirac's explanation for the prevalence[111] of the large numbers 10^{40} and 10^{80} amongst the dimensionless ratios involving atomic and cosmological quantities rests upon a radical assumption.[46] Rather than recourse to the mysterious combinatorical juggling of Eddington, Dirac chose to abandon one of the traditional constants of the physical world. He felt this step to be justified because of the huge gulf between the 'large numbers' and the more familiar second set of physical constants like m_N/m_e and $e^2/\hbar c$, which lie within a few orders of magnitude of unity. This dissimilarity suggested that some entirely different mode of explanation might be appropriate for each of these sets of constants.

Consider the following typical 'large numbers':

$$N_1 = \frac{t_0}{e^2/m_e c^3} \sim 6 \times 10^{39} = \frac{\text{age of Universe}}{\text{atomic light-crossing time}} \quad (4.30)$$

$$N_2 = \frac{e^2}{Gm_N m_e} \sim 2.3 \times 10^{39}$$

$$= \frac{\text{electric force between proton and electron}}{\text{gravitational force between proton and electron}} \quad (4.31)$$

The similarity between the magnitude of these superficially quite unrelated quantities suggested to Dirac that they might be *equal* (up to trivial numerical factors of order unity) due to some unfound law of Nature. To place this on a more formal basis he proposed the '*Large Numbers Hypothesis*' (LNH).[46]

Any two of the very large dimensionless numbers occurring in Nature are connected by a simple mathematical relation, in which the coefficients are of the order of magnitude unity.

Now, because Dirac chose to include a *time-dependent* factor—the Hubble age t_0, amongst his combinations of fundamental parameters, this simple hypothesis had a dramatic consequence: any large number $\sim 10^{40}$ equated with N_1 must *also* reflect this time variation.

The pay-off from this idea is that the time variation explains the enormity of the numbers: since all numbers of order $(10^{39})^n$ must now possess a time variation $\propto t^n$, they are large simply because the Universe is old.

There are now several routes along which to proceed. Incorporating the required time-dependence of N_2 into e^2, m_N or m_e would have overt and undesirable consequences for well-tried aspects of local quantum physics and so Dirac chose to confine the time variation within Newton's gravitational 'constant' G. For consistency with the LNH we see that gravity must weaken with the passage of cosmic time:

$$G \propto t^{-1} \quad (4.32)$$

Before following this road any further it is worth stressing that in this argument the variation of G (or any other 'constant') with time is not a consequence of the LNH *per se*. It has arisen because of a particular, subjective choice in the ranks of the large numbers. If one were to assume the Universe closed and finite in space-time then the proper time, t_{max}, taken by the Universe to expand to maximum volume is a fundamental cosmic time independent of the epoch at which we observe the Universe and list our large numbers. In our Universe, observation suggests t_{max} lies within an order of magnitude or so of the present time, t_0, and so if t_{max} replaces t_0 in the combination N_1 then the quantitative nature of the large number coincidence $N_1 \sim N_2$ remains. The qualitative change could not be greater: now the quantity $t_{max}m_ec^3/e^2$ possesses *no intrinsic time-variation* and so in conjunction with the LNH it can precipitate no time variation in other sets of traditional constants like N_2. In this form the LNH merely postulates exact equivalence between, otherwise causally unrelated, collections of natural constants. The conclusion that constants must vary in time can be spirited away if we believe the Universe to be closed (bounded in space and time).[47] A formulation along these lines appears implicit in a paper by Haas[48] published in 1938 sandwiched in time between the two initial contributions by Dirac. Instead of having Dirac's coincidences $N_1 \sim N_2 \sim N^{1/2}$ we have replaced N_1 by $N_1' = G(Nm_N)m_e/e^2 \sim 10^{40}$. Rather than three independent large numbers N_1, N_2 and N we now have only two because $N_1'N_2 = N$.

Other criticisms of Dirac's approach could be imagined: in the real world the Hubble age is a *local* construction. It changes from place to place because of variations in the density and dynamics or because of non-simultaneity in the big bang itself. If the age of the Universe is a spatial variable then the LNH implies that this spatial variation should be carried by the constants in N_2 just as surely as the temporal variation. To overcome this difficulty one would have to find some spatially-averaged Hubble age and employ that in the LNH as the fundamental cosmic time.

If spatial variation is introduced the possibility of an observational test of the hypothesis is considerably occluded. All our good tests of gravitation theories focus upon the behaviour of particular systems, for example the binary pulsar dynamics, and it is not clear how one would extricate the time and space variations in any particular case in order to test the theory against experiment. In 1963 when several second generation theories incorporating varying G were popular and viable theories of gravity, a criticism of this sort was put very strongly by Zeldovich[49]

> the local character of the general theory of relativity is not in agreement with the attempts of some authors to introduce an effect of the world as a whole on the phenomena occurring at a given point, and on the physical constants which appear in the laws of nature. From such an incorrect point of view one would have to expect ... the physical constants would change with time .... If we start from the Friedman model of the world, that state of the world can be characterized by the mean radius of curvature of space. The curvature of space is a local concept. One now assumes in the framework of local theory that a length contracted from

physical constants is proportional to the radius of curvature of space. Since in the Friedman world the radius changes in the course of time, the conclusion is drawn that the physical constants also change in the course of time. This pseudological view, however, cannot withstand criticism: the Friedman solution has a constant curvature of space only when one makes the approximation of a strictly uniform distribution of the matter density! ... a dependence of the constants on the local value of the curvature would lead to great differences in the constants at the earth's surface and near the sun, and so on, and hence is in complete contradiction with experience.

The novel course taken by Dirac leads to many unusual and testable predictions. If the Universe were finite then, because the number of particles contained within it is the square of a large number, this number must increase with time $N \propto t^2$. To avoid a violation of energy conservation Dirac concluded from this that the Universe must be infinite so N is not defined. Similar reasoning led to the conclusion that the cosmological constant, Λ, must vanish. Were this not the case, Eddington's large number involving Λ given in (4.6) would have to vary with epoch.

The earliest published reaction to Dirac's suggestion was that of Chandrasekhar[50] who pointed out that the LNH had a variety of consequences for the evolution of 'local' structures like stars and galaxies, whose sizes are governed by other large dimensionless numbers. He showed that if we form a set of masses out of the combination m_N, G, h and c then we can build a one-parameter family of masses:

$$m(\xi) = \left(\frac{hc}{G}\right)^{\xi} m_N^{1-2\xi} \tag{4.33}$$

Ranging through the values of ξ, members of this family are seen to lie remarkably close to the masses we observe in large aggregations of luminous material in the Universe. For instance, the Eddington number, N, is just $m(2)/m_N$ and,

$$m(1.5) = \left(\frac{hc}{G}\right)^{3/2} m_N^{-2} \sim 6 \times 10^{34}\ \text{gm} \sim M\ (\text{star}) \tag{4.34}$$

$$m(1.75) = \left(\frac{hc}{G}\right)^{7/4} m_N^{-5/2} \sim 1.7 \times 10^{11} M_\odot \sim M\ (\text{galaxy}) \tag{4.35}$$

$$m(2) = \left(\frac{hc}{G}\right)^{2} m_N^{-3} \sim 10^{21} M_\odot \sim M\ (\text{visible universe}) \tag{4.36}$$

These relations imply that the LNH should predict an increase in the 'number of particles in the galaxy' as $t^{3/2}$. Consequences of this sort were also outlined by Kothari[51] and discussed by Zwicky[52] who argued that these variations might alter the apparent brightness of stars in a systematic fashion that could be observationally checked.

Pascual Jordan[53] was another notable physicist attracted by the growing interest in a possible connection between large numbers and the time evolution of gravity. Like Chandrasekhar and Kothari, he noticed that a typical stellar mass is roughly $\sim 10^{60} m_N$ and so, according to Dirac's reasoning, should increase with time, $M_\odot \sim t^{3/2}$. Using (4.32) this indicated a relation of the form $M_\odot \propto G^{-3/2}$ would be anticipated to characterize the stellar mass scale. Since earlier theoretical work had provided good reasons for such a dependence of $M_\odot$ on G, Jordan interpreted this result as a confirmation of the idea of varying constants and its extension to time-varying stellar sizes; (there exists a straightforward explanation for the $M_\odot \propto G^{-3/2}$ dependence—see ref. 93 and Chapter 5, equation (5.121)).

In order to incorporate changing stellar masses in a consistent fashion, Jordan introduced the concept of continuous creation of material and even went so far as to argue that this creation will sometimes manifest itself by the spontaneous creation of entire stars *ex nihilo*. The rate of stellar creation per galaxy proceeds as $t^{1/4}$ in Dirac's scheme and Jordan wished to interpret supernovae as evidence for this style of stellar genesis. Unfortunately, such a connection required a supernovae rate ~ 1 supernova per galaxy per year—hundreds of times larger than the observed frequency.

These were the first 'continuous creation' theories of the Universe, although they have little in common with the spontaneous matter creation introduced by later proponents of the steady-state cosmology.[54] The latter philosophy introduced matter creation to *reduce* the number of degrees of freedom for evolutionary change in the Universe on a large scale, whereas Jordan admitted it as an *additional* degree of freedom.

Jordan made important contributions to the 'varying constants' debate in two areas: he showed how theories of gravity incorporating a gravitational coupling that is not constant in time can be derived from a well-posed variational principle and he brought a variety of geological and paleontological considerations to bear on the question of testing alternate theories of gravitation.

Before proceeding to examine the population explosion of cosmological theories and models which grew out of Dirac's LNH and the suggestion of 'varying constants' it is useful to place the idea of a time-varying G in some historical context. Just how heretical an idea would this have been in 1937?

Some perspective can be gained by examining a few early experimental and theoretical claims for time variation in fundamental 'constants'. Many of these claims were connected with cosmological questions.

Although Newton's gravitational constant[55] was the first fundamental constant to be identified it was not the first to have its constancy questioned. As early as 1874, Thomson [Lord Kelvin] and Tait[56] had claimed to observe a systematic decrease with time in the velocity of light, c, of 8 km s^{-1} century^{-1}. By the early 1930's several other authors[57] had claimed to measure a significant diminution in c over a span of fifty years. Variations of this type had been incorporated into unconventional cosmological models by Stewart,[34] Buc[58] and Wold[59] to create the first 'tired light' explanations for the redshifting of spectral lines in distant nebulae relative to their values on earth. A decay of the photon energy, $E = hc/\lambda$, as c changes during the transit time between emission and detection for light of wavelength λ was invoked to explain the dependence of redshift with distance from its source in preference to the standard explanation based upon Doppler recession. The same qualitative effect could be achieved superficially by supposing the magnitude of Planck's constant decreases in time, or a secular increase in the wavelength of radiation takes place. Various suggestions of that sort were made in the period 1935–7 by Chalmers,[60] Nernst,[61] and Sambursky.[62] In 1931, Sir James Jeans proposed an interesting scenario, wherein, in effect the atomic size decreases in time. This would, he claimed, give the appearance of an expanding Universe,[63]

> Another possibility . . . is that the Universe retains its size, while we and all material bodies shrink uniformly. The redshift we observe in the spectra of the nebulae is then due to the fact that the atoms which emitted the light millions of years ago were larger than the present-day atoms with which we measured the light—the shift is, of course, proportional to distance.

According to this view, the galaxies are not receding because of a universal Hubble expansion but rather, everything inside the galaxies, including us, is shrinking! If the atomic mass increases but the charge of the electron remains constant then the electrons will orbit closer to the nucleus (until they fall within the range of the nuclear force). This is similar to the electrons simply occupying a higher energy state and so radiation emitted by these 'smaller' atoms during an atomic transition would have higher frequency than that from a less tightly bound atom. The atoms we now see in remote galaxies would be larger than those here and so the light we receive from them would be redder than that seen in the corresponding local emission spectra.

Not surprisingly, dispute arose over the meaning of change in *dimensional* quantities like c or h. When examined critically it is clear that only a variation in dimensionless quantities possesses an invariant meaning. (We shall have more to say about this later). For this reason, supporters of the conventional expansion hypothesis were fairly critical of these heretical ideas. Lemaitre[64] contested Jean's suggestion (which has also been resurrected recently by Hoyle[65] incidentally). He remarks that

> it is clear that any artificial expansion could be provided by arbitrarily varying the units of length, time and mass . . . the expansion of the Universe is in some sense relative to the whole set of essential properties of matter being assumed to be constant.

Writing some years later, Eddington argued that all such alternative explanations for the redshift were without a logical foundation:[66]

> The ratio of the wavelength to the period of H_α light is the velocity of H_α light. Thus it follows from the definition of the ultimate standards of length and time that the velocity of light is constant everywhere and everywhen. Alleged experimental evidence for a rather large change of the velocity of light in the last 70 years has been put forward. From the nature of the case there can be no such evidence; if anything is put in doubt by the experimental results, it is the agreement of the standards used by the various observers. More baleful, because it has received more credence, is the speculation of various writers that the velocity of light has changed slowly in the long periods of cosmological time, which has seriously distracted the sane development of cosmological theory. The speculation is nonsensical because a change in the velocity of light is self-contradictory.

It is evident from these examples that the general idea of a slow time-variation in some of the traditional constants of Nature was not a new one. Indeed, the specific suggestion that G vary in time had also been prefigured. As Dirac himself pointed out, there were many features common to his theory and the earlier kinematic cosmological model of Milne.

Milne's gravitation theory[67] involved something more subtle than an explicit time-variation in G. He had built-up a theoretical structure of 'kinematic relativity' in which distinct physical clocks constructed upon different fundamental processes (say atomic or gravitational) would 'tick' at different rates as time passed. If one reckoned the passage of time using a system based upon atomic spectral frequencies, its rate might not remain constant relative to measurements of time carried out by some 'gravitational clock', like an oscillating pendulum or periodic planetary orbit. The problem with introducing this multiplicity of clocks is that there appears to exist a large number of possible time standards. Milne argued that two timescales were preferred in his kinematic theory. Electromagnetic and atomic phenomena should be reckoned in a preferred t-time in which the relative motion of fundamental cosmic observers is measured to possess no acceleration. Dynamical processes were associated with τ-time in which observers appeared to be relatively at rest. There exists a simple non-linear transformation between these two temporal graduation scales

$$\tau = \ln\left(\frac{t}{t_0}\right) + t_0 \tag{4.37}$$

They are coincident at $t_0 \equiv \tau_0$. The zero of 't' time occurs in the infinite past of τ-time. If one observer's subjective time were measured on the τ scale they could perceive an infinite time span in an interval that was perceived to be finite by observers experiencing t-time. (We shall return to discuss some ramifications of this in section 10.6.)

In Milne's picture the gravitation constant can be interpreted as possessing a time-variation when a dynamical process is measured in (the inappropriate) t-time. In the t-system observers must remain in causal contact with a mass that remains time-invariant, so $c^3G^{-1}t$ remains constant and $G \propto t$. This ensures that the gravitational radius of the Universe, $R_g \sim GM/c^2$, increases with t. Notice that the same relation results from Dirac's proposal because $R_g \propto GM \propto GNm_N \propto t^{-1}t^2 \propto t$.

Dirac[68] compared his ideas with those of Milne and many years afterwards updated his approach to incorporate the two measurement systems through a pair of conformally related metric intervals: one each for atomic and gravitational phenomena.[69,70] This idea does not seem very plausible. After all, why *two* metrics and why are atomic and gravitational phenomena taken as fundamental? We might ask which of the two metrics will govern a purely nuclear reaction or a weak interaction. There is nothing ultimately fundamental about 'atomic' phenomena.

4.5 Varying Constants

> Austrian trains are always late. A Prussian visitor asks the Austrian conductor why they bother to print timetables. The conductor replies: If we didn't, how would we know how late the trains are?
>
> V. Weisskopf

Dirac's innovation created a new sub-culture within gravitation physics. Theorists suggested all manner of possible time variations in quantities traditionally regarded as constant. Observational evidence gleaned from various disciplines was collated to limit the extent of these hypothetical variations and new theories of gravity were developed to place the possibility of varying G on a secure theoretical basis.[71] A complete survey of all these developments would take us too far afield and several comprehensive surveys already exist.[72] However, it is well to bear in mind that many of the 'experimental' limits on varying constants have a rather questionable status. No precise theory is yet available which allows any particular parameter one wishes to possess arbitrary space-time variations and still be consistent with other constraints and laws. As a consequence the hypothetical time-dependence of quantities other than G tends to be just 'written-into' the relations which hold when they are constant. This procedure is of doubtful physical meaning because it assumes that no compensatory changes are needed in either the laws of physics or other fundamental parameters to accommodate a varying 'constant'. (The nearest possibility to this is provided by Kaluza–Klein cosmological models which we shall introduce later in this chapter in connection with dimensionality questions).

When applying the LNH Dirac had assumed the electron-proton mass ratio and the fine structure constant both remain independent of cosmic epoch. Any change in these microscopic invariants would have generated problematic consequences for established aspects of 'local' physics—not least quantum mechanics. But later, the ramifications of the LNH for these other constants began to be examined more closely.

In 1948, whilst outlining some of the biological and geological consequences of a time-varying G, Edward Teller[101] pointed out the rough numerical coincidence between the value of the fine structure constant and the logarithm of Gm_N^2/hc. So according to the LNH we should anticipate an equality of the form,

$$\alpha^{-1} \sim \ln\left(\frac{hc}{Gm_N^2}\right) \propto \ln(t) \tag{4.38}$$

and hence the fine structure constant would fall-off at a logarithmic rate with the passage of cosmic time.[73]

Teller also argued that the simple incorporation of $G \propto t^{-1}$ evolution into the equations of stellar structure and planetary dynamics would have extremely unsavoury consequences for the terrestrial environment. The solar luminosity would have been considerably higher in past ages and, by angular momentum conservation, the radius of the Earth's orbit about the Sun, R_{orb}, much smaller. Precisely, the dependence of these quantities on Newton's constant is (where $L_\odot$ denotes the solar luminosity),

$$L_\odot \propto G^7; \qquad R_{orb} \propto G^{-1} \tag{4.39}$$

Any changes in G appear to alter the solar 'constant' and the temperature at the surface of the Earth, $T_\oplus$, dramatically

$$T_\oplus \sim \left(\frac{L_\odot}{R_{orb}^2}\right)^{1/4} \propto G^{9/4} \propto t^{-9/4} \tag{4.40}$$

This rapid evolution implies that, in the past, the Earth's surface would have been far hotter than it is today. Quantitatively, Teller estimated that oceans would have been actually boiling in the pre-Cambrian era. By

1967 a recalibration of the Universe's age had pushed the age of boiling oceans a little further into the past but the discovery of fossilized bacteria $\sim 3.1\times 10^9$ years old subsequently restored the contradictory evidence to its former strength.

In addition, Pochoda and Schwarzschild[74] (and independently, Gamow[75]), had shown that the higher past rate of thermonuclear fusion within the solar interior would have exhausted the supply of nuclear energy well before the present day. The Sun would not now be shining.

Motivated by these difficulties for the varying G hypothesis, Gamow[76] re-examined Dirac's coincidences and noticed that the LNH, (4.30), could equally well be interpreted as predicting an increase in the electron charge (or fine structure constant) with G constant,

$$e^2 \propto t \tag{4.41}$$

This change of strategy ameliorates the 'boiling ocean' problem. A gradual increase in the electron charge obviously leaves the Earth's orbit unaffected but the luminosity history of the Sun is still influenced; this time by a change in the Kramers opacity, κ, which determines the rate at which energy is transported to the surface of the Sun, its luminosity varies as $L_\odot \propto \kappa^{-1} \propto e^{-6}$. However, the total effect is milder than that given by (5.3); we now have

$$T_\oplus \propto L_\odot^{1/4} \propto t^{-3/4} \tag{4.42}$$

This moves the era of ocean boiling so far into the past that any conflict with biological evidence is avoided and the Sun's total lifetime now comfortably exceeds the age of the solar system. Gamow also noted that his hypothesis (4.41) would have observational consequences through redshifting the absorption spectra of distant quasi-stellar radio sources.

Gamow's paper provoked a flurry of articles[77] pointing out untoward consequences of the variation (4.41) and it soon lost its appeal as a viable hypothesis. Interestingly, in 1963, four years prior to Gamow's paper, Stanyukovich[78] had discussed a variation in e^2. He reformulated Dirac's hypothesis incorporating two additional constraints, namely, that the total energy and electric charge of the Universe both remain constant. This also predicted a variation in the electron charge with time:

$$Nm_Nc^2 = \text{constant} \tag{4.43}$$

$$Ne^2 = \text{constant} \tag{4.44}$$

This led to a 'self-consistent' set of variations in all the fundamental parameters (note that (4.44) is not the correct condition for total charge conservation though)

$$m_N \propto h \propto e^2 \propto G^{-2} \propto t^{-2} \tag{4.55}$$

After Gamow's suggestion, schemes like this were suggested by several authors[79] and subsequently, attempts made to incorporate variations in the weak and strong couplings into this framework and introduce observational constraints upon their magnitude.[80] Before leaving the history of specific variations in fundamental constants to trace the introduction of the modern Anthropic Principle it is worth making some remarks about the problem of formulating and interpreting tests of these theories involving varying 'constants' of Nature.

Consider first, a specific example: suppose, as many workers have done, one attempts to test by experiment a potential variation of a *dimensional* quantity. One should not expect such a variation—or such a test—to possess any invariant meaning because the constant can be altered by changing units arbitrarily at different space-time points.[81] Only variations of dimensionless quantities possess an absolute meaning. An example will illustrate the point:

Various attempts have been made to infer the time-invariance of the dimensional combination, hc, by examining the spectra of distant quasars.[82] Experiments measure the wavelength, λ, of incoming photons using a grating spectrometer and, independently, the photon energy E can be measured by a photomultiplier. The value of their product,

$$E\lambda = hc \tag{4.46}$$

can be calculated and is always found to be the same for local and distant sources possessing a wide span of redshifts ($z \leqslant 1.5$). This might be interpreted as experimental evidence that hc had the same value at the time of emission as it does at reception in our instruments. Is this supposition correct?

It is not. The argument has tacitly assumed that e and λ suffer the same amount of Doppler shifting in transit from source to the laboratory detector so the 'redshifting' of E exactly cancels the 'blueshifting' of λ and thus

$$(E\lambda)_{\text{lab}} = (E\lambda)_{\text{source}} \tag{4.47}$$

However, geometrical optics reveals that this is only true if the energy and momentum can both be propagated parallel to themselves. This requires that the constant of proportionality between e and λ, that is hc, remain constant along the entire space-time path of the photon from source to observer.[81] The invariance of hc has been implicitly imposed in the formulation of the problem.

The problem of measuring changes in dimensionless constants is also not without its subtleties. All the physical measurements we carry out are based upon a comparison between particular events and some reference standard, whether it be a wristwatch or the vibrations of a caesium atom.

Clearly, the particular reference standard we employ determines the class of changes we are able to discern. We could not, for example, determine whether all metre rods have invariant lengths by comparing them only with another metre rod.

If we measure time with a grandfather clock then the 'ticks' of the timepiece are fixed in duration by the frequency at which the pendulum swings in the Earth's gravitational field. This rate is determined in part by the mass and radius of the Earth, which are both subject to minute but unpredictable changes, together with a collection of fundamental constants like G. The simple electric clock found in most households is different; it uses the alternation cycle of the domestic A.C. supply as its 'pendulum'. Although the cycle frequency is fairly constant ~ 50 Hz it is clearly also subject to unpredictable changes due to random fluctuations, trade union action and so forth. Both these clocks are unfortunately influenced by parameters which are not fundamental constants of Nature. If examined closely they would be seen to vary slightly from place to place and time to time. This makes them useless as standards for defining what we mean by 1 unit of time in a way that would enable us to communicate this standard to someone in another galaxy requesting the time.

Our time standards must be based upon natural clocks whose frequencies are determined solely by the constants of Nature. One such chronometer uses the frequency of hyperfine transitions in caesium atoms as its 'pendulum'. This frequency is determined by atomic constants alone,

$$\nu_c = \frac{m_e^2 c^{-2} e^8}{h^5 m_N} \equiv T^{-1} \tag{4.48}$$

Analogously, the standard of length is defined, not by any artefact, but via the wavelength of krypton-86 near λ 6057.8. Therefore it is determined essentially by the Rydberg length, R_∞, which is given by fundamental constants as

$$4\pi R_\infty = \frac{m_e e^4}{ch^3} \equiv L \tag{4.49}$$

If we adopt L and T as our standards of length and time then they are *defined* as constant. We could not measure any change in fundamental constants which are functions of L and T. For example, no change in the quantity

$$LT^{-1} = c\left(\frac{e^2}{hc}\right)^2 \frac{m_e}{m_N} \tag{4.50}$$

could be measured against our standards.

We are, of course, at liberty to pick whatever standards of length and time we wish. If we possessed a means of measuring time by the Compton frequency of the electron and length in units of the Bohr radius of hydrogen then we would have different standards of length and time. In this case L' and T' would be defined as constants where

$$T' = \frac{h}{m_e c^2} \tag{4.51}$$

$$L' = \frac{h^2}{e^2 m_e} \tag{4.52}$$

Using these new standards it would be impossible to measure a change in, say, $T'/L' = e^2/hc^2$, but it would, in principle, be possible to measure a change in T/L.

The point of these model examples is simply this: one cannot merely postulate a completely arbitrary change in fundamental quantities and expect that any type of experiment will be able to measure its consequences (see ref. 83). To measure real variations one must compare the behaviour of one event against two or more independent standards. This was the technique used by van Flandern[84] to examine the constancy of G. He measured the deceleration of the lunar longitude in atomic time, defined by caesium transitions, and in ephemeris time which is determined by the motion of the Earth about the Sun. He claimed the deceleration was larger in atomic time. However, this particular comparison is extremely difficult to make because of the uncertainties in fixing ephemeris time and there has been much debate as to the reality and interpretation of his results.

In later Chapters 5 and 6 we shall not be concerned with examining Universes with varying constants but an ensemble of Universes, each member of which possesses different, but time-invariant, physical constants.

4.6 A New Perspective

> Imagine a game of Russian roulette played on a grand scale by many individuals using randomly distributed loaded and unloaded guns. At the conclusion of the deadly game a brilliant statistician after exhaustive statistical analyses concludes that there is a high probability of the randomly selected unloaded guns being drawn by the survivors of the game.
>
> R. Dicke and P. J. Peebles

The interest in cosmological coincidences involving the large dimensionless numbers of Eddington and Dirac together with the possibility of time variation in fundamental constants catalysed the development of a new cosmological perspective. A point of view that was to lead cosmologists directly to a modern form of the Anthropic Principle.

After Dirac's innovation, several scientists realized that our own existence was a relevant consideration in deciding upon viable theories of natural phenomena. Cosmological models, especially those with varying gravitational constant could have dramatic consequences for geology and protobiology. They could even predict the evolution of man to be impossible. J. B. S. Haldane was the first to stress that Milne's cosmological model, or any other theory incorporating dual chronology, could have profound consequences for evolutionary biology. In an article[85] written just after Dirac's first paper he pointed out that if one takes Milne's model at face value, then kinematic t-time, which has a finite past, appears to be appropriate for describing chemical processes, whilst dynamical processes like the Earth's motion in the solar system move on the τ-time which has an infinite past. This means that the conditions necessary for the existence of organisms cannot always have existed:

> ... a given mass of a particular mixture at a certain temperature and pressure generates so many ergs per second. Clearly such properities of matter must change with time if Milne is correct; and $1.5 \cdot 10^9$ years ago on the kinematic scale they may well have been sufficiently different to make life as we know it impossible. The conversion of chemical into kinetic energy is an essential feature of animal and even plant life; in terms of our ordinary dynamical units this was, according to Milne, less rapid in the remote past.

Haldane claimed that the energy available from atomic or chemical transitions, although constant on the t-scale, increases on the τ-scale.

A specific example of this 'two-timing' is radioactive decay: if N_0 radioactive nuclei exist at time t_0 then at a later time t there will remain only $N(t)$ where

$$N(t) = N_0 e^{-\lambda t}; \qquad t \geqslant t_0 \tag{4.53}$$

where λ is the decay constant on the t-scale. When viewed on the τ scale, (related to t via equation 4.37), the decays do not proceed at a constant rate because

$$\frac{dN}{d\tau} = \left(-\lambda N_0 \frac{t}{t_0}\right) e^{-\lambda t} \tag{4.54}$$

The decay 'constant' on the τ scale is $\lambda t/t_0$. Haldane points out that this could mean that elements now regarded as stable on the t-scale are really radioactive on the τ-scale.

In evolutionary biology this aspect of dual timescales appeared particularly intriguing to Haldane. The energy yield from oxidation processes or the exoergic breakdown of adenosine triphosphoric acid would vary in rate like $\sim t^2$ when measured on the τ-scale. At very early times these reactions would have proceeded too slowly to produce living organisms because they would have been unable to provide the energies necessary for cell division. So perhaps, Haldane suggests,[85]

> There was, in fact, a moment when life of any sort first became possible and the higher forms of life may only have become possible at a later date. Similarly a change in the properties of matter may account for some of the peculiarities of pre-Cambrian geology.

The gradual increase in the level of available activation energy could even lead to an explanation for the continual growth of biological complexity with time. At first all life would have been impossible but then, as the maximum potential energy yield from chemical processes grew, so different types of stable structure could arise and survive:[85]

> ... At a later time, life of a simple sort would have been impossible, but locomotion would have been very difficult, and large swimming or crawling animals could not have existed.

Looking to the future, this type of evolution clearly has a bearing upon the inevitable approach of ever-expanding Universes to degenerate thermal equilibrium. A model like Milne's might evade a 'heat death' at late times if the efficiency of energy production were to increase rapidly enough in the future organisms to exist indefinitely—other factors being ignored.

This remarkable scenario did not attract the imagination of other biologists although Haldane wrote about it on several occasions[86] in relation to the whole class of cosmological models incorporating varying G. It is an excellent example of the interplay that can arise between the dynamics of the Universe as a whole and the local conditions required for biological evolution to operate.

We have already discussed Teller's observation that Dirac's simple theory leads to an anomalously large temperature (>100°C) on the Earth's surface in the pre-Cambrian era with catastrophic consequences for land and water-based organisms. In the early 1960's Robert Dicke and Carl Brans[87] developed a rigorous self-consistent theory of gravitation which allowed the consequences of a varying G to be evaluated more precisely. The Brans–Dicke theory also had the attractive feature of approaching Einstein's theory in the limiting situation where the change in G tends asymptotically to zero. This enabled arguments like Teller's to be examined more rigorously, and the largest tolerable rate of change in G to be calculated.

Dicke and his colleagues had previously carried out a wide-ranging series of investigations[88] to examine the geological and astronomical evidence for any varying constants of Nature. In his 1957 review[89] of the theoretical and observational situation Dicke made his first remarks concerning the connection between biological factors and the 'large number coincidences'. Dicke realized that the observation of Dirac's coincidences between the Eddington number N and the other quantities not possessing a time-variation is 'not random but is conditioned by biological factors'.[90] This consideration led him to see a link between the large number coincidences and the type of Universe that could ever be expected to support observers. Seen in this light,[90]

> The problem of the large size of these numbers now has a ready explanation . . . there is a single large dimensionless number which is statistical in origin. This is the number of particles in the Universe. The age of the Universe 'now' is not random but is conditioned by biological factors. The radiation rate of a star varies as $\varepsilon^{-7.9}$ and for very much larger values of ε than the present value, all stars would be cold. This would preclude the existence of man to consider this problem . . . if [it] were presently very much larger, the very rapid production of radiation at earlier times would have converted all hydrogen into heavier elements, again precluding the existence of man.

Some years later, in 1961, Dicke presented these ideas in a more quantitative and cogent form specifically geared to explaining the large number coincidences.[91]

Life is built upon elements heavier than hydrogen and helium. These heavy elements are synthesized in the late stages of stellar evolution and are spread through the Universe by supernovae explosions which follow the main sequence evolution of stars. Dicke argued that only universes of roughly the main sequence stellar age could produce the heavy elements, like carbon, upon which life is based. Only those Universes could evolve 'observers'. Quantitatively, the argument shows that the main-sequence stellar lifetime is roughly

$$t_{ms} = \frac{\begin{pmatrix}\text{nuclear energy}\\ \text{available from}\\ \text{hydrogen fusion}\end{pmatrix} \times \begin{pmatrix}\text{time for radiation}\\ \text{to diffuse out of}\\ \text{the star}\end{pmatrix}}{\begin{pmatrix}\text{radiation energy trapped}\\ \text{within the star}\end{pmatrix}}$$

that is (see Chapter 5 for the proof):

$$t_{ms} \sim \left(\frac{hc}{Gm_N^2}\right)\left(\frac{h}{m_N c^2}\right) \tag{4.55}$$

'Observers' could not exist at times greatly in excess of t_{ms} because no hot stable stars would remain to support photochemical processes on planets; all stars would be white dwarfs, neutron stars or black holes. Living beings are therefore most likely to exist when the age of the Universe, t_0, is roughly equal to t_{ms} and so must inevitably observe Dirac's coincidence $N_1 \sim N_2$ to hold. It is a prerequisite for their existence and no hypothesis of varying constants is necessary to explain it. At a time t_{ms} after the beginning of the expansion of the Universe it is inevitable that we observe N_1 to have the value

$$N_1 \sim t_{ms} \frac{m_e c^3}{he^2} \sim \left(\frac{hc}{Gm_N^2}\right)\left(\frac{m_e}{m_N}\right)\left(\frac{hc}{e^2}\right) \sim N_2 \tag{4.56}$$

Two points are worth making at this stage. Although Dicke's argument explains the coincidence of N_1 and N_2 it does not explain why the coincident value is so large. Further considerations are necessary to resolve this question.

Also, Dicke made his 'anthropic' suggestion at a time when the cosmic microwave background radiation was undiscovered and the steady state universe remained a viable cosmological alternative to the Big Bang theory. However, a closer scrutiny of Dicke's argument at that time could have cast doubt upon the steady-state model. For, in the Big Bang model it is to be expected that we measure the Hubble age, H_0^{-1}, to lie close to a typical stellar lifetime, whereas in the steady-state theory it is a complete coincidence. In an infinitely old steady-state Universe manifesting 'continuous creation' there should exist no correlation between the time-scale on which the Universe is expanding and the main sequence lifetime. We should be surrounded by stars in all possible states of maturity.

There were others who had been thinking along similar lines. Whitrow had sought to explain why, on Anthropic grounds, we should expect to observe a world possessing precisely three spatial dimensions.[102] His ideas were also extended to consider the question of the size and age of the expanding Universe. In the 1956 Bampton Lectures, Mascall elaborated upon some of Whitrow's ideas concerning the relation between the size of the Universe and local environmental conditions. In effect, they anticipate why the *size* of the large numbers, $N^{1/2}$, N_1 and N_2 (rather than just their numerical coincidence) is likely to be conditioned by biological factors:[92]

> Nevertheless, if we are inclined to be intimidated by the mere size of the Universe, it is well to remember that on certain modern cosmological theories there is a direct connection between the quantity of matter in the Universe and the conditions in any limited portion of it, so that in fact it may be necessary for the Universe to have the enormous size and complexity which modern astronomy has revealed, in order for the earth to be a possible habitation for living beings.

These contributions by Dicke and Whitrow provide the first modern examples of a 'weak' anthropic principle; that the observation of certain, *a priori*, remarkable features of the Universe's structure are necessary for our own existence.

Having gone so far, it is inevitable that some would look at the existence of these features from another angle; one reminiscent of the traditional 'Design arguments'[102] that the Universe either must give rise to life or that it is specially engineered to support it. Carter[93] gave the name 'strong' Anthropic Principle to the idea that the Universe must be 'cognizable' and 'admit the creation of observers within it at some stage'. This approach can be employed to 'retrodict' certain features of any cognizable Universe.[102]

There is one obvious defect in this type of thinking as it now stands. We appear to be making statements of comparative reference and evaluating *a posteriori* the likelihood of the Universe—which is by definition unique—possessing certain structural features. Various suggestions have been made as to how one might generate an entire ensemble of possible worlds, each with different characteristics; some able to support life and some not. One might then examine the ensemble for structural features which are necessary to generate 'observers'. This scrutiny should eventually single out a cognizable subset from the metaspace of all possible worlds. We must inevitably inhabit a member of this subset in which living systems can evolve. Carter suggested[93] that a 'prediction made using this strong version of the Anthropic Principle could boil down to a demonstration that a particular feature of the world is common to all members of the cognizable subset'. Obviously, it would be desirable to have some sort of probability measure on this ensemble of worlds.

These speculations sound rather far-fetched, but there are several sources of such an ensemble of different worlds. If the Universe is finite and bounded in space and time, it will recollapse to a second singularity having many features in common with the initial big bang singularity. Wheeler[94] has speculated that the Universe may have a cyclic character, oscillating *ad infinitum* through a sequence of expanding and contracting phases. At each 'bounce' where contraction is exchanged for expansion, the singularity may introduce a permutation in the values of the physical 'constants' of Nature and of the form of the expansion dynamics. Only in those cycles in which the 'deal' is right will observers evolve. If there is a finite probability of a cognizable combination being selected then in the course of an infinite number of random oscillatory permutations those worlds allowing life to evolve must appear infinitely often. The problem with this idea is that it is far from being testable. At present, only the feasibility of a bounce which does not permute physical constants (although perhaps the expansion dynamics) is under scrutiny. Also, if the permutation at each singularity extends to the constants of Nature, why not to the space-time topology and curvature as well?[95] And if this were the case, sooner or later the geometry would be exchanged for a noncompact structure bound to expand for all future time. No future singularity would ensue and the constants of Nature would remain forever invariant. Such a scheme actually makes a testable prediction! The Universe should currently be 'open' destined to expand forever since this state will always be reached after a finite series of oscillations. However, why should this final permutation of the constants and topology just happen to be one which allows the evolution of observers?

A more attractive possibility, which employs no speculative notions regarding cyclic Universes, is one suggested by Ellis. If the Universe is randomly infinite in space-time then our ensemble already exists.[96] If there is a finite probability that a region the size of the visible Universe ($\sim 10^{10}$ light years in diameter) has a particular dynamical configuration, then this configuration must be realized infinitely often within the infinite Universe at any moment. This feature is more striking when viewed in the following fashion. In a randomly infinite Universe, any event occurring here and now with finite probability must be occurring simultaneously at an infinite number of other sites in the Universe. It is hard to evaluate this idea any further, but one thing is certain: if it is true then it is certainly not original!

Finally, a completely different motivation for the 'many worlds' idea comes from quantum theory. Everett,[97] in an attempt to overcome a number of deep paradoxes inherent in the interpretation of quantum theory and the theory of measurement, has argued that quantum mechanics requires the existence of a 'superspace' of worlds spanning the range of all possible observations. Through our acts of measurement we are imagined to trace a path through the mesh of possible outcomes. All the 'worlds' are causally disjoint and the uncertainty of quantum observation can be interpreted as an artefact of our access to such a limited portion of the 'superspace' of possible worlds. The evolution in the superspace as a whole is entirely deterministic. Detailed ramifications of this 'many worlds' interpretation of quantum mechanics will be explained later, in Chapter 7.

One other aspect of the ensemble picture is worth pointing out. There are two levels at which it can be used. On the one hand we can suppose the ensemble to be composed of 'theoretical' Universes in which the quantities we now regard as constants of Nature, e^2/hc, m_N/m_e and so forth, together with the dynamical features of the Universe; its expansion rate, rotation rate, entropy content etc. take on all possible values. On the other, we can consider only the latter class of variations. There is an obvious advantage to such a restricted ensemble. The second class of alternative worlds amount to considering only the consequences of varying the initial boundary conditions to solutions of Einstein's equations (which we assume here to provide a reliable cosmological theory). An examination of these alternatives does not require any changes in the known laws of physics or the status of physical parameters. Our Universe appears to be described very accurately by an extremely symmetrical solution of Einstein's cosmological equations; but there is no difficulty in finding other solutions to these equations which describe highly asymmetric Universes. One can then examine these 'other worlds' to decide how large a portion of the possible initial conditions gives rise to universes capable of, say, generating stars and planets.

A good example of considering this limited ensemble of universes defined by the set of solutions to Einstein's equations is given by Collins and Hawking.[98] Remarkably, they showed that the presently observed Universe may have evolved from very special initial conditions. The present Universe possesses features which are of infinitesimal probability amongst the entire range of possibilities. However, if one restricts this range by the stipulation that observers should be able to exist then the probability of the present dynamical configuration may become finite. The calculations that lead to these conclusions are quite extensive and are examined more critically elsewhere;[102] we shall discuss them in detail in Chapter 6.

It is also interesting to see the idea that our Universe may be a special point in some superspace containing all possible Universes is not a new one and a particularly clear statement of it was given by the British zoologist Charles Pantin[99] in 1951, long before the above-mentioned possibilities were recognized. By reasoning similar to Henderson's,[100] Pantin had argued that the Universe appears to combine a set of remarkable structural 'coincidences' upon which the possibility of our own existence crucially hinges.

> ...the properties of the material Universe are uniquely suitable for the evolution of living creatures. To be of scientific value any explanation must have predictable consequences. These do not seem to be attainable. If we could know that our own Universe was only one of an indefinite number with varying properties we could perhaps invoke a solution analogous to the principle of Natural Selection, that only in certain Universes, which happen to include ours, are the conditions suitable for the existence of life, and unless that condition is fulfilled there will be no observers to note the fact. But even if there were any conceivable way of testing such a hypothesis we should only have put off the problem of why, in all those Universes, our own should be possible?!

Another early subscriber to an ensemble picture, this time of the variety suggested by Ellis,[96] was Hoyle. His interest in the many possible worlds of the Anthropic Principle was provoked by his discovery of a remarkable series of coincidences concerning the nuclear resonance levels of biological elements.

Just as the electrons of an atom can be considered to reside in a variety of states according to their energy levels so it is with nucleons. Neutrons and protons possess an analogous spectrum of *nuclear* levels. If nucleons undergo a transition from a high to a low energy state then energy is emitted and conversely, the addition of radiant energy can effect an upward transition between nuclear levels. This nuclear chemistry is a crucial factor in the chain of nuclear reactions that power the stars.[107]

When two nuclei undergo fusion into a third nuclear state, energy may be emitted. One of the most striking aspects of low-energy nuclear reactions of this type is the discontinuous response of the interaction rate, or cross-section, as the energy of the participant nuclei changes; see Figure 4.1.

A sequence of sharp peaks, or *resonances*, arises in the production efficiency of some nuclei as the interaction energy changes. They will occur below some characteristic energy (typically $\sim$few$\times 10$ MeV) which depends on the particular nuclei involved in the reaction. Consider the schematic reaction

$$A+B \rightarrow C \tag{4.57}$$

We could make this reaction resonant by adjusting the kinetic energy of the A and B states so that when we add to it the intrinsic energy of the states in the nuclei A and B we obtain a total lying just above a possible energy level of the nucleus C. The interaction (4.57) would then be resonant. Although reactions can be made resonant in this way it may not always be possible to add the right amount of kinetic energy to obtain resonance. In stellar interiors the kinetic energy will be determined by the temperature of the star.

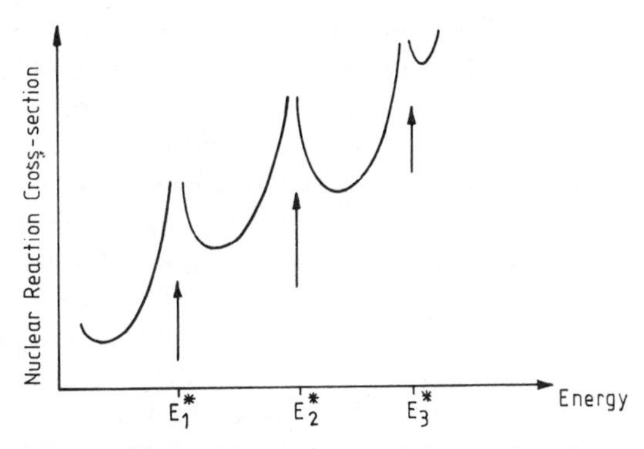

Figure 4.1. Schematic representation of the influence of nuclear resonances upon the cross-section for a particular nuclear reaction to occur. Typically, a series of energies, E_i^* will exist at which the reactions are maximally efficient, or resonant.

The primary mechanism whereby stars generate gas or radiation pressures to support themselves against gravitational collapse is exothermic fusion of hydrogen in helium-4. But, eventually a star will exhaust the supply of hydrogen in its core and its immediate source of pressure support disappears. The star possesses a built-in safety valve to resolve this temporary energy crisis: as soon as gravitational contraction begins to increase the average density at the stellar core the temperature rises sufficiently for the initiation of helium burning (at $T \sim 10^8$ K, $\rho \sim 10^{4.5}$ gm cm^{-3}), via

$$3He^4 \rightarrow C^{12} + 2\gamma \tag{4.58}$$

This sequence of events (fuel exhaustion → contraction → higher central temperature → new nuclear energy source) can be repeated several times but it is known that the nucleosynthesis of all the heavier elements essential to biology rests upon the step (4.58).

Prior to 1952 it was believed that the interaction (4.58) proceeded too slowly to be useful in stellar interiors. Then Salpeter[108] pointed out that it might be an 'autocatalytic' reaction, proceeding via an intermediate beryllium step,

$$\begin{aligned} 2He^4 + (99 \pm 6)\ \text{keV} &\rightarrow Be^8 \\ Be^8 + He^4 &\rightarrow C^{12} + 2\gamma \end{aligned} \tag{4.59}$$

Since the Be^8 lifetime ($\sim 10^{-17}$ s) is anomalously long compared to the $He^4 + He^4$ collision time ($\sim 10^{-21}$ s), the beryllium will co-exist with the He^4 for a significant time and allow reaction (4.59) to occur. However, in 1952 so little was known about the nuclear levels of C^{12} that it was hard to evaluate the influence of the channel (4.59) on the efficiency of (4.58).

Two years later Hoyle made a remarkable prediction: in the course of an extensive study of stellar nucleosynthesis he realized that unless reaction (4.58) proceeded *resonantly* the yield of carbon would be negligible. There would be neither carbon, nor carbon-based life in the Universe. The evident presence of carbon and the products of carbon chemistry led Hoyle to predict that (4.58) and (4.59) *must* be resonant, with the vital resonance level of the C^{12} nucleus lying near ~7.7 MeV. This prediction was soon verified by experiment. Dunbar *et al.*[109] discovered a state with the expected properties lying at 7.656 ± 0.008 MeV. If we examine the level structure of C^{12} in detail we find a remarkable 'coincidence' exists there. The 7.6549 MeV level in C^{12} lies just *above* the energy of Be^8 plus He^4 (=7.3667 MeV) and the acquisition of thermal energy by the C^{12} nucleus within a stellar interior allows a resonance to occur. Dunbar *et al.*'s discovery confirmed an Anthropic Principle prediction.

However, this is not the end of the story. The addition of another helium-4 nucleus to C^{12} could fuse it to oxygen. If this reaction were also resonant all the carbon would be rapidly burnt to O^{16}. However, by a further 'coincidence' the O^{16} nucleus has an energy level at 7.1187 MeV that lies just *below* the total energy of $C^{12}+He^4$ at 7.1616 MeV. Since kinetic energies are always positive, resonance *cannot* occur in the 7.1187 MeV state. Had the O^{16} level lain just below that of $C^{12}+He^4$, carbon would have been rapidly removed via the alpha capture

$$C^{12}+He^4 \rightarrow O^{16} \tag{4.60}$$

Hoyle realized that this remarkable chain of coincidences—the unusual stability of beryllium, the existence of an advantageous resonance level in C^{12} and the non-existence of a disadvantageous level in O^{16}—were necessary, and remarkably fine-tuned, conditions for our own existence and indeed the existence of any carbon-based life in the Universe.

These coincidences could, in principle, be traced back to their roots where they would reveal a meticulous fine-tuning between the strengths of the nuclear and electromagnetic interactions along with the relative masses of electrons and nucleons. Unfortunately no such back-track is practical because of the overwhelming complexity of the large quantum systems involved; such resonance levels can only by located by experiment in practice.

Hoyle's anthropic *prediction* is a natural successor to the examples of Henderson.[100] It exhibits further relationships between *invariants* of Nature which are necessary for our own existence. Writing and lecturing in 1965 Hoyle added some speculation as to the conditions in 'other worlds' where the properties of beryllium, carbon and oxygen might not be so favourably arranged. First[110] 'suppose that Be^8 ... had turned out to be moderately stable, say bound by a million electron volts. What would be the effect on astrophysics?' There would be many more explosive stars and supernovae and stellar evolution might well come to an end at the helium burning stage because helium would be a rather unstable nuclear fuel,[110]

> Had Be^8 been stable the helium burning reaction would have been so violent that stellar evolution with its consequent nucleosynthesis would have been very limited in scope, less interesting in its effects ... if there was little carbon in the world compared to oxygen, it is likely that living creatures could never have developed.

Hoyle chose not to regard these coincidences as absolute. Rather he favoured the idea that the so-called 'constants' of Nature possess a spatial variation. This he believed to be suggested by the additional coincidence that the dimensionless ratio or the gravitational and electric interaction strengths ($\sim 10^{-40}$) is numerically related to the total number of nucleons

($N \sim 10^{80}$) in the observable Universe by a $1/\sqrt{N}$ relation (4.14) that is suggestive of a statistical basis if the coupling constants have some Gaussian probability distribution in space.[111] If this were true (although there is no evidence for such a view) then the coincidences discussed above would not abide everywhere in the Universe but life could only evolve in regions where they did,[112]

> ... we can exist only in the portions of the universe where these levels happen to be correctly placed. In other places the level in O^{16} might be a little higher, so that the addition of α-particles to C^{12} was highly resonant. In such a place ... creatures like ourselves could not exist.

When it comes to assessing the consequences of making small changes in the dimensionless constants of Nature one is on shaky ground (even if we ignore the possibility of an all-encompassing unified theory that fixes the values of these constants uniquely). Although a small change in a dimensionless quantity, like Gm_N^2/hc or the resonance levels in C^{12} and O^{16}, might so alter the rate of cosmological or stellar evolution that life could not evolve, how do we know that compensatory changes could not be made in the values of other constants to recreate a set of favourable situations? Interestingly, one can say something quantitative and general about this difficulty. Suppose, for simplicity, we treat the laws of physics as a set of N ordinary differential equations governing various physical quantities $x_1, x_2, \ldots, x_N$ (allowing them to be partial differential equations would probably only reinforce the conclusion) that contain a set of constant parameters λ_i which we call the constants of physics

$$\dot{\mathbf{x}} = F(\mathbf{x}; \lambda_i); \qquad \mathbf{x} \in (x_1, x_2, \ldots, x_N) \tag{4.61}$$

The structure of our world is represented by the solutions of this system; let us call the particular realization of the constants that we observe, $\mathbf{x}^*$. It will depend upon the particular set of fundamental constants we observe, call these λ_i^*. We can ask if the solution $\mathbf{x}^*$ is stable with respect to small changes of the parameters λ_i^*. This is the type of question addressed recently by mathematicians.[155,156] Any solution of the system (4.61) corresponds to a trajectory in an N-dimensional phase space. In two dimensions, ($N = 2$), the qualitative behaviour of the possible trajectories is completely classified. Trajectories cannot cross in two dimensions without intersecting, and the property that they must not intersect in the phase plane ensures that the possible stable asymptotic behaviours are simple: after large times the trajectories either approach a 'focus' (which represents an oscillatory approach towards a stationary solution) or a 'limit cycle' (which represents oscillatory approach towards a periodic solution). However, when $N \geqslant 3$, trajectories can behave in a far more exotic fashion. Now, they are able to cross and develop complicated knotted configurations without actually intersecting. All the possible detailed behaviours are not known but when $N \geqslant 3$ it has been shown[155] that the generic behaviour of trajectories is approach to a '*strange attractor*'. This is a compact region of the phase space containing neither foci nor limit cycles and in which all neighbouring solution trajectories diverge from each other exponentially whether followed forwards or backwards in time; so there is sensitive dependence on starting conditions. An infinitesimal change in the starting position of a solution trajectory will soon develop into a huge difference in subsequent position. This tells us that in our case, so long as $N \geqslant 3$, (which it will certainly be in our model equations (4.61)), the solution $\boldsymbol{x}^*$ will become unstable to changes in λ_i away from λ_i^* when they exceed some critical (but small) value. If the original attractor at $\boldsymbol{x}^*$ was not 'strange' then our set of laws and constants are very special in the space of all choices for the set λ_i, and a small change in one of them will bring about a catastrophic change in Nature's equilibrium solutions $\boldsymbol{x}^*$. If the attractor at $\boldsymbol{x}^*$ is 'strange' then there may be many other similar sets in the λ_i parameter space. This might ensure that there were other permutations of the values of constants of Nature allowing life.

4.7 Are There Any Laws of Physics?

> There is no law except the law that
> there is no law.
>
> J. A. Wheeler

The ensembles of Worlds we have been outlining involve either hypothetical other possible universes possessing different sets of fundamental constants or different initial conditions. That is, they appeal to a potential non-uniqueness concerning both the laws of Nature and their associated initial conditions. A contrasting approach is to generate the ensemble of possibilities within a single Universe. One means of doing this can be found in the work of some particle physicists on so-called 'chaotic gauge theories'. Instead of assuming that Nature is described by gauge symmetries whose particular form then dictates which elementary particles can exist and how they interact, one might imagine there are no symmetries at high energies at all: in effect, that there are no laws of physics.

Human beings have a habit of perceiving in Nature more laws and symmetries than truly exist there. This is an understandable error in that science sets out to organize our knowledge of the world as well as increase it. However, during the last twenty years we have seen a gradual erosion of 'principles' and conserved quantities as Nature has revealed a deep, and previously unsuspected flexibility. Many quantities that traditionally were believed to be absolutely conserved—parity, charge conjugation, baryon and lepton number—all appear to be violated in elementary particle interactions. The neutrino was always believed to be a massless particle but recent experiments have provided evidence that it possesses a tiny rest mass $\sim$30 eV. Likewise, the long-held myth that the proton is an absolutely stable particle may be revised by recent theoretical arguments and tentative experimental evidence for its instability. Particle physicists have now adopted an extremely revolutionary spirit and it is reasonable to question other long-standing conservation laws and assumptions—is charge conserved, is the proton massless, is the electron stable, is Newton's law of gravity exact at low energy, is the neutron neutral . . . ?

The natural conclusion of this trend from more laws of Nature to less is to ask the overwhelming question: '*Are there any laws of Nature at all*?' Perhaps complete microscopic anarchy is the only law of Nature? If this were even partially true, it would provide an interesting twist to the traditional Anthropic arguments which appeal to the fortuitous coincidence of life-supporting laws of Nature and numerical values of the dimensionless constants of physics.

It is possible that the rules we now perceive governing the behaviour of matter and radiation have a purely random origin, and even gauge invariance may be an 'illusion': a selection effect of the low-energy world we necessarily inhabit. Some preliminary attempts[113] to flesh out this idea have shown that even if the underlying symmetry principles of Nature are random—a sort of chaotic combination of all possible symmetries—then it is possible that at low energies ($\ll 10^{32}$ K) the appearance of local gauge invariance is inevitable under certain circumstances. A form of 'natural' selection may occur wherein, as the temperature of the Universe falls, fewer and fewer of the entire gamut of 'almost symmetries' have a significant impact upon the behaviour of elementary particles, and orderliness arises. Conversely, as the Planck energy (which corresponds to a temperature of 10^{32} K) is approached, this picture would predict chaos. Our low-energy world may be necessary for physical symmetries as well as physicists.

Before mentioning some of the detailed, preliminary calculations that have been done in pursuance of this 'chaotic gauge theory' idea, let us recall a simpler example of what might be occurring: if you went out into the street and gathered information, say, on the heights of everyone passing-by over a long period of time, you would find the graph of the frequency of individuals versus height tending more and more closely towards a particular shape. This characteristic 'bell' shape is called the 'Normal' or 'Gaussian' distribution by statisticians. It is ubiquitous in Nature. The Gaussian is characteristic of the frequency distribution of all truly random processes regardless of their specific physical origin. As one goes from one random process to another the resulting Gaussians differ only by their width and the point about which they are centred. A universality of this sort might conceivably be associated with the laws of physics if they had a random origin.

Nielsen *et al.*[113,114] have shown that if the fundamental Lagrangian from which physical laws are derived is chosen at random then the existence of local gauge invariance at low energy can be a stable phenomenon in the space of all Lagrangian theories. It will not be generic. That is, the presence, say, of a massless photon is something that will emerge from an (but not every) open set of Lagrangians picked from the space of all possible functional forms. This will give the illusion of a local U(1) gauge symmetry at low energy and also of a massless photon.

Suppose that a programme of this sort could be substantiated and provide an explanation for the symmetries of Nature we currently observe—according to Nielsen,[115] it is even possible to estimate the order of magnitude of the fine structure constant in lattice models of random gauge theories; if so, then perhaps some of the values of fundamental constants might have a quasi-statistical character. In that case, the Anthropic interpretation of Nature must be slightly different. If the laws of Nature manifested at low energy are statistical in origin, then again, a *real* ensemble of different possible universes actually does exist. Our own Universe is one member of the ensemble. The question now is, are all the features of our Universe stable or generic aspects of the ensemble, or are they special? If unstable or non-generic, stochastic gauge theories require an Anthropic interpretation: they also allow, in principle, a precise mathematical calculation of the probabilities of seeing a particular aspect of the present world, and a means of evaluating the statistical significance of any cognizable Universe. In general, we can see that the crux of any analysis of this type, whatever its detailed character, is going to be the temperature of the Universe. Only in a relatively cool Universe, $T \ll 10^{32}$ K, will laws or symmetries of Nature be dominant and discernible over chaos; but, likewise, only in a cool Universe can life exist. The existence of physics and physicists may be more closely linked than we suspected.

Other physicists have adopted a point of view diametrically opposite to that of the stochastic gauge theorists: for instance, S. W. Hawking, B. S. DeWitt, and in the early 1960's J. A. Wheeler, have suggested that there is only one, unique law of physics, for the reason that only one law is logically possible! The main justification for this suggestion is scientific experience: it is exceedingly difficult to construct a mathematical theory which is fully self-consistent, universal, and in agreement with our rather extensive observations.

The self-consistency problem can manifest itself in many ways, but perhaps the most significant example in the last half-century is the problem of infinities in quantum field theory. Almost all quantum field theories one can write down are simply nonsensical, for they assert that most (or all) observable quantities are infinite. Only two very tiny classes of quantum field Lagrangians do not have this difficulty: finite quantum field theories and renormalizable quantum field theories. Thus, the mere requirement of mathematical consistency enormously restricts the class of acceptable field theories. S. Weinberg[163], in particulaı, has stressed how exceedingly restrictive the requirement of renormalization really is, and how important this restriction has been in finding accurate particle theories.

Furthermore, most theories which scientists have written down and developed are not universal; they can apply only to a limited number of possible observations. Most theories of gravity, for example, are incapable of describing both the gravitational field on the scale of the solar system and the gravitational field on the cosmological scale. Einstein's general theory of relativity is one of the few theories of gravity that can be applied on all scales.[164] Universality is a minimum requirement for a fundamental theory. Since, as Popper has shown, we cannot prove a theory, we can only falsify one, we can never know if in fact a universal theory is true. However, a universal theory *may* in principle be true; a non-universal theory we know to be false even before we test it experimentally.

Finally, our observations are now so extensive that it is exceedingly difficult to find a universal theory which is consistent with them all.

In the case of quantum gravity, these three requirements are discovered to be so restrictive that Wheeler[165] and DeWitt[166] have suggested that the correct quantum gravity theory equation (which is itself unique) can have only one unique solution!

We have discussed in sections 2.8 and 3.10 the philosophical attractiveness of this unique solution: it includes all logically possible physical universes (this is another reason for believing it to be unique, for what else could possibly exist?). The stochastic gauge theory also has this attractive feature of realizing all possibilities. The unique law theory may, however, allow a global evolution, whereas the stochastic gauge theory is likely to be globally static like Whitehead's cosmology (see section 3.10).

4.8 Dimensionality

> We see . . . what experimental facts lead us to ascribe three dimensions to space. As a consequence of these facts, it would be more convenient to attribute three dimensions to it than four or two, but the term 'convenient' is perhaps not strong enough; a being which had attributed two or four dimensions to space would be handicapped in a world like ours in the struggle for existence.
>
> H. Poincaré

The fact that we perceive the world to have three spatial dimensions is something so familiar to our experience of its structure that we seldom pause to consider the direct influence this special property has upon the laws of physics. Yet some have done so and there have been many intriguing attempts to deduce the expediency or inevitability of a three-dimensional world from the general structure of the physical laws themselves. The thrust of these investigations has been to search for any unique or unusual properties of three-dimensional systems which might render them naturally preferred. It transpires that the dimensionality of the World plays a key role in determining the form of the laws of physics and in fashioning the roles played by the constants of Nature.[118] Whatever one's view of such flights of rationalistic fancy they undeniably provide an explicit example of the use of an Anthropic Principle that pre-dates the applications of Dicke[1,91] and Carter.[93]

In 1955 Whitrow[116] suggested that a new resolution of the question '*Why do we observe the Universe to possess three dimensions*'? could be obtained by showing that observers could only exist in such universes:

> I suggest that a possible clue to the elucidation of this problem is provided by the fact that physical conditions of the Earth have been such that the evolution of Man has been possible ... this fundamental topological property of the world ... could be inferred as the unique natural concomitant of certain other contingent characteristics associated with the evolution of the higher forms of terrestrial life, in particular of Man, the formulator of the problem.

This anthropic approach to the dimensionality 'problem' was also taken in a later, but apparently independent, study of atomic stability in universes possessing an arbitrary dimension by the Soviet physicists Gurevich and Mostepanenko.[117] They envisaged an ensemble of universes ('metagalaxies') containing space-times of all possible dimensionality and enquired as to the nature of the habitable subset of worlds, and, as a result of their investigation of atomic stability they concluded that

> If we suppose that in the universe metagalaxies with various number of dimensions can appear it follows our postulates that atomic matter and therefore life are possible only in 3-dimensional space.

Interest in explaining why the world has three dimensions is by no means new. From the commentary of Simplicius and Eustratius,[118] Ptolemy is known to have written a study of the 3-D nature of space entitled '*On Dimensionality*' in which he argued that no more than three spatial dimensions are possible, but unfortunately this work has not survived. What does survive is evidence that the dramatic difference between systems identical in every respect but spatial dimension was discovered and appreciated by the early Greeks. The Platonic solids, first discovered by Theaitetos,[119] brought them face-to-face with a dilemma: why are there an infinite number of regular, convex, two-dimensional *polygons* but only five regular three-dimensional *polyhedra*? This mysterious property of physical space was later to spawn many mystical and metaphysical 'interpretations'—a veritable 'music of the spheres'.

In the modern period, mathematicians did not become actively involved in attempting a rigorous formulation of the concept of dimension until the early nineteenth century, although as early as 1685 Wallis[120] had speculated about the local existence of a fourth geometrical dimension. During the nineteenth century Möbius considered the problem of superimposing two enantiomorphic solids by a rotation through 4-space and later Cayley, Riemann, and others, developed the systematic study of N-dimensional geometry although the notion of dimension they employed was entirely intuitive.[121] It sufficed for them to regard dimension as the number of independent pieces information required for a unique specification of a point in some coordinate system. Gradually the need for something more precise was impressed upon mathematicians by a series of counter-examples and pathologies to their simple intuitive notions. For example, Cantor and Peano produced injective and continuous mappings of $\mathcal{R}$ into $\mathcal{R}^2$ to refute ideas that the unit square contained more points than the unit line. After unsuccessful attempts by Poincaré, it was Brouwer[122] who, in 1911, established the key result: he showed that there is no continuous injective mapping of $\mathcal{R}^N$ into $\mathcal{R}^M$ if $N \neq M$. The modern definition of dimension[123] due to Menger and Urysoln grew out of this fundamental result.

The question of the *physical* relevance of spatial dimension seems to arise first in the early work of Immanuel Kant.[124] He realized that there was an intimate connection between the inverse square law of gravitation and the existence of precisely three spatial dimensions, although he regards the three spatial dimensions as a consequence of Newton's inverse square law rather than *vice versa.* As we have already described in Chapter 2, William Paley[125] later spelt out the consequences of a change in the form of the law of gravitation for our existence. Many of the points he summarized in 1802 have been rediscovered by modern workers examining the manner in which the gravitational potential depends on spatial dimensions, which we shall discuss below.

In the twentieth century a number of outstanding physicists have sought to accumulate evidence for the unique character of physics in three dimensions. Ehrenfest's famous article[126] of 1917 was entitled '*In what way does it become manifest in the fundamental laws of physics that space has three dimensions*'? and it explained how the existence of stable planetary orbits, the stability of atoms and molecules, the unique properties of wave operators and axial vector quantities are all essential manifestations of the dimensionality of space. Soon afterwards, Hermann Weyl[127] pointed out that only in $(3+1)$ dimensional space-times can Maxwell's theory be founded upon an invariant, integral form of the action; only in $(3+1)$ dimensions is it conformally invariant, and this

...does not only lead to a deeper understanding of Maxwell's theory but the fact that the world is four dimensional, which has hitherto always been accepted as merely 'accidental', becomes intelligible through it.

In more recent times a number of novel ideas have been added to the store of examples provided by Ehrenfest and these form the basis of the anthropic arguments of Whitrow, Gurevich and Mostepanenko. These arguments, like most other anthropic deductions, rely on the knowledge of our ignorance being complete and assume a 'Principle of Similarity'—that alternative physical laws should mirror their actual form in three dimensions as closely as possible.

As we have already stressed, the development of the first quantitative theory of gravity by Newton brought with it the first universal constant of Nature and this in turn enabled scientific deductions of a very general nature to be made regarding the motions of the heavenly bodies. In his '*Natural Theology*' of 1802 William Paley[125] considered in some detail the consequences of a more general law of gravitational attraction than the inverse square law. What, he asks, would be the result if the gravitational force between bodies varied as an arbitrary power law of their separation; say as,

$$F \propto r^{-N+1} \tag{4.62}$$

Since he believed 'the permanency of our ellipse is a question of life and death to our whole sensitive world' he focused his attention upon the connection between the index N and the stability of elliptical planetary orbits about the Sun. He determined that unless $N<1$ or ≥ 4 no stable orbits are possible[128] and furthermore only in the cases $N=3$ and $N=0$ is Newton's theorem, which allows extended spherically symmetric bodies to be replaced by point masses at their centres of gravity, true. The case $N=0$ he regarded as unstable and so excluded and this provoked Paley to argue that the existence of an inverse square law in Nature was a piece of divine pre-programming with our continued existence in mind. Only in universes in which gravity abides by an inverse square law could the solar system remain in a stable state over long time-scales.

Following up earlier qualitative remarks of Kant and others, Ehrenfest[126] gave a quantitative demonstration of the connection between results of the sort publicized by Paley and the dimensionality question. He pointed out that the Poisson–Laplace equation for the gravitational field of force in an N-dimensional space has a power-law solution for the gravitational potential, ϕ, of the form

$$\phi \propto r^{2-N} \quad \text{if} \quad N \neq 2 \tag{4.63}$$

for a radial distribution of material. The inverse square law of Newton follows as an immediate consequence of the tri-dimensionality. A planetary motion can only describe a central elliptic orbit in a space without $N=3$ if its path is *circular*, but, as Paley also pointed out, such a configuration is unstable to small perturbations. In three dimensions, of course, stable *elliptical* orbits are possible. If hundreds of millions of years in stable orbit around the Sun are necessary for planetary life to develop then such life could only develop in a three-dimensional world. In general, the existence of stable, periodic orbits requires of the central force field $F(r)=-\partial\phi/\partial r$ that $r^3F(r)\to 0$ as $r\to 0$ and $r^3F(r)\to\infty$ as $r\to\infty$. Thus, by (4.62) we require $N<4$. In addition it can be shown that trajectories under a central force[129] field $F(r)$ are only closed when $F(r)\propto r$ or r^{-2}, that is when $N=0$ or 3.

These problems have also been treated in some detail by Tangherlini[130] in the context of general relativity. Specifically, he considers the equations of motion in the gravitational field of an $(N+1)$ dimensional Schwarzschild space-time geometry. By analogy with the Newtonian results it transpires that in general relativity no stable bound orbits are possible in the Schwarzschild geometry for $N>3$, as one would expect.

One of Newton's classic results was his proof that if two spheres attract each other under an inverse square law of force then they may both be replaced by points concentrated at the centre of each sphere, each with a mass equal to that of the associated sphere. We can ask what the general form of the gravitational potential with this property is. Consider a spherical shell of radius a whose surface density is σ and whose centre, at O, lies at distance r from some arbitrary point P outside its edge. If the gravitational potential at r is $\phi(r)$ then the potential at P due to the sphere will be the same as that due to some point mass $M(a)$ at O is

$$M(a)\phi(r)+2\pi\sigma a\lambda(a)=\frac{2\pi\sigma a}{r}\int_{r-a}^{r+a}x\phi(x)\,dx \tag{4.64}$$

where $\lambda(a)$ is a constant that we can always add to the potential without altering the associated force law. There are two classes of solution[131] to (4.64):

(a) *The Yukawa-type potentials:*

$$\phi(r)=\frac{Ae^{\mu r}+Be^{-\mu r}}{r}+E \tag{4.65}$$

with the equivalent point mass given by

$$M(a)=4\pi\sigma a\frac{sh(\mu a)}{\mu} \tag{4.66}$$

where A, B and E are arbitrary real constants, μ is a real or complex constant and $\lambda = 2Ea$. We notice that Newton's result is recovered with $E = 0$ as $\mu \to 0$ in which case $M(a) = 4\pi a^2\sigma$ and the equivalent point mass equals the mass of the sphere.

(b) *The algebraic potential:*

$$\phi = \frac{A}{r} + Br^2 + E \tag{4.67}$$

with the equivalent point mass again having the Newtonian value

$$M(a) = 4\pi\sigma a^2 \tag{4.68}$$

Again, A, B, and E are arbitrary real constants but now

$$\lambda(a) = 2Ea + 2Ba^2 \tag{4.69}$$

The form (4.67) is interesting because the Br^2 is the Newtonian equivalent of adding a cosmological constant to Einstein's equations, (see Chapter 6). We notice that the simple result (4.68) is associated with the specific algebraic form of the potential and hence the dimensionality, N, via (4.63). Recall that Newton delayed publication of his theory of gravitation until he had demonstrated rigorously that his inverse square law of gravitational attraction allowed a spherical shell to be replaced by a point of equal mass located at its centre. If Newton had asked what the general form of the force law could be which possessed this property, he would have calculated (4.67) and could have predicted the existence of the 'cosmological constant'. It often appears that the cosmological constant is inevitable in derivations of general relativity, but has no motivation in Newtonian theory. however, (4.63) and (4.67) show how it arises naturally in the Newtonian formulation.

In order to single out the Newtonian result, $\phi \propto r^{-1}$, which is associated with three-dimensional spaces, (4.63), we must appeal to another property of the inverse square law found by Newton. For the $\phi \propto r^{-1}$ potential the interior of a spherical shell is an *equipotential* region; in general, $\phi(r)$ will only have this property if,[132] for $r < a$,

$$\lambda(a)r = \int_{a-r}^{a+r} x\phi(x)\, dx \tag{4.70}$$

and this has the unique solution

$$\phi(r) = \frac{A}{r} + C \tag{4.71}$$

where C and A are arbitrary real constants and C can be set equal to zero without altering the force law.

These results show why gravitation physics is simplest in three spatial dimensions.[133] The inverse square law of force that is dictated by the three dimensions of space is unique in that it allows the local gravitational field within the spherical region we considered to be evaluated independently of the structure of the entire Universe beyond its outer boundary. Without this remarkable safeguard our local world would be at the mercy of changes in the gravitational field far away across our Galaxy and beyond.[167]

It is widely known that matter is stable: by this we mean that the ground state energy of an atom is finite. However, the common text-book argument which employs the Heisenberg Uncertainty Principle to demonstrate this is actually false. Although the energy equation for a single electron of mass m and charge $-e$ in circular orbit around a nuclear charge $+e$ gives a total atomic energy of

$$E = \frac{h^2}{2mr^2} - \frac{e^2}{r} \tag{4.72}$$

and this energy apparently has a finite minimum of $r_0 \sim h^2/2me^2$ where $E'(r_0) = 0$, it is, in principle, possible for the electron to be distributed in a number of widely separated wave packets. The packet close to the nucleus could then have an arbitrarily sharp momentum and position specification at the expense of huge uncertainty in the other packets. In this manner the ground-state energy might be made arbitrarily negative.

A much stronger, *non-linear* constraint is required in addition to the Heisenberg Uncertainty Principle if one is to rule out ground state energies becoming arbitrarily negative. The strongest result is supplied by the non-linear Sobolev inequality.[134] This supplies the required bound on the ground-state energy and shows that matter is indeed stable in quantum theory.

For these technical reasons analyses of atomic stability such as those of Ehrenfest[126] and Buchel[129] which use only the Uncertainty Principle must be regarded as only heuristic. However, their results are confirmed by an exact analysis of the Schrödinger equation in simple cases. In 1917, Ehrenfest considered only the simple Bohr theory of an N-dimensional hydrogen atom. He found the energy and radii of the energy levels and noted that when $N > 5$ the energy levels increase with quantum number whereas the radii of the Bohr orbits $r_\lambda(N) \sim (me^2\lambda^{-2}h^{-2})^{1/(N-4)}$ decrease with increasing quantum number λ, and electrons just fall into the nucleus. Alternatively, if we write down the total energy for the system and use the Uncertainty Principle to estimate the kinetic energy resisting localization we have (p is the momentum and V the potential energy)

$$E = \frac{p^2}{2m} + V \sim \frac{h^2}{2m}\frac{1}{r^2} - \frac{e^2}{r^{N-2}} \tag{4.73}$$

It can be seen that for $N>5$ there is no energy minimum. For $N=4$ the situation is ambiguous because there ceases to exist any characteristic length in the system. This also indicates that no minimum energy scale can exist. It is possible to demonstrate this more rigorously by including special relativistic effects in the energy equation (4.73). Thus, for $N=4$, the relativistic energy is, (where m_0 is the rest mass of the electron now),

$$E=(p^2c^2+m_0^2c^4)^{1/2}+V \qquad (4.74)$$
$$\sim\left(\frac{c^2h^2}{r^2}+m_0^2c^4\right)^{1/2}-\frac{e^2}{r^2}$$

and so as $r\to 0$, $E\to -1/r^2$ and E can become arbitrarily negative, hence no stable minimum can exist.

On the basis of these arguments it has been claimed that if we assume the structure of the laws of physics to be independent of the dimension, stable atoms, chemistry and life can only exist in $N<4$ dimensions. (Note that in two dimensions all energy levels are discrete and there exists a finite energy minimum together with a spectrum extending to infinity, the radius of the first orbit is huge $\sim$0.5 cm.)

These simple arguments were confirmed by Tangherlini[130] and Gurevich and Mostapenenko[117] when they solved the Schrödinger equation for generalized hydrogen in N-dimensions. Separating the hydrogen wave function into radial and angular parts a radial wave equation can be found in n dimensions, (′ denotes differentiation with respect to r),

$$R''+\left(\frac{N-1}{r}\right)R'+\frac{2m}{h^2}\left[E-\frac{h^2}{2m}\,l\,\frac{(l+N-2)}{r^2}+\frac{e^2}{(N-2)r^{N-2}}\right]R=0 \qquad (4.75)$$

where l is the eigenvalue of the angular momentum. If the effects of special relativity are incorporated this equation is generalized to

$$R''+\left(\frac{N-1}{r}\right)R'+\left[\frac{E^2+m_0^2}{h^2}-\frac{l(l+N+2)}{r^2}+\frac{2Ee^2}{h^2(N-2)r^{N-2}}+\frac{e^4}{h^2(N-2)^2r^{2(N-2)}}\right]R=0 \qquad (4.76)$$

where E is the energy of the electron.

Analysis of these equations indicates that there are no stable bound orbits for $N>3$. These conditions could also be established using the analytical techniques of Lieb.[134] Thus we see that the dimensionality of the Universe is a reason for the existence of chemistry and therefore, most probably, for chemists also.

The arguments cited above have been used to place an upper bound ($N\leqslant 3$) on the spatial dimension of life-supporting universes governed by dimension-independent physical laws. Whitrow[135] attempted to place a lower bound on N by considering the conditions necessary for some crude form of information-processing to exist:

. . . it seems to me that the solution to this problem lies in the geometrical structure of the human brain . . . . In three or more dimensions any number of cells can be connected with each other in pairs without intersection of the joins, but in two dimensions the maximum number of cells for which this is possible is only four.

He argues that if the spatial structure were of dimension two or less then nerve cells (or their analogues) would have to intersect when superimposed and a severe limitation on information-processing of any complexity would result. In effect, Whitrow is ruling out the existence of worlds in which the Jordan Curve Theorem is not true for all possible paths.[136] However, Tangherlini[130] claimed that with a little ingenuity it might be possible to evade this restriction by locating the cells on multiply connected surfaces. The possibility that by such a device intelligent beings could exist in a two-dimensional world of our own conception provokes us to examine the possibility of Abbott's[137] fictional '*Flatland*' a little more seriously.

The Canadian computer scientist A. K. Dewdney[138] has spent considerable effort developing detailed analogues of modern scientific theories and technological devices which would function in a two-dimensional world. In order to make his 'planiverse' viable Dewdney has to deal with biological objections of the type raised by Whitrow.

He counters these objections by claiming that one can construct a credible neural network based on a version of the McCullough–Pitts[139] model for an inter-connected grid of neurons. At each neural intersection a signal either can or cannot be transmitted—this creates a system of binary aritmemtic—and Dewdney imagines that some degree of fidelity may be possible in transmitting signals through the two-dimensional array rather like a grid of dodgem cars passing information at their collision points. However, the McCullough–Pitts neural network seems too dramatic a simplification to provide the basis for a real nervous system since one would like it to have the capacity to repair itself in cases of occasional malfunction.

Many authors[126,140,141] have drawn attention to the fact that the properties of wave equations are very strongly dependent upon the spatial dimension. Three-dimensional worlds appear to possess a unique combination of properties which enable information-processing and signal transmission to occur via electromagnetic wave phenomena. Since our Universe appears governed by the propagation of classical and quantum waves it is interesting to elucidate the nature of this connection with dimensionality and living systems.

Let us recall, as motivating examples, the solutions to the simple classical wave equation in one, two and three dimensions.

One dimension:

$$\frac{1}{c^2}\frac{\partial^2 u}{\partial t^2}=\frac{\partial^2 u}{\partial x^2}; \tag{4.77}$$

where c is the signal propagation speed and where initial conditions for $u(x, t)$ are set at $t=0$ as

$$\begin{aligned} u(x, 0) &= f(x) \\ \frac{\partial u}{\partial t}(x, 0) &= g(x) \end{aligned} \tag{4.78}$$

This has the solution of D'Alembert,

$$u(x, t)=\frac{f(x+ct)+f(x-ct)}{2}+\frac{1}{2c}\int_{x-ct}^{x+ct} g(y)\, dy \tag{4.79}$$

Two dimensions:

$$\frac{1}{c^2}\frac{\partial^2 u}{\partial t^2}=\frac{\partial^2 u}{\partial x^2}+\frac{\partial^2 u}{\partial y^2} \tag{4.80}$$

with initial conditions at $t=0$ for $u(x, y, t)$ of

$$\begin{aligned} u(x, y, 0) &= f(x, y) \\ \frac{\partial u}{\partial t}(x, y, 0) &= g(x, y) \end{aligned} \tag{4.81}$$

This has the solution of Poisson

$$u(x, y, t)=\frac{1}{2\pi c}\frac{\partial}{\partial t}\iint_{\rho\leq ct}\frac{f(\xi, \eta)\, d\xi\, d\eta}{(c^2t^2-\rho^2)^{1/2}}+\frac{1}{2\pi c}\iint_{\rho\leq ct}\frac{g(\xi, \eta)\, d\xi\, d\eta}{(c^2t^2-\rho^2)^{1/2}} \tag{4.82}$$

where $\rho^2=[(\xi-x)^2+(\eta-y)^2]$.

Three dimensions:

$$\frac{1}{c^2}\frac{\partial^2 u}{\partial t^2}=\frac{\partial^2 u}{\partial x^2}+\frac{\partial^2 u}{\partial y^2}+\frac{\partial^2 u}{\partial z^2} \tag{4.83}$$

with initial conditions at $t=0$ for $u(x, y, z, t)$ of

$$\begin{aligned} u(x, y, z, 0) &= f(x, y, z) \\ \frac{\partial u}{\partial t}(x, y, z, 0) &= g(x, y, z) \end{aligned} \tag{4.84}$$

This has the solution of Kirchhoff

$$u(x, y, z, t) = \frac{1}{4\pi c^2} \frac{\partial}{\partial t} \left(\frac{1}{t} \iint_{r=ct} f(\varepsilon, \eta, \zeta) \, dS \right) + \frac{1}{4\pi c^2 t} \iint_{r=ct} g(\xi, \eta, \zeta) \, dS \quad (4.85)$$

where $r^2 = (\xi - x)^2 + (\eta - y)^2 + (\zeta - z)^2$ and dS is the surface element with respect to (ξ, η, ζ) on the sphere $r = ct$ centred at $(x, y, z) = (0, 0, 0)$.

From these three solutions (4.77–4.85) something remarkable emerges. We see that in the one and two-dimensional cases, the domain of dependence which determines the solution $u(x, t)$ at point (x, t) is given by the closed interval $[x - ct, x + ct]$ and the disk (interior plus boundary) $r \leq ct$, respectively. Therefore in both cases the signals may propagate at any speed less than or equal to c. In complete contrast, the three-dimensional solution has a domain of dependence consisting only of the *surface* of the sphere of radius ct. All three-dimensional wave phenomena travel *only* at the wave velocity c.[142]

What this means in practice is that in two-dimensional spaces wave signals emitted at different times can be received simultaneously: signal reverberation occurs. It is impossible to transmit *sharply* defined signals in two dimensions, for example, by waves on a liquid surface. Now it has been shown[141] that in general the transmission of wave impulses in a reverberation-free fashion is impossible in spaces with an *even* number of spatial dimensions. The favourable odd-dimensional cases are said to obey Huygen's Principle.[143] This situation has led many to suppose that life could only exist in an odd-dimensional world because living organisms require high-fidelity information transmission at a neurological or mechanical level.

Interestingly, one can narrow down the number of reasonable odd-dimensional spaces even more dramatically by appealing to the need for wave signals to propagate *without distortion* if they are to be useful in a mechanical or neural network. Three-dimensional worlds allow spherical waves of the form

$$u(x_1, x_2, x_3; t) = h(r) f(r - ct) \quad (4.86)$$

with

$$r^2 = \sum_{i=1}^{3} x_i^2 \quad (4.87)$$

to propagate in *distortionless* fashion to large distances. But this is no longer the case for odd $N > 3$, For example, in seven dimensions, a solution of the spherical wave equation is,

$$u(x_1, \ldots, x_7; t) = \frac{A}{r^5} f(t - r/c) + \frac{Bf'(t - r/c)}{r^4} + \frac{Df''(t - r/c)}{r^3} \quad (4.88)$$

where A, B and D are constants. Thus, at time t there is no reverberation; only signals which were emitted at the time $(t-r/c)$ are received. However, these signals are now strongly distorted because at large r the terms in f'' and f' determine the form of the signal $u(x, t)$.

Only three-dimensional worlds appear to possess the 'nice' properties necessary for the transmission of high-fidelity signals because of the simultaneous realization of sharp and distortionless propagation. This situation led Courant and Hilbert[143] to conclude that

> ... our actual physical world, in which acoustic or electromagnetic signals are the basis of communication seems to be singled out among other mathematically conceivable models by simplicity and harmony.

If living systems require high-fidelity wave propagation for their existence to be possible, then we could not expect to observe the world to possess other than three spatial dimensions.

Some recent investigations by Nielsen[113,114] and collaborators as to the inevitability of local gauge invariance arising in low-energy phenomenology from a random Lagrangian have yielded some interesting claims regarding dimensionality. They point out that four (3 space plus 1 time) dimensional space-times are preferred in that they allow four linearly independent 2×2 Hermitian matrices and the two-component Weyl equations to exist in $3+1$ dimensions. A two-component Weyl equation in a World of more than $3+1$ dimensions would give rise to some directions in which fermions could only move with fixed velocity and along that direction they could not overtake each other. Observers composed of Weyl particles would only ever encounter a piece of three-dimensional space.

Our Universe appears to possess a collection of fundamental or 'natural' units of mass, length and time which can be constructed from the physical constants G, h and c, (see ref. 144). A dimensionless constant can only be constructed if the electron charge, e, is also admitted and then we obtain the dimensionless quantity e^2/hc first emphasized by Sommerfeld.[21] In a world with N dimensions the units of h and c remain ML^2T^{-1} and LT^{-1} in mass (M), length (L) and time (T), but the law of gravitation changes in accord with (4.61) and hence the units of G become $M^{-1}L^NT^{-2}$. Likewise, Gauss' theorem relates e to the spatial dimension and the units of e^2 are ML^NT^{-2}. Thus in N dimensions the dimensionless constant of Nature is proportional to

$$h^{2-N}e^{N-1}G^{(3-N)/2}c^{N-4} \tag{4.89}$$

It is interesting to notice that for $N=1$, 2, 3 and 4 the constants of electromagnetism, quantum theory, gravity and relativity are absent respectively. Only for $N>4$ are they all included in a single dimensionless unit. Only for $N=3$ is gravity the distinguished interaction.

The dimensionality of space also seems to shed light upon what we might term 'the unreasonable effectiveness of dimensional analysis'. The technique of estimating the rough magnitude of physical quantities using dimensional analysis, so beloved of professional physicists (see Chapter 5 for extensive examples) was first employed by Newton[145] and Fourier.[146] It enables one, for example, to estimate the period, T, of a simple pendulum of length l oscillating in a gravitational field with acceleration due to gravity g, as $T \sim (l/g)^{1/2}$; in good agreement with the exact formula

$$T = 2\pi \left(\frac{l}{g}\right)^{1/2} \tag{4.90}$$

But why, when we use this technique, do we find it accurate; why are the dimensionless factors of proportionality always of order unity? In discussing this problem Einstein remarked[147] concerning the fact that these dimensionless factors invariably turn out to be of order unity, that

> we cannot require this rigorously, for why should not a numerical factor like $(12\pi)^3$ appear in a mathematical-physical deduction? But without doubt such cases are rarities!

We would like to suggest that it is the low dimension of space that makes dimensional analysis so effective. The factors like 2π in (4.90) invariably have a geometrical origin (note we are not concerned with dimensionless combinations of fundamental physical constants here, only numerical factors). Most of the quantities appearing in physical formulae are linked to circumferences, areas or volumes in some way.[148] The purely arithmetic quantities that appear in physical formulae like (4.90) are usually, therefore, associated with the geometry of circles, shells and spheres.[149] They derive ultimately from the coefficients in the expressions for the circumference, C, and volume, V of N-dimensional balls of radius r:

$$C(N) = \frac{2\pi^{(N+1)/2} r^{N-1}}{N\Gamma\left(\frac{N+1}{2}\right)} \tag{4.91}$$

and

$$V(N) = \frac{2\pi^{(N+1)/2} r^{N}}{\Gamma\left(\frac{N+1}{2}\right)} \tag{4.92}$$

where $\Gamma(N)$ is the gamma function.

These formulae have interesting behaviours for $N > 3$ as can be seen in Figure 4.2,

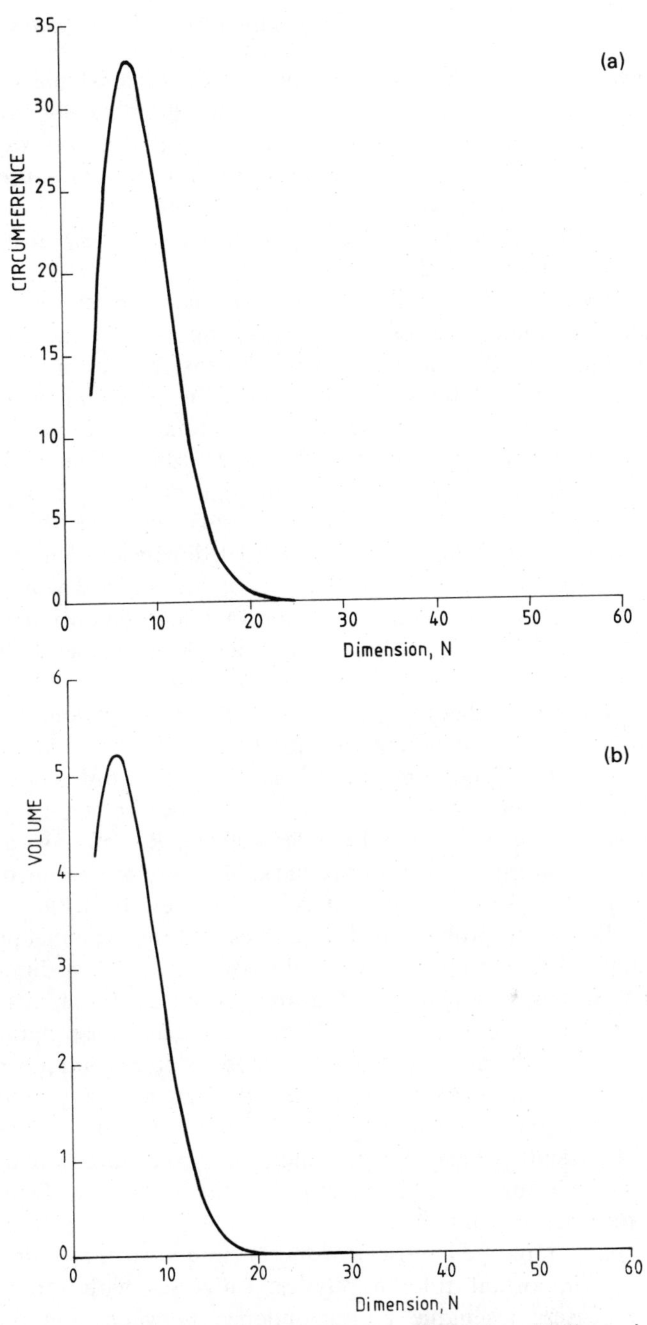

Figure 4.2. (a) The variation in circumference, $C(N)$, of an N-dimensional ball of unit radius as a function of N. (b) The variation in volume, $V(N)$, of an N-dimensional ball of unit radius as a function of N; see equations (4.91) and (4.92).

In particular, we note that the magnitude of the dimensionless geometrical factors pre-multiplying r^{N-1} and r^N depart dramatically from unity for large N as the gamma function $\Gamma(N) \propto (N/e)^N$ for large N. If we lived in a world with $N \gg 3$ dimensional analysis would not be a very useful tool for approximate analysis: the dimensionless geometrical factors would invariably be enormous. This is perhaps the nearest one can get to an answer to the problem posed by Einstein.[150]

So far, we have displayed a number of special features of physics in three dimensions under the assumption that the form of the underlying differential equations do not change with dimension. One might suspect the form of the laws of physics to be special in three dimensions only because they have been constructed solely from experience in three dimensions. If we could live in a world of seven dimensions perhaps we would end up formulating its laws in forms that made seven dimensions look special. One can test the strength of such an objection to some extent by examining whether or not 3 and 3+1 dimensions lead to special results in pure mathematics where the bias of the physical world should not enter.[151] Remarkably, it does appear that low-dimensional groups and manifolds do have anomalous properties. Many general theorems remain unproven or are untrue only in the case of $N=3$; a notable example is Poincaré's theorem that a smooth N-dimensional manifold with homotopy type S^N is homeomorphic to S^N. This theorem is known to be true if $N \neq 3$ and the homeomorphism is in fact a diffeomorphism if $N = 1$ 2, 5 or 6 (the $N=4$ case is open). It is still not known if Poincaré's conjecture is true for $N=3$. For Euclidean space, R^N, all have a unique differentiable structure if $N \neq 4$, but remarkably there are an uncountable number of differentiable structures if $N=4$ (see ref. 162). Other examples of this ilk are the problem of Schoenflies and the Annulus problem; each has unusual features when $N=3$. In addition, the low-dimensional groups possess many unexpected features because of the 'accidental' isomorphisms that arise between small groups. The twistor programme of Penrose,[152] takes advantage of some of these features unique to 3+1 dimensional space-times. As a general rule, the geometry and topology of two-dimensional spaces is simple, that of three and four dimensions is unusual and difficult, whilst that of dimensions exceeding four does not exhibit any significant dependence on the dimensionality. Dimensions three and four act as a threshold.

There is one simple geometrical property unique to three dimensions that plays an important role in physics: universes with three spatial dimensions possess a unique correspondence between rotational and translational degrees of freedom. Both are defined by only three components. In geometrical terms this dualism is reflected by the fact that the number of coordinate axes, N, is only equal to the number of planes through pairs of axes, $N(N-1)/2$, when $N=0$ or 3. These features are exploited in physics by the Maxwell field. In an $(N+1)$ dimensional space-time, electric, **E**, and magnetic, **B**, vectors can be derived from an $(N+1)$ dimensional potential **A**. The field **B** is derived from $N(N+1)/2$ components of curl **A**, whilst the **E** field derives from the N components of $\partial\mathbf{A}/\partial t$. Alternatively, we might say that in order to represent an antisymmetric second rank tensor as a vector, the $N(N-1)/2$ independent components of the tensor must equal the spatial dimension, N. So the existence of axial vector representations[153] of quantities like the magnetic vector **B** and the particular structure of electromagnetic fields is closely linked to the tri-dimensional nature of space.

There also exists an interesting property of Riemannian spaces which has physical relevance: in an $(N+1)$ dimensional manifold the number of independent components of the Weyl tensor is zero for $N \leq 2$ and so all the 1, 2 and 3 dimensional space-times will be conformally flat and they will not contain gravitational waves. The non-trivial conformal structure for $N=3$ leads to the properties of general relativity[154] in four-dimensional space-times.

As a final example where the mathematical consequences of dimensionality spill over into areas of physics we should mention the theory of dynamical systems, or ordinary differential equations,

$$\dot{\mathbf{x}} = F(\mathbf{x}); \qquad \mathbf{x} = (x_1, \ldots, x_N) \tag{4.93}$$

The solution of the system (4.93) corresponds to a trajectory in an N-dimensional phase space. We discussed earlier, why in two dimensions the qualitative behaviour of the possible trajectories is completely classified. As trajectories cannot cross without intersecting in two dimensions, the possible stable asymptotic behaviours are simple: after large times trajectories either approach a stable focus (stationary solution) or a limit cycle (periodic solution). However, when $N \geq 3$ trajectories can behave in a far more exotic fashion. They are now able to cross and develop complicated knotted configurations without intersecting. All the possible behaviours as $t \to \infty$ are not known for $N \geq 3$. When $N \geq 3$ it has been shown[155] that the *generic* behaviour of trajectories is to approach a *strange attractor*.

Before ending our investigation of how the dimensionality of the Universe enters into the structure of physics and those features of the world that allow life to exist, we should mention that we have assumed that our Universe does actually possess only three spatial dimensions. This may seem self-evident, but it may not in fact be true.

The idea that the Universe really does possess more than three spatial dimensions has a distinguished history.[157] Kaluza and Klein sought to associate an extra spatial dimension with the existence of electromagnetism. Under a particular symmetry assumption Einstein's equations in $4+1$ dimensions look like Maxwell's equations in $3+1$ dimensions together with an additional scalar field. Very roughly speaking one imagines uncharged particles as moving only in the $3+1$ dimensional subspace but charged particles move through $4+1$ dimensions. Their direction of motion determines the sign of their charge. Miraculously, local gauge invariances of $3+1$ dimensional space-time appear entirely as coordinate invariances in $4+1$ dimensions.

Supersymmetric gauge theories have rekindled interest in higher dimensional gauge theories that reduce to the $N=3$ theory by a particular process of dimensional reduction. A topical example is $10+1$ dimensional supergravity theory.[158] By analogy with the original Kaluza–Klein theories we would associate $3+1$ of these dimensions with our familiar space-time structure whose curvature is linked to gravitational fields while the additional dimensions correspond to a set of internal gauge symmetries. We perceive them as electromagnetic, weak and strong charges whose internal gauge invariances are just higher dimensional coordinate invariances. These extra dimensions are expected to be compactified to sizes of order

$$L \sim \alpha_*^{-1/2} L_P \tag{4.94}$$

where $L_P = (Gh/c^3)^{1/2} \sim 10^{-33}$ cm is the Planck length and $\alpha_* = 10^{-1} - 10^{-2}$ is the gauge coupling at the grand unification energy (see Chapter 5). Thus, according to such theories the Universe will be fully N-dimensional (with $N > 3$) when the Big Bang is hotter then $\sim 10^{17}$ GeV but all except three spatial dimensions will become confined to microscopic extent when it cools below this temperature after about 10^{-40} s of expansion. Only three dimensions will be perceived by living beings.

Kaluza–Klein cosmologies of this type have two exciting consequences that may allow them to be experimentally tested and which bring us around full circle to some of the questions concerning fundamental constants that motivated the introduction of the Anthropic Principles by Dicke and Carter. First, it has been shown[159] that they allow, in principle, the exact numerical calculation of certain fundamental constants of Nature, like $e^2/\hbar c$, in terms of combinatorical factors. Second, the time-variation of the extra compactified dimensions can lead to time-evolution of what otherwise we would regard as time-independent constants in our three space dimensions.[160]

Suppose there exist an additional D spatial dimensions to the Universe, and the distances between points in these extra dimensions change in time by a scale factor $R_D(t)$. The cosmological evolution of such a world with $D+3$ spatial dimensions can be studied, using higher dimensional extensions of general relativity. The gravitational constant for all the dimensions, $\tilde{G}$, will be related to the usual Newtonian gravitational constant we observe in three dimensions, G, by $G = \tilde{G}R_D^{-D}$, and the coupling strengths of the other interactions vary inversely with the geometric mean radius of the extra dimensions.[161] For example, if these extra dimensions do exist, then the fine structure constant and the gravitational constant would be seen to vary with R_D as

$$G \propto R_D^{-D}; \qquad \alpha \propto R_D^{-2} \tag{4.95}$$

Analogous variations in the strong and weak interaction strengths would be seen also. These predictions are rather dramatic and can only be reconciled with our observations of the time-invariance of 'constants' of Nature, like G and α, if R_D is essentially unchanging with time today. At present, it is not known what could keep all the additional dimensions of the Universe static whilst the three we exist in expand cosmologically. Such a demarcation seems slightly unnatural. Perhaps effects associated with the quantum character of gravitation, which only become strongly evident on length-scales smaller than the Planck length, keep the extra D dimensions confined to sizes close to the Planck length, whilst the remaining three dimensions expand. This may not be the best way of describing this problem, however. It could be said that the extra dimensions have the naturally expected dimension if they are all of the Planck length in extent. The real mystery is why three of them are about 10^{60} times larger than the Planck length. No answer is known, although one might hope that an answer could be provided by showing that the three dimensions inflate along the lines to be described in Chapter 6. Of course, there are Weak Anthropic reasons why we are observing a Universe which has a three-dimensional size of this enormous magnitude.

At present there is no theoretical understanding of why just *three* dimensions have expanded to a large size if the others are indeed confined to minute extent. However, the Anthropic arguments we gave concerning the special properties of three-dimensional space and four-dimensional space-time show that there would be a Weak Anthropic explanation for this observation also; but, for all we know, it may also be a consequence of the unique topological properties that four-dimensional manifolds have recently been found to possess. The fact that only they admit more than one distinct differentiable structure may well turn out to have something to do with the fact that observable space-time has four dimensions.

In this chapter we have traced some aspects of the history of coincidences in the physical sciences, concentrating upon the famous large number coincidences of Weyl and Dirac. We have tried to show that the recognition of coincidences often precedes the development of rigorous new physical explanations. Dirac's coincidences stimulated a vast outpouring of effort to invent and investigate new theories of gravity in which the strength of gravity decreased with cosmic time. Eventually, Dirac's arguments for such a variation were undermined by Dicke's use of the Anthropic Principle. The Dirac coincidence was shown to be a necessary property of an expanding universe containing carbon-based observers. Earlier, Hoyle had been able to use Anthropic reasoning to predict successfully the presence of a new resonance level in the carbon nucleus. We found that other scientists had developed Anthropic arguments independently, to explain why we must find the observed universe to possess three dimensions. An analysis of this question sheds light on many aspects of physics and reveals the extent to which the form of the laws of Nature are conditioned by the dimensionality of space. Finally, we saw how attempts to explain gauge invariance in Nature lead to new theories, one in which there are essentially no laws of physics at all and another in which the Universe is required to possess additional spatial dimensions. Both have fascinating interconnections with the Anthropic Principle and the questions of coincidences and varying constants which provoked its resuscitation in the 1960's.

In order to follow the detailed examples put forward by Carter and others during this more recent period, we must first examine the spectrum of structures we find around us in Nature, and attempt to ascertain which of their characteristics are determined by constants of Nature. This will enable us to separate the invariant aspects of the World from those which, being due to chance, could have been found to be otherwise arranged today. In short, we must separate coincidence from consequence.

References

1. R. H. Dicke, *Rev. Mod. Phys.* **29,** 375 (1957); *Nature* **192,** 440 (1961). B. J. Carr and M. J. Rees, *Nature* **278,** 605 (1979).
2. Archimedes, *The sand reckoner*, in *The works of Archimedes*, ed. T. L. Heath, (Dover, NY, 1953), pp. 221–32.
3. G. Sarton, *History of science*, Vol. 2 (Norton, NY, 1959).
4. E. R. Harrison, *Physics Today* **25,** 30 Dec. (1972).
5. H. A. M. Snelders, *Janus* **60,** 25 (1973); *Studies in Romanticism* **9,** 193 (1970).
6. It appears that early Buddhist cosmology had some preoccupation with large numbers $\sim 10^{51}$; see W. M. McGovern, *Manual of Buddhist philosophy*, Vol. 1 (Kegan Paul, Trench, Trübner London 1923), pp. 39f.
7. C. S. Weiss, *Abh. Physik.*, Classe der Berlin Akad. d. Wissen (1820).
8. J. D. Titius v. w., *Betrachtung uber die Natur vorn Herrn Karl Bonnet* (Johann Friedrich Junius, Leipzig, 1766).

9. C. Bonnet, *Contemplation de la Nature* (Marc-Michel Rey, Amsterdam, 1764).
10. The note in question occurs on p. 8 of ref. 9 and p. 7 of ref. 8.
11. For a comprehensive historical account of the Titius–Bode Law see M. N. Nieto, *The Titius–Bode law of planetary distances: its history and theory* (Pergamon, Oxford, 1972).
12. J. E. Bode, *Anleitung zur Kenntniss des gestirnten Himmels*, 2nd edn (Hamburg, 1772), p. 462.
13. *The scientific papers of Sir William Herschel*, Vol. 1 (Royal Society, London, 1912) p. xxiv.
14. M. W. Ovenden, *Nature* **239,** 508 (1972); M. Lecar, *Nature* **242,** 318 (1973); S. F. Dermott, *Nature* **244,** 17 (1973).
15. J. Balmer, *Basel. Verh.* **7,** 548; 750 (1885) and see ref. 22.
16. J. Rydberg, *Phil. Mag.* **29,** 331 (1890).
17. Cited in A. Hermann, *The genesis of quantum theory* (MIT Press, Cambridge, Mass., 1971). See also N. Bohr, *On the constitution of atoms and molecules*, ed. L. Rosenfeld (W. A. Benjamin, Copenhagen, and NY, 1963), repr. from *Phil. Mag.* **26,** 2 (1913).
18. M. Planck, letter to P. Ehrenfest, Rijksmuseum Leiden, Ehrenfest collection (accession 1964), July 1905. See Hermann, ref. 17.
19. A. Einstein, *Phys. Z.* **10,** 192 (1909).
20. A. Haas, see Herman, ref. 17, p. 148.
21. A. Sommerfield, *Phys. Z.* **12,** 1057 (1911).
22. M. Weber and R. Kohlrausch, see M. Mason and W. Weaver. *The electromagnetic field* (Dover, NY, 1929), p. x, and E. T. Whittaker, *History of the theories of aether and electricity* (Thomas Nelson, London, 1953), pp. 110, 266; see also *The scientist speculates*, ed. J. Good (Basic Books, NY, 1962), p. 315.
23. For a detailed survey see *Black holes, Les Houches lectures*, ed. C. de Witt and B. S. de Witt (Gordon & Breach, NY, 1973); and for a personal account by someone extensively involved in these developments, J. D. Bekenstein, *Physics Today* **33** (1), 24 (1980).
24. S. W. Hawking, *Nature* **248,** 30 (1974); *Commun. Math. Phys.* **43,** 199 (1975); *Scient. Am.* **236,** 34 (1977). D. W. Sciama, *Vistas in Astron.* **19,** 385 (1976); V. Frolov, *Sov. Phys. Usp.* **19,** 244 (1976).
25. H. Weyl, *Ann. Physik.* **59,** 129 (1919) with our translation of the German original. A more general discussion of large number coincidences is given by Weyl in *Naturwiss.* **22,** 145 (1934). It is interesting that the hypothesis of equality between large numbers of similar order was called 'Weyl's hypothesis' by F. Zwicky, *Phys. Rev.* **55,** 726 (1939).
26. A. S. Eddington, *The mathematical theory of relativity* (Cambridge University Press, London, 1923), p. 167.
27. E. Whittaker, *Eddington's principle in the philosophy of science* (Cambridge University Press, London, 1951), see also *Am. Scient.* **40,** 45 (1952); *Math. Gaz.* **29,** 137 (1945); *From Euclid to Eddington* (Dover, NY 1958): H. Dingle, *The sources of Eddington's philosophy* (Cambridge University Press, Cambridge, 1954); N. B. Slater, *Development and meaning of Eddington's fundamental theory* (Cambridge University Press, Cambridge, 1957); C. W.

Kilmister and B. O. J. Tupper, *Eddington's statistical theory* (Oxford University Press, Oxford, 1962).

28. A. S. Eddington, *New pathways in science* (Cambridge University Press, London, 1935); *Fundamental theory* (Cambridge University Press, London, 1946); *The nature of the physical world* (Cambridge University Press, London, 1929); *The philosophy of physical science* (Cambridge University Press, London, 1939).
29. An interesting discussion of attempts to derive a fundamental mass in terms of N is given by E. Schrödinger in *Scientific papers presented to Max Born* (Oliver & Boyd, Edinburgh, 1953), although this article was originally written about 1940. If the lowest momentum state that can be measured in a finite Universe is $p \sim h/R_g$ where $R_g \sim GM/c^2$ is the gravitational radius of Universe, then we have $M_* \sim h/R_g c \sim 10^{-64}$ gm for the fundamental mass when $R_g \sim ct_0 \sim 10^{27}$ cm. Schrödinger derives another fundamental mass, $m'_* \sim 10^{-38}$ gm which has no known significance outside of this calculation. Other coincidences of a similar nature are described by E. J. Zimmerman, *Am. J. Phys.* **23,** 136 (1955).
30. A clear discussion of this later work is given by J. Singh, *Great ideas of modern cosmology* (Penguin, London, 1970).
31. G. Beck, H. Bethe and W. Riezler, *Naturwiss.* **19,** 39 (1931). For an amusing commentary on the consequences of this hoax see the article by M. Delbrück in *Cosmology, fusion and other matters*, ed. F. Reines (A. Hilger, Bristol, 1972).
32. M. Born, *Experiment and theory in physics* (Cambridge, University Press, Cambridge, 1944), p. 37. A more objective difficulty was suggested by B. Podolsky and H. Branson, *Phys. Rev.* **57,** 494 (1940). They studied the behaviour of the Dirac equation for the electron in an expanding Universe and claimed this investigation lent no support to the idea that the mass of the electron is determined by the radius of the Universe.
33. A. Haas, *Anz. Akad. Wiss. Wien*, No. 11 (1932); *Naturwiss.* **20,** 906 (1934); *Die Kosmologischen Probleme der Physik* (Akad. Verlagsgesselschaft, Leipzig, 1934); *Phys. Rev.* A **48,** 973, (1935); *Phys. Rev.* A **51,** 65 (1937); *Phys. Rev.* A **51,** 1000 (1937); *Phys. Rev.* A **53,** 207 (1938); *Proc. natn. Acad. Sci., U.S.A.* **24,** 274 (1938); *Science* **87,** 195 (1938); see also K. Sitte and W. Glaser, *Z. Physik* **88,** 103 (1934).
34. J. Q. Stewart, *Phys. Rev.* **38,** 2071 (1931).
35. S. Weinberg, *Gravitation and cosmology* (Wiley, NY 1972), p. 619.
36. S. Hayakawa and H. Tanaka, *Prog. Theor. Phys.* **25,** 858; S. Hayakawa, *Prog. Theor. Phys. Suppl.* **25,** 100 (1963); *Prog. Theor. Phys. Suppl.* **33,** 532 (1965); N. Hokkyo, *Prog. Theor. Phys.* **34,** 856 (1965); *Prog. Theor. Phys.* **39,** 1078 (1968); *Prog. Theor. Phys.* **40,** 1439 (1968). These authors also attempt to extend Eddington's approach to calculate the weak and strong coupling constants. Aspects of these investigations have been rediscovered by J. K. Laurence and G. Szamosi, *Nature* **252,** 538 (1974). An attempt to incorporate the gravitational potential and the Yukawa nuclear potential was made by K. C. Wang and H. L. Tsao, *Phys. Rev.* **66,** 155 (1944).
37. E. Teller, *Phys. Rev.* **73,** 801 (1948).
38. C. Isham, A. Salam and J. Strathdee, *Phys. Rev.* D **3,** 1805 (1971).

39. L. Landau, in *Niels Bohr and the development of physics*, ed. W. Pauli (McGraw-Hill, NY 1955), p. 52.
40. P. J. E. Peebles, R. H. Dicke, *Phys. Rev.* **128,** 2006 (1962).
41. F. Lenz, *Phys. Rev.* **82,** 554 (1951).
42. R. Muradyan, *Sov. J. Part. Nucl.* **8,** 78 (1977).
43. Y. Tomozawa, *Prog. Theor. Phys.* **52,** 1715 (1974).
44. M. H. MacGregor, *Phys. Rev.* D **9,** 1259 (1974); *Nuovo Cim. Lett.* **1,** 759 (1971).
45. For example, should one ascribe significance to empirical relations like $m_e/m_\mu \sim 2\alpha/3$, $m_e/m_\pi \sim \alpha/2$, $(3\alpha \ln 2\pi)^{-1} m_e = 105.794\,\text{MeV} \sim m_\mu$, $m_\eta - 2m_{\pi^\pm} = 2m_{\pi^0}$, or the suggestion that perhaps elementary particle masses are related to the zeros of an appropriate special functions? A. O. Barut has given a simple formula for predicting the masses of charged leptons which yielded a good predition of the τ mass, *Phys. Rev. Lett.* **42,** 1251 (1979); *Phys. Lett.* B **73,** 310 (1978), see also W. M. Lipmanov, *Sov. J. Nucl. Phys.* **30,** 227 (1980). Using the Hagedorn statistical bootstrap M. Alexanian, *Phys. Rev.* D **11,** 722 (1975), has used the large number of coincidences to derive a relation for the fine structure constant as

$$\alpha = \frac{3\pi m_e m_N}{32 m_\pi^2} = (137.94 \pm 0.07)^{-1}$$

46. P. A. M. Dirac, *Nature* **139,** 323 (1937); *Proc. R. Soc.* A **165,** 199 (1938); *Pon. Acad. Comment.* **11,** 1 (1973); *Proc. R. Soc.* A **333,** 403 (1973), **338,** 439 (1974); *Nature* **254,** 273 (1975). For a commentary on the development of ideas see P. C. W. Davies, *Nature* **250,** 460 (1974). Soon after the first of Dirac's articles, a resumé of his theory together with an analysis of its rationale was given by P. Jordan, *Naturwiss.* **25,** 513 (1937) and **26,** 417 (1938).
47. For further discussion of this interpretation see the comments by J. A. Wheeler following Dirac's contribution to *The physicist's conception of Nature*, ed. J. Mehra (Reidel, Dordrecht, 1973). Dirac claims this formulation does not explain the size of the 'large numbers', only their coincidental values. However, if the Universe is closed, the hypothesis of varying constants does not explain them either. See also J. D. Barrow, *Nature* **282,** 698 (1979), where a new large number coincidence is pointed out; namely, that the ratio of the proton half-life (predicted by Grand Unified gauge theories) to the Planck time is $\sim 10^{80}$, see also J. D. Barrow, *Nature* **287,** 566 (1980).
48. A. Haas, *Phys. Rev.* **53,** 207 (1938) and *Proc. natn. Acad. Sci., U.S.A.* **24,** 274 (1938), considered a coincidence of the form $(R_g/a)^2 = (e^2/Gm_N m_e)^2$ where $R_g = 2Nm_N G/c^2$ is the gravitational radius of the Universe.
49. Y. B. Zeldovich, *Sov. Phys. Usp.* **5,** 931 (1963).
50. S. Chandrasekhar, *Nature* **139,** 757 (1937); see also the remarks by P. Dirac in *Nature* **139,** 1001 (1937). It is interesting to note that E. R. Harrision, *Phys. Rev.* D **1,** 2726 (1970), has argued that with the assumption of a Hagedorn state for high density matter in the early stages of the Universe, and the absence of preferred mass scales among the primordial in-

homogeneities, a galaxy might have a mass $\sim m_\pi(hc/Gm_\pi^2)^{3/2} \sim 10^{11}\ M_\odot$, where m_π is the pion mass. However, this argument is based upon the assumption that the pion is an elementary particle without substructure. Modern theories indicate they have constituents with a pointlike character, and the reason for regarding m_π as fundamental disappears. Another estimate, with a firmer physical basis, is the argument by J. Silk. *Nature* **265,** 710 (1977), *Astrophys. J.* **211,** 638 (1977), J. J. Binney, *Astrophys. J.* **215,** 483 (1977) and M. J. Rees and J. P. Ostriker, *Mon. Not. R. astron. Soc.* **183,** 341 (1978), who have all shown that only masses smaller than $\sim(hc/Gm_N^2)^2\alpha^5(m_N/m_e)^{1/2}m_N \sim 10^{12}M_\odot$ are able to cool, fragment, and develop substructure during the lifetime of the Universe (see Chapter 6).

51. D. S. Kothari, *Nature* **142,** 354 (1938) and see also *Proc. R. Soc.* A **165,** 486 (1938), argued that the Chandrasekhar mass [M_0] (S. Chandrasekhar, *Mon. Not. R. astron. Soc.* **91,** 456 (1931)) must be connected to the large numbers N_1 and N_2, and so it must mirror their time-dependence and '... according to the ideas of Dirac and Milne will be reflected in a corresponding time-dependence of M_0 and R_{max} and other astrophysical magnitudes'.
52. F. Zwicky, *Phys. Rev.* **55,** 726 (1937); see ref. 25, *Phys. Rev.* **43,** 1031 (1933) and **53,** 315 (1938); *Phil. Sci.* **1,** 353 (1934); *Proc. natn. Acad. Sci., U.S.A.* **23,** 106 (1937).
53. P. Jordan, *Ann. Phys.* **36,** 64 (1939); *Nature* **164,** 637 (1949); *Schwerkraft und Weltall* (2 vols, Vieweg, Braunschweig, 1955); *Z. Physik.* **157,** 112, *Rev. Mod. Phys.* **34,** 596; *Z. Physik.* **201,** 394 (1967); *Z. Astrophys.* **68,** 201 (1968); *The expanding earth* (Pergamon, NY, 1971). Jordan rejected the idea of extending Dirac's hypothesis to give time-variation of the weak coupling constant (manifested in radioactivity) or proton-electron mass ratio; *Z. Physik.* **113,** 660 (1939), *Naturwiss.* **25,** 513 (1937).
54. F. Hoyle, *Mon. Not. R. astron. Soc.* **108,** 372 (1948); H. Bondi and T. Gold, *Mon. Not. R. astron. Soc.* **108,** 253 (1948); see also H. Bondi, *Cosmology* (Cambridge University Press, London, 1952). Bondi's book contains an excellent chapter on 'large numbers' which influenced several later workers, notably Carter.
55. I. Newton, *Philosophiae naturalis principia mathematica* (1713), transl. A. Motte (revised F. Cajori), (University of California Press, Berkeley 1946).
56. W. Thomson, and P. G. Tait, *Natural philosophy*, Vol. 1, p. 403 (1874).
57. G. E. J. Gheury de Bray, *Astron. Nachr.* **230,** 449 (1927); *Nature* **127,** 892 (1931); V. S. Vrklyan, *Z. Physik*, **63.** 688 (1930).
58. H. E. Buc, *J. Franklin Inst.* **214,** 197 (1932), suggested a 'tired light' hypothesis for the cosmological redshift.
59. P. I. Wold, *Phys. Rev.* **47,** 217 (1935) suggested the speed of light varies linearly with time; $c = c_0(1 - kt)$; where c_0, k are constant.
60. J. A. and B. Chalmers, *Phil. Mag.* **19,** 436 (1935) Suppl.
61. W. Nernst, *Ann. Physik.* **32,** 38 (1938).
62. S. Sambursky, *Phys. Rev.* **52,** 335 (1937).
63. J. Jeans, *Nature* **128,** 703 (1931).
64. G. Lemaitre, *Nature* **128,** 704 (1931).
65. F. Hoyle, *Astrophys. J.* **196,** 661 (1975).
66. A. S. Eddington, *Fundamental theory* (Cambridge University Press, Cam-

bridge, 1946), p. 8. An interesting discussion of the measurement process is given in *New pathways in science*, ref. 28.

67. E. A. Milne, *Relativity, gravitation and world structure* (Clarendon, Oxford, 1935); *Kinematic relativity* (Clarendon, Oxford, 1948); *Proc. R. Soc.* A **165,** 351 (1938); *Mon. Not. R. astron. Soc.* **106,** 180 (1946). An interesting description of Milne's ideas is given at a semi-popular level by J. Singh, ref. 30.
68. See refs. 46.
69. Conformal transformations, $(d\bar{s}^2 \rightarrow \Omega^2(\boldsymbol{x})\, ds^2)$, leave the light-cone structure of space-time invariant; but because the metric is transformed to $\tilde{g} = \Omega^2(x)g$ they do not leave the Einstein equations invariant unless Ω is constant. (Note: conformal transformations are not merely coordinate transformations, because the latter leave ds^2 invariant). Thus, Einstein's equations are not scale invariant, whereas Maxwell's equations for free electromagnetic fields are.
70. F. Hoyle and J. V. Narlikar, *Proc. R. Soc.* A **277,** 178, 184 (1966); A **290,** 143, 162, 177 (1966); A **294,** 138 (1966); *Nature* **233,** 41 (1971); *Mon. Not. R. astron. Soc.* **155,** 305, 323 (1972); *Action at a distance in physics and cosmology* (Freeman, San Francisco, 1974).
71. For a review of Jordan's theory and related work by Thierry and Fierz see D. R. Brill, *Proc. XX Course Enrico Fermi Int. School of Physics* (Academic Press, NY, 1962) p. 50. See also M. Fierz, *Helv. Phys. Acta.* **29,** 128 (1956). Other earlier developments were made by C. Gilbert, *Mon. Not. R. astron. Soc.* **116,** 678, 684 (1960); and in *The application of modern physics to the earth and planetary interiors*, ed. S. K. Runcorn (Wiley, NY, 1969), p. 9–18.
72. P. S. Wesson, *Cosmology and geophysics* (Adam Hilger, Bristol, 1978). F. J. Dyson in *Aspects of quantum theory*, ed. A. Salam and E. P. Wigner (Cambridge University Press, London, 1972).
73. The self-consistent form of all possible variations in the parameters of N, N_1, and N_2 was considered by W. Eichendorf and M. Reinhardt, *Z. Naturf.* **28,** 529 (1973).
74. P. Pochoda, M. Schwarzschild, *Astrophys. J.* **139,** 587 (1964).
75. G. Gamow, *Proc. natn. Acad. Sci., U.S.A.* **57,** 187 (1967).
76. G. Gamow, *Phys. Rev. Lett.* **19,** 757, 913 (1967). For an account of Gamow's contribution to the 'large number' problem, see R. Alpher, *Am. Scient.* **61,** 52 (1973).
77. F. Dyson, *Phys. Rev. Lett.* **19,** 1291 (1967); A. Peres, *Phys. Rev. Lett.* **19,** 1293 (1967); R. Gold, *Nature* **175,** 526 (1967); T. Gold, *Nature* **175,** 526 (1967); S. M. Chitre and Y. Pal, *Phys. Rev. Lett.* **20,** 2781 (1967); J. N. Bahcall and M. Schmidt, *Phys. Rev. Lett.* **19,** 1294 (1967).
78. K. P. Stanyukovich, *Sov. Phys. Dokl.* **7,** 1150 (1963); see also D. Kurdgelaidze, *Sov. Phys. JETP* **20,** 1546 (1965).
79. J. O'Hanlon and K. H. Tam, *Prog. Theor. Phys.* **41,** 1596 (1969); Y. M. Kramarovskii and V. P. Chechev, *Sov. Phys. Usp.* **13,** 628 (1971), V. P. Chechev, L. E. Gurevich, and Y. M. Kramarovsky [*sic*], *Phys. Lett.* B **42,** 261 (1972).
80. P. C. W. Davies, *J. Phys.* A **5,** 1296 (1972) and reference therein. E. Schrödinger had discussed 'large number' coincidences involving the strong coupling, *Nature* **141,** 410 (1938). The best limits claimed for the constancy

of the weak, strong and electromagnetic couplings are those of I. Shlyakhter, *Nature* **264,** 340 (1976), and are based upon an interpretation of data from the Oklo Uranium mine in Gabon (see M. Maurette, *Ann. Rev. Nucl. Sci.* **26,** 319 (1976). Shlyakhter bases his argument upon the abundance ratio of the two light samarium isotopes, Sm-149: Sm-148. In ordinary samarium the natural ratio of these isotopes is ~ 0.9, but in the Oklo sample it is ~ 0.02. This depletion is due to the bombardment received from thermal neutrons over a period of many millions of years during the running of the natural 'reactor'. The capture cross-section for thermal neutrons on samarium-149 can be measured in the laboratory at ~ 55 kb and is dominated by a strong capture resonance when the neutron source has energy ~ 0.1 eV. The Oklo samples imply the cross-section could not have exceeded 63 kb two billion years ago (all other things being equal), and the capture resonance cannot have shifted by as much as 0.02 eV over the same period. The position of this resonance sensitively determines the relative binding energies of different samarium isotopes in conjunction with the weak, α_w, strong α_s, and electromagnetic, α, couplings. The allowed time-variations are constrained by $\dot{\alpha}/\alpha \lesssim 10^{-17}\,\text{yr}^{-1}$, $\dot{\alpha}_w/\alpha_w \lesssim 2\times 10^{-12}\,\text{yr}^{-1}$, $\dot{\alpha}_s/\alpha_s \lesssim 5\times 10^{-19}\,\text{yr}^{-1}$.

81. J. D. Bekenstein, *Comm. Astrophys.* **8,** 89 (1979); M. Harwit, *Bull. Astron. Inst. Czech.* **22,** 22 (1971).
82. W. A. Baum and R. Florentin-Nielson, *Astrophys. J.* **209,** 319 (1976); J. E. Solheim, T. G. Barnes III, and H. J. Smith, *Astrophys. J.* **209,** 330 (1967); L. Infeld, *Z. Physik.* **171,** 34 (1963).
83. R. d'E. Atkinson, *Phys. Rev.* **170,** 1193n (1968); H. C. Ohanian, *Found. Phys.* **7,** 391 (1977).
84. T. C. van Flandern, in *On the measurement of cosmological variations of the gravitational constant*, ed. L. Halpern (University of Florida, 1978), p. 21, concludes that $\dot{G}/G \simeq -6\times 10^{11}\,\text{yr}^{-1}$. The latest Viking Lander data yield a limit $|\dot{G}/G| < 3\times 10^{-11}\,\text{yr}^{-1}$; see R. D. Reasenberg, *Phil. Trans. R. Soc.* A **310,** 227 (1983).
85. J. B. S. Haldane, *Nature* **139,** 1002 (1937). The ideas broached in this note were developed in more detail in *Nature* **158,** 555 (1944).
86. See *New biology*, No. 16 (1955), ed. M. L. Johnson, M. Abercrombie, and G. E. Fogg (Penguin, London, 1955), p. 23.
87. C. Brans and R. Dicke, *Phys. Rev.* **124,** 924 (1961); C. Brans, Ph.D. Thesis, 1961 (Princeton University, NJ).
88. R. Dicke, *The theoretical significance of experimental relativity* (Gordon & Breach, NY, 1964) contains early references; R. Dicke and P. J. E. Peebles, *Space Sci. Rev.* **4,** 419 (1965); R. Dicke, *Gravitation and the universe* (America Philosophical Society, Philadelphia, 1969).
89. R. H. Dicke, *Rev. Mod. Phys.* **29,** 355, 363 (1957).
90. Ref. 89, p. 375. Here ε is the dielectric constant, and so the fine structure constant is $\alpha = e^2/\varepsilon_0 \hbar c$. This is introduced because Maxwell's equations imply charge conservation, so a change in α is most conveniently interpreted as being due to a change in the permittivity or permeability of free space. See also K. Greer, *Nature* **205,** 539 (1965), and *Discovery* **26,** 34 (1965).
91. R. H. Dicke, *Nature* **192,** 440 (1961). An accompanying reply by P. A. M. Dirac, *Nature* **192,** 441 (1961), argues that 'On Dicke's assumption habit-

able planets could exist only for a limited period of time. With my assumption they could exist indefinitely in the future and life need never end'.

92. E. Mascall, *Christian theology and natural science* (Longmans, London, 1956); also private communication from G. J. Whitrow (August 1979).
93. B. Carter, in *Confrontation of cosmological theories with observation*, ed. M. S. Longair (Reidel, Dordrecht, 1974), p. 291; see also 'Large numbers in astrophysics and cosmology', paper presented at Clifford Centennial Meet, Princeton (1970).
94. J. A. Wheeler, in *Foundational problems in the special sciences* (Reidel, Dordrecht, 1977) pp. 3–33; see also the final chapter of C. W. Misner, K. S. Thorne, and J. A. Wheeler, *Gravitation* (Freeman, San Francisco, 1973), also C. M. Patton and J. A. Wheeler, in *Quantum gravity: an Oxford symposium*, ed. C. J. Isham, R. Penrose and D. W. Sciama (Clarendon, Oxford) pp. 538–605.
95. For the consequences of topology change, see R. P. Geroch, *J. Math. Phys.* **8,** 782 (1967) and **11,** 437 (1970); F. J. Tipler, *Ann. Phys.* **108,** 1 (1977); C. W. Lee, *Proc. R. Soc.* A **364,** 295 (1978), R. H. Gowdy, *J. Math. Phys.* **18,** 1798 (1977). They show that a closed universe must undergo a singularity to admit a topology change. Since some topologies require parity violation and others do not, we can see a way in which topology change could act to alter conservation laws of physics.
96. G. F. R. Ellis, R. Maartens and S. D. Nel, *Mon. Not. R. astron. Soc.* **184,** 439 (1978); *Gen. Rel. Gravn.* **9,** 87 (1978) and **11,** 281 (1979); G. F. R. Ellis and G. B. Brundrit, *Quart. J. R. astron. Soc.* **20,** 37 (1979).
97. H. Everett, *Rev. Mod. Phys.* **29,** 454 (1957); for other expositions see especially B. de Witt, *Physics Today*, Sept., p. 30 (1970); P. C. W. Davies, *Other worlds* (Dent, London, 1980), J. D. Barrow, *The Times Higher Educ. Suppl.* No. 408, p. 11 (22 Aug. 1980), and Chapter 7.
98. C. B. Collins and S. W. Hawking, *Astrophys. J.* **180,** 137 (1973); S. W. Hawking in *Confrontation of cosmological theories with observation*, ed. M. S. Longair (Reidel, Dordrecht, 1974). A discussion of the possible conclusions that may be drawn from these papers is given in J. D. Barrow and F. J. Tipler, *Nature* **276,** 453 (1978); J. D. Barrow in *Problems of the cosmos* (Einstein Centenary Volume, publ. 1981, Enciclopedia Italiana, in Italian). A detailed discussion is given in Chapter 6.
99. C. F. A. Pantin, *Adv. Sci.* **8,** 138 (1951); and in *Biology and personality*, ed. I. T. Ramsey (Blackwell, Oxford, 1965), pp. 83–106; *The relation between the sciences* (Cambridge University Press, Cambridge, 1968).
100. L. J. Henderson, *The fitness of the environment* (Smith, Gloucester, Mass. 1913), repr. 1970, and *The order of Nature* (Harvard University, Cambridge, Mass. 1917).
101. E. Teller, *Phys. Rev.* **73,** 801 (1948), and in *Cosmology, fusion and other matters*, ed. F. Reines (Hilger, Bristol, 1972), p. 60.
102. J. D. Barrow, *Quart. J. R. astron. Soc.* **23,** 344 (1982).
103. P. C. W. Davies, *Other worlds: space, superspace and the quantum universe* (Dent, London, 1980).
104. A. Peres, *Physics Today*, Nov., **24,** 9 (1971).
105. A Wyler, *C.r. Acad. Sci. Paris* A **269,** 743 (1969).

106. F. Hoyle, D. N. F. Dunbar, W. A. Wensel and W. Whaling (1953) *Phys. Rev.* **92,** 1095.
107. J. P. Cox and R. T. Giuli, *Principles of stellar structure*, Vol. 1 (Gordon & Breach, NY, 1968). D. D. Clayton, *Principles of stellar evolution and nucleosynthesis* (McGraw-Hill, NY, 1968).
108. E. E. Salpeter, *Astrophys. J.* **115,** 326 (1965); *Phys. Rev.* **107,** 516 (1967).
109. F. Hoyle, D. N. F. Dunbar, W. A. Wensel, and W. Whaling, *Phys. Rev.* **92,** 649 (1953).
110. F. Hoyle, *Galaxies, nuclei and quasars* (Heinemann, London, 1965), p. 146.
111. It is amusing to note the coincidence that Erno Rubik's 'Hungarian' cube, now available in any toyshop or mathematics department has $\sim 10^{20}$ distinct configurations; for details see D. R. Hofstadter, *Science. Am.* **244** (3), 20 (1981). Also, if neutrinos possess a small non-zero rest mass $m_\nu \sim 10$–30 eV, as indicated by recent experiments, then neutrino clusters surviving the radiation-dominated phase of the Universe have a characteristic scale which encompasses $\sim(m_P/m_\nu)^3 \sim 10^{80}$ neutrinos, where m_P is the Planck mass.
112. Ref. 110, p. 159.
113. H. Nielsen, in *Particle physics 1980*, ed. I. Andric, I. Dadic, and N. Zovko (North-Holland, 1981), pp. 125–42, and *Phil. Trans. R. Soc.* A **310,** 261 (1983); J. Iliopoulos, D. V. Nanopoulos, and T. N. Tamaros, *Phys. Lett.* B **94,** 141 (1983); J. D. Barrow, *Quart. J. R. astron. Soc.* **24,** 146 (1983); J. D. Barrow and A. C. Ottewill, *J. Phys.* A **16,** 2757 (1983).
114. D. Foerster, H. B. Nielsen, and M. Ninomiya, *Phys.* Lett. B **94,** 135 (1980); H. B. Nielsen and M. Ninomiya, *Nucl. Phys.* B **141,** 153 (1978); M. Lehto M. Ninomiya, and H. B. Nielsen, *Phys. Lett.* B **94,** 135 (1980). Generally speaking, it is found that small symmetry groups tend to be stable attractors at low energy whilst larger groups are repellers. As an example of the type of analysis performed by Nielsen *et al.*, consider a simple Yang–Mills gauge theory that is Lorentz invariant and has one dimensionless coupling constant. If Lorentz invariance is not demanded of the theory then the Lagrangian can be for more general and up to 20 independent couplings are admitted (when they are all equal the Lorentz invariant theory is obtained). One then seeks to show that the couplings evolve to equality as energy falls. A similar strategy is employed to check for the stability of gauge invariance at low energy.
115. N. Brene and H. B. Nielsen, Niels Bohr Inst. preprint NBI-HE-8242, (1983).
116. G. J. Whitrow, *Br. J. Phil. Sci.* **6,** 13 (1955).
117. L. Gurevich and V. Mostepanenko, *Phys. Lett.* A **35,** 201.
118. O. Neugabauer, *A history of ancient mathematical astronomy*, pt. 2 (Springer, NY 1975), p. 848; C. Ptolemy, *Opera* II, 265, ed. J. L. Heiberg (Teubner, Leipzig 1907).
119. G. Sarton, *History of science*, Vol. 1 (Norton, NY, 1959), pp. 438–9.
120. J. Wallis, *A treatise of algebra; both historical and practical* (London, 1685), p. 126.
121. M. Jammer, *Concepts of space* (Harper & Row, NY, 1960).
122. L. E. J. Brouwer, *Math. Annalen* **70,** 161 (1911); *J. Math.* **142,** 146 (1913).

123. K. Menger, *Dimensions Theorie* (Leipzig, 1928). W. Hurewicz and H. Wallman, *Dimension theory* (Princeton University Press, NJ, 1941).

124. I. Kant, 'Thoughts on the true estimation of living forces', in J. Handyside (transl.), *Kant's inaugural dissertation and early writings on space* (University of Chicago Press, Chicago, 1929).

125. W. Paley, *Natural theology* (London, 1802). It is interesting to note that Paley was Senior Wrangler at Cambridge.

126. P. Ehrenfest, *Proc. Amst. Acad.* **20,** 200 (1917); *Ann. Physik* **61,** 440 (1920).

127. H. Weyl, *Space, time and matter* (Dover, NY, 1922), p. 284.

128. A derivation of these standard results can be found in almost any text on classical dynamics although they are not discussed as consequences of worlds of different dimension, merely as examples of different possible central force laws. See, for example, H. Lamb, *Dynamics* (Cambridge University Press, Cambridge 1914), pp. 256–8; J. Bertrand, *Compt. rend.* **77,** 849 (1873).

129. For other derivations of these results and a discussion of their relevance for spatial dimension see; I. M. Freeman, *Am. J. Phys.* **37,** 1222; W. Buchel, *Physik. Blätter* **19,** 547 (1963); appendix I of W. Buchel, *Philosophische Probleme der Physik* (Herder, Freiburg, 1965); E. Stenius, *Acta. phil. fennica* **18,** 227 (1965); K. Schafer, *Studium generale* **20,** 1 (1967); R. Weitzenbock, *Der vierdimensionale Raum* (Braunschweig, 1929).

130. F. R. Tangherlini, *Nuovo Cim.* **27,** 636 (1963).

131. For a partial solution of this problem see I. N. Sneddon and C. K. Thornhill, *Proc. Camb. Phil. Soc.* **45,** 318 (1949). These authors do not find the solution (b).

132. A Barnes and C. K. Keogh, *Math. Gaz.* **68,** 138 (1984).

133. Note, however, that the so-called gravitational paradox that $\int \phi d^3r$ is infinite for $\phi \propto r^{-1}$ does not arise for $\phi \propto r^{-1}\exp(-\mu r)$ for real μ, but this paradox disappears in the general relativistic theory of gravitation, which is able to deal with infinite spaces consistently.

134. E. Lieb, *Rev. Mod. Phys.* **48,** 553; F. J. Dyson, *J. Math. Phys.* **8,** 1538 (1967), J. Lenard, *J. Math. Phys.* **9,** 698 (1968).

135. G. Whitrow, *The structure and evolution of the universe* (Harper & Row, NY, 1959).

136. Recall Eddington's remark during his 1918 Royal Institution Lecture: 'In two dimensions any two lines are almost bound to meet sooner or later; but in three dimensions, and still more in four dimensions, two lines can and usually do miss one another altogether, and the observation that they do meet is a genuine addition to knowledge.'

137. E. A. Abbott, *Flatland* (Dover, NY, 1952). For a modern version see D. Burger, *Sphereland* (Thomas Y. Crowell Co., NY, 1965).

138. A. K. Dewdney, *Two dimensional science and technology* (1980), pre-print, Dept. of Computer Science, University of Western Ontario; *J. Recreation. Math* **12,** 16 (1979). For a commentary see M. Gardner, *Scient. Am.*, Dec. 1962, p. 144–52, *Scient. Am.*, July 1980, and for a full-blooded fantasy see A. K. Dewdney's *Planiverse* (Pan Books, London, 1984).

139. W. McCullough and W. Pitts, *Bull. Math. Biophys.* **5,** 115 (1943).

140. H. Poincaré, *Dernières pensées* (Flammarion, Paris, 1917).

141. J. Hadamard, *Lectures on Cauchy's problem in linear partial differential equations* (Yale University Press, New Haven, 1923).

142. We are ignoring the effects of dispersion here.

143. R. Courant and D. Hilbert, *Methods of mathematical physics* (Interscience, NY, 1962).

144. J. D. Barrow, *Quart J. R. astron. Soc.* **24,** 24 (1983).

145. I. Newton, *Principia* II, prop. 32, see ref. 55.

146. J. B. Fourier, *Theoria de la chaleur,* Chapter 2, § 9 (1822).

147. A. Einstein, *Ann. Physik* **35,** 687 (1911).

148. For example, surface is a force per unit length, pressure a force per unit area and density a mass per unit volume, and so on.

149. There is a general tendency for spherical configurations to dominate because of the prevalence of spherically symmetric force laws in Nature, but our reasoning does not depend upon this; we consider spherical symmetry for simplicity of explanation only.

150. There are obviously other factors that play a role: the simple local topology of space-time avoids the introduction of dimensionless parameters describing the identification of length scales; for example, if we make identifications $(x, y, z) \longleftrightarrow (x+\alpha_x, y+\alpha_y, z+\alpha_z)$, then α_x, α_y and α_z will feature as numerical factors (scaled by an appropriate length) in physical formulae.

151. We do not enter here into a discussion of whether mathematics is something that exists independently of mathematicians, as Platonists would claim.

152. R. Penrose, *Rep. Math. Phys.* **12,** 65 (1977).

153. J. J. Faris, *Am. J. Phys.* **38,** 1265 (1970).

154. There would be no way of propagating the gravitational interaction in a $2+1$ dimensional space-time in general relativity. However, we do not know whether other theories of gravity can avoid this problem.

155. D. Ruelle and F. Takens, *Commun. Math. Phys.* **20,** 167 (1971); R. Plykin, *Sb. Math.* **23,** 333 (1974); S. Newhouse, D. Ruelle and F. Takens, *Commun. Math. Phys.* **64,** 35 (1978).

156. Strange attractors have 'non-integral dimension' (F. Hausdorff, *Math. Annalen.* **79,** 157 (1918), B. Mandelbrot, *Fractals: form, chance and dimension* (W. H. Freeman, San Francisco, 1977), but this is not a *topological* dimension of the sort we have been discussing here. Rather, it is a measure of the amount of information necessary to specify the structure of the attractor in phase space. For simple attractors, like foci and limit cycles, this is the same as the topological dimension, but in more complicated situations it is not (see J. D. Barrow, *Phil. Trans. R. Soc.* A **310,** 337 (1983)).

157. T. Kaluza, *Sber. preuss. Akad. Wiss.* Phys. Math. Kl. 966 (1921); O. Klein, *Z. Physik* **37,** 895 (1926); A. Einstein and P. Bergmann, *Ann. Math.* **39,** 683 (1938). For a recent review, see A. Salam and J. Strathdee, *Ann. Phys.* (NY) **141,** 316 (1982). The general idea actually predates both Kaluza, Klein, and even general relativity—see G. Nordström, *Phys. Z.* **15,** 504 (1914).

158. E. Witten, *Nucl. Phys.* B **186,** 412 (1981); P. G. O. Freund and M. A. Rubin, *Phys. Lett.* B **97,** 233 (1980). $10+1$ dimensions are the smallest that

can accommodate the full SU(3)×SU(2)×U(1) gauge symmetry of strong and electro-weak interactions.

159. P. Candelas and S. Weinberg, *Nucl. Phys.* B **237,** 397 (1984); D. Bailin, A. Love, and C. Vayonakis, *Phys. Lett.* B **142,** 344 (1984).
160. W. J. Marciano, *Phys. Rev. Lett.* **52,** 489 (1984), but note that the observational limits quoted in this paper are probably not necessarily meaningful in connection with testing the proposed time variations, because they are derived by interpreting observations using theories with time-independent constants. The author is also aware of this problem.
161. The dimensional dependence of the gravitational coupling and the electron charge can be seen by examining eqn (4.89) and the discussion preceeding it.
162. D. S. Freed and K. K. Uhlenbeck, *Instantons and four-manifolds* (Springer-Verlag, NY, 1984); S. K. Donaldson, *Bull. Am. Math. Soc.* **8,** 81 (1983).
163. S. Weinberg, *1979 Nobel Prize Lecture* (Nobel Foundation, Stockholm, 1980).
164. C. M. Will, *Theory and experiment in gravitation physics* (Cambridge University Press, Cambridge, 1981).
165. J. A. Wheeler, *Monist* **47,** 40 (1962).
166. B. S. DeWitt, *Phys. Rev.* **160,** 1113 (1967).
167. Note, however, that solutions of Poisson's equation in other dimensions possess the Newtonian spherical property. For example, in two dimensions the logarithmic potential allows a disc to be replaced by a point of equal mass at its centre. The disc can be replaced by a point at its centre when the potential is a sum of zero order Bessel functions in general, but the point mass differs from that of the disc. The spherical property is not unique to three-dimensions but it is unique to the inverse square law in three dimensions.

5 The Weak Anthropic Principle in Physics and Astrophysics

'Any coincidence', said Miss Marple to herself, 'is always worth noticing. You can throw it away later if it *is* only a coincidence.'

Agatha Christie

5.1 Prologue

'The time has come', the Walrus said,
'To talk of many things: Of Shoes—and ships,—and sealing wax—Of cabbages —and kings—And why the sea is boiling hot—And whether pigs have wings'.

Lewis Carroll

There has grown up, even amongst many educated persons a view that everything in Nature, every fabrication of its laws, is determined by the local environment in which it was nurtured—that natural selection and the Darwinian revolution have advanced to the boundaries of every scientific discipline. Yet, in reality, this is far from the truth. Twentieth-century physics has discovered there exist invariant properties of the natural world and its elementary components which render inevitable the gross size and structure of almost all its composite objects. The size of bodies like stars, planets and even people are neither random nor the result of any progressive selection process, but simply manifestations of the different strengths of the various forces of Nature. They are examples of possible equilibrium states between competing forces of attraction and repulsion.

A study of how these equilibrium states are set up and how their form is determined reveals that the structure of the admissible stable states is determined, aside from geometrical factors like 2π, by those parameters we have come to call the fundamental constants of Nature; for example, quantities like the electric charge of the electron, the ratio of the electron and proton masses, the strength of the strong force between nucleons and so forth. This approach typifies the modern reaction to the facts that fuelled the Design Arguments of past centuries: but whereas the ancients might regard it as a consequence solely of divine favour that the Earth possesses a life-supporting atmosphere whilst the Moon does not, now it would be more immediately attributable to the fact that only bodies exceeding a particular critical size will exert sufficient gravitational pull to prevent gas molecules escaping.[1] The presence or absence of an atmosphere is seen to be most immediately related to the size of the planetary body alone. But, it would be superficial to claim that our understanding of the absence of atmospheres around some objects entirely replaces the old Design Arguments for, as scientists like Henderson[2] clearly appreciated, this type of explanation for the nature of things reduces, in the end, to explaining the existence of everything in terms of a number of 'fundamental' parameters, the values of which we are, at present, only able to determine by experiments. If their values had differed in this Universe (or do differ in other possible universes) then the overall characteristics of the World could have been remarkably different. If one could in some way argue that all alternative 'worlds' in which the fundamental constants differ slightly from those in our own could not arise then the result would be a version of the Design Argument remarkably similar to that proposed by Paley[3] in the early nineteenth century.

Some scientists and philosophers have been keen to develop such an approach and in order to pursue it they have attempted to trace the manner in which the large scale features of Nature are contingent upon the precise values of unchanging fundamental constants of physics. The aim of these investigators was to uncover striking coincidences between *prima facie* independent constants of Nature whose existence made our own possible. The most interesting feature of these investigations is the simple qualitative picture they give of the way the world is structured. A picture that all scientists can appreciate whether or not they are interested in the Weak Anthropic Principle which has motivated it. Here, we shall present a number of the arguments and derivations that have been used in support of an Anthropic ansatz that the subset of cognizable universes, amongst a collection in which the constants of Nature take on all possible permutations of all possible values, is very small.[4] We can show that the order of magnitude of the key features of astronomy and physics can be deduced as inevitable once the constants of Nature are specified. Unfortunately, derivations of this sort rarely find their way into physics texts in any extensive way simply because, more often than not, much more precise numerical answers to such questions are required.

In order to present these derivations, and the interrelationships they display, we shall base many arguments upon dimensional analysis. This technique, first used by Newton[5] and Fourier,[6] enables the magnitude of physical quantities to be ascertained to within purely arithmetic factors. In practice, exact calculations reveal that these numerical factors are very close to unity. For example, the volume of a sphere must, on dimensional grounds, be proportional to the cube of its radius. The constant of proportionality, $4\pi/3$, cannot be deduced by dimensional analysis but is close to unity. Clearly, there is an interesting article of faith hidden away here that believes all such purely numerical factors will be small. In practice they all seem to be, but no rigorous proof of the fact exists.[7] However, some explanation of the fact was given in our discussion of dimensionality in Chapter 4.

Suppose we were to commission a survey of all the different types of object in the Universe from the scale of elementary particles to the largest clusters of galaxies. A picture could be prepared that plotted all these objects according to their masses and sizes, or average dimension.

The result would look like[8] Figure 5.1,

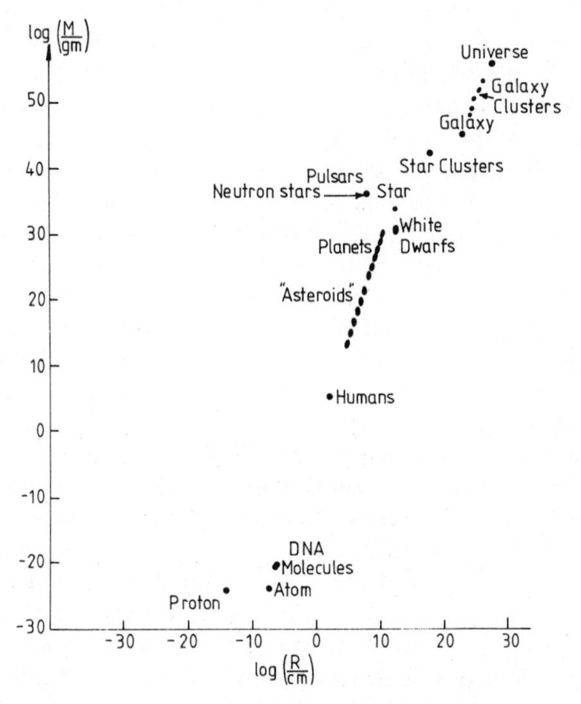

Figure 5.1. Observed objects in the Universe plotted in a size-mass diagram.[31] Note the accumulation of points in particular regions of the plane and the absence of bodies populating large regions of the plane.

A priori we might have expected our graph to be covered by points in a fairly haphazard fashion: but this is clearly not the case. Some regions of the diagram are heavily and systematically populated whilst others remain very obviously empty. One of our goals will be to understand the reasons for the particular distribution we see and show how it lies at the foundation of modern Anthropic arguments.

When you were first shown Figure 5.1, various ideas might occur to explain the distribution of points:

(i) They are completely random—any preference for a particular region is purely statistical. All the apparent correlations are real coincidences.
(ii) We are the victim of a 'selection effect'—the whole plane is fairly evenly populated but some types of structure are not detectable by us and this explains the apparent depopulation of certain areas. They are unobservable.
(iii) Stability—the 'rules' of Nature allow only certain types of structure to exist for long periods of time. The portions of the M–R diagram containing the observed structures are the portions that describe the stable equilibria that can exist between different fundamental forces.

In order to see clearly why an amalgam of the second and third of these possibilities is the correct interpretation we need to set up some absolute system of description—a set of units. Conventionally physicists and engineers employ all manner of units—centimetres, feet and inches, pints, years, solar masses *et cetera*. When we describe the dimension of subatomic systems we are at liberty to employ more appropriate yardsticks like the fermi ($=10^{-13}$ cm, close to the radius of atomic nuclei) or measure energies in electron-volts.

In 1874 the Irish physicist G. Johnstone Stoney[9] first discussed the possibility that there exist particular systems of units picked out by Nature herself, what we might term 'Natural Units'. In order to determine them, he wrote,[10] 'we must select phenomena that prevail throughout the whole of Nature, and are not specially associated with individual bodies'. The appropriate candidates he claims to be the velocity of light, c, because of the manner in which it connects all systems of electrostatic and electromagnetic units; Newton's gravitation constant, G, because of its universal character and lastly, e, the unit of electric charge deduced from Faraday's Law. From these quantities a length, a mass and a time can be constructed. The values of the units Johnstone Stoney evaluated from these three standards were,

$$L_J = \left(\frac{Ge^2}{c^4}\right)^{1/2} \sim 10^{-35}\ \text{cm} \tag{5.1}$$

$$T_J = \left(\frac{Ge^2}{c^6}\right)^{1/2} \sim 3\times 10^{-46}\ \text{s} \tag{5.2}$$

$$M_J = \left(\frac{e^2}{G}\right)^{1/2} \sim 10^{-7}\ \text{gm} \tag{5.3}$$

Some years later, in 1906 and apparently not knowing of Johnstone

Stoney's work, Planck also considered the question of 'natural units'.[11] He chose the velocity of light and the Newtonian gravitation constant as components like his predecessor but his third choice was the newly introduced quantum of action, h, which we now call Planck's constant. From these three universal constants it is also possible to determine units of mass, length and time which in Planck's words[11] 'are independent of special bodies or substances, which necessarily retain their significance for all times and for all environments, terrestrial and human or otherwise, and which may, therefore, be described as "natural" units'. The Planck units were

$$l_p = \left(\frac{G\hbar}{c^3}\right)^{1/2} \sim 10^{-33}\,\text{cm} \tag{5.4}$$

$$t_p = \left(\frac{G\hbar}{c^5}\right)^{1/2} \sim 5\times 10^{-44}\,\text{s} \tag{5.5}$$

$$m_p = \left(\frac{c\hbar}{G}\right)^{1/2} \sim 10^{-5}\,\text{gm} \tag{5.6}$$

we notice that the Planck units differ from Johnstone Stoney units only in employing Planck's constant h instead of the electric charge e as one of the basic elements. Since the ratio $e^2/\hbar c$ is dimensionless and of order 1/137 we see that each Johnstone Stoney unit just differs from the corresponding Planck unit by a numerical factor $\sim\sqrt{137}$. Hereafter, we shall use the Planck quantities and note that various other secondary quantities can be derived from them: for example, a Planck temperature $T_p \sim k_B^{-1} c^{5/2} \hbar^{1/2} G^{-1/2} \sim 10^{32}$ K is defined. The Boltzmann constant k_B is merely a conversion factor between energy and temperature units and not a fundamental constant having the same[12] status as G, $\hbar$ or c.

It is interesting to note that the Planck length and time are both extremely small, many orders of magnitude smaller even than nuclear sizes and times ($\sim 10^{-13}$ cm and $\sim 10^{-23}$ s respectively), whereas the Planck mass is macroscopic and roughly equal to the mass of a small grain of sand.

We are at liberty to employ any unit system we please to describe physical quantities. The Planck or Johnstone Stoney units can be thought of as the system 'chosen' by Nature herself. In the following sections we shall set $\hbar = c = 1$ so energy, frequency and mass are all inverse lengths and inverse times. We are free to set G equal to unity also but we shall not do so in order to retain a characteristic scale and we notice that when $\hbar = c = 1$, the Planck mass is $m_p = G^{-1/2}$. The only meaningful quantities are *dimensionless* ones and the strength of the *gravitational* force will be described by α_G, an analogue of the fine structure constant, where

$$\alpha_G = \frac{Gm^2}{\hbar c} \tag{5.7}$$

and m is any mass scale. Conventionally, the proton mass, m_N, is chosen as the reference mass scale so in (5.7) it yields a numerical value of

$$\alpha_G = \frac{Gm_N^2}{\hbar c} \sim 10^{-39} \tag{5.8}$$

The observed natural phenomena appear to be controlled by three other distinct force fields besides gravity. Unlike the gravitational interactions these other fields do not act on every type of particle. The familiar *electromagnetic interaction* governs the interaction of electrically charged particles and light. It determines the structure of atoms, molecules and solid materials together with the behaviour of light. The strength of the interaction is conveniently described by the fine structure constant of Sommerfeld.[13] It is the ratio of the electrostatic energy of repulsion between two elementary charges, e, separated by one Compton wavelength, to the rest energy of a single charge,

$$\alpha = \frac{e^2/(\hbar/mc)}{mc^2} = \frac{e^2}{\hbar c} = (7.29720 \pm 0.00003) \times 10^{-3} \tag{5.9}$$

Roughly, we have $\alpha \sim (137)^{-1}$ and because this coupling is independent of the mass of the electric charge, it describes the coupling to the electromagnetic field of any elementary particle carrying electric charge e. The ratio of the electron to proton mass, β, is found to be

$$\beta = \frac{m_e}{m_N} = (1836.12)^{-1} \tag{5.10}$$

and is the only other dimensionless parameter necessary to formulate quantum electrodynamics. The gross physical properties of atoms, molecules and solids can, in principle, be determined as functions of the pure numbers α and β. The detailed aspects of atomic structure also depend on numerical factors like the atomic number, Z, or atomic weight A, the eigenvalues of particular operators and geometrical factors like 2π, but the principal dependence on 'physics' is given by the α and β dependence alone.

The *weak interaction* is felt by both leptons and hadrons and its most familiar manifestation is radioactivity (β-decays),

$$n \to p + e^- + \bar{\nu}_e \tag{5.11}$$

If this interaction is modelled as a point interaction, as done originally by Fermi[14] in 1934, then the relation between the half-life of the neutron and the energy difference, E_0, of the initial and final nucleon states is,

$$\tau_N \sim 60\pi^3 E_0^{-5} G_F^{-2} \tag{5.12}$$

From observations of τ_N, the Fermi coupling constant, G_F is determined to be

$$G_F \simeq 1.435 \times 10^{-49} \text{ erg cm}^3 \tag{5.13}$$

The dimension of G_F is of mass-squared if $\hbar = c = 1$ so, more conveniently, it can be represented formally in terms of the proton mass

$$G_F \simeq \frac{1.0 \times 10^{-5}}{m_N^2} \tag{5.14}$$

If the *weak interaction* is modelled, not as a point interaction of strength G_F, but through the exchange of a massive gauge boson of mass m_W and coupling strength g_W then by analogy with the electromagnetic interaction we can introduce a weak 'charge' g_W, and express the weak interaction strength in dimensionless form, $g_W^2/\hbar c$. If the weak force is mediated by the exchange of a massive W boson then the Uncertainty Principle indicates that the range of the force, r_W, is linked to the W mass by

$$r_W \sim \frac{\hbar}{m_W c} \tag{5.15}$$

Experiment shows $r_W \leqslant 0.1 \times 10^{-13}$ cm so the W must be very massive. The connection between the Fermi coupling and the coupling g_W must be[15]

$$G_F = \sqrt{2}\, g_W^2 r_W^2 = \frac{\sqrt{2}}{\pi}\left(\frac{\hbar}{m_W c}\right)^2 g_W^2 \tag{5.16}$$

and so the 'weak structure constant' is

$$\frac{g_W^2}{\hbar c} = \frac{1}{\sqrt{2}} \frac{1}{\hbar}\left(\frac{m_W c}{\hbar}\right)^2 G_F \tag{5.17}$$

Conventionally, the ratio of e to g_W is specified by the Weinberg angle, θ_W, first introduced by Glashow, via

$$g_W = \frac{e}{\sin \theta_W} \tag{5.18}$$

Experiment shows[17] $\sin^2 \theta_W \sim 0.21$ so $m_W \sim 80$ GeV is expected as recently measured at CERN.

The most complicated of the microscopic interactions is the *strong* (or nuclear) interaction. It is responsible for the structure of nuclei, fusion reactions between light nuclei, the structure of the stars and the phenomena of elementary particle physics.

Prior to the hypothesis that hadrons contained quarks as internal constituents, the picture of the strong interaction resembled the original suggestion of Yukawa that the nuclear forces were mediated by meson exchange. The theory gives the potential energy of a pair of nucleons separated by a distance r as $g_s^2 r^{-1} \exp(-m_\pi r)$ where g_s is the Yukawa (scalar) coupling and m_π the mass of the exchange particle. If we consider the binding of the deuteron state of a single proton and neutron then the role of the fine structure constant in atomic binding is played by the constant $g_s^2/\hbar c \sim 0.1$. If we take into account that the pion must be made with an extra energy of at least $m_\pi \sim 139$ MeV (which reduces the probability of pion emission) then the analogue of the fine structure constant is

$$\frac{g_s^2}{\hbar c}\left(\frac{m_\pi}{2m_N}\right) \sim 0.08 \tag{5.19}$$

and so

$$\alpha_s = \frac{g_s^2}{\hbar c} \sim 15 \tag{5.20}$$

This dimensionless quantity also determines the interaction between two quarks and, like α and α_W is not a constant but a function of the energy at which it is measured. We have given the values at low energy. Later, in section 5.10, we shall see how the effective strength of these interactions scales with the energy of the environment in which the forces are measured.

5.2 Atoms and Molecules

> What is a man, that the electron
> is mindful of him?
>
> Carl Becker

Let us now turn to consider the nature of the elementary objects of which we are composed—atoms and molecules. From our study we would like to gain an understanding of why, in rough terms, things 'are as they are'; why atoms are so small ($\sim 10^{-8}$ cm), why light and matter interact with their characteristic strength, why aggregates of atoms are stable and why biological structures are forced to inhabit a particular niche of the environment with a temperature close to what we conventionally call 'room temperature'. Remarkably, it transpires that the gross properties of all atomic and molecular systems are controlled by only two dimensionless *physical* parameters—the fine structure constant, $\alpha = e^2 \sim (137)^{-1}$, and the electron to proton mass ratio, $\beta \equiv m_e/m_N \sim (1836)^{-1}$. No physical theory has yet been able to explain the numerical values of these two pure numbers that determine, to within an order of magnitude or so, all the qualitative features of bound states of the electromagnetic interaction. The difference between simple order of magnitude estimates for physical parameters involving powers of α and β and the exact calculations of the quantum theory (which agree remarkably with observation) consists only of geometrical factors like $4\pi/3$ and integral quantum numbers. We shall see also that the particular values of α and β are responsible for various 'coincidences' of Nature on which the possibility of our own existence is contingent. The first physicist to stress the all-encompassing role of α and β in determining the inevitable structure of atomic systems seems to have been Max Born.[18] In 1935 he delivered a lecture to the Indian Scientific Association entitled '*The Mysterious Number 137*' which highlighted the importance of the fine structure constant in atomic physics.

Rather than enlist the aid of the full-blooded quantum theory, we shall derive the results we require by simple arguments based on Bohr's[19] model for the atom: electrons moving in orbits of quantized angular momentum about a central nucleus. The centripedal force required to sustain the rotational motion is supplied by the electromagnetic attraction between the positively charged nucleus and the negatively charged electron(s).

The hydrogen atom can be modelled by two particles: a nuclear proton bound by the Coulomb force to an orbiting electron. The electron has potential energy $\sim e^2/r$ in an orbit of radius r, and kinetic energy $\sim p^2/2m_e$ if p is its linear momentum. However, quantum mechanics teaches us that the electron possesses a wave-like character and its position is 'spread-out' over a de Broglie wavelength $\lambda \sim p^{-1}$. Since most quantum waves will lie around the classical position $\lambda \sim r$, the kinetic energy is $\sim 1/2m_e r^2$ and the total energy of the system is

$$E = -\frac{\alpha}{r} + \frac{1}{2m_e r^2} \tag{5.21}$$

Unlike in classical physics, the minimum energy state of the system does not correspond to the minimum of r. A concentration of the electron wave-packet at small r leads to an increase in its momentum and to an increase in its kinetic energy which resists localization in accordance with the Uncertainty Principle of Heisenberg. The energy reaches a minimum at a radius a_0 where $E'(r) = 0$ and this defines the characteristic atomic size—this is displayed in Figure 5.4. Minimizing E with respect to r, (essentially we are employing a Virial Theorem here) the value of a_0 is found to be $\sim (\alpha m_e)^{-1}$. If the nucleus contains Z protons then we have

$$a_0 \sim (Z\alpha m_e)^{-1} \sim 0.5 \times 10^{-8} Z^{-1}\ \text{cm} \tag{5.22}$$

or equivalently, with $Z=1$,

$$a_0 \sim \alpha^{-2}\frac{e^2}{m_e} \sim \alpha^{-2} \times (\text{classical electron radius}) \tag{5.23}$$

The Compton wavelength of the nucleus extends out to $\sim(Zm_N)^{-1}$ and so the ratio of the range of influence of the nuclear constituents to the atomic radius is small $\sim(\alpha\beta)^{-1}$. Atoms are relatively large, distended structures, full of open space with well-defined central nuclei because α and β are so much less than unity. This is one of the basic reasons for the existence of chemistry—(and also chemists). A similar calculation works for larger, more complex atoms containing several shells of electrons. However, when many electrons are present they exert small repulsive forces upon their neighbours and this tends to screen the nucleus from exerting the full force of the nuclear charge Ze on the electrons. This detail can be accounted for by evaluating an effective atomic number Z_{eff}, (which will be less than Z because electron repulsion lowers the binding energy), felt by the electrons. For example, in the case of neon with $Z=10$ the system of 2 K-shell and 8 L-shell electrons are well described by the simple model if $Z_{\text{eff}} \sim 8$. Of course, quantum mechanics replaces the naive idea of a precise electron orbit with a probability distribution for its position—an 'orbital'. The ground state of hydrogen ($Z=1$) will have its electron distributed with a probability $\sim\exp(-r/a_0)$ which is also governed by α and m_e.

Equations (5.21) and (5.22) also tell us the magnitude of the atomic binding energy, or the ionization energy necessary to remove the electron from the atom,

$$E \sim \tfrac{1}{2}m_e(Z\alpha)^2(1+\beta)^{-1} \sim Z^2 \times 13.6\ \text{eV} \sim Z^2 \times 10^5\ \text{K} \tag{5.24}$$

where we have taken into account the detail that the electron will not orbit a fixed nucleus but rather both will orbit around their common centre of gravity. The smallness of β makes this detail superfluous to rough estimates like (5.24) but illustrates one of the consequences of β being so much less than unity.

The simple relation (5.24) displays several interesting points. In small atoms, $(Z \ll \alpha^{-1})$, the electron motion does not possess sufficient kinetic energy to allow the spontaneous production of electron-positron pairs—this would require kinetic energy $\sim 2m_e$. Large $(Z \gtrsim \alpha^{-1})$ systems appear unstable to pair production, but in practice a susceptibility to fission arises at a lower value of Z. This feature also indicates why atoms of arbitrarily large size do not exist and why atomic masses roughly span the range m_N to $m_N\alpha^{-1}$.

The velocities of the orbital electrons are relatively small compared with that of light,

$$v \sim Z\alpha \sim Z \times 2.3.10^8\ \text{cm s}^{-1} \tag{5.25}$$

Small atoms are non-relativistic, (relativistic corrections are of order $v^2 \sim Z^2\alpha^2$), slowly moving systems. The time, τ, taken for an electron to orbit the central nucleus—the 'atomic year'—is $2\pi a_0 v^{-1}$ and so

$$\tau \sim m_e^{-1}(Z\alpha)^{-2} \sim Z^2 \times 10^{-15}\,\text{s}. \tag{5.26}$$

This motion is equivalent to the flow of an electric current in a loop of wire and therefore possesses a magnetic moment. The ratio of the magnetic moment to the angular momentum is $\sim\alpha^{1/2}m_e^{-1}$ and the magnetic effects of the electron motion are small because they are governed by the combination $ev \sim Z\alpha^{3/2}$. The mild magnetic properties of materials, reflect the weak electromagnetic coupling and the slow motion of the atomic electrons. Paramagnetic susceptibilities are controlled by the combination $\alpha^{3/2}m_e^{-3}$.

As the number of nuclear constituents becomes large the atomic binding energy grows as Z^2 according to (5.24), (although when Z becomes very large it grows only as $Z^{4/3}$, while a_0, the radius of the inner electron shell, falls as Z^{-1}. The increasing nuclear charge pulls the inner electrons closer to the nucleus but the outer orbits remain of order a_0 in radius so all atoms have virtually the same size.

The fine structure constant also determines the relative permanence of atoms. An electron wave has to 'oscillate' $\sim\alpha Z^2$ times before having a significant chance of photon emission or absorption. This is why atomic and molecular states are fairly stable with lifetimes of order

$$T \sim \alpha^{-1}m_e^{-1}(Z\alpha)^{-4} \sim Z^{-4} \times 10^{-9}\,\text{s} \tag{5.27}$$

and T is $\sim\alpha^{-3}Z^{-2}$ 'atomic years'. The emission of photons is inhibited by the coupling to the velocity which arises because they must always carry away whole quanta of angular momentum. Just carrying one quantum of angular momentum they would need to appear at a radius $\sim\alpha Z^{-1}a_0$ and electrons rarely move that far from the nucleus.[20] The probability of double photon emission is of order the square of that for single emission. The probability of electrons spontaneously annihilating with positrons into light is $\sim\alpha$ and the smallness of α ensures the distinguishability of matter and radiation. An electron can be thought of as spending a fraction α of its time as an electromagnetic wave. The characteristic frequency of atomic electron transitions is $\sim Eh^{-1}$ and the corresponding wavelength of emitted light is $\sim 2\pi\alpha^{-1}a_0 \sim 10^{-7}\,\text{cm}$ which lies in the ultraviolet.

Some familiar parameters governing the interaction between matter and light are easily estimated: if radiation is scattered from atomic electrons then the spread in their velocities produces a spectral broadening with a Doppler form

$$\frac{\Delta\lambda}{\lambda} \sim v \sim Z\alpha \qquad (5.28)$$

In an electric field $\mathbf{E}$ the electron is endowed with an acceleration $e\mathbf{E}m_e^{-1}$ and it radiates energy at a rate controlled by $(e\mathbf{E}m_e^{-1})^2$. As the incident energy flux per unit area is $\mathbf{E}^2$ the cross-section for this (Compton) scattering is

$$\sigma_c \sim (\mathbf{E}\alpha m_e^{-1})^2 \sim 10^{-25}\ \text{cm}^2 \qquad (5.29)$$

and represents the 'area' an electron present to an incoming photon for electromagnetic interaction and the associated opacity is $\kappa = \sigma_c/m_N \sim 0.4\ \text{cm}^2\ \text{gm}^{-1}$.

The cross-section for bremsstrahlung (free-free emission) must incorporate the probability of photon emission during scattering. For fast ($v \sim 1$) particles the bremsstrahlung cross-section, σ_{ff}, will be

$$\sigma_{ff} \sim \sigma_c \times (\text{emission probability}) \sim \alpha\left(\frac{Z\alpha}{m_e}\right)^2 \sim Z^2 \times 10^{-28}\ \text{cm}^2 \qquad (5.30)$$

and allows about 0.5 cm of propagation in heavy metals like lead.

Other, higher order, effects can be calculated similarly; for example the energy of the Lamb shift is $\sim m_e\alpha^5 \ln \alpha$. The characteristics of positronium—a bound state in which the nuclear protons are replaced by positrons is somewhat similar to hydrogen in size but the probability that an e^-e^+ pair get close enough for annihilation is of order the 'atomic' volume $\sim\alpha^3$ and so positronium spends a fraction $\sim\alpha^4$ of its time as light.

Before moving on to consider the characteristics of aggregates of atoms and the structure of solids it is good to have in mind the average density of an atomic system since it should be a close approximation to the density of structures built from large numbers of atoms. For the simple ($Z = 1$) hydrogen system we have, where A is the atomic weight,

$$\rho_{\text{AT}} \sim \frac{Am_N}{(4\pi/3)a_0^3} \sim Am_N^3 m_e^3 \sim 0.04A\ \text{gm cm}^{-3} \sim 1.5\ \text{gm cm}^{-3} \qquad (5.31)$$

which is about right (in reality the dimensionless factor depends on $A^{-1/5}$ and $\rho \sim 0.8A^{2/5}\ \text{gm cm}^{-3}$). An increase in atomic number Z usually goes hand-in-hand with an increase in atomic weight. Since the outer atomic radius changes very little in this sequence of events, the density of materials grows (albeit erratically) with atomic number and the ratio of the density of uranium-238 to that of hydrogen is very close to 238 to 1.

A molecule is a stable configuration of nuclei and electrons. The simplest contain just two nuclei. Since bonds between atoms will require only a reorientation of the electron distributions around the atoms, the binding energies of molecules will be less than those of atoms; for example, only about 5 eV is necessary to break the covalent bond between two hydrogen atoms. As two atoms approach, their electron shells begin to overlap and each electron feels the attraction of the other nucleus which results in further attraction into a larger cloud, or molecular orbital. The electrons now feel the electrostatic attraction of both nuclei. The repulsive forces between the two nuclei fix their distance of closest approach, the attraction stops and the system resides in a configuration that is more stable than that of two isolated atoms.

The simplest inter-atomic forces arise because of electron transfer in which there is little or no electron sharing and give rise to the ionic bond. One of the participating atoms sheds an electron which the other atom has an affinity for. The atoms then exchange an electron and the ions attract each other to form a bond. Complete interpenetration is excluded by the Pauli Principle. The force falls as the inverse square of the ion separation, does not saturate, and is spherically symmetric. It tends to create rigid and extensive crystal lattices with definite packing patterns as for example in sodium chloride. The covalent bond falls off as the inverse cube of the ion separation in examples like the hydrogen molecule and saturates. Unlike the ionic bond, it is not an isotropic force but depends on the ability of an atom to provide electrons between itself and a bonding neighbour. The final structure is determined largely by the electronic configurations of the binding atoms rather than their packing properties. Finally, the weaker van der Waals bonds are isotropic and unsaturated with a binding energy $\sim\alpha(a_0/x)^7$ which falls off very rapidly with the ionic separation x.

The bonds between molecules are similar to those between atoms but include a metallic bond.[21] This bond depends on the presence of mobile, free electrons in solids. Its strength varies greatly from one material to another and is reflected by the huge range of melting points that metals display. In water an important force is that exerted between the dipolar electric charges of the molecules. The hydrogen and oxygen atoms exert unequal forces of attraction on the valence electrons and so the oxygen atom tends to be more negatively charged and the hydrogen atoms more positively charged than 'average'. Therefore, when dipolar molecules encounter each other, the negative portion of one molecule attracts the positive portion of the other. These forces tend to be ~10–20 times smaller than those binding covalently. This is an example of van der Waals bonding. Although these are the weakest of the chemical bonds they appear to be the most universal.

Hydrogen bonding occurs with a hydrogen atom in a covalent bond when an extremely electronegative atom, like oxygen, approaches another atom. The hydrogen atom is attracted by both electronegative atoms and

Bond Type	Example	Energy per Bond (Joules)	Effects	Configuration
Ionic	Sodium Chloride	3×10^{-19}	Strong enough to allows solids at room temperature	⊖ ⊕
Covalent	Diamond	6×10^{-19}		⊕ – – ⊕
Van der Waals	Solid Methane	3×10^{-21}	Cohesion of substances that are liquid at or below room temperature	± ±
Metallic	Iron	5×10^{-19}	Depends on free electrons in solids: very variable strength, hence wide range of metallic melting points	– ⊕ – ⊕ – ⊕ – ⊕ – ⊕ – ⊕ – ⊕ –

Figure 5.2. A summary of some characteristics of the four types of inter-atomic bond. The energy per bond is for zero temperature. The diagrams in column four summarize the character of the bonding interaction; the positive signs denote positive ions, negative signs electrons and, in the case of Van der Waals bonds, the symbols represent phase-correlated oscillating dipoles.[22]

provides a link between them. These bonds are generally weak (~25 times weaker than covalent bonds) because they arise from small electrostatic perturbations of chemical structures. However, in complicated systems like water, there exists a whole network of hydrogen bonds and therefore the binding is tighter. Intermolecular bonds are strongest when material is in the solid state. The following Figure 5.2 summarizes[22] the basic properties. Note that ionic and covalent bonds are strong enough to ensure solidity at room temperature whilst van der Waals bound states tend to be liquids below room temperature.

A dimensionless measure of the strength of materials is provided by the chemical binding energy fraction which compares the binding energy to the total mass of the system $\sim N_A m_N$, where N_A is the number of atoms in the molecule. This fraction is of order $5\times10^{-2}\alpha^2\beta N_A \sim 10^{-9}-10^{-11}$ as one goes from light to heavy elements. This indicates the efficiency of energy release one can expect from chemical processes like non-nuclear explosions, burning firewood or eating.

Molecules possess excited states just like atoms: suppose a molecule is diatomic and the two-component nuclei possess atomic numbers Z_1 and Z_2 respectively. Then the vibrational energy is $\sim m_e\alpha^2(Z_1Z_2)^{1/2}(m_e/\mu)^{1/2}$ where μ is the reduced mass of the system. This energy is typically $\sim\beta^{1/2}\sim10^{-2}$ times smaller than the analogous atomic levels. The energies associated with rigid rotation of molecules are typically $\sim(m_e/\mu)m_e\alpha^2$—about 10^{-4} times smaller than their atomic counterparts. As the environmental temperature rises, constituents of atoms and molecules will become increasingly energetic and eventually they will become ionized or dissociated. If a molecule has N bonding neighbours and the intermolecular binding energy is ε then the energy required to break the bonds between one molecule and all its neighbours is $\sim 0.5N\varepsilon$. Atoms and molecules cannot exist in an environment with temperature (which is measured in the same units as energy when Boltzmann's constant is set equal to unity, $k_B = 1$) exceeding $\sim 0.5\alpha^2 m_e$ because atoms will become completely ionized. In fact, significant ionization and opacity already occurs at $\sim 0.05\alpha^2 m_e$ because the Maxwell distribution about this energy contains enough high energy particles in its tail to cause significant collisional ionization. The hydrogen bond will snap when the temperature is about $\sim 10^{-2}\alpha^2 m_e$ and molecules will be completely dissociated at energies of $\sim 0.05\alpha^2 m_e$. All biological structures will disintegrate if the temperature rises above $\sim 10^{-3}\alpha^2 m_e$. Proteins then lose their internal mobility and denaturize—this, incidentally, defines the cooking temperature in our kitchens; for example, proteins in egg-white become denaturized and solidify at $\sim 10^{-3}\alpha^2 m_e \sim 470$ K. Thus biological molecules and living organisms inevitably inhabit environments with temperature below T_B where[86]

$$T_B \sim 10^{-3}\alpha^2 m_e \sim 470 \text{ K} \tag{5.32}$$

This defines what we generally term 'room temperature'. At temperatures far below T_B the hydrogen bond becomes very rigid and the flexibility of atomic configurations is debilitated. Most substances are liquid or solid below $\sim T_B$, although some unusual ones, like helium, remain gaseous down to $\sim 10^{-2}T_B$. Biology occurs in environments with ambient temperature within an order or magnitude or so of T_B.

In our discussion of atomic and molecular stability we have so far avoided mention of the most important ingredient responsible for the stability of all matter—the Pauli Exclusion Principle. Although it appears that atoms, molecules and solids achieve stable configurations by setting up a balance between attractive electrostatic forces and other inter-ionic repulsive forces of quantum mechanical origin, this is not the whole story. Since the systems involved contain positive and negative charges in equal numbers they could orient themselves so that interactions between nearest neighbours predominated. According to our understanding of chemistry the entire system would then rapidly collapse to enormous density greatly in excess of ρ_{AT}. However, because electrons are fermions this catastrophe is averted by a principle of quantum mechanics. The question seems to have been considered first by Ehrenfest.[23]

The Exclusion Principle can be formulated by stating that no more than one particle of a particular kind and spin is permitted in a single quantum state. If this restriction did not hold for atomic electrons they would all occupy the state with the lowest available energy—the ground state. The Exclusion Principle can also be formulated as follows: if N identical particles occupy a volume $V = 4\pi R^3/3$ then the minimum energy state of each particle exceeds that of a single electron. The minimum electron energy is of order of the kinetic energy $\sim m_e^{-1}R^{-2}$ that the Uncertainty Principle ensures will resist confinement; but the minimum particle energies are larger $\sim m_e\lambda^{-2}$, where $\lambda = (R^3N^{-1})^{1/3}$. So each particle behaves as though confined to a dimension $RN^{-1/3}$, rather than just R, and this determines its minimum energy. This effect also means that the pressure resisting the confinement of N particles within the volume R^3 is $N^{5/3}$ times larger than that resisting the confinement of a single particle. The 'exclusion energy' is thus $\sim m_e^{-1}N^{5/3}R^{-2}$. Technically, only those particles with identical spin orientations are excluded from occupying the same quantum state by this effect and the accurate value for the exclusion energy is,

$$E_F \sim 0.2(3\pi^2)^{2/3} m_e^{-1} N^{5/3} R^{-2} \sim 1.9 m_e^{-1} N^{5/3} R^{-2} \tag{5.33}$$

This repulsive effect is usually referred to as the *degeneracy pressure.* In small objects this pressure can be counter-balanced by the electrostatic attraction of nuclei and electrons when $n \sim (\alpha m_e)^3$.

The Exclusion Principle plays a key role in Nature. Aside from guaranteeing the stability of matter and the 'large' size of atomic and molecular structures, it creates the shell structure of atomic electrons. These electronic heirarchies are responsible for the existence and enormous diversity of chemical properties. One could imagine a world in which the Exclusion Principle did not exist or one in which electrons were bosons but it would be a world of compact, superdense bodies with little scope for complex structures or living organisms and any two molecules that encountered one another would release huge quantities of binding energy.

The familiar solid materials around us are composed of lattices of atoms and molecules. Their average densities reflect the average densities of atomic rather than nuclear systems and their characteristic 'hardness' is a reflection of the large bulk modulus of a single atom, or collection of atoms. To appreciably deform a solid, one must apply at least one Rydberg of energy $\sim(Z\alpha)^2 m_e(1+\beta)^{-1}$ to an atomic volume $\sim a_0^3$, thus the bulk modulus B, is of order

$$B \sim \frac{Z^5\alpha^5 m_e^4}{(1+\beta)} \sim 10^{11} - 10^{12} \text{ erg cm}^{-3} \tag{5.34}$$

and so about a million atmospheres of pressure is necessary for the deformation of a solid. The sound-speed in the solid is given dimensionally by $v_s \sim (B/\rho)^{1/2}$ and so by (5.11) and (5.14)

$$v_s \sim \left(\frac{Z^5}{A}\right)^{1/2} \alpha\beta^{1/2} \sim 1.7\times10^{-4}\left(\frac{Z^5}{A}\right)^{1/2} \tag{5.35}$$

This is roughly correct for metals like copper or lead where $v_s \sim 3\times10^5\ \text{cm s}^{-1}$. The coefficient of thermal expansion is also an expression of internal binding energy and will be $\sim E^{-1}$ with E given by (5.24) which ensures values $\sim 6\times10^{-6}\ \text{K}^{-1}$.

Although the Exclusion Principle provides for the overall stability of solid bodies it is not responsible for the comparatively fixed properties of ions within solids. Consider a lattice of ions: it is more realistic to think of the electrons moving through it as a sea amid the islands of fixed ions. Every ion behaves as an independent harmonic oscillator with mass $\sim m_N$ and vibrates with a frequency ω. If an ion is displaced a distance x from its equilibrium position then it will gain a potential energy $\sim 0.5 m_N \omega^2 x^2$ which must become $\sim \alpha a_0^{-1}$ when x is of order a_0 since the bonds then break. For these ionic oscillators the mean-square velocity is $\langle v^2\rangle \sim 2\omega^2 x^2$ and so the mean-square momentum is

$$\langle p^2\rangle \sim \alpha m_N (\alpha m_e)^3 \langle x^2\rangle\,. \tag{5.36}$$

The Uncertainty Principle ensures that the momentum resisting localization is $\langle p^2\rangle \sim \langle x^2\rangle^{-1}$ and the uncertainty in the electron's position, λ, will be of order a_0 so the relative fluctuation in the ion location relative to that of the passing electrons in the lattice is[24]

$$\sqrt{\langle x^2\rangle/\lambda^2} \sim \beta^{1/4} \ll 1 \tag{5.37}$$

and so the nuclei are accurately and rigidly located in the solid. The uncertainty in the position of an atom is $\sim \beta^{1/4}$ of the inter-atomic separation. If ions tried to move further afield than this they would push the electrons into such a small region that their momentum would grow to resist localization and force them back. The dependence in (5.37) reveals the key role that β plays in Nature. It ensures that nuclei have well-defined, relatively invariant, locations. When a substance is heated, the positional uncertainty of the ions rises. If atoms stray $\sim a_0$ from their locations the material will melt or, if molecules stray, dissociate. If one tried to build up ordered materials built upon the strong nuclear force one would not have this important property since neutrons and protons have similar masses so neither are located with precision in nuclei and from the outside nuclei appear fairly spherically symmetric. It appears that well-ordered structures rely heavily upon the small value of β. The specific application to DNA replication fidelity was highlighted by Regge,[25] who remarks that

It might as well be that a whole set of perfectly reasonable S-matrices exist for any choice of these m_e/m_N parameters, all of them yielding rather weird, self-consistent universes, all but one of them existing only in the sense of Plato. Our universe would be determined by the fact that only the choice $m_N/m_e = 1837$ guarantees that there are long chain molecules of the right kinds and size as to make biological phenomena possible. It could be for instance that the slightest variation in these parameters would change critically the size and length of the rings in the DNA helix as to invalidate its typical way of replicating itself. In this sense we could say that $m_N/m_e = 1837$ just because we are here. Other universes do exist as well but nobody is around to see them. I am describing this somewhat paradoxical mechanism just in order to warn that we may expect quite exotic criteria to come into play in fixing the fundamental constants.

There is one additional aspect of physics that is essential for the existence and stability of atomic systems—*quantization*. In 1913 Niels Bohr proposed the radical revision of the naïve atomic models that imagined the electrons to orbit a nucleus in the manner of a mini solar system. The quantization principle he used restricted the energy of the orbital electrons to certain discrete values: multiples of a universal energy quantum fixed by Planck's constant. In the non-quantum atom, electrons can possess all possible energies. They can reside at any orbital radius so long as their velocity is sufficient to establish an equilibrium between centrifugal and Coulomb forces. All atoms would be different under these circumstances and, worse still, the continuous buffeting of electrons by photons and other particles would cause a steady change in electron orbit (and hence chemistry). The quantum principle avoids this: if one electron is added to a proton there is only one orbital radius available to it in quantum theory and consequently all hydrogen atoms are identical. This could not be the case in a non-quantum theory. Likewise, tiny environmental perturbations do not upset the structure of the atom because an entire quantum of energy must be added before the electron orbital is altered. Thus, despite its traditional reputation as the harbinger of chance and indeterminism, quantum theory is the basis for the fidelity and large-scale stability of Nature.

5.3 Planets and Asteroids

> Then it is reasonable to think
> that one can see, by looking in
> a microscope, what is going on
> in another planet.
>
> Johann Strindberg

If the temperature falls below $\sim 0.05 - 0.2\alpha^2 m_e$ materials can exist in liquid form and if it drops below $\sim 0.2\alpha^2 m_e$ then stable solid matter is possible. Since the ambient temperature of the Universe is comfortably below these levels, large solid bodies can and evidently do exist in Nature. We are familiar with some of the more notable examples: people, planets, asteroids, comets and so forth. Although these various manifestations differ vastly in size they all possess a roughly similar density close to the atomic density, ρ_{AT}. Ordinary 'small' objects like this book are consequences of the equilibrium state that can exist between the repulsive degeneracy pressure of atomic electrons and the electrostatic attraction. This equilibrium is reflected by the average density of collections of atoms. On the scale of this page the gravitational attraction of the constituents is utterly negligible, smaller than the electromagnetic and exclusion forces by a factor $\sim\alpha(Gm_N^2)^{-1}\sim 10^{37}$. However, as we consider larger and larger collections of atoms the gravitational effects do not saturate but accumulate in an additive fashion because the gravitational 'charge'—mass—can only be positive. Large aggregates of electromagnetic charge cannot exist in stable equilibrium because they involve positive and negative electric charges which are unstable against mutual annihilation. Thus, eventually the intrinsic weakness of the gravitational force is made up for by the large number of electrically neutral particles that can be accumulated into stable bodies. One might therefore anticipate that it should be possible to create large, cold, stable bodies in which the repulsive exclusion pressure resisting the confinement of electrons balances the attractive force of gravity. How large will these structures be?

If a spherical body of mass M and radius R is built of N molecules each of molecular weight A then its gravitational binding energy E_g will be given by Newton's law of gravitation as

$$E_g \sim \frac{GM^2}{R} \sim \frac{GA^2 m_N^2 N^2}{R}. \tag{5.38}$$

A rocket would require kinetic energy of order E_g to escape from the gravitational field of this body.

The body will remain in stable equilibrium when the gravitational binding energy, E_g, causing collapse is balanced by the electrostatic and degeneracy pressure $\sim N\alpha^2 m_e$ resisting gravitational collapse. This occurs when the radius of the body is equal to R_+ where[26]

$$R_+ \sim NA^2 \frac{\alpha}{\alpha_G} a_0 \tag{5.39}$$

but if the body is composed of atoms then $N \sim R_+^3 a_0^{-3}$ and the radius R_+ is found to be of order that of planetary bodies

$$R_+ \sim \left(\frac{\alpha}{\alpha_G}\right)^{1/2} \frac{a_0}{A} Z^{2/3} \sim \frac{0.7}{A} Z^{2/3} 10^6 \text{ km} \tag{5.40}$$

The Earth is composed primarily of quartz (silicon dioxide) hence to a good approximation $A = 28 + (2 \times 16) = 60$ and in this case our rough calculation gives $R \sim 10^4 - 10^5$ km which is roughly correct. The corresponding planetary mass contained in a spherical body of radius R_+ at atomic density is

$$M_+ = \frac{4\pi}{3} R_+^3 \rho_{AT} \approx \left(\frac{\alpha}{\alpha_G}\right)^{3/2} \frac{m_N}{A^2} Z^3 \sim 10^{31}\ \text{gm}\left(\frac{Z^3}{A^2}\right) \qquad (5.41)$$

and $Z^3/A^2 \sim 0.5 - 15$ for the relevant materials, (note $\alpha/\alpha_G = 1.2 \times 10^{38}$).

This is in excess of the mass of the Earth ($\sim 6 \times 10^{27}$ gm) but is of order the mass of the major planets whose densities vary from $\sim 0.17 - 1.7$ gm cm^{-3} along the sequence Saturn, Uranus, Neptune, Jupiter.[27] The mass of Jupiter is $\sim 1.9 \times 10^{30}$ gm. Bodies appreciably larger than Jupiter would have such a large central pressure that electrons would be squeezed closer together than in conventional solids where inter-particle separations are $\sim a_0$. In fact, a body just a little bigger than Jupiter would have a high enough central pressure and temperature to initiate nuclear burning and so become a *star*. The expressions (5.40) and (5.41) give the upper size and mass limits for solid or liquid planets. Again, these calculations indicate that the observed range of planetary masses is not an accident, nor is it surprising that planets populate[28] the region of the mass-radius diagram (Figure 5.1) that they do. It is an inevitable consequence of the relative strengths of the electromagnetic and gravitational interactions. Finally, since we shall often refer to it, we note that the force of gravitational attraction exerted by a planetary body can be calculated and the acceleration experienced by a body dropped close to the surface is independent of its mass and equal to GM_+/R_+^2. This is called the 'acceleration due to gravity', g_+, of the planet and is calculated from (5.41–42) to be of order

$$g_+ \sim Z^{2/3}\left(\frac{\alpha_G}{\alpha}\right)^{1/2} \alpha^3 \left(\frac{m_e}{m_N}\right)^2 m_N \qquad (5.42)$$

A *lower* size limit for planetary bodies can be calculated by considering the maximum size, H, of surface undulations (i.e. mountains) and demanding that spherical planets are those for which $H \ll R_+$. All planets place a limit on the extent of mountain-building above their average radii because the higher a mountain becomes so the greater the pressure on the planetary crust beneath it. If the pressure at the mountain base should exceed the strength of atomic bonding in the crustaceous material then the latter would liquify without any appreciable change in volume.[26] Plastic flow would result; see Figure 5.3.

If the mountain sinks by an amount H then it will do so until its loss in gravitational potential energy, mg_+H, equals the amount of work that must be done to liquify a mass of rock of height H. The liquefaction

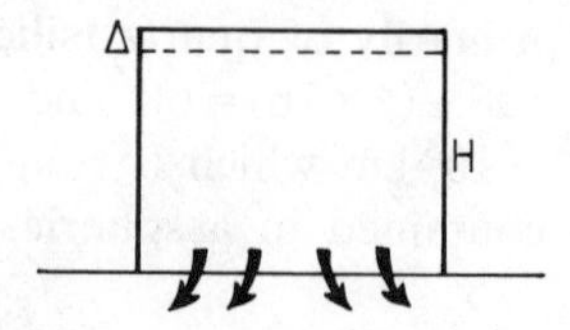

Figure 5.3. The sinking of a mountain of height H by Δ corresponds to the displacement of a layer of thickness Δ at the summit into the ground, and leads to a plastic flow of a comparable volume in the ground.[26]

energy is $E_L \sim \eta m_e Z^2 \alpha^2$ where $\eta \sim 10^{-2}$ and if the cross-sectional area of the mountain is $A \sim \pi R^2$, then stable mountains must be bounded in height by

$$H \leqslant \frac{E_L}{A m_N g_+}. \tag{5.43}$$

So, on the Earth, a quartz mountain must be less than ~30 km in height. Using equation (5.42) and the expression for E_L we have[30], with $R \sim R_+$

$$H \leqslant \eta \frac{Z^{4/3}}{A} \left(\frac{\alpha}{\alpha_G}\right)^{1/2} a_0 \tag{5.44}$$

and, if we compare this with the radius of the planet, R_+, we have

$$\frac{H}{R_+} \leqslant \eta Z^{2/3} \tag{5.45}$$

Planets possessing mountains of height $H \sim R_+$ will be non-spherical and this criterion ($H \sim R_+$) fixes the sizes of asteroids. For example, the Martian moon Phobos has a mass $\sim 10^{-6}$ that of a major planet whereas equation (5.45) predicts that this ratio should be

$$\frac{\text{Mass of Phobos}}{\text{Mass of Mars}} \sim \left(\frac{H}{R_+}\right)^3 \sim \eta^3 Z^2 \sim 6 \times 10^{-4} \tag{5.46}$$

in reasonable agreement with observation given the crudeness of the treatment. The major reason for disagreement is likely to be the poor estimate for the hardness of rock and the extreme sensitivity of liquefaction to such factors as the temperature and water content of rock. It is interesting to consider that terrestrial continents have a density slightly less than that of the Earth's mantle and sub-oceanic structure and so can be thought of as 'floating' on the Earth's surface. Perhaps the existence of continental structures on a planetary surface arises only for particular range of values for H/R_+? Finally, what about planets with spherical form and significant mountain building? On the Earth, the Moon and on Mars the ratio of the highest mountain to the planetary radius is 1.3×10^{-3},

2.8×10^{-3} and 8.8×10^{-3} respectively and is again close to our rough estimate (5.45).

Habitable planets require considerably more than a spherical form. The evolution of sophisticated organisms and their continued survival requires an atmosphere and the accompanying photochemistry. In 1897 G. Johnstone Stoney[1] showed that the atmospheric composition of a planetary body was constrained by its mass. Small planets are unable to retain atmospheres because their escape velocities are smaller than the root-mean-square velocity of gas molecules near their surface. This is why the Moon has no gaseous atmosphere, for if the terrestrial atmosphere were added to the lunar surface it would rapidly disperse into space to be recaptured by the Earth or the Sun. How big must a planet be in order to possess a gaseous atmosphere and the photochemistry that accompanies it?

A planet which can support biological structures must have a temperature close to $T_B \sim \varepsilon\alpha^2 m_e$ where $\varepsilon \sim 10^{-2}$ if we require hydrogen bonding or van der Waals bonding of molecules to be possible. If the atmosphere is gaseous then the mean energy of the gas molecules will be of order T_B and in order that they not have sufficient energy to escape the gravitational field of the planet we must have

$$G\frac{Mm_N}{R^2} \geqslant \varepsilon\alpha^2 m_e \tag{5.47}$$

Planets of this size will be supported by Exclusion forces against collapse under gravity and so they will possess approximately atomic density on average. Therefore $M/R^3 \sim Am_N/a_0^3$ and so their masses must be of order,

$$M \sim \varepsilon^{3/2} Z^{-3/2} A^{-1/2} \left(\frac{\alpha}{\alpha_G}\right)^{3/2} m_N \tag{5.48}$$

This is the minimum mass for a 'habitable' planet (at least 'habitable' by gas-breathing beings like ourselves).

It is interesting to notice that, although the approximate size and mass of planetary and other solar system bodies are seen to be quite inevitable by the reasoning of this section, they do not guarantee life-supporting environments. Hart has shown that the conditions necessary to support life on a planetary surface depend very sensitively on the radius of the planet[32], see section 8.7. Dimensional analysis alone is not sufficient to determine the necessary size of the habitable planet. Later, in Chapter 8, we shall consider some of these more detailed constraints on habitable planets by considering the evolution of their atmospheres.

Lightman[40,41] has shown that if we note that most planets in the solar system have an angular velocity that lies within an order of magnitude of that required for centrifugal break-up (the others are tidally entrained to a close companion), then since the angular velocity at break-up is $\sim(G\rho)^{1/2}$ and planets have atomic densities, (5.31), the length of a 'day' can be calculated (1 day = orbital period of planet around its central star) as $\sim 2\pi(G\rho_{AT})^{-1/2}$, so we have the rough estimate

$$t_{\text{day}} \sim 4\pi m_e \left(\frac{m_N}{m_e}\right)^{3/2} \alpha^{-1/2} \alpha_G^{-1/2} \sim 6\text{–}10 \text{ hours.} \quad (5.49)$$

5.4 Planetary Life

From a drop of water a logician could infer the possibility of an Atlantic or a Niagara without having seen or heard of one or the other.

A. Conan Doyle

Many authors[33–38] have realized that size is a crucial parameter which determines the qualitative nature of living organisms. If only because brain size appears[39] to be proportional to the 9/5 power of the linear size of most birds and vertebrates, it can be argued that total size is fundamental in controlling the qualitative and quantitative features of living creatures—after all, there exist thousands of varieties of insect and rodent but only two types of elephant. Size will determine whether creatures can move easily without risk of internal damage, wield weapons with sufficient kinetic energy to hunt other creatures or use fire to smelt metal and so forth. The possible evolutionary histories and strategies available to any creature are strongly, if not absolutely, circumscribed by its size.

In the last section we showed how the size of planetary bodies is determined by the equilibrium state between gravity and electrostatic exclusion forces at the atomic level. In this section we shall show that the same type of reasoning actually fixes the inevitable size of a human being to within dimensionless factors. This type of argument cannot tell us why humans tend to be five or six rather than eight or nine feet tall but it does explain why they are not one hundred feet or one inch tall! The aim of this section is to display this and other calculations of structural features in living creatures in terms of the fundamental constants of Nature. This enables us to develop considerable insight into the factors upon which the existence of living systems hinges. More light-heartedly, we can determine whether or not Bunyan's Giant Despair or Swift's Lilliputians are physiologically viable. The techniques we shall employ are familiar to biologists and were known to Galileo.[42] They are based upon scaling arguments. However, biologists do not seem to have realized that the constants of proportionality in these scaling laws can be determined by fundamental constants and one need not simply evaluate them from observational data (unless of course a very precise result is required).

From our previous calculations regarding the size of habitable planets—that is, planets with atmospheres—we can roughly estimate the maximum size of a living being like man or woman whose composition relies on chemical bonding.

The reason why organisms cannot be constructed with arbitrarily large size was first spelt out by Galileo in 1638. He realized that the *strength* and the *size* of bodies scale at different rates. If you double the size, or mass, of a body you do not double its strength. If the linear scale of a body is L then its mass grows like L^3 but its strength grows only as the cross-sectional area across its bones, hence as L^2. There exists a maximum mammalian size and if all animals were slowly inflated in dimension they would eventually be unable to support themselves under their own weight, Galileo writes:[42]

> Nor could Nature make trees of immeasurable size because their branches would eventually fall of their own weight, and likewise it would be impossible to fashion skeletons for men, horses, or other animals which could exist and carry out their functions proportionability when such animals were increased to immense weight. ... It follows that when bodies are diminished, their strengths do not proportionally diminish; rather, in very small bodies the strength grows in greater ratio, and I believe that a little dog might carry on his back two or three dogs the same size, whereas I doubt if a horse could carry even one horse his size...

If a human has mass M_H and size L_H, (assume for the present that the being is spherical), then it will contain $\sim M_H/m_N$ atoms. To fracture the bonds between the molecules of the body only requires breakage along a two-dimensional surface which should contain $\sim(M_H/m_N)^{2/3}$ atoms. Therefore, the energy required for fracture, E_{fr}, is

$$E_{fr} \sim \left(\frac{M_H}{m_N}\right)^{2/3} \times (\text{molecular binding energy}). \tag{5.50}$$

If our human lives on a planetary body we know that the acceleration due to gravity at its surface will be roughly of order, (from (5.42)),

$$g_+ \sim Z^{2/3}\left(\frac{\alpha_G}{\alpha}\right)^{1/2} \alpha^3 \left(\frac{m_e}{m_N}\right)^2 m_N \tag{5.51}$$

The surface of the planet will only be safe for humans if, when they fall, the potential energy lost in falling $\sim 0.5L_H$ is insufficient to fracture molecular bonds. Alternatively we might say that the compression energy at the being's base must be insufficient to rupture its bonds. Now the molecular binding energy will be $\sim \varepsilon\alpha^2 m_e$ where $\varepsilon \sim 10^{-3}$, so the human existence criterion reduces to

$$\begin{pmatrix}\text{Gravitational potential}\\ \text{on planet surface}\end{pmatrix} < (\text{Energy required for fracture}) \tag{5.52}$$

that is

$$M_H L_H g_+ < \varepsilon \left(\frac{M_H}{m_N}\right)^{2/3} \alpha^2 m_e \tag{5.53}$$

but, since the human composition is atomic we know

$$M_H \sim \rho_{AT} L_H^3 \tag{5.54}$$

and equation (5.53) produces the intriguing constraint on size,[43]

$$L_H \leqslant Z^{2/3} \varepsilon^{1/4} \left(\frac{\alpha}{\alpha_G}\right)^{1/4} a_0 \sim 10 \left(\frac{\alpha}{10^{-3}}\right) \text{cm} \tag{5.55}$$

So, by (5.54), its maximum mass is of order

$$M_H \sim AZ^2 \left(\frac{\alpha}{\alpha_G}\right)^{3/4} m_N \sim 10^5 \text{ gm} \tag{5.56}$$

These estimates are extremely close to the actual size and mass of humans and animals, although they do underestimate their sizes slightly because they are not as brittle as the simple molecular bond energy would suggest.[44] (They also explain why small children rarely break bones when they fall.) The polymeric form of human molecular structure distributes stresses over large areas and increases the overall length. Tendons possess a many-stranded fibrous structure, reminiscent of suspension bridges, which prevent the propagation of cracks in their structure.[45] Incidentally, Man is virtually the largest two-legged animal; larger creatures tend to be quadrupeds. This posture relieves pressure on the base of the skeleton and also minimizes the risk of falling and breaking.

A few points are worth making in regard of the results (5.55) and (5.56): the first is, do we really fracture ourselves by compression? No, we tend to break bones because we accidently apply too great a bending moment and for this very reason large animals tend to stand on straight legs with as little flexure as possible. Bending moments are forces which must be applied over a cross-section of bone and therefore scale in similar fashion to fracture energies, as L_H^2. The rough size of an object able to withstand bending stresses will depend on fundamental constants in a fashion identical to (5.55) and (5.56) but the dimensionless factors involved will be slightly different. Galileo seems also to have been aware of the bending moment restriction and after him it was studied quantitatively by Euler[46] and others[47] who were interested in the maximum height to which a tree could grow before bending under its own weight after a slight displacement from the vertical by wind. This[37] effect can be observed in cats: a kitten's tail can be supported erect but the longer tail of a fully grown cat bends under its own weight.

Of course, greater stability to bending stresses is possible if an object is tapered, or if its height is not equal to its width. Suppose an object, or even a creature's leg, has a cylindrical form with height L but cross-sectional area d^2. Then the area of bone feeling the breaking stress is $A \propto d^2$ and the criterion for it not to break under its own weight when displaced from the vertical is $L \propto d^{2/3}$. Now since the mass $M \propto Ld^2 \propto d^{8/3} \propto A^{4/3}$ (which differs from the relation between mass and area in the spherical model where $M \propto A^{3/2}$), we see the strength to mass ratio scales not as L but as $d^{2/3}$ in the cylindrical structures.[48]

Lastly, we should consider some practical consequences of (5.55) and (5.56): Large animals must possess slightly stronger bones than small ones because they are composed of the same molecular materials with the same stress tolerance. Indeed, one finds that very large animals tend to have thicker bones and muscles than smaller ones. Another interesting consequence of (5.55) and (5.56) is the precarious evolutionary niche of the largest animals. The largest animals and trees that now exist are fairly close to the maximum allowed by the limits (5.55) and (5.56) and extinct species like the dinosaurs must have lain extremely close to the boundaries defined by the laws of physics. Their lack of scope for adaptation and poor mobility is clearly something that rendered them susceptible to extinction.

One obvious way to alleviate the pressures of size is to live in water, as Galileo was also the first to realize:[49]

> What happens in aquatic animals is the opposite of the case with land animals; in the latter, it is the task of the skeleton to sustain its own weight and that of the flesh, while in the former the flesh supports its own weight and that of the bones. And here the marvel ceases that there can be very vast animals in the water but not on the earth—that is to say, in the air!

The buoyancy makes the effective weight in water less than that in air by a factor

$$1-\left(\frac{\text{density of water}}{\text{density of animal}}\right) \sim 0.5\text{–}0.7. \tag{5.57}$$

This is borne out by experience: the largest whales grow to $\sim$130 tons, larger than any land-going animal that has ever existed; the largest dinosaurs were Brachiosaurii $\sim$80 tons, whilst a typical elephant weighs-in at only $\sim$7 tons. Interestingly, it seems that sea-going creatures have a real advantage in being large because they must work to overcome a frictional resistance, R say, and their muscles can supply energy in proportional to their mass $\propto L^3$ to overcome the resistive force RV^2 which acts on them when they swim at speed V. Therefore the speed attainable scales *directly* with size, $V \propto L^{1/2}$, and large sea-going creatures will be the most rapid movers and thereby are able to exploit the resources of their environment more efficiently.

The consideration of very small land and sea-going creatures involves quite different factors.[50] Consider some of the smallest ciliate protozoa that are less than a millimetre in length but propel themselves at speeds of order 0.4–2 mm s^{-1}. Why are they so small? Suppose an animal is covered by cilia and it has length L and cross-sectional area A, then to propel itself at speed v the cilia must generate power $\sim \eta A v^2 L^{-1}$ where η is the viscosity of water. The power available is directly proportional to the number of cilia on its surface $\propto A$ and also to its length. So the maximum speed—at which the available power balances the frictional resistance is given by $V_\star$ when

$$V_\star \propto L\eta^{-1/2} \tag{5.58}$$

and is independent of the number of cilia that cover the animal because it is independent of area, A. There exists an absolute limit to the speed of propulsion by cilia because the force exerted by a single cilia is $\propto vL^{-1}$ and the accompanying bending moment at its base is $\propto L(vL^{-1}) \propto v$. From (5.58) we see that this implies an upper limit on cilia size and explains why only very small creatures employ ciliatic propulsion. This shows also how smallness is advantageous to a cellular organism—the shearing forces and stress moments that it must withstand are smaller across small cells and so it can exist with very thin and fragile cell walls.

Let us turn from aqueous life at low Reynolds number to see how the constants of physics limit the nature of flying creatures.[52] Observationally, there appears to be a clear upper size limit for flying creatures. This is not surprising. Far more energy must be expanded when flying or hovering than standing still, supported by the Earth's surface. As flying is a complicated procedure, we restrict our example to calculating the size of the largest hovering bird. A hovering creature must be able to generate enough kinetic energy by its wing motion to overcome the pull of gravity, therefore the power needed to hover is of order

$$P_{\text{hov}} \sim \frac{m^3 g_+^3}{\rho_a \lambda^2} \tag{5.59}$$

where m is the mass of the 'bird', λ its wing-span, and ρ_a is the density of the air in which it flies. We would anticipate that $\lambda \propto m^{1/3}$ and so $P_{\text{hov}} \propto m^{1.17}$, (in fact data[39,53] indicate that $\lambda \propto m^{0.4}$ and $P_{\text{hov}} \propto m^{1.11}$, quite close to the crude scaling estimates we are using). As the bird grows bigger the power needed to support itself grows more rapidly than the power that its muscles can exert and there is a maximum size beyond which hovering flight is impossible. If we assume the maximum power the bird can exert must not rupture his atomic bonds then

$$P_{\max} \sim \varepsilon \alpha^4 m_e^2 \tag{5.60}$$

where $\varepsilon \sim 10^{-2}$ or so. Now the density of air is close to atomic density, $\rho_{AT} \sim m_N \alpha^3 m_e^3$ and the acceleration due to gravity is given by (5.42). Since $m \sim \lambda^3 \rho_{AT}$ we have, very roughly, for the maximum wing-span

$$\lambda \leq \left(\frac{m_N}{m_e}\right)^{1/7} \varepsilon^{2/7} \left(\frac{\alpha}{\alpha_G}\right)^{3/14} Z a_0 \tag{5.61}$$

So, numerically

$$\lambda \leq 1.3Z \text{ cm} \tag{5.62}$$

This is a surprisingly good estimate for the size of the largest hovering birds—these are the largest humming-birds which are ~20 gm in mass. Birds, like kestrels, that are substantially larger than this (~400 gm) do hover, but only for brief moments and they do so by exploiting air currents not by wing-power alone.

There also seems to exist a lower bound on the size of birds, ($\gtrsim$2 gm), and indeed to all mammals. Can the particular values of the fundamental constants shed some light on the order of this limit as well? The basic limit to the existence of increasingly small mammals (rather than insects) arises from a consideration of their circulatory systems. A warm-blooded creature substantially smaller than a shrew simply could not ingest the amount of food necessary to maintain a constant body temperature. Thus, mammals do not predominantly populate the ranks of the tiniest creatures. In addition, small creatures lose heat very rapidly compared to their potential for heat generation. Mice do not exist in Arctic climates; rather, large polar bears are favoured, and the average size of birds grows as one moves to the far North or South away from the Equator. The heat loss from an animal's body scales as his area $\propto L^2$ whilst its heart size is $\propto L$ and its stroke volume $\propto L^3$, thus

$$\frac{\text{heat loss}}{\text{heat generation}} \propto \frac{L^2}{L^3} \propto \frac{1}{L} \tag{5.63}$$

The total amount of blood pumped by the heart of the animal or human in a fixed time is $\propto L^3 p$ where p is the pulse rate. To balance the heat loss $\propto L^2$ it is necessary that $p \propto L^{-1}$ and so large animals have slower, but stronger hearts. This is the reason for the decrease in human pulse rate as one grows from childhood into an adult. Biologists have also observed that[54] the lifespan of animals is roughly proportional to their size L and so the total number of heart beats that occur in an animal or human lifetime is independent of size, a constant that observation indicates to be 3×10^9. It is possible to understand this in the following way: the ambient temperature on a solid life-supporting planet must be $\sim T_B$ and the flux of solar energy incident on the Earth is

$$F \sim \sigma T_B^4 \sim 1.4 \times 10^6 \text{ erg cm}^{-3} \text{ s}^{-1} \tag{5.64}$$

where σ is the Stefan–Boltzmann constant ($=\pi^2/15$ in our units) and $T_B \sim \varepsilon\alpha^2 m_e$. The quantity F is usually called the 'solar constant' and the numerical value in (5.64) is the observed value. One can obtain a natural human or animal timescale, τ_H, from the total energy of their chemical bonds and the rate of solar energy incidence on an area $\sim L_H^2$ of the Earth's surface. If Man were a tree this might correspond to the time of tree-ring growth, if he were a plant it would represent a characteristic growth time. Since Man's food chain is ultimately linked to flora this is a reasonable criterion. It yields an estimate for his characteristic lifetime of[43]

$$\tau_H \sim \sigma^{-1} \rho_{AT} m_N^{-1} L_H T_B^{-3} \tag{5.65}$$

Since $T_B \sim \varepsilon\alpha^2 m_e$ we have, using (5.55) for L_H, that

$$\tau_H \sim 10\varepsilon^{-2.75} \alpha^{15/4} \alpha_G^{1/4} m_e^{-1} \tag{5.66}$$

$$\tau_H \sim 14\left(\frac{\varepsilon}{3\times 10^{-3}}\right)^{-2.7} \text{hours} \tag{5.67}$$

This result is very sensitive to the dimensionless factor in the bonding strength ε and is extremely heuristic. However, what (5.44) does indicate is that $\tau_H \propto L$.

Since the pulse rates scales as $p \sim L_H^{-1} \sim \alpha_G^{1/4} a_0^{-1} \varepsilon^{-1/4}$ we have an estimate for the total number of heart beats, $p\tau_H$, as

$$p\tau_H \sim \varepsilon^{-3} \alpha^{-11/4} \tag{5.68}$$

$$\sim \left(\frac{10^{-2}}{\varepsilon}\right)^3 10^{11} \tag{5.69}$$

and the order of magnitude is almost correct.[54] This derivation shows why all mammals experience roughly the same number of heart beats during their lifetimes.

The work rates that humans and animals can sustain tells us something about general mobility as well. The work that must be done to take a stride of length $d \propto L$ will be proportional to L^3 and the energy imparted to the limb will scale as $mv^2 \propto L^3 v^2$ if motion is at speed v. So if the energy available from the metabolic activity is employed in running at speed v, the equilibrium state requires $L^3 \propto L^3 v^2$ and the maximum running speed is size-independent. Interestingly, this conclusion is borne out by the data.[55] Men and bears can run at roughly the same speed on flat ground. However, if you run uphill the situation changes: the heavier animal must then exert enough power to overcome the vertical component of his weight. The power required will be proportional to the velocity times the weight, $\propto L^3 v$, but the power available is still $\propto L^3$. Thus $v \propto L^{-1}$ and small animals run faster uphill. This makes the strategy for escaping from Grizzly bears very clear!

We can estimate the typical universal running speed on the flat (for humans and mammals) by noting[41] that the power of working, P, will be roughly given by T_B multiplied by the human size and the skin conductivity so

$$P \sim \left(\frac{\alpha}{\alpha_G}\right)^{1/4} \frac{\alpha^2}{a_0^2} \left(\frac{m_e}{m_N}\right)^{5/8} \sim 200 \text{ watts} \tag{5.70}$$

The power needed to run at the velocity v is $\sim mv^3 \div$(human size) and so

$$v \sim \left(\frac{\alpha_G}{\alpha}\right)^{1/12} \alpha \left(\frac{m_e}{m_N}\right)^{11/24} \sim 10 \text{ ms}^{-1} \tag{5.71}$$

Some animals obtain the oxygen they require directly from the gaseous atmosphere whilst others extract it from water after it has been dissolved. The distribution of the oxygenated blood to the body tissue is performed by a circulatory system in large animals but in small organisms this can be done by the process of diffusion. Humans are far too large for this slow diffusive action to be relevant and one can calculate that the largest diffusive organism should be only about 10^{-2} cm in size.[56]

Finally, we make a few general remarks about the influence of size on life. In his famous book *What is Life*?, Erwin Schrödinger[57] pointed out a fundamental property of 'large' systems: they are extremely stable to statistical fluctuations. The amplitude of purely statistical fluctuations in a collection of N bodies is of order $N^{-1/2}$. Living things must be relatively large and contain a huge number of atoms if they are to avoid statistical fluctuations reaching a lethal level. A gene might contain $\sim 10^6$ atoms and has a large statistical fluctution $\sim 0.1\%$. A system must be a good deal larger than this to maintain a stable organized structure. One would expect that this type of statistical limit is very important on the cellular scale and probably determines the characteristic size of a simple cell.[58] This is not the sole criterion though: no organism is known that does not contain enzymes and all enzymes contain proteins. If one could evaluate how many enzymes are necessary for the cell to function then this would set a lower limit to its size. Some rough estimates indicate that ~ 45 may be the order of this minimum.[59] Organisms smaller than the tiniest known mucoplasm, $\sim 3 \times 10^{-5}$ mm in diameter, would find it impossible to accommodate the ribosomes needed to generate their proteinous content and would be living very dangerously with a minimum of essential enzymes and a maximum of statistical fluctuation. The probability of lethal errors accumulating quickly would be very large indeed.

This completes our precarious digression from astro to bio-physics. Our aim was to be neither exhaustive nor trivial but rather to indicate the all-encompassing role played by the fundamental constants of nature and the manner in which everyday facts depend upon them. We know too little biology to speculate upon the ingenuity of natural selection in overcoming deficiencies of strength that govern the evolution of size. Yet, one can be reasonably certain that certain basic components and properties are necessary before a living organism can exist—atoms and heavy elements. To the origin of these fundamental building-bricks we now turn. Astrophysicists have discovered that the answers to the question 'where did the building bricks of life come from?' lies in the stars.

5.5 Nuclear Forces

> As we look out into the Universe and identify the many accidents of physics and astronomy that have worked together to our benefit, it almost seems as if the Universe must in some sense have known that we were coming.
>
> F. Dyson

In order to discuss the structure of stars we need to introduce the strong interaction which is responsible for inter-nucleonic forces. Besides binding the nuclear constituents together this interaction also determines the energetics of interactions between nuclei and thereby the thermonuclear properties of the stars. Unlike the electromagnetic interaction needed to understand atoms, planets and people, the strong nuclear forces are not well understood although the theory of quantum chromodynamics offers an elegant and testable candidate for their explanation. For our present purposes it is sufficient to consider a particular model of the nuclear interaction—basically that proposed by Yukawa in 1936—which allows a simple analysis analogous in many respects to that of the electromagnetic forces. We shall examine the basis of this 'secondary' model in more detail later on when we come to consider the internal structure of nuclei, quarks and grand unified gauge theories.

The strong interaction contrasts sharply with the electromagnetic interaction: it has finite range and is repulsive at very short range, a typical form of the potential is shown[26] in Figure 5.4.

The nuclear force appears, in some ways, analogous to the van der Waals intermolecular forces. The model of Yukawa[60] assumed that nuclear forces arose from the exchange of some fields or particles between the nucleons—specifically by the exchange of a π-meson. If this view were completely correct the nucleon system would manifest a primary interaction. However, since that early proposal the nucleus has been found to exhibit internal structure and the detailed structure of the interaction is enormously complex. It is believed that the origin of the nuclear force lies in the so called 'colour' force between elementary quarks and gluons. However, since we are interested in order of magnitude estimates, the Yukawa picture will be a useful heuristic and any

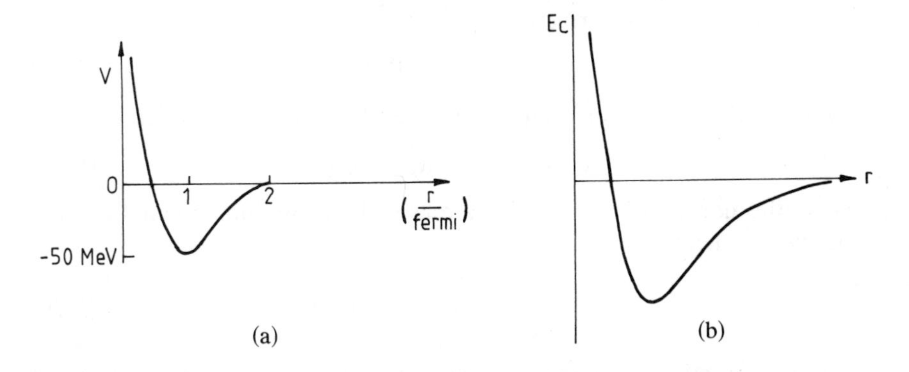

Figure 5.4. (a) The nuclear potential of the deuteron as a function of the radius r. Roughly, the potential is repulsive for $r \leqslant 0.5$ fermi and attractive for $r \gtrsim 1$ fermi with a finite range of about 2 fermi (1 fermi = 10^{-13} cm). (b) The electromagnetic Coulomb potential in an atomic system.

result obtained in terms of its defining coupling parameters can be transformed into relations between more fundamental quantities describing the coupling of quarks and gluons when necessary.

The dominant portion of the inter-nucleonic force can be viewed as arising from pion exchange with a pseudo-scalar coupling governed by the strong coupling constant, g_s, or an analogue of the fine structure constant, α_s defined by

$$\alpha_s \equiv \frac{g_s^2}{4\pi} \sim 0.2 \tag{5.72}$$

According to the Yukawa theory the potential energy of interaction between a pair of nucleons separated by a distance r is

$$V \sim -\alpha_s r^{-1} \exp(-r/r_0) \tag{5.73}$$

where

$$r_0 = m_\pi^{-1} \sim 1.4 \times 10^{-13} \text{ cm} \tag{5.74}$$

is the Compton wavelength of the pion. Thus at short range (r small) we can treat the two-nucleon interaction in a similar fashion to the hydrogen atom if relativistic effects are negligible. When r is small enough for the exponential cut-off in (5.73) to be insignificant the potential has the r^{-1} form of the Coulomb force.[61] The size of the two-nucleon state, the nuclear radius, a_N, is given by

$$a_N \sim (\alpha_s m_N)^{-1} \sim 2 \times 10^{-13} \text{ cm} \tag{5.75}$$

whilst the nucleon binding energy is

$$E_N \sim \alpha_s^2 m_N \sim 5 \text{ MeV} \tag{5.76}$$

and the binding energy fraction ε_N is

$$\varepsilon_N \sim \alpha_s^2 \sim 10^{-2} \tag{5.77}$$

From (5.53) we see that a typical nucleon 'velocity' is of order

$$v_N \sim \alpha_s \tag{5.78}$$

and so the non-relativistic model is reasonable. The fundamental nuclear time is τ_N where

$$\tau_N \sim \frac{a_N}{v_N} \sim (\alpha_s^2 m_N)^{-1} \sim 10^{-22}\ \mathrm{s} \tag{5.79}$$

It is instructive to compare some of these estimates with those of atomic systems to emphasize the differences of scale. The ratio of nuclear to atomic dimensions is

$$\frac{a_N}{a_0} \sim \left(\frac{\alpha}{\alpha_s}\right)\left(\frac{m_e}{m_N}\right) \sim 3 \times 10^{-5} \tag{5.80}$$

whilst the relative energies of nucleons and atomic electrons are

$$\frac{E_N}{E_0} \sim \left(\frac{\alpha_s}{\alpha}\right)^2 \left(\frac{m_N}{m_e}\right) \sim 4 \times 10^5 \tag{5.81}$$

This illustrates why nuclear energy sources are so much more efficient than chemical ones (roughly E_N/E_0 tons of conventional chemical explosive is necessary to match the explosive capacity of one ton of nuclear explosive). The density of nucleon material is also readily estimated,

$$\rho_N \sim \frac{m_N}{a_N^3} \sim m_N^4 \alpha_s^3 \sim 3 \times 10^{14}\ \mathrm{gm\ cm}^{-3} \tag{5.82}$$

and

$$\frac{\rho_N}{\rho_{AT}} \sim \left(\frac{m_N^3}{m_e}\right)\left(\frac{\alpha_s^3}{\alpha}\right) \sim 10^{14} \tag{5.83}$$

Unlike in atomic structure, the constituent nucleons have roughly similar mass and so no fixed directional structures are possible as in atomic solids where $m_e \ll m_N$. This ensures that the mass distribution in nuclei is roughly spherically symmetric.

Whereas atomic transitions occur and involve the emission of the massless quanta (photons) which mediate the electromagnetic interaction, similar nuclear transitions are not possible. The available binding energy $\sim E_N$ is much smaller than that required to create a pair of exchange particles $\sim 2m_\pi$ because of the 'coincidence',[62]

$$\alpha_s^2 m_N \leqslant 2m_\pi \tag{5.84}$$

As Weisskopf[63] has pointed out, the atomic nucleus is well-shielded from perturbative influences by the electron shells and consequently a nucleus can spend many hours in excited states if the angular momentum law excludes its spontaneous de-excitation.[64] An excited atomic state, on the other hand, is very short-lived because it is not shielded from interaction with neighbouring particles.

The simplest compound systems involve just two nucleons: the deuteron (proton + neutron) and diproton (proton + proton) and the dineutron (neutron + neutron). We shall see in what follows that the existence and non-existence of various two-nucleon bound states is crucial to the overall evolution of the Universe in general and to the burning rates within the stars in particular.[65]

Consider first the deuteron, which can be formed in reactions like

$$\begin{aligned} p+p &\rightarrow D+e^{+}+\nu \\ p+n &\rightarrow D+\gamma \end{aligned} \tag{5.85}$$

A simple model[66] is provided by solving the Schrödinger equation for two particles—each of reduced mass $0.5m_N$ in a three-dimensional square well potential of depth $-V_0$ and width b. It transpires that the deuteron can exist in a triplet state of zero angular momentum which has $V=38.5$ MeV and $b=1.93\times10^{-13}$ cm

$$(Vb^2)_D = 7.3\times10^{-14}\ \text{cm} \tag{5.86}$$

but the condition for two nuclei to form a bound system in the well is that the potential be greater than the zero-point kinetic energy of the nucleus $\sim 0.25\pi^2 m_N b^2$; that is

$$Vb^2 > \frac{\pi^2}{4m_N} \sim 5.2\times10^{-14}\ \text{cm} \tag{5.87}$$

If the potential falls off faster than b^{-2} there is no bound state. Thus (5.86) and (5.87) show that the deuteron is a bound state of the proton and neutron—but only just. Its binding energy is very small; (5.86) and (5.87) indicate it is roughly 1 MeV and experiment gives $\sim 2.2\ \text{MeV} = 1.1\times10^{11}\ \text{cm}^{-1}$. No states with non-zero angular momentum are possible; the binding is too loose to overcome the centrifugal force associated with these states.

Now the diproton (He^2) and dineutron can only exist in the singlet states of zero angular momentum. For the singlet state, one has

$$V = 13.3\ \text{MeV}; \qquad b = 2.58\times10^{-13}\ \text{cm} \tag{5.88}$$

$$(Vb^2)_{\substack{nn\\pp}} \simeq 4.5\times10^{-14}\ \text{cm} \tag{5.89}$$

and they just *fail* to be bound. Experiment indicates the diproton fails to be bound by a mere ~92 keV.

The existence of deuterium and the non-existence of the diproton therefore hinge precariously on the precise strength of the nuclear force. If the strong interaction were a little stronger the diproton would be a stable bound state with catastrophic consequences—all the hydrogen in the Universe would have been burnt to He^2 during the early stages of the Big Bang and no hydrogen compounds or long-lived stable stars would exist today. If the di-proton existed *we* would not! Also, if the nuclear force were a little weaker the deuteron would be unbound with other adverse consequences for the nucleosynthesis of biological elements because a key link in the chain of nucleosynthesis would be removed. Elements heavier than hydrogen would not form. In our potential approximation for the deuteron the dependence on α_s is roughly linear

$$V \propto \alpha_s \tag{5.90}$$

A decrease in α_s of about 9% is sufficient to unbind the deuteron whilst an increase in α_s of 3.4% is sufficient to bind the diproton. In the case of the dineutron, only a 0.3% increase suffices for binding.

The precise dependence of the nuclear binding energy on α_s becomes very complicated when one examines large nuclei because each nucleon moves in the average potential of all its neighbours. However, these larger nuclei are essential to living systems. Hydrogen and helium exhibit insufficient diversity to provide a basis for living organisms—heavier elements must exist for any form of life based upon chemistry to be possible. Before we can evaluate the stability of heavier elements, we must recall some basic facts about the nuclear force. Firstly, it is charge independent: removing the electromagnetic contributions, the nuclear forces between n–n, n–p and p–p are all the same. Secondly, the nuclear force saturates. If every one of A nucleons in a nucleus were to attract its neighbours then there would exist $A(A-1)/2$ interacting nucleon pairs and we would find the nuclear binding energy growing as $A(A-1) \sim A^2$. All nuclei would have diameters of order the range of the nuclear force and possess a constant volume. This is not what is observed—rather the nuclear radius R scales as the nucleon number,

$$R \simeq r_0 A^{1/3}; \qquad r_0 = 1.2 \times 10^{-13}\ \text{cm} \tag{5.91}$$

The volume of the nucleus thus varies linearly with A and so the density of all nuclei is roughly constant and they appear more reminiscent of liquids than solids. We recall that the radius of a liquid drop is also proportional to the cube of the number of molecular constituents. The nuclear force saturates: each nucleon attracts only a small number of others. This is reminiscent of chemical bonds where exchange forces lead to saturation. The suggestive analogy between liquids and nuclear material led to the development of a liquid-drop model of the nucleus[67] which enables an analysis of its stability to be made by accounting for the various factors that contribute attractively and repulsively inside nuclei. The liquid-drop model represents the nuclear binding energy, B, as a sum of five terms[68]

$$B = a_v A - a_s A^{2/3} - a_c Z^2 A^{-1/3} - a_{\text{sym}}(Z - 0.5A)^2 A^{-1} + \delta \qquad (5.92)$$

The first (*volume*) term is the contribution of the total number of nucleons in the nucleus. This contribution is reduced by the second (*surface*) term because nuclei at the surface of the nucleus have fewer bonding partners. Since the radius of the nucleus is proportional to $A^{1/3}$ this surface energy is proportional to the area, $A^{2/3}$, and is analogous to the effect of surface tension on a liquid where the surface molecules are more loosely bound. This dependence on $A^{2/3}$ indicates that a fraction $\sim 4A^{-1/3}$ of all nuclei are at the nuclear surface and so light nuclei have nearly all their constituents at the surface. The volume term is reduced still further by the third (*Coulomb*) repulsion which is the repulsive electromagnetic force acting between any two protons. If we assume the protons are distributed spherically symmetrically throughout a nucleus of radius R then the loss of binding energy is $\sim -0.6\alpha Z^2/R \equiv -a_c Z^2 A^{-1/3}$. Clearly, this effect becomes important for large Z. The fourth (*asymmetric*) contribution to the binding energy arises because the Exclusion Principle makes it energetically more economical for nuclei to be built with equal numbers of neutrons and protons—as, for example, in carbon or oxygen. If Z protons and $(A - Z)$ neutrons are present in a nucleus they will be able to occupy the lowest Z energy states. But, if N more neutrons are added, only $N - Z$ of them will be allowed in the lowest energy states, the rest will have to occupy states of higher kinetic and lower potential energy. Therefore, these neutrons will have less binding energy than the first Z protons and Z neutrons and the reduction will vary as $(Z - A/2)$. Obviously the protons and neutrons can be exchanged and the effect must therefore be independent of the sign of $(Z - A/2)$. A detailed calculation based on the Fermi-gas model gives a contribution to E_N of[69]

$$\Delta E_N \simeq -\frac{1}{6}\left(\frac{9\pi}{8}\right)^{1/3} \frac{(Z - A/2)^2}{m_N r_0^2 A} \sim 11 \frac{(Z - A/2)^2}{A} \text{MeV} \qquad (5.93)$$

assuming nucleons have the same mass. Large nuclei always contain more neutrons than protons because equal numbers would lead to a huge Coulomb energy; a neutron excess is necessary to prevent Coulomb disruption. This effect is entirely quantum mechanical in origin and has no analogue in classical liquids. Finally, there exists a small pairing energy

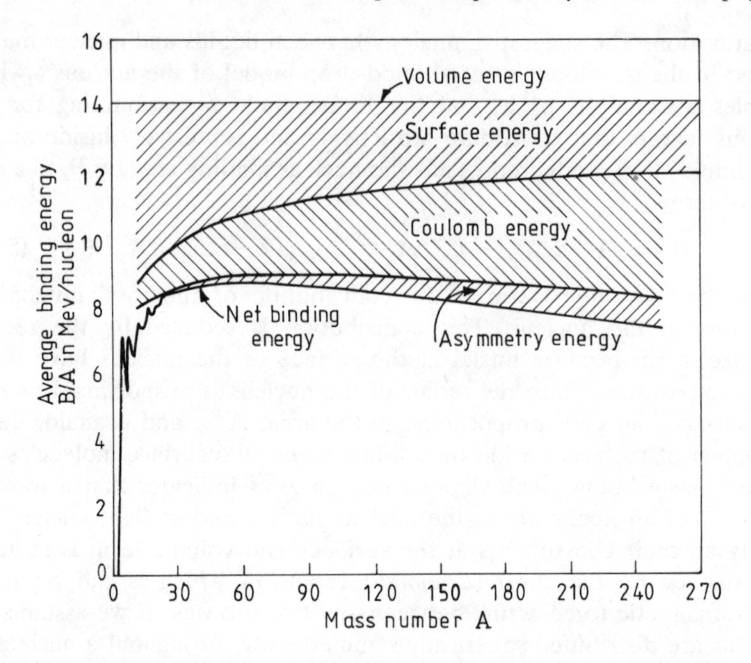

Figure 5.5. The relative contributions of the different components of the binding energy per nucleon versus mass number according to the liquid drop model discussed in the text.[68]

term, δ in (5.92), arising because of the intrinsic spin of the nucleons (it is zero for nuclei with odd A and otherwise falls off as $\delta \sim 12A^{-1/2}$ MeV) and we shall neglect it. Figure 5.5[70] shows the contributions made by the various terms in (5.92) and illustrates how the decrease in free surface energy along with an increase in the Coulomb repulsion produces a maximum of the binding energy per nucleon at $A \sim 60$. To extract binding energy from nuclei with $A \gtrsim 60$ they must be split (fission) but to extract it from nuclei with $A \lesssim 60$ they must be fused. The unknown constants in (5.92) which enable Figure 5.5 to be plotted are determined from data-fitting as $a_v = 16$ MeV, $a_{\text{sym}} = 50$ MeV and $a_c = 0.7$ if $r_0 = 1.24 \times 10^{-13}$ cm. On dimensional grounds they must all be of order $E_N \sim \alpha_s^2 m_N$ and in principle they can be calculated. The relation (5.92) now enables us to decide how the strong interaction strength decides which stable nuclei can exist.

It is energetically favourable for a nucleus to disintegrate into two equal parts of constitution ($Z/2$, $A/2$) if the binding energy change ΔB is positive, where

$$\Delta B \equiv B(Z, A) - 2B(Z/2, A/2) \tag{5.94}$$

ΔB is the energy released by the *fission* of the nucleus (Z, A); for the fission of uranium $\Delta B \sim 180$ MeV. Using (5.92), with the experimental values for a_v, a_s, a_{sym} and a_c, the binding energy change is just

$$\Delta B = a_s A^{2/3}(1-2^{1/3}) + a_c Z^2 A^{-1/3}(1-2^{-2/3})$$
$$= -4.5A^{2/3} + 0.26 Z^2 A^{-1/3}\ (\text{MeV}) \qquad (5.95)$$

The susceptibility to fission is determined by competition between the surface forces of nuclear origin and the electromagnetic Coulomb interaction between the charged nuclei. The Coulomb force tends to deform the nucleus away from a spherical configuration whilst the surface tension tries to maintain it.[72] If the Coulomb forces win then the nucleus can fission and $\Delta B > 0$ gives the instability criterion as $Z^2/A \gtrsim 18$. However, this does not describe the inevitable change in the Coulomb and surface forces as the nucleus is gradually deformed away from sphericity; it is only a static criterion. If the nucleus possesses axial symmetry when deformed, with major axis $R(1+\varepsilon)$ and minor axes $R(1-0.5\varepsilon)$, the surface energy deforms to

$$E_s = -a_s A^{2/3}(1+0.4\varepsilon^2+\ldots) \qquad (5.96)$$

while the Coulomb energy becomes

$$E_c = -a_c Z^2 A^{-1/3}(1-0.2\varepsilon^2+\ldots) \qquad (5.97)$$

So, when the deformations are small, $(\varepsilon \ll 1)$ the total energy change after deformation is

$$\Delta B = \frac{\varepsilon^2}{5}(a_c Z^2 A^{-1/3} - 2a_s A^{2/3}) \qquad (5.98)$$

and this is only positive if

$$\frac{Z^2}{A} > \frac{2a_s}{a_c} \sim 49 \qquad (5.99)$$

Any nuclei satisfying (5.99) splits into two parts. This is one reason why we do not observe very heavy elements in Nature. Uranium-238, one of the heaviest nuclei, has $Z^2/A \sim 35.5$ and is close to the limit. If a nucleus is very close to the fission limit, the addition of small amounts of energy can render it unstable to fission. For example, when U^{235} captures a slow-moving neutron the binding energy of the neutron becomes available to the nuclear system. This extra energy is ~6 MeV and ensures that the new state of U^{236} is formed in a highly energetic state from which it is much easier to deform the nucleus and fission.

The criterion (5.99) shows that the dividing line between those nuclei which are stable and those which are not is drawn by the strong and electromagnetic interactions. Their relative strengths determine the susceptibility to fission. The condition for a nucleus (Z, A) to be *stable* is roughly that

$$\frac{Z^2}{A} \leqslant 49\left(\frac{\alpha_s}{10^{-1}}\right)^2\left(\frac{1/137}{\alpha}\right) \tag{5.100}$$

Thus, if the electromagnetic interaction were stronger (increased α) or the strong interaction a little weaker (decreased α_s), or both, then biologically essential nuclei like carbon would not exist in Nature. For example, if the electron charge were increased by a factor ~3 no nuclei with $Z > 5$ would exist and no living organisms would be possible. The existence of carbon-based organisms hinges upon a 'coincidence' regarding the relative strengths of the strong and electric forces, namely that

$$\left(\frac{\alpha}{1/137}\right) \leqslant 16.3\left(\frac{\alpha_s}{10^{-1}}\right)^2 \tag{5.101}$$

If one assumes the electromagnetic force strength is fixed, then the effect of small variations is α_s for the stability of nuclei is shown[75] in Figure 5.6.

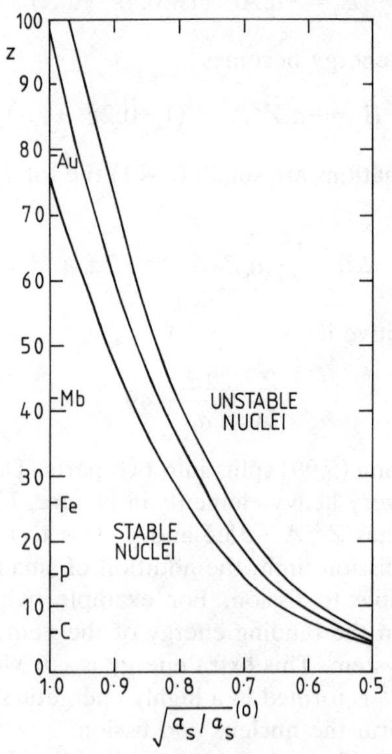

Figure 5.6. Nuclear stability as a function of the strong coupling, α_s, variation away from the observed value, $\alpha_s(0)$, with Coulomb forces constant.[66]

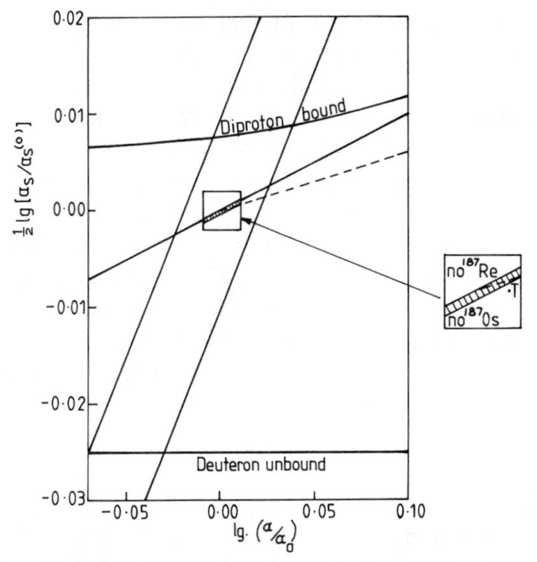

Figure 5.7. Consequences of simultaneous variations in the nuclear and electromagnetic coupling strengths.[66]

A 50% decrease in the strength of the nuclear force ($\alpha_s \leqslant 0.025$) would adversely affect the stability of all the elements essential to living organisms and biological systems. Similarly, holding the strong force constant, we see that the stability of carbon requires the fine structure constant α to be less than ~ 0.1.

In Figure 5.7[75] are plotted the effects of varying the nuclear and electromagnetic couplings simultaneously.

We shall see later that other constraints exist to limit these interactions if Nature exhibits a grand unification of fundamental forces.

5.6 The Stars

> Twinkle, twinkle little star
> I don't wonder what you are,
> For by spectroscopic ken,
> I know that you are hydrogen.
> Ian D. Bush

Any body of mass M and average radius R possesses a gravitational potential energy E_g of order[76]

$$E_g \sim -\frac{GM^2}{R} \tag{5.102}$$

If no other forces existed in Nature this attractive gravitational force would cause all bodies to collapse indefinitely. However, as we have already seen, there do exist other physical forces which can support small bodies against gravitational collapse. The characteristic sizes of planets and asteroids result from a stable balance between gravity and the quantum mechanical exclusion forces. However, not all systems need appeal to pressures of quantum mechanical origin to support themselves against gravity. Whereas we have regarded planetary material as possessing zero temperature, it is obvious that matter could exist in large quantities with a finite temperature. In that case the object would possess an 'ordinary' gas pressure by virtue of the thermal motion of its constituents. If the motions are non-relativistic (root mean square gas velocities much less than the speed of light) the body could be termed 'cool'. Then, the thermal pressure is given in terms of the temperature, T, and volume of the gas by Boyle's law.

$$P \sim \frac{NT}{R^3} \tag{5.103}$$

(where N is the total number of nucleons in a volume $\sim R^3$). Clearly, as the material is compressed isothermally, E_g falls and the pressure rises. If the body has an average density ρ, then the condition for an equilibrium to exist between gravity and thermal pressure is that the central pressure of the body, $P_c \sim \rho GMR^{-1}$ equal the thermal pressure $P \sim \rho T m_N^{-1}$. This criterion yields the relation

$$NT \sim \frac{GM^2}{R} \tag{5.104}$$

that is, simply that

$$\text{(total thermal energy)} \sim \text{(gravitational potential energy)} \tag{5.105}$$

If the average inter-nucleon separation inside the star is represented by d where, by definition,

$$d^3 N \sim R^3 \tag{5.106}$$

then (5.104) implies the temperature to be

$$T \sim \frac{Gm_N^2 N^{2/3}}{d} \tag{5.107}$$

This contribution to the pressure does not involve the electrons because they are so light (their contribution to the thermal pressure is smaller than that of the nucleons by a factor $\sim m_N^2 m_e^{-2} \sim 3 \times 10^6$); however, if the body continues to shrink to a higher density state, it will begin to squeeze the electrons into regions small enough for their degeneracy pressure to be significant. In that event, the thermal pressure of the nucleons becomes augmented by the degeneracy pressure of the electrons. Recall that the

Exclusion Principle reduces to the imposition that electrons of average separation d possess a *minimum* kinetic energy $\sim m_e^{-1}d^{-2}$. (The corresponding contribution from nucleon degeneracy is clearly negligible because $m_N \gg m_e$.) The equation of energy balance now looks like equation (5.107) plus the electron degeneracy term:

$$T+\frac{1}{m_e d^2}\simeq\frac{Gm_N^2N^{2/3}}{d^2}. \tag{5.108}$$

When the body is large and the density quite low, the degeneracy term ($\propto d^2$) is the least significant term in (5.108) and the temperature will just increase according to the ideal gas law (5.103) as the body shrinks under gravity, ($T \propto R^{-2}$). However, this shrinkage ensures the degeneracy pressure must eventually intervene and guarantees a temperature maximum when the combination $(Gm_N^2N^{2/3}d^{-1}-m_e^{-1}d^2)$ attains its maximum value. This occurs when d equals $d_\star$ where

$$d_\star = 2(\alpha_G m_e N^{2/3})^{-1} \tag{5.109}$$

which corresponds to a maximum central temperature of

$$T_\star \sim \alpha_G^2 N^{1/3} m_e \tag{5.110}$$

Figure 5.8 shows the variation of temperature T versus density d. Incidentally, the form of the 'potential' closely resembles that in nuclei and molecules because in these systems stable states also arise from a competition between r^{-1} and r^{-2} forces (see Figure 5.4).

The defining characteristic which turns our 'warm' body into a *star* is that the central temperature, $T_\star$, be high enough to initiate and sustain

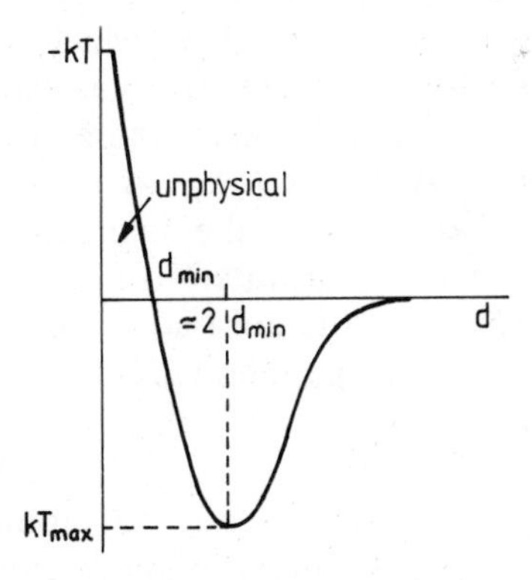

Figure 5.8. Temperature versus inter-particle separation, d, for a star. We require T_{max} to be great enough for nuclear reactions to occur in order to produce a star. If the temperature is always low then the system collapses, heats up and then cools down over a period of about 10^6 years whereas stars that initiate nuclear burning last for more than 10^9 years.[26]

thermonuclear reactions.[77] In order to establish the size of such stars we must determine the threshold for nuclear fusion reactions.

When the ambient temperature is low, two light nuclei will not have enough kinetic energy to break through the Coulomb barrier of superficial electrostatic repulsion that exists between them. This barrier height varies as Z_1Z_2, where Z_i are the atomic numbers of the interacting nuclei; so clearly, light nuclei will be able to interact more readily than heavy ones. This is also advantageous because the fusion of light nuclei is exothermic. However, nuclei can undergo nuclear burning when their mean kinetic energies are significantly lower than the Coulomb barrier $\sim 1\,\text{MeV} \sim 10^{10}\,\text{K}$. The reasons are twofold: the energies of nuclei participating in a nuclear interaction will possess a Maxwellian number distribution $N(E) \propto \exp(-E/T)$, so although the mean energy may sit below the Coulomb threshold, there will still be many nuclei in 'the tail' of the distribution with energies high enough to surmount the potential barrier. Also, there is a help from quantum mechanics: nuclei with energies *less* than that of the Coulomb barrier can still penetrate it by quantum tunnelling. Ignoring angular momentum, the probability of tunnelling through the barrier E_c by particles with energy E is

$$(\text{Tunnelling Probability}) \sim \exp\left(-\int_{R_n}^{r_0} (E_c - E)^{1/2}\, dr\right) \tag{5.111}$$

where r_0 is the distance of closest approach 'classically' which is given by

$$r_0 = Z_1Z_2\alpha E^{-1} \tag{5.112}$$

and R_n is the nuclear radius. The reaction rate is controlled by competition between the Maxwell factor $\exp(-E/T)$ which tends to zero for *large* E, and the tunnelling probability which varies as $\exp(-bE^{1/2})$ and goes to zero for *small* E; here $b \sim Z_1Z_2\alpha A^{1/2}m_N^{1/2}$ where A is the reduced atomic weight of the reactants, $A \equiv A_1A_2/(A_1 + A_2)$. There exists an intermediate energy $\sim$15–30 keV where the interaction probability is optimized. This 'Gamow peak' is illustrated in Figure 5.9.[79]

The energy $E_0 \sim (0.5bT)^{2/3}$ is the most advantageous for nuclear burning and corresponds to an average thermal energy of

$$T_{\text{NUC}} \sim \eta\alpha^2 m_N \sim \eta 5.7 \times 10^7\,\text{K} \tag{5.113}$$

where η incorporates small factors due to atomic weights, intrinsic nuclear properties and so forth. For hydrogen burning ($\sim 1.5 \times 10^7$ K) we have $\eta(\text{H}) \sim 0.025$; helium burning ($\sim 2 \times 10^8$ K) has $\eta(\text{He}) \sim 3.5$ whilst $\eta(\text{C}) \sim 14$, $\eta(\text{Ne}) \sim \eta(\text{O}) \sim 30$ and $\eta(\text{Si}) \sim 60$.

Returning to (5.110) and (5.113) we see that hydrogen ignition is

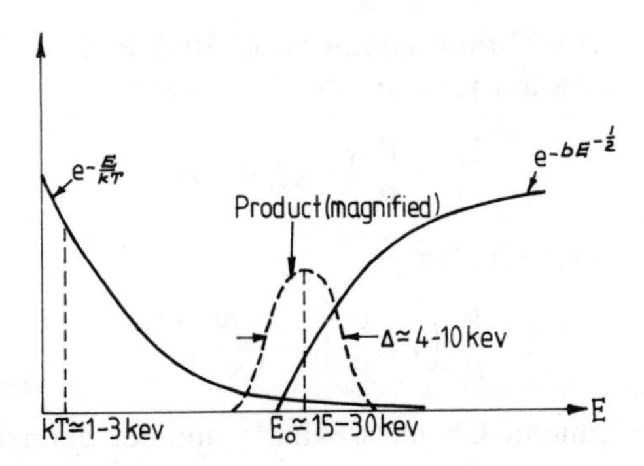

Figure 5.9. The Gamow Peak:[79] the dominant energy-dependent factors in thermonuclear reactions. Most reactions occur in the high-energy tail of the Maxwellian distribution which introduces a thermal factor, $\exp(-E/kT)$. The path through the Coulomb barrier introduces a factor $\exp(-bE^{1/2})$. The product of these factors has a sharp (Gamow) peak at E_0.

possible if $T_\star > T_{NUC}$, that is, if the body is larger than $M_\star$ where[26]

$$M_\star \gtrsim \eta^{3/4}\left(\frac{\alpha}{\alpha_G}\right)^{3/2}\left(\frac{m_N}{m_e}\right)^{3/4} m_N \sim 10^{33}\text{ gm} \tag{5.114}$$

This simple argument explains why stars contain no less than about $M_\star m_N^{-1}$ nucleons and shows that the largest planet in our solar system Jupiter—is fairly close to fulfilling the condition for nuclear ignition in its interior. It was almost a star (as a consequence we expect planets to exist over a mass range of $\sim(m_e/m_N)^{-3/4} \sim 300$). The rough lower size limit corresponding to the mass constraint (5.114) is

$$R_\star \sim \alpha_G^{-1/2}\alpha^2 m_e \sim 10^{10}\text{ cm} \tag{5.115}$$

In order to ascertain whether there is also a *maximum* stellar size we must consider a third source of pressure support within the interior—*radiation pressure.* Equilibrium radiation will possess a pressure, P_γ, given by the black body law, which in our units is,

$$P_\gamma = \frac{\pi^2}{45} T^4 \tag{5.116}$$

From (5.103), we see that the relative importance of gas and radiation pressure in a stellar interior is given by the ratio,

$$\frac{P_\gamma}{P_g} = \frac{2\pi^3}{135} N^{-1}(TR)^3 = \frac{2\pi^3}{135}\left(\frac{T}{d^{-1}}\right)^3 \tag{5.117}$$

If we consider large bodies, so the electron degeneracy is smaller than the gas pressure, the equilibrium condition (5.107) is now modified by the inclusion of the radiation pressure and becomes

$$T\left(1+\frac{P_\gamma}{P_g}\right)\sim\alpha_G^2N^{4/3}m_e \tag{5.118}$$

or equivalently, using (5.117),

$$\frac{P_\gamma}{P_g}\left(1+\frac{P_\gamma}{P_g}\right)^3\sim\left(\frac{N}{N_\star}\right)^2 \tag{5.119}$$

where $N_\star$ is the Landau–Chandrasekhar[80] number defined by

$$N_\star\equiv\alpha_G^{-3/2}=2.2\times10^{57} \tag{5.120}$$

This relation shows that the relative importance of radiation pressure grows with the size of the star as N^2. However, if P_γ becomes significantly greater than P_g, a star will become pulsationally unstable and break up.[81] Therefore (5.119) provides an upper bound on the number of nucleons in a stable hydrogen burning star, $N\lesssim 50\alpha_G^{-3/2}$ and, in combination (5.114), we see that simple physical considerations pin down the allowed range of stellar sizes very closely as

$$\alpha_G^{-3/2}\left(\frac{\alpha^2m_N}{m_e}\right)^{3/4}m_N\lesssim M_\star\lesssim 50\alpha_G^{-3/2}m_N \tag{5.121}$$

A stable, non-relativistic star must inevitably contain $\sim\alpha_G^{-3/2}\sim10^{57}$ nucleons.

The most obvious outward characteristic of a star, besides its mass, is its *luminosity*—the rate of energy production. In the case of the Sun, it is this property that determines the ambient temperature one astronomical unit away, on the Earth's surface.

Photons produced near the stellar centre do not simply leave the star after a time $\sim cR_\star$ of flight. Rather, they undergo a whole series of quasi-random scatterings from electrons and charged ions which results in a much slower diffusive exit from the stellar interior. This path is called a 'random walk'; see Figure 5.10.

Consider first the effect of electron scattering, for which the (Thomson) cross-section σ_T is[82]

$$\sigma_T\sim\alpha^2m_e^{-2} \tag{5.122}$$

This mean free path λ gives the average distance travelled by photons between collisions by electrons and is[83]

$$\lambda\sim(\sigma_Tn_e)^{-1} \tag{5.123}$$

where the electron number density is $n_e\sim NR^{-3}$. The time to traverse a

Figure 5.10. Absorption and emission processes together with scattering allow radiation to leak out of a star by a random-walk path as shown rather than to free-stream.

linear distance R from the centre to the boundary of the star by a random walk is the escape time ($c \equiv 1$)

$$t_{ex} \sim \left(\frac{R}{\lambda}\right) \times R \tag{5.124}$$

and the luminosity, L, of the star is defined as

$$L = \frac{\text{Nuclear energy available}}{\text{Escape time from centre}} \tag{5.125}$$

so

$$L \sim \frac{T_\star^4 R^3}{R^2/\lambda} \sim f\left(\frac{\alpha}{m_e^2}\right)^{-1}\left(\frac{N}{N_\star}\right)^3 \alpha_G^{-1/2} \tag{5.126}$$

where the dimensionless factor f accounts for deviations from exact Thomson scattering which result at low temperature or high density. The estimate (5.126) gives a reasonably accurate estimate of

$$L \sim 5 \times 10^{34}\left(\frac{M}{M_\star}\right)^3 \text{erg s}^{-1} \tag{5.127}$$

which is independent of the stellar radius and temperature. We can also deduce the lifetime of a star burning its hydrogen at this rate. This gives the 'main sequence' lifetime, $t_\star$, as

$$t_\star \sim \frac{\text{(Nuclear Energy from Hydrogen Fusion)}}{L} \tag{5.128}$$

$$\sim \frac{\varepsilon \alpha \alpha_G^{-1}}{m_e^2}\left(\frac{M_\star}{M}\right)^2 \sim 10^{10}\left(\frac{M_\star}{M}\right)^2 \text{yr} \tag{5.129}$$

Massive stars have short lifetimes because they are able to attain high internal temperatures and luminosities. They burn their nuclear fuel very rapidly. A star of $\sim 30M_\odot$ has a hydrogen-burning lifetime of only ten million years whereas the Sun can continue burning hydrogen for more than ten billion years.

The fact that $t_\star$ can be determined by the fundamental constants of Nature has many far-reaching consequences. It means that we can understand why we observe the Universe to be so old and hence so large, and it also provides a point of contact with the timescales that biologists estimate for evolutionary change and development.[84] To these questions we shall return in Chapter 6.

Our estimates of stellar luminosities and lifetimes have assumed that the opacity controlling the transport of energy in the star's interior is entirely due to Thomson scattering. However, when matter becomes denser the nuclei can begin to affect the electrons through free-free and bound-free transitions. For hydrogen the free-free and bound-free opacities—or Kramers opacities—are roughly the same but, unlike the Thomson opacity, they are temperature dependent. Thus, whereas the Thomson opacity per nucleon per unit volume is

$$\kappa_T \sim \alpha^2 m_e^{-2}, \tag{5.130}$$

the Kramers opacity is

$$\kappa_K \sim \alpha^3 m_e^{-2}\left(\frac{m_e}{T}\right)^{1/2}. \tag{5.131}$$

When the Kramers opacity is significant, the luminosity differs slightly from the form (5.126) and is

$$L_K \sim 10^{-4}\left(\frac{\alpha}{m_e^2}\right)^{-2}\left(\frac{T}{m_e}\right)^{1/2}\alpha^{-1}\alpha_G^{-1/2}\left(\frac{M}{M_\star}\right)^5. \tag{5.132}$$

In practice, one uses the formula which gives the lowest luminosity of (5.126) and (5.132). We can simplify (5.132) further because we know the relevant central temperature to consider is $T_{\text{NUC}} \sim \eta\alpha^2 m_N$ and this gives

$$L_K \sim 10^{-4}\eta^{1/2}\alpha^{-2}\left(\frac{m_e}{m_N}\right)^{1/2}. \tag{5.133}$$

The luminosities (5.133) and (5.126) become equal when $M \sim 3\eta^{-1/4}M_\star$ and the Sun is thus controlled by Kramers opacity.

So far, we have only discussed the central temperature of stars, $T_\star$, but we are also interested in the surface temperature of a star. In the solar case it is this parameter which determines the energy flux incident on the Earth's surface. The surface temperature T_s should be simply related to the luminosity by an inverse square law, so

$$L \sim 0.5R^2T_s^4 \tag{5.134}$$

where T_s^4 is the radiant energy at the surface. Applying this result, we obtain

$$T_s^4 \sim 0.1\eta^2\alpha^2 m_N^2 m_e^2 \alpha_G^{1/2}\left(\frac{M}{M_\star}\right) \tag{5.135}$$

with Thomson opactiy and

$$T_s^4 \sim 10^{-3}\eta^{5/2}\alpha_G^{1/2}\alpha^2\left(\frac{m_e}{m_N}\right)^{3/2} m_N^4\left(\frac{M}{M_\star}\right)^3 \tag{5.136}$$

with Kramers opacity.

However, these results implicitly assume that Thomson or Kramers scattering maintains a large opacity right out to the boundary of the star. This will only be possible if material is ionized near the surface. If the temperature near the stellar surface falls below the dissociation temperature of molecules, T_I, where

$$T_I \sim 10^{-2}\alpha^2 m_e \tag{5.137}$$

the matter will cease to be opaque there. What then happens if the values of T_s calculated in (5.135) and (5.136) fall below T_I? In order to remain in equilibrium, the star must have other means of transporting heat to its surface and it is believed that convection is responsible for maintaining the surface temperature at T_I if the radiative transport described by (5.135) or (5.136) is inadequate. Inside the boundary of a star whose surface temperature lies close to T_I there should exist a thin convection layer associated with the atomic and molecular transitions. If the temperature at the surface falls below T_I the convective layer will spread into the star until it increases the heat flux sufficiently for T_s to attain the value T_I. Convection should therefore extend far enough into the star to maintain the surface temperature close to T_I. Thus if the formulae (5.135) and (5.136) predict a value for T_s lower than T_I, that value should be replaced by T_I. For main sequence stars, this leads to an interesting result;[86] we see that

$$\left(\frac{T_s}{T_I}\right)^4 \sim \eta^2\frac{\alpha_G^{1/2}}{\alpha^6}\left(\frac{m_N}{m_e}\right)^2\left(\frac{M}{M_\star}\right) \tag{5.138}$$

when Thomson scattering dominates the opacity within the central regions and

$$\left(\frac{T_s}{T_I}\right)^4 \sim \eta^{5/2}\frac{\alpha_G^{1/2}}{\alpha^6}\left(\frac{m_N}{m_e}\right)^{5/2}\left(\frac{M}{M_\star}\right)^3 \tag{5.139}$$

when Kramers scattering dominates.[87] These two formulae reveal a striking 'coincidence' of Nature, first recognized by Carter: the surface temperature only neighbours the ionization temperature T_I of stars with mass $M \sim M_\star$ because of the numerical 'coincidence' that[88]

$$\alpha^{12}\left(\frac{m_e}{m_N}\right)^4 \sim \alpha_G \tag{5.140}$$

which reduces numerically to the relation

$$2.2\times 10^{-39} \sim 5.9\times 10^{-39} \tag{5.141}$$

The existence of this unusual numerical coincidence (5.140) ensures that the typical stellar mass $M_\star$ is a dividing line between convective and radiative stars. Carter argues that the relation (5.140) therefore has strong Anthropic implications: the fact that α_G is *just* bigger than $\alpha^{12}(m_e/m_N)^4$ ensures that the more massive main sequence stars are radiative but the smaller members of the main sequence, which are controlled by Kramers opacity, are almost all convective. If α_G had been slightly greater all stars would have been convective red dwarfs; if α_G had been slightly smaller the main sequence would consist entirely of radiative blue stars. This, Carter claims,[86]

suggests a conceivable world ensemble explanation of the weakness of the gravitational constant. It may well be that the formation of planets is dependent on the existence of a highly convective Hayashi track phase on the approach to the main sequence. (Such an idea is of course highly speculative, since planetary formation theory is not yet on a sound footing, but it may be correlated with the empirical fact that the larger stars—which leave the Hayashi track well before arriving at the main sequence—retain much more of their angular momentum than those which remain convective.) If this is correct, then a stronger gravitational constant would be incompatible with the formation of planets and hence, presumably of observers.

This argument is hard to investigate more closely because of lack of evidence. It is maintaining that planetary formation is associated with convective stars and their relatively low angular momentum relative to blue giants makes it conceivable that stellar angular momentum was lost during the process of planet formation and now resides in the orbital motion of planetary systems around them.

Finally, we note that the classic means[90] of classifying stars and tracing their evolutionary history is via the Hertzsprung–Russell diagram which plots the position of stars according to their surface temperature and luminosity, (Figure 5.11),

An extremely crude determination of its main branch is possible using (5.133) and (5.139) or (5.126) and (5.134) which give fundamental relations between L and T_s. For Thomson scattering opacity these formulae give, omitting the small numerical constants, a dependence

$$T_s \propto L^{1/12} \tag{5.142}$$

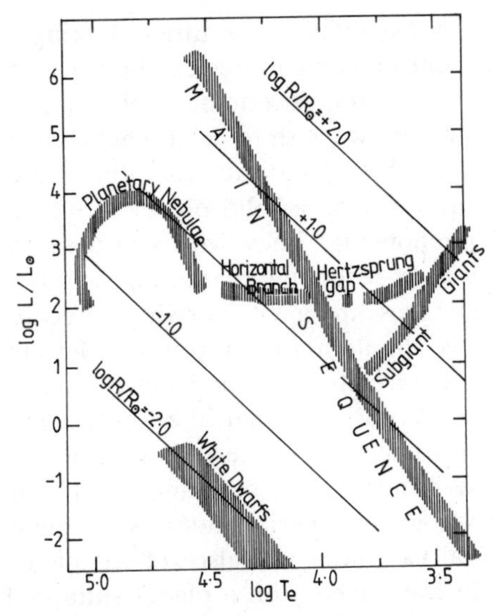

Figure 5.11. Schematic Hertzsprung–Russell diagram plotting luminosity (in solar units) versus effective temperature. The lines of constant slope represents stars having identical radii (see ref. 90).

whereas for Kramers opacity

$$T_s \propto L^{3/20} \tag{5.143}$$

remarkably close to the observational situation of $T_s \propto L^{0.13}$ in Figure 5.11.

Finally, we note that if we take the typical stellar mass as $M_\star \sim 0.1 m_N \alpha_G^{-3/2}$ then the distance at which a habitable planet will reside in orbit is given by requiring that it be in thermal equilibrium at the biological temperature (5.32) necessary for life. Therefore we can calculate the 'astronomical unit' which gives the distance of a habitable planet from its parent star (assuming that its orbit is not too eccentric) as[40,41]

$$d \sim m_e^{-1} \alpha^{-5} \alpha_G^{-1/4} \left(\frac{m_N}{m_e}\right)^{1/2} \sim 5 \times 10^{13}\,\text{cm} \tag{5.144}$$

If we now use Kepler's laws of motion, which follow from Newton's second law of motion, we can calculate the typical orbital period of such a planet. This determines[40,41] what we call a 'year' to be

$$t_{\text{year}} \sim \left(\frac{m_N}{m_e}\right)^{1/2} \alpha^{-15/2} \alpha_G^{-1/8} m_e^{-1} \tag{5.145}$$

This result, together with (5.49), may have a deeper significance than the purely astronomical. It has been argued by some historians of science[85] that the homogeneity of the thread linking so many mythological elements in ancient human cultures can be traced to an origin in their shared experience of striking astronomical phenomena. If this were true (and it is not an issue that we wish to debate here) then the results (5.145) and (5.49) for t_{day} and t_{year} indicate that there are Weak Anthropic reasons why any life-form on a solid planet should experience basically similar heavenly phenomena. They will record seasonal variations and develop systems of time-reckoning that are closely related to our own. If astronomical experiences are a vital driving force in primitive cultural development then we should not be surprised to find that planetary-based life-forms possess some cultural homogeneity. This homogeneity would be a consequence of the fact that the timescales t_{day} and t_{year} are strongly constrained to lie close to the values we observe because they are determined by the fundamental constants of Nature. Any biological phenomenon whose growth cycle and development is influenced by seasonal and diurnal variations will also reflect this universality.

The fact that life can develop on a planet suitably positioned in orbit about a stable, long-lived star relies on the close proximity of the spectral temperature of starlight to the molecular binding energy ~1 Rydberg. Were it to greatly exceed this value, living organisms would be either sterilized or destroyed; were it far below it, the delicate photochemical reactions necessary for biology to flourish would proceed too slowly. A good example is the human eye: the eye is receptive only to that narrow wave-band of electromagnetic radiation between 4000–8000 Å which we call the 'visible' region. Outside this wave-band electromagnetic radiation is either so energetic that the rhodopsin molecules in the retina are destroyed or so unenergetic that these molecules are not stimulated to undergo the quantum transitions necessary to signal the reception of light to the central nervous system.

Press and Lightman[41] have shown that the relation between the biological temperature, T_B, and the spectral temperature (that is, the surface temperature of the Sun) is due to a real coincidence, that

$$\frac{T_s}{e^2/2a_0} \sim \left(\frac{m_N}{m_e}\right)^{1/2} \alpha_G^{1/8} \alpha^{-3/2} \sim 0.9 \tag{5.146}$$

where T_s is given by (5.135) or (5.136).

We can even deduce something about the weather systems on habitable planets.[41] The typical gas velocity in an atmosphere will be set by the sound speed at the biologically habitable temperature T_B. This is just

$$v_s \sim \left(\frac{T_B}{m_N}\right)^{1/2} \sim \alpha \left(\frac{m_e}{m_N}\right)^{3/4} \sim 2\times 10^5 \text{ cm s}^{-1} \tag{5.147}$$

5.7 Star Formation

He made the stars also.
Genesis 1 : 16

Our discussion of stellar structure implicitly assumes that one begins with some spectrum of massive bodies some with initial mass far in excess of $\alpha_G^{-3/2}m_N \sim 1M_\odot$ and, perhaps, some much smaller. Only those with mass close to $\alpha_G^{-3/2}m_N$ will evolve into main-sequence stars because only bodies with mass close to this value get hot enough to initiate nuclear burning and yet remain stable against disruption by radiation pressure. However, what if some prior mechanism were to ensure that no protostars could exist with masses close to $\alpha_G^{-3/2}m_N$? This brings us face to face with the problem of star formation—a problem that is complicated by the possible influence of strong magnetic or rotational properties of the protostellar clouds.[91] One clear-cut consideration has been brought to bear on the problem by Rees.[92] His idea develops a previous suggestion of Hoyle,[93] that stars are formed by the hierarchical fragmentation of gaseous clouds.

A collapsing cloud will continue to fragment while it is able to cool in the time it takes to gravitationally collapse. If the fragments radiate energy at a rate per unit area close to that of a true black-body then they will be sufficiently opaque to prevent radiation leaking out from the interior and cooling will be significantly inhibited. Once the fragments begin to be heated up by the trapped radiation the pressure builds up sufficiently to support the cloud against gravity and a protostar can form. These simple physical considerations enable the size of protostellar fragments to be estimated: at any stage during the process of fragmentation, the smallest possible fragment size is given by the Jeans mass (the scale over which pressure forces balance gravitational attraction). If the first opaque fragments to form have temperature T then, since they must behave like black bodies they will be cooling at rate $\sim T^4R_J^{-1}$; where R_J is the Jeans length—the depth from which radiation escapes.

The cooling time in the cloud is given by the ratio of the thermal energy density to the radiative cooling rate,

$$t_{\text{cool}} \sim \frac{nT}{T^4/R_J} \tag{5.148}$$

where n is the particle number density in the cloud. In order for cooling to occur, the cooling time must be shorter than the time for gravitational collapse, t_g, where

$$t_g \sim (Gnm_N)^{-1/2} \tag{5.149}$$

This is the case if, by (5.148) and (5.149),

$$n \lesssim T^{5/2}m_N^{1/2}$$

and so the collapsing cloud must cease to fragment when the average mass of the fragment is

$$m_{fr} \sim \left(\frac{T}{m_N}\right)^{1/4} \alpha_G^{-3/2} m_N \sim \left(\frac{T}{m_N}\right)^{1/4} M_\odot \tag{5.150}$$

The inevitable size of protostellar fragments is relatively insensitive to temperature over the range of conditions expected in such clouds $T \sim 10^2 - 10^4$ K. Further fragmentation is not possible because the fragments have reached the maximum rate of energy disposal. It is interesting that the oldest stars must therefore have masses $\lesssim \alpha^{-3/2} m_N$.

5.8 White Dwarfs and Neutron Stars

> For a body of density 10^{12} gm/cc—which must be the maximum possible density, and its particles would be then all jammed together,—the radius need only be 400 kilometres. This is the size of the most consolidated body.
>
> Sir Oliver Lodge (1921)

The picture of a star we have sketched above cannot be sustained indefinitely. Eventually the sources of thermonuclear energy within the star will be exhausted, all elements will be systematically burnt to iron by nuclear fusion and no means of pressure support remains available to the dying star. What is its fate? We have already said enough to provide a partial answer. According to the energy equation (5.108) it should evolve towards a configuration wherein the electron degeneracy pressure balances the inward attraction of gravity. This, we recall, was the criterion for the existence of a planet. However, planets are cold bodies, that is, their thermal energies are far smaller than the rest of the mass energies of the electrons that contribute degeneracy pressure. However, if a body is warm enough for the electrons to be *relativistic* ($T \gtrsim m_e$), then the electron degeneracy energy is no longer given by $\sim p^2 m_e^{-1} \sim d^{-2} m_e^{-1}$ but rather by the relativistic value $\sim d^{-1}$. The equilibrium state that results is called a *white dwarf* and has a mass and radius given by,[94]

$$M_{WD} \sim \alpha^{-3/2} m_N \tag{5.151}$$

$$R_{WD} \sim \alpha^{-1/2} m_e^{-1} \tag{5.152}$$

Thus, although they are of similar mass[95] to main sequence stars, white dwarfs have considerably smaller radii. They are roughly the size of planets but a million times heavier:

$$\frac{R_{WD}}{R_\star} \sim \alpha \tag{5.153}$$

Therefore, they are denser than ordinary stars by a factor $\sim \alpha^{-2} \sim 10^6$ and the density of a white dwarf is roughly

$$\rho_{WD} \sim m_N m_e^3 \sim 10^6 \text{ gm cm}^{-3} \tag{5.154}$$

Figure 5.12[31] illustrates the details of the mass-size plane in the neighbourhood that includes stars, planets and white dwarfs.

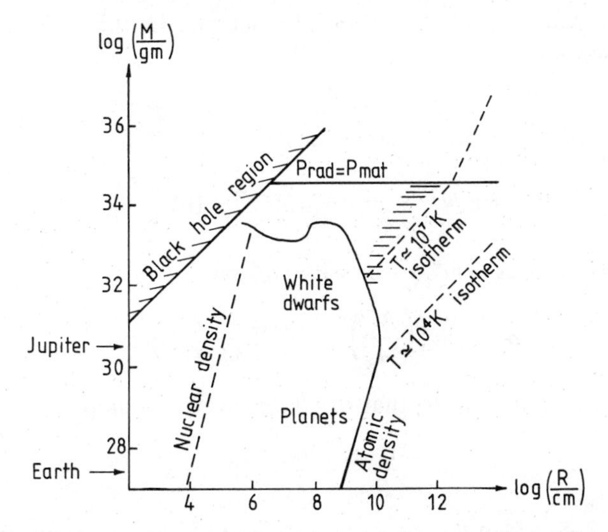

Figure 5.12. Detailed view of the mass-size diagram in the region containing planetary and white dwarf masses.[38]

Although these objects appear bizarre, they do not involve general relativistic considerations because their binding energy per unit mass is $\sim m_e/m_N$ and thus is much less than unity.

Now, as Chandrasekhar first discovered,[86,94] the mass M_{WD} represents an upper limit to the mass which can be supported by electron degeneracy pressure. Heavier bodies will continue to collapse to densities in excess of $\rho_{WD} \sim 10^6 \text{ gm cm}^{-3}$. In that situation it becomes energetically favourable for the degenerate electrons to combine with nuclear protons to form neutrons (because of the 'coincidence' that $m_{pr} - m_n \sim m_e$) when $E_e \sim$ 1 MeV so

$$e^- + p \to n + \nu - 0.8 \text{ MeV} \tag{5.155}$$

The electron number density therefore drops and, along with it, the electron degeneracy pressure. But, eventually the neutrons will become so closely packed that their degeneracy pressure becomes significant because they are initially non-relativistic. The fluid, or perhaps solid, of degenerate neutrons will have a degeneracy energy given by the Exclusion Principle as $\sim r_0^{-2} m_N^{-1}$ where r_0 is the mean inter-nucleon separation.

The balance between gravity and neutron degeneracy creates a new equilibrium state that is called a *neutron star*. For equilibrium we require that,

$$\frac{N}{r_0^2 m_N} \sim \frac{GM^2}{R} \tag{5.156}$$

where $N = Mm_N^{-1}$ is the number of nucleons in the neutron star and $r_0 = N^{1/3}R^{-1}$ so

$$r_0 \sim m_N^{-1}\alpha_G^{-1}N^{-1}. \tag{5.157}$$

The radius of the neutron star is thus

$$R_{NS} = r_0 N^{1/3} \sim m_N^{-1}\alpha_G^{-1}N^{-1/3} \sim 10\left(\frac{M}{M_\star}\right)^{1/3} \text{km} \tag{5.158}$$

and until ρ reaches $\sim m_N^4$ its density will be

$$\rho_{NS} \sim m_N^4\left(\frac{M}{M_\star}\right)^2 \sim 10^{14}\left(\frac{M}{M_\star}\right)^2 \text{gm cm}^{-3} \tag{5.159}$$

and the ratio of its size to that of white dwarfs is simply

$$\frac{R_{WD}}{R_{NS}} \sim \frac{m_N}{m_e} \sim 10^3. \tag{5.160}$$

If $N \sim \alpha_G^{-3/2}$ as it will be for typical stars, then we see that neutron stars are much larger than their gravitational radii, $R_{NS} \sim m_N^{-1}\alpha_N^{-1/2} \gg \alpha_G M_{NS} m_N^{-2}$, and so they are objects in which general relativity is unimportant.[96]

If a neutron star is only slightly larger than $M \sim 3M_\odot$, the neutrons within it become relativistic and are again unstable to gravitational collapse. When this stage is reached no known means of pressure support is available to the star and it must collapse catastrophically. This dynamic state, inevitable for all bodies more massive than a few solar masses, leads to what is called a *black hole*.

If we assume that a neutron star has evolved from a typical main sequence star with $R_\star \sim R_\odot \sim 10^{11}$ cm and $M_\star \sim M_\odot \sim 10^{33}$ gm and if both mass and angular momentum were conserved during its evolution (which is rather unlikely), then the frequency of rotation of the neutron star will be related to that of the original star $\nu_\star$ by

$$\nu_{NS} \sim \left(\frac{R_\star}{R_{NS}}\right)^2 \nu_\star \sim \left(\frac{\alpha m_e}{m_N}\right)^{-2} \nu_\star \tag{5.161}$$

The sun rotates roughly once a month and, if typical of main sequence stars, this suggests $\nu_\star \sim 5\times10^7\ \text{s}^{-1}$ and $\nu_{NS} \sim 10^{-4}\ \text{s}^{-1}$. The stipulation that centrifugal forces not be so large that equatorial regions become unbound places an upper bound on ν_{NS} of[98]

$$\nu_{NS} < \frac{1}{2\pi}\left(\frac{GM_{NS}}{R_{NS}^3}\right)^{1/2} \sim 0.1\alpha_G^{1/2} m_N \sim 10^4\,\mathrm{s}^{-1} \qquad (5.162)$$

The neutron star introduces a qualitatively different type of astronomical object from those discussed up until now—an object whose average density is close to that of the atomic nucleus and in whose interior nuclear timescales determine events of physical interest. For these reasons many scientists and science fiction writers have speculated that if living systems could be built upon the strong rather than the electromagnetic interaction, then neutron stars might for them play the role that planets play for us. Freeman Dyson and others[99] have suggested that intelligent 'systems' which rely upon the strong interaction for their organization might reside near or on the surface of neutron stars. It appears that no quantitative investigations have been made to follow up this intriguing speculation and so we shall sketch some results that give some feel for the type of systems that are allowed by the laws of physics.

Analysing the surface conditions likely on a neutron star is a formidable problem, principally because of the huge magnetic fields anticipated there. Just as the rotation frequency spins up during the contraction of main-sequence stars into neutron stars, so the magnetic field, B, amplifies with radius, R, as $B \propto R^{-2}$, and fields as large as $\sim 10^{13}$ gauss could result from an initial magnetic field close to the solar value ~ 1 gauss, (a magnetic field of $\sim 10^{12}$ gauss on the neutron star would contribute an energy $\sim 10^{42}$ erg, far smaller than the gravitational energy $\sim 10^{53}$ erg and possible rotational energy $\sim 2\times 10^{53}$ erg). However, for the moment, let us ignore the magnetic field. The neutron star will possess a density and composition gradient varying from the centre to the boundary.[100] The general form[97] of this variation is probably like that shown in Figure 5.13.

In the outer region where the density is less than $\sim 10^4\,\mathrm{gm\,cm^{-3}}$, electrons are still bound to nuclei, the majority of which are iron. A little deeper into the crust there should exist a sea of free electrons alongside the lattice of nuclei. The estimated surface temperature is $\sim 5\times 10^6$ K and much less than the melting temperature of the nuclei there. Above the outer crust there will exist a thin atmosphere of charged and neutral particles. This atmosphere is characterized by a scale height h over which temperatures and pressures vary significantly and which is defined by

$$h_{NS} \sim \frac{T_s}{m_N g_{NS}} \qquad (5.163)$$

where g_{NS} is the acceleration due to gravity on the surface (so, for example, in the Earth's atmosphere with $T \sim 290$ K and $g \sim 980\,\mathrm{cm\,s^{-2}}$,

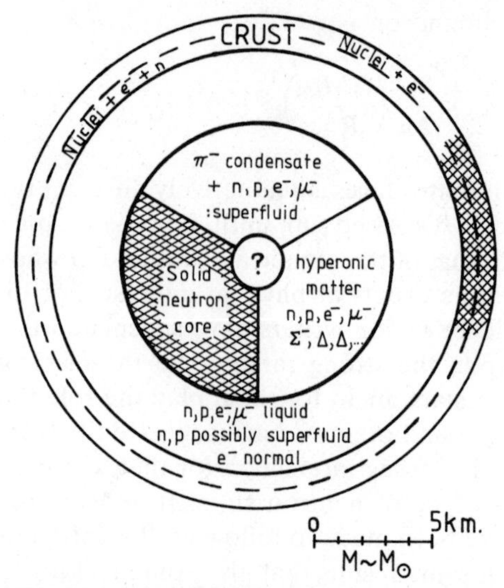

Figure 5.13. Schematic slice through a neutron star displaying the outer crust, the liquid interior and the various theoretical alternative suggested for the core (solid neutrons or pion condensate or hyperons).[97] (Reproduced, with permission, from the *Annual Review of Nuclear and Particle Science*, Vol. 25, copyright 1975 by Annual Reviews Inc.)

one has $h \sim 50$–100 km). On the neutron star surface $T_s \sim 10^6$ K and

$$g_{\rm NS} \sim \frac{GM_{\rm NS}}{R_{\rm NS}^2} \sim 5 \times 10^{13}\ \text{cm s}^{-2} \tag{5.164}$$

and so

$$h_{\rm NS} \sim \frac{T_s}{m_N^2} \sim \frac{\varepsilon m_e}{m_N^2} \sim 1\ \text{cm} \tag{5.165}$$

with $T_s \sim \varepsilon m_e$ and $\varepsilon \sim 1.5 \times 10^{-4}$.

Just as we were able to calculate the height of mountains on planetary surfaces by considering the maximum stress that can be supported by solid atomic material ($\rho_{\rm AT} \sim 1$ gm cm^3) at their bases, so we can estimate the largest 'mountains' that could exist on a neutron star.[101] The yield stress, Y, or bulk modulus at the surface will be

$$Y \sim \frac{\eta Z^2 \alpha}{a_{\rm N}^4} \sim 10^{12} \rho^{4/3}\ \text{dyne cm}^{-2} \tag{5.166}$$

with $\eta \sim 0.01$ and $a_{\rm N}$ the average inter-nucleon separation. The maximum height of a mountain strong enough to withstand the gravitational force at its base is therefore

$$h \sim \frac{Y}{\rho g} \sim 20 \left(\frac{\rho}{10^9\ \text{gm cm}^{-3}} \right)^{1/3} \text{cm} \tag{5.167}$$

If we assume that neutron star 'inhabitants' are subject to analogous constraints as are atomic systems on planetary surfaces—that is, they do not grow so tall that on falling they break their atomic bonds or make themselves susceptible to unacceptable bending moments when slightly displaced from the vertical—then their maximum height is calculated to be

$$L_{NS} \sim \varepsilon^{1/2} \alpha \alpha_G^{-1/4} m_N^{-1} \sim 10^{-6} \text{ cm} \tag{5.168}$$

if the energy of their bonding is $\varepsilon\alpha^2 m_e$. Note that on the surface of the neutron star *nuclear* 'life' based on the strong interaction is not likely. Only in the deep interior where densities approach $\sim 10^{14}$ gm cm^{-3} would such a possibility be realized. The mildest conditions allowing it might be those just about 1 km from the boundary at a radius $\sim 0.9 R_{NS}$ where $\rho \sim 10^{14}$ gm cm^{-3}. Suppose, for amusement's sake, nuclear life existed there with bonding—or communication networks—that would be destroyed by stresses which exceed the nuclear binding energy $\sim \alpha_s^2 m_N$. By equating the gravitational potential on a nuclear system of size λ situated at a radius $\sim \eta R_{NS}$ from the centre, bound by a bond energy of $\sim \varepsilon \alpha_s^2 m_N$ we find its maximum size to be

$$\lambda \lesssim \left(\frac{\varepsilon}{\eta}\right)^{1/2} \alpha_G^{-1/4} \alpha_S^{-1} m_N^{-1} \sim 10^{-3} \text{ cm} \tag{5.169}$$

smaller than an atomic being on the surface by a factor $\eta^{-1/2}\alpha\alpha_s$. If a nuclear 'civilization' formed a shell in the neutron star interior of thickness $\sim\lambda$ it would enclose a total mass $M_{civ} \sim \rho_N \lambda (\eta R_{NS})^2$ where

$$M_{civ} \sim \varepsilon^{1/2} \eta^{3/2} \alpha_s^2 \alpha_G^{-5/4} m_N \sim 10^{21} - 10^{22} \text{ gm} \tag{5.170}$$

This is smaller than the Earth by a factor of a million or so. The human race collectively has a mass of about 10^{14} gm. Thus a nuclear 'civilization' would have rather little mass in which to code information.

These simple arguments must be changed considerably if the role of magnetic fields in neutron stars is taken into account. Detailed models of pulsars and X-ray emission[102] indicate that fields $\sim 10^{12}$ gauss probably exist in neutron stars. Fields of such enormous strength considerably alter the properties of atomic matter near the neutron star surface.

An electron in a strong magnetic field, B, assumes a helical motion around a cylinder of radius $\tilde{r} \sim \alpha^{-1/4} B^{-1/2} \sim 2.6 \times 10^{-4} B^{-1/2}$ cm (where B is in gauss). If $\tilde{r}$ is smaller than the smallest Bohr orbital radius $\sim Z^{-1} a_0$, then the magnetic field will determine the behaviour of electrons moving perpendicular to the magnetic field.[103] The atom will assume a compact cylindrical shape with its axis of symmetry aligned along the field. So if $Z^{-1} a_0 > \tilde{r}$, atomic structure will be radically altered. This is the case if

$$B \gg 5Z^2 \times 10^9 \text{ gauss} \tag{5.171}$$

and for iron nuclei the field required is thus in excess of $\sim 5 \times 10^{11}$ gauss.

The electron distribution in magnetized hydrogen will appear as a cylinder of radius $\tilde{r}$ and length L (for heavier elements the cylinder width grows as $\sqrt{Z}$). The electron energy will possess two pieces: the usual kinetic energy demanded by the Uncertainty Principle (which is associated with motion parallel to B) as in equation (5.21) and the electrostatic energy for the *two*-dimensional motion with $L \gg \tilde{r}$. The latter contributes a logarithmic potential, so[104]

$$E \sim \frac{1}{2m_e L^2} - \frac{\alpha}{L} \ln\left(\frac{L}{\tilde{r}}\right). \tag{5.172}$$

The minimum stable energy state occurs with

$$L \sim a_0\left(\ln\left(\frac{a_0}{\tilde{r}}\right)\right)^{-1} \ll a. \tag{5.173}$$

and

$$E \sim \frac{1}{ma_0^2} \ln\left(\frac{a_0}{\tilde{r}}\right)^2 \propto \ln^2 B. \tag{5.174}$$

So when the magnetic field strength is high, the atomic binding energy increases above its usual value $\sim\alpha^2 m_e$.

These unusually strong magnetized atoms also form anomalously strong atomic bonds.[97] Long, covalently bonded chains of atoms can exist in which the binding energy per atom is very large, $\sim 0.5(Z^3\alpha/a_0)(a_0/Z\tilde{r})^{4/5} \sim 10^5$ eV, and where the distance between neighbouring nuclei is $\sim 2.5a_0 Z^{-1}(a_0/Z\tilde{r})^{4/5} \sim 10^{-10}$ cm. Ruderman[103] points out that these chains are extraordinarily strong: a chain of single atoms can support nearly one gram! Neighbouring chains are also very strongly bound together by van der Waals forces, and at zero pressure the average density of matter arranged in these magnetic chains is

$$\rho_{\text{mag}} \sim \frac{AZ^3 m_N}{12a_0^3}\left(\frac{a_0 Z^{1/2}}{\tilde{r}}\right)^{12/5} \sim 4\times 10^3\left(\frac{A}{56}\right)\left(\frac{26}{Z}\right)^{3/5} \times \left(\frac{B}{10^{12}\text{ gauss}}\right)^{6/5} \text{gm cm}^{-3} \tag{5.175}$$

where A is the atomic weight. These chains would be expected to protrude vertically from the neutron star surface like 'whiskers'.

Organisms that choose to exploit these unusual properties of atomic matter in huge magnetic fields would clearly be structurally dissimilar to those constrained solely by the competition between gravity and conventional atomic binding forces. In this peculiar environment, atomic bonds could be more than one thousand times stronger than our own but their information-processing would also have to cope with the ultra-strong magnetic effects and it is not clear what their capabilities might be. Further speculation is left as an exercise for the reader.

5.9 Black Holes

> The black holes of nature are the most perfect macroscopic objects there are in the universe: the only elements in their construction are our concepts of space and time.
>
> S. Chandrasekhar

Stars with considerably more than a few solar masses of material within them will be unable to overcome gravitational collapse by neutron degeneracy pressure. Eventually, the gravitational field of these bodies will become so intense that nothing, not even light, will be able to escape from the gravitational field—a *black hole* will have been formed.[105]

Although the concept of a black hole is non-Newtonian, the idea of such structures occurred first to the English clergyman John Michell as early as 1783. Michell[106] examined the consequences of assuming that gravity acts upon light in similar fashion to its action on matter. He stated what we would now call the Equivalence Principle,

> Let us now suppose the particles of light to be attracted in the same manner as all other bodies with which we are aquainted, . . .

and went on to calculate the characteristics of a massive body with the same density as the sun ($\sim 1\ \text{gm cm}^{-3}$) but with a gravitational pull of sufficient magnitude to ensure that 'all light emitted from such a body would be made to return towards it, by its own proper gravity'. In a remarkable paragraph, Michell described the dimensions of a 'black hole' with solar density, pointed out the possibility of there existing large quantities of hidden material in the Universe and suggested that black holes could be detected by searching for their effects in binary star systems where one member was visible whilst the other was a black hole. All three suggestions are now focal points of research in astrophysics and cosmology; he wrote

> If there should really exist in nature any bodies, whose density is not less than that of the sun, and whose diameters are more than 500 times the diameter of the sun, since their light could not arrive at us; or if there should exist any other bodies of somewhat smaller size, which are not naturally luminous; of the existence of bodies under either of these circumstances, we could have no information from sight; yet if any other luminous bodies should happen to revolve about them we might still perhaps, from the motions of these revolving bodies, infer the existence of the central ones with some degree of probability, as this might afford a clue to some of the apparent irregularities of the revolving bodies, which would not be easily applicable on any other hypothesis!.

Some of these ideas were independently rediscovered by Laplace[107] a few years later in 1796. In effect Michell and Laplace show that the radius of an object with an escape velocity equal to the speed of light ($c = 1$ here) is linearly proportional to its mass

$$R = 2GM \tag{5.176}$$

So, if the density of this object is $\rho = 3M/4\pi R^3$, we have

$$\rho = 3 \times (8\pi GR^2)^{-1} \tag{5.177}$$

If it has solar density $\rho_\odot \sim \rho_{AT} \sim \alpha^3 m_e^3 m_N$, then it must have a radius of

$$R_M \sim (G\rho_{AT})^{-1/2} \sim \alpha_G^{-1/2} \alpha^{-3/2} m_N^{1/2} m_e^{3/2} \sim 250 R_\odot \tag{5.178}$$

and a mass

$$M_M = R_M/2G \sim \alpha_G^{-3/2} \alpha^{-3/2} \left(\frac{m_N}{m_e}\right) m_N \sim 10^7 M_\odot. \tag{5.179}$$

We shall refer to these quantities as the Michell mass and radii, respectively; they give the size and mass of a black hole possessing atomic density.

The existence of black holes in general relativity was uncovered after the discovery of a coordinate singularity in the metric for the space-time external to a static, spherically symmetric and isolated body of mass M. In 1916 Karl Schwarzschild found the following metric as a solution of Einstein's equations[108]

$$ds^2 = -\left(1 - \frac{2GM}{r}\right) dt^2 + \left(1 - \frac{2GM}{r}\right)^{-1} dr^2 + r^2(d\theta^2 + \sin^2\theta \, d\phi^2) \tag{5.180}$$

The metric has infinities or 'singularities' at two values of the radial coordinate $r = 0$ and $r = 2MG$. Physicists were not bothered by the singularity at $r = 0$—Newtonian theory has a similar singularity in the gravitational potential (5.102)—but they were puzzled by the singularity at $r = 2MG \equiv R_s$. However, early investigators convinced themselves that the singularity at R_s was unphysical by the following argument: since the metric (5.180) is a description of the gravitational field *exterior* to a static spherically symmetric matter distribution, one can calculate the associated *interior* metric assuming the matter density to be static and constant inside the matter. The radius of such a distribution cannot be less than $9R_s/8$. As (5.180) is valid only in the matter-free region, it is concluded that the singularity at R_s would never exist in the actual universe, for the matter-free region could not be extended to values of r less than $9R_s/8$.

The Schwarzschild singularity at $r = R_s$ was first considered to have physical significance by the English physicist Sir Oliver Lodge in 1921. Lodge[109] contended that 'if light is subject to gravity, if in any real sense light has weight, it is natural to trace the consequence of such a fact. One of the consequences would be that a sufficiently massive and concentrated body would be able to retain light and prevent its escaping'. Independently of Michell and Laplace, he pointed out that this was true in Newtonian mechanics. He also realized that it could occur in Einstein's theory of gravitation:

So from Schwarzschild's metric [(5.180)] we see it is possible for the speed of light to be in . . . the neighbourhood of a mass so great that $2M/R = c^2/G$ and in that case light cannot altogether escape from the body.

This is the first explicit prediction of a black hole in general relativity. Interestingly, Lodge's reaction to his result was far more conservative than that of Michell and Laplace over a century before him; he argues that,

we find that a system able to control and retain its light must have a density and size comparable

$$\rho R^2 = 1.6 \times 10^{27} \text{ ergs}$$

It is hardly feasible for any single mass to satisfy this condition; either the density or the size is too enormous.

The largest density Lodge is willing to admit is what we would now call nuclear density (actually he uses ~10% of this value) and he calculates the size of what we would now call a neutron star[110]

For a body of density 10^{12}[gm cm^{-3}],—which must be the maximum possible density, as its particles would be than all jammed together,—this radius need only be 400 kilometres. This is the size of the most consolidated body.

Despite these explicit predictions astronomers did not take the possibility of black holes very seriously until 1969, when Lynden-Bell[111] suggested that they might exist in the nuclei of galaxies and provide viable explanations for spectacular high-energy emission by a number of X-ray objects. More recently, Hills[112] has shown that supermassive black holes can be formed by the process of accretion. Hills showed that a black hole can swallow whole stars if it is large enough. This process, if it can occur, enables a small black hole to grow rapidly to the size envisaged for a galactic nucleus by Lynden-Bell. What fixes the size of a black hole that can do this is the magnitude of its tidal gravitational field. Black holes *smaller* than a critical size will tidally disrupt infalling stars before they enter the horizon. Most of the initial stellar mass will avoid capture by the hole when this happens, and instead, escape as radiation or a 'wind'. The criterion that disruption be avoided is roughly that the density of a star be less than that of the black hole. Since stars have roughly atomic density this condition is (see equation (5.31)),

$$\rho_\star < \rho_\odot \sim \rho_{\text{AT}} \tag{5.181}$$

This gives the size required before a black hole can swallow an entire star whole as

$$M_M \sim \alpha^{-3/2}\left(\frac{m_N^{3/2}}{m_e}\right)\alpha_G^{-3/2} m_N \sim 10^7 M_\odot \tag{5.182}$$

and was the characteristic size calculated by both Michell and Laplace. Black holes larger than $M \sim M_M$ should remain invisible even during collisions with stars whereas smaller black holes would show signs of stellar disruption and radiative emission.[112] Black holes close to the mass M_M are potentially interesting to astrophysicists because they may be the power-source of all manner of high-energy phenomena in the Universe—galactic nuclei, quasars, bursting X-ray sources and Seyfert galaxies. The ubiquity of such exotic astrophysical behaviour may be associated with the number of evolutionary pathways that end in the black hole state; this situation is illustrated by Figure 5.14.[113]

Black holes smaller than M_M have a characteristic growth rate because they grow by accreting diffuse gas rather than whole stars and the infalling gaseous material can be resisted by an outgoing flux of radiation. The possibility of equilibrium between these two opposing fluxes was discovered by Eddington. He pointed out that a flux $F \sim Lr^{-2}$ of energetic photons of luminosity L—in practice, optical to X-ray—will exert a force per nucleon on ionized matter equal to

$$f_e \sim \frac{\alpha^2}{m_e^2} \times \frac{L}{r^2} \tag{5.183}$$

at a distance r from the central energy source (unless conditions are so dramatic near $r \approx 0$ that they do not allow the flux to fall as r^{-2} by the radius r). The gravitational attraction on each nucleon exerted by the accreting mass M is

$$f_g \sim \frac{GMm_N}{r^2} \tag{5.184}$$

So the condition that the attractive force of gravity exceed the outgoing radiative pressure force $f_g \geq f_e$, leads to an upper-bound on the luminosity of an object which is called the Eddington luminosity, L_E, and we usually expect astronomical energy sources to have luminosities satisfying

$$L \leq L_E \sim \frac{\alpha_G m_e^2}{\alpha^2 m_N} M \sim \frac{m_e^2}{\alpha_G^{1/2}\alpha^2}\left(\frac{M}{M_\odot}\right) \tag{5.185}$$

So if radiation makes up a fraction ε of the accreted mass, then the mass increases at rate

$$\dot{M} \sim \varepsilon^{-1} L < \varepsilon^{-1} L_E \sim M\tau^{-1} \tag{5.186}$$

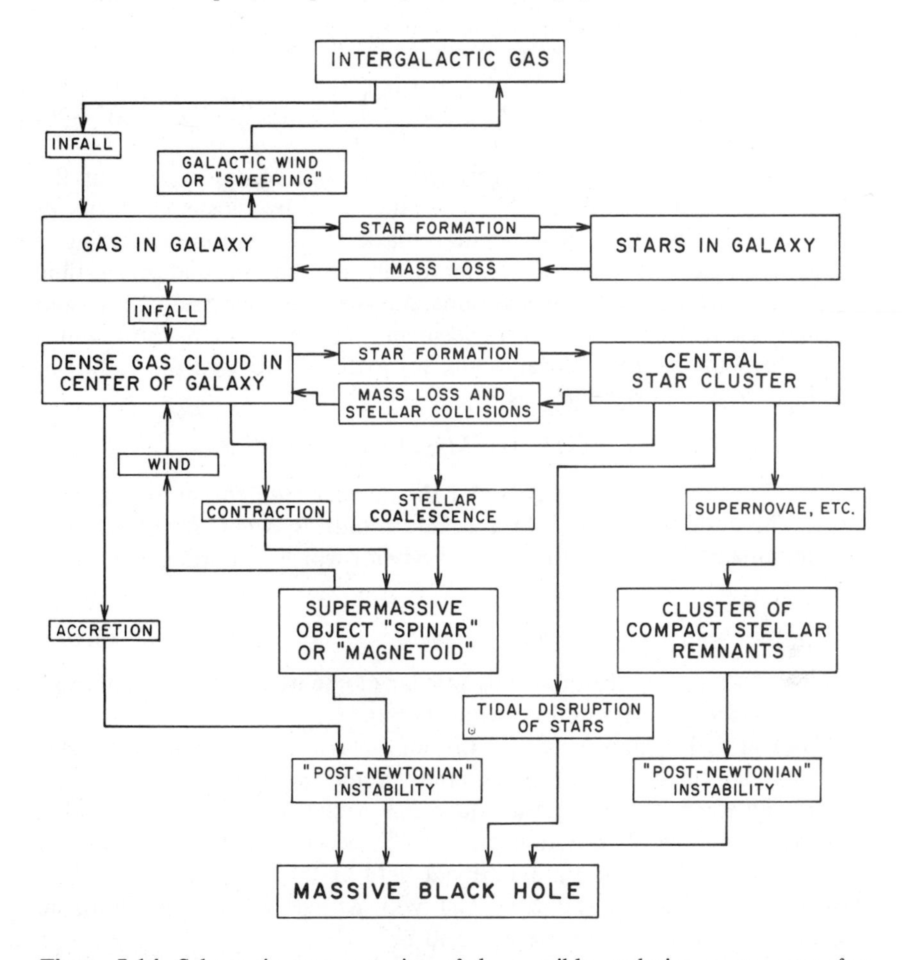

Figure 5.14. Schematic representation of the possible evolutionary sequence for massive gas clouds. Almost all tracks lead to the black hole state.[113]

where the characteristic time for e-folding of the black hole mass τ is given from (5.185) as

$$\tau \sim \varepsilon\alpha^2 m_N \alpha_G^{-1} m_e^{-2} \tag{5.187}$$

which is roughly the same as the main-sequence nuclear burning time.

In addition, we note that a black hole that began with $M \ll M_M$ and which gradually grew by accretion to M_M would be most luminous in the final stages when it would have a luminosity equal to the Eddington Luminosity[114] of a Michell mass black hole, that is, by (5.183) and

(5.185),

$$L_{\max} \sim L_E(M_M) \sim \alpha_G^{-1/2}\alpha^{-7/2}m_e^2\left(\frac{m_N}{m_e}\right)^{3/2} \sim 3\times 10^{45}\ \mathrm{erg\,s^{-1}} \quad (5.188)$$

Let us leave astrophysical problems now and consider some of the characteristics of general black holes as described by Einstein's theory of general relativity: the most general black hole solution describing a stationary black hole in an asymptotically flat background space-time possesses only three defining parameters—its mass, electric charge and angular momentum. The Kerr–Newman metric[115] has a form which generalizes the simple Schwarzschild form (5.180) to give the radius of the black hole's horizon R_K as

$$R_K = GM + (G^2M^2 - J^2M^{-2} - Q^2)^{1/2} \quad (5.189)$$

where Q is the electric charge and J the angular momentum of the black hole. This reduces to the Schwarzschild radius R_s when $J \equiv Q \equiv 0$. The maximum angular momentum of the body for which an horizon still forms is

$$J_{\max} = GM^2 \quad (5.190)$$

A black hole having this maximal angular momentum is called 'extremal' or the 'extreme Kerr solution' when $Q = 0$.

Hawking[116] has shown that, in the semi-classical approximation, black holes are actually black bodies—they obey the laws of equilibrium thermodynamics. When analysed quantum mechanically they are found to emit a thermal distribution of particles. This process can be viewed as pair production in the strong gravitational field at the horizon of the black hole. Using $\Delta E\,\Delta t \sim \hbar$ and $\Delta E \sim k_B T$ with $\Delta t \sim c^{-1}R_s$ near the horizon, we find the black hole temperature to be

$$T_{bh} \sim \frac{\hbar c^3}{k_B GM} \sim \alpha_G^{-1}\left(\frac{m_N}{M}\right)m_N c^2 k_B \sim 10^{-7}\ \mathrm{K}\left(\frac{M_\odot}{M}\right) \quad (5.191)$$

where we have left the fundamental constants $\hbar$, k_B, c present to illustrate the simultaneous quantum, thermodynamic and relativistic character of this gravitational phenomenon.[116–117] The time taken to radiate away the mass M of the black hole by this black-body emission is just

$$t_{bh} \sim \alpha_G^2\left(\frac{M}{m_N}\right)^3 m_N^{-1} \sim 10^{10}\ \mathrm{yrs}\left(\frac{M}{10^{14}\ \mathrm{gm}}\right)^3 \quad (5.192)$$

Thus, black holes in the mass range $\sim 10^{14}$–10^{15} gm would have a lifetime of order the current age of the universe ($\sim$15 billion years) and the effects of these black holes could be observable.[120] It is an interesting 'coincidence' of physics to note that the quantum mechanical entropy of a black hole, $S_{bh} \sim M^2$, is roughly equal to the classical entropy of the black hole mass S_c, (S_c is just the number of nucleons the mass M contains), for a 10^{14}–10^{15} gm black hole, and also that the Schwarzschild radius of such a black hole is equal to the Compton wavelength of the proton ($r_{pr} \sim m_N^{-1}$)

$$\frac{S_{bh}}{S_c} \sim \left(\frac{R_s}{r_{pr}}\right) \tag{5.193}$$

The coincidence that black holes with equal quantum and classical entropies evaporate in the present age of the universe is equivalent to the Anthropic prerequisite for observers: that the present age of universe be at least of order the main sequence stellar lifetime, (see equations (5.128–5.129)).

Hawking[118] has also developed a picture of space-time as a 'foam' of black holes of mass $\sim m_p$. Since they can only form when the Universe has aged $\sim t_p$ and their evaporation time t_{bh} is also of order t_p this leads to a space-time structure composed ultimately of continually forming and dissolving mini black holes of mass close to $m_p \sim 10^{-5}$ gm. None of these black holes with masses $\ll 1 M_\odot$ can form from the gravitational collapse of an ordinary massive star in the usual way. As we have already seen, objects of such low mass would just end up supported by degeneracy pressure and would not continue to collapse inexorably. The only site we know of where black holes with masses $\ll 1 M_\odot$ can form is in the early stages of the Universe where the external radiation pressure can force material inside its Schwarzschild radius if it began with a density sufficiently in excess of the ambient average density.[119] In the early stages of the universe the mass of radiation in causal contact after a time t is given by

$$M_H = \frac{4\pi}{3} t^3 \rho_\gamma \tag{5.194}$$

where the radiation density is

$$\rho_\gamma = \frac{3}{32\pi G t^2} \tag{5.195}$$

so

$$M_H = \frac{t}{8G} \tag{5.196}$$

and black holes of mass M_H can form only at times given by

$$t \leqslant 8 G M_H \sim \alpha_G \left(\frac{M_H}{m_N}\right) m_N^{-1} \tag{5.197}$$

5.10 Grand Unified Gauge Theories

> When you follow two separate chains of thought, Watson, you will find some point of intersection which should approximate the truth.
>
> A. Conan Doyle

So far we have treated the principal interactions of Nature, and the fundamental parameters that describe them, as independent forces. It has long been a goal of theoretical physics to economize on the profusion of free parameters: Newton[5] showed that terrestrial and celestial gravitation were of the same origin; Maxwell that electricity, magnetism and optics could be united. More ambitious attempts to create a completely unified theory of Nature were made by many outstanding physicists, notably by Eddington[121] and Einstein.[122] Both were unsuccessful. However, during the last ten years more specific schemes of unification have been formulated and partially tested by experiment. The first successful theory of this type, the Weinberg–Salam model,[123] based on the $SU(2)\times U(1)$ gauge group, provides a partially unified model for the electromagnetic and weak interactions. Strictly, it is not a truly unified theory; there still exist two independent coupling parameters, α and α_W which are related by an arbitrary parameter. (The dimensionless weak coupling α_W is related to the Fermi[14] constant G_F by $\alpha_W = G_F m_e^2$.) This arbitrary parameter (see equation (5.18)), the Weinberg angle, θ_W, can be measured by experiment but may be uniquely determined by more ambitious extensions of the Weinberg–Salam model which incorporate the gauge theory of the strong interaction (quantum chromodynamics based on the SU(3) colour group) within some simple gauge group,[124] G, so

$$SU(3)\times SU(2)\times U(1)\subset G \qquad (5.198)$$

These larger unified theories of the strong and electro-weak interactions are called '*grand unified theories*' (GUT's). They do not, as yet, indicate how the gravitational interaction can be incorporated into a unified description of Nature but suggest the three microscopic interactions of Nature to be different manifestations of a single interaction. The incorporation of gravity into this edifice will require an understanding of the quantum behaviour of gravitational fields and must await a major extension of known physics.[125]

The first obstacle to any notion of unifying different interactions of Nature appears to be their quite different strengths. However, this is not as formidable an obstacle as it first appears because the *effective* coupling strengths of the interactions depend on the energy at which they are measured. In the case of the electromagnetic interaction there is an increase in strength, but the strong interaction decreases in strength, at high energy. To understand why this is so, consider first the electromagnetic interaction. Quantum field theory shows an electron can emit virtual photons which then turn into virtual e^+e^- pairs. The created positron gets pulled towards the electron while the other electrons are repelled and the original, or 'bare', electron charge is screened. Suppose two electrons scatter from each other when the momentum Q is large. This scattering occurs more strongly when Q is large because the electrons can then penetrate their screening electric fields and get close to each other. However, the target electron will be screened by a collection of electron-positron pairs. This phenomenon is called *vacuum polarization.* Now, when Q is very large the incoming electron partially penetrates the 'screen' and feels the stronger 'bare' electron charge. The fine structure constant α is equal to the square of the completely unscreened charge but as Q rises the interaction will have a strength determined by the square of the partially screened charge, $\alpha(Q^2)$. This is the qualitative reason for the increase in strength of the electromagnetic coupling 'α' at high energy and the resultant $\alpha(Q^2)$ is called the *effective* or *running* coupling strength. When $Q^2 \gg m_e^2$ it has the form[17]

$$\alpha(Q^2) = \alpha\left[1 + \frac{\alpha}{3\pi}\ln\left(\frac{Q^2}{m_e^2}\right) + \dots\right] \tag{5.199}$$

The sum on the right hand side can be more conveniently approximated by the form

$$\alpha(Q^2) = \frac{\alpha}{1 - \left(\frac{\alpha}{3\pi}\right)\ln\left(\frac{Q^2}{m_e^2} + 1\right)} \tag{5.200}$$

which has the property $\alpha(Q^2 = 0) = \alpha$ and $\alpha(Q^2) > \alpha$ for $Q^2 > 0$. For example, at $Q^2 = (10\text{ GeV})^2$ we find $\alpha(10\text{ GeV})^2 = 0.0074 = 1/135.1$. The perturbation analysis used to derive (5.200) breaks down when the denominator vanishes; that is, when $Q^2 \sim m_e^2 \exp(3\pi/\alpha)$. This corresponds to extraordinarily high energies where neglect of gravity is unwarranted and the theory used to derive (5.200) is invalid.

In the case of the strong interaction, although a quark will have its bare colour charge screened by quark-antiquark pairs this is not the only consideration. Indeed, if it were, the strong coupling $\alpha_s(Q)$ would increase above α_s at high energy also and we would be no nearer unification with the electromagnetic force. However, whereas the photons which mediate the electromagnetic interaction do not carry the electromagnetic charge, the gluons mediating the strong force do carry the colour quantum charge. Therefore the gluons, unlike the photons, are *self-interacting.* This enables the gluon field to create a colour deficit near a quark and so there can exist anti-screening of the quark's bare colour charge when that charge is smeared out. A quark can emit gluons which carry colour and so the quark colour is spread or smeared out over a much larger volume and *decreases* the effective coupling as Q increases. Incoming quarks will then see a colour field that is stronger outside than within the local smeared-out colour field. Thus, although the production of $q\bar{q}$ pairs strengthens the strong interactions at high Q because the interaction distance is then smaller, the production of gluon pairs acts in the opposite sense to disperse colour and weaken the effective interaction at high Q. The winner of these two trends is determined, not surprisingly, by the population of coloured gluons relative to that of quark flavours, f. The gluons will dominate if the number of quark flavours is less than 17. If, as we believe, $f<17$ then the strong interaction will *weaken* at high Q—this is called *asymptotic* (or ultraviolet) *freedom*. The effective strong coupling felt at momentum Q is given by $\alpha_s(Q^2)$ relative to its value $\alpha_s(\mu^2)$ at some arbitrary, smaller momentum scale, μ^2, as

$$\alpha_s(Q^2)=\frac{\alpha_s(\mu^2)}{1+\dfrac{\alpha_s(\mu^2)}{12\pi}(33-2f)\ln\left(\dfrac{Q^2}{\mu^2}\right)} \tag{5.201}$$

and we see $d\alpha_s(Q^2)/dQ^2<0$ if $f\leq 16$.

Although it appears that (5.201) contains two arbitrary parameters, μ and $\alpha^2(\mu_s)$, they are in fact related. It is conventional to replace them by the single QCD energy parameter Λ where

$$\ln \Lambda^2 \equiv \ln \mu^2-\frac{12\pi}{(33-2f)\alpha_s(\mu)^2} \tag{5.202}$$

so that (5.201) becomes

$$\alpha_s(Q^2)=\frac{12\pi}{(33-2f)\ln(Q^2/\Lambda^2)}, \qquad Q^2\gg\Lambda^2 \tag{5.203}$$

Experiment indicates that $\Lambda\sim 200$–500 MeV. With a choice $\Lambda=400$ MeV we see that ($\Lambda=400$ MeV corresponds to a characteristic quark-antiquark separation $\sim\Lambda^{-1}\sim 0.5\times10^{-13}$ cm) sample values of the strong coupling at 1 and 10 GeV are predicted from (5.203) to be

$$\begin{aligned}\alpha_s((1\,\text{GeV})^2)&=0.76\\ \alpha_s((10\,\text{GeV})^2)&=0.25\end{aligned} \tag{5.204}$$

where we have $f=3$ at 1 GeV but rising to $f=5$ at 10 GeV (additional quarks appear when the energy rises above their mass).

The coupling $\alpha_s(Q^2)$ depends only on Q^2/Λ^2 and the hope is that other aspects of nuclear physics, for example the proton mass, would be given by Λ and some purely numerical coefficient; Λ sets the energy scale for

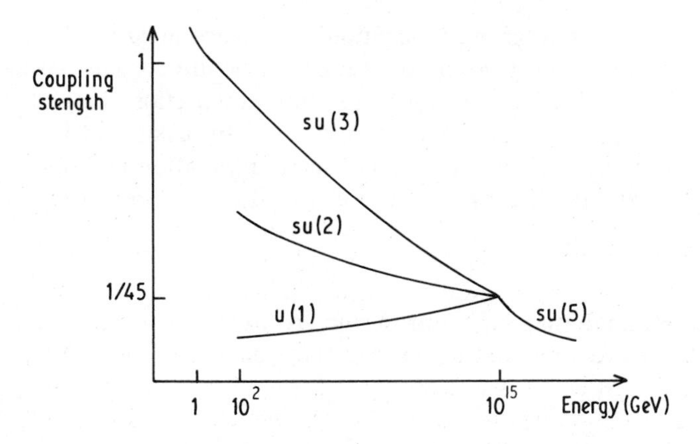

Figure 5.15. Proposed evolution of the effective U(1), SU(2) and SU(3) coupling strengths in the SU(5) scheme of grand unification.[126]

strong interactions. In Figure 5.15 the calculated scaling of the weak, electromagnetic and strong couplings is shown with energy.[126]
A coincidence of interaction strengths is possible at extremely high energy $\sim 10^{15}$ GeV. This is the energy scale at which 'grand unification' is suspected to take place.

This overcomes the 'quantitative' obstacle to unification—how can three interactions of such different strengths be unified? The other hurdle is what we might call the 'qualitative' obstacle—how can we unify interactions that act upon such different and disjoint classes of elementary particle? The electromagnetic and weak interactions govern the interactions of a class of particles called leptons and their mutual interactions. Leptons conserve a quantity called lepton number. The strong interaction governs a class of particles (quarks and gluons) which possess a conserved quality called baryon number but quarks and gluons do not possess lepton number. The electro-weak and strong interactions cannot be unified unless quarks can turn into leptons in elementary particle interactions and *vice versa*. In GUT's this qualitative hurdle is overcome by the appearance of a new heavy X boson, with mass m_X. Its mass is naturally of order the unification energy, $m_X \sim 10^{15}$ GeV. This boson can mediate lepton and baryon number violating interactions between quarks and leptons. These will be prevalent at energies $\sim m_X$, at which X and $\bar{X}$ boson pairs will be produced in thermal equilibrium. At everyday temperatures $\sim T_B$, (see (5.32)), X bosons will be rare and the quark and lepton interactions will not appear manifestly unified and both baryon and lepton numbers will be conserved to high precision.

If we demand that 'grand unification' take place at $Q<m_p\sim 10^{19}$ GeV then this places a lower bound on the fine structure constant at habitable temperatures and it leads to the very strong restriction that $\alpha \gtrsim 1/180$.

We can also obtain an upper bound on α by using the fact that the grand unified theories predict baryon number violation and hence proton decay.[127] The lifetime, τ_N, has the four-Fermi form, (see equation (5.121)),

$$\tau_N \sim \alpha^{-2} m_X^4 m_N^{-5} \tag{5.205}$$

If we demand that an anthropically constructed universe must have stars then we require that τ_N exceed the stellar lifetime (5.128, 5.129)

$$\tau_N > t_\star \tag{5.206}$$

So, in order to contain stars and hence life we require that Nature obeys

$$\alpha^{-2} m_X^4 m_N^{-5} > \alpha^2 \alpha_G^{-1} \left(\frac{m_N}{m_e}\right)^2 m_N^{-1} \tag{5.207}$$

that is,

$$\left(\frac{m_X}{m_N}\right)^4 > \frac{\alpha^4}{\alpha_G} \left(\frac{m_N}{m_e}\right)^2 \tag{5.208}$$

Since at the unification energy scale $m_{GUT} \sim m_X$ we have[128]

$$\alpha \sim \frac{0.25}{\ln(m_X/m_N)} \tag{5.209}$$

so (5.205) becomes

$$\tau_N \sim (\alpha^2 m_N)^{-1} \exp(\alpha^{-1}) \tag{5.210}$$

and (5.208) translates into the constraint

$$\exp(\alpha^{-1}) > \alpha^4 \alpha_G^{-1} \left(\frac{m_N}{m_e}\right)^2 \tag{5.211}$$

For example, with the actual values for α_G and m_N/m_e one obtains

$$\alpha \leq \frac{1}{85} \tag{5.212}$$

Rozenthal[129] has pointed out that if one takes a closed universe of mass M, so that its present mass can, using Dirac's observation,[130] be written $M \sim \alpha_G^{-2} m_N$ then in conjugation with $\tau_N > t_0$ where the present age of the Universe is $t_0 \sim (\alpha_G m_e)^{-1}$, we have (compare equation (4.23)),

$$\alpha < -(\ln \alpha_G)^{-1} \tag{5.213}$$

Another interesting coincidence exists.[131] The ratio of τ_N to t_p is $\sim 10^{80}$

which is also roughly the number of nucleons in a closed universe (so long as it is not extraordinarily close to spatial flatness). This is equivalent to the coincidence

$$\alpha_G^{-2} \sim (\alpha^{-2} m_X^4 m_N^{-5})(\alpha_G^{1/2} m_N^{-1}) \tag{5.214}$$

that is,

$$\alpha_G^{-3/2} \sim \alpha^{-2} \left(\frac{m_X}{m_N}\right)^4 \tag{5.215}$$

now since $m_X/m_N \sim \exp(-4\alpha)$ we can write this coincidence as

$$\alpha_G^{-1} \sim \alpha^{-4/3} \exp\left(\frac{2\alpha}{3}\right) \sim 3 \times 10^{-43} \tag{5.216}$$

In summary, grand unified theories allow very sharp limits to be placed on the possible values of the fine structure constant in a cognizable universe. The possibility of doing physics on a background space-time at the unification energy and the existence of stars made of protons and neutrons enclose α in the niche

$$\frac{1}{180} \lesssim \alpha \lesssim \frac{1}{85} \tag{5.217}$$

These unified theories also show us why we observe the World to be governed by a variety of 'fundamental' forces of apparently differing strengths: inevitably we must inhabit a low-temperature world with $T < T_B \lesssim \alpha^2 m_e$, and at these low energies the underlying symmetry of the World is hidden; instead we observe only its spontaneously-broken forms.

There are further consequences of grand unified theories for cosmology.[132] Most notably, the simultaneous presence of baryon number, *CP* and *C* violating interactions makes it possible for us to explain the observed baryon asymmetry of the Universe—the overt propensity for matter rather than antimatter in the Universe. This leads us to consider next what we know of cosmology.

In this chapter we have shown how it is possible to construct the gross features of the natural world around us from the knowledge of a few invariant constants of Nature. The sizes of atoms, people, and planets are not accidental, nor are they the inevitable result of natural selection. Rather, they are consequences of inevitable equilibrium states between competing natural forces of attraction and repulsion. Our study has shown us, in a rough way, where natural selection stops. It has enabled us to separate those aspects of Nature which we should regard as coincidences, from those which are inevitable consequences of fundamental forces and the values of the constants of Nature. We have also been able to ascertain which invariant combinations of physical constants play a key role in making the existence of intelligence possible. This possibility appears to hinge upon a number of unrelated coincidences whose existence may or may not be inevitable. In our survey we have ranged from the scale of elementary particles to stars. We stopped there for a reason; beyond the scale of individual stars it is known that cosmological coincidences and initial conditions may also play a major role in rendering the Universe habitable by intelligent observers. In the next chapter we shall investigate these interconnections in some detail.

References

1. G. Johnstone Stoney, *Phil. Mag.* (ser. 5) **11,** 381 (1881); *Trans. R. Dublin Soc.* **6** (ser. 2) Pt xiii, 305 (1900).
2. L. J. Henderson, *The fitness of the environment* (Harvard University Press, Mass., 1913).
3. W. Paley, *Natural theology*, Vol. 3 of *The complete works of William Paley* (Cowie, London, 1825).
4. J. D. Barrow, *Quart. J. R. astron. Soc.* **22,** 388 (1981).
5. I. Newton, *Philosphiae naturalis, principia mathematica* II, prop 32 (1713), ansl. A. Motte (University of California Press, Berkeley, 1946).
6. J. B. Fourier, *Theoria de la chaleur* (1822), Chapter 2, §9. For a detailed account of modern dimensional methods see R. Kurth, *Dimensional analysis and group theory in astrophysics* (Pergamon, Oxford, 1972).
7. An interesting discussion of this was given by A. Einstein, *Ann. Physik* **35,** 687 (1911). For further discussion see section 4.8 of this book for a possible anthropic explanation.
8. Adapted from J. Kleczek, *The universe* (Reidel, Dordrecht, 1976), p. 218.
9. G. Johnstone Stoney, *Phil Mag.* (ser. 5) **11,** 381 (1881). This work was presented earlier at the Belfast meeting of the British Association in 1874.
10. op. cit., p. 384.
11. M. Planck, *The theory of heat radiation*, transl. M. Masius (Dover, NY, 1959); based on lectures delivered in 1906–7 in Berlin, p. 174.
12. The initials of the celebrated Mr. C. G. H. Tompkins, a bank clerk with an irrepressible interest in modern science, were given by these constants. For an explanation see *Mr. Tompkins in paperback* by G. Gamow (Cambridge University Press, Cambridge, 1965) p. vii.
13. A. Sommerfeld, *Phys. Z.* **12,** 1057 (1911).
14. E. Fermi, *Z. Physik* **88,** 161 (1934) transl., in *The development of weak interaction theory,* ed. P. K. Kabir (Gordon & Breach, NY, 1963).
15. The factor $2^{-1/2}c^2$ is purely conventional; for details see D. C. Cheng and G. K. O'Neill, *Elementary particle physics: an introduction* (Addison-Wesley, Mass., 1979).
16. These expressions are in rationalized units, $g^2(\text{rat}) = 4\pi g^2(\text{unrat})$.
17. P. Langacker, *Phys. Rep.* **72,** 185 (1981).
18. M. Born, *Proc. Indian Acad. Sci* A **2,** 533 (1935).

19. For a good overview see S. Gasiorowicz, *The structure of matter; a survey of modern physics* (Addison-Wesley, Mass., 1979). For historical background to the Bohr theory see M. Hammer, *The conceptual development of quantum mechanics* (McGraw-Hill, NY, 1966), and *Sources of quantum mechanics*, ed. B. L. van der Waerden (Dover, NY, 1967).
20. W. E. Thirring, *Principles of quantum electrodynamics* (Academic Press, NY, 1958).
21. For a discussion of the structure of materials, see D. Tabor, *Gases, liquids and solids*, 2nd edn (Cambridge University Press, Cambridge, 1979).
22. A. Holden, *Bonds between atoms* (Oxford University Press, Oxford, 1977), p. 15.
23. F. Dyson quotes Ehrenfest: '... why are atoms themselves so big? ... Answer: only the Pauli Principle, 'No two electrons in the same state? That is why atoms are so unnecessarily big, and why metal and stone are so bulky'. *J. Math. Phys.* **8,** 1538 (1967).
24. F. Kahn, in *The emerging universe*, ed. W. C. Saslaw and K. C. Jacobs (University of Virginia Press, Charlottesville, 1972).
25. T. Regge, in *Atti de convegus Mendeleeviano*, Acad. del Sci. de Torino (1971), p. 398.
26. V. F. Weisskopf, *Science* **187,** 605 (1975).
27. J. M. Pasachoff and M. L. Kutner, *University astronomy* (Saunders, Philadelphia, 1978).
28. H. Dehnen, *Umschau* **23,** 734 (1973); Konstanz Universitätsreden No. 45 (1972).
29. The height allowed will be slightly less than ~30 km because the rock is not initially at zero temperature and so does require so much energy to liquify.
30. The melting temperature of quartz is 1968 K according to D. W. Hyndman, *Petrology of igneous and metamorphic rocks* (McGraw-Hill, NY, 1972).
31. B. J. Carr and M. J. Rees, *Nature* **278,** 605 (1979).
21. M. H. Hart, *Icarus* **33,** 23 (1978).
33. F. W. Went, *Am. Scient.* **56,** 400 (1968).
34. A. V. Hill, *Science Prog.* **38,** 209 (1950).
35. J. B. S. Haldane, in *Possible worlds* (Hugh & Bros., NY, 1928).
36. L. J. Henderson, Proc. natn. Acad. Sci., *U.S.A.* **2,** 645 (1916).
37. W. D'A. Thompson, *On growth and form* (Cambridge University Press, London, 1917).
38. F. Moog, *Scient. Am.* **179,** 5 (1948); C. J. v. d. Klaauw, *Arch. neerl. Zool.* **9,** 1 (1948).
39. R. M. Alexander, *Size and shape* (E. Arnold, Southampton, 1975).
40. A. Lightman, *Am. J. Phys.* **52,** 211 (1984).
41. W. H. Press and A. Lightman, *Phil. Trans. R. Soc.* A **310,** 323 (1983).
42. G. Galileo, *Two new sciences*, English transl., S. Drake (University of Wisconsin Press, Madison, 1974); the first edition was published in Italian as *Discorsi e dimostrazioni matematiche, intorno a due nouve scienze atteneti alla mecanica ed ai muovimenti locali* (1638); the quotation is from p. 127.
43. W. Press, *Am. J. Phys.* **48,** 597 (1980). The size estimates given by Press are a better estimate of the size of a creature able to support itself against

gravity by the surface tension of water which is some fraction of the intermolecular binding energy, say $\varepsilon\alpha^2 m_e$ per unit area, and Press's size limits, $\sim$1 cm, more realistically correspond to the maximum dimension of pond-skaters rather than people.

44. A. Rauber showed that elephants are quite close to the maximum size allowed for a land-going animal in *Morph. Jb.* **7,** 327 (1882). Notice that some ingenious organisms (sponges) have evolved means of increasing their surface areas without inflating their masses by the full factor $\sim$(area)$^{3/2}$. The bathroom towel exploits this design feature.
45. J. Woodhead-Galloway, *Collagen: the anatomy of a protein* (Arnold, Southampton, 1981). However, it appears that, in general, good resistance to crack and compression tend to be mutually exclusive features of structures.
46. L. Euler, *Acta acad. sci. imp. petropol.* (1778), p. 163.
47. W. Walton, *Quart. J. Math.* **9,** 179 (1868). A. G. Greenhill, *Proc. Camb. Phil. Soc.* **4** (Pt II), 5 (1881).
48. H. Lin, *Am. J. Phys.* **50,** 72 (1982). A. Herschmann, *Am. J. Phys.* **42,** 778 (1974), E. D. Yorke, *Am. J. Phys.* **41,** 1286 (1973). T. McMahon, *Science* **179,** 1201 (1973). H. J. Metcalf, *Topics in biophysics* (Prentice-Hall, NJ, 1980).
49. Ref. 42, p. 129.
50. E. M. Purcell, *Am. J. Phys.* **45,** 3 (1977).
51. Note that the resistive drag force, $F_d \propto$ (cross-sectional area) $\times$ (density) $\times$ (velocity)^2, is only independent of the viscosity of the ambient medium when the velocities are large. When they are small the familiar Stokes law holds with $F_d \propto$ (radius) $\times$ (velocity) $\times$ (viscosity) and this is exploited in centrifuges: since macromolecules have mass α (radius)^3 they will sediment out at rates proportional to their size.
52. H.-C. Berg, *Nature* **254,** 389 (1975); *Ann. Rev. Biophys. Biol.* **4,** 119 (1975): *Scient. Am.* **233,** 36 (Aug. 1975). J. M. Smith, *Mathematical ideas in biology* (Cambridge University Press, Cambridge, 1980).
53. C. J. Pennycuik, Col. livia. *J. Exp. Biol.* **49,** 527 (1968).
54. Actually only 1% of the work done by the human heart is deployed to accelerate blood. Most of the rest overcomes viscous resistance to the blood flow through small blood vessels.
55. See M. Kleiber, *Physiol. Rev.* **27,** 511 (1947); *Scale effects in animal locomotion*, ed. T. Pedley (Academic Press, NY, 1927); P. Altman and D. Dittmer, *Biology data book* 2nd edn (Federation of American Societies for Experimental Biology, Bethseda, Maryland, 1974). There is excellent agreement with the data over a mass range of $\sim 10^{-1}$–10^5 kg encompassing mice, birds, rabbits, dogs, physicists, and elephants.
56. Jellyfish are an obvious exception to this statement. They grow hundreds of times larger but are exceptionally constructed with all their cells superficially situated within $\sim 10^{-2}$ cm of the water from which they extract oxygen.
57. E. Schrödinger, *What is life*? (Cambridge University Press, Cambridge, 1944).
58. Despite the huge range of aniamal and plant sizes these organisms all possess cells of roughly the same size. The difference in their gross size is due to variations in cell number. A typical cell has a volume of about a thousand cubic microns.

59. N. W. Pirie, *Ann. Rev. Microbiol.* **27,** 119 (1973); W. R. Stahl, *J. Theor. Biol.* **8,** 371 (1965); H. J. Morowitz, *Prog. Theor. Biol.* **1,** 35 (1967).
60. H. Yukawa, *Proc. Phys. Math. Soc. Japan,* **17,** 48 (1935). The analogy between chemical and nuclear forces was evident to Heisenberg prior to the work of Yukawa. The Yukawa model can only explain a few aspects of the nuclear force. There exist many mesons besides the π. Other facts must be taken into account to describe the short range ($<2\times10^{-13}$ cm) part of the interaction.
61. The mean inter-nucleon-separation will be $\sim 2\alpha_s m_N^{-1}$ and the mean value of the exponent in (5.49) is $\sim 2\alpha_s^{-1} m_\pi m_N^{-1}$, so neglecting the exponent overestimates the binding energy. If one considers states other than the ground level, its inclusion is essential and the fall-off is fast enough to prevent higher bound-states existing.
62. Actually, $m_\pi/m_N \sim 1/7$; much slower emission of e^+ is possible.
63. V. Weisskopf, in *Scientific endeavour* (Rockefeller Institute Press, Centennial Volume of the National Academy of Sciences, 1965).
64. Here the comparison with molecular systems is interesting: many molecules can exist in rotationally excited states of very large angular momentum.
65. F. Dyson, *Scient. Am.* **225,** 25 (Sept. 1971).
66. P. C. W. Davies, *J. Phys.* A **5,** 1296 (1972). L. Okamoto and C. Pask, *Ann. Phys.* **68,** 18 (1971).
67. N. Bohr, *Nature* **137,** 344 (1936); C. F. von Weiszäcker, *Z. Physik* **96,** 431 (1933); H. A. Bethe and R. F. Bacher, *Rev. Mod. Phys.* **8,** 82 (1936).
68. R. D. Evans, *The atomic nucleus* (McGraw-Hill, NY, 1955). E. Segre, *Nuclei and particles*, 2nd edn (Benjamin, California, 1977).
69. To see this, ignore the nucleon–nucleon interaction and consider protons and neutrons in a box of volume $4\pi R^3 = 4\pi r_0^3 A$. For any collection of non-interacting fermions in a box of volume V the energy is given in terms of their number density, n, and mass m, as $E = V\pi^3(3n/\pi)^{5/3}/10m$. For the protons $n = 3Z/4\pi r_0^3 A$ and so $E(\text{protons}) \propto (Z/A)^{5/3}$, and similarly $E(\text{neutrons}) \propto (A-Z)^{5/3}$, where the coefficients of proportionality are equal if we neglect the very small proton–neutron mass difference. Therefore, the total energy can be expanded in terms of $(0.5A - Z)$, and the leading terms are $E(\text{protons}) + E(\text{neutrons}) = \text{const} + 0(0.5A - Z)^2$, where the constant is just the volume term is (5.68). Finally, we note that the liquid-drop model ignores the local energy variations arising from nuclear energy-shell structure.
70. From Evans, ref. 68, p. 382.
71. The term 'fission' was abstracted from biology by L. Meitner and O. R. Frisch, *Nature* **143,** 471 (1939) and was suggested by the division of a liquid drop into smaller droplets. A major theoretical development of the liquid drop model was made by N. Bohr and J. A. Wheeler, *Phys. Rev.* **56,** 426 (1939), and the details are considerably more complicated than our short sketch of the salient features implies.
72. The value of Z which minimizes E for fixed A is given by $\partial E/\partial Z = 0$; that is, when

$$Z/A = 101.3/(200 + 1.4A^{2/3})$$

so, $Z/A \sim 0.5$ except for very heavy nuclei where it drops off slightly to 0.4 at $A \sim 216$.

73. Nuclei with $Z^2/A < 49$ can fission by quantum tunnelling, and asymmetric fission, like α-decay, can occur with Z^2/A as low as ~ 33.
74. The nuclear 'surface' energy is taken proportional to α_s^2.
75. Adapted from Davies, ref. 66.
76. The precise value depends on the mass distribution. For a homogeneous sphere $E_g = -3GM^2/5R$.
77. At the end of the nineteenth century it was believed that gravity might be the sole force acting in the interior of stars. As the stars collapsed, the loss in gravitational potential was taken up as heat by the internal constituents, to provide a temporary means of pressure support. However, if this were the sole consideration necessary to support our Sun, its total lifetime would be given roughly by the ratio of the gravitational energy, (E_g in (5.102)) to the rate of energy loss (or luminosity). For the Sun, $M_\odot \sim 2 \times 10^{33}$ gm, $R_\odot \sim 7 \times 10^{10}$ cm and $L_\odot \sim 4 \times 10^{33}$ erg s^{-1}, which predicts a lifetime of a little over 10^7 years. However, since the oldest micro-fossils exceed $\sim 2 \times 10^9$ years in age, another source of energy must be present in the Sun (see section 3.6 for the early debate about this issue).
78. For nuclei with $Z_1 = Z_2$ that approach to within $r \sim 10^{-13}$ cm, the Coulomb barrier has energy $\sim Z_1 Z_2 \alpha r^{-1} \sim 1$ MeV. Thermal energies in the Sun are $\sim 10^7$ K, and they alone would only allow nuclei to approach to within $\sim 10^{-10}$ cm, about a thousand times greater than the range of nuclear forces.
79. D. Clayton, *Principles of stellar evolution and nucleosynthesis* (McGraw-Hill, NY, 1968), p. 302.
80. L. D. Landau, *Phys. Abh. Soviet Union* **1,** 285 (1932); L. D. Landau and E. M. Lifshitz, *Statistical physics*, 2nd edn (Pergamon Press, Oxford, 1977); S. Chandrasekhar, *Mon. Not. R. astron. soc.* **95,** 207 (1935).
81. In the Sun radiation pressure is relatively unimportant, since $N_\odot \sim 0.6\alpha_G^{-3/2}$.
82. Note that similar scattering from ions has a much smaller cross-section $\sim \alpha^2 m_N^{-2}$.
83. We assume heat transport by radiation alone. Convection is neglected along with the small variations associated with the fact that not all photons originate at the centre of the star.
84. R. H. Dicke, *Nature* **192,** 440 (1961).
85. G. de Santillana and H. Von Dechend, *Hamlet's mill* (Gambit, Boston, 1969).
86. B. Carter, 'The significance of numerical coincidences in nature', unpublished preprint (University of Cambridge, 1967); in *Confrontation of cosmological theories with observational data*, IAU Symposium No. 63, ed. M. S. Longair (Reidel, Dordrecht, 1974), p. 291.
87. These formulae are only applicable when $M \leqslant 10M_\star$. If $M \geqslant 10M_\star$, then $(T_S/T_I)^4 \sim (\eta/10^{-3.5})^2 \alpha_G^{1/2} \alpha^{-6}(m_N/m_e)$.
88. We have used the Thomson formula (5.114) here. If the Kramers form (5.139) is used, the required coincidence is similar, $\alpha_G \sim \alpha^{1/2}(m_N/m_e)^5$.
89. Taken from Carter (unpublished), ref. 86.
90. M. Harwit, *Astrophysical concepts* (Wiley, NY, 1973).
91. L. Mestel, IAU Symposium No. 75, ed. T. de Jong and A. Maeder (Reidel, Dordrecht, 1977); I. Appenzeller, J. Lequeux, and J. Silk, *Star formation*, 10th Advanced Course Swiss Soc. Astron. Astrophys, Saas-Fee (Geneva

Observatory, 1980); V. C. Reddish, *Stellar formation* (Pergamon Press, Oxford, 1978).

92. M. J. Rees, *Mon. Not. R. astron. Soc.* **176,** 483 (1976); see also J. Silk, *Astrophys. J.* **211,** 638 (1976).
93. F. Hoyle, *Astrophys. J.* **118,** 513 (1953).
94. S. Chandrasekhar, in *Physics and astrophysics of neutron stars and black holes*, ed. R. Giaconni and R. Ruffini (Enrico Fermi School; Academic Press, NY, 1978).
95. More detailed analysis shows, $M_{WD} \sim 5.8\mu^2 M_\odot$ where μ is the number of free electrons per nucleon; see W. B. Hubbard, *Rev. Geophys. and Space Phys.* **18,** 1 (1980).
96. The detailed structure of neutron stars is considerable more complex than indicated here. An understanding of the internal structure involves, as yet, uncertain aspects of nuclear physics; see ref. 97.
97. G. Baym and C. Pethick, *Ann. Rev. Nucl. Part. Sci.* **25,** 27 (1975).
98. Pulsars like the Crab gave periods $\sim 0.033s$ and appear to be rapidly rotating neutron stars.
99. R. Forward, *The dragon's egg* (Ballantine, NY, 1980); D. Goldsmith and T. Owen, *The search for life in the universe* (Benjamin, California, 1980). p. 219; *Communication with extraterrestrial intelligence CETI*, ed. C. Sagan (MIT Press, Cambridge, Mass., 1973), p. 199; F. Drake, *Astronomy* **1,** 5 (No. 5), (Dec. 1973).
100. C. G. Källman, *Fund. Cosmic Phys.* **4,** 167 (1979).
101. F. Dyson, *Nature* **223,** 486 (1969); V. Canuto, in *Selected topics in physics, astrophysics and biophysics*, ed. A. de Laredo (Reidel, Dordrecht, 1973).
102. G. R. Blumenthal and H. W. Tucker, *Ann. Rev. Astron. Astrophys.* **12,** 23 (1974).
103. M. A. Ruderman, *Neutron stars*, Scuola di Fisica Cosmica, Ettore Majorana Center reprint (1971); *Phys. Rev. Lett.* **27,** 1306 (1971).
104. The Coulomb potential is $\sim \ln r$ in two dimensions compared with r^{-1} in three dimensions.
105. The term 'black hole' was first used in print by J. A. Wheeler, *Am. Scient.* **56,** 1 (1968). For a guide to the subsequent literature, see S. Detweiler, *Am. J. Phys.* **49,** 394 (1981).
106. J. Mitchell, *Phil. Trans.* **74,** 35 (1783). For some historical background, see S. Schaffer, *J. Hist. Astron.* **10,** 42 (1979). Notice that the Newtonian 'black hole' is not really like the general relativistic version. In the former, an escape velocity of c still allows light to travel farther than a distance $2GM/c^2$ from the centre.
107. P. S. Laplace, *Exposition du systeme du monde* (Cercle-Social, Paris, 1796) Vol. 2, p. 305. For a translation see S. W. Hawking and G. F. R. Ellis, *The large scale structure of space-time* (Cambridge University Press, Cambridge, 1973). Laplace's article was read and extended by the German astronomer Johann Soldner who drew attention to the effects of light-bending by massive objects in *Astronomiches Jahrbuch* (1804), ed. J. E. Bode. Soldner's paper was rediscovered and republished in *Ann. Physik* **65,** 593 (1921). Sadly, the motivation for its republication was Lenard's pathetic attempt to discredit Einstein and 'Jewish science'.

108. K. Schwarzschild, *Sber. preuss. Akad. Wiss.* 189 (1916); *Sber. preuss. Akad. Wiss.* 424 (1916); see also S. Chandrasekhar, *Nature* **252,** 15 (1974).
109. O. Lodge, *Phil. Mag.* (ser. 6) **41,** 549 (1921).
110. Nuclear density is closer to $\sim 10^{14}$ gm cm^{-3} and so by using 10^{12} gm cm^{-3} Lodge obtains a size ~400 km rather than the value ~1–10 km accepted today.
111. D. Lynden-Bell, *Nature* **233,** 690 (1969).
112. J. G. Hills, *Nature* **254,** 295 (1975); B. Carter in C. Hazard and S. Mitton, *Active galactic nuclei* (Cambridge University Press, Cambridge, 1980).
113. M. Begelman, R. Blandford, and M. J. Rees, *Rev. Mod. Phys.* **56,** 297 (1984).
114. A. S. Eddington, *The internal constitution of the stars* (Cambridge University Press, Cambridge, 1926).
115. R. Kerr, *Phys. Rev. Lett.* **11,** 237 (1963); S. Chandrasekhar, *The mathematical theory of black holes* (Oxford University Press, Oxford, 1983).
116. S. W. Hawking, *Commun. Math. Phys.* **43,** 189 (1975).
117. D. W. Sciama, *Vistas in Astron.* **19,** 385 (1976); *Ann. NY. Acad. Sci.* **302,** 161 (1977).
118. S. W. Hawking, *Nucl. Phys.* B **144,** 349 (1978).
119. B. J. Carr, *Astrophys. J.* **201,** 1 (1975); S. W. Hawking, *Nature* **243,** 30 (1974).
120. M. J. Rees, *Nature* **266,** 333 (1977); J. D. Barrow and G. Ross, *Nucl. Phys.* B **181,** 461 (1981).
121. A. S. Eddington, *Fundamental theory* (Cambridge University Press, Cambridge, 1946).
122. A. Einstein, *The meaning of relativity*, 5th rev. edn (Chapman & Hall, London, 1967).
123. S. L. Glashow, *Nucl. Phys.* **22,** 579 (1961); S. Weinberg, *Phys. Rev. Lett.* **19,** 1264 (1967); A. Salam, *Proc. 8th Nobel Symposium, Stockholm* 1968, ed. N. Svartholm (Almquist & Wiksell, Stockholm, 1968), p. 367.
124. H. Georgi and S. L. Glashow, *Phys. Rev. Lett.* **33,** 451 (1974).
125. N. Birrell and P. C. W. Davies, *Quantum fields in curved space* (Cambridge University Press, Cambridge, 1982).
126. J. Ellis, *Phil. Trans. R. Soc.* A **310,** 279 (1983).
127. H. Georgi, H. Quinn, and S. Weinberg, *Phys. Rev. Lett.* **33,** 451 (1974).
128. J. Ellis and D. V. Nanopoulos, *Nature* **292,** 436 (1981).
129. I. L. Rozenthal, *Sov. Phys. JETP Lett.* **31,** 490 (1980).
130. Dirac's coincidence can be expressed as the statement that the present-day observable universe contains a mass $\sim 10^{78} m_N \sim \alpha_G^{-2} m_N$.
131. J. D. Barrow, *Nature* **282,** 698 (1979).
132. J. D. Barrow, *Fund. Cosmic Phys.* **8,** 83 (1983).

6 The Anthropic Principles in Classical Cosmology

There was a most ingenious architect who had contrived a new method for building houses, by beginning at the roof, and working downwards to the foundation.

Jonathan Swift

6.1 Introduction

All science is cosmology, I believe.

K. Popper

In the last chapter we saw why the Weak Anthropic Principle enters as a significant factor when assessing the existence of physical structures. The coarse-grained features of equilibrium states in Nature arise principally as a result of the numerical values of a small number of dimensionless constants of Nature. These constants also determine whether or not biological structures are possible and thereby an invariant link is forged between the existence of observers and the variety of phenomena they can ever expect to observe. We have seen that, from the scale of nuclei to stars, our World is conditioned principally by the values of the fundamental constants α, m_N/m_e, α_G, α_W and α_s. There is also a more speculative 'strong anthropic' aspect. For, were the values of these fundamental constants significantly different, then life as we know it would be impossible.

In order to complete our survey of the Universe's structure and the manner in which it is tied to absolute, invariant parameters we must turn to cosmology. This will enable us to extend our discussion of structures beyond that of stars. We shall find, not unexpectedly, that to give descriptions of the possible character, size and age of galaxies, galaxy clusters and the whole observable universe we require some additional 'cosmological' parameters. Our local dimensionless constants α, m_N/m_e, α_G, α_W and α_s will be joined by three cosmological parameters. It is one of the mysterious features of our Universe that there could, as far as we understand things, exist a large number of additional parameters associated with the wide range of geometrical configurations available to the universe. Nevertheless, these parameters are found to have values so small that they have yet to be positively discriminated from zero by astronomical measurements.

Our survey of cosmology will again stress two aspects: we shall try to unravel those features of the Universe which are inevitable consequences of the values of fundamental constants from those that are apparent 'coincidences'; we shall also examine how these features are associated with the necessary prerequisites for the existence of observers. We begin by giving a rapid resumé of the Big Bang cosmological model that avoids mathematical details.[1] We shall then examine the observational evidence for the Big Bang model in more quantitative detail.

The 'Big Bang' theory of the 'origin' and evolution of the Universe is the paradigm of modern cosmology. Within its theoretical and observational framework the insights of many areas of physics and astronomy can be coordinated to build up an understanding of the large-scale structure and evolution of the cosmos in which we live. The Big Bang cosmological model is a prediction of Einstein's general theory of relativity.[2] This theory of gravitation supersedes that of Newton and predicts that the Universe must be in a state of dynamic change. When, in 1929, Hubble first interpreted the 'redshifting' in the spectra of distant galaxies as a manifestation of the Doppler effect, he discovered this dynamic state to be one of overall expansion rather than contraction.[3] The whole Universe, everything that is, is in a state of dynamic inflation and evolutionary change.

The Big Bang theory received remarkable confirmation with the discovery of the microwave background radiation in 1965 by Penzias and Wilson.[4] It had been predicted by Alpher and Herman[5] in 1948 that the hot fireball of the Big Bang should leave an 'echo', a glimmer of its former self, in the present-day Universe. They calculated that the adiabatic expansion of the Universe should have cooled the heat radiation from the hot initial state down to a level ~ 5 K or thereabouts by the present, some fifteen billion years after the initial 'Bang'. While they were calibrating a communications satellite at the Bell Laboratories in New Jersey, Penzias and Wilson stumbled upon this primordial radiation. It manifested itself in their microwave antenna as an isotropic source of background noise that remained ubiquitous in the face of every conceivable experimental check for instrumental malfunctions and local sources of radiation. Eventually, the theoretical prediction of Alpher and Herman was recalled, following discussions with Dicke and his colleagues[6] at nearby Princeton, and seen to be extraordinarily close to what was discovered—radiation at 3 K. Later measurements over a variety of wavelengths have determined the spectral form of the 3 K radiation and shown it to possess the characteristic Planckian spectrum of heat radiation. The other observational cornerstone of the Big Bang model is its successful prediction that the Universe should contain particular abundances of the lightest elements: hydrogen, deuterium and helium. Gamow and his students were also the first to realize that the early stages of a Big Bang cosmology experienced temperatures and densities so high that the entire Universe could behave as a nuclear fusion reactor, synthesizing light nuclei from primeval neutrons and protons. Subsequent, detailed calculations[7] predicted that the present Universe should contain about 75% of its mass in the form of hydrogen and 25% as helium-4 with about one part in a million ending up in the form of all the other elements—including those of which we are composed (although such biological elements are predominantly fused in stars rather than in the inferno of the Big Bang). These predictions have been strikingly confirmed by observations.[8] Even in sites where the heavy elements like carbon, produced locally by the stars, have anomalous abundancies reflecting peculiarities in the local environment, the abundance of helium is universal, witnessing to its origin in the early stages of the Universe and also to the essential accuracy of our cosmological model right back to merely one second after the expansion began.

Edwin Hubble's discovery that the Universe is in a state of expansion has an equally dramatic corollary. If, in our mind's eye, we reverse the sense of the expansion and follow it backwards in time to earlier and earlier stages of cosmic history we encounter conditions of higher and higher density, and continuously increasing temperature. Eventually we reach a state of (apparent) infinite density at a finite proper time in the past—a space-time singularity where our theory breaks down. Using the present observations of the Universe's rate of expansion and deceleration this singular state existed about fifteen billion years ago.[9] To orient ourselves, suppose we begin at that singularity fifteen billion years ago and follow the Universe forwards in time.

What follows is the history of the Universe in a nut-shell as we currently imagine it.[10] A standard model, if you like, which cosmologists use as a benchmark. Prior to the Planck time 10^{-43} s we know nothing of the state of space and time nor even if such familiar entities existed; neither quantum theory nor general relativity are valid before this time, the entire Universe is a quantum phenomenon, and a major extension of physics—the theory of quantum gravity—will be necessary before anything sensible can be said of these moments. When we reach times later than 10^{-43} s things become a little simpler, but only a little. The notions of space and time are unambiguous, the force of gravity is independent of the other forces of Nature and the quantum aspect of gravity becomes unimportant. Nevertheless, the force of gravity is still so strong that the random differences in the gravitational field that exist from place to place because matter is not smoothly distributed create forces that are large enough to produce elementary particles spontaneously. Energy is conserved in this process, the mass-energy needed to create the particles is extracted from the spatial gradient in the gravitational field strength, leaving it a little smoother than it was before the particles were 'created'. During the interval between 10^{-43} s and about 10^{-35} s the Universe is so hot that enough ambient radiation energy exists to create extremely massive analogues of the photon. In particular, the X particles, which we discussed in Chapter 5, are created and are able to interact with both quarks and leptons. The X particles carry both the electric and colour charges that precipitate electro-weak and strong interactions. Thus, X particles are able to mediate transmutations between particles that feel the strong interaction and those that feel only the electromagnetic and weak forces. They ensure that during this fleeting era there is a complete symmetry between all these interactions; all have effectively the same strength—a completely different state of affairs to that at more everyday 'biological' energies where X particles virtually never appear. The X particles mediate interactions that can change the value of quantities traditionally believed to be immutable like the baryon and lepton numbers of the Universe. Because the X particles and their antiparticles are so energetic they can produce both particles and also antiparticles and rapidly erase any favouritism for one over the other that may have existed initially. During this period there is thus complete symmetry between matter and antimatter. However, at the end of this period something remarkable happens: the X particles and their equally populous antiparticles decay into quarks and leptons and antiquarks and antileptons, but they do so at different rates. The result is that the Universe inherits a preponderance of quarks over antiquarks. The imbalance we observe now between matter and antimatter was built into the Universe at these early moments—roughly ten billion and one quarks for every ten billion antiquarks.[1] When the X particles disappear nothing remains to mediate transmutations between strongly interacting quarks and the leptons which feel only the electromagnetic and weak interactions. The result is that the strong and electro-weak interactions now become effectively different in strength and distinct in the spectrum of particles they act upon. The electromagnetic and the weak forces still remain of effectively equal strength because other fairly massive W and Z particles still exist to transmute particles that only feel the electromagnetic force into those that feel the weak force and *vice versa*. The W and Z bosons continue to be produced and the two forces remain of equal strength until the Universe cools down to about 10^{15} K which is its temperature when it has been expanding for a little less than a billionth of a second. After that the weak and electromagnetic forces become distinct in both range and strength and in the spectrum of particles on which they act. Shortly after that moment a number of important changes in the state of the Universe occur in quick succession: first, after a microsecond, the density drops to such an extent that the local forces which quarks exert on their nearest neighbours become strong enough to bind them irretrievably together in triplets to make hadrons like the proton and neutron and in pairs to make some of the many mesons. The vast bulk of the quarks end up as protons and neutrons and, because there are slightly more (roughly $1+10^9$ to 10^9) quarks than antiquarks, we end up with roughly one surviving proton or neutron for every 2×10^9 photons that are produced when the 10^9 quarks annihilate upon encountering 10^9 antiquarks. This magic ratio of about $\sim10^9$ photons per baryon is seen in the Universe today, it is a witness to the asymmetrical behaviour of the X particles when the Universe was just 10^{-35} seconds old. This 'phase transition' which transforms matter from the quark state to the nucleon state is completed after a billionth of a second of cosmic history, and until the Universe is one second old the protons and neutrons are transformed very rapidly into each other as a result of weak interactions with neutrinos, electrons and positrons. These interactions keep the number of protons equal to the number of neutrons. When the Universe is about one second old there is a vital change. The neutrinos find that the density of the Universe is now too low for them ever to encounter protons and neutrons, and the transmutations between neutrons and protons cease. Because the neutron is very slightly heavier than the proton (and so requires a little more energy to make) we are left with fractionally fewer neutrons than protons; about 13% neutrons for 87% protons to be precise. The neutrinos, meanwhile, have ceased to encounter other particles; they will remain in the Universe as a ghostly presence right until the present, about five hundred neutrinos in every cubic centimetre of space—in fact, many of them are passing through the reader's head at this very moment.

During the next three minutes of the Universe's life the lightest elements form. The temperature falls-off to a billion degrees and light nuclei can survive once they are formed by the nuclear fusion of free protons and neutrons. The population of neutrons and protons quickly gets burnt into deuterium nuclei and these are rapidly bound together in pairs by the strong nuclear force to form helium-4. Almost all the neutrons end up in helium-4 and the final fraction of the mass of the Universe in the form of helium is about $2\times13\%=26\%$ (since each helium nucleus contains two protons and two neutrons). The rest is almost entirely hydrogen nuclei—protons (the temperature is still too high for atoms to exist) with just a few traces of deuterium, and helium-3: these other nuclei comprise about $10^{-3}\%$ of the primordial nuclei. Although minute, the abundances of these nuclei have been measured by astronomers today.[8] The measurements confirm the predictions of the standard model.

From the time it is three minutes old until about a million years after its apparent beginning the Universe is dominated by the influence of the photons. They interact rapidly with the electrons and charged nuclei (it is still far too hot for any atoms to form) and the huge pressure they exert rapidly disperses any protostars or protogalaxies that might begin to form under the attractive forces of gravity. When the Universe has aged a million years or so, the photon pressure finally ceases to be important and the density of the nuclei becomes predominant; the radiation era is over and the Universe is no longer in a plasma state. At about the same time the photons experience the same fate as did the neutrinos after one second of expansion: they cease to interact with the particles of matter because the cosmic medium has become too rarefied. They propagate towards us through space and time experiencing occasional absorption and emission *en route* and arrive with a temperature of ~ 3 K and a spectral structure that reflects the state of thermal equilibrium in the Universe when they were last scattered in the Universe's youth. The temperature is now low enough for atoms to form. At first only simple hydrogen atoms form, but then the first molecules of hydrogen appear, followed by heavier and more complex states as chemistry begins in the cosmic medium. After several billion years, great agglomerations of matter begin to be noticeable, drawn together by gravitational attraction. They collapse and fragment, creating protogalaxies, globular clusters and intergalactic debris. Finally, after ten billion years the galaxies condense, quasars form and embryonic stars begin to shine within the dusty regions of galaxies so creating the heavier elements of which living creatures are made. Our parent interstellar cloud forms, and within it a protosolar nebula condenses to form the planetary system we call the solar system orbiting a star called the Sun.

We shall now back-track and describe in more detail the simplest possible model of the expanding universe. Fortunately, it also gives a picture that very closely resembles the observed Universe. The model is 'simple' because it assumes at the outset that the Universe is isotropic and spatially homogeneous.[11] The astronomical evidence confirms that this is an extremely good approximation to reality—why it is so is one of the outstanding problems of modern cosmology to which we shall return later. We shall also discuss the observational evidence for this type of Big Bang cosmological model.

6.2 The Hot Big Bang Cosmology

> The evolution of the world may be compared to a display of fireworks that has just ended: some few red wisps, ashes and smoke. Standing on a cooled cinder, we see the slow fading of the suns, and we try to recall the vanished brilliance of the origin of the worlds.
>
> G. Lemaître

The expansion of the Universe can be described by a simple Newtonian argument. Consider a mass m at radius r of a homogeneous and isotropic

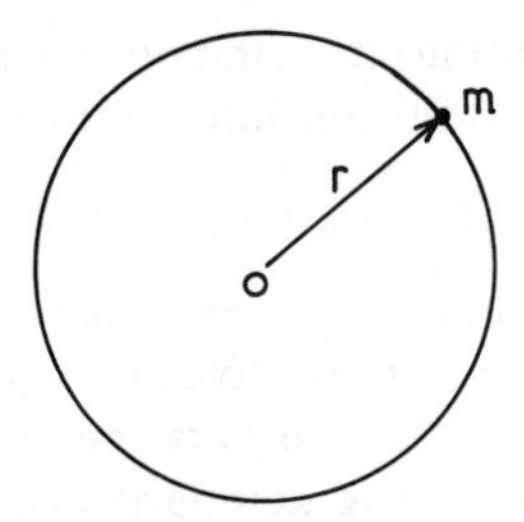

Figure 6.1. Radial notion of receding mass point, m. The sphere of radius r centred on O contains a homogeneous and isotropic mass density $\rho(t)$.

sphere, Figure 6.1. Now expand all length scales by a scale factor $R(t)$; the initial radius vector r_0 will transform to r where

$$r = R(t)r_0 \tag{6.1}$$

Therefore the velocity of expansion, $v = \dot{r}$, is just

$$v = \dot{r} = \frac{\dot{R}}{R}r \equiv Hr \tag{6.2}$$

where $H \equiv \dot{R}/R$ is the Hubble parameter.[3] If the spherical ball represents our Universe then the present value of the expansion rate, H_0, is called Hubble's constant. We can write down an energy conservation equation for the motion of m, (kinetic energy + potential energy = constant). If ρ is the average density of the ball of matter, this yields,[12,13]

$$\frac{\dot{R}^2}{2} - \frac{4\pi}{3}G\rho R^2 = \frac{-k}{2} = \text{constant} \tag{6.3}$$

This is just Friedman's equation[2] for the expansion of a homogeneous and isotropic expanding universe filled with pressure-free material. The corresponding general relativistic space-time metric for a homogeneous and isotropic space is:

$$ds^2 = -dt^2 + R^2(t)\left[\frac{dr^2}{1-kr^2} + r^2(d\theta^2 + \sin^2\theta\, d\phi^2)\right] \tag{6.4}$$

where (r, θ, ϕ) are comoving coordinates (that is, an observer expanding with the Universe maintains constant (r, θ, ϕ) coordinates) and the constant k is equal to 0 or ± 1.

The model of the expansion given by (6.1) and (6.2) assumes the expansion of the Universe to be isotropic and homogeneous. If it were not symmetric in this way, then we would require a 3×3 matrix of 'Hubble constants' to describe the velocity gradients

$$v_i = H_{ij}r_j, \qquad i, j = 1, 2, 3 \tag{6.5}$$

The matrix H_{ij} can be split into its symmetric and antisymmetric parts and the symmetric part analysed further into its trace and tracefree parts, so

$$H_{ij} = \sigma_{ij} + \omega_{ij} + \tfrac{1}{3}H\,\delta_{ij} \tag{6.6}$$

where the antisymmetric part ($\omega_{ij} = -\omega_{ji}$) represents rotation of the expanding medium, the symmetric tracefree part ($\sigma_{ij} = \sigma_{ji}$, $\sigma^i_i = 0$) represents shear and the scalar trace, H, represents the volumetric expansion; δ_{ij} is the Kronecker delta symbol. The standard Friedman model of the Universe and the Hubble law (6.2) assume σ_{ij} and ω_{ij} are much less than $H\,\delta_{ij}$; an assumption borne-out by observations.[14]

The fluid continuity equations describe the evolution of the pressure, p, and density, ρ, of the expanding Universe,

$$\frac{d}{dt}\left(\frac{4\pi}{3}\rho R^3\right) = -p\frac{d}{dt}\left(\frac{4\pi}{3}R^3\right) \tag{6.7}$$

so

$$\dot{\rho} = -3(\rho + p)\frac{\dot{R}}{R} \tag{6.8}$$

To solve this equation we require an equation of state for the medium, $\rho(p, T)$ where T is the temperature. The most important equation of state is that of non-relativistic matter or 'dust', with[15] $p = nk_BT$. For galaxies, $v \sim 250$ km s$^{-1} \ll c$, therefore $p \sim \rho v^2 \ll \rho c^2$ and $p = 0$ is an excellent approximation, so by (6.8)

$$p = 0 \rightarrow \rho \propto R^{-3} \tag{6.9}$$

In the early, hot phase of the Universe the cosmic medium is an equilibrium radiation gas with $\rho \sim T^4$, so

$$p = \rho/3 \propto T^4 \propto R^{-4} \tag{6.10}$$

In general, with a perfect fluid equation of state

$$p = (\gamma - 1)\rho; \qquad 1 \leqslant \gamma \leqslant 2 \tag{6.11}$$

we have

$$\rho \propto R^{-3\gamma} \tag{6.12}$$

It is useful to define a number of parameters that help to quantify the observed expansion of the Universe. The present value of the Hubble constant is inferred to lie in the range.

$$H_0 = 75 \pm 25 \text{ km s}^{-1}\text{ Mpc}^{-1} \tag{6.13}$$

We shall measure H_0 in dimensionless form,

$$h_0 = H_0/100 \text{ km s}^{-1}\text{ Mpc}^{-1} \tag{6.14}$$

The effect of gravity on the Universe is to decelerate the expansion. The deceleration parameter, q_0, is defined by.[12]

$$q_0 = -\left(\frac{\ddot{R}R}{\dot{R}^2}\right)_0 \tag{6.15}$$

And observations indicate[16]

$$0 \lesssim q_0 \lesssim 2 \tag{6.16}$$

If the cosmological constant is zero then (6.2), (6.3) and (6.15) give, when evaluated today

$$H_0^2(2q_0 - 1) = kR_0^{-2} \tag{6.17}$$

and from the time-derivative of (6.3) evaluated today,

$$2q_0 = \frac{8\pi G}{3H_0^2}\rho_0 \equiv \frac{\rho_0}{\rho_c} \tag{6.18}$$

where the critical density, ρ_c, is the largest density the Universe can possess and still expand for all future time

$$\rho_c \equiv \frac{3H_0^2}{8\pi G} = 2\times 10^{-29}\, h_0^2 \text{ gm cm}^{-3} \tag{6.19}$$

We can measure ρ_0 in units of the critical density by defining

$$\Omega_0 \equiv \frac{\rho_0}{\rho_c} = \frac{\text{Potential energy of Universe}}{\text{Kinetic energy of Expansion}} \tag{6.20}$$

There are three simple Friedman solutions of (6.3) when $p = 0$ according as $k = 0, \pm 1$:

(a) *Flat model*, ($k = 0$, $q_0 = 0.5$, $\Omega_0 = 1$):

$$R(t) = \left(\frac{9GM}{2}\right)^{1/2} t^{2/3}, \qquad H = \frac{\dot{R}}{R} = \frac{2}{3t} \tag{6.21}$$

$$\rho(t) = (6\pi G t^2)^{-1} \tag{6.22}$$

where the quantity $4\pi\rho R^3/3 \equiv M$ is constant.

(b) *Open model*, ($k = -1$, $q_0 < 0.5$, $\Omega > 1$):

$$R(\eta) = GM(\cosh\eta - 1); \qquad t = GM(\sinh\eta - \eta) \tag{6.23}$$

$$H > \frac{2}{3t} \tag{6.24}$$

(c) *Closed model*, ($k=+1$, $q_0>0.5$, $\Omega_0>1$):

$$R(\eta)=GM(1-\cos\eta); \qquad t(\eta)=GM(\eta-\sin\eta) \tag{6.25}$$

$$H<\frac{2}{3t} \tag{6.26}$$

where in both cases η is a time parameter. In the closed model, H changes sign at $\eta=\pi$ when the expansion reaches a maximum radius. Recollapse to a second singularity occurs at $\eta_\star=2\pi$ when $t_\star=2\pi GM$. The total lifetime of the closed model is

$$t_\star=\frac{2\pi q_0}{H_0(2q_0-1)^{3/2}}=\frac{\pi\Omega_0}{H_0(\Omega_0-1)^{3/2}} \tag{6.27}$$

and, although always finite, can be arbitrarily long if Ω_0 is close enough to one. In Figure 6.2 the three alternative pictures of the Universe's evolutionary history are illustrated according to the solutions (a), (b) and (c).

The wavelength of radiation, along with all other lengths unaffected by other forces of Nature, scales as $\lambda\propto R$ under the expansion (6.1) and so its energy $E=\lambda^{-1}\propto R^{-1}$ falls, or 'reddens', doing work against the expansion of the Universe. Suppose we define two wavelengths: λ_e, the wavelength of a particular spectral line in the rest frame of its source and λ_0, the wavelength of that spectral line measured by an observer moving with velocity v relative to the source along the line of sight. The observed spectral shift, z, is defined as

$$z\equiv\frac{\lambda_0-\lambda_e}{\lambda_e} \tag{6.28}$$

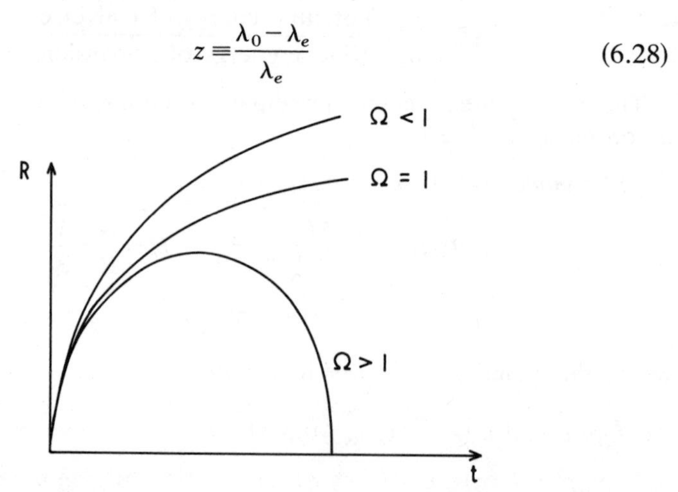

Figure 6.2. Time-evolution of the expansion scale factor in open ($\Omega_0<1$), flat ($\Omega_0=1$), and closed ($\Omega_0>1$), Friedman model universes with zero cosmological constant.

If $z>0$ then it is termed a 'redshift'; if $z<0$, a 'blueshift'. Beyond the Local Group of nearby galaxies we always observe $z>0$ and the most distant observed quasars have $z \sim 3.7$.

The age of the Friedman universes can be calculated exactly in terms of H_0 and Ω_0 (or equivalently q_0) to give[12]

$$t_u = H_0^{-1} f(\Omega_0) \tag{6.29}$$

where f is a messy function that can be found exactly, but for our purposes it is sufficient to note that $f'(\Omega)<0$ and $f(0)=1$, $f(1)=2/3$ and $f(2)=0.57$; we find (6.13) and (6.16) give $t_u \sim 13$–20 Gyr which can be compared with lower bounds provided by dating of the oldest terrestrial rocks ($\sim$3.9 Gyr), lunar and meteoritic material ($\sim$4.6 Gyr), and stars in globular clusters ($\sim$8–15 Gyr).[17]

The density of matter inferred to reside in the Universe shows a strong correlation with the scale surveyed. The following are typical measurements[18] of Ω_0,

$$\begin{aligned} &\Omega \text{ (solar neighbourhood)} &&= 0.004 \pm 0.007 \\ &\Omega \text{ (galaxies)} &&= 0.006 - 0.014 \\ &\Omega \text{ (binary galaxies and groups)} &&= 0.04 - 0.13 \\ &\Omega \text{ (clusters)} &&= 0.2 - 0.7 \end{aligned} \tag{6.30}$$

The trend is clear and if we look at a tabulation of the mass to luminosity ratios in different objects we see that $M/l \propto r$ implying that considerable quantities of dark material are associated with groups and clusters of galaxies,[19] but there could be large contributions from undetected hot gas in these structures.

In this connection we recall the evidence from galactic rotation curves which exhibit $v_{\text{rot}} \to$ constant at large distance, r, from the centre of spiral galaxies. For a system in rotational equilibrium

$$\frac{GM}{r} \sim v_{\text{rot}}^2 \tag{6.31}$$

and so the observations indicate $\rho \propto r^{-2}$ at large r. Thus some sort of dark material must be present. The form of this material is still a mystery, although we shall see later that various species of elementary particle may provide natural explanations for such a 'halo' of dark material.

The rough value of the density contrast of particular structures compared with the average density of the Universe is shown in Figure 6.3 along with their length and mass scales.

Because the luminous matter content of the Universe exhibits such overt structure it is useful to employ some quantitative measure of the clustering. The simplest statistic to use is the two-point correlation function $\xi(r)$ which gives the *excess* probability, over random, of finding a

Object	$\rho/\bar{\rho}$	λ	$M/M_\odot$
Globular star cluster	10^{10}	0.1 Kpc	10^5
Galaxy	10^6	30 Kpc	10^{11}
Galaxy cluster	10^3	1 Mpc	10^{13}
Supercluster	2–3	30 Mpc	10^{17}

Figure 6.3. Typical density enhancement, ρ, over the mean background density of the Universe, $\bar{\rho}$, displayed in observed astronomical aggregates of varying mass (in solar units) and of average linear dimension, λ.

galaxy at a distance r from another galaxy picked at random.[20–21] If δP is the infinitesimal probability of finding that galaxy in a volume δV in a field of mean number density $\bar{n}$, then

$$\delta P = \bar{n}(1+\xi(r))\,\delta V \tag{6.32}$$

Thus $\xi(r)\in[-1,\infty)$ and $\xi>0$ implies clustering; $\xi=0$ is a purely random, or Poissonian distribution, and $\xi<0$ implies anticlustering, that is, 'voids' in the distribution. In practice, we usually observe the angular correlation function on the celestial sphere and use the luminosity function to deproject and infer the three-dimensional distribution $\xi(r)$ although new surveys now exist which include measured redshifts and hence distances can be inferred using Hubble's Law (6.2); Peebles,[21] finds

$$\xi(r) = \left(\frac{4h_0^{-1}\,\text{Mpc}}{r}\right)^{1.8};\ 10\ \text{kpc}<r<10\ \text{Mpc} \tag{6.33}$$

The clustering is well-developed. $\xi\gtrsim 1$, on a scale corresponding to masses below

$$M_\star \sim 5\times 10^{14}\Omega_0\, h_0^{-1} M_\odot \tag{6.34}$$

where $M_\odot$ is the mass of the sun. This is the characteristic scale that separates weak clustering ($M\gtrsim M_\star$) from well-developed clustering ($M\lesssim M_\star$). One of the aims of any theory of galaxy and cluster formation should be an explanation of the magnitude of $M_\star$. The observations (6.33) imply that the Universe can be considered homogeneous for $r\gtrsim 50$ Mpc and in fact counts of radio sources indicate that fluctuations from homogeneity in their number density over volumes $\sim(1\ \text{Gpc})^3$ are less than one per cent.[22]

From the cosmologist's point of view the most important ingredients of the cosmic matter density are the abundances of light elements (helium, deuterium and lithium) relative to hydrogen since these particular light elements are synthesized by nuclear reactions in the early ($t\sim 1$–10^3 s) stages of the Universe rather than in stars. The results of the synthesis process depend upon the dynamics and content of the universe during these early moments in a very sensitive way and so provide us with a unique observational probe of the Big Bang.

Typical observations of helium-4 are those in globular star clusters where a mass fraction $Y = 0.22 \pm 0.04$ is observed, galactic HII regions with low metal abundances ($Z < 0.02$; 'metals' are elements heavier than helium-4) where $Y = 0.23 \pm 0.02$ is observed, and galactic HII regions with normal ($Z > 0.02$) metals yielding $Y = 0.30 \pm 0.02$. Since we are interested in deducing the *primordial* Y value we must slightly renormalize these values to account for the increase in the primordial value Y_p due to stellar processing. The above-quoted data illustrate the trend; where metals are low there has been little stellar processing and only a small contribution to Y_p by stellar processes; however, where Z is higher so is Y. Peimbert[23] suggests a decomposition of Y into the sum of a primordial part Y_p and an increment ΔY due to processing. A plot of Y versus Z indicates $\Delta Y / \Delta Z = 2 \pm 1$ which leads to a deduction of the primordial Y_p value

$$Y_p = 0.23 \pm 0.02 \tag{6.35}$$

The most reliable measurements made of deuterium are those made by the Copernicus satellite which observed the Lyman absorption lines of atomic deuterium and hydrogen in light from hot OB stars near the Sun. The importance of these interstellar measurements is their detection of deuterium in *extra-molecular* form. The preferential incorporation of deuterium into molecules and the complexities of the associated chemistry make deductions of the deuterium abundance from that of deuterated molecules (for example in sea water or on Jupiter's surface) very uncertain. The Copernicus measurement gives a mass fraction[8,24] of $X(D) = (2.5 \pm 1.5) \times 10^{-5}$. In view of the possibility of the destruction of deuterium in stars following primordial nucleosynthesis, it is safe to regard its primordial value as laying somewhere in the interval

$$X(D) = 2 \times 10^{-4} - 2 \times 10^{-5} \tag{6.36}$$

Before leaving the question of the matter content of the Universe we should remark that there is no evidence for any primary sources of cosmic antimatter.[10] The only antiparticles detected directly are cosmic ray antiprotons ($\bar{p}$). At energies ~5–12 Gev we observe $\bar{p}/p \sim 5 \times 10^{-4}$ as would be expected from secondary interactions of cosmic ray protons ($p + p \rightarrow p + \bar{p} + p + p$) *en route* from source to detector. However, there have been recent claims to observe a $\bar{p}/p$ flux of about the same level at low energies ~130–320 MeV. This flux is hundreds of times larger than would be expected with such energies. The interpretation (and perhaps even the correctness) of this measurement is still the subject of some debate but it is quite likely that the primary cosmic rays have been significantly decelerated *en route* towards us by the Galactic magnetic field. The apparent absence of cosmic antimatter is a striking feature of the Universe.[25]

The discovery of the microwave background radiation by Penzias and Wilson in 1965 began a new era of cosmological thinking. The background radiation has turned out to be a sort of cosmic 'Rosetta stone' on which is inscribed the record of the Universe's past history in space and time. By interpreting the spectral structure of the radiation we can learn of violent events in the Universe's distant past.

The microwave contribution is far and away the dominant contribution to the radiation background but its mass density is about one millionth that in the luminous matter today.

The initial measurement of Penzias and Wilson provided just a single data point of the spectrum, but since then experiments of ever-increasing precision have shown the microwave spectrum to be Planckian to a very high degree of accuracy.[26] This tells us that the radiation must originate in the distant past ($z \gg 200$) when the cosmic density was high enough to relax it to an equilibrium form over the observed waveband. The best map of the spectrum has been obtained by Woody and Richards using a liquid-cooled, balloon-borne spectrophotometer, (see Figure 6.4). The best fit thermal spectrum has a present temperature

$$T_{\gamma 0} = 2.9 \pm 0.1 \text{ K} \tag{6.37}$$

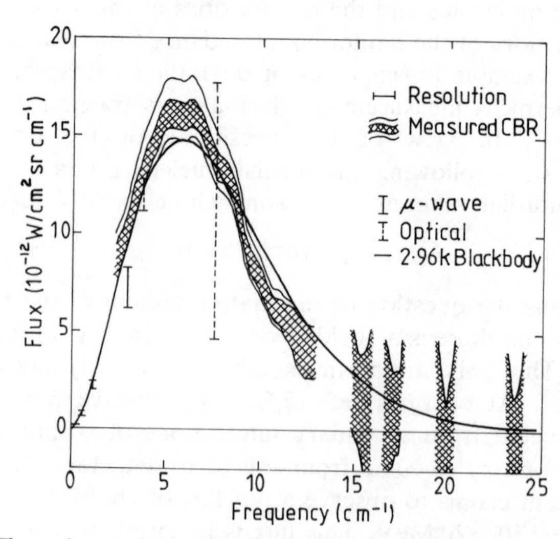

Figure 6.4. The microwave background spectrum showing the data taken by Woody and Richards.[26] The cross-hatched areas and solid lines bound the experimental errors. A 2.96 K black body spectrum is superimposed.

which corresponds to a black-body density (our units have $k_B = \hbar = c = 1$) of

$$\rho_{\gamma 0} = \frac{\pi^2}{15} T_{\gamma 0}^4 = 5.9 \times 10^{-34} \left(\frac{T_{\gamma 0}}{2.9 \text{ K}}\right) \text{gm cm}^{-3} \tag{6.38}$$

This is equivalent to a density parameter, (6.20), of

$$\Omega_{\gamma 0} = 3.0 \times 10^{-5}\, h_0^{-2} \left(\frac{T_{\gamma 0}}{2.9 \text{ K}}\right)^4 \tag{6.39}$$

and the number density of black-body photons is

$$n_\gamma = 2\frac{\zeta(3)}{\pi^2} T^3 = \frac{2.404}{\pi^2} T^3 \sim 20 T^3 \text{ cm}^{-3}\text{ K}^{-3} \tag{6.40}$$

where ζ is the Riemann zeta function. Since the present nucleon density in the Universe is, (6.20),

$$n_B = 1.1 \times 10^{-5} \Omega h_0^2 \text{ cm}^{-3} \tag{6.41}$$

we see that the number of baryons per photon in the Universe is

$$\left(\frac{n_B}{n_\gamma}\right)_0 = 2.2 \times 10^{-8} \Omega h_0^2 \tag{6.42}$$

The baryon to photon ratio given by (6.42) should be regarded as a new fundamental parameter which ought to be explained by any complete cosmological theory. The quantity n_B/n_γ is not quite constant during the expansion history of a homogeneous and isotropic Universe. It changes slightly when elementary particles annihilate into photons—for example, when electron-positron pairs annihilate after ~25 s. However, a closely related quantity—which is conserved in the Friedman universe, is the entropy per baryon (or, specific entropy), S. The photon entropy $\mathbf{s}_\gamma$ is, (recall $k_B \equiv 1$),

$$\mathbf{s}_\gamma = \tfrac{4}{3}\rho_\gamma T^{-1} = \frac{4\pi^2}{45} T^3 \tag{6.43}$$

and therefore, defining

$$S \equiv \frac{\mathbf{s}_\gamma}{n_B} \tag{6.44}$$

we see that

$$\frac{n_\gamma}{n_B} \simeq 0.28 S \tag{6.45}$$

If non-equilibrium phenomena of a violent nature, for example, shock waves, black hole formation or anisotropy damping, occur during the history of the Universe then S will increase (unless for some reason n_B decreases) since the Second Law of thermodynamics ensures $\dot{S} \geq 0$.

Although the present density of matter $\sim 10^{-30}$ gm cm^{-3} greatly exceeds that in the microwave background radiation $\sim 10^{-34}$ gm cm^{-3}, this domination by matter could not always have been the case. When the universe expands by a scale factor $R(t)$ the matter density falls as R^{-3} but the radiation density falls more rapidly because the energy of each quantum is inversely proportional to its wavelength and that also scales as $R(t)$. Therefore, $\rho_\gamma \propto R^{-4}$, and at some time in the past the value of ρ_γ must have exceeded ρ_B. This occurs at a moment t_{eq}; from the present value of ρ_γ and ρ_B given in (6.30) and (6.38) we can calculate the temperature, age, and density of the Universe at this moment when the densities of matter and radiation are equal. We obtain

$$t_{eq} \simeq 2.2 \times 10^{10} \Omega^{-2}\,\mathrm{s} \tag{6.46}$$

$$T_{eq} \simeq 1.2 \times 10^{5} \Omega\,\mathrm{K} \tag{6.47}$$

Another important cosmological epoch is the time before which the radiation temperature is high enough to ionize all atoms. The ionization temperature is ~ 3000 K and this temperature is attained at a 'recombination' time $t_{rec} \sim 10^{13}$ s.

The value of S is responsible for the gross pattern of cosmic history. For example, the end of the radiation-dominated phase of the early Universe, determined numerically in (6.46) and (6.47), is fixed by S as

$$t_{eq} \sim 2 \times 10^{-4} \left(\frac{m_p}{m_N}\right)^2 t_p S^2 \sim 2 \times 10^{-5} S^2\,\mathrm{sec} \tag{6.48}$$

and the epoch, t_{rec}, when the primeval plasma combines into atoms, leaving the photons collisionless is given by

$$t_{rec} \sim 10^8\, S^{1/2}\,\mathrm{sec} \tag{6.49}$$

The 'coincidence' that, in our Universe $t_{rec} \sim t_{eq} \sim 10^{13}$ s is a consequence of the fact that $S \sim 10^9$. Lastly, we note that during the radiation era of the Universe ($t < t_{eq}$) the number of particles N_D, in a Debye volume of the Universe, is

$$N_D \sim 8 \times 10^3\, S^{1/2} \gg 1 \tag{6.50}$$

Therefore, there is good, quasi-neutral, collective behaviour and the early Universe behaves as a good plasma before t_{rec}.

We shall refer to the period $t < t_{eq}$ when $T > T_{eq}$ as the *radiation era* and it is in this period that the most interesting interconnections between cosmology and elementary particle physics lie. At times prior to

t_{eq} the curvature parameter k is negligible in the Friedman equation and the expansion of an isotropic, homogeneous Universe filled with radiation has the simple solution[10,12]

$$R(t) \propto t^{1/2}; \qquad \rho_\gamma = \frac{3}{32\pi G t^2}; \qquad H = \frac{1}{2t} \tag{6.51}$$

The energy density in the radiation-dominated phase of the early universe is dominated by black-body radiation. There may exist several different equilibrium species of elementary particles (either interacting or non-interacting) and in general we write

$$\rho = \frac{g(T)}{2} \qquad \rho_\gamma = \frac{g\pi^2}{30} T^4 = 3p \tag{6.52}$$

where g is the number of helicity states—the effective number of degrees of freedom—so since in general this counts bosons and fermions,

$$g = g_b + \tfrac{7}{8} g_f \tag{6.53}$$

where $b \equiv$ bosons and $f \equiv$ fermions.

During the radiation era (6.52), (6.3) and (6.8) yield a solution which, when combined with

$$T \propto R^{-1} \tag{6.54}$$

gives the temperature-time adiabat as

$$\frac{t}{1s} = 2.42 g^{-1/2} \left(\frac{1\ \text{MeV}}{T}\right)^2 \tag{6.55}$$

In Planck units ($c = \hbar = 1$, $m_p = G^{-1/2} \simeq 10^{-5}$ gm $\simeq 10^{19}$ GeV, $k_B = 1$) the temperature-time adiabat is

$$t \sim 0.3 m_p g^{-1/2} T^{-2} \tag{6.56}$$

This establishes the essential quantitative features of the 'standard' hot Big Bang model. Some further pieces of observational evidence that support it will be introduced later. For the moment we stress its special character: it is homogeneous and isotropic, has an entropy per baryon close to 10^9 and is expanding at a rate that is irresolvably close to the critical divide that separates an infinite future from a finite one. We now turn to examine some of these key properties of the Universe with a view to determining which of them are important for the process of local biological evolution. Thus will enable us to identify those aspects of the Universe, our discovery of which may in some sense be necessary consequences of the fact that we are observers of it.

6.3 The Size of The Universe

I don't pretend, to understand the
Universe—its a great deal bigger than
I am.

T. Carlyle

In several other places we have used the fact of the Universe's size as a striking example of how the Weak Anthropic Principle connects aspects of the Universe that appear, at first sight, totally unrelated.[27] The meaning of the Universe's large size has provided a focus of attention for philosophers over the centuries. We find a typical discussion in *Paradise Lost* where Milton evokes Adam's dilemma: why should the Universe serve the Earth with such a vast number of stars, all[28]

. . . merely to officiate light
Round this opacious earth, this punctual spot
One day and night, in all their vast array
Useless besides?

Perplexed, he tells Raphael that he cannot understand

How nature, wise and frugal, could commit
Such disproportions, with superflous hand
So many nobler bodies to create?

The archangel replies only that the 'Heaven's wide circuit' is evidence of 'The Maker's high magnificence'.

Adam's concern was shared by an entourage of philosophers, ancient and modern: if life and mind are important, or unique, why does their appearance on a single minor planet require a further 10^{22} stars as a supporting cast? In the past, as we saw in Chapter 2, this consideration provided strong circumstantial evidence against naïve Design Arguments. However, the modern picture of the expanding universe that we have just introduced renders such a line of argument, at best, irrelevant to the question of Design.

Einstein's special theory of relativity unified the concepts of space and time into a single amalgam: *space-time.*[29] The existence of an invariant quantity in Nature with the dimensions of a velocity, (the velocity of light, *in vacuo*, c) places space and time on an equal footing. The size of the observable universe, λ, is inextricably bound-up with its age, t_u, through the simple relation

$$\lambda = ct_u \tag{6.57}$$

The expanding Big Bang model, (6.22), allows us to calculate the total mass contained in this observable universe,

$$M_u = \frac{4\pi}{3}\rho\lambda^3 \sim c^3G^{-1}t_u \tag{6.58}$$

which yields,

$$M_u \sim 10^5 \left(\frac{t_u}{1s}\right) M_\odot \tag{6.59}$$

These relations display explicitly the connection between the size, mass and age of an expanding universe. If our Universe were to contain just a single galaxy like the Milky Way, containing 10^{11} stars, instead of 10^{12} such galaxies, we might regard this a sensible cosmic economy with little consequence for life. But, a universe of mass $10^{11}M_\odot$ would, according to (6.59) have expanded for only about a month. No observers could have evolved to witness such an economy-sized universe.

An argument of this sort, which exploits the connection between the age of the Universe, t_u, and the global density of matter within it, was first framed by Idlis and Whitrow. Later, it was stressed by Dicke and Wheeler as an explanation for Dirac's famous 'Large number coincidences',[27,30] (see Chapter 4).

A minimum time is necessary to evolve astronomers by natural evolutionary pathways and stars require billions of years, $(\sim\alpha_G^{-1}m_N^{-1})$, to transform primordial hydrogen and helium into the heavier elements of which astronomers are principally constructed. Thus, only in a universe that is sufficiently mature, and hence sufficiently large, can 'observers' evolve. In answer to Adam's question we would have to respond that the vastness of 'Heavens' wide circuit' is necessary for his existence on Earth.

Later, we shall see that the use of (6.58) in this way relies upon particular properties of our Universe like small anisotropy, close proximity to the critical density and simple space-time topology.

It is also interesting to recall that even in 1930 Eddington[31] entertained an Anthropic interpretation of cosmological models possessing long-lasting static phases due to the presence of a non-zero cosmological constant. He pointed out that if a period of $\sim 10^{10}$ years had elapsed from the static state, astronomers would have to 'count themselves extraordinarily fortunate that they are just in time to observe this interesting but evanescent feature of the sky [the dimming of the stars]'.

6.4 Key Cosmic Times

> Since the universe is on a one-way slide towards a state of final death in which energy is maximally degraded, how does it manage, like King Charles, to take such an unconscionably long time a-dying.
>
> F. Dyson

The hot Big Bang cosmological model contains seven times whose relative sizes determine whether life can develop and continue. The first six are all determined by microscopic interactions:

(a) t_{ev}: the minimum time necessary for life to evolve by random mutation and natural selection. We cannot, as yet, calculate t_{ev} from first principles. (See Section 8.7 for further discussion of this time-scale.)

(b) $t_\star$: the main-sequence stellar lifetime, necessary to evolve stable, long-lived, hydrogen-burning stars like the Sun and $t_\star \sim \alpha^2(m_N/m_e)^2\alpha_G^{-1}m_N^{-1} \sim 10^{10}$ yr.

(c) t_{eq}: the time before which the expansion dynamics of the expanding universe are determined by the radiation, rather than the matter content of the Universe. It depends on the observed entropy per baryon, S, and thus

$$t_{eq} \sim S^2\alpha_G^{-1/2}m_N^{-1} \sim 10^{12}\text{ s}$$

(d) t_{rec}: the time after which the expanding Universe is cool enough for atoms and molecules to form,

$$t_{rec} \sim S^{1/2}\alpha^{-3}\alpha_G^{-1/2}(m_N/m_e)^{1/2}m_e^{-1} \sim 10^{12}\text{ s}$$

(e) τ_N: the time for protons to decay; according to grand unified gauge theories this is $\tau_N \sim \alpha^{-2}m_X^4 m_N^{-5} \sim 10^{31}$ yr

(f) t_p: the Planck time, determined by the unique combination of fundamental constants $G, \hbar, c$ having dimensions of a time, $t_p = (G\hbar/c^5)^{1/2} \sim 10^{-43}$ s

(g) t_u: the present age of the Universe, $t_u \simeq (15 \pm 3) \times 10^9$ yr.

Of these fundamental times, only two are not expressed in terms of constants of Nature—the current age, t_u, and the biological evolution time, t_{ev}. From the list (a)–(g) we can deduce a variety of simple constraints that must be satisfied by any cognizable universe. If life requires nuclei and stellar energy sources then we must have

$$t_u > \tau_N > t_{ev} > t_\star > t_{rec} \tag{6.60}$$

We shall see that in order for galaxies to form—and perhaps therefore, stars—we require $t_\star > t_{eq}$. We notice, incidentally, that

$$\frac{t_{eq}}{t_{rec}} \sim S^{3/2}\alpha^3\left(\frac{m_e}{m_N}\right)^{3/2} \tag{6.61}$$

and the fact that $t_{eq} \sim t_{rec} \sim 10^{12}$ s in our Universe is an immediate consequence of the fact that we have

$$S \sim \alpha^{-2}\left(\frac{m_N}{m_e}\right) \sim 10^9 \tag{6.62}$$

The condition that atoms and chemistry exist before all stars burn out requires $t_\star > t_{rec}$, and leads to an upper bound on the value of S of

$$S \lesssim \alpha^{10}\left(\frac{m_N}{m_e}\right)\alpha_G^{-1} \tag{6.63}$$

whilst the condition that stellar lifetimes exceed the radiation-dominated phase of the Universe during which galaxy and star formation is suppressed yields the requirement

$$S \lesssim \alpha\left(\frac{m_N}{m_e}\right)\alpha_G^{-1/4} \tag{6.64}$$

The most powerful constraint, which was also derived in Chapter 5, arises if the proton is unstable with a lifetime of order that predicted by grand unified theories. In order that the proton lifetime exceed that of stars, $t_\star$, we require

$$S \lesssim \left(\frac{m_N}{m_e}\right)^{1/2} \exp(0.25\alpha^{-1}) \tag{6.65}$$

Again, we find the ubiquitous trio of dimensionless quantities, m_N/m_e, α_G and α appearing; however, on this occasion it is a property of the entire Universe that they place constraints upon rather than the existence of local structures, as was their role in Chapter 5. So far, the parameter S giving the number of photons per baryon in the Universe has been treated as a free parameter that is an initial condition of the Universe and whose numerical value can only be determined by observation. Later, we shall see that grand unified gauge theories offer some hope that this quantity can be calculated explicitly in terms of other fundamental parameters like α and α_G.

6.5 Galaxies

> If galaxies did not exist we would have
> no difficulty in explaining the fact.
>
> W. Saslaw

We have already shown that the gross character of planetary and stellar bodies is neither accidental nor providential, but an inevitable consequence of the relative strengths of strong, electromagnetic and gravitational forces at low energies. It would be nice if a similar explanation could be provided for the existence and structure of galaxies and galaxy clusters. Unfortunately, this is not so easily done. Whereas the structure of celestial bodies up to the size of stars is well understood—aided by the convenient fact that we live on a planet close by a typical star—the nature of galaxies is not so clear-cut. It is still not known whether galaxies owe their sizes and shapes to special conditions at or near the beginning of the Universe (if such there was) or whether these features are conditioned by physical processes in the recent past. To complicate matters further, it is now suspected that the large quantities of non-luminous material in and around galaxies is probably non-baryonic in form. If the electron neutrino were found to possess a non-zero rest mass ~30 eV as claimed by recent experiments[32] then our whole view of galaxy formation and clustering would be affected. For simplicity, let us first describe the simplest situation wherein we assume that no significant density of non-baryonic material exists.[21]

We imagine that in the early stages of the Big Bang some spectrum of density irregularities arises which we describe by the deviation of the density ρ from the mean $\bar{\rho}$ using

$$\frac{\delta\rho}{\rho} \equiv \frac{\rho - \bar{\rho}}{\bar{\rho}} \tag{6.66}$$

In general, we would expect $\delta\rho/\rho$ to vary as a power-law in mass so no mass scale is specially picked out, say as

$$\frac{\delta\rho}{\rho} \propto M^{-n}; \qquad n > 0 \tag{6.67}$$

Cosmologists now ask whether some damping process will smooth out the smallest irregularities up to some particular mass, M_D. If this occurs the mass scale M_D might show up observationally in the Universe as a special one dividing large from moderate non-unofirmity.

If the initial irregularities involve only non-uniformities in the matter content of the universe, but not in the radiation, they are called *isothermal* and isothermal irregularities will survive above a mass determined by the distance sound waves can travel whilst the Universe is dominated by radiation,[33] ($t \leq t_{eq}$). This gives a mass close to that of globular clusters $\sim 10^6 M_\odot$.

$$M_{Di} \approx S^{1/2} \alpha_G^{-3/2} m_N \tag{6.68}$$

Another type of density non-uniformity arises if both the matter and radiation vary from place to place isentropically. These fluctuations are called *adiabatic*. The survival of adiabatic inhomogeneities is determined by the mass scale which is large enough to prevent radiation diffusing away[34] during the period up to t_{eq}. This yields

$$M_{Da} \sim S^{5/4} \alpha^{-21/2} \alpha_G^{-3/4} \left(\frac{m_e}{m_N}\right)^{3/4} m_N \tag{6.69}$$

This can be compared with the maximum extent of the Jeans mass, M_J, which is the largest mass of a gas cloud which can avoid gravitational collapse by means of pressure support during the Universe's history.[35] This maximum arises at t_{eq} and since $M_J \sim G^{-3/2}p^{3/2}\rho^{-2}$, where p is the pressure, we have

$$(M_J)_{max} \sim G^{-1}t_{eq} \sim \alpha_G^{-3/2}S^{1/2}\left(\frac{m_N}{m_e}\right)^{3/2}\alpha^{-3}m_N \quad (6.70)$$

If inhomogeneities were of the isothermal variety then the first structures to condense out of the smoothly expanding universe would have a mass $\sim M_{Di}$ and would have to be associated with either globular clusters or dwarf galaxies. Galaxies could, in principle, be formed by the gravitational clustering of these building-blocks; subsequent clustering of galaxies would be the source of galaxy clusters. The extent of galaxy clusters would reflect the time interval from t_{rec} until $\sim\Omega_0 t_u$ when gravitational clustering stops because gravity ceases to be cosmologically significant after a time $\Omega_0 t_u$ in universes with $\Omega_0 < 1$.

By way of contrast, if homogeneities were initially adiabatic then we can argue a little further. The first structures to condense out of the expanding universe and become gravitationally bound should have a mass $\sim M_{Da}$, close to the observed mass of galaxy clusters. It is then inevitable that these proto-clusters will contract asymmetrically under their own self-gravity and fragment. Some simple arguments allow us[36] to estimate the masses and radii of typical fragments. The condition that a gravitating cloud be able to fragment is that it be able to cool and, hence, radiate away its binding energy. After the cosmic recombination time, t_{rec}, the dominant cooling mechanism will be bremsstrahlung on a time-scale dictated by the Thomson cross-section, σ_T, so the cooling time is

$$t_c \sim (n\alpha\sigma_T)^{-1}\left(\frac{T}{m_e}\right) \sim m_e^2\alpha^{-3}n^{-1}\left(\frac{T}{m_e}\right)^{1/2} \quad (6.71)$$

where n is the particle number density within the cloud, and T its temperature. The cloud will only cool efficiently if t_c is less than the gravitational contraction time, t_f; for a cloud of mass M and radius R, this is

$$t_f \sim \left(\frac{GM}{R^3}\right)^{-1/2} \quad (6.72)$$

The cloud will only cool and fragment, therefore, if

$$R \lesssim R_g \equiv \alpha^4\alpha_G^{-1}m_e^{-1}\left(\frac{m_N}{m_e}\right)^{1/2} \sim 75\ \text{kpc} \quad (6.73)$$

When $R > R_g$ the cloud contracts slowly without fragmenting and thus the characteristic dimension R_g divides frozen-in primordial structure from well-developed fragmentation. This argument will only hold so long as the temperature within the cloud stays below the ionization temperature $\sim\alpha^2 m_e$ before the cloud contracts to a radius $\sim R_g$. This condition requires that the cloud mass satisfy

$$M \gtrsim M_g \equiv \alpha_G^{-2}\alpha^5\left(\frac{m_N}{m_e}\right)^{1/2} m_N \sim 10^{12} M_\odot \tag{6.74}$$

Clouds with masses less than M_g will cool very efficiently by atomic recombination radiation and will never be pressure-supported. This singles out M_g as the mass-scale dividing well-developed, fragmented cosmic structure from quasi-static, under-developed clustering. The fact that M_g and R_g are so close to the masses and sizes of real galaxies is very suggestive. If irregularities that arise in the early universe are of adiabatic type (and the latest ideas[37] in elementary particle physics suggest that this will be the case) and if the arguments leading to (6.73) and (6.74) hold then the characteristic dimensions of galaxies are, like those of stars and planets, determined by the fundamental constants α_G, α and m_N/m_e independent of cosmological parameters. The only condition of a cosmological nature that is implicit in these deductions is that the maximum Jeans mass of (6.70) exceed M_g in order that galaxies can form from fragments of a larger surviving inhomogeneity; this implies[35,38]

$$S \gtrsim \alpha_G^{-1/4}\alpha^{9/2}\left(\frac{m_N}{m_e}\right)^{5/4} \sim 10^6 \tag{6.75}$$

In the past few years there has been growing interest in the possibility that the predominant form of matter in the Universe might be non-baryonic. There are a variety of non-baryonic candidates supplied by supersymmetric gauge theories. The most attractive would be a light massive electron neutrino since its mass can be (and may already have been) measured[32] in the laboratory. Others, like the axion, gravitino or photino,[39] do not as yet readily offer prospects for direct experimental detection. Cosmologists find the possibility that the bulk of the Universe exists in non-luminous, weakly interacting particles a fascinating possibility because it might offer a natural explanation for the large quantities of dark material inferred to reside in the outer regions of spiral galaxies and within clusters.[40] If this is indeed the case then the masses of these elementary particles will play a role in determining the scale and mass of galaxies and galaxy clusters. By way of illustration we show how, in the case of a massive neutrino, this connection arises.[41]

If a neutrino possesses a rest mass less than 1 MeV and is stable then it will become collisionless after the Universe has expanded for about one second and will always have a number density of order the photon number density, n_γ. The mass density of light neutrinos in the present

Universe is then given by

$$\rho_{\nu_0} = \frac{3}{22} g_\nu m_\nu n_\gamma \tag{6.76}$$

where m_ν is the neutrino mass, and g_ν is the number of neutrino spin states (for the total collection of known neutrinos $\nu_e, \bar{\nu}_e, \nu_\mu, \bar{\nu}_\mu$ we have $g_\nu = 4$); hence, today,

$$\rho_{\nu_0} \sim 10^{-31} g_\nu \left(\frac{m_\nu}{1 \text{ eV}}\right) \text{gm cm}^{-3} \tag{6.77}$$

If $m_\nu \gtrsim 3.5$ eV then the neutrino density will exceed that of luminous matter.

Neutrinos are also a natural candidate for galaxy or cluster halos because their distribution remains far more extended than that of baryons. Whereas baryonic material can radiate away its binding energy through the collisional excitation and de-excitation of atomic levels the neutrinos, being collisionless, cannot. One might therefore expect luminous baryonic material to condense within extended halos of neutrinos. If neutrinos are to provide the dominant density within these systems we can derive an interesting limit on the neutrino mass. Since neutrinos are fermions they must obey the Pauli Exclusion Principle. If neutrinos within a spherical region of mass M and radius r have an average speed σ and momentum p, then the volume of phase space they occupy is[42]

$$V_{\mathbf{p}} \sim \int d^3p \int d^3x \sim (m_\nu \sigma)^3 r^3 \tag{6.78}$$

Since $V_{\mathbf{p}}$ cannot exceed unity the total mass of the neutrino sphere is at most $M \sim m_\nu V_{\mathbf{p}} \sim m_\nu^4 \sigma^r r^3$. If the system is in virial equilibrium then we must have

$$\langle \sigma^2 \rangle \sim \frac{GM}{r} \tag{6.79}$$

Therefore there exists a lower limit on the neutrino mass of

$$m_\nu \gtrsim G^{-1/4} \sigma^{-1/4} r^{-1/2} \tag{6.80}$$

This corresponds to $m_\nu \gtrsim 20$ eV for galaxies and $m_\nu \gtrsim 5$ eV for clusters if we use appropriate values for σ and r. If we identify M with the characteristic baryon fragment size in (6.74) then (6.80) corresponds to

$$m_\nu \gtrsim \alpha_G^{1/4} \left(\frac{m_e}{m_N}\right)^{1/2} \alpha^{-17/8} m_N \tag{6.81}$$

If massive neutrinos determine the content of galaxies they may also determine the scale of their clustering. During the radiation-dominated period of the early universe inhomogeneous clumps of neutrinos will be dispersed if their size is smaller than the distance neutrinos can travel in the age of the Universe. Neutrinos will just escape from these small irregularities which will thence be erased, leaving a characteristic size in the hierarchy of irregularities in the Universe. The mass encompassed by this characteristic dimension is given by m_ν and the Planck mass, m_p, as[43]

$$M_\nu \sim m_p^3 m_\nu^{-2} \sim \alpha_G^{-3/2} m_\nu^{-2} m_N^3 \sim 10^{15} M_\odot \tag{6.82}$$

This is similar to the extent of large galaxy clusters. If the mass-scale (6.82) is associated with the large scale structure of the Universe it illustrates how an additional dimensionless parameter, m_ν/m_N, can enter into the invariant relations determining the inevitable sizes of large scale structures. In this picture of galaxy formation, which is 'adiabatic', galaxies must form by fragmentation of clusters of mass M_ν. The arguments leading to (6.74) should still apply and we would require M_ν to exceed M_g, hence

$$\alpha_G > \alpha^{10}\left(\frac{m_\nu}{m_N}\right)^4\left(\frac{m_N}{m_e}\right) \tag{6.83}$$

There are two further interesting coincidences in the case when $m_\nu \sim$ 30 eV as has been claimed by one recent experiment.[32] Not only is such a neutrino mass sufficient to ensure neutrinos dominate the Universe, (6.77); it also ensures that the cosmic time, t_ν, when the radiation temperature falls to m_ν, and the neutrinos become non-relativistic, is of order t_{rec} and t_{eq}. In general $\rho_\nu \sim G^{-1}t^{-2}$ and so as $\rho_\nu \sim T^4$ we find that the time t_ν, when $T \sim m_\nu$, is $t_\nu \sim \alpha_G^{-1/2} m_\nu^{-2} m_N$ and this is only of order $t_{eq} \sim S^2 \alpha_G^{-1/2} m_N^{-1}$ if

$$S \sim \frac{m_N}{m_\nu} \sim 10^8\left(\frac{30 \text{ eV}}{m_\nu}\right) \tag{6.84}$$

In addition, we have $t_\nu \sim t_{rec} \sim S^{1/2}\alpha^{-3}\alpha_G^{-1/2}(m_N/m_e)^{1/2} m_e^{-1}$ if

$$S \sim \alpha^6\left(\frac{m_e}{m_N}\right)^3\left(\frac{m_N}{m_\nu}\right)^4 \tag{6.85}$$

and combining (6.84) and (6.85) leads to the suggestive relation

$$m_\nu \sim \alpha^2 m_e \tag{6.86}$$

In fact, this formula may turn out to have some deeper theoretical basis as a prediction of the electron neutrino rest-mass since we notice that $\alpha^2 m_e$ is 27 eV, within the error bars of the reported measurements by Lyubimov *et al.*[32]

Galaxy formation in the presence of massive neutrinos is a version of the *adiabatic* theory outlined above in which clusters form first and then break up into subcomponents of galactic size. It appears that if this is to be the route to galaxy formation then a high density of neutrinos must exist (exceeding that of baryons) otherwise the level of density fluctuation required in the early universe would exceed that allowed by observational limits on the fine scale temperature fluctuations in the microwave background over minutes of arc[44]—this is the typical angular scale subtended by a galaxy cluster when the radiation was last scattered to us at high redshift. However, recent numerical simulations of galaxy clustering in the presence of massive neutrinos carried out on fast computers[45] reveal that the clustering of the ordinary luminous matter in the presence of 30 eV neutrinos has statistical properties not shared by the real universe; (see Figures 6.5(a) and 6.5(b)).

Neutrinos are not the only non-baryonic candidates for the non-luminous material that apparently dominates the present structure of the Universe. Elementary particle physicists have predicted and speculated about the existence of an entire 'zoo' of weakly interacting particles like axions, photinos and gravitinos. These particles should, if they exist, behave in many ways like massive neutrinos, for they do not have electromagnetic interactions with baryons and leptons during the early radiation era of the Universe but respond to gravity. Yet, unlike the neutrino, these more exotic particles are predicted to possess negligible velocities relative to the overall systematic expansion of the universe today, either because of their greater mass or, in the case of the axion, because they were formed with negligible motion.[39,40] This means that only very small clouds of these particles get dispersed by free-streaming during the first few thousand years of cosmic expansion. In contrast to the neutrino model, in which no irregularities survive having mass less than $\sim 10^{15} M_\odot$, (see equation (6.82)), non-uniform distributions of these exotic particles are only erased over dimensions smaller than $-10^6 M_\odot$. In effect, the characteristic survival mass is still given by (6.82) but the mass of a gravitino or photino necessary to generate all the required missing matter is ~ 1 GeV hence the analogue of M_ν is close to $10^6 M_\odot$. In this picture of cosmogony, events follow those of the *isothermal* scenario outlined earlier, with star clusters forming first and then aggregating into galaxies which in turn cluster in hierarchical fashion into great clusters of galaxies. Remarkably, computer simulations of these events[45] in the presence of axions or photinos predict patterns of galaxy clustering with statistical features matching those observed if the total density of the universe satisfies $\Omega_0 \sim 0.2$, but unfortunately the velocities predicted for the luminous galaxies do not agree with observation; see Figure 6.6.

This completes our attempt to extend the successes of the last chapter

(a)

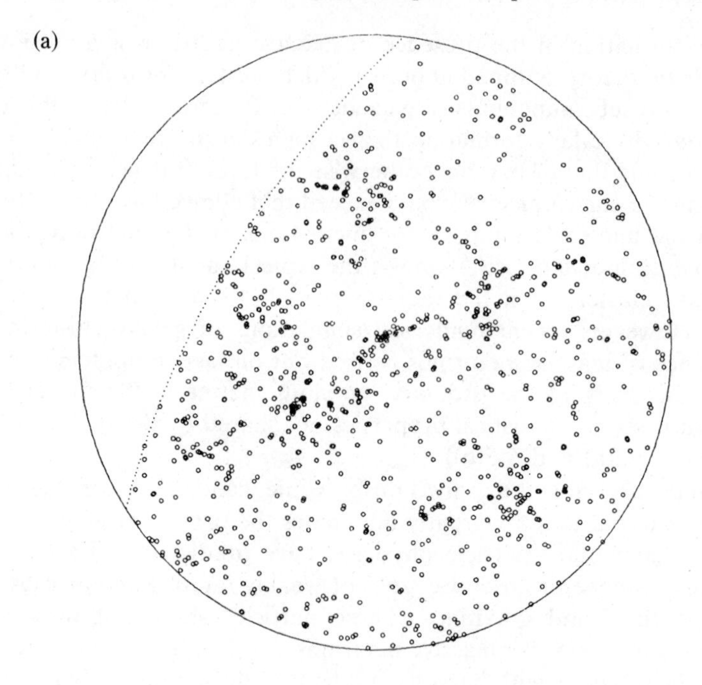

(b)

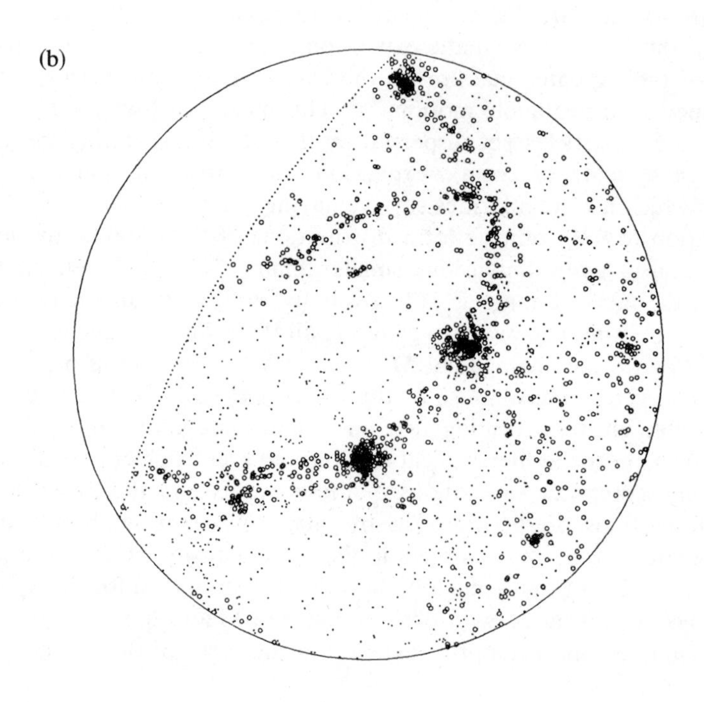

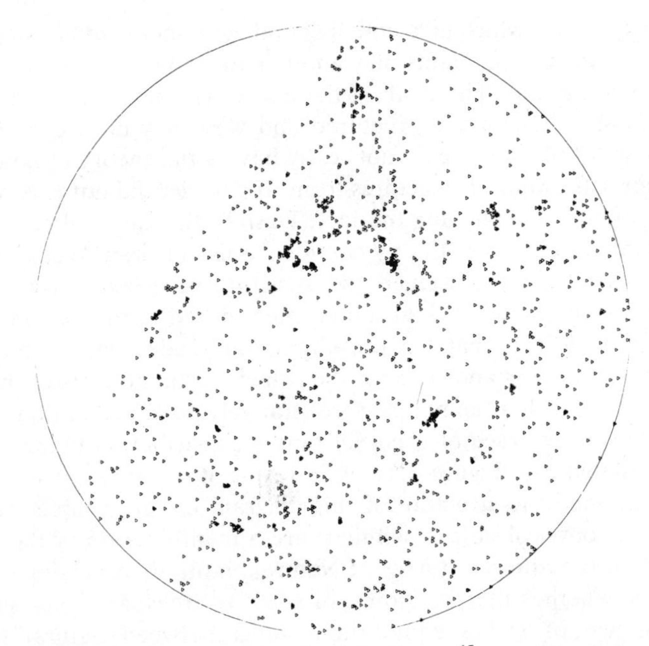

Figure 6.6. As Figure 6.5(b), but for a model universe[45] containing axions, one of the exotic elementary particle species that may exist in the Universe, with a total density equal to $\Omega_0 = 0.2$. There is little evidence for filamentary structures forming and the axions and baryons are clustered in identical fashion with no segregation of mass and light. This model offers a better match to the observed clustering of galaxies shown in Figure 6.5(a) than does the neutrino-dominated model 6.5(b) but the distribution of velocities predicted for the luminous matter is at variance with observation.

into the extragalactic realm. Here we have encountered awkward uncertainties and unknown factors that prevent us ascribing the structures we observe to the values of the constants of Nature alone. Although we can think of theories of galaxy formation in which galaxy masses are determined by fundamental constants alone, (as in (6.74)), we can also think

---

Figure 6.5. (a) The semi-volume-limited distribution of galaxies fainter than 14.5 mag. with recession velocities less than 10,000 km s^{-1} observed out to a distance of about 100 Mpc found by M. Davis, J. Huchra, D. Latham and J. Tonry.[141] (b) The clustering of galaxies predicted by a computer simulation of the Universe.[45] The computed cosmological model contains a critical density ($\Omega_0 = 1$) of neutrinos ($m_\nu = 30$ eV). The circles trace the distribution of luminous (baryonic) material, whilst the dots trace the neutrino distribution. Notice the segregation of luminous from non-luminous material and the filaments and chains of luminous matter created by 'pancake' collapse. The luminous material is predicted to reside in far more concentrated form than.is observed in the sample (a).

up other theories, which give equally good agreement with observation, in which fundamental constants play a minor role compared with cosmological initial conditions. The truth of the matter is simple: whereas we know how stars and planets are structured and why they must exist given the known laws of physics, we do not really have a full theory of how galaxies and larger astronomical structures form. If galaxies did not exist we would have no difficulty explaining the fact! Despite this incompleteness, which means that we cannot with any confidence as yet draw Weak Anthropic conclusions from the existence and structure of galaxies, this is a good point to take a second look at the problem posed at the beginning of the last chapter. Recall that we presented the reader with a plot of the characteristic masses and sizes for the principal components of the natural world.[38] We saw that the points were strangely polarized in their positions and there was no trace of a purely random distribution filling the entire plane available (see Figure 5.1). As a result of our investigations we can now understand the structure of this diagram in very simple terms. The positions of physical objects within it are a manifestation of the invariant strengths of the different forces of Nature. Naturally occurring composite structures, whether they be atoms, or stars, or trees, are consequences of the existence of stable equilibrium states between natural forces of attraction and repulsion. If we review the detailed analysis of the last chapter and the present one, the structure of the diagram can be unravelled (see Figure 6.7). There are two large empty regions: one covers the area occupied by black holes:

$$R \leqslant 2GM \tag{6.87}$$

Nothing residing within this region would be visible to external observers like ourselves. The other vacant region is also a domain of unobservable phenomena, made so by the Uncertainty Principle of quantum mechanics, which in natural units reads,

$$\Delta R \, \Delta M \geqslant 1 \tag{6.88}$$

All the familiar objects like atoms, molecules, solids, people, asteroids, planets and stars are atomic systems held in equilibrium by the competing pressures of quantum exclusion and either gravity or electromagnetism. They all have what we termed atomic density, ρ_{AT}, which is roughly constant at one proton mass per atomic volume. Thus all these atomic bodies lie along a line of constant *atomic* density; hence for these objects

$$M \propto R^3 \tag{6.89}$$

Likewise the atomic nuclei, protons and neutrons all lie along a line of constant *nuclear* density which they share with neutron stars. As we go beyond the scale of solid, stellar bodies and enter the realm of star

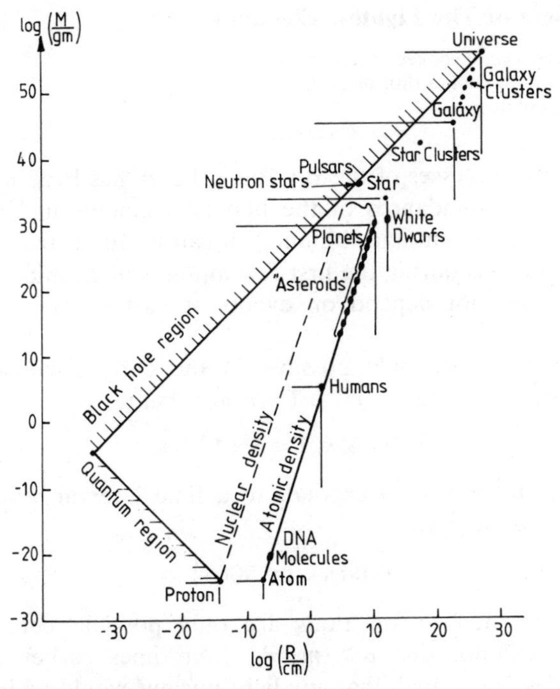

Figure 6.7. A revised version of Figure 5.1 in which the particular distribution of cosmic objects in the mass-size plane is shown to be conditioned by the existence of regions excluded from direct observation by the existence of black holes and quantum mechanical uncertainty and structured by the lines of constant atomic and nuclear densities. The latter pick out ranges of possible equilibrium states for solid bodies (based on ref. 38).

systems—globular clusters, galaxies, galaxy clusters and superclusters, we stray from the line of constant density. These systems are supported by a balance between the inward attraction of gravity and the outward centrifugal forces generated by the rotation of their components about their common centres of gravity. Finally, off at the top corner of the diagram we see the point marking the entire visible universe. Its exact mass we do not yet know because of our uncertainties regarding the extent of non-baryonic matter and dead stars in space, but if it lies a little below the black hole line so $R_u > 2GM_u$ then the Universe will continue to expand forever. However, if the final value of the cosmological density yields $\Omega_0 > 1$ then we will lie in the region $R_u < 2GM_u$ and the Universe will recollapse to a second singularity of high density. We would, in this case, be living inside a very large black hole.

6.6 The Origin of The Lightest Elements

The elements were cooked in less time
than it takes to cook a dish of duck
and roast potatoes.

G. Gamow

One of the great successes of the Big Bang theory has been its successful prediction of the abundances of the lightest elements in Nature: hydrogen, helium, deuterium and lithium. All can be fused from primordial protons and neutrons during the first few minutes of cosmic expansion in quantities that do not depend on events at earlier, more exotic moments.[10]

Nuclear reactions are only possible in the early universe during a narrow temperature niche, $0.1m_e \leq T \leq \alpha m_N$; that is

$$5\times 10^8\ \text{K} \leq T \leq 5\times 10^{10}\ \text{K} \tag{6.90}$$

This, according to (6.56) corresponds to a time interval between about $m_p m_e^{-2} \geq t \geq \alpha^{-2} m_p m_N^{-2}$, that is

$$0.04\ \text{s} \leq t \leq 500\ \text{s} \tag{6.91}$$

Thus, primordial nuclear reactions are only possible because of the Anthropic coincidence that $\alpha > (m_e/m_N)$. At times earlier than 0.04 s thermal energies are so high that any light nucleus would be immediately photodisintegrated, whilst after ~500 sec the energies of nucleons are too low to allow them to surmount the Coulomb barriers and come within range of the strong nuclear force

One might have thought that the final abundances of light nuclei, all of which are composed solely of neutrons and protons, would have been unpredictable, depending on the relative initial abundances of protons and neutrons at the Big Bang. Fortunately, this is not the case. When the temperature exceeds $\sim(G_F m_N)^{-2/3}(m_N/m_p)^{1/3} m_N^{-1} \sim 1\ \text{MeV}$ there arise weak interactions[46] involving nucleons which proceed more rapidly than the local cosmic expansion rate. These reactions are,

$$\begin{gathered} p+e^- \leftrightarrow n+\nu_e \\ n+e^+ \leftrightarrow p+\nu_e \\ n \leftrightarrow p+e^- +\bar{\nu}_e \end{gathered} \tag{6.92}$$

Whilst proceeding faster than the expansion rate they maintain an *equilibrium* abundance of neutrons relative to protons, so at temperature T,

$$\frac{n}{p} = \exp\left(-\frac{\Delta m}{T}\right) \tag{6.93}$$

where $\Delta m \equiv m(n) - m(p) \sim 1.293$ MeV is the neutron-proton mass difference. When T falls to $\sim \Delta m$ the ratio (n/p) falls below unity, reflecting the fact that slightly less energy is required to make a proton than a neutron. This rapid fall of (n/p) continues as the temperature drops until the weak interaction rates of (6.92) equal the cosmological expansion rate when $T_f \sim 1$ MeV. The neutron-proton ratio is then fixed at $n/p = \exp(-\Delta m/T_f) \sim 0.2$. Eventually, beta decays would reduce this to zero, but nuclear reactions intervene when the temperature of the universe falls to $T \sim 10^9$ K at $t \sim 100$ s. Proton capture proceeds slowly at first, via $p + n \rightarrow D + \gamma$, to be followed rapidly by fast nuclear chain-reactions[47] $p + D \rightarrow {}^3He + \gamma$, $n + D \rightarrow {}^3H + \gamma$, $p + {}^3H \rightarrow {}^4He + \gamma$, $n + {}^3He \rightarrow {}^4He + \gamma$, $D + D \rightarrow {}^4He + \gamma$. Here the reactions essentially stop; helium-4 is tightly bound and there is no stable nucleus with mass number 5. Virtually all the original neutrons left at T_f wind-up in helium-4 nuclei hence the number of helium-4 nuclei to hydrogen nuclei will be roughly $0.5 \times 0.2 = 0.1$, there being two neutrons per helium-4 nucleus. This corresponds to a helium-4 mass fraction of ~22–25%, as observed, (6.35). If the baryon density of the present universe equals that observed, $\Omega_B = 0.03$, then this process successfully predicts the observed cosmic abundances of helium-3, deuterium and lithium-7 also.[8]

The fact that the early universe gives rise to an 'interesting' abundance of helium-4, that is, neither zero nor 100%, is a consequence of a delicate coincidence between the gravitational and weak interactions. It arises because we have $T_f \sim \Delta m \sim m_e$, so the exponent in (6.93) is neither very large nor very small, and because the temperature T_f is suitable for electron and neutrino production. This coincidence is equivalent to the coincidence[38]

$$G_F m_e^2 \sim (G m_e^2)^{1/4} \tag{6.94}$$

Were this not the case then we would either have 100% hydrogen emerging from the Big Bang or 100% helium-4. The latter would likely preclude the possibility of life evolving. There would be no hydrogen available for key biological solvents like water and carbonic acid, and all the stars would be helium-burning and hence short-lived. Almost certainly, helium stars would not have the long-lived nuclear burning phase necessary to encourage the gradual evolution of biological life-forms in planetary systems. However, there appears no 'anthropic' reason why a universe containing 100% hydrogen initially would not be hospitable to life.

Carr and Rees[38] have pointed out that the coincidence (6.94) may be associated with another one that probably is closely tied to the conditions necessary for the existence and distribution of carbon in space following its production in stellar interiors (see section 5.2). It may be that the envelope of a supernova is ejected into space by the pressure of neutrinos generated in the core of the stellar explosion. If this is indeed the way the stellar envelope is ejected, then the timescale for interactions between nuclei in the envelope and free neutrinos must be close to the dynamical timescale $\sim(G\rho)^{-1/2}$ of the stellar explosion if the debris has density ρ. This ensures that the neutrinos have enough time to reach the envelope before dumping their energy and momentum but not so much time that they escape beyond the envelope. This would allow the envelope to be expelled. This condition requires the delicate balance[38]

$$G_F^2 n T^2 \sim (G n m_N)^{1/2} \tag{6.95}$$

where n is the nucleon number density and T the temperature. Now in order that the supernova be hot enough to produce neutrinos by $e^+ + e^- \to \nu_e + \bar{\nu}_e$ we must have $T \sim m_e$. The density expected when the core explodes is close to the nucleon degeneracy density found within neutron stars. This is roughly the nuclear density $n \sim m_N^3$. Using these relations we have the Carr–Rees coincidence[38]

$$G_F m_e^2 \sim (G m_e^2)^{1/4} (m_e/m_N)^{1/2} \tag{6.96}$$

which differs from the primordial nucleosynthesis coincidence (6.94) only by a factor $(m_e/m_N)^{1/2} \sim 0.02$ and suggests[38] a fundamental relationship between the weak and gravitational couplings of the form $\alpha_w \sim \alpha_G^{1/4}(m_N/m_e)^{3/2}$.

The other part of the nucleosynthesis coincidence (6.94) arises because[48] the neutron-proton mass difference is $\Delta m \sim m_e$. In fact, this is only part of a very delicate coincidence that is crucial for the existence of a life-supporting environment in the present-day Universe. We find that

$$\Delta m - m_e = 1.394\ \text{MeV} - 0.511\ \text{MeV} = 0.883\ \text{MeV} \tag{6.97}$$

Thus, since $m(n)$ and $m(p)$ are of order 1 GeV the relation is a one part in a thousand coincidence. If instead of (6.97) we found $\Delta m - m_e \lesssim 0$ then we would not find the beta decay $n \to p + e^- + \bar{\nu}_e$ occurring naturally. Rather, we would find the decay $p + e^- \to n + \bar{\nu}_e$. This would lead to a World in which stars and planets could not exist. These structures, if formed, would decay into neutrons by pe^- annihilation. Without electrostatic forces to support them, solid bodies would collapse rapidly into neutron stars (if smaller than about $3M_\odot$) or black holes. Thus, the coincidence that allows protons to partake in nuclear reactions in the early universe also prevents them decaying by weak interactions. It also, of course, prevents the 75% of the Universe which emerges from nucleosynthesis in the form of protons from simply decaying away into neutrons. If that were to happen no atoms would ever have formed and we would not be here to know it.

6.7 The Value of S

> God created two acts of folly. First,
> He created the Universe in a Big Bang.
> Second, He was negligent enough to
> leave behind evidence for this act, in
> the form of the microwave radiation.
>
> P. Erdös

In our discussion of cosmology so far we have confined our discussion to events and effects that are independent of cosmological initial conditions. They have, like the structures discussed in the previous chapter, been conditioned by various unalterable coupling constants and mass ratios α_G, α_w, α_s and m_N/m_e. But we have seen these dimensionless parameters joined by one further parameter introduced in equations (6.42)–(6.45): the entropy per baryon of the Universe, S. This quantity arose from the discovery of the microwave background radiation and was first discussed as a dimensionless parameter characterizing possible hot Big Bang models by Zeldovich and Novikov, and by Alpher, Gamow and Herman.[49] It is interesting to note how fundamental advances in our understanding of Nature are usually accompanied by the discovery of another fundamental constant and in this case it was Penzias and Wilson's serendipitous discovery[4] of the 3 K background radiation which introduced the parameter S.

We have seen already that the observed numerical value of $S \sim 10^9$ determines the key cosmic times t_{eq} and t_{rec}, (see equations (6.48) and (6.49)), and hence plays a role in various coincidences that are necessary for the evolution of life, (6.60–6.65). Furthermore, it is possible that S controls the characteristic sizes of galaxies and clusters in our Universe (6.68–6.75), (6.85).

The appelation 'hot' is often used of the Big Bang model of the Universe. This is partially because the observed value of $S \sim 10^9$ is so large. Indeed, over the period since the discovery of the microwave background radiation in 1965 cosmologists have repeatedly tried to explain why[50] the value of S is not, like many other dimensionless constants of physics, of order unity say, or, like many cosmological parameters, of order $10^{20} \sim 10^{40}$. It is clear from (6.60–6.85) that the requirement that galaxies exist and that the Universe is not dominated by radiation today (a situation that would prevent the growth and condensation of small irregularities into fully-fledged galaxies by the process of gravitational instability) we must[35,38,51] have $S \lesssim 10^{11}$.

One approach to explaining why $S \gg 1$ is to recognize that, since the photon entropy, $\mathbf{s}_\gamma$, which defines S, (6.42), is monotonic non-decreasing with time, by the Second Law of thermodynamics, so also is S if the baryon number is unchanging. Hence $\dot{S} \geq 0$ and if the Universe were extremely anisotropic or inhomogeneous during its early stages it might be possible for dissipation of non-uniformities to smooth the universe out into the observed state of virtual isotropy and homogeneity whilst boosting an initial entropy per baryon of $S \sim 1$ to the large observed value of order 10^9. Unfortunately, a detailed investigation[52] revealed that this dissipation inevitably results in a catastrophic overproduction of photon entropy from anisotropies in the cosmological expansion. A universe dominated by anisotropy close to the Planck time in which baryon number was conserved would produce a present-day value of $S \sim \alpha_G^{-1} \sim 10^{39}$ and conditions would never be cool enough to allow the formation of living cells at the vital moments of cosmic history (6.60–6.65).

Another variation of this idea appealed not to the irregularity of the very early universe to produce a large value of S but to the recent activity of explosive stars. Rees[53] has argued that if a population of supermassive stars formed prior to the emergence of galaxies (and there are reasons why this might be an appealing idea) then they might naturally account for the observed value of $S \sim 10^9$. These objects would radiate their mass in a Salpeter time[54] $t_s \sim \sigma_T G^{-1} m_N^{-1} \sim 4 \times 10^8$ yr and if, when they formed, we had $S \ll 10^9$ the expansion dynamics would possess a density-time relation (6.22). If a fraction f of the Universe condenses into these stars and they radiate with the optimal 'Eddington' Luminosity, L_E, of (5.185) then the entropy per baryon that results is uniquely prescribed by

$$S \sim \alpha_G^{1/4} \alpha \left(\frac{m_N}{m_e}\right) f^{3/4} \tag{6.98}$$

Three of the terms in this expression are collectively of $O(1)$ and so we have simply $S \sim \alpha_G^{1/4} \sim 10^9$–$10^{10}$ as observed. The same result would arise if, instead of being generated by stellar exothermia, the radiaton arose by accretion of cosmic material onto large primordial black holes.[55] However, despite the superficial attraction of these predictions, one pays a high price for them. If $S \ll 10^9$ prior to the generation of entropy by supermassive stars at $t_s \sim 10^8$ yrs then the beautiful agreement between the predictions of the standard model (having $S \sim 10^9$ for all times after $\sim 10^{-6}$ s) and observations of the present helium and deuterium abundances is ruined.

This type of non-primordial explanation of the observed $S \sim 10^9$ by recent events in an initially cold or tepid ($S < 10^8$ initially) universe illustrates an important point of principle: one which separates the above idea from more appealing explanations that arise within the standard hot Big Bang model and new discoveries in elementary particle physics.

As we follow the evolution of the Universe backwards in time we see that S is a consequence of a baryon asymmetry in the Universe. At times less than $\sim 10^{-6}$ s, when temperatures exceed the proton rest mass, the observed S must arise from an asymmetry between baryons and antibaryons.[10] The antibaryons annihilate with baryons to create two photons per annihilation event for every 'extra' baryon. If this is not the case and one imagines the Universe to be baryon symmetric at the moment when protons become non-relativistic, $\sim 10^{-6}$ s, then the final number of protons per photon (equal to the number of anti-protons per photon) can be calculated exactly and we would today observe[56]

$$S \sim \alpha_G^{-1/2} \sim 10^{19} \tag{6.99}$$

Clearly, we do not live in such a universe; nor could we, for the present matter density would be over 10^{10} times smaller that what it is today. No stars or galaxies could form. This is just as well perhaps, for if they did the matter-antimatter symmetry would create catastrophic annihilation. Our observation of $S \sim 10^9$ is telling us that when $T \sim m_N \sim 1$ GeV there already existed an asymmetry in the Universe between the number density of baryons, n_B, and antibaryons, $n_{\bar{B}}$, of order

$$\frac{n_B - n_{\bar{B}}}{n_B + n_{\bar{B}}} \sim 10^{-9} \equiv S^{-1} \tag{6.100}$$

Since baryons are composed of quarks and when $t < 10^{-6}$ the cosmic medium is a soup of free quarks and leptons, of which only quarks carry baryon number, the asymmetry (6.100) arises from an inbuilt bias for quarks over antiquarks in the Universe prior to 10^{-6} s; that is,

$$\frac{n_q - n_{\bar{q}}}{n_q + n_{\bar{q}}} \sim 10^{-9} \sim S^{-1} \tag{6.101}$$

Therefore, we see that any explanation for the observed value of S that appeals to events prior to the first microsecond of the Universe's life also provides an explanation for the observed matter-asymmetry of the Universe. Those attempts to explain the value of S by recent events like that of Rees[53] described above, do not automatically provide an explanation for the baryon asymmetry because the S-generating events involve processes which occur after nucleons become non-relativistic.

In the last chapter we introduced grand unified gauge theories (GUT's) which predict that baryon number is not conserved and which offer the decay of protons as a direct experimental test of that prediction. This property opens the door for a theory which could explain the origin of the asymmetry (6.100), or equivalently (6.101). If baryon number is conserved then $(n_q - n_{\bar{q}})$ in (6.101) cannot be altered after the cosmic expansion begins and (6.100) and (6.101) must be manifestations of cosmological initial conditions with S independent of other constants of Nature like α, α_s, and α_w. On the other hand, if baryon number is *not* conserved it might be possible to show that $S \sim 10^9$ arises naturally from arbitrary initial conditions. This would have the effect of shifting the responsibility for our fortuitous value of $S \sim 10^9$, which allows the evolution of life, away from the initial conditions and onto the laws of evolution themselves. However, the presence of baryon non-conserving interactions in a grand unified theory is not a sufficient condition to generate a non-zero baryon asymmetry in the Universe;[57] for every interaction generating a net baryon number there might also exist its inverse reaction destroying baryon number at the same rate.

If we assume unitarity (all probabilities of interactions add up to 1) holds and *CPT* is a good symmetry, then a non-zero baryon asymmetry can be generated if the following conditions hold (where the operators *C*, *P* and *T* denote $C \equiv$ charge conjugation so C(particle) = antiparticle; $P \equiv$ parity reversal so P (right hand) = left hand, and $T \equiv$ time-reversal.);

(i) baryon number is not conserved

(ii) *C* and *CP* are not conserved

(iii) there is departure from thermal equilibrium.

Condition (i) is necessary if we are to pass from a state with baryon number (B), zero to $B \neq 0$; however, it is not a sufficient condition. The C operator changes n_q into $n_{\bar{q}}$, so if C is conserved we must have $n_q = n_{\bar{q}}$ in the system and hence $B = 0$. Since the P operator leaves both n_q and $n_{\bar{q}}$ unchanged, the *CP* conservation also requires $n_q = n_{\bar{q}}$ and hence $B = 0$. Therefore, the condition (ii) is necessary. Finally, if thermal equilibrium obtains then T is a good symmetry and so *CPT* symmetry would imply *CP* symmetry and $B = 0$ by (ii): therefore, we require condition (iii).

The general picture for the generation of a non-zero baryon number in the early Universe goes back to the work of Sakharov who in 1967 proposed a model in which hypothetical heavy bosons mediated interactions between quarks and leptons. He also pin-pointed[57] the necessary and sufficient conditions for baryon number generation: (i), (ii) and (iii). Later, in 1970, Kuzmin[58] proposed an *ad hoc* model which added an arbitrary baryon non-conserving part to the lagrangian for *CP*-violating $K^0\,\bar{K}^0$ decay.

With the advent of GUT's it is now possible to produce a more complete and compelling picture of baryon-number generation because particular theories, like SU(5), predict interactions that violate *C*, *CP*, and *B* and which occur out of equilibrium during the early stages of the Universe.[59] These interactions are mediated by the superheavy *X* or *Y* bosons which mediate the unification of electroweak and strong interactions in the grand unified theory (see Chapter 5 section 10). There has emerged a simple general picture of how baryon and *CP* conserving interactions can be forced out of equilibrium by the rapid expansion of the Universe when it is just 10^{-35} s old and the temperature exceeds the rest mass energy of the *X* bosons.

When $T \gg m_X \sim 10^{14}$ GeV, equal numbers of X and $\bar{X}$ bosons should exist (let us assume for the sake of argument that this is the case), and the Universe will be matter-antimatter symmetric. After a characteristic time, $\sim 10^{-35}$ s, these heavy bosons will decay asymmetrically (due to C and CP violation) into *unequal* numbers of quarks and antiquarks. These decays will be out of equilibrium if the X-bosons are heavy enough. For simplicity, suppose the X bosons (although they could equally be Y or Higgs bosons), decay into two channels with baryon number B_1 and $B_2 \neq B_1$ having branching ratios r and $(1-r)$ respectively. The $\bar{X}$ bosons will decay into similar channels with some branching ratios $\bar{r}$ and $(1-\bar{r})$. Because C and CP are not conserved we will have $r \neq \bar{r}$, and decays into quarks (q) and leptons (l) will go as follows:

$$X \xrightarrow{r} \bar{q}\bar{q}(B_1 = -\tfrac{2}{3}), \quad X \xrightarrow{1-r} ql(B_2 = \tfrac{1}{3}); \qquad \bar{X} \xrightarrow{\bar{r}} qq(B_1 = -\tfrac{2}{3}), \quad \bar{X} \xrightarrow{1-\bar{r}} \bar{q}\bar{l}(B_2 = \tfrac{1}{3}) \tag{6.102}$$

The mean net baryon number created by the $X\bar{X}$ decays are respectively

$$\begin{aligned} B_X &= rB_1 + (1-r)B_2 \\ B_{\bar{X}} &= -\bar{r}B_1 - (1-\bar{r})B_2 \end{aligned} \tag{6.103}$$

So, the decay of an $X\bar{X}$ pair produces, on average, a baryon number ε

$$\varepsilon \equiv B_{\bar{X}} + B_X = (r-\bar{r})(B_1 - B_2) \tag{6.104}$$

If C and CP are not conserved then $r \neq \bar{r}$; (note that baryon minus lepton number is conserved in these decays which is a feature of many simple GUT's).

The X and $\bar{X}$ bosons decay at a rate

$$\Gamma_d \sim g_d \alpha_X m_X^2 (T^2 + m_X^2)^{-1/2} \tag{6.105}$$

where g_d is the number of decay channels ($=2 \times$ no. of lepton generations $= 6$) and α_X is the strength of interactions between X particles and fermions. The $X\bar{X}$ decay proceeds efficiently until the cosmological expansion rate

$$H = \frac{1}{2t} = 1.66 g_\star^{1/2} T^2 m_p^{-1} \tag{6.106}$$

equals Γ_d. This occurs at a temperature, T_D, where

$$T_D \sim (\alpha_X g_d^{1/2} m_X m_p)^{1/2} \tag{6.107}$$

and a time,

$$t_D \sim (\alpha_X g_d m_X)^{-1} \sim 10^{-35} \text{ s} \tag{6.108}$$

To generate the baryon number (6.104) these decays must occur out of equilibrium. Originally Weinberg[59] gave the criterion for non-equilibrium decay as

$$m_X > T_D \tag{6.109}$$

which leads to,[60]

$$m_X \geqslant 2\times 10^{-2} g_d \bar{\alpha} g_\star^{-1/2} m_p \tag{6.110}$$

that is

$$m_X \geqslant 0.3\bar{\alpha} m_p \equiv m_\star \tag{6.111}$$

For gauge vector bosons (X, Y), $\bar{\alpha} \equiv \alpha_X \sim 1/45$ so $m_X > 10^{17}$ GeV; whereas for scalar (Higgs) bosons, $\bar{\alpha} \equiv \alpha_H \sim \alpha_X(m_f/m_W)^2 \sim 10^{-4}$–$10^{-6}$, (where m_f and m_W are light fermion and W boson masses), we have a more acceptable requirement, $m_H \geqslant 10^{12}$–10^{14} GeV.

The number density of heavy bosons before decay is thermal, so the number density is

$$n_X \sim \frac{\zeta(3)}{\pi^2} g_X T_D^3 \tag{6.112}$$

and the total entropy is

$$\mathbf{s} \sim \frac{4\pi^2}{45} g_\star(T_D) T_D^3 \tag{6.113}$$

Therefore, if ε is the mean net baryon number produced in an $X\bar{X}$ decay, we end up generating as specific entropy,[59]

$$S^{-1} \equiv \frac{n_B}{\mathbf{s}} \sim \frac{45\zeta(3)}{g_\star} \varepsilon \sim 10^{-2}\,\varepsilon \tag{6.114}$$

A *CP* violation of order $\varepsilon \sim 10^{-7}$ is necessary to explain the observed specific entropy $S^{-1} \sim 10^{-9}$. In the standard three-generation SU(5) model, calculations show $\varepsilon < 10^{-16}$ and extra fermionic families need to be added to[61] yield $\varepsilon \sim 10^{-8}$. In general one knows only that $|\varepsilon| < \alpha$ and in fact the sign of ε is also unknown because a 'definition' of matter in $X\bar{X}$ decay may not necessarily agree with the convention adopted in other *CP*-violating systems—for example neutral kaon decay.

Since the original paper of Weinberg it has been realized that condition (6.109), although *sufficient* for non-equilibrium decay, is by no means necessary. Accordingly, the result (6.114) gives the *maximum* value of $n_B/\mathbf{s}$ that can be generated in GUT's. When $m_X < T_D$ there is still only partial equilibrium because the equilibrium temperature is constantly falling as the Universe expands; the Boltzmann exponential cut-off is moderated into a power-law by the expansion and we can gauge the effectiveness of decays by a parameter,[62]

$$K \equiv \left(\frac{\Gamma_d}{H}\right)_{T=m_X} = \frac{m_\star}{m_X} \tag{6.115}$$

$$\sim \frac{3\times 10^{17}\alpha}{m_X}\,\text{GeV} \tag{6.116}$$

K gives the effectiveness of the $X\bar{X}$ (or $Y\bar{Y}$ or $H\bar{H}$) decays. When $K<1$ the non-equilibrium is strong, condition (6.109) holds, and the maximal baryon asymmetry (6.114) is generated. When $1 \leqslant K \leqslant K_c$, where K_c is a critical value $\sim 10^2$–10^4, then the non-equilibrium is partial, and numerical studies reveal a simple prediction[62]

$$\frac{n_B}{s} \sim \frac{g_X}{g_\star}\,\varepsilon(1+K)^{-1/3} \tag{6.117}$$

When $K > K_c$ the baryon number is damped exponentially[60,62] by baryon non-conserving scatterings of quarks and leptons; their rate is

$$\Gamma_{22} \sim g_d\alpha^2 T^5 (T^2 + m_X^2)^{-2} \tag{6.118}$$

and thus becomes equal to the cosmological expansion rate, H, when

$$K \equiv K_c \sim \frac{g_d}{g_\star}\,\alpha^{-1} \tag{6.119}$$

Thus when $K > K_c$; we have,[60,62]

$$\frac{n_B}{s} \sim \varepsilon\frac{g_X}{g_\star}\,K^{5/6}\alpha^{5/6}\exp(-\alpha^{1/3}K^{1/3}) \tag{6.120}$$

The calculations that have been performed to determine the value of m_X from the energy at which all interactions have the same effective strength yield a value $m_X \sim 5.5\times 10^{14}$ GeV, which corresponds to a K value for $X\bar{X}$ decays of,

$$K_X \sim 10 \tag{6.121}$$

If the explanation of grand unified theories for the value of $S \sim 10^9$ is correct then we can see from (6.114) and (6.120) that everything hinges upon the magnitude (and sign) of the *CP* violation ε in heavy boson decays, like (6.102). Since, as yet, there appears no hope of calculating ε precisely, (although it is possible in principle), we seem to have simply replaced an initial condition for S by an initial condition for ε. However, ε is an invariant and some restrictions on its value are known: we must have $|\varepsilon| < \alpha$. The weak anthropic limits on S that we examined earlier, and which are necessary for the evolution of life in the late stages of the

Universe, provide[63] circumstantial evidence that we may find a theory in which $S \sim \alpha^{-4}$. Furthermore, Barrow and Turner[10,64] have shown that if baryon number is generated in the manner described above it has strong implications for the nature of the density irregularities that can arise in the early universe and adiabatic perturbations are preferred over those of isothermal type.

Before leaving this topic it is worth recalling that our extrapolation back to the early moments $\sim 10^{-35}$ s rests upon the belief that elementary particle interactions become *asymptotically free* at high energy. Were this not the case, interaction strengths would grow in complexity as $T \to \infty$ and the very early universe would resemble an intractable strongly interacting state, not amenable to analysis employing the ideal gas laws (which apply to systems with particle interaction times much shorter than the typical time between interactions). We have already seen that the phenomenon of asymptotic freedom arises in quantum chromodynamics because the smearing of colour charge by gluons beats the vacuum polarization effects of virtual quark-antiquark pairs, see (5.200–5.202). This only happens if the number of quark flavours arising in Nature is less than 17. Since quarks and leptons appear in a paired generation structure, the number of neutrino species, N_ν, must be bounded by half the number of quark flavours, hence $N_\nu < 8$. As yet there is no understanding of the observed value of N_ν. Primordial nucleosynthesis requires $N_\nu \leqslant 4$ and if we had $N_\nu \sim 6$–8 virtually all the Universe would be burnt to helium during primordial nucleosynthesis and the present Universe would be devoid of hydrogen, water and life.

This completes our brief discussion of the parameters of a physical origin that characterize the structures emerging within the Big Bang Universe. We now turn to consider the anthropic consequences of a different collection of parameters that are unique to the cosmological problem but which are also the major prerequisites for the existence of a 'user-friendly' Universe.

6.8 Initial Conditions

> 'Things are as they are because they were as they were.
>
> T. Gold

The scientific theories that prove to be the most effective descriptions of the physical world are invariably mathematical. It is an interesting question, although not one that concerns us here, as to why this should be the case. Furthermore, the relevant mathematical laws are based upon differential equations. Such representations consists of two components: a set of evolution and constraint equations, which usually arise from an Action Principle, together with a set of boundary conditions to denote the starting (or finishing) state of the evolution. In the cosmological case the evolution equations are assumed (at least for cosmic times exceeding 10^{-43} s) to be Einstein's equations of general relativity. By wishing to specify boundary conditions for these evolution equations we are not pre-empting the issue of whether or not there ever was a beginning to the evolution. We can consider initial conditions to be specified at any past time. There are philosophical reasons for wanting to prescribe them at the earliest conceivable moment. There have been attempts to reconcile the huge apparent age ($\sim 15\times 10^9$ years) of the Universe evidenced by the fossil record and astronomical motions, with prejudices for a very young ($\sim$6000 years) Earth and Universe by appealing to very special boundary conditions. If one imagines the Universe to have come into being 6000 years ago with boundary conditions giving the appearance of a 15×10^9 year age, then such a Universe is observationally indistinguishable from one with different boundary conditions that really is 15×10^9 years old. The young universe model does not, of course, *explain* anything and is rejected as of no intrinsic interest to scientists who are interested only in studying the *appearance* of reality. They have no practical interest in any 'absolute physical reality' for how could it ever be known what it is? The real quantitative difference between the 'old' and the 'young' universe boundary conditions is in the amount of information that must be built into the initial conditions. The 'young' universe has all the information in the initial conditions and the subsequent evolution creates essentially no more. The 'old' universe has a minimum of information in the initial conditions but is able to generate a high information content for its description of Nature today because the evolutionary laws are unstable: complex final states can develop from simple initial states. This is a manifestation of epistemological indeterminism arising from ontologically deterministic evolution equations (See section 3.2).

In general relativity the specification of initial conditions at, say for the sake of argument, an initial singularity, is non-trivial. If this initial state is *timelike* then some points of this initial state will lie in the causal future of others. We shall assume here that this is not the case and that the initial data are prescribed on a spacelike hypersurface of constant cosmic time so we can formulate a conventional Cauchy problem.[65] In the cosmological context these initial conditions will determine (for a given topology) the present expansion rate, rotation, shear, curvature and density of the Universe.

So far we have been examining the 'standard' hot Big Bang model which is isotropic and homogeneous. We shall begin by examining the role of initial conditions in these very special models (which remarkably *do* describe the present-day universe very accurately[66]) before examining the anthropic status of the 'coincidences' that allow the homogeneous and isotropic models to provide such an accurate description of the Universe.

In the absence of a cosmological constant term the isotropic and homogeneous cosmological models arising as solutions to Friedman's equation, (6.3), require two arbitrary constants to be specified on a spacelike surface of constant time. In practice, these two parameters are usually taken as H_0 and Ω_0 defined by (6.2) and (6.20) and are specified today when they are measured. The total lifetime of the closed Universe is related simply to these parameters by (6.27). We showed in section 6.2 that there is a simple necessary criterion for life to evolve in a closed Universe: that $t_\star$ defined by (6.27) exceed the main sequence stellar lifetime.

It has been appreciated for some time that one of the two parameters which define the Friedman model that best describes our own Universe has a very unusual value. As we saw in (6.30) the present-day universe is extremely close to the critical state with $\Omega_0 = 1$; so close, in fact, that our observations cannot tell with certainty whether $\Omega_0 > 1$ or $\Omega_0 < 1$ (amusingly, if $\Omega_0 = 1$ exactly our measurements will never be accurate enough to demonstrate the fact). This is a remarkable state of affairs and is equivalent to the requirement that in the Friedman equation (6.3) the constant term on the right is roughly of the same order as the $4\pi G\rho R^2/3$ term. Since this latter term falls off as R^{-2} during the radiation era and as R^{-1} during the dust-dominated era we can see that, in order to be of similar magnitude after $\sim 10^{10}$ years of expansion, these terms must have had a very special relative size when the expansion first began at $t \lesssim 10^{-43}$ s. If we define ρ_c as the density of a Universe with $\Omega_0 = 1$ (see (6.19)) then our present observations require that at the Planck time, 10^{-43} s, the Universe must have been expanding at a fantastically special rate with a total density close to the critical value then of $3H^2(t_p)/8\pi G$, where $H(t_p) \sim t_p^{-1}$,

$$\left|\frac{\rho - \rho_c}{\rho_c}\right|_{t_p} \lesssim 10^{-57} \tag{6.122}$$

This extraordinary relation regarding the initial conditions has been called the *flatness problem* by Alan Guth.[67] This name arises because the cosmological models that have $\rho = \rho_c$ are those with zero spatial curvature, (6.4), and hence possess flat, Euclidean spatial geometry. A more physical version of the coincidence (6.122) (and one which is, in fact, roughly the square root of (6.122)) involves $\mathcal{R}$, the ratio of the present radius of curvature of the Friedman Universe relative to the scale that the Planck length would have freely expanded to after a time equal to the present age of the Universe,[68] $t_0 \sim 10^{10}$ yr. Thus,

$$\mathcal{R} = \left(\frac{\text{Friedman curvature radius}}{\text{Planck scale at } t_0}\right) = \left(\frac{t_0}{t_p}\right)^{1/2}\left(\frac{t_{eq}}{t_0}\right)^{1/6} \times \frac{1}{|\Omega_0 - 1|^{1/2}} \sim \frac{10^{30}}{|\Omega_0 - 1|^{1/2}} \tag{6.123}$$

where t_{eq} appears because we allow for the change-over from radiation to dust-dominated expansion after t_{eq}, (6.46). This relation can be expressed in terms of fundamental constants and S as

$$\mathcal{R} \sim \left(\frac{Sm_N t_0}{\alpha_G}\right)^{1/3} \frac{1}{|\Omega_0 - 1|^{1/2}} \tag{6.124}$$

If t_0 is to exceed the time required to produce stable stars, so $t_0 > t_\star \sim \alpha^2 \alpha_G^{-1} m_e^{-1}$, then we have a Weak Anthropic constraint on a cognizable Universe

$$R \gtrsim \left(\frac{S\alpha^2 m_N}{\alpha_G^2 m_e}\right)^{1/3} \frac{1}{|\Omega_0 - 1|^{1/2}} \tag{6.125}$$

Another way of stating this problem is to formulate it as an '*oldness*' *problem*. The laws of physics create one natural timescale for cosmological models, $t_p = (G\hbar/c^5)^{1/2} \sim 10^{-43}$ s. The fact that our Universe has existed for at least $\sim 10^{60} t_p$ suggests there is something very unusual and improbable about the initial conditions that gave rise to our Universe. (But see Chapter 7.) This situation was first stressed by Collins and Hawking[69] in 1973 and it is one that has striking anthropic implications. We can see from (6.4) and (6.27) that when $|\ln \Omega| \gg 1$ the expansion timescale of the Friedman models is altered and we have $t_0 \propto \Omega_0^{-1/2}$ approximately. Models with $\Omega_0 \gg 1$ would have recollapsed before stars ever had a chance to form or life to evolve. Models with $\Omega_0 \ll 1$ would expand so rapidly that material would never be able to condense into galaxies and stars. Only for a very narrow range of $\Omega_0 \sim 10^{-3}$–10 corresponding to a range $\sim 10^{-56}$–10^{-60} in (6.122) does it appear that life can evolve, (see Figure 6.8). Why did the initial conditions lie in this peculiar and special range that allows observers to exist?

One approach to resolving the *flatness problem*, which is in accord with the Weak Anthropic Principle, is to imagine that the Universe is inhomogeneous and infinite, (so $\Omega_0 < 1$ if we assume that the topology of space is simple). Since life can only evolve in regions where (6.122) is roughly satisfied, and because in an infinite Universe there is a finite probability that an arbitrarily large region obeying (6.122) will occur somewhere if the initial conditions are *random*, we would expect to

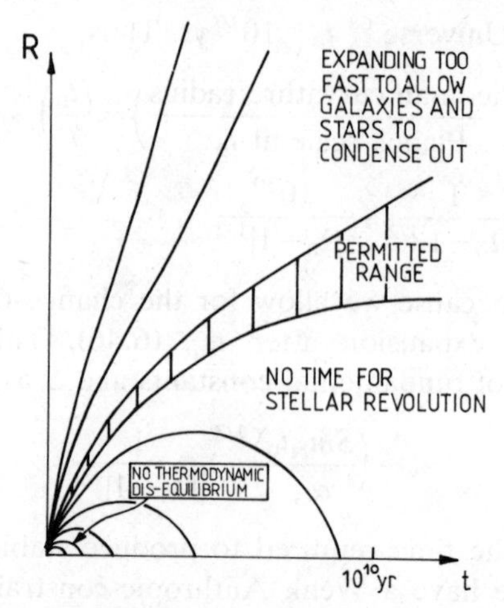

Figure 6.8. In universes that are expanding much more slowly than the rate which allows them to lie close to the critical, $\Omega_0 = 1$, state, the universe will evolve to a second singularity too soon for stars to form and evolve ($\gtrsim 10^9$ yr) or even for conditions to cool off sufficiently for non-equilibrium structures like atoms to form ($\gtrsim 10^6$ yr). If the expansion is much faster than the critical rate, material will recede with so high a velocity that gravitational condensations like stars and galaxies will not form. Only for a range of initial conditions lying close to $\Omega_0 = 1$ will conditions be conducive to the evolution of life in the universe after billions of years.[70]

observe (6.122). A strong Anthropic explanation would have to argue for a special choice of initial condition to produce (6.122) but would allow Ω_0 to be greater than unity. In fact, we shall see below that there may be an explanation for the special state (6.122) as a result of the unusual behaviour of elementary particles during the first 10^{-35} s of the Universe's expansion.

6.9 The Cosmological Constant

> I am a detective in search of a criminal—the cosmical constant. I know he exists, but I do not know his appearance; for instance I do not know if he is a little man or a tall man.
>
> A. S. Eddington

When Einstein first applied his field equations of general relativity to the cosmological problem he discovered that static solutions were impossible.[2] Since there was at that time no observational evidence to suggest the Universe was in a non-static state and the philosophic prejudices of centuries underpinned the notion of a changeless background universe,

Einstein altered his field equations to include the cosmological constant, Λ. The Einstein equations with cosmological constant have a static cosmological solution: the Einstein static universe. The addition of the cosmological constant to general relativity produces a theory which, in the case of weak, non-relativistic gravity fields alters Poisson's equation for the Newtonian gravitational potential, ϕ, to

$$\nabla^2\phi + \Lambda = 4\pi G\rho \tag{6.126}$$

Friedman's equation (6.3) is generalized to

$$\frac{\dot{R}^2}{R^2} = \frac{8\pi G\rho}{3} - \frac{k}{R^2} + \frac{\Lambda}{3} \tag{6.127}$$

In order that at present the term $|\Lambda|/3$ not greatly exceed $8\pi G\rho/3$, the value of $|\Lambda|$ must be very small. In dimensionless form we see this implies (since $\rho_0 < 10^{-29}$ gm cm$^{-3} \sim 10^{-123} m_p^4$) from observations of universal expansion that,[71]

$$\left|\frac{\Lambda}{m_p^2}\right| \lesssim 10^{-120} \tag{6.128}$$

To get an idea of how small this limit is, consider Λ_{min} the smallest value of the parameter Λ that could be measured in $t_0 \sim 10^{10}$ yrs (the age of the Universe) according to the Uncertainty Principle of Heisenberg (which yields $\Lambda_{min}^{1/2} t_0 > \hbar$). This minimum value is larger than the limit (6.120) by nearly 65 orders of magnitude!

$$\frac{\Lambda_{min}}{m_p^2} \gtrsim 10^{-56} \tag{6.129}$$

Indeed, the limit (6.128) is the smallest dimensionless number arising naturally anywhere in physics. It has led to determined efforts to demonstrate that there is some deep underlying principle that requires Λ to be *precisely* zero.[72] Some of these ideas appear promising, but as yet there is no convincing explanation for the smallness of the observational limit on the possible magnitude of Λ. If we express the gravitational lagrangian of general relativity as a constant plus a linear four-curvature term in the standard way then

$$L_g = \Lambda + a_1 R \tag{6.130}$$

and the limit (6.128) implies $\Lambda/a_1 \lesssim 10^{-120}$. However, this limit and its equivalent, (6.128), have great significance for the possibility of life evolving in the Universe. If $|\Lambda|$ exceeds $8\pi G\rho_0$ today then the expansion dynamics are dominated by the Λ term. In the case of $\Lambda < 0$ and $|\Lambda|$ large the Universe will collapse to a second singularity after a time t_s where

$$-\Lambda < \frac{3\pi^2}{(2t_s)^2} \tag{6.131}$$

In order for the Universe to have a total lifetime great enough to evolve stars, produce heavy elements and hence biochemistry, the Universe must have $t_s > t_\star \sim \alpha^2\alpha_G^{-1}(m_N/m_e)^2 m_N^{-1}$ and so we have the Anthropic limit,

$$\frac{|\Lambda|}{m_p^2} \leqslant \alpha^{-4}\left(\frac{m_e}{m_N}\right)^4\left(\frac{m_N}{m_p}\right)^6 \tag{6.132}$$

The same limit applies to Λ/m_p^2 in the case when $\Lambda > 0$ because in this case a violation of (6.132) creates expansion dynamics that are dominated by the positive cosmological constant term at times $t_\star \geqslant t$; hence, by (6.127) with $\dot{R}^2/R^2 \simeq \Lambda/3$

$$R \propto \exp\left(t\sqrt{\frac{\Lambda}{3}}\right) \tag{6.133}$$

and expansion takes place too rapidly for galaxy and subsequent star formation to occur. Gravitational instability is quenched in a medium undergoing rapid expansion like (6.133) and over-densities behave as[73] $\delta\rho/\rho \to$ constant (this is intuitively obvious since Jeans' instability amplifies at a rate $\delta\rho/\rho \propto e^t$ in a static medium and exponential expansion of that medium will exactly cancel the growth rate of the Jeans instability).

There have been various attempts to calculate the constant Λ in terms of other known constants of Nature.[74] These amount to nothing more than dimensional analysis except in one case which we shall examine in detail below. It rests upon the fact that the Λ term in general relativity appears to have a physical interpretation as the energy density, ρ_V, of a Lorentz-invariant quantum vacuum state

$$\Lambda = \frac{8\pi}{m_p^2}\langle\rho_V\rangle \tag{6.134}$$

Unfortunately, it appears that quantum effects arising in the Universe at $t_p \sim 10^{-43}$ s should create $\langle\rho_V\rangle \sim m_p^4$ and $\Lambda \sim m_p^2$ which violates the observational bound and the anthropic limit (6.128) by almost 120 orders of magnitude. How this conclusion is to be avoided is not yet known.

6.10 Inhomogeneity

> Homogeneity is a cosmic undergarment
> and the frills and furbelows required
> to express individuality can be readily
> tacked onto this basic undergarment!
>
> H. Robertson

The accuracy of the Friedman models as a description of our Universe is a consequence of the Universe's homogeneity and isotropy. Only two constants (or three if $\Lambda \neq 0$) are necessary to completely determine the dynamics. The homogeneous and isotropic universes containing matter and radiation are uniquely defined at all times by adding the value of S. But, fortunately for us, the Universe is not perfectly homogeneous. The density distribution is non-uniform with evident clustering of luminous matter into stars, galaxies and clusters. The statistical properties of this clustering hierarchy were outlined in (6.30)–(6.32). Roughly speaking, the level of inhomogeneity in the observable Universe is small and the matter distribution becomes increasingly homogeneous in sample volumes encompassing more than about $10^{15}\,M_{\odot}$. The constant of proportionality and the spectral index n of (6.67) are two further parameters that appear to be specified by the initial data of the Universe, either directly or indirectly.

The modern theory of the development of inhomogeneity in the Universe[21] rests upon the idea that the existing large scale structure that manifests itself in the form of galaxies and clusters did not always exist. Rather, it grew by the mechanism of gravitational instability from small beginnings.[75] Some (statistical?) graininess must have existed in the earliest stages of the Universe and regions of size x would contain a density $\rho(x)$ that exceeds the smooth average density of the universe, $\bar{\rho}$. The amplitude of this inhomogeneity is measured by the density contrast

$$\frac{\delta\rho}{\rho} \equiv \frac{\rho(x)-\bar{\rho}}{\rho} \tag{6.135}$$

As the Universe expands and ages, density inhomogeneities that were once very small ($\delta\rho/\rho \ll 1$) can amplify by gravitational instability until they become gravitationally bound ($\delta\rho/\rho \gtrsim 1$) and then condense into discrete structures resembling galaxies and clusters.

Suppose our Universe to be well-described by a flat or open Friedman model. If the present age of the Universe is denoted by t_0 then all the Friedman models resemble the flat model early on when $t < \Omega_0 t_0$. At such times, and at all times in the flat model, the density inhomogeneities enhance at a rate directly proportional to the expansion scale factor when the pressure is negligible ($p = 0$)

$$\frac{\delta\rho}{\rho} \propto R(t) \propto t^{2/3}, \qquad t \lesssim \Omega_0 t_0 \tag{6.136}$$

However, when a Friedman model with $\Omega_0 < 1$ ages beyond $\sim\Omega_0 t_0$ it becomes dominated by the effect of its negative spatial curvature and approaches free expansion at a rate $R \propto t$. In this regime, distant reference points in the Universe are receding at the speed of light and linear gravitational inhomogeneities cannot grow. They are 'frozen-in' with constant amplitude if they are still linear, $\delta\rho/\rho \leqslant 1$, so

$$\frac{\delta\rho}{\rho} \simeq \text{constant}; \qquad \Omega t_0 \leqslant t; \qquad \Omega_0 < 1 \tag{6.137}$$

This result can be viewed in two ways: if protogalaxies have not grown to a sizeable amplitude, $\delta\rho/\rho \gtrsim 1$, by a cosmic time $t_\star \sim \Omega_0 t_0$ they will never form bound condensed structures like galaxies. Alternatively, given equal initial amplitudes set by some, as yet unknown, aspect of fundamental physics, universes in which Ω_0 is very small allow less time for gravitational instability to occur and so have a lower probability for galaxy formation than those with $\Omega_0 = 1$, where conditions are optimal.

Similar considerations apply to universes that are 'closed' ($\Omega_0 > 1$): In a matter-dominated closed universe the time to the expansion maxim is t_m where

$$t_m = \frac{\Omega_0 \pi}{2H_0(\Omega_0 - 1)^{3/2}} \tag{6.138}$$

If we parametrize the time by the so-called conformal time τ, (which will be discussed in more detail in Chapter 7) the Friedman equation for $p = 0$ admits a parametric solution for $R(t)$ with[76]

$$1 - \cos\tau = 2\left(\frac{\Omega_0 - 1}{\Omega_0}\right)\frac{R(t)}{R(t_0)} \tag{6.139}$$

$$t = H_0^{-1}(\Omega_0 - 1)^{-3/2}(\tau - \sin\tau) \tag{6.140}$$

and small pressureless density inhomogeneities are found to grow as

$$\frac{\delta\rho}{\rho} \propto \frac{5 + \cos\tau}{1 - \cos\tau} - \frac{3\tau\sin\tau}{(1 - \cos\tau)^2} \tag{6.141}$$

These solutions reduce to (6.136) at early times since $t \to 0$ when $\tau \to 0$. However, the larger the value of Ω_0, the shorter the age of the universe at maximum expansion ($\tau_m = \pi$), and the faster the amplification of $\delta\rho/\rho$. Since the total age of the universe is $2t_m$, and this is $\sim 10^{10}\Omega_0^{-1/2}$ yr when $\Omega_0 \gg 1$ we see that main-sequence stellar evolution and biological evolution would not have time to occur if $\Omega_0 > 10^4$. If $\Omega_0 \gg 1$ and the initial value of $\delta\rho/\rho$ were the same as in the flat model ($\Omega_0 = 1$), then the density inhomogeneities would rapidly evolve into condensations of high density or black holes. Equation (6.141) shows that $\delta\rho/\rho$ grows at a faster rate than $t^{2/3}$ when $\Omega_0 > 1$.

In order to produce gravitationally bound structures resembling galaxies and clusters, the density contrast $\delta\rho/\rho$ must have attained a value ~ 5 in the recent past. The above equations[77] allow the following general conclusions to be arrived at:

(a) if the initial conditions are such that $\delta\rho/\rho$ exceeds a 'critical value' equal to $(1+z_i)^{-1}(1-\Omega_0)\Omega_0^{-1}$ at a redshift $z_i \sim 10^3$ then density inhomogeneities will collapse and form galaxies prior to a redshift $z > \Omega_0^{-1}-1$.

(b) if initial conditions are such that $\delta\rho/\rho$ is roughly equal to the 'critical' value of (a) at z_i, then by the present it will have attained a fixed value $\sim 4\Omega_0^{-1}/9$ and galaxies and clusters will not condense out of the overall expansion.

(c) if initial conditions are such that $\delta\rho/\rho$ is significantly less than the 'critical' value at z_i then the present density contrast approaches a steady asymptotic value of order $1.5\left(\frac{\delta\rho}{\rho}\right)_{z=z_i}(1+z_i)\Omega_0(1-\Omega_0)^{-1} \ll 1$ and, again, no galaxies or clusters will condense out.

Thus, we see there is a narrow range of initial conditions for the inhomogeneity level of the Universe with $\delta\rho/\rho \sim 10^{-3}-10^{-4}$ at the recombination redshift, $z_i \sim 10^3$, which allows galaxies to form. If $(\delta\rho/\rho)_i \gtrsim 10^{-2}$ then non-uniformities condense prematurely into black holes before any stars form. If $(\delta\rho/\rho)_i \lesssim 10^{-5}$ then inhomogeneities are unable to amplify sufficiently for galaxies and clusters to form. The conditions at the initial singularity (or in its vicinity) of our Universe have given rise to a level of inhomogeneity of order 10^{-4}. Of course, the level of inhomogeneity may vary with length scale in the initial conditions. In Figure 6.9 we show[68] the observational upper limits on the inhomogeneity amplitude $\delta\rho/\rho$ over various length scales together with the minimum level necessary to form galaxies and clusters under various assumptions for the composition of the non-luminous matter content discussed earlier.

These results show that the amplitude and spatial variation of the density non-uniformity is crucial to the evolution of galaxies and stars. A very narrow range of initial amplitudes allow these structures to originate and life to form within them. This adds a further parameter to the list of cosmological invariants (H_0, Ω_0, Λ, S) that we have already delineated as underpinning the Universe's structure. Later, we shall discuss the possibility that this dimensionless amplitude might be expressed in terms of other constants of Nature. Finally, we should remark that our discussion of the particular inhomogeneity level necessary to make the Universe habitable has assumed, for simplicity, that the other cosmological parameter S is fixed with its observed value $\sim 10^9$. However, if S is regarded as a free parameter, along with δ, the inhomogeneity amplitude, so defining a plane of possible universes, then the relative role of radiation in the history of density perturbations is altered and the combination (δ, S) controls the development of gravitationally bound structures. The situation is summarized in Figure 6.10, drawn[78] for the case of $\Omega_0 = 1$. The

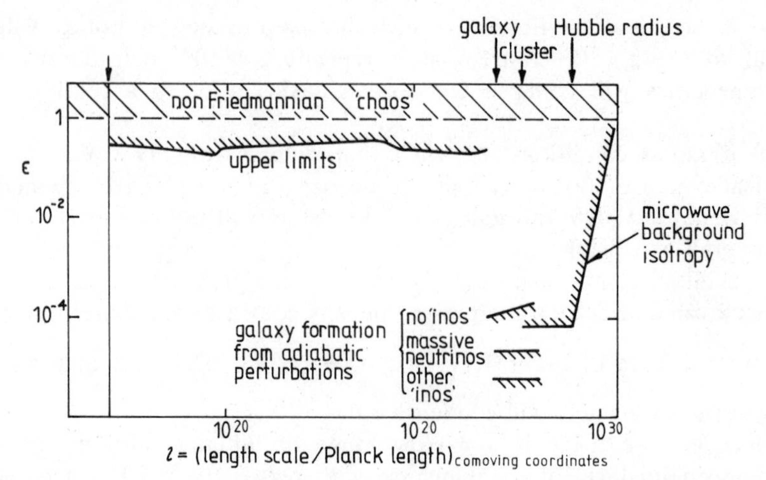

Figure 6.9. Limits on the amplitude of adiabatic density perturbations, ε, from uniformity on various cosmic length scales. On large scales, strong upper limits are imposed by the isotropy of the microwave background radiation. If galaxies and clusters are to form from these over-densities then certain lower limits on ε are required. They differ in the cases of universes with massless neutrinos, massive neutrinos, and heavier neutrinos or axion-like particles as discussed in the text. If ε is to be approximately scale independent then we see that the amplitude must be of order 10^{-4}–10^{-5}, (based on ref. 68).

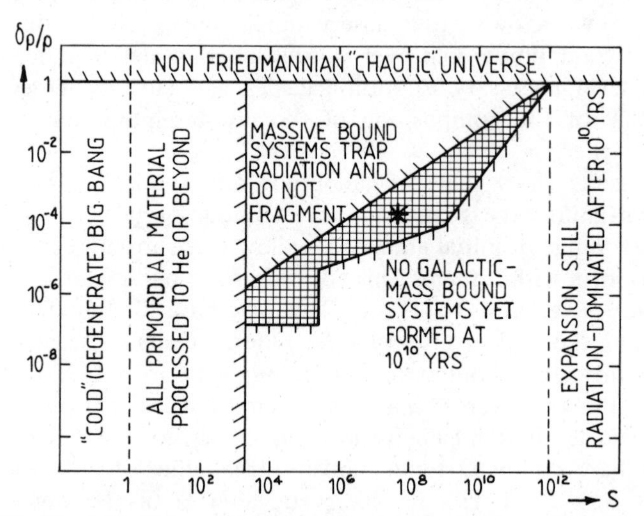

Figure 6.10. The behaviour of universes characterized by arbitrary combinations of dimensionless inhomogeneity amplitude, ε, (see Figure 6.9) and entropy per baryon S. In our Universe (marked by the cross when $\Omega_0 = 1$) we have $(\varepsilon, S) \simeq (10^{-4}, 10^9)$. Only the hatched region of the ε–S plane leads to the formation of galaxies which can fragment into stars. It is likely that this region describes cognizable universes. If the cosmological model has $\Omega_0 < 1$, the allowed, hatched region decreases in size (based on ref. 78).

hatched region of the (δ, S) plane allows massive systems like galaxies to condense. If $\Omega_0 < 1$ the 'habitable' region decreases in size.

6.11 Isotropy

> The fact that we have observed the universe to be isotropic is only a consequence of our existence.
>
> S. W. Hawking

The remaining large-scale properties of the Universe, in addition to H_0, Ω_0, S and $\delta\rho/\rho$, are associated with its isotropy. In theory, it would appear that the Universe could possess significant large-scale shear, rotation and curvature anisotropy (see equations (6.5) and (6.6)). These properties are determined by the cosmological initial conditions. The fact that no large-scale anisotropy of this sort has ever been detected witnesses to something very special about either the cosmological initial conditions or the early history of the Universe. It has also been investigated to what extent the extreme isotropy of the Universe (in effect its accurate description by the Friedman metric (6.4)) is connected with the existence of observers within it.

No result has been more frequently exhibited as an example of the Anthropic Cosmological Principle than the remarkable conclusion of Collins and Hawking that[69]

> '... the isotropy of the Universe and our existence are both results of the fact that the Universe is expanding at just about the critical rate. Since we could not observe the Universe to be different if we were not here, one can say, in a sense, that the isotropy of the Universe is a consequence of our existence.'

This striking statement emerges from the results of a technical paper entitled '*Why is the Universe Isotropic*?' To understand why this question was asked how it was answered and what the answer has to do with ourselves we must provide a thumb-nail sketch of an unusual cosmological conundrum.

Since its discovery in 1964 the 3 K microwave background radiation has provided a wealth of information about the past and present structure of the Universe. Its most striking property is a temperature isotropy over a wide range of angular scales, from seconds to degrees of arc on the sky.[10] The fact that the temperature and intensity of the microwave background radiation appear independent of direction in the sky to within one part in a thousand is telling us something extraordinary about the Universe. Over the largest observable dimensions, ~6000 Mpc, the expansion dynamics of the Universe are meticulously coordinated and isotropic. This singular state of affairs is revealed to be even more baffling when we recall that the expanding universe has a 'horizon' structure[79]; that is, the finite velocity of light partitions the Universe into causally coherent volumes which have not had time to send light signals to each other since the beginning of the Universe (for example, if the age of the Universe is t_u then, roughly speaking, regions separated by a distance $x > ct_u$ will have had insufficient time to communicate internally and enter each others' horizons. Thus, if the standard model is true, regions of the microwave background separated by more than $\sim 30°$ on the sky could not have been in causal communication with each other at or before the time when the radiation was last scattered to us. How, then, did these widely separated and apparently independent regions conspire to have the same temperature and radiation density today to better than one part in a thousand?

Traditionally this question has not been asked. Prior to the discovery of the microwave background radiation the only direct test of the isotropy of the Universe was the counting of galaxies in different solid angles around the sky and was first done by Edwin Hubble. Because the evidence gleaned by this technique was so meagre and non-uniform solutions of general relativity so hard to find, cosmologists concentrated their early theoretical and observational studies upon the simple, isotropic Friedman universes. And with good reason: these simple models provide a remarkably accurate description of the present Universe only because the Universe is so close to isotropy and homogeneity. The question of why this large scale isotropy and uniformity exists was never asked by cosmologists prior to 1967. Rather, they were interested in explaining the presence of the small deviations from perfect homogeneity: the heterogeneities that grew into galaxies, stars, planets and ultimately, ourselves.

Soon after the isotropy of the microwave background was first measured over large angular scales, the American physicist Charles Misner[80] realized that the question of the origin of the large scale regularity of the Universe was potentially more important than the long-studied problem of the origin of the small-scale irregularities. Misner made a new proposal which sought to transform general relativistic cosmology from simply being a way to describe and correlate observations of the Universe into a predictive theory. Misner's 'chaotic cosmology' programme was proposed with the hope that,[81]

> ideally one might try to show that almost all solutions of the Einstein equations which lead to star formation also have many other properties compatible (or incompatible!) with observation. More modest but more feasible approaches would attempt to survey much more limited classes of solutions of the Einstein equations to see whether some presently observable properties of the Universe may be largely independent of the initial conditions admitted for study.

The goal was to show that the present large-scale structure of the

Universe is largely independent of the initial conditions at the Big Bang. Almost every set of physically realistic initial conditions might be shown to evolve inevitably towards regularity as the Universe expands and ages. If this could be established it would prove the Universe unique in theory as well as in practice. It appears that this sort of idea was first considered by Hoyle and Narlikar in 1963. They showed that the steady-state cosmological model was stable against perturbations to its isotropy and inhomogeneity and concluded that[73]

> ...provided the continuous creation of matter is allowed, the creation acts in such a way as to smooth out an initial anisotropy or inhomogeneity over any specified volume... In other words, any finite portion of the universe gradually loses its "memory" of an initially imposed anisotropy or inhomogeneity... the universe attains the observed regularity irrespective of initial boundary conditions.

Let us consider a more everyday example of this loss of memory of initial conditions which illustrates the philosophy of the 'chaotic cosmology' programme: suppose you place a colleague on a cliff-top, armed with stones. You then blindfold yourself and ask him to begin throwing the stones down onto the beach. Could you predict the speed with which the stones will hit the beach? Yes you could, so long as the cliff is high enough. The reason for your clairvoyance is the action of air resistance on the falling stones. It ensures that after they have fallen for a sufficient time the downward force of gravity will be balanced by the upward force of resistance; no net force now acts on the falling stone and, in accord with Newton's third law of motion, it falls at a *constant* 'terminal velocity'. If the resisting force of the air opposing motion is mkv, then after time t the velocity v of the stone of mass m falling from the cliff-top with some unknown initial velocity v_0 can be determined from Newton's second law of motion as

$$v(t) = v_\infty(1 - e^{-kt}) + v_0 e^{-kt} \tag{6.142}$$

where the *terminal velocity*, v_∞, is derived from the acceleration due to gravity, g, and the air friction, k, as

$$v_\infty = g/k \tag{6.143}$$

For air $k \sim 0.1\ \mathrm{s}^{-1}$, $g \sim 9.81\ \mathrm{ms}^{-2}$ and so v_∞ is $\sim 98\ \mathrm{ms}^{-1}$. The frictional resistance causes an exponential ($\propto e^{-kt}$) decrease in the relevance of the unknown condition v_0 for the determination of the stone's velocity at a later time. Chaotic cosmology is a more grandiose application of this simple idea: it envisages that however non-uniform and chaotic the cosmological initial conditions were, as the Universe expands and ages so there might arise natural frictional processes that cause dissipation of the initial non-uniformities, and, after a sufficiently long time, ensure the

Universe would inevitably appear isotropic and smooth. If this scenario were true one could 'predict' the isotropy of the microwave background radiation as an inevitable consequence of gravitation alone.

The appeal of this type of evolutionary explanation is obvious: it makes knowledge of the (unknowable!) initial conditions at the 'origin' of the Universe largely superfluous to our present understanding of its large scale character. In complete contrast, the alternative 'quiescent cosmology'[82] pictures the present state of regularity as a reflection of an even more meticulous order in the initial state.

Unfortunately, it transpired that Misner's programme did not possess the panaceatic properties he had hoped. Viscous processes can only smooth out anisotropies in the initial state if these anisotropies are not too large in magnitude and spatial extent.[83] If the anisotropies over-step a certain level the overall expansion rate of the Universe proceeds too rapidly for inter-particle collisions to mediate viscous transport processes. In this rapidly expanding, non-equilibrium environment the Einstein equations possess an important property: the present structure of the Universe is a unique and continuous function of initial conditions and a counter-example to the chaotic cosmology scheme is now easy to construct: pick any model for the present-day Universe which is in conflict with the isotropy measurements on the microwave background. Evolve it backwards and it will generate a set of initial conditions to the Einstein equations which do not tend to regularity by the present, irrespective of the level of dissipation. In the context of our example described by equations (6.142) and (6.143), if we make observations at some predetermined time T then the measured velocity, $v(T)$, can be made arbitrarily large by picking enormous values of v_0, and we could avoid the inevitable asymptotic result $v(T) \approx v_\infty$. Stones thrown with huge initial velocity could confound our predictions that $v \to v_\infty$ inevitably because they need not have attained a speed close to v_∞ by time T. On the other hand, if we pick v_0 first, then there will always be a T such that $v(T)$ is as close as one wishes to v_∞. In cosmology we, in effect, observe a $v(T)$ while v_0 is given at the initial singularity.

This type of objection to the chaotic cosmology programme might not worry us too greatly if it could be shown that the set of counter-examples is of measure zero amongst all the possible initial states for the Universe. This is where the Collins and Hawking paper enters the story. It attempts to discover just how large the set of cosmological initial conditions which do not lead to isotropic Universes really is.

Collins and Hawking sought to demonstrate that the chaotic cosmological principle is false and that the *generic* behaviour of physically realistic solutions to Einstein's equations is to approach *irregularity* at late times. To establish this they singled-out for investigation the set of cosmological models that are spatially homogeneous but *anisotropic*. This set is finite in size and is divided into ten equivalence classes according to the particular spatial geometry of the Universe. This classification into ten equivalence classes is called the Bianchi classification and it has a hierarchical structure.[84] The most general members, which contain all the others as special cases, are those labelled Bianchi types VI_h, VII_h, VIII and IX. The Cauchy data for the vacuum cosmological models of these Bianchi types are specified by four arbitrary constants,[85] of which the subscript h marking types VI_h and VII_h is one. Not all of these four general classes contain the isotropic Friedman models though; types VI_h and VIII do not and therefore cannot isotropize completely (although they could, in principle, come arbitrarily close to isotropy). However, the VII_h class contains the isotropic, ever-expanding ('open') Friedman universes and the type IX models include the 'closed' Friedman models which recollapse in the future. Collins and Hawking first investigated the properties of the universes in the VII_h class, and we can view this choice as an examination of the *stability* of the open, isotropic Universe with respect to spatially homogeneous distortions.

Before that examination can be made a definition of *isotropization* must be decided upon. The following criteria were chosen by Collins and Hawking to establish that a cosmological model tends to isotropy:

I1: The model must expand for all future time; $V \to \infty$ where V is the comoving volume.

I2: The energy density in the Universe, ρ, must be positive and the peculiar velocities of the material relative to the surfaces of homogeneity must tend to zero as $t \to \infty$. Precisely, we require $T^{0\mu}/T^{00} \to 0$ as $t \to \infty$ where $T^{\mu\nu}$ is the energy-momentum tensor (the indices μ, ν run over the values 1, 2, 3) and $\rho \equiv T^{00}$.

I3. If σ is the shear in the expansion and if $H \equiv \dot{V}/3V$ is the volumetric expansion rate, then the *distortion* σ/H must approach zero as $t \to \infty$.

I4. If the *cumulative distortion* in the dynamics is defined by $\beta = \int^t \sigma\, dt$, then β must approach a constant[86] as $t \to \infty$.

If the conditions I1–4 are satisfied, the cosmological model was said by Collins and Hawking to *isotropize.*

In order to use these criteria, two further physical restrictions on the properties of matter are required;

M1: *The Dominant Energy Condition* requires that $T^{00} > |T^{\alpha\beta}|$ and says that negative pressures ('tensions') cannot arise to such an extent that they dominate the energy density of the fluid.

M2: *The Positive Pressure Criterion* stipulates that the sum of the principal pressures in the stress-energy tensor must be non-negative: $\sum_{\kappa=0}^{3} T_{\kappa\kappa} \geq 0.$

The conditions M1 and M2 are satisfied by all known classical materials but might be violated microscopically in the Universe if quantum black holes evaporate via the Hawking process or if particle creation occurs in empty space in the presence of strong gravitational fields.[87] However, even in this case the violations would be confined to small regions $\leqslant 10^{-33}$ cm and M1, 2 should still be valid on the average over large spatial scales, late in the Universe.[88] Notice that these conditions on the matter tensor exclude a positive cosmological constant, Λ, and a negative cosmological constant is excluded by I1.

Collins and Hawking then write down the Einstein equations of the VII_h model. They are an autonomous system of non-linear ordinary differential equations[89] of the general form

$$\dot{\mathbf{x}} = F(\mathbf{x}); \qquad \mathbf{x} = (x_1, x_2 \ldots x_n) \tag{6.144}$$

Suppose the isotropic Friedman Universe is the solution of (6.144) given by the null solution (this can always be arranged by a coordinate transfomration of the x_i),

$$\mathbf{x} = 0 \tag{6.145}$$

then it is a necessary condition of the chaotic cosmology programme that this solution be stable. The usual way of deciding whether or not (6.145) is a stable solution to (6.144) we linearize (6.144) about the solution (6.145) to obtain,

$$\dot{\mathbf{x}} = A\mathbf{x} \qquad A : R^n \to R^n \tag{6.146}$$

where A is a constant matrix. Now we determine the eigenvalues of A and if any have positive real part then the Friedman solution (6.145) is *unstable*; that is, neighbouring cosmological solutions that start close to isotropy continuously deviate from it with the passage of time. The situation Collins and Hawking discovered was not so clear-cut. They found one of the eigenvalues of A was purely imaginary and so the stability could not be decided by the linear terms[146] alone. However, they were able to decide the stability by separating out the variable with the imaginary eigenvalue and performing a second order stability analysis on it. The open Friedman universe was shown to be unstable, but the deviations from it grow slowly like $\ln t$ rather than a power of t. More precisely: *If M1 and M2 are satisfied, then the set of cosmological initial data giving rise to models which approach isotropy as $t \to \infty$ is of measure zero in the space of all spatially homogeneous initial data.*

A closer examination[90] of the Bianchi VII_h universe reveals that it fails to isotropize because conditions I3 and I4 are not met. As $t \to \infty$ the ratio of the shear to the expansion rate, σ/H, approaches a constant and $\beta \propto t$. This result tells us that almost every ever-expanding homogeneous Universe which can isotropize will not do so regardless of the presence of dissipative stresses (so long as they obey the conditions M1 and M2).

A detailed investigation of the Bianchi VII_h universe has been made by Barrow and Siklos[91] who have shown that there exists a special solution of Bianchi type VII_h which is stable, but not asymptotically stable, in the space of VII_h initial data. This particular solution, which was found some years ago by Lukash, contains two arbitrary parameters which, when chosen appropriately, can make the expansion arbitrarily isotropic. This result considerably weakens the Collins and Hawking conclusions: it shows that isotropic open universes are stable in the same sense that our solar system is stable. As $t \to \infty$ there exist spatially homogeneous perturbations with $\sigma/H \to$ constant but there are none with $\sigma/H \to \infty$. The demand for asymptotic stability is too strong a requirement. However, despite this we shall assume that the Collins and Hawking theorem retains its force because its interpretation in connection with the Anthropic Principle will transpire to be non-trivial.

Next, Collins and Hawking focused their attention upon a special subclass of the VII_h universes—those of type VII_0. These specialize to the 'flat', Einstein–de Sitter universe when isotropic. These models have the minimum of kinetic energy necessary to undergo expansion to infinity and Euclidean space sections, and are of measure zero amongst all the ever-expanding type VII universes. The stability properties of these universes turn out to differ radically from those in the larger VII_h class. If the matter content of the universe is dominated by fluid with zero pressure—as seems to be the case in our Universe today since galaxies exert negligible pressure upon each other—then flat, isotropic Universes are stable. More precisely: *If matter has zero pressure to first order and M*1 *holds, then there exists an open neighbourhood of the flat* ($k = 0$) *Friedman initial data in the type* VII_0 *subspace of all homogeneous initial data such that all data in this neighbourhood give rise to models which isotropize.*

If the Universe is close to the 'flat' state of zero energy then, regardless of its initial state, it will eventually approach isotropy when it is old enough for the pressure-free material to dominate over radiation. Finally, we should add that if this type of analysis is applied to *closed* homogeneous universes which can isotropize—the type IX models—then one sees that in general they will *not* approach isotropy. A slightly different criterion of isotropization is necessary in this case because $\sigma/H \to \infty$ when the universe approaches maximum volume because $H \to 0$ there even if the universe is almost isotropic; as an alternative criterion, one might require the spatial three-curvature to become isotropic at the time of maximum expansion although it is not clear that the type IX universe model can recollapse unless this occurs.[92]

From these results two conclusions might be drawn; either: (A) *The*

Universe is 'young' and it is not of zero measure amongst all the ever-expanding models and is growing increasingly anisotropic due to the influence of generic homogeneous distortions which have had, as yet, insufficient time to create a noticeable effect upon the microwave radiation isotropy.

Or: (B) *The Universe is a member of the zero measure set of flat, zero binding-energy models. The most general homogeneous distortions admitted by its geometry are of Bianchi type VII_0 and all decay at late times. The Universe is isotropizing but is of zero measure in the metaspace of all possible cosmological initial data sets.*

The stance taken by Collins and Hawking is to support option (B) by invoking the Weak Anthropic Principle in the following manner. We saw in section 6.8 that our astronomical observations show the Universe to be remarkably close to 'flatness', (6.122); indeed, this is one of the reasons it has proven so difficult to determine whether the Universe is expanding fast enough for infinite future expansion or whether it will recollapse to a second and final space-time singularity. Collins and Hawking conclude that the reason for not observing the Universe to be strongly anisotropic is its proximity to the particular expansion rate required to expand forever. And there is a way we can explain our proximity to this very special state of expansion[69] if

> ... there is not one universe but a whole infinite ensemble of universes with all possible initial conditions. From the existence of the unstable anisotropic mode it follows that nearly all of the universes become highly anisotropic. However, these universes would not be expected to contain galaxies, since condensations can grow only in universes in which the rate of expansion is just sufficient to avoid recollapse. The existence of galaxies would seem to be a necessary precondition for the development of any form of intelligent life.

In the last section we saw how the probability of galaxy formation is closely related to the proximity of Ω_0 to unity. In universes that are now extremely 'open', $\Omega_0 \ll 1$, density inhomogeneities do not condense into self-gravitating units like galaxies, whereas if $\Omega_0 \gg 1$ they do so very rapidly and all regions of above average density would evolve into supermassive black holes before life-supporting biochemistry could arise. Conditions for galaxy formation are optimal in Universes that are flat, $\Omega_0 = 1$. We would not have expected humans to have evolved in a universe that was not close to flatness and because flat universes are stable against anisotropic distortions 'the answer to the question "why is the universe isotropic"? is "because we are here"'.[69]

Striking as the previous argument appears, it is open to criticism in a variety of places. We have already mentioned that Collins and Hawking could simply have concluded that the universe is relatively young, open, and tending towards anisotropy but they felt 'rather unhappy about believing that the universe had managed to remain nearly isotropic up to the present day but was destined to be anisotropic eventually'. However, the Anthropic Principle provides just as good a basis for this interpretation as it does for the sequential argument that observers require heavy elements which require stars and galaxies and these require spatial flatness which, in turn, ensures isotropy at late times. There are good reasons why we should be observing the Universe when it is relatively youthful and close to the main-sequence stellar lifetime $\sim 10^{10}$ yr. All stars will have exhausted their nuclear fuel in $\sim 10^{12}$ yrs and galaxies will collapse catastrophically to black holes after $\sim 10^{18}$ yr; all nuclear matter[93] may have decayed after $\sim 10^{31}$ years. Planet-based beings like ourselves could not expect to observe the Universe in the far future when the effects of anisotropy had grown significant and when any deviation from flatness becomes unmistakable because life-supporting environments like ours would, in all probability, no longer exist for carbon-based life. If we scrutinize the calculations which demonstrate the isotropic open universes to be unstable we shall see we have to take this 'young universe' option (A) more seriously than the Collins–Hawking interpretation (B).

The criteria I1–4 adopted for isotropization are *asymptotic* conditions that are concerned only with the cosmological behaviour as $t \to \infty$. In the case of open universes a powerful result was possible without making assumptions like M1 or M2 about the matter content of the universe. This is a reflection of the fact that as $t \to \infty$ 'matter ceases to matter' in open universes. The dynamical evolution becomes entirely dominated by the three-curvature of the space-time. Thus, the proof that open, isotropic universes are unstable is only a statement about the late *vacuum* stage of their evolution. It would be quite consistent with Collins and Hawking's result if almost every open universe *tended to isotropy* up until the time when it became vacuum-dominated ($t_\star \sim \Omega_0 t_0$) and then tended towards anisotropy thereafter. Since we are living fairly close to $t_\star$, ($\Omega_0 \sim 10^{-1}$ in our World), the presence of comparative isotropy in our Universe has little or nothing to do with a proof that open universes become increasingly anisotropic in their vacuum stages. Perhaps open universes also become increasingly anisotropic during temporary radiation or matter-dominated phases but this is not yet known (although recent analyses[94] indicate they do not). The Universe could be open, have begun in a very anisotropic state and have evolved towards the present state of high isotropy without in any way conflicting with the theorems of ref. 69. In such a situation the present level of isotropy does not require close proximity to flatness for its explanation and the Anthropic interpretation of Collins and Hawking become superfluous.

The proof that flat anisotropic models approach isotropy requires a condition on the matter content of the Universe (M1 and M2). This is not surprising since flat models, by definition, contain sufficient matter to influence the expansion dynamics at all times. Their stability depends crucially upon the matter content and would not exist if the flat universe were filled with radiation ($p = \rho/3$) rather than dust ($p = 0$). Yet, the bulk of the Universe's history has seen it dominated by the effects of radiation. Only comparatively recently, after $t_{eq} \sim 10^{12}$ s, has the influence of pressureless matter predominated. So the theorem that flat, isotropic universes are stable tells us nothing about their behaviour during the entire period of classical evolution from the Planck time, $t_p \sim 10^{-43}$ s, until the end of the radiation era at $t_{eq} \sim 10^{12}$ s. It tells us only that anisotropies must decay after t_{eq} up until the present, $t_0 \sim 10^{17}$ s, if the Universe is flat. The Universe could have begun in an extremely irregular state (or even a comparatively regular one) and grown increasingly irregular throughout its evolution during the radiation era until t_{eq}. The anisotropy could then have fallen slightly during the short period of evolution from t_{eq} to t_0 yet leave the present microwave background anisotropy greatly in excess of the observed level. Again, a flat, dust-dominated universe could be highly anisotropic today without in any way contradicting the theorems of Collins and Hawking and without in any way invoking the Anthropic Principle.

Another weakness of the Anthropic argument for the isotropy of the Universe is that it is based upon an unconfirmed theory for the origin of protogalaxies. For, we might claim, the outstanding problem in explaining the presence of galaxies from the action of gravitational instability on small fluctuations from homogeneity in *any* cosmological model is the size and nature of the initial fluctuations. In existing theories, these initial amplitudes are just chosen to grow the observed structure in the time allowed. Thus the cosmological model begins with the protogalaxies embedded in it at 10^{-43} s, and they are given just the right appearance of age to grow galaxies by now $\sim 10^{17}$ s. The is amusingly similar to the theory of Philip Gosse[95] who, in 1857, suggested a resolution of the conflict between fossils of enormous age and religious prejudice for a very young Earth might be provided by a scheme in which the Universe was of recent origin but was created with ready-made fossils of great apparent age already in it!

So, in practice, *ad hoc* initial amplitudes are chosen to allow *flat* Friedman universes to produce galaxies by the present. If these amplitudes were chosen significantly smaller or larger in the flat model the theory would predict no galaxies, or entirely black holes, respectively. By the same token, initial amplitudes might be chosen correspondingly larger (or smaller) in very open (or closed) models to compensate for the slower (or faster) amplification up to the present. Such a procedure would be no less *ad hoc* than that actually adopted for the flat models. It is therefore hard to sustain an argument that galaxies grow too quickly or too slowly to allow the evolution of observers in universes deviating greatly from flatness.

It could also be argued that to establish whether or not isotropy is a stable property of cosmological models one must examine general *inhomogeneous* cosmologies close to the Friedman model. Strictly speaking, spatially homogeneous models are of measure zero amongst the set of *all* solutions to Einstein's equations. This may not be as strong an objection as it sounds, probably not as strong as those arguments given against the Anthropic Principle explanation above. The instability of open universes could only be exacerbated by the presence of inhomogeneities, but it is possible that flat universes might turn out to be unstable to inhomogenous gravitational wave perturbations. A resolution of this more difficult question virtually requires a knowledge of the general solution to the Einstein equations and this is not likely to be found in the very near future. A few investigations of the late-time behaviour of inhomogeneous models[96] do exist but are not helpful since they examine very special models that are far from representative of the general case. It could be argued that the real Pandora's box opened by the inclusion of inhomogeneous universes is the possibility of an *infinite inhomogeneous universe.*

Our observations of the 'Universe' are actually just observations on and inside our past light-cone, which is defined by that set of signals able to reach us over the age of the universe. The structure of our past light-cone appears homogeneous and isotropic but the grander conclusion that the entire universe possesses this property can only be sustained by appeal to an unverifiable philosophical principle, for example, the 'Copernican' principle—which maintains that our position in the Universe is typical. As Ellis has stressed,[97] it is quite consistent with all cosmological observations so far made to believe that we inhabit an infinite universe possessing bizarre large scale properties outside our past light-cone (and so are unobservable by us), but which is comparatively isotropic and homogeneous on and inside that light-cone. This reflects the fact that we can observe only a finite portion of space-time.[98] If the Universe is 'closed'— bounded in space and time—this finite observable portion may comprise a significant fraction of the entire Universe if Ω_0 is not very close to unity and will allow conclusions to be drawn from it which are representative of the whole Universe. However, if the Universe is 'open' or 'flat' and infinite in spatial extent, our observational data has sampled (and will only ever sample) an *infinitesimal* portion of it and will never provide an adequate basis for deductions about its overall structure unless augmented by unverifiable assumptions about uniformity. If the Universe is infinite and significantly inhomogeneous the Collins and Hawking analysis would not even provide an answer to the question 'why is our past light-cone isotropic' unless one could find general inhomogeneous solutions to Einstein's equations which resembled the VII_h and VII_0 models locally. But perhaps in this infinite, inhomogeneous universe the Anthropic explanation could re-emerge. Only some places within such a universe will be conducive to the presence of life, and only in those places would we expect to find it. Perhaps observers at those places necessarily see isotropic expansion; perhaps only world-lines with isotropic past light-cones eventually trace the paths of intelligent beings through space and time.

6.12 Inflation

> It is therefore clear that from the direct data of observation we can derive neither the sign nor the value of the curvature, and the question arises whether it is possible to represent the observed facts without introducing a curvature at all.
>
> A. Einstein and W. de Sitter

We have seen that our Universe possesses a collection of unusual properties—a particular, small level of inhomogeneity, a high degree of isotropy and a close proximity to the 'critical' density required for 'flatness'. All of these properties play an important role in underwriting the cosmological conditions necessary to evolve galaxies and stars and observers. Each has, until recently, been regarded as an independent cosmic conundrum requiring a separate solution. We can always appeal to very special starting conditions at the Big Bang to explain any puzzling collection of current observations but, in the spirit of the 'chaotic cosmologists' mentioned in the last section, it is more appealing to find physical principles that require the Universe to possess its present properties or, less ambitiously, to show that some of its unusual properties are dependent upon the others.

A new approach to explaining some of these fundamental cosmological problems began in 1981 with the work of Sato[99] and Guth.[67] Subsequently this package of ideas, dubbed the 'inflationary universe' by Guth, has undergone a series of revisions and extensions.[100] We shall focus upon general points of principle desired of any working model of the inflationary type.

During the first 10^{-35} s of cosmic expansion the sea of elementary particles and radiation that fill the Universe can reside in a variety of physical states that physicists call 'phases'. At a more elementary level, recall that ordinary water exists in three phases of gaseous, liquid or solid type which we call steam, water and ice respectively. These 'phases' correspond to different equilibrium states of the molecules. Steam is the most energetic state whilst ice is the least energetic. If we pass from a high to a low energy state then excess heat will be given out. This is why your hand will be scalded when steam condenses upon it.

If changes of phase occur between the different elementary particle states in the early universe then dramatic events can ensue. The energy difference between the two phases can accelerate the expansion of the Universe for a finite period of time. This brief period of 'inflation' can produce a series of remarkable consequences.

Alternative phases can exist for the scalar Higgs fields, ϕ, associated with the supermassive X and Y bosons that we discussed earlier in connection with the baryon asymmetry of the Universe. They will possess some potential energy of interaction, $V(\phi)$, that will in general depend upon the temperature. The ϕ field evolution in the expanding Universe will be governed by a wave equation

$$\ddot{\phi}+\frac{3\dot{R}}{R}\dot{\phi}+V'(\phi)=0 \tag{6.147}$$

where $R(t)$ is the usual scale factor of the Friedman universe. The ϕ field has a total energy (kinetic $\dot{\phi}^2/2$ and potential V) given by

$$\rho_\phi=\tfrac{1}{2}\dot{\phi}^2+V(\phi) \tag{6.148}$$

and so the cosmological dynamics will be governed by a Friedman equation

$$\frac{\dot{R}^2}{R^2}=\frac{8\pi G\rho}{3}-\frac{k}{R^2} \tag{6.149}$$

where the density, ρ, is composed of ρ_ϕ and the radiation density, ρ_γ

$$\rho=\rho_\phi+\phi_\gamma \tag{6.150}$$

The popular 'new inflationary universe' model of[100] Linde, Albrecht and Steinhardt, and Hawking and Moss assumes that the potential energy $V(\phi)$ possesses the particular shape and temperature behaviour exhibited by a class of potentials first investigated for other reasons by Coleman and Weinberg.[101] Their character is exhibited in Figure 6.11:

When the temperature is very high the potential has a single minimum (or 'vacuum' state) at $\phi=0$. As the temperature approaches the unification energy, $m_X\sim10^{14}$ GeV, a second local minimum appears in $V(\phi)$ at $\phi=\sigma\neq0$. As the temperature falls below m_X this asymmetric local minimum becomes the global minimum (or 'true vacuum') with $V(\sigma)\ll V(0)$. Thus if ϕ were to evolve from $\phi=0$ down the potential to $\phi=\sigma$

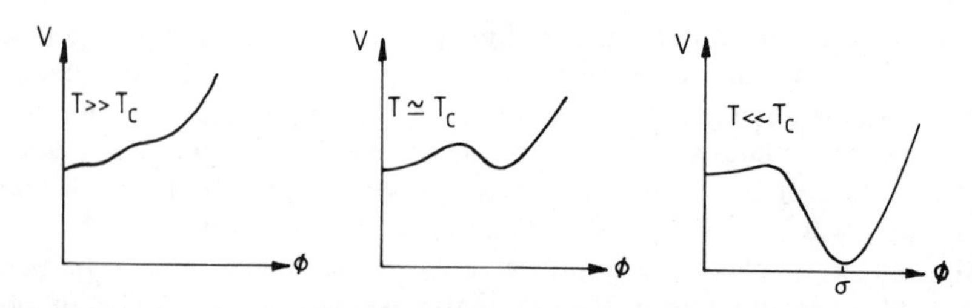

Figure 6.11. The Coleman–Weinberg form of the effective Higgs field potential, $V(\phi)$: (a) when T exceeds the unification energy, T_c, it displays a single, symmetric global minimum; (b) when T equals T_c and there exist two minima, and (c) when T falls below T_c and the asymmetric minimum at $\phi=\sigma$ becomes the true minimum of $V(\phi)$ and the symmetric minimum at $\phi=0$ is now metastable.

the energy difference ΔV could be liberated. Besides the multiple vacuum states the Coleman–Weinberg interaction potentials have one other vital property. Near $\phi=0$ the potential is extremely shallow, $V(\phi)\sim\phi^3\ln(\phi^2/\sigma^2)$ and if the ϕ field tunnels through the low barrier near $\phi=0$, (or is thermally excited over it), it will roll very *slowly* down the potential towards the true vacuum at $\phi=\sigma$. This 'slow roll-over' enables the vacuum energy $V(\phi)$ to dominate the Friedman equation (6.149) during the transition. 'Slow-rolling' means $\dot{\phi}/\phi\ll\dot{R}/R$; that is, the ϕ field evolves more slowly than the Universe is expanding and hence (6.147) gives $3\dot{R}\dot{\phi}R^{-1}\approx-V'(\phi)$ and $\rho_\phi\approx V(\phi)\approx$constant. The characteristic 'rolling' time is just determined by ρ_ϕ, and the Friedman equation (6.149) yields an exponentially expanding solution[99]

$$R(t)\propto\exp(t/t_I)\qquad(6.151)$$

where 'e-folding' time of the expansion is

$$t_I\equiv\left(\frac{8\pi G}{3}\rho_\phi\right)^{-1/2}\simeq\left(\frac{8\pi G}{3}V(0)\right)^{-1/2}\qquad(6.152)$$

When ϕ reaches $\phi\sim\sigma$ it evolves more rapidly ($\dot{\phi}/\phi\gtrsim\dot{R}/R$) as the potential steepens and the vacuum energy is converted into radiation in particle decays and production which can reheat the universe to a temperature exceeding that necessary for baryon number generation. After the transition has been completed the expansion will resume its usual radiation-dominated behaviour with $R(t)\propto t^{1/2}$ again. However, the brief period of exponential inflation has specific results:[67]

(a) It offers an explanation for the present proximity of the Universe to the critical density, $\Omega_0\sim1$, which we termed the 'flatness' problem above. Any period of exponential expansion (6.151) during which $\rho\approx$constant because it is dominated by ρ_ϕ in (6.150) will result in the curvature term

kR^{-2} becoming negligible with respect to $8\pi G\rho/3$ by a huge factor $\sim\exp(2t/t_I)$, even if inflation lasts only for $t\sim 100t_I$. This small period of inflationary expansion would also explain how the Universe has contrived to become nearly sixty orders of magnitude greater than the natural length scale of a self-gravitating, general relativistic system, $l_p\sim 10^{-33}$ cm or equivalently, why its age is of order $10^{60}t_p$.

(b) The exponential expansion leads to a rapid increase in the cosmological horizon size over which causal signals can propagate. Regions of size $\sim m_X^{-1}\sim 10^{-25}$ cm when the phase transition begins at $T\sim m_X$ would be correlated by quantum fluctuations. The region destined to expand out to encompass the presently observable Universe is, at this time, compressed into ~ 10 cm and is causally disjoint. Now if inflation were to ensue for a period $\sim Nt_I$ then the causally related regions would be inflated to encompass $\sim e^N m_X^{-1}$. With $N\gtrsim 70$ this exceeds ~ 10 cm. Thus the present regularity of our Universe could be explained: the distance over which causality can correlate properties has been exponentially increased at an early moment.

(c) The premature enlargement of the size of causally related regions at $t\sim 10^{-35}$ s can resolve the dilemma of magnetic monopoles in grand unified theories[67] that had been pointed out by Khlopov and Zeldovich[102] and Preskill.[103] Monopoles arise because of misalignments in the Higgs fields between one point of space and another.[104] The mismatches can only be erased by causal processes and so they remain over dimensions exceeding $\sim 10^{-25}$ cm at the time of grand unification. The monopoles are stable particles of large mass $\sim m_X$ and are predicted to contribute more than 10^{10} times the observed density in galaxies on the basis of this argument! Inflation removes this problem. Each causally coherent region of scale $\sim m_X^{-1}$ can inflate to encompass our entire observable Universe. Within that region, Higgs fields would have been aligned by causal process and at most one monopole would be generated within the region that expands to become our observable Universe.

(d) The fact that the right-hand side of (6.149) is dominated for a period by the 'vacuum' energy, $\rho_\phi\approx$constant, ensures that the Universe evolves rapidly towards the de Sitter space-time[105]

$$ds^2=-dt^2+e^{2Ht}(dx^2+dy^2+dz^2) \tag{6.153}$$

All small anisotropies die away on approach to this homogeneous space-time and a geodesic observer will observe the universe becoming more and more isotropic.[73] However, the inflated ϕ field will not be perfectly smooth within the region of size $\sim m_X^{-1}$ that inflates to encompass what we observe today. It will inevitably possess quantum fluctuations and these will be amplified by the inflation process to create a definite spectrum of density inhomogeneities which may subsequently evolve into fully fledged galaxies and clusters.[106] The spectral index of the density perturbations (see (6.67)) can be deduced in a straightforward way. The de Sitter metric (6.153) is a homogeneous space-time: we see that it is time-translation invariant, for if we put $t \to t+T$ with T constant then the metric remains invariant so long as we just re-label spatial coordinates (x, y, z) by

$$e^{HT}(x, y, z) \tag{6.154}$$

This means that no length scale can be identified by the magnitude of density irregularity it receives. The *metric* perturbation must be the same on every length scale as it enters the horizon; this requires

$$\frac{\delta\rho}{\rho} \propto M^{-2/3} \tag{6.155}$$

This spectrum is known as the 'Harrison–Zeldovich'[107] or 'constant curvature' spectrum and is consistent with observation (see Figure 6.9) so long as the amplitude of proportionality is of order 10^{-4}. Inflation can make a definite prediction of this amplitude yielding a result $\sim H^2\dot{\phi}^{-1}$ where $H \sim t_I^{-1}$ in (6.152) and (6.153) and $\dot{\phi} \sim$ constant during the slow roll-over. The first calculations of this amplitude yielded answers nearly 10^5 times too large; subsequently, alternative potentials have been tried and it is a topic of detailed research at present to find a working model that naturally yields the correct magnitude of order 10^{-4}. At present all we can conclude is that inflation can, in principle, predict this amplitude in terms of the grand unified coupling constant $\alpha_X \sim \alpha(10^{14}\text{ GeV})$ via a formula like[108]

$$\frac{\delta\rho}{\rho} \sim \alpha_X \ln^{3/2}\left(\frac{\tau}{t_I}\right) \tag{6.156}$$

where τ is the time when a particular wavelength of inhomogeneity re-enters the horizon after inflation. For a cluster of galaxies we have $\tau \sim 10^{23} t_I$ and (6.156) can only give a sensible amplitude $\sim 10^{-3}$–10^{-4} if α_X is chosen to be unnaturally small $\sim 10^{-7}$.

6.13 Inflation and The Anthropic Principle

> We are unable to obtain a model of the Universe without some specifically cosmological assumptions which are completely unverifiable.
>
> G. F. R. Ellis

The inflationary idea is especially interesting when viewed from an anthropic perspective. Formerly, one could identify a number of *indepen-*

dent properties of the Universe each of which plays a key role in creating a cosmic environment that is life-supporting; but inflation is able to link together a large number of these properties. We also see that these properties are made to depend on fundamental parameters with exponential sensitivity. If the phase transition occurs at time t_1 and lasts for a time t_I in the early universe when its scale is $R(t_1)$, then inflation will create a universe with a size at least of order[109]

$$R^* = R(t_1)\exp(t_1/t_I) \sim t_I \exp(2t_1/t_I) \tag{6.157}$$

in which the radiation temperature eventually will fall below $T_p \exp(-t_1/t_I)$. In order to create a universe large enough, and hence old enough for life to evolve, the ratio t_1/t_I must be bounded below by a number ~ 65. Furthermore, we have seen that the new inflationary picture relies upon the special properties of the Coleman–Weinberg potential, in particular the slow evolution of the ϕ field away from the symmetric minimum.

It is clear from what we have said so far that because our whole observable universe of size $\sim H_0^{-1} \sim 10^{27}$ cm derives from an inflation of the properties of a single quantum fluctuation $\sim m_X^{-1}$, there may exist a non-trivial probabilistic element in the whole picture. In order for a region to inflate it is necessary for it to get trapped in the symmetric $\phi = 0$ vacuum as the Universe cools to $\sim m_X$. But, it is not necessary for the entire Universe to do so. Just one single region of size $\sim m_X^{-1}$ will do. For all we know, regions beyond our horizon may have undergone different amounts of inflation (or none at all). They may be very different in density to the region we observe despite being smooth within their own horizons (Figure 6.12). If the Universe is open and has simple topology (so is infinite) then inflation creates for us an ensemble of causally disjoint universes each of which may possess different size, level of density inhomogeneity, and so on, reflecting the stochastic behaviour of the duration of inflation over space. For example, if the constant term H defining the de Sitter space-time (6.153) were regarded as a Gaussian random variable in space then we would have a stochastic Friedman equation of the form

$$\frac{\dot{R}}{R} \sim H. \tag{6.158}$$

where $H = t_1/t_I$ is the length of inflation. The resulting size of the inflated universe R^* will then be lognormally distributed with a time-dependent variance. Of course, if the universe is *closed* (and although inflation predicts Ω_0 should be very close to unity today, it does not predict whether it will be greater or less than unity) the number of inflated regions comprising the entire Universe will be finite and, even if the

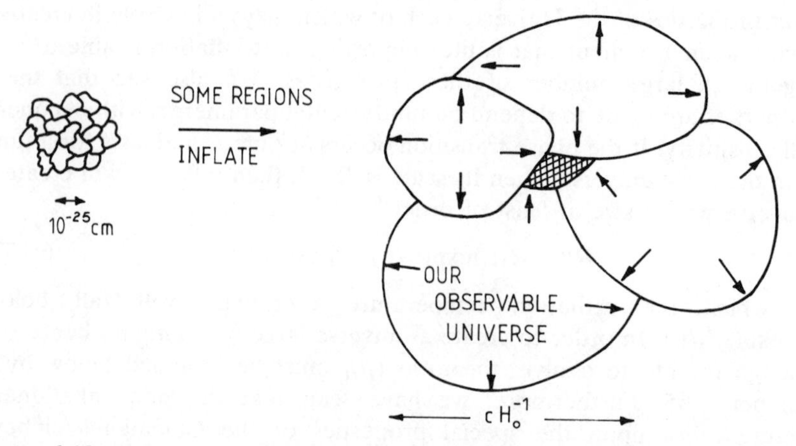

Figure 6.12. Different causally disjoint regions each of the size roughly equal to $m_X^{-1} \sim 10^{-25}$ cm at the epoch of inflation. They undergo different amounts of inflation and are subsequently causally disjoint. We could inhabit a single bubble yet be unable to observe the additional bubbles until the far future. Although each inflated bubble should be very smooth and inhomogeneous, conditions may vary greatly from bubble to bubble (based on ref. 144).

properties were random, all possible configurations would not be realized. It can be argued that in an infinite inflationary universe life can only arise in a region that is significantly inflated and this 'explains' why we observe the Universe to possess the large size, spatial flatness and high isotropy that we do. Other bubbles (Fig. 6.12) that either did not inflate at all, or did so more moderately, would not possess the conditions needed to create long-lived stable stars.

This type of anthropic inflation idea is best illustrated by a suggestion of Linde's[110] which he calls 'chaotic inflation'. It is much simpler in concept than the new inflationary universe model described above because it does not involve any phase transitions or specially chosen potentials. Suppose there exists in Nature a scalar field ϕ with potential energy of interaction, $V(\phi)$, where

$$V(\phi) = \lambda\phi^4; \qquad \lambda < 1 \tag{6.159}$$

with some arbitrary coupling contant λ. This is pictured in Figure 6.13. The precise form of $V(\phi)$ is not crucial here; it is only required that it have a shallow, slowly varying slope near $\phi \sim 0$. The evolution of this field in the early universe will be described by (6.147)–(6.150) as before. Suppose that the Planck time, $t_p \sim 10^{-43}$ s, when the temperature was $\sim 10^{19}$ GeV, the Universe was dominated by chaotic fluctuations of a thermal and quantum nature. We would expect to find the ϕ field having

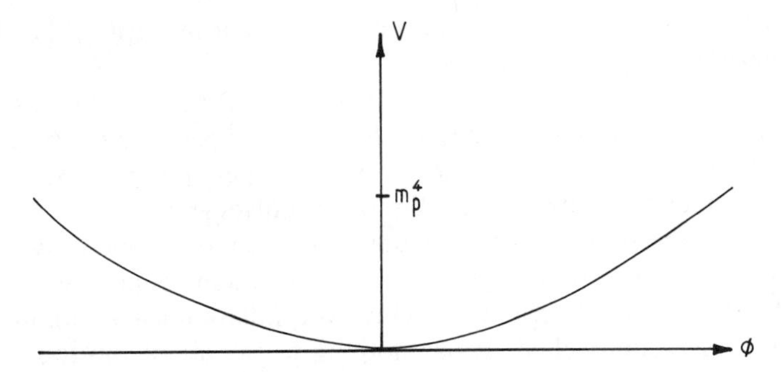

Figure 6.13. A typical shallow potential of the type required in the chaotic inflationary universe picture; here $V(\phi)=\lambda\phi^4$ with λ a dimensionless coupling constant.

some random energy value in the range $-m_p\lambda^{-1/4}$ to $m_p\lambda^{-1/4}$ since the only constraint is that $V(\phi)\leq m_p^4$. If there exists a region larger than $\sim m_p^{-1}$, over which the ϕ field is almost constant (that is, in which $\dot{R}/R \gg V''(\phi)$) then ρ_ϕ will be essentially constant for a time $\sim \dot{R}/R(V'')^2$. This region will inflate if the potential is sufficiently flat and the Universe would expand with[111] $R \propto \exp(Ht)$ where $H=(8\pi V(\phi)m_p^{-2})^{1/2}$. Therefore the typical inflation factor would be $\sim\exp(4\pi/9\lambda^{1/2})$ and it is sufficient to resolve the flatness and horizon problems if the potential is very shallow, $\lambda \leq 10^{-4}$; however, λ may need to be much smaller, $\lambda \sim 10^{-12}$, to generate low level density fluctuations and the correct level of reheating to enable baryosynthesis to occur after inflation.[112] It remains to be established how such a weakly coupled scalar field arises in Nature. The anthropic aspect enters in this picture because it is argued that *somewhere* in an infinite universe that begins in a randomly chaotic state there must exist regions larger than $\sim m_p^{-1}$ over which the ϕ field is roughly constant, as required for the exponential expansion. Only these particular regions will inflate and evolve observers. For many, this type of cosmological theory would be unsavoury. After all, if we wish to appeal to a particular property occurring inevitably in a random initial state,[113] why should we bother conjuring up the $\phi \approx$ constant state necessary for inflation to occur? Why not argue that in a chaotically random, infinite initial data set there *must* exist a large, virtually homogeneous and isotropic region, expanding sufficiently close to flatness ($\Omega_0=1$) so that after fifteen billion years it looks like our universe? In fact, special initial data of this type must arise infinitely often in causally disjoint portions of an infinite random initial data set. If we argue that life can only evolve in causal futures of these special initial data sets we have no need to invoke inflation or any further theory of cosmology at all: all is statistics. Only if such a scenario could make a testable prediction would cosmologists take it seriously. However, we note that it is very similar in spirit to the chaotic inflationary philosophy.

If inflation does occur during the early stages of the Universe then many, apparently disjoint aspects of the Universe's structure can be linked together and the number of free independent parameters that could characterize a long-lived Big Bang universe is considerably reduced.[114] A long-lived inflationary universe must simultaneously possess high isotropy, no global rotation, $\Omega_0 \sim 1$ and an inhomogeneity level determined by the scalar field couplings. However, inflation leaves a number of important issues untouched. It has nothing to say about the topology of the Universe, nor can it predict whether the Universe is open or closed: these properties depend on initial conditions. One would like, also, some observational test capable of confirming the inflationary hypothesis in a convincing way, because the predictions that it does make (flatness and constant curvature inhomogeneity spectrum) are the sort of general predictions that could be imagined to arise from any number of plausible quantum gravity theories since they single out the simplest models, determined by a minimum of parameters.[115] However, the biggest problem with the inflationary picture is the puzzle of the cosmological constant. In order for inflation to work we must tune the absolute scale of V in Figure 6.11 to have a very special value. Today, our observations of the deceleration of cosmic expansion tell us (see section 6.9) that the cosmological constant must now be very small. This is equivalent to requiring the value of V in the true asymmetric vacuum we now find ourselves inhabiting, ($\phi = \sigma$ in Fig. 6.11), to be less than about $10^{-46}\,\text{GeV}^4$. Since the *change* in energy brought about by the transition from the symmetric vacuum state at $\phi = 0$ to the asymmetric final minimum at $\phi = \sigma$ is huge, roughly $T_c^4 \sim (10^{14}\,\text{GeV})^4$, we require an extraordinarily special choice of the absolute scale $V(0)$ to achieve the 'right' final state; in fact we require,

$$\frac{V(\text{true vacuum})}{V(0)} \lesssim 10^{-102} \tag{6.160}$$

At present we have no understanding as to why this ratio has to be chosen to be so fantastically small. It is just a restatment of our ignorance as to why the cosmological constant is so very small. And, we recall from section 6.9, that the very small value that the cosmological constant can possess in our Universe is a vital factor in allowing the Universe to survive long enough or expand slowly enough for life to evolve.[116] Thus requirement (6.160) is reinforced as an anthropic coincidence of a most extraordinary type if the picture of inflation sketched-out above is really true.

We should also point out that the value of S, the entropy per baryon that we argued could be predicted by grand unified theories in section 6.7 is essentially unconstrained by the simple inflationary models described above. They all assume that the vacuum energy difference $\Delta V \sim T_c^4$ which is thermalized at the end of the inflationary phase is large enough to reheat the Universe above the grand unification energy $m_X \sim 10^{14}$ GeV so that baryon nonconserving interactions occur in equilibrium and drive the cosmic baryon number to zero[117], independent of what happened to it prior to, and during, inflation. Then, as the universe resumes its standard expansion history, cooling as $T \propto t^{-1/2}$, baryosynthesis will occur along the lines laid out in section 6.9. Thus, this merely places a lower bound on the reheating caused by inflationary models. The actual value of any subsequent baryon asymmetry produced is determined by the CP violation.

Before leaving the inflationary universe we should comment upon its one concrete prediction: that the present expansion should exhibit flatness to the extent of $\Omega_0 = 1 \pm \varepsilon$ with $\varepsilon \lesssim 10^{-6}$ if inflation was sufficient to extend to a distance $\gtrsim H_0^{-1} \sim 10^{27}$ cm today. Our observations of the luminous matter content of the Universe, (6.30), reveal only $\Omega_{\text{Lum}} \lesssim 0.014$ and the quantity of dark or non-baryonic material necessary to explain the flat rotation curves of spiral galaxies and the virial equilibrium of large clusters, (6.31), and groups of galaxies requires at most $\Omega_0 \sim 0.2$. Dynamical arguments show that the total density of *clustered* matter in the Universe, whether luminous or dark, cannot contribute more than $\Omega_0 \sim 0.2$ to the total density without creating larger random velocities between galaxies and clusters relative to the Hubble flow than are observed. The only way $\Omega_0 = 1$ can be reconciled with observation is by the existence of a smooth (unclustered) sea of non-baryonic weakly interacting particles like gravitons or massive neutrinos. However, numerical simulations of how the clustering of luminous matter would proceed in the presence of a sea of these particles (see Figures 6.5 and 6.6) reveal that the particle species that create the correct pattern of galaxy clustering in numerical simulations do so only when $\Omega_0 \lesssim 0.2$ unless luminous matter is not a good indicator of the total mass distribution. So far there has been found no observational reason to suppose our universe does possess $\Omega_0 = 1 \pm \varepsilon$ with ε very small, but there may exist new types or combinations of non-baryonic material able to produce the correct clustering patterns. One escape route which has been suggested as a rather unsatisfactory way of rescuing inflation in the face of evidence that $\Omega_0 \lesssim 0.2$ is to suppose that the cosmological constant associated with the final asymmetric vacuum state today is non-zero. In this case the results of inflation can be deduced from (6.127) in which Λ is the surviving cosmological constant term. Inflation will drive the early Universe to a state in which

$kR^{-2} \ll (8\pi G\rho + \Lambda)/3$ and so today we should observe

$$\Omega_0 = 1 - \frac{\Lambda}{3H_0^2} \tag{6.161}$$

which, for $\Lambda < 0$, can be arranged to explain a present observation of $\Omega_0 \sim 0.2$. Another recent idea is that the Universe does indeed have $\Omega_0 \simeq 1$, but its material content resides principally in the form of relativistic particles, like neutrinos, that result from the recent non-radiative decays of an earlier population of more massive unstable neutrinos.

6.14 Creation *ex nihilo*

> In the beginning, the world has nothing at all
> heaven was not, nor earth, nor space
> Because it has not, it bethought itself:
> I will be. It emitted heat.
>
> Ancient Egyptian text

The success of the inflationary universe picture has stimulated some far-reaching speculation about the origin of the Universe. In particular, there have arisen mathematical discussions of the problem of the creation of the Universe out of 'nothing'.[118] Roughly speaking, these ideas envisage the whole universe to be a giant, quantum mechanical virtual fluctuation of the vacuum. First, recall that the modern picture of the quantum vacuum differs radically from the classical and everyday meaning of a vacuum—nothing. The Uncertainty Principle of quantum mechanics forbids us the complete information we would need to make such a dogmatic statement about a region of space-time. The quantum vacuum (or vacua, as there can exist many) states we have been discussing in the preceding sections are defined simply as local, or global, energy minima ($V'(\phi) = 0$, $V''(\phi) > 0$).

The microstructure of the quantum vacuum is envisaged to be a sea of continually creating and annihilating particle-antiparticle pairs that exist for times Δt. If the energy for pair creation that is 'borrowed' from the vacuum is $\Delta E = 2mc^2$ then, according to the energy-time form of the Uncertainly Principle these 'virtual' particle pairs will be unobservable as individual events so long as

$$\Delta E \, \Delta t \lesssim \hbar \tag{6.162}$$

Despite the metaphysical sound of this conception, the existence of virtual particle pairs does have observable consequences for atomic structure and its detailed predictions are confirmed by experiment to one part in a billion.

Long before the emergence of the inflationary universe idea, Tryon[119] suggested that such an idea might be applied to the Universe as a whole.

If we wish to interpret the Universe (which must be finite in this case) as a long-lived vacuum fluctuation ($\Delta t \gtrsim 15\times 10^9$ yr) then we must have $\Delta E \simeq 0$ to high precision and it might be possible to associate this feature with the proximity of the expansion to the state of zero binding energy, ($\Omega_0 = 1$). Unfortunately, this simple picture does not hold up well when examined more carefully. The 'energy' of the Universe is not a well-defined concept[120] and if we think of what is occurring in the 'wave' rather than the 'particle' picture, then the Uncertainly Principle, (6.161), is equivalent to the spontaneous existence of an acausal fluctuation; that is, a fluctuation with wavelength exceeding the time required for light to travel across it.

Attempts to make this idea more rigorous have been made by Vilenkin and others.[118,142] Vilenkin has pointed out there exist formulae which have a natural interpretation as probabilities for the quantum mechanical tunneling from 'nothing' (in the quantum mechanical sense of 'nothing') to a closed, expanding de Sitter Universe. It remains to be seen whether any real physical meaning can be associated with these results. If this does turn out to be fruitful line of enquiry one would imagine that it will be found that small universes of size $\sim m_p^{-1}$ are overwhelmingly more probable to be created from the vacuum than ones with our huge size $\sim 10^{60} m_p^{-1}$, but small fluctuations may undergo inflation out to $\gtrsim 10^{60} m_p^{-1}$ if they contain suitably coupled scalar fields of the type discussed above in section 6.8. Not all fluctuations would experience the right scalar field evolution and again the Weak Anthropic Principle would need to be invoked to 'explain' our own existence in one of the fluctuations that did get inflated. We shall develop in the next chapter another way of using WAP to explain the size of the Universe.

Before leaving this topic it is worth noting that we know of no non-zero conserved quantities that the universe must possess when it comes into being. Vacuum fluctuations would be matter-antimatter symmetric but subsequent baryon non-conserving interactions could arise to endow the Universe with its observed asymmetry.

It is, of course, somewhat inappropriate to call the origin of a bubble universe in a fluctuation of the vacuum 'creation *ex nihilo*', for the quantum mechanical vacuum is not truly 'nothing'; rather, the vacuum state has a rich structure which resides in a previously existing substratum of space-time, either Minkowski or de Sitter space-time. Clearly, a true 'creation *ex nihilo*' would be the spontaneous generation of everything—space-time, the quantum mechanical vacuum, matter—at some time in the past.

Such a true creation *ex nihilo* has been discussed by cosmologists in both classical and quantum gravity. In classical general relativity, it can be shown[9] that if the Universe is deterministic, if gravity is always attractive, and if the universe is expanding on the average (this last condition is observed to be true, at least in our past light-cone), then all timelike curves have a proper time length less than some universal constant T (which is roughly 30 billion years). In simple models such as the Friedman universe, the finite length of all timelike curves is caused by an all-encompassing singularity, a finite time in the past, as we discussed in section 6.2. At this singularity, space and time came into existence; literally nothing existed before the singularity, so, if the Universe originated at such a singularity, we would truly have a creation *ex nihilo*. The singularity is to be regarded as being on the 'boundary' of space-time (in a sense that will be made precise in section 10.3).

Since the singularity occurs at a finite time in the past, there is a temptation to ask the question 'what happened before the singularity?'. However, this question makes as much sense as the question 'What happened before the Universe began, assuming it has existed forever?' In both cases, the answer is that nothing happened 'before', because there is no 'before' in either case. To see this, compare the expanding closed Friedman universe with the Einstein static universe. Both have the metric (6.4) with $k = \pm 1$, but the Friedman universe has $R(t) = 0$ at some finite time t in the past, while the Einstein static universe has $R(t) = \text{constant}$. In other words, the expanding Friedman universe begins a finite time in the past, while the Einstein static universe exists for infinite past time; it is eternal.

But this respectively finite or infinite time is proper time, and one can always choose a measure of time in which the expanding Friedman universe exists for an infinite period, and another time in which the Einstein static universe exists for only a finite period. For example, we could choose to measure time in the expanding universe by the negative of its Hubble parameter: $\tau = -H = (-1/R)\, dR/dt$. Since $R \to 0$ and $dR/dt \to +\infty$ as the initial singularity is approached, the singularity occurs at $\tau = -\infty$, i.e. an infinite time in the past. In Hubble time, the time of maximum expansion is $\tau = 0$, and the final singularity occurs at $\tau = +\infty$, so the Friedman universe exists forever in Hubble time. Similarily, we could measure time in the Einstein universe by $\tau = \arctan(t)$, where t is the proper time. In this new time-variable (which is called 'conformal time'), the Einstein static universe exists for only a finite time. The 'beginning' of the eternal Einstein universe occurs at $\tau = -\pi/2$, and it 'ends' at $\tau = +\pi/2$. The key point to note is that neither in the expanding Friedman universe, nor in the Einstein universe, is the 'beginning' (in t or τ time) included in space-time. Both 'beginnings' are on the boundary of space-time, and that by suitable choice of the measure of time we can bring this boundary of space-time into a finite distance, or push it out to an infinite distance. In a sense, both the expanding Friedman universe and the Einstein universe are created *ex nihilo*, though it is usual to apply this term only to those cosmologies which have existed for only a finite proper time. However, proper time may not be the best measure of physical time near a singularity. We shall discuss this point in section 10.6.

It is sometimes objected that the Universe cannot have originated a finite proper time ago in the past, as the Friedman universe does, because this would violate the law of conservation of mass-energy. This objection is invalid. At every instant of time in the Friedman universe the general relativity stress-energy conservation law $T^{ij}_{;j}=0$ holds. The law does not hold at the singularity, but the singularity is not in time. If we use Hubble time as our time variable in the Friedman universe, then the conservation law would hold 'forever'; that is, for infinite past time. Similarly, if we use conformal time as our time variable in the Einstein universe, then the stress-energy conservation law would hold everywhere in time, but only for a finite time. Again, the conservation law would not hold on the boundary, but the boundary is not in time. Thus, we see it makes no more sense to wonder at the 'spontaneous' creation of matter in the Friedman universe than it would to wonder at the spontaneous creation of matter in the eternal Einstein universe. Mass-energy is never created or destroyed in either universe.

It is open to serious question whether classical gravity can provide an accurate model of creation *ex nihilo*, because, as we have discussed earlier in this chapter, we would expect quantum gravity effects to be very important near a singularity. A number of quantum gravity models of creation *ex nihilo* of the entire Universe, including space and time, have been constructed. The idea common to all of them is that the Universe is envisaged as originating out of a 'point' in the past.

The best known model of quantum creation *ex nihilo* is due to Hartle and Hawking.[127] In the Hartle–Hawking picture, the wave function of the Universe $\Psi[h_{\mu\nu}]$ is viewed as a function of spatial three-geometries $h_{\mu\nu}$ alone; there is no functional dependence on time. They argue that there is a 'natural' choice for the quantum state of the Universe, defined by

$$\Psi_0[h_{\mu\nu}]=N\int D[g_{ij}]\exp(-I_E[g_{ij}]) \tag{6.163}$$

where N is a normalization constant and the integral is a path integral over all the Riemannian four-geometries which have the three-geometry $h_{\mu\nu}$ as a boundary. The function $I_E[g_{ij}]$ is the Euclidean Einstein action obtained by replacing the time t with $(-i\tau)$ in the usual Einstein action and adjusting the sign so that I_E is positive. As in standard quantum mechanics, $|\Psi_0[h_{\mu\nu}]|^2$ is the probability that we will observe the Universe to have the three-geometry $h_{\mu\nu}$. The reason the above wave function is the 'natural' choice is that Lorentzian path integrals (those with iI_E rather than I_E) are not well-defined mathematically; the Euclidean path integral, however, is well-defined (probably), and it is one of the few path integrals in quantum gravity which is.

Hartle and Hawking argue by analogy with standard quantum mechanics that the above expression is the amplitude for the universe with spatial geometry $h_{\mu\nu}$ to arise from a zero three-geometry, i.e., from 'nothing'. In standard quantum mechanics, the amplitude for a particle to go to the event (x, t) from the event (x', t') is given by the propagator $\langle x, t \mid x', t' \rangle$. If we set $t=0$ and $x'=0$ and expand the resulting propagator in a complete set of energy states, we obtain

$$\langle x, 0 \mid 0, t' \rangle = \sum_n \Psi_n(x)\bar{\Psi}_n(0)\exp(\mathrm{i}E_n t') = \int Dx(t')\exp(\mathrm{i}S[x(t')]) \quad (6.164)$$

where $S[x(t')]$ is the action. If we replace t' by $(-\mathrm{i}\tau')$ and take the limit as $\tau' \to -\infty$, the path integral becomes

$$\int Dx \exp[-I(x)] \quad (6.165)$$

which is the same in form as the Euclidean Einstein action given above. Only the $n=0$ state with $E_0=0$ (the ground state) survives in the sum, and the propagator tells us that this wave function is the amplitude for arriving at the spatial point x from the point $x=0$. By analogy, the above expression for the wave function of the Universe gives the amplitude for the Universe to go to its present state $h_{\mu\nu}$ from 'nothing' ($h_{\mu\nu}=0$). There is no reference to time in the Hartle–Hawking quantum model, so it is not a meaningful question to ask when the Universe originated. Time direction is not even defined near $h_{\mu\nu}=0$.

In Chapter 7, we shall develop a model quantum universe in which the wave function does depend on a time parameter explicitly, and in which the Universe originates out of the point $R(t)=0$—that is, out of nothing—at a finite time in the past as measured by this time-parameter.

6.15 Boundary Conditions

> There ought to be something very special about the boundary conditions of the universe and what can be more special than the condition that there is no boundary.
>
> S. W. Hawking

The inflationary Universe and the ideas associated with it are natural successors to the chaotic cosmology programme discussed in section 6.6. These theories all attempt to produce cosmological explanations of the present structure of the Universe that are independent of the initial conditions. This is an attractive idea, if only because we can be quite confident that we will never know what the initial conditions were like. However, there are still questions we would like to ask concerning the structure of the Universe prior to an epoch of inflation at, say, the time[121] when it has cooled to $\sim 10^{15}$ GeV. We would like to know whether there was an initial singularity,[122] whether the universe is open or closed and, indeed, if there are more compelling explanations of the observational problems that inflation addresses. Last, but not least, we would like to understand the present value of the cosmological constant. Problems of this sort will require some discussion of the cosmological initial conditions. The antithesis of the chaotic cosmology programme—quiescent cosmology[123]—would seek to find compelling reasons for supposing the Universe to possess very special initial conditions. We have already seen that the present high isotropy of the Universe is neither a generic nor a stable property of cosmological initial data sets. Indeed, the most general homogeneous cosmological models behave in a chaotically unpredictable fashion during their initial stages.[124]

There have been two quite distinct suggestions regarding how to approach the problem of the boundary conditions of the Universe. The first, stressed by Penrose,[125] encourages a thermodynamic approach to the gravitational field, whilst the other, promoted by Hawking[126] and collaborators,[127] seeks to dispense with the need for boundary conditions altogether. We discuss first the proposal of Penrose: that there exists a gravitional entropy obeying the Second Law of thermodynamics.

Entropy has proved to be one of the most fertile concepts ever introduced into the physical sciences. First employed by Rudolf Clausius to describe thermal behaviour, it has steadily extended its domain of applicability to encompass a vast array of natural phenomena. Very soon after its inception, Boltzmann and Gibbs developed its interconnections with probability theory, mechanics and physical chemistry and in the post-war era Shannon has used it as the guiding principle at the heart of deep theory of communication and information channelling. More recently, Hawking has discovered that *stationary* gravitational fields possess an entropy and display an intrinsic thermodynamic structure.[128]

Hawking's discovery is one of the most striking in gravitation physics because it reveals a *ménage à trois* between quantum theory, general relativity and thermodynamics in the mechanics of black hole processes.[129] More specifically, black holes are seen to be *black bodies.* They obey the laws of equilibrium thermodynamics and possess a thermal entropy which is determined by a purely geometric entity—the area of the event horizon.

Even prior to Hawking's discovery that black hole thermodynamics has a rigorous quantum mechanical basis, Penrose had conjectured[130] that the striking analogy between the relations governing black hole mechanics and the laws of equilibrium thermodynamics might extend into cosmology. He suggested that the Weyl curvature could be intimately related to the gravitational entropy of space-time and so parametrize the global evolution of the Universe. This Weyl portion of the space-time curvature governs 'tidal' forces which, while stretching bodies in one direction, squashes them in others so no overall change in volume results. By contrast, the Ricci portion of the curvature is associated with a simultaneous crushing in all directions. The Friedman model universe displays only the Ricci part of the curvature whilst the famous vacuum solution of Kasner[131] is an example of a universe model that contains only Weyl curvature.

The motivation for establishing a link between entropy and a geometrical aspect of space-time like Weyl curvature is obvious. It would provide us with a very natural thermodynamic boundary condition to place on the initial structure of the big bang. The Universe would necessarily expand away from a virtually homogeneous and isotropic low entropy singularity where the Ricci curvature dominates the Weyl curvature. If the space-time were to evolve towards a second singularity this would be a maximum entropy state, more disordered and totally dominated by the anisotropizing effect of the Weyl curvature. Clearly, if successful, such a theoretical scenario could lead to an understanding of the present large scale structure of the Universe: a structure that is mysterious for its large scale quiescence, isotropy and uniformity in the midst of all the dynamically more favourable degrees of freedom apparently available for non-uniform expansion.

To take the notion of gravitational entropy seriously we must imagine[132] that there exists a gravitational entropy, S_g, which measures the deviation of the Universal space-time geometry from homogeneity and isotropy. By the Second Law of thermodynamics, the total entropy, which presumably would be the sum of the matter entropy, S_M, and the gravitational entropy, must be monotonically non-decreasing in proper time, so $\dot{S}_M + \dot{S}_g \geq 0$.

By way of contrast, we note that the existence of S_g implies the antithesis of the once popular 'chaotic cosmology' programme[80–82] which sought to show that however the Universe begins, it inevitably becomes isotropic and homogeneous with the passage of time through the action of dissipative processes. Although it is philosophically attractive to dispense with the need for initial conditions in this way the chaotic cosmology programme cannot be tested by observations. We have no way of telling the difference between a Universe which began expanding isotropically with high thermal entropy and one which did not, but which subsequently underwent dissipative smoothing.

Unfortunately, as yet there is no obvious candidate to use as a gravitational entropy S_g. Penrose has suggested[125] that we might relate S_g to 'some suitably integrated measure of the size of the Weyl curvature'. Such a direct line of attack encounters immediate difficulties because many cosmological models evolve at late times towards the space-time of a plane gravitational wave.[91] All polynomial invariants of the Weyl tensor vanish identically in these space-times regardless of the degree of expansion anisotropy. A gravitational entropy directly related to any linear combination of curvature invariants is clearly unsatisfactory. It could decrease monotonically in a Univese which expanded for all time towards an increasingly non-Friedmanian but plane wave asymptote. Cosmological evolution is not completely described by the behaviour of the curvature invariants.

As an illustration of what a working gravitational entropy might look like which side-steps these technical difficulties with the Weyl curvature we might explore an axiomatic approach which does not identify S_g explicitly. We can simply assume that it does exist and can be defined in the following rough operational fashion:

(a) $S_g = 0$ if and only if the space-time is isotropic and homogeneous; that is, if and only if it is Friedmanian.
(b) $\dot{S}_g \geqslant 0$, in vacuum where '·' denotes differentiation with respect to proper time.
(c) S_g increases with the deviation of the space-time from isotropy and homogeneity.

Property (c) could be made more rigorous by defining an appropriate distance function on the spacelike hypersurfaces foliating the space-time and various technical conditions and caveats could be introduced to make (a)–(c) precise and unambiguous.

One way of identifying the consequences of (a), (b) and (c) might be to exploit global results on the behaviour of Ricci curvature and shear stresses. The following theorem[133] and its corollary[134] place general limits on the maximum growth of the Ricci curvature on approach to a singularity and the minimum rate of shear anisotropy decay to the future of an ever-expanding Universe.

Theorem: *Let* (M, g) *be a space-time foliated by spatially homogeneous Cauchy surfaces. If* $R_{ab}k^ak^b \geqslant 0$ *for any time-like vectors* k^a *then, if the foliation is future complete,*

$$\liminf_{t\to 0} t^2 R_{ab}k^ak^b \leqslant \tfrac{3}{4} \tag{6.166}$$

Corollary: *If* σ^2 *is the shear of the timelike geodesic congruence normal to the surfaces of homogeneity then*

$$\liminf_{t\to +\infty} t^2\sigma^2 \leqslant \tfrac{3}{8} \tag{6.167}$$

A natural scenario in line with that originally envisaged by Penrose would start the Universe at $t \simeq 0$ in a state that possessed the largest Ricci curvature allowed by (6.166) and allow it to evolve towards[135] a non-uniform state with the largest anisotropy permitted by (6.167). In this scheme of things the present high isotropy of the Universe would be seen as a consequence of its relative youth. We may be constrained to observe the Universe in the early phase of its evolution before all stars have died. If so, the isotropy could be a selection effect of our very existence. One can show that the Ricci curvature bound is only attained by an isotropic radiation gas in a Friedman Universe. This prediction regarding the equation of state is in line with current ideas regarding the behaviour of elementary particles at ultra-high energies and the results of quantum processes close to the initial singularity.[136]

The corollary (6.167) provides us with the intriguing possibility of characterizing the future stage of 'gravitational equilibrium'—a geometrical analogue of the famous 'Heat Death' of the Universe once popularized by Jeans and Eddington.[137] Space-times which are thermodynamically general should tend to attain the upper bound on anisotropy as $t \to \infty$. They become increasingly irregular, asymptotically approaching the largest anisotropy level (6.167) compatible with the strong energy condition. This is a state of gravitational equilibrium for ever-expanding cosmologies.

Penrose has employed his idea of gravitational entropy to argue against the Strong Anthropic Principle. He points out that the present entropy content of the observable Universe could be $\sim 10^{30}$ times larger than is observed if the material were gathered into black holes (assuming that we just calculate this new entropy by using Hawking's formulae (5.192–193)). Since entropy is proportional to the logarithm of the probability of that entropic state emerging, this implies that the present arrangement of cosmic matter has only a one in $10^{10^{30}}$ chance of emerging from the Big Bang. There are so many far more probable ways for the Big Bang to generate black holes than smoothly distributed material. Penrose goes on to argue that if life were the goal of the Universe it would be far more economical to bring it about by setting up a small number of meticulously organized particle collisions to bring about the *local* life-supporting environment we require than to produce it as a subset of a vastly bigger entropy fluctuation the size of the entire observable Universe. However, it is hard to evaluate such a claim without knowledge of whether there exist couplings between the local and global structure of the Universe.

The other suggestion regarding the boundary conditions of the Universe is quite different in spirit to that of Penrose. It seeks to find formulations of the quantum cosmological problem in which there are no boundary conditions at all. The most straightforward, but disconcerting, way to achieve this state of affairs with the classical Einstein equations might be to search for solutions all of whose timelike curves periodic. A number of such cases are known[139] and, in the absence of a complete classification, a number of existence theorems have been proven concerning them. If the minimum length of all closed timelike curves were extremely long, say greater than $\sim 10^{32}$ years, in a cosmological solution of this type it would behave for all practical purposes like an ordinary cosmology with non-periodic time. It is known[139] that the existence of closed timelike curves is a stable property of space-time, in the sense that such curves persist if the 'size' of the light-cone is slightly changed at every event. It is also known[139] that the property of having all timelike curves periodic is stable in the same sense. However, it is far from clear that this is the appropriate physical notion of stability. The more conventional measure of stability, which we used in section 6.11, assumes at the start that closed timelike curves do not exist, and so it cannot be applied when they do.

Hawking[126–127,143] has approached the removal of boundary conditions in a different way by adopting the Euclidean approach to quantizing gravity. By transforming Lorentzian space-times into Euclidean space-time manifolds by means of a complex transformation of the time coordinate, $\tau \rightarrow -it$, it is possible to exclude singularities from the resulting Euclidean region. Path integrals have nice properties in this Euclidean region and Hawking claims that by integrating the path integral only over compact metrics the need for any boundary condition at all disappears. Hence, Hawking, suggests that the quantum wave function of the Universe is defined by a path integral over compact metrics without boundary. Hawking has argued that, in the classical limit, the quantum state derived from this condition has desirable cosmological properties:[140] it must be almost isotropic and homogeneous and be very close to the $\Omega_0 \sim 1$ state. This quantum state can be regarded as a sort of superposition of Friedman universes with these classical properties.

The type of boundary condition proposed by Hawking, unlike that of Penrose, explicitly involves quantum gravitation and, in particular, must come to grips with the problem of what is meant by the 'wave function of the Universe' after it has been written down. In the next Chapter we move on to consider this complex problem which brings together the roles of observer and observed in Nature in an intimate and intricate fashion.

In this chapter we have discussed the ideas of modern theoretical and observational cosmology in some detail. We have completed the study, begun in Chapter 5, of the size spectrum of objects in Nature and have shown how properties of the Universe as a whole, perhaps endowed at its inception, may be crucial if the existence of observers is to ever be possible within it.

References

1. For popular accounts of modern cosmology, see S. Weinberg, *The first three minutes* (Deutsch, London, 1977), and J. D. Barrow and J. Silk, *The left hand of creation* (Basic Books, NY, 1983, and Heinemann, London, 1984).
2. This prediction was not made by Einstein himself, who sought to suppress the expansion of the Universe emerging from his original formulation by introducing another, mathematically admissible, parameter into his gravitational field equations. This type of static cosmological solution with a uniform non-zero density was in line with the prevailing philosophical view. These solutions proved to be unstable, and the first expanding universe solutions without cosmological constant were found by A. Friedman, *Z. Physik.* **10,** 377 (1922). For more detailed history see A. Pais, *Subtle is the Lord* (Oxford University Press, Oxford, 1983), J. North, *The measure of the universe* (Oxford University Press, 1965), and F. J. Tipler, C. J. S. Clarke, and G. F. R. Ellis, in *General relativity and gravitation: an Einstein centenary volume,* edited by A. Held (Pergamon Press, NY, 1980).
3. E. Hubble, *Proc. natn. Acad. Sci., U.S.A.* **15,** 169 (1929). Technically speaking, the Hubble redshift is not a true Doppler effect, since in a non-Euclidean geometry we cannot invariantly characterize relative velocities of recession, and the effect arises because of light propagation through a curved space-time, although it is described by identical formulae as the Doppler effect.
4. A. A. Penzias and R. A. Wilson, *Astrophys. J.* **142,** 419 (1965).
5. R. Alpher and R. Herman, *Nature* **162,** 774 (1948), and see also S. Weinberg, ref. 1 for same historical background.
6. R. H. Dicke, P. J. E. Pebbles, P. G. Roll, and D. T. Wilkinson, *Astrophys. J.* **142,** 414 (1965).
7. R. V. Wagoner, W. A. Fowler and F. Hoyle, *Astrophys. J.* **148,** 3 (1967).
8. B. Pagel, *Phil. Trans. R. Soc.* A **307,** 19 (1982).
9. S. W. Hawking and G. F. R. Ellis, *The large scale structure of space-time* (Cambridge University Press, Cambridge, 1973).
10. For a detailed overview see J. D. Barrow, *Fund. Cosmic Phys.* **8,** 83 (1983), and for popular accounts see ref. 1.
11. This assumption is sometimes called 'The Cosmological Principle', following E. A. Milne.
12. S. Weinberg, *Gravitation and cosmology* (Wiley, NY, 1972). We assume $\Lambda = 0$ here.
13. D. W. Sciama, *Modern cosmology* (Cambridge University Press, Cambridge, 1975).
14. J. D. Barrow, *Mon. Not. R. astron. Soc.* **175,** 359 (1976).
15. Here k_B is Boltzmann's constant; henceforth it will be set equal to unity.
16. A. Sandage and E. Hardy, *Astrophys. J.* **183,** 743 (1973).
17. J. Audouze, in *Physical cosmology*, ed. R. Balian, J. Audouze, and D. N. Schramm (North-Holland, Amsterdam, 1979); S. van den Bergh, *Quart. J. R. astron. Soc.* **25,** 137 (1984).
18. S. M. Faber and J. S. Gallagher, *Ann. Rev. Astron. Astrophys.* **17,** 135 (1979).

19. P. J. E. Peebles, in *Physical cosmology*, op. cit., M. Davis, J. Tonry, J. Huchra, and D. W. Latham, *Astrophys. J. Lett.* **238,** 113 (1980).
20. H. Totsuji and T. Kihara, *Publ. Astron. Soc. Japan* **21,** 221 (1969); S. M. Fall. *Rev. Mod. Phys.* **51,** 21 (1979).
21. P. J. E. Peebles, *The large scale structure of the universe* (Princeton University Press, NJ, 1980).
22. A. Webster, *Mon. Not. R. astron. Soc.* **175,** 61; **175,** 71 (1976).
23. M. Peimbert, *Ann. Rev. Astron. Astrophys.* **13,** 113 (1975).
24. C. Laurent, A. Vidal-Madjar, and D. G. York, *Astrophys. J.* **229,** 923 (1979).
25. F. Stecker, *Nature* **273,** 493 (1978); G. Steigman, *Ann. Rev. Astron. Astrophys.* **14,** 339 (1976).
26. D. P. Woody and P. L. Richards, *Phys. Rev. Lett.* **42,** 925 (1979).
27. R. Dicke, *Nature* **192,** 440 (1961). The fact that cosmological descriptions of the expanding Universe link local conditions and habitability with global facets like the size of the Universe was first stressed by G. Whitrow, see E. L. Mascall, *Christian theology and natural science* (Longmans, London, 1955) and G. M. Idlis, *Izv. Astrofiz. Inst. Kazakh. SSR* **7,** 39 (1958), (in Russian), whose paper was entitled 'Basic features of the observed astronomical universe as characteristic properties of a habitable cosmic system'.
28. J. Milton, *Paradise Lost,* Book 8 (1667).
29. H. Minkowski introduced this concept in a lecture entitled 'Space and Time' delivered in Cologne, 1908.
30. J. A. Wheeler, in *Essays in general relativity*, ed. F. J. Tipler (Academic Press, NY, 1980), and see also L. C. Shepley, in this volume. These authors investigate the fact that anisotropic universes of Galactic mass can have expanded for 10^{10} yrs compared with only a few months in the isotropic case.
31. A. S. Eddington, *Proc. natn. Acad. Sci., U.S.A.* **16,** 677 (1930).
32. V. A. Lyubimov, E. G. Novikov, V. Z. Nozik, E. F. Tret'yakov, V. S. Kozik, and N. F. Myasoedov, *Sov. Phys. JETP* **54,** 616 (1981).
33. J. D. Barrow, *Phil. Trans. R. Soc.* A **296,** 273 (1980).
34. J. Silk, *Astrophys. J.* **151,** 459 (1968).
35. J. Silk, *Nature* **265,** 710 (1977).
36. M. J. Rees and J. Ostriker, *Mon. Not. R. astron. Soc.* **179,** 541; J. Silk, *Astrophys. J.* **211,** 638 (1976); J. Binney, D.Phil. thesis, Oxford University (1977).
37. J. D. Barrow and J. S. Turner, *Nature* **291,** 469 (1981).
38. B. J. Carr and M. J. Rees, *Nature* **278,** 605 (1979).
39. M. Dine, W. Fischler and M. Srednicki, *Phys. Lett.* B **104,** 199 (1981); M. B. Wise, H. Georgi, and S. L. Glashow, *Phys. Rev. Lett.* **47,** 402 (1981).
40. G. R. Blumenthal, S. M. Faber, J. R. Primack, and M. J. Rees, *Nature* **311,** 517 (1984).
41. J. E. Gunn, B. W. Lee, I. Lerche, D. N. Schramm, and G. Steigman, *Astrophys. J.* **223,** 1015 (1978).

42. R. Cowsik and J. McClelland, *Phys. Rev. Lett.* **29,** 669 (1972); *Astrophys. J.* **180,** 7 (1973); J. E. Gunn and S. Tremaine, *Phys. Rev. Lett.* **42,** 407 (1979).
43. G. Bisnovathy-Kogan and I. D. Novikov, *Soc. Astron. Lett.* **24,** 516 (1981); J. Bond, G. Efstathiou, and J. Silk, *Phys. Rev. Lett.* **45,** 1980 (1981).
44. A. G. Doroshkevich, M. Y. Khlopov, A. S. Szalay, and Y. B. Zeldovich, *Ann. NY Acad. Sci.* **375,** 32 (1980).
45. M. Davis, G. Efstathiou, C. Frenk and S. D. M. White, *Astrophys. J.* **292,** 371. (1985). For a popular description, see J. D. Barrow and J. Silk, *New Scient.*, 30 Aug. (1984).
46. C. Hayashi, *Prog. Theor. Phys.* **5,** 224 (1950), *Prog. Theor. Phys. Suppl.* **49,** 248 (1971).
47. F. Hoyle and R. J. Tayler, *Nature* **203,** 1108 (1964); P. J. E. Peebles, *Phys. Rev. Lett.* **43,** 1365; R. V. Wagoner, in *Confrontation of cosmological theories with observation*, ed. M. Longair (Reidel, Dordrecht 1974).
48. B. Carter, unpublished manuscript 'The significance of numerical coincidences in Nature' (DAMTP preprint, University of Cambridge, 1967): our belief that Carter's work should appear in print provided the original motivation for writing this book, in fact. F. Hoyle, *Astronomy and cosmology: a modern course* (Freeman, San Francisco, 1975).
49. Y. B. Zeldovich and I. D. Novikov, *Sov. Phys. JETP Lett.* **4,** 117 (1966); R. Alpher and G. Gamow, *Proc. natn. Acad. Sci., U.S.A.* **61,** 363 (1968).
50. M. J. Rees, *Phys. Rev. Lett.* **28,** 1969 (1972); Y. B. Zeldovich, *Mon. Not. R. astron. Soc.* **160,** 1p (1972); E. P. T. Liang, *Mon. Not. R. astron. Soc.* **171,** 551 (1975); J. D. Barrow, *Nature* **267,** 117 (1977); J. D. Barrow and R. A. Matzner, *Mon. Not. R. astron. Soc.* **181,** 719 (1977); B. J. Carr, *Acta cosmologica* **11,** 113 (1983).
51. M. Clutton-Brock, *Astrophys. Space Sci.* **47,** 423 (1977).
52. J. D. Barrow and R. A. Matzner, *Mon. Not. R. astron. Soc.* **181,** 719 (1977).
53. M. J. Rees, *Nature* **275,** 35 (1978); B. J. Carr, *Acta cosmologica* **11,** 131 (1982).
54. E. Salpeter, *Astrophys. J.* **211,** 161 (1955).
55. B. J. Carr, *Mon. Not. R. astron. Soc.* **181,** 293 (1977), and **189,** 123 (1978).
56. H. Y. Chiu, *Phys. Rev. Lett.* **17,** 712 (1965); Y. B. Zeldovich, *Adv. Astron. Astrophys.* **3,** 242 (1965); G. Steigman, *Ann. Rev. Nucl. Part Sci.* **29,** 313 (1979). The result changes slightly if the expansion is anisotropic but is not enough to affect the general conclusions, J. D. Barrow, *Nucl. Phys.* B **208,** 501 (1982).
57. A. D. Sakharov, *Sov. Phys. JETP Lett.* **5,** 24 (1967); E. Kolb, and S. Wolfram, *Nucl. Phys.* B **172,** 224.
58. V. A. Kuzman, *Sov. Phys. JETP Lett.* **12,** 228 (1970). Note that an interesting example of a system violating baryon conservation but conserving both C and CP is provided by the gravitational interaction. This is manifested during the collapse of a cloud of material to the black hole state and its subsequent evaporation via the Hawking effect. The final state will always be baryon symmetric so long as no particles which can undergo CP violating decays are evaporated (for this case see J. D. Barrow, *Mon. Not. R. astron. Soc.* **192,** 427 (1980), and J. D. Barrow and G. Ross, *Nucl. Phys.* B **181,** 461 (1981)).

59. M. Yoshimura, *Phys. Rev. Lett.* **41,** 281 (1978); A. Y. Ignatiev, N. Krasinokov, V. Kuzmin, and A. Tavkhelkize, *Phys. Lett.* B **76,** 436 (1978); S. Dimopoulos and L. Susskind, *Phys. Rev.* D **19,** 1036 (1979); J. Ellis, M. K. Gaillard, D. V. Nanopoulos, and S. Rudaz, *Phys. Lett.* B **99,** 101 (1981); S. Weinberg, *Phys. Rev. Lett.* **42,** 850 (1979); A. D. Dolgov, *Sov. J. Nucl. Phys.* **32,** 831 (1980).
60. J. D. Barrow, *Mon. Not. R. astron. Soc.* **192,** 19p (1980).
61. E. Kolb and M. S. Turner, *Ann. Rev. Nucl. Part. Sci.* **33,** 645 (1983).
62. J. N. Fry, K. A. Olive, and M. S. Turner, *Phys. Rev.* D **22,** 2953 (1980). For other numerical results see Kolb and Wolfram, ref. 57.
63. D. V. Nanopoulos, *Phys. Lett.* B **91,** 67 (1980).
64. J. D. Barrow and M. S. Turner, *Nature* **291,** 469 (1981).
65. S. W. Hawking and G. F. R. Ellis, *The large scale structure of space-time* (Cambridge University Press, Cambridge, 1973).
66. C. B. Collins and S. W. Hawking, *Mon. Not. R. astron. Soc.* **162,** 307 (1973); J. D. Barrow, *Mon. Not. R. astron. Soc.* **175,** 359 (1976) and *Quart. J. R. astron. Soc.* **23,** 344 (1982).
67. A. Guth, *Phys. Rev.* D **23,** 347 (1981).
68. M. J. Rees, *Phil. Trans. R. Soc.* A **310,** 311 (1983).
69. C. B. Collins and S. W. Hawking, *Astrophys. J.* **180,** 317.
70. M. J. Rees, *Quart. J. R. astron. Soc.* **22,** 109 (1981).
71. Observations of the deceleration parameter lead to this limit. See also D. Tytler, *Nature* **291,** 289 (1981) and J. D. Barrow, *Phys. Lett.* B **107,** 358.
72. S. W. Hawking, *Phil. Trans. R. Soc.* A **310,** 303 (1983).
73. F. Hoyle and J. V. Narlikar, *Proc. R. Soc.* A **273,** 1 (1963); also articles by J. D. Barrow and by W. Boucher and G. Gibbons, in *The very early universe,* ed. G. Gibbons, S. W. Hawking, and S. T. C. Siklos (Cambridge University Press, Cambridge, 1983).
74. Y. B. Zeldovich, *Sov. Phys. JETP Lett.* **6,** 1050 (1967); *Sov. Phys. Usp.* **24,** 216 (1982). This interpretation of the cosmological constant was pioneered by W. H. McCrea, *Proc. R. Soc.* A **206,** 562 (1951).
75. E. M. Lifshitz, *Sov. Phys. JETP* **10,** 116 (1946); E. M. Lifshitz and I. Khalatnikov, *Adv. Phys.* **12,** 185 (1963).
76. S. Weinberg, *Gravitation and cosmology* (Wiley, NY, 1972).
77. A. A. Kurskov and L. Ozernoi, *Sov. Astron.* **19,** 937 (1975).
78. M. J. Rees, in *Physical cosmology*, ed. R. Balian, J. Audouze, and D. N. Schramm (North-Holland, Amsterdam, 1979).
79. W. Rindler, *Mon. Not. R. astron. Soc.* **116,** 662 (1955).
80. C. W. Misner, *Nature* **214,** 30 (1967); *Phys. Rev. Lett.* **19,** 533 (1967).
81. C. W. Misner, *Astrophys. J.* **151,** 431 (1968).
82. J. D. Barrow, *Nature* **272,** 211 (1978).
83. J. M. Stewart, *Mon. Not. R. astron. Soc.* **145,** 347 (1969); C. B. Collins and J. M. Stewart, *Mon. Not. R. astron. Soc.* **153,** 419 (1971); A. G. Doroshkevich, Y. B. Zeldovich, and I. D. Novikov, *Sov. Phys. JETP* **26,** 408 (1968).
84. L. Bianchi, *Mem. Soc. It.* **11,** 267 (1898), repr. in Opere IX, ed. A. Maxia

(Editizoni Crenonese, Rome, 1952); M. Ryan and L. C. Shepley, *Homogeneous relativistic cosmologies* (Princeton University Press, New Jersey, 1975); D. Kramer, H. Stephani, E. Herlt, and M. A. H. MacCallum, *Exact solutions of Einstein's field equations* (Cambridge University Press, 1980).

85. S. T. C. Siklos, in *Relativistic astrophysics and cosmology*, ed. E. Verdaguer and X. Fustero (World Publ., Singapore, 1984).
86. The reason for this condition is that, even though the shear σ may be decaying, it is still possible for the integrated effect of the shear down our past null cone to be large (for example, if $\sigma \propto t^{-1}$ then the cumulative microwave anisotropy would grow logarithmically in time).
87. S. W. Hawking, *Commun. Math. Phys.* **43,** 189 (1975).
88. F. J. Tipler, *Phys. Rev.* D **17,** 2521 (1978).
89. See, for example, V. I. Arnold, *Ordinary differential equations* (MIT Press, Cambridge, 1978).
90. A. G. Doroshkevich, V. N. Lukash, and I. D. Novikov, *Sov. Phys. JETP* **37,** 739 (1974); J. D. Barrow and F. J. Tipler, *Nature* **276,** 453 (1978); J. D. Barrow, ref. 66.
91. J. D. Barrow and S. T. C. Siklos, in preparation (1984); for a summary of results and the techniques employed, see J. D. Barrow and D. H. Sonoda, *Gen. Rel. Gravn.* **17,** 409 (1985) and *Phys. Rep.* (In press.)
92. S. P. Novikov, *Sov. Phys. JETP* **35,** 1031 (1977).
93. P. Langacker, *Phys. Rep.* **72,** 185 (1981).
94. V. N. Lukash, *Nuovo Cim.* B **35,** 268 (1976).
95. P. Gosse, *Omphalos: an attempt to untie the geological knot* (J. van Voorst, London, 1857).
96. W. B. Bonnor, *Mon. Not. R. astron. Soc.* **167,** 55 (1974); W. B. Bonnor and N. Tomimura, *Mon. Not. R. astron. Soc.* **175,** 85 (1976). The models considered in these papers do not contain any gravitational wave modes, and do not belong to the Bianchi classification in the spatially homogeneous limit. They possess special geometrical properties and are of measure zero in initial data space.
97. G. F. R. Ellis, *Gen. Rel. Gravn.* **9,** 87 (1978) and **11,** 281 (1979); *Quart. J. R. astron. Soc.* **16,** 245 (1975).
98. Note, however, that closed universes can be made arbitrarily large by choosing Ω_0 infinitesimally close to unity. This seemingly artificial situation turns out to be extremely relevant if current ideas in elementary particle physics turn out to be correct, see §6.7 below. Also, 'open' universes can be made finite in volume by a suitable choice of topology.
99. K. Sato, *Mon. Not. R. astron. Soc.* **195,** 467 (1981); *Phys. Lett.* B **99,** 66 (1981).
100. A. Albrecht and P. I. Steinhardt, *Phys. Rev. Lett.* **48,** 1220 (1982); A. Linde, *Phys. Lett.* B **108,** 389 (1982) and B **114,** 431 (1982); S. W. Hawking and I. G. Moss, *Phys. Lett.* B **110,** 35 (1982); see also *The very early universe*, cited in ref. 73.
101. S. Coleman and E. Weinberg, *Phys. Rev.* D **7,** 1888 (1973).
102. Y. B. Zeldovich and M. Y. Khlopov, *Phys. Lett.* B **79,** 239 (1978).
103. J. P. Preskill, *Phys. Rev. Lett.* **43,** 1365 (1979).

104. A. M. Polyakov, *Sov. Phys. JETP Lett.* **20,** 194 (1974); G. t'Hooft, *Nucl. Phys.* B **79,** 176 (1974).
105. W. de Sitter, *Mon. Not. R. astron. Soc.* **90,** 3 (1917).
106. S. W. Hawking, *Phys. Lett.* B **115,** 195 (1982); A. A. Starobinskii, *Phys. Lett.* B **117,** 175 (1982); A. H. Guth and S.-Y. Pi, *Phys. Rev. Lett.* **49,** 1110 (1982); C. Vayonnakis, *Phys. Lett.* B **123,** 396; J. M. Bardeen, P. J. Steinhardt, and M. S. Turner, *Phys. Rev.* D **28,** 679 (1983).
107. E. R. Harrison, *Phys. Rev.* **1,** 2726 (1969); Y. B. Zeldovich, *Mon. Not. R. astron. Soc.* **160,** 1p (1970). Since the metric perturbation $\delta g/g$ is related to the density perturbation $\delta\rho/\rho$ over a length scale $\lambda \propto M^{1/3}$, the dependence (6.155) yields scale-independent metric fluctuations. This means that every mass scale has the same density perturbation amplitude when it enters the cosmological particle horizon, because $M_H \propto t \propto \delta\rho/\rho$.
108. For different types of inflationary model which claim to obtain a more acceptable prediction of $\delta\rho/\rho$, see J. Ellis, D. V. Nanopoulos, K. A. Olive, and K. Tamvakis, *Nucl. Phys.* B **221,** 524 (1983); G. B. Gelmini, D. V. Nanopoulos, and K. A. Olive, *Phys. Lett.* B **131,** 53 (1983).
109. Y. B. Zeldovich, *Sov. Astron. Lett.* **7,** 323 (1981).
110. A. D. Linde, *Phys. Lett.* B **129,** 177 (1983), *Sov. Phys. JETP Lett.* **38,** 176 (1983).
111. In effect, the right-hand side of the Friedman equation is governed by a constant stress (up to logarithmic accuracy) and as a consequence has a solution of the exponentially expanding, de Sitter form.
112. The precise quartic form of the potential (6.159) is obviously not essential to this argument. Any symmetric $V(\phi)$ with shallow slope near its minimum would work, (for example, $V(\phi) = \mu\phi^2$), although the constraints on the coupling constant, μ, would differ accordingly.
113. Note that for these probabilistic arguments to work it is not sufficient merely to have an infinite number of possibilities to choose from; they must also be exhaustive of all possibilities and hence a *random* infinity is necessary for the argument to work.
114. This cannot be done with any confidence yet, because there is still no working model of inflation that produces all the advantages results simultaneously without a special *ad hoc* choice of the free parameters involved.
115. For example, the simple versions of Dirac's Large Numbers Hypothesis discussed in Chapter 4.1 require that $\Omega_0 = 1$ and $\Lambda = 0$ identically in order that dimensionless numbers cannot be associated with the curvature radius and Λ respectively. The simple oscillating universe model studied by P. Landsberg and D. Park, *Proc. R. Soc.* A **346,** 485 (1976), increases in size with each oscillation due to entropy production, and should therefore be infinitesimally close to $\Omega_0 = 1$ after a past eternity of oscillations. However, the Landsberg–Park model is unphysical because it contains no mechanism which would allow evolution through the singularity which occurs at the end of each oscillation. Some quantum gravitational theories also naturally predict $\Omega_0 \simeq 1$, see J. V. Narlikar and T. Padmanabhan, *Phys. Rep.* **100,** 151 (1983), T. Padmanabhan *Phys. Lett.* A **96,** 110 (1983), and Chapter 7 of this book.
116. S. W. Hawking, in *Quantum structure of space and time*, ed. H. Duff and C. Isham (Cambridge University Press, Cambridge, 1982), p. 423.

117. This will occur so long as there does not exist some other conserved combination of baryon and lepton numbers, as there does in the simplest GUT's, like SU(5).
118. L. Grishchuk and Y. B. Zeldovich, in ref. 116; D. Atkatz and H. Pagels, *Phys. Rev.* D **25,** 2065 (1982); A. Vilenkin, *Phys. Lett.* B **117,** 25 (1982).
119. E. Tryon, *Nature* **246,** 396 (1973); see also P. I. Fomin, *Dokl. Ukran. Acad. Sci.* A **9,** 831 (1975).
120. In the closed Friedman model the rest mass energy of the material content exactly equals its potential energy; see Y. B. Zeldovich, *Adv. Astron. Astrophys.* **3,** 242 (1965).
121. If inflation occurs at the energy of grand unified then initial conditions must be such as to allow the temperature to fall to that level.
122. It is possible to avoid the initial singularity in space-time with a cosmological constant, or where there is a self-interacting scalar field of the type described in section 6.8 (because $V(\phi)>0$ allows a violation of the strong energy condition; see ref. 88 and J. D. Barrow and R. A. Matzner, *Phys. Rev.* D **21,** 336 (1980))—although it has been argued that this may not prevent a singularity in the future, see S. Bludman, *Nature* **308,** 319 (1984).
123. J. D. Barrow, *Nature* **272,** 211 (1978).
124. J. D. Barrow, *Phys. Rep.* **85,** 1 (1982); D. Chernoff and J. D. Barrow, *Phys. Rev. Lett.* **50,** 134 (1983) and Gravity Essay (1982); Y. Elskens, *Phys. Rev.* D **28,** 1033; V. A. Belinskii, E. M. Lifshitz, and I. M. Khalatnikov, *Sov. Phys. Usp.* **13,** 745 (1971); E. M. Lifshitz, I. M. Lifshitz, and I. M. Khalatnikov, *Sov. Phys. JETP* **32,** 173; J. D. Barrow, in *Classical general relativity*, ed. W. Bonnor, J. Islam, and M. A. H. MacCallum (Cambridge University Press, Cambridge, 1984).
125. R. Penrose, in *Theoretical principles in astrophysics and relativity*, ed. N. Lebovitz, W. H. Reid, and P. O. Vandervoort (University of Chicago Press, 1978) and in *Proc. First Marcel Grossman meeting on general relativity*, ed. R. Ruffini (North-Holland, Amsterdam, 1977), and in *Physics and Contemporary needs*, ed. Riazuddin (Plenum, NY, 1977).
126. S. W. Hawking in *Astrophysical cosmology, Pont. Acad. Scient. Scripta Varia.* **48,** 563 (Pontificia Acad. Scient., Vatican City, 1982); *Nucl. Phys.* B **239,** 257 (1984).
127. J. Hartle and S. W. Hawking, *Phys. Rev.* D **28,** 2906 (1983).
128. S. W. Hawking, *Commun. Math. Phys.* **43,** 199 (1975).
129. P. Candelas and D. W. Sciama, *Phys. Lett.* **38,** 1372 (1977).
130. R. Penrose, in *Confrontation of cosmological theories with observational data*, ed. M. S. Longair (Reidel, Dordrecht, 1974), p. 263.
131. E. Kasner, *Ann. J. Math.* **43,** 217 (1921).
132. J. D. Barrow, unpublished Gravity Essay (1980).
133. F. J. Tipler, *Phys. Rev.* D **15,** 9423 (1977).
134. F. J. Tipler, *Gen. Rel. Gravn* **10,** 1005 (1979).
135. In fact, the stable late-time asymptotes of homogeneous universes investigated in ref. 91 do appear to attain the bound (6.166) as $t\rightarrow\infty$.
136. J. Collins and M. J. Perry, *Phys. Rev. Lett.* **34,** 1353 (1975); B. L. Hu in *Recent developments in general relativity* (North-Holland, Amsterdam, 1980).

137. J. D. Barrow and F. J. Tipler, *Nature* **276,** 453 (1978).
138. R. Penrose, in *General relativity: an Einstein centenary survey*, ed. S. W. Hawking and W. Israel (Cambridge University Press, Cambridge, 1979), and in *Progress in cosmology*, ed. A. Wofendale (Reidel, Dordrecht, 1982), p. 87.
139. K. Gödel, *Rev. Mod. Phys.* **21,** 447 (1949). For a detailed survey see F. J. Tipler, *Ann. Phys.* **108,** 1 (1977), *Phys. Rev.* D **9,** 2203 (1974), and *Phys. Rev. Lett.* **37,** 879 (1976). The question of the stability of closed timelike curves is investigated in detail in S. W. Hawking, *Gen. Rel. Gravn* **1,** 393 (1971); and in F. J. Tipler, *J. Math. Phys.* **18,** 1568 (1977).
140. S. W. Hawking and J. C. Luttrell, *Phys. Lett.* B **143,** 83 (1984).
141. M. Davis, J. Huchra, D. Latham and J. Tonry, *Astrophys. J.* **270,** 20 (1983).
142. A. D. Linde, *Rep. Prog. Phys.* **47,** 925 (1984); *Lett. Nuovo Cim.* **39,** 401 (1984).
143. J. B. Hartle, 'Initial conditions' (Lecture delivered at the Fermi-Lab Inner-Space/Outer-Space Conference, 4 May 1984).
144. H. Kodama, 'Comments on Chaotic Inflation', KEK Report 84–12, ed. K. Odaka and A. Sugamoto (1984).

7 Quantum Mechanics and the Anthropic Principle

Nothing ever becomes real till
it is experienced.

John Keats

7.1 The Interpretations of Quantum Mechanics

When I hear of
Schrödinger's cat, I reach
for my gun.

S. W. Hawking

In classical physics Man seemed entirely superfluous to the Universe. He was only a cog—and a rather small cog at that—in the Newtonian world-machine. However, his role in the Cosmos appears greatly enhanced in quantum mechanics. According to the so-called Copenhagen Interpretation of the quantum mechanical formalism—and this interpretation is the most widely accepted interpretation among contemporary physicists—Man, in his capacity as the observer of an experiment, is an essential and irreducible feature of physics.

The historians S. G. Brush[1] and P. Forman[2] have claimed that the idea of the observer playing an important role in a physical measurement can be traced back to the nineteenth century, but in quantum mechanics this idea was first put forward only in 1926 by Born[3] in his 'probability interpretation' of Schrödinger's wave function. This function and the Schrödinger equation which it satisfies were very successful in solving many outstanding problems in atomic physics and spectroscopy, but in the interpretation which Schrödinger himself gave to the wave function—that of measuring the charge density of the electron in a system described by the Schrödinger equation[4]—there were a number of difficulties. For instance, the wave function which described a beam of electrons incident upon a photographic plate was greatly extended in space, yet each electron actually impinged upon the plate at a localized point. In the Schrödinger interpretation this was interpreted as a sudden instantaneous 'collapse' of the charge spread out over a wide area down to a point on the plate. It was hard to see how such a collapse could be consistent with the requirement of special relativity that no information be transmitted faster than the velocity of light. In the probabilistic interpretation put forward by Born,[5] the wave function $\psi(x)$ was a measure of the probability that the electron, viewed as always remaining a point particle, was at the point

x in space. More precisely, since ψ is a complex number, and a probability must be a non-negative real number $|\psi(x)|^2$ is the probability that the electron is at the point x. This interpretation removed the difficulty presented by causality. For before the electron hits the plate, it has a small probability of being at many points over a wide area, and hence the wave function is spread out over a wide area. When the electron actually hits the photographic plate, and hence is measured to be at a particular spot on that plate, the probability that the electron is at that particular spot suddenly becomes one at that point, and zero at all other points. This means that the wave function must suddenly collapse if it is to measure the sudden change in the probabilities. What changes is not something physical, but as Born put it, rather 'our knowledge of the system suddenly changes'. Thus with this interpretation, a property of Man in the role as observer of the physical universe enters the formalism of physics in an essential way.

The Born interpretation of the Schrödinger wave function was extended by the great Danish physicist Niels Bohr, who turned it into the so-called Copenhagen interpretation, and it is this interpretation of the quantum mechanical formalism which is most widely accepted among physicists today, at least in some form. Bohr first defended the essential role played by the observer in quantum mechanics in his Como Lecture of 1927:[6]

> On one hand, the definition of the state of a physical system, as ordinarily understood, claims the elimination of all external disturbances. But in that case, according to the quantum postulate, any observation will be impossible, and above all, the concepts of space and time lose their immediate sense. On the other hand, if in order to make observation possible we permit certain interactions with suitable agencies of measurement, not belonging to the system, an unambiguous definition of the state of the system is naturally no longer possible, and there can be no question of causality in the ordinary sense of the word. The very nature of the quantum theory thus forces us to regard the space-time coordination and the claim of causality, the union of which characterizes the classical theories, as complementary but exclusive features of the description, symbolizing the idealization of observation and definition respectively.

Both in the above passage and in his later writings on the quantum theory of measurement and the philosophical significance of quantum mechanics, Bohr goes far beyond the bare bones of a probabilistic interpretation of the wave function. For a probabilistic or a statistical interpretation of the wave function is perfectly consistent with the notion of the world as a deterministic system in which both causality and a space-time description are valid simultaneously. However, in this case one must admit that the statistical quantum theory is not the ultimate theory of the world. If the world is deterministic in the classical mechanical sense and its properties exist independently of human observation, then it must be that quantum theory is statistical only because it contains no reference to some of the classical variables which are actually governing the behaviour of atomic particles. These unknown factors not considered by quantum theory are termed 'hidden variables' by those physicists who support such a deterministic world view.[7] This view is not inherently unreasonable because classical statistical mechanics was pictured in precisely this way during the nineteenth century.[8] The atoms of a gas were pictured as governed by deterministic Newtownian laws of motion. However, because there is such an enormous number of atoms in a macroscopic volume of gas—10^{23} atoms in a cubic centimetre being a typical number—it is a practical impossibility to take into account all of the variables—6 variables per atom—which would have to be considered in a deterministic classical description of the system of atoms comprising the gas. Therefore, the description of the system was vastly over-simplified by taking certain statistical averages of these variables, thereby reducing the number of independent variables to a tractable number. It is, of course, impossible to give an absolutely precise deterministic description of the time evolution of the gas in terms of these new and fewer variables, but the average behaviour of the new variables can be predicted, and this is sufficient for most practical purposes.

Bohr denied that there could exist hidden variables which would ultimately replace the probabilistic description of the world by quantum theory with a deterministic description. He based his position upon the essential role played by the observer in quantum physics. The observer and the world were so inextricably connected that 'an independent reality in the ordinary physical sense can neither be ascribed to the phenomena nor to the agencies of observations'.[9] In other words, many physical properties of atomic particles did not even exist before the act of observation, the act of observation was necessary to bring these properties into existence.

Bohr argued that ascribing simultaneous independent physical reality to all properties which an electron could possess would contradict the formalism of quantum theory. For, from this formalism one could derive the Heisenberg uncertainty relations, which for the position of the electron in the x-direction and momentum p_x in the x-direction can be written

$$\Delta x \, \Delta p_x \geqslant \hbar/2 \tag{7.1}$$

where $\hbar$ is Planck's constant divided by 2π, and $\Delta x, \Delta p_x$ are to be interpreted according to the Copenhagen view as the uncertainty in the measurement of the position of the electron in the x-direction, and Δp_x the uncertainty in the measurement of its momentum component in this direction, respectively. By applying this relation and the other uncertainty relations to a number of idealized experiments, Bohr showed they implied that a precise measurement of the position would so affect the electron as to make the momentum unknown. On the other hand, had we chosen to measure the momentum of the electron precisely, the experimental arrangement necessary to make this measurement would, through its interaction with the electron, make the electron's position completely unknown. Adopting the empiricist principle that what cannot be measured, even in principle, cannot be said to exist, Bohr therefore denied reality to the notions of electron position and electron momentum prior to their measurement. The electron's position and momentum would be determined by the particular experimental arrangement which the observer chose to interact with it, and quantum mechanics shows no experimental apparatus can be constructed which would determine both properties absolutely precisely in a single measurement. Thus after any measurement, the electron's position and momentum must be partially undetermined; these properties are 'real' only within the limits allowed by the uncertainty relations and the experimental apparatus chosen by the observer to measure them.

To a realist like Einstein, who held that a physical reality existed independently of Man the Observer, Bohr's view was anathema. Over a period of a decade Einstein tried repeatedly to contrive an idealized experiment in which precise measurement of a system's complementary properties—those orthogonal properties of a system which, like the position and momentum of an electron, have complementary uncertainties according to Bohr's interpretation of the uncertainty relations—could be made simultaneously. He failed in this endeavour, and was forced to admit that the uncertainty relations did, as Bohr claimed, restrict the precise simultaneous measurement of complementary properties. Nevertheless, he continued to feel that these properties possessed independent simultaneous reality even if they could not be simultaneously measured. To justify this point of view, he and two of his colleagues, Nathan Rosen and Boris Podolsky, proposed what has become known as the Einstein–Podolsky–Rosen (EPR) experiment.

This experiment was presented in a paper entitled, 'Can Quantum-Mechanical Description of Physical Reality be Considered Complete?'[10] The paper began with the authors' definition of 'physical reality:'

> If, without in any way disturbing a system, we can predict with certainty (i.e., with probability equal to unity) the value of a physical quantity, then there exists an element of physical reality corresponding to this physical quantity.[10]

The EPR experiment is an experimental arrangement to measure the complementary variables of a physical system. In their original paper, Einstein, Podolsky and Rosen chose position and momentum, but following Bohm[11] most modern discussions of the experiment have used the spin of an electron as measured in a given direction. The z-component, S_z, of the spin—that is, the value of the component of spin of the electron in the direction of the z-coordinate axis—can be shown to be complementary to the component of the spin in a direction perpendicular to the z-axis. Quantum theory tells us that a component of the electron spin can take on only one of two values: S_z can only be $\pm\hbar/2$. Similarly, S_x, which is the component of the spin in the x-direction, can only take on the values $\pm\hbar/2$. However, since the variables S_z and S_x are complementary, if it is known that S_z is equal to $\hbar/2$, then the value of S_x is completely undefined by quantum theory; in Bohr's view S_x has *no* value in this situation. It takes on a value if it is measured, but if it is measured, the very process of measurement simultaneously destroys the reality of the value of S_z.

The EPR experiment considers a system of two electrons, coupled so that the total spin of the system is zero; if S_z of the first electron is $+\hbar/2$, then S_z of the second electron is $-\hbar/2$, and similarly for a measurement of S_x. Now it can be shown[11] that the uncertainty relations allow an absolutely precise simultaneous measurement of the total spin S_T of the two-electron system and *either* S_z or S_x. That is, the pair of variables (S_T, S_z) or the pair (S_T, S_x) can be measured. Suppose the two-electron system is constructed so that $S_T = 0$ and the two electrons are moving apart very rapidly. Then after a very long time, S_T will still be zero and the electrons will be far apart—one light year, say. After the electrons have become widely separated, we perform a measurement of S_z on electron #1, with the result $S_z = +\hbar/2$, say. Then since $S_T = 0$, it follows that S_z of electron #2 must equal $-\hbar/2$. So, we know with certainty the value of S_z for electron #2 even though we have performed no measurement on electron #2. Thus by the EPR definition of physical reality, S_z of electron #2 has physical reality. On the other hand, we could have decided to measure S_x at electron #1. If S_x of electron #1 were $\hbar/2$, then (as before) the value of S_x of electron #2 must be $-\hbar/2$, and so according to the EPR definition, the value of S_x of electron #2 must have physical reality.

According to the EPR definition, both S_z and S_x of electron #2 must possess an element of physical reality independent of any observation, since all observations were performed on electron #1, not electron #2. The electrons were a light year apart when the measurement on electron #1 was performed, so because the speed of light is finite, there is no question of a measurement on electron #1 affecting the state of electron #2. Einstein contended that since the observation affects electron #1 and not electron #2, it is impossible for the measurement on electron #1 to bring into existence the properties of electron #2, as Bohr's Copenhagen Interpretation would claim.

The EPR experiment highlights the 'contrary-to-common-sense' nature of the Copenhagen Interpretation. The idea that the act of observation must have a non-negligible effect on the object being observed is certainly plausible, and this happens all the time in the social sciences. If the government announces that its economists have found the inflation rate has changed drastically, then people change their buying and saving habits accordingly. However, the EPR experiment shows the interaction of the observer with the observed in quantum mechanics can have *non-local* effects. If we grant, following Bohr, that the observation of electron #1 brings into existence some property of this electron, say, the z- or x-component of the spin—then this observation brings into existence the same property of electron #2 which is a light year away. Furthermore, this property of electron #2 is brought into existence at the instant the measurement is performed on electron #1, even though no information about the measurement, no forces and no influence of any kind can reach electron #2 for at least a year. There appears to be instantaneous action at a distance.

This non-local effect of the measurement process in quantum mechanics makes it possible to test for a certain class of hidden variables, those hidden variables which act locally like any of the known forms of physical interaction. The physicist J. S. Bell showed[12,13] that if the spin of the electrons in the EPR experiment were indeed controlled by local hidden variables, then the determination of the spin of electron #2 by a measurement on electron #1 could not take place instantaneously as it does in quantum mechanics. Thus by performing the EPR experiment one could test for the existence of local hidden variables. In the past few years the EPR experiment has actually been performed by a number of groups, with the result that the predictions of quantum theory are confirmed and the existence of local hidden variables ruled out,[13] (at least those local hidden variable theories in which the measuring process is assumed not to effect the distribution of the hidden variables[4] are ruled out). This is generally (see, however, refs 14 and 15) regarded as confirmation of the Copenhagen Interpretation, in which the act of observation is responsible for bringing properties of physical systems into existence. As John Clauser and Abner Shimony graphically put it:

> Physical systems cannot be said to have definite properties independent of our observations; perhaps an unheard tree falling in the forest makes no sound after all.[13]

Bohr's response to the EPR experimental proposal was to emphasize even more strongly the essential role of the observer in the measurement of a quantum system.[16,17] He denied the validity of the EPR criterion of reality. In his view, it was meaningless to say that a property of a quantum mechanical system existed without referring to the observer, or more precisely, to the observer's experimental arrangement which measured this property:

As a more appropriate way of expression I advocated the application of the word *phenomenon* exclusively to refer to the observations obtained under specified circumstances, including an account of the whole experimental arrangement [ref. 18, p. 238].... As regards the specification of the conditions for any well-defined application of the [quantum mechanical] formalism, it is moreover essential that the *whole experimental arrangement* be taken into account [ref. 18, p. 222].... As repeatedly stressed, the principal point here is that such measurements [of S_z or S_x] demand mutually exclusive experimental arrangements [that is, an apparatus which could measure S_z could not measure S_x, and vice versa] [ref. 18, p. 233].... of course, there is in [the case of the EPR experiment] no question of a mechanical disturbance of the system under investigation during the last critical stage of the measuring procedure. But even at this stage there is essentially the question of *an influence on the very conditions which define the possible types of predictions regarding the future behaviour of the system.* Since these conditions constitute an inherent element of the description of any phenomenon to which the term 'physical reality' can be properly attached, we see that the argumentation of [EPR]... does not justify their conclusion that the quantum-mechanical description is essentially incomplete.[18]

In other words, physical reality does not exist independently of the observer and his experimental apparatus. Even though there is no direct interaction between electrons #1 and #2 during the measurement, they are bound together by the observer's decision to obtain information about electron #2 by measuring a property of electron #1.

The Copenhagen Interpretation of quantum mechanics was first given a rigorous, axiomatic formulation by the mathematician John von Neumann in 1932.[19] Von Neumann's axioms represent a quantum state by a wave function which can change with time in one of two ways: first, it can evolve continuously as a solution to the Schrödinger equation; or second, it can undergo a discontinuous change as a result of a measurement. In the latter case, after a measurement the quantum state will be an eigenstate of the variable which is measured by the experimental apparatus. Since it is the observer who ultimately defines which experimental apparatus is employed, in effect the necessary presence of the observer in quantum physics is recognized by an explicit axiom. Von Neumann regarded the two processes of time evolution as mutually irreducible. He did, however, point out that there was no hard and fast dividing-line between the two. We might choose to say that the second process, the collapse of the wave function, occurs somewhere in the experimental apparatus itself, or we might want to say that the apparatus is part of the quantum system and that the collapse of the wave function occurs in the consciousness of the human observer. The last possibility was favoured by

London and Bauer,[20] who published a simplified discussion of the von Neumann theory of measurement, which made this theory widely known to physicists.[4]

This lack of a sharp dividing-line between the two types of basic quantum processes was felt to be very unsatisfactory by a number of physicists. Schrödinger proposed a famous experiment, canonized as 'Schrödinger's Cat Paradox' to illustrate the difficulties:

A cat is penned up in a steel chamber, along with the following diabolical device (which must be secured against direct interference by the cat): in a Geiger counter there is a tiny bit of radio-active substance, so small, that perhaps in the course of one hour one of the atoms decays, but also, with equal probability, perhaps none; if it happens, the counter tube discharges and through a relay releases a hammer which shatters a small flask of hydrocyanic acid. If one has left this entire system to itself for an hour, one would say that the cat still lives *if* meanwhile no atom has decayed. The first atomic decay would have poisoned it. The ψ-function of the entire system would express this by having in it the living and the dead cat (pardon the expression) mixed or smeared out in equal parts.[21]

The situation after one hour is pictured in Figure 7.1.

After one hour the wave function is a superposition of two states:

$$\psi = \frac{1}{\sqrt{2}}(\psi_{\text{dead}} + \psi_{\text{alive}}) \tag{7.2}$$

where ψ_{dead} is the quantum state of the cat being dead, and ψ_{alive} is the quantum state of the cat being alive. If the cat *were* dead it would be in

Figure 7.1. The Schrödinger's Cat Paradox. A cat is sealed in a steel chamber. If a radioactive atom decays within an hour, a hammer shatters a flask of hydrocyanic acid and the cat dies. If no atom decays, the flask is not shattered and the cat lives. Ater one hour the cat's wave function is a superposition of two states, given by equation (7.2); the cat is both alive *and* dead.[43]

the state

$$\psi = \psi_{\text{dead}} \tag{7.3}$$

while if the cat *were* alive it would be in the state

$$\psi = \psi_{\text{alive}} \tag{7.4}$$

Both state (7.3) and state (7.4) are quite different from the superposed state (7.2) which the cat quantum system must be in before the measurement is made according to quantum mechanics. During the measurement, wave function (7.2) of the cat quantum system must collapse into either wave function (7.3) or wave function (7.4). The question is, just where does this collapse occur? If we follow London and Bauer and say the collapse occurs when a human observer actually observes the system, then this means the cat is neither dead nor alive, but rather is a superposition (7.2) of both states, until the human observer opens the steel chamber. This seems absolutely contrary to common sense. Should then the cat be regarded as the observer who collapses the wave function? Most working physicists would probably take this view.[22,4] On the other hand, perhaps the Geiger counter tube, the device which irreversibly amplifies the atomic decay to macroscopic dimensions, should be regarded as the true 'observer'. This is the view defended by Wheeler,[23,24] and it has some experimental support,[24,25] if it is granted that the wave function is collapsed during the measurement process by *some* agency.

As shown by Arthur Fine from unpublished papers of Einstein,[26] the objection to quantum mechanics in its Copenhagen interpretation which Einstein was trying to express in the EPR experiment was actually the same problem that led Born to introduce the probabilistic interpretation of the wave function in the first place: collapse of the wave function during a measurement is inconsistent with the principle of separation—that information cannot be sent faster than light. Thus the reality of a particle property cannot depend on the result of a measurement made on another particle far away from it. Einstein's own simplified version of the EPR experiment is strikingly similar to the Schrödinger's Cat experiment. This simplified version is as follows:

Suppose a ball is in one of two closed boxes, with equal probability; if we know that there is exactly one ball in the system, then we can determine whether the ball is in box #2 by simply looking in box #1. If the boxes are sufficiently far apart, then according to the principle of separation the ball *really* was (or was not) in box #2 before the observer looked in box #1. However, according to quantum mechanics the ball is, so to speak, half in one box and half in the other—just as Schrödinger's Cat is a mixture of dead and alive states before the chamber is opened—and suddenly 'materializes' in one or the other box at the instant of measurement—the instant the wave function is collapsed by opening the first box.[27]

This ambiguity of just *where* the wave function collapses leads to further difficulties. In the case of Schrödinger's Cat, we have seen how it is unclear who should be called the observer: is it the Geiger counter, the cat, or the human observer? Why should even the human observer be regarded as responsible for the wave function collapse? Indeed, if one analyses the measurement process according to the laws of quantum mechanics without the axiom of wave function collapse, one finds that the state of the cat–human system is

$$\psi_{\substack{\text{cat-human}\\\text{system}}} = \left(\psi_{\text{dead}} \times \psi_{\substack{\text{human sees}\\\text{cat dead}}}\right) + \left(\psi_{\text{alive}} \times \psi_{\substack{\text{human sees}\\\text{cat alive}}}\right) \tag{7.5}$$

The state $\psi_{\text{dead}} \times \psi_{\substack{\text{human sees}\\\text{cat dead}}}$ means that the cat is dead and that the observer sees the cat dead. In other words, the human observer is in a mixed state just like the cat! He is in a state of having seen the cat dead and having seen the cat alive. The best that quantum mechanics can predict without the collapse axiom is that there will be a correlation between the observations of the human and the state of the cat. Quantum mechanics without wave function collapse cannot say either state of the cat has a claim to reality over the other, even after the observation of the cat by a human.

Eugene Wigner has emphasized this fact in order to argue that, in spite of the ambiguities, the wave function does collapse during a measurement, and that it is the interaction of human consciousness with the physical system that is responsible for the collapse.[29,30] His argument, which has become known as the 'Wigner's Friend Paradox', goes as follows[28]: suppose a Friend of the observer in the cat-observer system described by equation (7.5) makes an observation on the system himself to determine the state of the cat. The observation will take the form of simply asking the original observer whether the cat is alive or dead. Now equation (7.5) assumes that the observation of the cat by the first observer did not collapse the wave function of the cat; it merely associated a cat-state with a human-state, thereby extending the superposition of states to the observer. If we assume that the consciousness of the observer's Friend collapses the cat-human system, then after the Friend asks the question, the system wave function collapses into either

$$\psi = \psi_{\text{dead}} \times \psi_{\substack{\text{human sees}\\\text{cat dead}}} \tag{7.6}$$

or,

$$\psi = \psi_{\text{alive}} \times \psi_{\substack{\text{human sees}\\\text{cat alive}}} \tag{7.7}$$

Were we to replace the cat and the human in the above description by atomic systems there would be no doubt experimentally that the superposition (7.5) correctly describes these systems and that (7.3) or (7.4), or (7.6) or (7.7) do not. If, however, the observer's Friend asks the observer, 'What was the state of the cat before I asked you?', the observer would reply 'I just told you. The cat was alive (dead)'.

According to Wigner, this reply indicates that the cat–man system was already in either state (7.6) or in state (7.7) *before* the Friend asked the question. The alternative (7.5)

... appears absurd because it implies that [the observer] was in a state of suspended animation before he answered [the] question. It follows that the being with a consciousness must have a different role in quantum mechanics than the inanimate measuring device.... In particular, the quantum mechanical equations cannot be linear if the preceding argument is accepted. This argument implies that 'my friend' has the same types of impressions and sensations as I—in particular, that, after interacting with the object, he is not in that state of suspended animation which corresponds to the wave function [ψ]. It is not necessary to see a contradiction here from the point of view of orthodox quantum mechanics, and there is none if we believe that the alternative is meaningless, whether my friend's consciousness contains either the impression of having seen [a live cat] or of having seen a [dead cat]. However, to deny the existence of the consciousness of a friend to this extent is surely an unnatural attitude, approaching solipism, and few people, in their hearts, will go along with it.[28]

Wigner's Friend Paradox was extended by Hugh Everett III into what we might call the 'Everett Friends Paradox'. Everett pointed out[31] that if it is considered problematic whether a single observer A with consciousness can collapse a wave function of a quantum system Q, then it is equally problematic whether another observer B can collapse the wave function corresponding to the system A+Q, and whether a *third* observer C can collapse the wave function of the system B+C+Q, and so on for an infinite series of observers. There seem to be only five ways to avoid this quandary. First, solipsism, which, as Wigner emphasizes, any physicist would reject out of hand. Second, any being with consciousness can collapse wave functions by observations. Third, a 'community' of such beings can collectively collapse wave functions. Fourth, there is some sort of Ultimate Observer who is responsible for the collapse of wave functions. Fifth, wave functions *never* collapse.

The second possibility is the one defended by Wigner, von Neumann, and London and Bauer. The last-mentioned authors attempted to truncate the infinite sequence of observers by arguing that the key aspect of consciousness which causes wave function collapse is introspection:

The observer has an entirely different point of view [from inanimate objects]. For him it is only the object x and the apparatus y which belong to the external world—to that which he calls "objective". By contrast, he has *with himself* some relations of a completely special character: he has at his disposal a characteristic and quite familiar faculty, which we may call the "faculty of introspection". He can thus give an account of his own state in an immediate manner. It is in virtue of this "immanent knowledge" that he claims the right to create for himself his own objectivity, that is to say, to cut the chain of statistical coordinations expressed by [equation (7.5)] by certifying: "I am in the state [$\psi_{\substack{\text{human sees}\\ \text{live cat}}}$]" or more simply "I see [the cat alive]...[20]

We might mention that one appealing feature of using self-reference as the essential feature of consciousness which reduces wave functions is that such self-reference is often given as the defining characteristic of the word 'consciousness'.[33,34,35]

The problem with this argument is that it cannot account for the agreement which different observers find when they make the same measurement. As Abner Shimony points out,[36] from quantum theory (e.g., equation (7.5)) we can only infer that two observers of Schrödinger's cat would make the same statistical predictions about an ensemble of Schrödinger's cats. The agreement of two observers in a *specific* observation of the cat would be a coincidence unless the first observer to look collapses the wave function. But which observer is actually the first to look? Suppose they decide to observe the system simultaneously taking a photograph of the cat, but moving a light year apart before developing the film. If they develop their respective films at events which are outside each others' light-cones, then according to relativity of simultaneity there will be some reference frame in which observer one developed the film first, and so collapsed the cat's wave function, and another frame in which observer two developed the film first. (There is an analogue of this difficulty in the EPR experiment, so we can refer to this experiment if we want to avoid the question of whether the cat is conscious or not.) It would seem that one could justify this coincidence of wave function reduction to the same state by two different observers only by reference to something like Leibniz' idea of 'pre-established harmony'.

The third and fourth possibilities have not been explored to any extent. John Wheeler has been intrigued by the notion of collapse by inter-subjective agreement,[20] but he confesses[37] that he does not see any way to make this idea mathematically precise. What really interests him about possibility three is that it would be a mechanism of bringing the entire Universe into existence! In the opinion of Heisenberg[38] and von Weizsäcker,[39] the Copenhagen Interpretation implies that properties of objects do not exist until they are observed; the properties are 'latent' but are not 'actual' before the observation. Wheeler himself defends this Copenhagen Interpretation by reference to a class of experiments which he calls 'delayed choice' experiments.[84] In a delayed choice experiment, the experimenter makes the decision of the property he wishes to measure *after* the interaction which one would think determines the property has occurred. The EPR experiment is a good example of a delayed choice experiment, for in the EPR experiment the experimenter decides which spin component to measure long after the electrons have interacted and separated. One would expect that the spin component of the electron would be determined at the time of the interaction rather than at the much later time when the experimenter sets his apparatus to measure the spin component, if it were the interaction rather than the observation which actually determines the spin component of the electron. Thus it appears that the spin component of the electron is only a 'latent' property before the observation: the observation makes the spin component 'actual'. These latent properties are generally such properties as the electron's spin, position, and momentum, while the electron's mass and charge are pictured as 'actual' before the observation. But with the recent success of the unified gauge theory of the weak and electromagnetic interaction in which the electron mass is dependent on the details of symmetry breaking which apparently occurred in the early universe, there is no reason why we should not regard *all* electron properties as contingent in principle on some sort of observation. Wheeler points out[24] that according to the Copenhagen interpretation, we can regard some restricted properties of distant galaxies, which we now see as they were billions of years ago, as brought into existence now. Perhaps *all* properties—and hence the entire Universe is brought into existence by observations made at some point in time by conscious beings. The order in the Universe is brought about in some way by the manner in which these observations are made self-consistent. Wheeler calls a Cosmos arising in this Anthropic manner a 'Participatory Universe', hence our definition of Participatory Anthropic Principle in Chapter 1. However, we ourselves can bring into existence only very small-scale properties like the spin of the electron. Might it require intelligent beings 'more conscious' than ourselves to bring into existence the electrons and other particles?

This line of speculation leads naturally to the fourth possibility, that there is some Ultimate Observer who is in the end responsible for coordinating the separate observations of the lesser observers and is thus responsible for bringing the entire Universe into existence. (The *Scientific American* columnist Martin Gardner was the first to mention[40] that John Wheeler's train of reasoning might lead to this Berkelian *Weltbild*.) The sequence of observers in the Everett Friends Paradox could continue to run until an observer O_j is reached in the future who, by his observation, coordinates two such sequences of observers. But O_j himself is part of another sequence which is joined further in the future by observer O_{j+1}. This joining of sequences of observers continues—and even includes the observations made by different intelligent species elsewhere in the Universe—until *all* sequences of observations by all observers of all intelligent species that have ever existed and ever will exist, of *all* events that have ever occurred and will ever occur are finally joined together by the Final Observation by the Ultimate Observer. He must be located at the final singularity in a closed universe, or at future time-like infinity[41] in an open universe. Since no further observations are possible past this Final observation—there is no future time past the final singularity or past future time-like infinity—the infinite sequence of observations

$$\psi = \prod_{n} \sum_{i,k} \psi_{ik} O_{ink} \tag{7.8}$$

must come to an 'end' in the sense that no further terms can be added to (7.8); ((7.8) is an infinite sequence with no last term.) Here i labels the possible results of the kth experiment, and n labels the sequence of observers observing k. If the observer is assumed to be separable from the Universe, then eqn (7.8) is the ultimate generalization of equation (7.5). The Final Observer can be regarded as 'collapsing' (7.8) into one possibility—one value of the i label—without invoking the von Neumann collapse axiom because the Final Observer is not in the Universe to which quantum theory applies: the Final Observer is either at the final singularity or at future timelike infinity. In effect, the selection of which i actually occurs is not made until the Final State is 'reached'. Not until 'then' is the Universe actualized. (The words 'reached' and 'then' placed in quotes since the Final State is not in space-time.) The above scenario is admittedly very vague; it is intended just as an outline of how an interpretation of quantum theory based on the fourth possibility could be developed. This explanation is interesting because it asserts that intelligent life, or rather consciousness, is essential to bring the entire Cosmos into existence, and since this is not done until the Final State, consciousness must continue to exist as long as the Universe does. Furthermore, this explanation requires intelligence to eventually become coextensive with the Cosmos so that *all* events can eventually be observed (assuming that past and present measurements cannot collapse wave functions in the future, and this assumption is made in the Copenhagen interpretation.) If all sequences of observers are to be brought together at the Final state, the global causal structure must allow this—in particular, there must be no event horizons. In short, this interpretation of Wheeler's Participatory Anthropic Principle would imply the Final Anthropic Principle (FAP). We will delay additional comments on this Principle until the last chapter, where FAP is discussed together with the various scenarios for the final state of the Universe.

7.2 The Many-Worlds Interpretation

> This is often the way it is in physics—our mistake is not that we take our theories too seriously, but that we do not take them seriously enough.
>
> S. Weinberg

We will now consider the fifth possibility, that the wave function *never* collapses. Such an idea forms the basis of an interpretation of the quantum mechanical formalism, the *Many-Worlds Interpretation*, (MWI) which was first proposed by Everett in 1957[42,31,32] to solve the above-mentioned difficulties of the Copenhagen Interpretation. The Many-Worlds Interpretation is often[1] classed as a 'realist' interpretation of quantum mechanics, as opposed to the idealist Copenhagen Interpretation, which brings the observer into physics in an essential way. (The *Statistical Interpretation* is intermediate between realism and idealism. It claims that the formalism provides us with information about the statistical properties of ensembles of physical systems, but no information about individual systems. The extent to which the properties of individual systems are determined by the observer is undecided in this interpretation.)

We shall begin the presentation of the Many-Worlds theory with a simple example which illustrates the salient features without any unnecessary mathematical complication. The simplest quantum mechanical system of interest is a system with two possible states. For definiteness let us think of these states as the two possible states of the vertical component of the spin of the electron, and they will be represented as $|\uparrow\rangle$ and $|\downarrow\rangle$, the former denoting spin up, and the latter spin down. According to the postulates of quantum mechanics, these two states form a basis for the state space of the electron-spin system and so any state $|\psi\rangle$ of this system can be written as a linear superposition

$$|\psi\rangle = a\,|\uparrow\rangle + b\,|\downarrow\rangle \tag{7.9}$$

For reasons that will be apparent later, we shall not impose an *a priori* normalization condition on the constants a and b.

According to quantum theory, it is necessary to include some physics of the observer or measuring apparatus in the analysis if one wishes to talk about the result of a measurement on a system. A moment's reflection will show that the essential feature of a measuring apparatus is the ability to record the result of a measurement. The essence of a successful measurement is the transfer of information about the system being measured to the memory of the measuring apparatus, as Everett[42] and DeWitt[43] (see also refs 44 and 45) were the first to show. Since in our simple system the spin of the electron can be spin up or spin down, we need an apparatus whose memory is sufficiently complex to record either possibility. We should also have an apparatus neutral state, corresponding to no measurement having been performed. The minimal apparatus required to measure the spin of the electron should have three memory states $|u\rangle$, $|d\rangle$, $|n\rangle$, which represent the memory recording the electron spin to be up, the electron spin to be down, and no recording yet having been made, respectively. (The neutral state is not strictly speaking necessary, but it simplifies the analysis enormously.) The laws of quantum mechanics must be regarded as universal if we are to apply them to cosmology, so they must apply with equal force to the measuring apparatus as to the system being measured. Thus a general state $|\Phi\rangle$ of the measuring apparatus must also be a linear superposition of the three basis states. The general state $|\text{Cosmos}\rangle$ of the universe, which is defined to be everything—all systems and apparata—considered in the analysis, is a sum of a tensor products of the basis states of the system with the basis states of the apparatus:

$$|\text{Cosmos}\rangle = \sum_i a_i\,|\psi_i\rangle\,|\Phi_i\rangle \tag{7.10}$$

It is of course not obvious *a priori* that the universe as defined above can be sufficiently divided into systems and apparata to permit writing the state $|\text{Cosmos}\rangle$ as a sum of tensor products. However, such a division must be possible if the term 'measurement' is to be meaningful, so we must assume the universe is sufficiently inhomogeneous to allow such a division. This assumption is called The Postulate of Complexity by DeWitt.[46,47] The measurement corresponds to a change of state of the universe. According to laws of quantum mechanics, all changes of state are accomplished by linear unitary operators acting on the state. In our example, the appropriate state is $|\text{Cosmos}\rangle$, so the measurement must be represented as

$$M\,|\text{Cosmos(before)}\rangle = |\text{Cosmos(after)}\rangle \tag{7.11}$$

where M is the linear unitary operator, and $|\text{Cosmos(before)}\rangle$, $|\text{Cosmos(after)}\rangle$ are the states of the universe before and after the measurement is performed, respectively. It is very important to note we have eliminated all possibility of wave function reduction by our assumption that all apparata and systems are equally governed by the same quantum mechanical laws. This assumption is essential if standard quantum mechanics is to be applied to the Universe as a whole. Since M is a linear operator, its action on $|\text{Cosmos}\rangle$ can be completely determined by its action on the tensor products of the system and apparatus basis states. For simplicity, we shall choose M to represent what DeWitt calls a 'von Neumann measurement',[48] which is a measurement that has no effect on the system if the system is in an eigenstate of the observable measured by the particular apparatus used. In the electron spin example, if the apparatus is set to measure spin up or down, and the electron spin happens to be either up or down, then a von Neumann measurement is performed if the apparatus records spin up or down, respectively, and further the state of the electron is not changed by the measurement interaction. This measurement can be represented formally as

$$M\,|\uparrow\rangle\,|n\rangle = |\uparrow\rangle\,|u\rangle \tag{7.12a}$$

$$M\,|\downarrow\rangle\,|n\rangle = |\downarrow\rangle\,|d\rangle \tag{7.12b}$$

Thus if the system is in an eigenstate of the system variable to be measured by the apparatus a von Neumann measurement does not disturb the system. Ever since Heisenberg used his gamma-ray microscope thought-experiment to demonstrate the Uncertainty Principle for the position and momentum of an electron, many have believed that a measurement on a system necessarily disturbs the system, and this disturbance is the cause of the Uncertainty Principle. This is not true. The operator defined in (7.12) does not disturb the system (provided the system happens to be in an eigenstate of the component of spin measured by the apparatus.) For any variable, measurement operators can be defined which have the effect of recording the state of the system in the memory of the measuring apparatus without disturbing the system. In our simple two-spin-state electron example, the Stern-Gerlach apparatus can be regarded as a physical realization of such a von Neumann measuring apparatus, provided the vertical component of momentum of the atom is considered to be the memory trace of the apparatus, and spin precession is ignored. (See ref. 43 for a fuller discussion.)

The effect of a von Neumann measurement operator M acting on any state (7.10) with $|\psi\rangle$ given by (7.9) and $|\Phi\rangle = |n\rangle$ is then

$$\begin{aligned} M\,|\text{Cosmos(before)}\rangle &= M(a\,|\uparrow\rangle + b\,|\downarrow\rangle)\,|n\rangle \\ &= M(a\,|\uparrow\rangle\,|n\rangle) + M(b\,|\downarrow\rangle\,|n\rangle) \\ &= a\,|\uparrow\rangle\,|u\rangle + b\,|\downarrow\rangle\,|d\rangle \\ &= |\text{Cosmos(after)}\rangle \end{aligned} \tag{7.13}$$

We can assume that $\{|\uparrow\rangle\,|n\rangle, |\downarrow\rangle\,|n\rangle\}$ span the initial state space, for we shall assume the apparatus is always initially in the neutral position.

The fundamental problem in the quantum theory of measurement is deciding what the linear superposition of universe states in the third line of equation (7.13) means. The advocates of the Many-Worlds Interpretation decide this question by arguing as follows. It is obvious that each element in the two cases (7.12) corresponds to a real physical state of some actual entity either associated with the system or the apparatus. If we grant that the state (7.9) also corresponds to an actual physical state—and we can justify this by reference either to innumerable experiments or to the superposition principle of quantum mechanics—and we grant that quantum evolution of everything in existence occurs via linear operators, then we are led necessarily to the conclusion that each term in (7.13) corresponds to an actual physical state. We are forced to say that the universe 'splits' into two 'worlds'. In the first world, represented by the first term in (7.13), the electron has spin up, and its spin is measured to be spin up. In the second world, represented by the second term in (7.13), the electron has spin down, and its spin is measured to be spin down. Another way to express this is to say that all a quantum measurement does, or indeed can do, is establish a unique correlation between states of the system being measured and states of the measuring apparatus. In the above discussion, we qualified the statement that the operator (7.12) did not disturb the system with the proviso that the system be in a certain eigenstate. If the system is not in an eigenstate—as it is not in (7.13)—then the operator does affect the system. What the operator (7.12) does when the system is in a general state is establish correlations between the apparatus basis states and those system basis states which are selected by the choice of apparatus basis states. The existence of these correlations can be detected if the {system}+{apparatus} is measured by a second apparatus. For example, a short calculation would show that a measurement of the system by an apparatus with basis states corresponding to a measurement of spin in the horizontal rather than the vertical direction would give a different result if the system were measured by the second apparatus before the system interacts with the first apparatus, than the result which the second apparatus would obtain were it to measure the system after the system has been measured by the first apparatus. Needless to say, the practical importance of these correlations will depend on the size of the system, and the measuring apparatus, relative to Planck's constant, and in the situation where the system and the apparatus are both macroscopic objects (which is the case when humans make a measurement on the Universe), the correlations can be effectively ignored.

There is a misconception in popular accounts about the MWI which must be cleared up before the MWI can be applied to cosmology. The misconception arises because the word 'universe' is used in one sense in technical discussions about the MWI, and in another sense in non-technical discussions. We have said in our interpretation of (7.13), which is the state of the universe after the measurement, that the universe is split by the measurement. This is the standard terminology in the technical literature, but it is important to note this split is to be associated more with the measuring apparatus rather than with the system being measured. In the case of a von Neumann measurement, the system is not affected (again, with the exception of the correlations) by the measurement, so it is completely misleading to describe the system as splitting as a result of the measurement. On the other hand, as is obvious from (7.12), the measuring apparatus undergoes a tremendous change: it goes from $|n\rangle$ to either $|u\rangle$ or $|d\rangle$ (or both). Of course, in measurements which are not of the von Neumann type, the system variables and not just the system/apparatus correlations will be changed by the measurement, but for macroscopic systems the change of the system variables are very small; measurements of such systems can be regarded as essentially von Neumann measurements. In particular, a measurement of the radius of the Universe can be considered a von Neumann measurement, and it would thus be more appropriate to regard the recording apparatus rather than the Universe as splitting, although the 'universe' in the technical sense defined above does split. The 'universe' in the technical sense includes just the system and the measuring apparatus, whereas the Universe in the non-technical sense includes these two entities, plus everything else in existence. We have made a distinction between the two uses of the word 'universe' by capitalizing the word when it refers to the totality of everything in existence, and left it uncapitalized when it refers to just the system and the apparatus: i.e., to everything being considered in the analysis of the measurement. The other things in the Universe, those things which are not considered in the analysis of the measurement—the planets, stars, and galaxies—are coupled only very weakly to the measuring apparatus. Thus these other items do not split when the apparatus does. Looking at the split from this point of view obviates one of the major objections to the MWI, which is that the MWI seems to require if not an actual infinity, then at least a large number of 'Universes' (in the popular sense) to explain a measurement of some microscopic phenomena, and this is contrary to Ockham's Razor.[65] In the explanation of the MWI given above, there is only one Universe, but small parts of it—measuring apparata—split into several pieces. They split—or more precisely, they undergo a drastic change—upon the act of measurement because they are designed to do so. If they were not capable of registering changes on a macroscopic level they would be quite useless as measuring devies. This fact plus the linearity of quantum mechanical operators requires them to split.

Everett himself realized that it is more appropriate to think of the measuring apparatus rather than the Universe as splitting. In reply to a criticism by Einstein against quantum mechanics, to the effect that he [Einstein] '... could not believe ... a mouse could bring about drastic changes in the Universe simply by looking at it', Everett said, '... it is not so much the system which is affected by an observation as the observer...[49]...The mouse does not affect the Universe—only the mouse is affected'.[50]

We can see this formally by simply putting the non-interacting remainder of the universe in equation (7.13):

$$\begin{aligned} M\,|\text{Universe(before)}\rangle &= M(a\,|\uparrow\rangle + b\,|\downarrow\rangle)\,|n\rangle\,|\text{everything else}\rangle \\ &= a\,|\uparrow\rangle\,|u\rangle\,|\text{everything else}\rangle + b\,|\downarrow\rangle\,|d\rangle\,|\text{everything else}\rangle \\ &= (a\,|\uparrow\rangle\,|u\rangle + b\,|\downarrow\rangle\,|d\rangle)\,|\text{everything else}\rangle \end{aligned} \tag{7.14}$$

It is clear from (7.14) that 'everything else' does not split.

A human being, or indeed any measuring apparatus, would be unaware of, or in the case of an inanimate apparatus, could not detect, those splits which they do undergo. To detect the split would entail introducing a second observing apparatus into the universe which is capable of recording in its memory both worlds $|u\rangle$ and $|d\rangle$ of the split first apparatus. In the case of a human being, the two apparata could in principle be two sections of the human memory, the second of which observes the first. It is impossible to construct such a second apparatus if it is reasonably required that this second apparatus definitely record the first apparatus to be in the state $|u\rangle$ if in fact it is, or in the state $|d\rangle$ if in fact it is. We may as well let the second apparatus perform a von Neumann measurement on the system simultaneously with measuring the first apparatus, as a check. We require only that the second apparatus record the system as being in the state $|\uparrow\rangle$ if in fact it is in this state, and as being in the state $|\downarrow\rangle$ if in fact it is in this state. The state of the second apparatus, $|A_2\rangle$, can thus be expanded in terms of basis states of the form $|a_1, a_2\rangle$, where a_1 records the value of the system variable and a_2 records the content of the first apparatus' memory. Both a_1 and a_2 can have the values n, u, or d. Before the interaction between the second apparatus and the rest of the universe, we shall require the second apparatus to be in the state $|n, n\rangle$.

The above restrictions on what the second apparatus must record uniquely define the second apparatus interaction operator M_2 acting on the basis states of the universe. We have

$$M_2\,|\uparrow\rangle\,|u\rangle\,|n, n\rangle = |\uparrow\rangle\,|u\rangle\,|u, u\rangle \tag{7.15a}$$

$$M_2\,|\downarrow\rangle\,|d\rangle\,|n, n\rangle = |\downarrow\rangle\,|d\rangle\,|d, d\rangle \tag{7.15b}$$

$$M_2\,|\uparrow\rangle\,|n\rangle\,|n, n\rangle = |\uparrow\rangle\,|n\rangle\,|u, n\rangle \tag{7.15c}$$

$$M_2\,|\downarrow\rangle\,|n\rangle\,|n, n\rangle = |\downarrow\rangle\,|n\rangle\,|d, n\rangle \tag{7.15d}$$

The last two entries in (7.15) are effective only if we were to interact the second apparatus with the rest of the universe before the first apparatus has measured the state of the system. Before any measurements by any apparatus are performed, the state of the universe is

$$|\text{Cosmos(before)}\rangle = |\psi\rangle\,|n\rangle\,|n, n\rangle \tag{7.16}$$

A measurement of the state of the system by the first apparatus, followed by measurements of the state of the system and the state of the first apparatus is thus represented as:

$$M_2M_1\,|\text{Cosmos(before)}\rangle = M_2M_1(a\,|\uparrow\rangle + b\,|\downarrow\rangle)\,|n\rangle\,|n, n\rangle$$

$$= M_2(a\,|\uparrow\rangle\,|u\rangle\,|n, n\rangle + b\,|\downarrow\rangle\,|d\rangle\,|n, n\rangle) \tag{7.17a}$$

$$= a\,|\uparrow\rangle\,|u\rangle\,|u, u\rangle + b\,|\downarrow\rangle\,|d\rangle\,|d, d\rangle \tag{7.17b}$$

It is clear from (7.17) that the first apparatus is the apparatus responsible for the splitting of the universe. More precisely, it is the first apparatus that is responsible for splitting itself and the second apparatus. The second apparatus splits, but the split just follows the original split of the first apparatus, as is apparent in (7.17b). As a consequence, the second apparatus does not detect the splitting of the first apparatus. Again, the impossibility of split detection is a consequence of two assumptions: first, the linearity of the quantum operators M_2 and M_1; second, the requirement that M_2 measure the appropriate basis states of the system and the apparatus correctly. The second requirement is formalized by (7.15). Again, in words, this requirement says that if the system and first apparatus are in eigenstates, then the second apparatus had better record this fact correctly.

It is possible, of course, to construct a machine which would *not* record correctly. However, it is essential for the sensory apparatus of a living organism to record appropriate eigenstates correctly if the organism is to survive. If there is definitely a tiger in place A, (the tiger wave function is non-zero only in place A), then a human's senses had better record this correctly, or the results will be disastrous. Similarly for the tiger. But if the senses of *both* the tiger and the human correctly record approximate position eigenfunctions, then the linearity of quantum mechanical operators necessarily requires that if either of them are *not* in a position eigenstate, then an interaction between them will split them *both* into two worlds, in each of which they both act appropriately. Ultimately, it is natural selection that determines not only that the senses will record that an object is in an eigenstate if in fact it is. Natural selection even determines what eigenstates are the appropriate ones to measure; i.e., which measuring operators are to correspond to the senses. The laws of quantum mechanics cannot determine the appropriate operators; they are given. A different measuring operator will split the observed object into different worlds. But the WAP selection of operators will ensure that the class of eigenfunctions we can measure, and hence the measuring operators, will be appropriate. The self-selection of measuring operators is the most important role WAP plays in quantum mechanics.

Our ultimate goal is to develop a formalism which will tell us what we will actually observe when we measure an observable of a system while the system state is changing with time. One lesson from the above analysis of quantum mechanics from the Many-Worlds point of view is that to measure anything it is necessary to set up an apparatus which will record the result of that measurement. To have the possibility of observing a change of some observable with time requires an apparatus which can record the results of measuring that observable at sequential times. To make n sequential measurements requires an apparatus with n sequential memory slots in its state representation. At first we will just consider the simple system (7.9) that we have analysed before, so the time evolution measurement apparatus has the state $|E\rangle$, which can be written as a linear superposition of basis states of the form

$$|a_1, a_2, \ldots, a_n\rangle \tag{7.18}$$

where each entry a_j can have the value n, u, or d, as before. The jth measurement of the system state is represented by the operator M_j, defined by

$$M_j\,|\uparrow\rangle\,|a_1, a_2, \ldots, a_j, \ldots, a_n\rangle = |\uparrow\rangle\,|a_1, a_2, \ldots, u, \ldots a_n\rangle \quad \text{(7.19a)}$$

$$M_j\,|\downarrow\rangle\,|a_1, a_2, \ldots, a_j, \ldots, a_n\rangle = |\downarrow\rangle\,|a_1, a_2, \ldots, d, \ldots a_n\rangle \quad \text{(7.19b)}$$

As before, the initial state of the apparatus will be assumed to be $|n, n, \ldots, n\rangle$. The measurement is a von Neumann measurement.

Time evolution will be generated by a time evolution operator $T(t)$. It is a crucial assumption that $T(t)$ act only on the system, and not have any effect on the apparatus that will measure the time evolution. In other words, we shall assume the basis states (7.18) are not affected by the operator $T(t)$. This is a standard and indeed an essential requirement imposed on instruments that measure changes in time. If the record of the values of some observable changed on timescales comparable with the rate of change of the observable, it would be impossible to disentangle the change of the observable from the change of the record of the change. When we measure the motion of a planet, we record its positions from day to day, assuming (with justification!) that our records of its position at various times are not changing. If we write the apparatus state as $|\Phi\rangle$, the effect of a general time evolution operator $T(t)$ on the basis states of the system can be written as

$$T(t)\,|\uparrow\rangle\,|\Phi\rangle = (a_{11}(t)\,|\uparrow\rangle + a_{12}(t)\,|\downarrow\rangle)\,|\Phi\rangle \quad \text{(7.20a)}$$

$$T(t)\,|\downarrow\rangle\,|\Phi\rangle = (a_{21}(t)\,|\uparrow\rangle + a_{22}(t)\,|\downarrow\rangle)\,|\Phi\rangle \quad \text{(7.20b)}$$

Unitarity of $T(t)$ imposes some restrictions on the a_{ij}'s, but we do not have to worry about these. Interpreting the result of a measurement on the system in an initially arbitrary state after an arbitrary amount of time has passed would require knowing how to the interpret the a_{ij}'s, and as yet we have not outlined the meaning of these in the MWI. So let us for the moment analyse a very simplified type of time evolution. Suppose that we measure the state of the system every unit amount of time; that is, at $t = 1, 2, 3, \ldots,$ etc. Since time operators satisfy $T(t)T(t') = T(t+t')$, the evolution of the system from $t = 0$ to $t = n$ is given by $[T(1)]^n$. Again for simplicity, we shall assume $a_{11}(1) = a_{22}(1) = 0$, $a_{12}(1) = a_{21}(1) = 1$. This choice will give a unitary $T(t)$. We have

$$T(1)\,|\uparrow\rangle\,|\Phi\rangle = |\downarrow\rangle\,|\Phi\rangle \tag{7.21a}$$

$$T(1)\,|\downarrow\rangle\,|\Phi\rangle = |\uparrow\rangle\,|\Phi\rangle \tag{7.21b}$$

All that happens is that if the electron spin happens to be in an eigenstate, that spin is flipped from one unit of time to the next, with $[T(1)]^{2n} = I$, the identity operator.

After every unit of time we shall measure the state of the system. The time evolution and measurement processes together will be represented by a multiplicative sequence of operators acting on the universe as follows:

$$M_nT(1)M_{n-1}T(1)\ldots M_2T(1)M_1\,|\psi\rangle\,|n, n, \ldots, n\rangle \tag{7.22a}$$

$$= M_nT(1)M_{n-1}T(1)\ldots M_2T(1)[M_1(a\,|\uparrow\rangle + b\,|\downarrow\rangle)]\,|n, n, \ldots, n\rangle \tag{7.22b}$$

$$= M_nT(1)M_{n-1}T(1)\ldots M_2T(1)(a\,|\uparrow\rangle\,|u, n, \ldots, n\rangle + b\,|\downarrow\rangle\,|d, n, \ldots, n\rangle) \tag{7.22c}$$

$$= M_nT(1)M_{n-1}T(1)\ldots M_2(a\,|\downarrow\rangle\,|u, n, \ldots, n\rangle + b\,|\uparrow\rangle\,|d, n, \ldots, n\rangle) \tag{7.22d}$$

$$= M_nT(1)M_{n-1}T(1)\ldots M_3T(1)(a\,|\downarrow\rangle\,|u, d, n, \ldots\rangle + b\,|\uparrow\rangle\,|d, u, n, \ldots\rangle) \tag{7.22e}$$

and so on.

The particularly interesting steps in the above algebra are (7.22c) and (7.22e). The first measurement of the state of the system splits the universe (or more precisely, the apparatus) into two worlds. In each world, the evolution proceeds as if the other world did not exist. The first measurement, M_1, splits the apparatus into the world in which the spin is initially up and the world in which the spin is initially down. Thereafter each world evolves as if the spin of the entire system were initially up or down respectively.

If we were to choose $a = b$, then $T(1)\,|\psi\rangle\,|\Phi\rangle = |\psi\rangle\,|\Phi\rangle$; so the state of the system in the absence of a measurement would not change with time. It would be a stationary state. If the system were macroscopic—for instance, if it were the Universe—then even after the measurement the Universe would be almost stationary; the very small change in the state of a macroscopic system can be ignored. Nevertheless, the worlds would change with time. An observer who was capable of distinguishing the basis states would see a considerable amount of time evolution even though the actual, total state of the macroscopic system would be essentially stationary. Whether or not time evolution will be observed depends more on the details of the interaction between the system and the observer trying to see if the change occurs in the system, than on what changes are actually occurring in the system.

In order to interpret the constants a, b in (7.9), or the a_{ij}'s in (7.20), it is necessary to use an apparatus which makes repeated measurements on not merely a single state of a system, but rather on an ensemble of identical systems. The initial ensemble state has the form:

$$|\text{Cosmos(before)}\rangle = (|\psi\rangle)^m \,|n, n, n, \ldots, n\rangle \tag{7.23}$$

where there are m slots in the apparatus memory state $|n, n, \ldots n\rangle$. The kth slot records the measured state of the kth system in $(|\psi\rangle)^m$. The kth slot is changed by the measuring apparatus operator M_k, which acts as follows on the basis states of the kth $|\psi\rangle$:

$$\begin{aligned} &M_k\,|\psi\rangle\ldots|\psi\rangle\,|u\rangle\,|\psi\rangle\ldots|\psi\rangle\,|n, n, \ldots, n\rangle \\ &\quad = |\psi\rangle\ldots|\psi\rangle\,|u\rangle\,|\psi\rangle\ldots|\psi\rangle\,|n, \ldots, n, u, n, \ldots, n\rangle \end{aligned} \tag{7.24a}$$

$$\begin{aligned} &M_k\,|\psi\rangle\ldots|\psi\rangle\,|d\rangle\,|\psi\rangle\ldots|\psi\rangle\,|n, n, \ldots, n\rangle \\ &\quad = |\psi\rangle\ldots|\psi\rangle\,|d\rangle\,|\psi\rangle\ldots|\psi\rangle\,|n, n, \ldots, n, d, n, \ldots, n\rangle \end{aligned} \tag{7.24b}$$

The M_k operator effects only the kth slot of the apparatus memory. It has no other effect on either the system ensemble or the other memory slots.

If we perform m state measurements on the ensemble $(|\psi\rangle)^m$, an operation which would be carried out by the operator $M_mM_{m-1}\ldots M_2M_1$, the result is

$$\begin{aligned} &M_mM_{m-1}\ldots M_2[M_1(a\,|\uparrow\rangle + b\,|\downarrow\rangle)](|\psi\rangle)^{m-1}\,|n, n, \ldots, n\rangle \\ &\quad = M_mM_{m-1}\ldots M_3M_2(|\psi\rangle)^{m-1}(a\,|\uparrow\rangle\,|u, n, \ldots, n\rangle + b\,|\downarrow\rangle\,|d, n, \ldots, n\rangle) \\ &\quad = M_mM_{m-1}\ldots M_3(|\psi\rangle)^{m-2}(aM_2\,|\psi\rangle\,|\uparrow\rangle\,|u, n, \ldots, n\rangle \\ &\qquad\qquad + bM_2\,|\psi\rangle\,|\downarrow\rangle\,|d, n, \ldots, n\rangle \\ &\quad = M_m\ldots M_4(|\psi\rangle)^{m-3}(M_3\,|\psi\rangle)(a^2\,|\uparrow\rangle\,|\uparrow\rangle\,|u, u, n, \ldots, n\rangle \\ &\qquad + ab\,|\uparrow\rangle\,|\downarrow\rangle\,|u, d, n, \ldots, n\rangle + ba\,|\downarrow\rangle\,|\uparrow\rangle\,|d, u, n, \ldots, n\rangle \\ &\qquad + b^2\,|\downarrow\rangle\,|\downarrow\rangle\,|d, d, n, \ldots, n\rangle) \\ &\quad = \Sigma a^j b^{m-j}(|\uparrow\rangle)^j(|\downarrow\rangle)^{m-j}\,|s_1, s_2, \ldots, s_m\rangle \end{aligned} \tag{7.24c}$$

where the s_i's represent either u or d, and the final sum is over all possible permutations of u's and d's in the memory basis state $|s_1, s_2, \ldots, s_m\rangle$. All possible sequences of u's and d's are represented in the sum. The measurement operator $M_m \ldots M_1$ splits the apparatus into 2^m worlds. In this situation we have m systems rather than one, so each measurement splits the apparatus (or equivalently, the universe). Each measurement splits each previous world in two.

In each world, we now calculate the relative frequency of the u's and d's. Hartle,[51] Finkelstein,[52] and Graham[53] have shown that if a, b are defined by $a = \langle\psi \mid \uparrow\rangle$ and $b = \langle\psi \mid \downarrow\rangle$, then as m approaches infinity, the relative frequency of the u's approaches $|a|^2/(|a|^2+|b|^2)$, and the relative frequency of the d's approaches $|b|^2/(|a|^2+|b|^2)$ in the Hilbert space for which the scalar product defines $\langle\psi \mid \uparrow\rangle$ and $\langle\psi \mid \downarrow\rangle$, except for a set of worlds of measure zero in the Hilbert space. It is only at this stage, where a and b are to be interpreted, that it is necessary to assume $|\psi\rangle$ is a vector in a Hilbert space. For the discussion of universe splitting, it is sufficient to regard $|\psi\rangle$ as a vector in a linear space with $|\psi\rangle$ and $c\,|\psi\rangle$, for any complex constant c, being physically equivalent. If we impose the normalization condition $|a|^2+|b|^2=1$, then $|a|^2$ and $|b|^2$ will be the usual probabilities of measuring the state $|\psi\rangle$ in the state $|\uparrow\rangle$ or $|\downarrow\rangle$, respectively. It is not essential to impose the normalization condition even to interpret a and b. For example, $|a|^2/(|a|^2+|b|^2)$ would represent the relative probability of the subspace $\langle\psi \mid \uparrow\rangle$ as opposed to $\langle\psi \mid \downarrow\rangle$ even if we expanded $|\psi\rangle$ to include other states, enough to make $|\psi\rangle$ itself non-normalizable.

One key point should be noted: since there is only *one* Universe represented by only *one* unique wave function $|\Psi\rangle$, the ensemble necessary to measure $|\langle a \mid \Psi\rangle|^2$ cannot exist for *any* state $|a\rangle$. Thus, being unmeasurable, the quantities $|\langle a \mid \Psi\rangle|^2$ have no direct physical meaning. We can at best *assume* $|a|^2/(|a|^2+|b|^2)$ measures relative probability. But there is still absolutely no reason to assume that $|\Psi\rangle$ is normalizable.

Even in laboratory experiments, where we *can* form a finite-sized ensemble of identically prepared states, it is not certain that $|a|^2/(|a|^2+|b|^2)$ will actually be the measured relative frequency of observing u. All we know from quantum theory is that as the ensemble size approaches infinity, the relative frequency approaches $|a|^2/(|a|^2+|b|^2)$ in all worlds except for a set of measure zero in the Hilbert space. There will always be worlds in which the square of the wave function is *not* the observed relative frequency, and the likelihood that we are in such a world is greater the smaller the ensemble. As is well known, we are apparently not in such a world, and the question is, why not? DeWitt suggests that perhaps a WAP selection effect is acting:

It should be stressed that no element of the superposition is, in the end, excluded.

All the worlds are there, even those in which everything goes wrong and all the statistical laws break down. The situation is similar to that which we face in ordinary statistical mechanics. If the initial conditions were right the universe-as-we-see-it *could* be a place in which heat sometimes flows from cold bodies to hot. We can perhaps argue that in those branches in which the universe makes a habit of misbehaving in this way, life fails to evolve, so no intelligent automata are around to be amazed by it.[85]

We will now consider wave-packet-spreading from the Many-Worlds point of view. A simple system which will show the essential features has four degrees of freedom, labeled by the basis states $|\uparrow\rangle$, $|\downarrow\rangle$, $|\rightarrow\rangle$, and $|\leftarrow\rangle$. As before, we shall need a measuring apparatus to record the state of the system if we are to say anything about the state of the system. Since we are interested in measuring time evolution, say at m separate times (which will be assumed to be multiples of unit time, as before), we shall need an apparatus state with m slots: $|n, n, \ldots, n\rangle$, where the n denotes the initial 'no record' recording. The kth measurement of the system state will be carried out by the operator M_k, which changes the kth slot from n to u, d, r, or l, depending on whether the state of the system is $|\uparrow\rangle$, $|\downarrow\rangle$, $|\rightarrow\rangle$, or $|\leftarrow\rangle$, respectively. The time evolution operator $T(t)$ will not effect the apparatus state, and its effect on the system basis states is as follows:

$$T(1)\,|\uparrow\rangle = a_{\uparrow\rightarrow}\,|\rightarrow\rangle + a_{\uparrow\downarrow}\,|\downarrow\rangle \tag{7.25a}$$

$$T(1)\,|\downarrow\rangle = a_{\downarrow\leftarrow}\,|\leftarrow\rangle + a_{\downarrow\uparrow}\,|\uparrow\rangle \tag{7.25b}$$

$$T(1)\,|\leftarrow\rangle = a_{\leftarrow\uparrow}\,|\uparrow\rangle + a_{\leftarrow\rightarrow}\,|\rightarrow\rangle \tag{7.25c}$$

$$T(1)\,|\rightarrow\rangle = a_{\rightarrow\downarrow}\,|\downarrow\rangle + a_{\rightarrow\leftarrow}\,|\leftarrow\rangle \tag{7.25d}$$

The effect of the time evolution operator is easily visualized by regarding the arrow which labels the four basis states of the system as a hand of a clock. If the hand is initially at 12 o'clock, (basis state $|\uparrow\rangle$) the operator $T(1)$ carries the hand clockwise to 3 o'clock (basis state $|\rightarrow\rangle$), and to 6 o'clock (basis state $|\downarrow\rangle$). More generally for any basis state, the operator $T(1)$ carries the basis state (thought of a clock hand at 12, 3, 6, or 9 o'clock) clockwise one quarter and one half the way around the clock. We shall imagine that

$$|a_{ij}|^2 \gg |a_{ik}|^2 \tag{7.26}$$

if $j = i+1$, and $k = i+2$, where $i+n$ means carrying the arrow clockwise around n quarters from the ith clock hand position. The condition (7.26) means roughly that 'most' of the wave packet initially at one definite clock position is carried to the immediately adjacent position in the clockwise direction, with a small amount of spreading into the position halfway around the clock. In addition to satisfying (7.26), the constants a_{ij} must be chosen to preserve the unitarity of $T(t)$. The measured time evolution of the state $|\uparrow\rangle$ through three time units is then

$$M_4T_3M_3T_2M_2T_1M_1\,|\uparrow\rangle\,|m, n, n, n\rangle = M_4T_3M_3T_2M_2T_1\,|\uparrow\rangle\,|u, n, n, n\rangle \tag{7.27a}$$

$$= M_4T_3M_3T_2M_2(a_{\uparrow\rightarrow}\,|\rightarrow\rangle + a_{\uparrow\downarrow}\,|\downarrow\rangle)\,|u, n, n, n\rangle \tag{7.27b}$$

$$= M_4T_3M_3T_2(a_{\uparrow\rightarrow}\,|\rightarrow\rangle\,|u, r, n, n\rangle + a_{\uparrow\downarrow}\,|\downarrow\rangle\,|u, d, n, n\rangle) \tag{7.27c}$$

$$\begin{aligned} &= M_4T_3M_3[a_{\uparrow\rightarrow}(a_{\rightarrow\downarrow}\,|\downarrow\rangle + a_{\rightarrow\leftarrow}\,|\leftarrow\rangle)\,|u, r, n, n\rangle \\ &\quad + a_{\uparrow\downarrow}(a_{\downarrow\leftarrow}\,|\leftarrow\rangle + a_{\downarrow\uparrow}\,|\uparrow\rangle)\,|u, d, n, n\rangle] \end{aligned} \tag{7.27d}$$

$$\begin{aligned} &= M_4T_3[a_{\uparrow\rightarrow}a_{\rightarrow\downarrow}\,|\downarrow\rangle\,|u, r, d, n\rangle + a_{\uparrow\rightarrow}a_{\rightarrow\leftarrow}\,|\leftarrow\rangle\,|u, r, l, n\rangle \\ &\quad + a_{\uparrow\downarrow}a_{\downarrow\leftarrow}\,|\leftarrow\rangle\,|u, d, l, n\rangle + a_{\upuparrows\downarrow}a_{\downarrow\uparrow}\,|\uparrow\rangle\,|u, d, u, n\rangle] \end{aligned} \tag{7.27e}$$

$$\begin{aligned} &= M_4[a_{\uparrow\rightarrow}a_{\rightarrow\downarrow}(a_{\downarrow\leftarrow}\,|\leftarrow\rangle + a_{\downarrow\uparrow}\,|\uparrow\rangle)\,|u, r, d, n\rangle \\ &\quad + a_{\uparrow\rightarrow}a_{\rightarrow\leftarrow}(a_{\leftarrow\uparrow}\,|\uparrow\rangle + a_{\leftarrow\rightarrow}\,|\rightarrow\rangle)\,|u, r, l, n\rangle \\ &\quad + a_{\uparrow\downarrow}a_{\downarrow\leftarrow}(a_{\leftarrow\uparrow}\,|\uparrow\rangle + a_{\leftarrow\rightarrow}\,|\rightarrow\rangle)\,|u, d, l, n\rangle \\ &\quad + a_{\uparrow\downarrow}a_{\downarrow\uparrow}(a_{\uparrow\rightarrow}\,|\rightarrow\rangle + a_{\uparrow\downarrow}\,|\downarrow\rangle)\,|u, d, u, n\rangle] \end{aligned} \tag{7.27f}$$

$$\begin{aligned} &= a_{\uparrow\rightarrow}a_{\rightarrow\downarrow}a_{\downarrow\leftarrow}\,|\leftarrow\rangle\,|u, r, d, l\rangle + a_{\uparrow\rightarrow}a_{\rightarrow\downarrow}a_{\downarrow\uparrow}\,|\uparrow\rangle\,|u, r, d, u\rangle \\ &\quad + \text{six other terms} \end{aligned} \tag{7.27g}$$

Thus each measurement splits each previous world into two; the number of branches of the universe doubles with each measurement. The time evolution however, does not split worlds, for only measurements can do that. Each world is defined by a definite sequence of measured system basis states in the apparatus memory. (This is another indication that it is more appropriate to regard the apparatus as splitting rather than the system.) Every possible sequence of records allowed by the time evolution operator is represented in the universe after each of the three measurements. However, because of condition (7.26) and the probability interpretation of the constants a_{ij}, some of the worlds are much more probable than others. The first world in the list in (7.27g), the world $|\leftarrow\rangle\,|u, r, d, l\rangle$ is the most probable world to be in at the end of three time periods and four measurements, since the coefficient of this world has the largest relative modulus. When condition (7.26) is imposed, the time evolution operator is most likely to carry the system state into the clockwise adjacent state, and indeed this is what is recorded in the memory sequence of the most probable final state of the universe. We might regard this sequence as the 'classical' evolutionary sequence, because it is both the sequence of the peak of the wave packet initially in the state $|\uparrow\rangle$, and, as a consequence, the most probable final state. It is possible, of course, to have a memory sequence corresponding to a 'non-classical' world: one in which the observed motion is not from i to $i+1$. The most probable of the 'non-classical' worlds are those which have only one memory slot entry out of the classical sequence, so if one did not observe a purely 'classical' evolution, the most likely one to see is one of the ones which are as close to 'classical' as possible.

For all worlds—memory sequences—there is no overlap between the worlds, even though by the second time period the wave packets of the system have begun to overlap one another. This is a general property which is a consequence only of the linearity of the operators, the assumption that the time evolution does not effect the apparatus memory, and the assumption that the measurement is a von Neumann measurement.

If we had evolved and measured the time evolution of a general system state $|\psi\rangle$, the results would have been broadly speaking the same. For example, if we had chosen $|\psi\rangle = |\uparrow\rangle + |\rightarrow\rangle + |\downarrow\rangle + |\leftarrow\rangle$, then if the a_{ij}'s were chosen as in (7.25) but with the added proviso that $T(1)|\psi\rangle = |\psi\rangle$, then there would have been four maximum probability worlds, each of which would be observed to evolve 'classically', as we have defined it. The 'classical' worlds would be defined by the initial state recorded in the first memory slot: for each 'classical' world, the recorded value i in the first slot would be u, r, d, or l, and the value recorded in the kth slot would be $i+k$. The overlap between the system wave packets would be enormous, but the overlap would not be seen by the measurement apparatus.

We have hitherto assumed in our analysis that the eigenspectra of both the apparatus and the system are discrete. This is a convenient but not essential assumption. If one wishes to have the universe be split cleanly by a measurement into distinct, non-overlapping worlds then it will be necessary to assume that at least one of the system and apparatus have a discrete spectrum. It need not be both that have a discrete spectrum, but one of them must.

A particularly instructive example of a good measurement of a continuous variable by a discrete variable apparatus is the Wilson cloud chamber experiment, which was first analysed quantum mechanically by Mott[44] and Heisenberg.[45] In this experiment, the system variable is the position of a charged particle, an alpha particle, say, and this position is measured by exciting a series of atoms in a three-dimensional array. Since an atom has a non-zero size a, the apparatus will not be able to measure the location of the alpha particle at any given time closer than a. This limitation is essentially the same as pointed out in the simple model above.

The alpha-particle wave function will be a spherical wave outgoing from the nucleus from which it is emitted. By the time it reaches the cloud chamber, it can be approximated very accurately by a plane wave. The theory of measurement must explain how a plane wave function, which is spread out all over space, can give the localized straight line motion actually observed.

The explanation was given by Mott and Heisenberg (we shall follow the presentation of J. S. Bell).[54] The initial wave function of the alpha particle is

$$\psi(\vec{r}) = \exp(ik\,|\vec{r}-\vec{r}_0|) \tag{7.28}$$

and ϕ_0 will denote the ground state of the array of atoms. Let

$$\phi(n_1, n_2, \ldots) \tag{7.29}$$

denote a state of the array in which atoms $n_1, n_2, \ldots$ are excited. If no alpha particle were present, the universal state would be the product of (7.28) and (7.29). Because of the interaction between the alpha particle and the atoms of the array, the universal wave function will be the sum of this product and the scattered waves produced by the interaction. In a multiple scattering approximation the scattered waves are

$$\begin{aligned}\sum_{N}\sum_{n_1,n_2,\ldots n_N} &[\phi(n_1, n_2, \ldots, n_N)\exp(ik_N\,|\vec{r}-\vec{r}_N|)f_N(\theta_N)/|\vec{r}-\vec{r}_N|]\\ &\times[\exp(ik_{N-1}\,|\vec{r}_N-\vec{r}_{N-1}|)f_{N-1}(\theta_{N-1})/|\vec{r}_N-\vec{r}_{N-1}|]\\ &\times[\exp(ik_0\,|\vec{r}_1-\vec{r}_0|/|\vec{r}_1-\vec{r}_0|]\end{aligned} \tag{7.30}$$

The general term in (7.30) is a sum over all possible sequences of N atoms in the three-dimensional array. The position of the n_j atom is denoted by $\vec{r}_j$; $k_j = (k_{j-1}^2/2m - \varepsilon)^{1/2}$, where ε is the atomic excitation energy; θ_j is the angle between $\vec{r}_j - \vec{r}_{j-1}$, and $\vec{r}_{j+1} - \vec{r}_j$ (or $\vec{r} - \vec{r}_N$ for $n = N$); $f_j(\theta)$ is the inelastic scattering amplitude for an alpha particle of momentum k_{j-1} incident on a single atom.

An explicit formula for $f(\theta)$ can be calculated in the Born approximation in terms of atomic wave functions, and it is found that $f(\theta)$ peaks in the direction of the incident alpha-particle momentum k, with angular spread $\Delta\theta = (ka)^{-1}$. This means that the relative probability of observing a sequence of excited atoms $n_1, n_2, \ldots$ will be greatest if these atoms lie essentially in a straight line, or rather in a cone of opening angle $\Delta\theta$. For an alpha particle of energy ~ 1 MeV, and with a typical atomic size of $\sim 10^{-8}$ cm, we have $\Delta\theta \sim 10^{-5}$ radians, so it is easy to see why we see the alpha-particle track as a straight line.

However, it is not just a single straight line we should see. The relative probabilities for observing a sequence of atoms $n_1, n_2, \ldots$ are given by the squares of the moduli of the coefficients of $\phi(n_1, n_2, \ldots)$, and there are many straight line sequences of atoms in the sum (7.30), each having approximately the same probability. It is clear from our previous discussion of the MWI how to interpret this: the universe is split by the first stack of atoms in the array, and subsequent excitations respect the original split. Any other measuring apparatus we could bring in to measure the excitations (e.g., ourselves) would also respect the split, as discussed above, and so we see a single straight line alpha-particle track in the cloud chamber. The first atom in the array to be excited could be any atom, located at any point in the array, so there will be an enormous number of worlds in the universe. The split of the universe into clearly distinct straight lines will occur only if a, the atomic radius, is non-zero, for were a to be small in comparison to the alpha-particle momentum, the opening angle would be so large that no single particle track would be apparent. This illustrates our previous assertion that a continuous variable can be measured only by a discrete variable if one wants a clean split between the worlds.

The above analysis is static since it is concerned with the spatial shape of the alpha-particle tracks. However, a dynamical analysis[54] shows just what one would expect: the straight lines develop in time. It is worth considering in some detail the quantum dynamics for a one-dimensional array of atoms and an alpha particle moving in one dimension, for this situation is very closely analogous to the problem of measuring the radius of the Universe in the Friedman universe. The static wave function for the array and the alpha particle in one dimension is the same as (7.30), except that the factors $|\vec{r}_j - \vec{r}_{j-1}|$ in the denominator are removed. The array wave function $\phi(n_1, n_2, \ldots)$ now refers to a sequence of atoms whose positions are given by a single coordinate x. It will be useful to distinguish unexcited and excited atoms in the sequence, so we shall denote an unexcited atom in jth position in the array by 0_j, and an excited atom in the jth position by ε_j. For example, with four atoms in the array the wave function for the second and fourth atoms unexcited and the other atoms excited would be $\phi(\varepsilon_1, 0_2, \varepsilon_3, 0_4)$. Initially, the universal wave function is

$$\Psi_i = \phi(0_1, 0_2, \ldots, 0_N)\exp(i[kr - \varepsilon t]) \tag{7.31}$$

The interaction will be turned on at $t = t_0$, after which the wave function of the universe becomes

$$\begin{aligned}\Psi_i + \sum_N \sum_{n_1, n_2, \ldots} &\phi(n_1, n_2, \ldots)\exp(i[k_N |x - x_N| - k_N^2 t/2m])f_N(x, t)\\ &\times \exp(i[k_{N-1} |x_N - x_{N-1}| - k_{N-1}^2 t/2m])f_{N-1}(x, t)\\ &\times \exp(i[k_0 |x_1 - x_0| - k_0^2 t/2m])\end{aligned} \tag{7.32}$$

where the $f_j(x, t)$ are the time-dependent inelastic scattering amplitudes, m is the mass of the alpha particle, and the other symbols are defined as in (7.30). As in the static case the scattering amplitudes can be calculated in time-dependent perturbation theory, and the most probable atomic states at any time $t > t_0$ are illustrated in Figure 7.2.

Since, initially, the alpha-particle wave function is spread out equally over all space—that is, its squared modulus is independent of the position

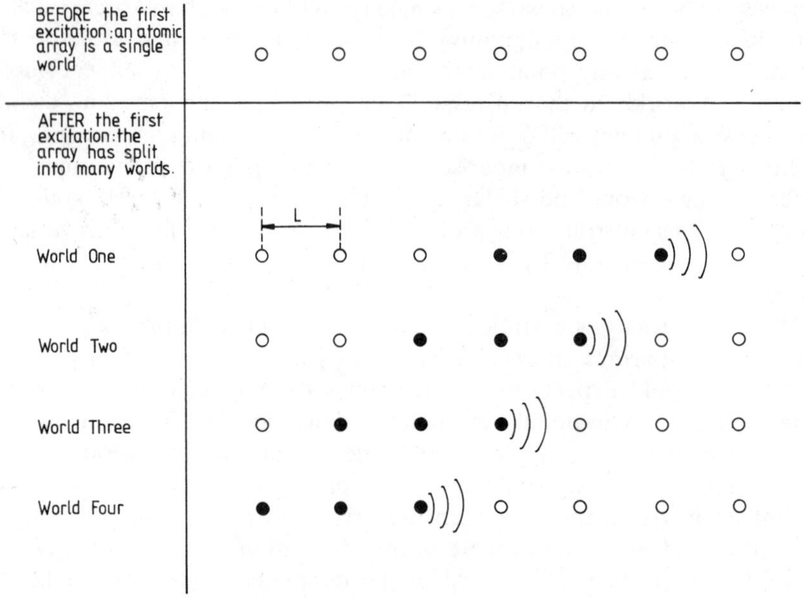

BEFORE the first excitation: an atomic array is a single world

AFTER the first excitation: the array has split into many worlds.

L

World One

World Two

World Three

World Four

Figure 7.2. The splitting of an apparatus designed to measure the position of an alpha particle as a function of time. The apparatus consists of a one-dimensional array of atoms which become excited by the passage of an alpha particle with definite energy. A darkened circle denotes an atom that has become excited, while an empty circle denotes an unexcited atom. The alpha particle momentum points from left to right. Each world is defined by the first atom to the left to become existed. Before the excitation of the first atom, the atomic array defines only one world, denoted by the single unexcited array at the top of the figure. The universe is split into a large number of worlds by the first excitation. In the figure, four such worlds are shown. Each world pictured is the most probable world defined by the leftmost excited atom, wherein the next atom to be excited is the adjacent atom which is excited at time $mL/\hbar k_i$, after the excitation of the first atom. (There would actually be one such most probable world defined by each atom in the array, and many worlds of lesser probability. The worlds of lesser probability are those in which excited atoms are interspersed with unexcited atoms.) In each world an outgoing wave packet is pictured moving to the right. The unexcited atom immediately to the right of the last excited atom is most likely to be excited when the packet reaches it.

coordinate x—the first atom to become excited is equally likely to be anywhere in the one-dimensional array. This first atom to be excited defines a branch of the universe for all succeeding times, and each atom in the array defines such a world. In each such world—the world defined by the ith atom being the first to be excited, say—the atom most likely to become excited next is in the $i+1$ position and the most likely time of its excitation is $t = mL/k_i + t_i$, where L is the spacing between the array atoms, and t_i is the time at which the ith atom becomes excited. Figure 7.2 shows four such worlds, in which the first atoms to be excited are adjacent atoms, and the time is such that three further atoms along the line have become excited. The direction of the propagation of the excitation is pictured in Figure 7.2 by the direction of the outgoing waves from the third atom in each world. Again, we must emphasize that only one of the four worlds pictured in Figure 7.2 would be seen by a human observer, because he himself would split into four branches if he were to try to measure the state of the array of atoms at the given time.

It is important to note that the most probable time for the $i+1$ atom to be excited in the branch defined by the ith atom is given by the time a wave packet of energy $k_i^2/2m$ would take to travel between the two atoms. This means that to investigate the most probable time evolution of a single branch (which is all we are physically capable of doing when we try to determine the time evolution of the entire Universe), it is sufficient to study the time evolution of a single wave packet of the appropriate characteristics outgoing from the first interaction centre which measures the radius of the Universe.

We have assumed in the above analysis that the alpha particle had a single definite energy, which means a wave function spread out over all space. However, the essential features would remain if we were to analyse the measurement of an alpha-particle wave packet that is localized in a region of physical space, and hence spread out in momentum space; i.e., being a super-position of plane wave functions with a range of energies. The splitting into worlds would be a bit more complicated, as the energy of a particle is determined by two position measurements: one at one time and another at some later time. Two position measurements, in other words, would determine a world, rather than a single position measurement in the single energy universe.

An incoming alpha-particle wave packet would cause the atomic array to be split by the first two atomic excitations into worlds with all energies consistent with the support of the alpha-particle wave packet and the discrete energy resolution of the atomic array.

The time evolution seen by the atomic array is in all essentials the same as that seen in our discrete time evolution model developed at length above. In both cases the splits occur for the dominant worlds in the first few interactions. There is, however, subsequent splitting into improbable branches at every measurement interaction. The most probable worlds will be those which evolve classically, as the Heisenberg–Mott analysis shows, and so in what follows we shall focus attention on them. When in doubt about what is going on in the more realistic continuous variable models, return to the toy discrete model.

7.3 The Friedman Universe from the Many-Worlds Point of View

> I'm afraid... that the Question and the Answer are mutually exclusive. Knowledge of one logically precludes knowledge of the other. It is impossible that both can ever be known about the same Universe.
>
> Douglas Adams

As we pointed out in Chapter 3, the deepest insight into the significance of a physical theory is obtained by expressing it in terms of an Action Principle. The full action in Einstein's gravitational theory is

$$S=\frac{1}{16\pi G}\int_M \mathbf{R}\sqrt{-g}\,d^4x+\frac{1}{8\pi G}\int_{\partial M} trK\sqrt{\gamma}\,d^3x+\int_M L_m\sqrt{-g}\,d^4x+C \tag{7.33}$$

where ∂M is the boundary of the four-dimensional region M, having extrinsic curvature K and intrinsic metric γ. The space-time metric is g, from which the Ricci curvature $\mathbf{R}$ is obtained. The matter Lagrangian is L_m. The constant C is a boundary term which must be considered in open universes, including asymptotically flat space-times,[55] but may be set to zero in closed universes. The action (7.33) is a global object and is well-defined only if the global topology is fixed (for example, as in refs 56, 57, 58; see however, refs 59 and 60). We shall consider here only the standard Friedman closed universe, so the global topology is $\mathbf{S}^3\times R^1$. The form of the action (7.33) makes it clear why theorists from Einstein[61] to Misner, Thorne, and Wheeler[62] have regarded closed universes as more physically reasonable than open universes. The latter have the boundary term C to deal with, and the surface integral in (7.33) is also more complicated in open universes, since in this case it must contain timelike portions. For closed universes, the surface integral can be taken over two disjoint spacelike hypersurfaces, or removed altogether by collapsing the boundary ∂M onto the initial and final singularities. For closed universes one would have only the two volume integrals in (7.33), and these terms would be finite even if M were chosen to be the entire spacetime (for suitable choices of the matter Lagrangians). In the open universe case the boundary terms enter the theory in a fundamental way, and there is no good way to decide what these terms should be for arbitrary open universes.[61,62]

We shall consider therefore only the closed Friedman universes, which means we shall consider only variations in the action (7.33) which preserve isotropy and homogeneity. Taking the path integral view of quantum mechanics, this means we shall consider only those paths in which the radius of the universe varies; paths in which the homogeneity or isotropy varies will be omitted from the Feynman sum. It is well-known that the metric for such a universe can be written

$$ds^2 = R^2(\tau)[-d\tau^2 + \sin^2\chi(d\theta^2 + \sin^2\theta\, d\phi^2)] \tag{7.34}$$

where the spatial variables have ranges $0 \leq \chi \leq \pi$, $0 \leq \theta \leq \pi$, $0 \leq \phi \leq 2\pi$, and τ is the 'conformal' time whose range must be determined from Einstein's equations.

For isentropic perfect fluids with pressure $\equiv p = (\gamma - 1)\rho$, where ρ is the matter density, the total action is, ($'$ is $d/d\tau$),

$$S = -\frac{3\pi}{4G}\int_{\tau_1}^{\tau_2} [(R')^2 - R^2 + CR^{-3\gamma+4}]\, d\tau \tag{7.35}$$

where C is a constant. The total Lagrangian will be quadratic only in three particular cases:

$$-3\gamma + 4 = \begin{cases} 0 \Rightarrow \gamma = 4/3, & \text{radiation gas} \\ 1 \Rightarrow \gamma = 1, & \text{dust} \\ 2 \Rightarrow \gamma = 2/3, & \text{unphysical, since it implies a} \\ & \text{negative pressure} \end{cases}$$

For the radiation gas, varying with respect to the metric gives the Lagrange equation as that of the simple harmonic oscillator (SHO),

$$\frac{d^2R}{d\tau^2} + R = 0 \tag{7.36}$$

since the constant term in the Lagrangian can be omitted. The general solution to (7.36) is of course

$$R(\tau) = R_0 \sin(\tau + \delta) \tag{7.37}$$

The two integration constants in (7.37) can be evaluated in the following way. It is clear that all solutions (7.37) have zeros with the same period π. Since it is physically meaningless to continue a solution through a singularity which occurs at every zero, all solutions exist only over a τ-interval of length π. Thus for all solutions we can choose the phase δ so that for all solutions the zero of τ-time occurs at the beginning of the universe, at $R = 0$. This implies $\delta = 0$ for all solutions, in which case the remaining constant R_0 is seen to be the radius of the universe at maximum expansion:

$$R(\tau) = R_{\max} \sin \tau \tag{7.38}$$

In the radiation gas case, all solutions are parameterized by a single number $R_{\max}$, the radius of the universe at maximum expansion. It is important to note we have obtained the standard result (7.38) without having to refer to the Friedman constraint equation. Indeed, we obtained the dynamical equation (7.36) by an unconstrained variation of the Lagrangian (7.35); we obtained the correct dynamical equation and the correct solution even though we ignored the constraint. The constraint equation contained no information that was not available in the dynamical equation obtained by unconstrained variation, except for the tacit assumption that $\rho \neq 0$. From the point of view of the dynamical equation, the vacuum 'radiation gas' (that is, $\rho = 0$) is an acceptable 'solution'. For a true ($\rho \neq 0$) radiation gas at least, ignoring the constraints is a legitimate procedure. It is well this is so, for we have precluded any possibility of obtaining the Friedman constraint equation by fixing the lapse N before carrying out the variation (in effect choosing $N = R(\tau)$). The fact that the constraint can be ignored in the radiation case is important because quantizing a constrained system is loaded with ambiguities[63,64]; indeed, the problem of quantizing Einstein's equations is mainly the problem of deciding what to do with the constraint equations,[64] and these ambiguities do not arise in the unconstrained case (see ref. 62, for a discussion of the relationship between the lapse and the Einstein constraint equations).

The constraint equation in the radiation gas case actually tells us two things: the density cannot be zero, and the solutions hit the singularity. Thus as long as these implications of the constraints are duly taken into account in some manner in the quantum theory, quantizing an unconstrained system should be a legitimate procedure, at least for a radiation gas. For simplicity, we will consider only the quantization of a radiation gas.

For a radiation gas, the Hamiltonian that is generated from the Einstein Lagrangian (7.35) is just the Hamiltonian, $\hat{H}$, for a simple harmonic oscillator (SHO), which is easy to quantize: the wave function of the Universe will be required to satisfy the equation

$$i \, \partial\Psi/\partial\tau = \hat{H}\Psi \tag{7.39}$$

There are other ways of quantizing the Einstein equations. The various quantization techniques differ mainly in the way in which the Einstein constraint equations are handled. It is an open question which way is correct. Consequently, we must attempt to pose only those questions which are independent of the quantization technique. The Friedman universe quantized via (7.38) will then illustrate the conclusions. After deriving the conclusions using our quantization technique, we shall state the corresponding results obtaining using the Hartle–Hawking[59] technique. The results obtained via these two techniques are identical.

Whatever the wave function of the Universe, the MWI implies that it should represent a collection of many universes. We would expect the physical interpretation of the time evolution of the Universal wave function Ψ coupled to some entity in the Universe which measures the radius R of the Universe, to be essentially the same as the physical interpretation of time-evolution of the alpha-particle wave function coupled to an atomic array. The first two measurements of the radius would split the Universe into the various branch universes—or more precisely, the observing system would split—and in each branch the evolution would be seen to be very close to the classical evolution expected from the classical analogue of the quantum Hamiltonian. Since the Hamiltonian is the SHO, the classical motion that will be seen by observers in a given branch universe will be sinusoidal, which is consistent with the motion predicted by the classical evolution equation (7.36).

If we assume that the collection of branch universes can be grouped together so that they all begin at the singularity at $R=0$ when $\tau=0$, then the Universe—the collection of all branch universes—will be as shown in Figure 7.3. Before the first radius measurement is performed, the Universe cannot be said to have a radius, for the Universe has not split into branches. After the first two radius measurements, the Universe has all radii consistent with the support of the Universal wave function and the support of the measurement apparatus.

The MWI imposes a number of restrictions on the quantization procedure. For example, the time parameter in equation (7.38) must be such as to treat all classical universes on an equal footing, so that all the classical universes can be subsumed into a single wave function. It is for this reason that the Einstein action (7.34) has been written in terms of the conformal time τ, for this time parameter orders all the classical closed Friedman universes in the same way: the initial singularity occurs at $\tau=0$, the maximum radius is reached at $\tau=\pi/2$, and the final singularity occurs when $\tau=\pi$. In contrast, a true physical time variable, which is the time an observer in one of the branch universes would measure, does of course depend on the particular branch one happens to be in. An example of such a physical time is the proper time. The proper time at which the maximum radius is reached depends on the value of the maximum radius, which is to say on the branch universe. Thus proper time is not an appropriate quantization time parameter according to the MWI.

The MWI also suggests certain constraints on the boundary conditions to be imposed on the Universal wave functions, constraints which are not natural in other interpretations. The other interpretations suggest that the Universe is at present a single branch which has been generated far in the past by whatever forces cause wave-function reduction. Consequently, in these non-MWI theories the effect of quantum gravity, at least at present, is to generate small fluctuations around an essentially classical universe. This view of quantum cosmology has been developed at length by J. V. Narlikar and his students,[66] and it leads to a cosmological model which is

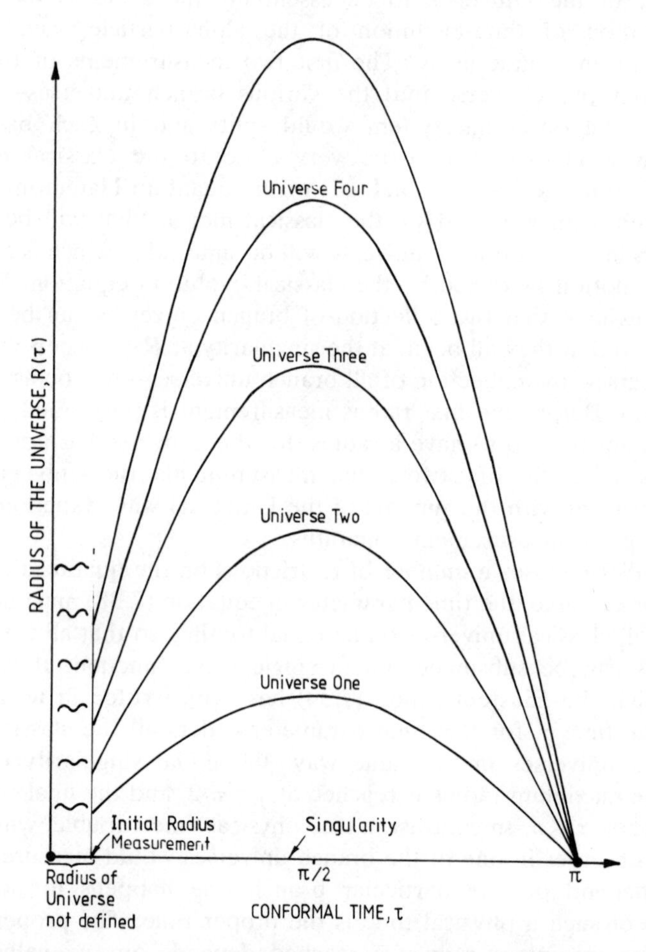

Figure 7.3. The branching of a quantum universe. Before the first interaction occurs that can encode a scale measurement, the Universe, represented before this interaction occurs as a series of wavy lines, has no radius. After the first two scaled interactions have occurred, the Universe has been split by the interactions into a large number of branches, in each of which an essentially classical evolution is seen. These branches are represented in the figure by the sine curves, each of which goes through the final singularity at $\tau = \pi$. The collection of all sine curves are all the classical radiation gas-filled Friedman models. Each curve is defined by R_{max}, the radius of the universe at maximum expansion. In the quantum Universe, all the classical universes are present, one classical universe defining a single branch. The classical universes are equally probable. Five such classical universes are pictured.

physically distinct from the models suggested by the MWI. A detailed analysis of what an observer would see would show a difference between the MWI models and the Narlikar models, although to a very good approximation the evolution would be the classical Friedman evolution in the present epoch. The two models would differ enormously very close to the initial singularity, and this could lead to experimentally testable differences between the MWI on the one hand, and the wave function reduction models on the other. Other experimentally distinguishable differences between the MWI and the other interpretations have been pointed out by Deutsch.[67]

These experimentally distinguishable differences between the MWI and the other interpretations obviate the most powerful argument which opponents bring against the MWI. This argument was succinctly stated by Shimony:

> From the standpoint of any observer (or more accurately, from the standpoint of any 'branch' of an observer) the branch of the world which he sees evolves stochastically. Since all other branches are observationally inaccessible to the observer, the empirical content (in the broadest sense possible) of Everett's interpretation is precisely the same as the empirical content of a modified quantum theory in which isolated systems of suitable kinds occasionally undergo 'quantum jumps' in violation of the Schrödinger equation. Thus the continuous evolution of the total quantum state is obtained by Everett at the price of an extreme violation of Ockham's principle, the entities being entire universes.[15]

But Ockham's principle is not violated by the MWI. Note that when the system being observed is small, the Universe in the usual sense of being everything that exists, does not split. Only the measuring apparatus splits, and it splits because it is designed to split. When the system being observed is the entire Universe it is meaningful to think of the Universe as splitting, but strictly speaking even here it is the observing apparatus that splits. If we chose to regard the Universe as splitting, then we have the Universe consisting of all classical universes consistent with the support of the Universal wave function, as in Figure 7.3. This is a violation of Ockham's principle only in appearance, for one of the problems at the classical level is accounting for the apparent fact that only a single point in the initial data space of Einstein's equations has reality. Why this single point out of the aleph-one points in initial data space? *Any* classical theory will have this problem. It is necessary to raise the Universal initial conditions to the status of physical laws to resolve this problem on the classical level. We also have to allow additional physical laws to account for wave function reduction. No additional laws need be invoked if we adopt the MWI, for here all the points in initial data space—classical universes—actually exist. The question of why does this universe rather than that universe exist is answered by saying that *all* logically possible universes do exist. What else could there possibly be? The MWI cosmology enlarges the ontology in order to economize on physical laws.

The ontological enlargement required by the MWI is precisely analogous to the spatial enlargement of the Universe which was an implication of the Copernican Theory. Indeed, philosophers in Galileo's time used Ockham's principle to support the Ptolemaic and Tychonic Systems *against* the Copernican system. For example, the philosopher Giovanni Agucchi[68] argued in a letter to Galileo that one of the three most powerful arguments against the Copernican system was the existence of the vast useless void which the Copernican system required.

In 1610 there were three interpretations of the planetary motions, the Ptolemaic, the Tychonic, and the Copernican systems, all of which were empirically equivalent and entirely viable,[69] and two of which—the Tychonic and the Copernican—were actually mathematically equivalent if applied to circular orbits.[69] The Ptolemaic system was just made the most implausible by Galileo's observations with the telescope which he announced in 1610, just as the Statistical Interpretation of quantum mechanics has been rendered implausible in the opinion of most physicists by the experiments to test local hidden variables theories. What finally convinced Galileo[70] of the truth of the Copernican system as opposed to the Tychonic system was the fact that astronomers who would not regard the Earth's motion as real were under a great handicap in understanding the motions they observed, regardless of 'mathematical equivalence'. This was also the major factor in convincing other physicists and astronomers of the truth of the Copernican System.[69] Furthermore, the Tychonic system was dynamically ridiculous and almost impossible to apply other than to those particular problems of planetary orbits which it had been designed to analyse. Similarly, the wave function collapse postulated by the Copenhagen Interpretation is dynamically ridiculous, and this interpretation is difficult if not impossible to apply in quantum cosmology. We suggest that the Many-Worlds Interpretation may well eventually replace the Statistical and Copenhagen Interpretations just as the Copernican system replaced the Ptolemaic and Tychonic. Physicists who think in terms of the Copenhagen Interpretation may become handicapped in thinking about quantum cosmology.

The different versions of the Anthropic Principle will themselves differ according to the boundary conditions that are imposed on the Universal wave function even in the MWI, and since different boundary conditions imply different physics, it is possible, at least in principle, to determine experimentally which of the different versions of the Anthropic Principle actually applies to the real Universe.

7.4 Weak Anthropic Boundary Conditions in Quantum Cosmology

Listen, there's a hell of a
good universe next door: let's go!
E. E. Cummings

From the viewpoint of the Weak Anthropic Principle, the particular classical branch of the Universe we happen to live in is selected by the fact that only a few of the classical branches which were illustrated in Figure 7.2 can permit the evolution of intelligent life. The branches which have a very small R_{max},—a few light years, say—will not exist long enough for intelligent life to evolve in them. Nevertheless, according to WAP these other branches exist; they are merely empty of intelligent life. Therefore, if WAP is the only restriction on the Universal wave function, the spatial domain of the Universal wave function $\Psi(R, \tau)$ must be $(0, +\infty)$, for any positive universal radius R is permitted by WAP. The origin must be omitted from the domain because $R = 0$ is the singularity, while negative values of R have no physical meaning.

The key problem one faces on the domain $(0, +\infty)$ is the problem of which boundary conditions to impose at the singularity $R = 0$. A straightforward calculation[71,72,73] shows that in order for the operator $-d^2/dR^2 + V(R)$ to be self-adjoint on $(0, +\infty)$, where the time-independent potential is regular at the origin and the operator acts on functions which make it L^2 on $(0, +\infty)$, the operator must be restricted to those functions which satisfy one of the following boundary conditions: either

$$\Psi(R = 0, \tau) = 0 \tag{7.40}$$

or

$$\Psi'(R = 0, \tau) + \alpha\Psi(R = 0, \tau) = 0 \tag{7.41}$$

Condition (7.40) is a boundary condition which Bryce DeWitt[63] argued must be imposed on the wave function of the Universe, for it has the effect of keeping wave packets away from the singularity. We shall therefore call condition (7.40) the DeWitt boundary condition. Condition (7.41) has an arbitrary constant α. Were condition (7.41) the appropriate boundary condition to impose on the Universal wave function, then this constant would be a new fundamental physical constant. We could then avoid introducing a new physical constant only by requiring it to be zero; i.e., by imposing the boundary condition

$$\Psi'(R = 0, \tau) = 0 \tag{7.42}$$

Both the DeWitt boundary condition and (7.41) tell us what happens to wave packets when they hit the singularity at $R = 0$. It should be emphasized that in either case, the singularity is a real entity which influences the evolution of the Universe (or more precisely, its wave function) at all times via the boundary condition at the origin. In the classical universe, the singularity is present only at the end and at the beginning of time, so in a sense the singularity is even more noticeable in quantum cosmology than in classical cosmology.

Because they are the only boundary conditions which do not introduce a new physical constant, the DeWitt boundary condition or (7.42) are the most natural boundary conditions to impose. We shall henceforth restrict attention to these conditions only. The wave function of the Universe, $\Psi(R, \tau)$, can be expressed in terms of the boundary conditions $\Psi(R, \tau=0)$ imposed at the beginning of time and the Green's function $G(R, \tilde{R}, \tau)$ via

$$\Psi(R, \tau) = \int_0^{+\infty} d\tilde{R}\, \Psi(\tilde{R}, \tau=0) G(R, \tilde{R}, \tau) \qquad (7.43)$$

The initial conditions $\Psi(\tilde{R}, \tau=0)$ are determined by the hitherto ignored constraint equations. As pointed out in section 7.3, the effect of the constraint equations in the classical case was to require all classical solutions to pass through the singularity $R=0$ when $\tau=0$. It is natural to include the constraints in the quantum model by requiring all quantum universes to do the same. The only way this can occur is if

$$\Psi(\tilde{R}, \tau=0) = f(\tilde{R})\delta(\tilde{R}) \qquad (7.44)$$

From the properties of the delta function, the functional form of $f(\tilde{R})$ is irrelevant since only $f(0)$ gives a contribution. The value of the constant $f(0)$ cannot be measured, even in principle, for it is normalization constant for the Universal wave function, and we pointed out in section 7.2 that such a constant is not measurable. Therefore for mathematical simplicity we shall set $f(0)=1$.

The Green's function for the SHO on the domain $(-\infty, +\infty)$ can be found in many textbooks.[74] If the boundary conditions imposed at $R=0$ are (7.40) or (7.42), then the Green's function for the SHO on $(0, +\infty)$ can be obtained from the Green's function on $(-\infty, +\infty)$ by linear superposition. If $\tilde{G}(R, \tilde{R}, \tau)$ denotes the Green's function on $(-\infty, +\infty)$, and $G(R, \tilde{R}, \tau)$ is the Green's function satisfying the appropriate boundary conditions at the origin, then for $\psi(\tilde{R}, \tau=0)$ being an $L^2[0, +\infty]$ function:

$$G(R, \tilde{R}, \tau) = \tilde{G}(R, \tilde{R}, \tau) - \tilde{G}(R, -\tilde{R}, \tau) \qquad (7.45)$$

if the boundary condition at the singularity is (7.40), and

$$G(R, \tilde{R}, \tau) = \tilde{G}(R, \tilde{R}, \tau) + \tilde{G}(R, -\tilde{R}, \tau) \qquad (7.46)$$

if the boundary condition at the singularity is (7.42).

If the boundary condition $\Psi(\tilde{R}, \tau=0)$ is not a smooth function but a distribution with support at $R=0$—the situation we wish to consider—the appropriate Green's function for boundary condition (7.42) is

$$G(R, \tilde{R}, \tau)=\{\tilde{G}(R, \tilde{R}, \tau)+\tilde{G}(R, -\tilde{R}, \tau)\}/2 \tag{7.47}$$

The DeWitt boundary condition at the singularity is inconsistent with the initial boundary condition (7.44), as a simple calculation using (7.40) and (7.45) (or (7.45) times some constant) will show. Therefore, (7.42) is the appropriate singularity boundary condition to use to obtain the Universal Green's function.

Putting (7.47) into (7.40) and using the Hamiltonian obtained from (7.34) to generate the Green's function $\tilde{G}(R, \tilde{R}, \tau)$, we get for the wave function of a radiation-dominated Friedman universe:

$$\Psi(R, \tau)=[3i/4L_P \sin \tau]^{1/2} \exp[(3\pi/4i)(\cot \tau)(R/L_P)^2] \tag{7.48}$$

where we have put the units back in to show the scale dependence: L_P is the Planck length. The wave function (7.48) is actually just the Green's function $\tilde{G}(R, 0, \tau)$.

The wave function (7.48) not only begins as a delta function at $\tau=0$, it recombines into a second delta function $\delta(R)$ when $\tau=\pi$; in other words, all quantum worlds terminate in a second singularity at $\tau=\pi$, just as all the classical closed Friedman universes do. This shows that the initial boundary condition (7.44) is consistent, for the logic used to derive (7.44) requires that all the quantum universes terminate in a final singularity at $\tau=\pi$.

Although the wave function is scaled by the Planck length, as a quantum cosomology should be scaled, the scale only affects the wave function phase. The wave function modulus is independent of the radius of the Universe R, except at $\tau=0$ and $\tau=\pi$. Since at the initial instant the Universal wave function (or more precisely, wave functional) is concentrated entirely at $R=0$, it has all values of momenta initially. These momenta cause the wave function to explosively spread out from the singularity to $+\infty$ the instant after $\tau=0$.

The physical interpretation of this Universal wave function is essentially the same as that given to a highly localized alpha-particle wave packet in section 7.2. The first two measurements of the radius of the Universe will split the Universe into a large number of worlds, in each of which an almost classical motion will be observed. As we have shown at length in section 7.2, the measurement of any variable requires a physical variable wherein the measurement is recorded, and our simple Friedman model contains no such variable. In the actual Universe the 'measuring device' would be some non-gravitational field in the early universe which could define a scale length. The radiation gas is conformally invariant and so defines no intrinsic length, but a conformally invariant field can be used to define a non-intrinsic length: an electromagnetic wave Gaussian packet has its standard deviation as a length scale. The first such field to couple to the radius of the Universe, and which retains the result of the coupling on timescales long compared to the expansion of a given branch of the Universe was the actual 'apparatus' that initially split the Universe. Whatever generated the perturbation spectrum that eventually gave rise to the galaxies would be a candidate for such a field, or it could be that the spectrum is a relic of the initial split. As with the localized alpha-particle, the first two such field interactions would define the branches of the Universe. At the present time the galaxies themselves would serve as benchmarks for the radius measurements, as they do in classical cosmology for measurements of the radius of the Universe.[62] The observed motion of the alpha particle is approximately classical in each branch and is determined by the motion of the wave packets scattered from each atom in the array. Similarly, the observed motion of the Universe in each branch would be approximately classical (so long as the observer is far from the final singularity), but the scattered wave packets will be evolving in an harmonic oscillator potential. This will give the usual sinusoidal motion of a radiation universe in each branch, for the motion of $\langle R \rangle$ of wave packets in such a potential satisfies the harmonic oscillator equation. This interpretation of the Universal wave function is the one pictured in Figure 7.2.

The Universe is split into branches by the first two measurements. Say this first measurement occurred at $\tau = \tau_{\text{initial}}$. Since the wave function modulus is independent of R, this means the probability of being in a world with radius R at τ_{initial} is independent of R: all classical universes are equally likely. In particular, there is no tendency for the worlds to be typically a Planck length in size at the time of maximum expansion, in this quantum cosmology at least. Some relativists have argued on dimensional grounds that such a tendency should be a generic property of quantum cosmologies. The Planck length is indeed a scale, but here it scales the phase of the wave function, not the overall size of the Universe.

Another consequence of $|\Psi(R, \tau_{\text{initial}})|^2$ being independent of R is that it is overwhelmingly probable the particular world we happen to be in will have an enormous radius of maximum expansion. That is, the probability is $1-\varepsilon$ that the density parameter Ω in our particular branch of the Universe equals $1+\delta$, where ε and δ are true infinitesimals. To see this, we need only recall the discussion in Section 7.2 about the meaning of relative probabilities calculated from non-normalizable wave functions: the probability of $|A\rangle$ relative to $|B\rangle$ is given by $|\langle\psi \mid A\rangle^2/(|\langle\psi \mid A\rangle|^2 + |\langle\psi \mid B\rangle|^2)$ even if $|\psi\rangle$ is not normalizable. In the case of the radius of the Universe at τ_{initial}, the probability that the radius is smaller than a given radius R_1 relative to the probability that it is larger than R_1 is

$$\int_0^{R_1} |\Psi(R, \tau_{\text{initial}})|^2 \, dR \Big/ \left(\int_0^{R_1} |\Psi(R, \tau_{\text{initial}})|^2 \, dR + \int_{R_1}^{+\infty} |\Psi(R, \tau_{\text{initial}})|^2 \, dR \right) \tag{7.49}$$

But this is zero, which gives the result claimed. To put it simply, if we pick a single integer (to represent the radius of Universe to the nearest parsec) at random from the set of all positive integers, and if all integers are equally probable, then it is overwhelmingly probable that the integer we pick will be extremely large.

We do not actually need the various worlds to be equally probable in order for the value of Ω we would measure to be infinitesimally close to 1. The expression (7.49) would be zero for any non-normalizable wave function, which is regular at the origin, since with such a wave function the second term in the denominator would be infinite. Thus whatever the actual probability distribution as a function of R, any non-normalizable wave function would yield an overwhelmingly most probable value of Ω of $1+\delta$. Narlikar and Padmanabham[66] were the first to suggest that quantum gravity might naturally lead to the prediction $\Omega = 1$.

It is a general working principle in physics that what is not forbidden is compulsory, and we showed in section 7.2 that there is no physical reason which requires the wave function of the Universe to be normalizable. Therefore, we would expect that the correct quantum gravity theory would yield a non-normalizable Universal wave function. The general Universal wave function (not the special Friedman model) advanced by Hartle and Hawking[59] probably has this property. The non-normalizability of the Hartle–Hawking wave function arises from the desire of its creators to include all possibilities in the Feynman sum-over-histories.

Another general guiding rule in cosmology is a *Copernican Principle*: our place in the Universe is typical. In standard cosmology the word 'place' is interpreted to mean position in space: the Universe on a sufficiently large scale ought to have the same properties independently of position. But even in classical cosmology there is another possible meaning to the word 'place'—position in initial data space. In the case of the radiation-dominated Friedman universe, the initial data space is parametrized by one variable, which can be chosen to be the radius of the Universe at a set time τ_1 (it is conventional in classical cosmology to pick τ_1 to be $\pi/2$, the time of maximum expansion). The Flatness Problem is essentially the problem of explaining why, out of all possible points in the one-dimensional initial data space of the Friedman universe, we happen to live at a very special point corresponding to a very large radius at maximum expansion. (This is equivalent to asking why Ω_0 is extremely close to 1). In classical cosmology the only possible answer to this question is to say that we have been misinformed as to the size of the initial data space: there are more forces governing the expansion of the Universe than a radiation gas coupled to gravity, and these other forces restrict the actual Friedman initial data space to a narrow range around $\Omega = 1$, at least in our neighbourhood in space. Such an answer to the Flatness Problem is the one provided by the inflationary universe model.[75] However, the inflationary model does not provide a unique value for the Universal initial condition. Although the initial data space is reduced in size, it is still not reduced to a single point, and so the question of why we happen to live in a very special Universe defined by a definite particular value of the radius at maximum expansion is left unanswered by the inflationary model. Indeed, any classical cosmological model must leave this question unanswered. It would also remain unanswered in any interpretation of quantum mechanics that has some force responsible for wave function reduction.

But it has an answer in quantum mechanics if we accept the Many Worlds Interpretation, for here we have the possibility of having many universes, each defined by a different radius at any given time, existing simultaneously. The whole of initial data space can be spanned by the various universes. Each point in the initial data space would be as real as the points in the sensible three-dimensional physical space. Thus we should expect the Copernican Principle to apply to the initial data space as it applies to three-dimensional physical space.

A *Quantum Copernican Principle* would require all the points in classical initial data space to be equally probable; we would be no more likely to find ourselves in one classical universe than another. Such a Principle is suggested by WAP, but it is not actually required by it. We have seen that the quantum cosmological model defined above has this property, but we would expect that any accurate model of the Universe would have this property if the Quantum Copernican Principle were true. A consequence of the Quantum Copernican Principle is a non-normalizable wave function if the wave function domain is $(0, +\infty)$, as required by WAP, and this leads to an $\Omega = 1 + \delta$ prediction if the Universe is closed, and $\Omega = 1 - \delta$ if it is open with δ very small.

The WAP/Copernican prediction $\Omega = 1 \pm \delta$ will be falsified if it is discovered that δ is not an infinitesimal. In fact, the current data suggest $\Omega_0 \approx 0.1 - 0.3$,[76,77] so if we accept both WAP and the current data, we must have a normalizable Universal wave function. WAP by itself restricts only the spatial domain, which means it imposes the boundary conditions (7.39) and (7.40). We should emphasize again that these boundary conditions are largely independent of the quantization procedure one uses to obtain the Universal wave function. A different procedure would not in general restrict the initial conditions on the wave function to be (7.43), in which case tests of WAP would involve calculations of how observer wave functions (which are of course included in the full Universal wave function) evolve when they are subject to (7.39) or (7.40). Note, however, that since these conditions are just the conditions necessary to make the Universal Hamiltonian self-adjoint, WAP without the Copernican Principle is logically indistinguishable from a conventional quantum cosmology theory with a Many-Worlds Interpretation. Nevertheless, even without the Copernican Principle, WAP can be experimentally distinguished from SAP, as we shall see in section 7.5.

In principle, $\Omega_0 = 1$ arising from quantum gravity could be distinguished experimentally from that arising from inflation, for the effective cosmological constant in inflation does not quite inflate the Universe completely down to $\Omega_0 = 1$; in the new inflationary universe, for example, the density parameter would actually be 1 ± 10^{-6}. In practice, it is quite impossible to distinguish these two numbers by direct astronomical observation of the universal mass density; however, see ref. 83 for an example where it is possible. However, the inflationary universe has yet to overcome all its technical problems and it is not at all clear why the spontaneous symmetry-breaking scalar field which leads to inflation should exist in Nature.[78] Thus quantum gravity may be the only way of explaining $\Omega_0 = 1$, if in the end the observational data do imply such an Ω_0 value.

7.5 Strong Anthropic Boundary Conditions in Quantum Cosmology

> The whole is always more, is capable of
> a much greater variety of wave states,
> than the combination of its parts.
>
> H. Weyl

In contrast to WAP, SAP requires a universe branch which does not contain intelligent life to be non-existent; that is, branches without intelligent life cannot appear in the Universal wave function. Since short-lived universes cannot evolve intelligent life, there is a radius R_{maxmin} such that all closed classical universe branches in the Universal wave function have a radius of maximum expansion greater than R_{maxmin}. If $R_{i\,\text{min}}$ is the radius of the R_{maxmin} branch at the time of the first radius 'measurement,' then SAP requires

$$\Psi(R < R_{i\text{min}}(\tau), \tau > \tau_i) = 0 \tag{7.50}$$

where $R_{i\,\text{min}}(\tau)$ is the trajectory for all $\tau > \tau_i$ of the classical universe with $R_{\text{max}} = R_{\text{maxmin}}$, and as before τ_i is the time of this first radius measurement. The effect of the boundary condition (7.50) is to restrict the domain of the Universal wave function to $(R_{i\text{min}}(\tau), +\infty)$. The WAP only requires (7.50) with $R_{i\text{min}}(\tau) = 0$, for all τ. The identical WAP boundary condition is obtained by Hartle and Hawking[59] using their quantization technique. A boundary condition identical to (7.50) would make their quantum cosmological model consistent with SAP.

The boundary condition (7.50) is far more restrictive than the boundary conditions (7.40) and (7.41) permitted by WAP. For $\tau > \tau_i$, the Universal wave function must satisfy conditions analogous to (7.40) or (7.41) at the lower bound of the domain if the Hamiltonian $\hat{H} = -d^2/dR^2 + V(R)$, where $V(R)$ is regular at $R = R_{i\min}(\tau)$, is to be self-adjoint. These conditions are: either

$$\Psi(R = R_{i\min}(\tau), \tau) = 0 \tag{7.51}$$

or

$$\Psi'(R = R_{i\min}(\tau), \tau) + \alpha(\tau)\Psi(R = R_{i\min}(\tau), \tau) = 0 \tag{7.52}$$

where the function $\alpha(\tau) = \Psi^{*\prime}(R = R_{i\min}(\tau), \tau)/\Psi^*(R = R_{i\min}(\tau), \tau)$, the prime means partial differentiation taken with respect to R before R is set equal to $R_{i\min}(\tau)$, and the asterisk denotes complex conjugation. Equations (7.51) and (7.52) are obtained in the same manner as (7.40) and (7.41). First, one finds the boundary conditions which must be imposed on the square-integrable functions at $R_{i\min}(\tau)$ in order for the operator $\hat{H}$ to be Hermitian on the domain $(R_{i\min}(\tau), +\infty)$. ($\hat{H}$ will be Hermitian if $\int \psi^* \hat{H} \phi \, dR = \int (\hat{H}\psi)^* \phi \, dR$ where ψ and ϕ are admissible functions.) Then one verifies that either (7.51) or (7.52) is sufficient to make $\hat{H} = -d^2/dR^2 + V(R)$ self-adjoint on the C^2 functions of compact support on $(R_{i\min}(\tau), +\infty)$. (See refs 72 or 79 for a discussion of the detailed procedure).

In addition to the boundary conditions (7.51) and (7.52), further conditions must be imposed for $\tau < \tau_i$ in order to ensure that there is no wave function seepage into the region $R(\tau > \tau_i) < R_{i\min}(\tau > \tau_i)$. Calculating necessary conditions to prevent such seepage would require knowledge of the non-gravitational matter Hamiltonian at τ_i, and this is not known. A sufficient condition to prevent seepage is

$$\Psi(R = R_{i\min}(\tau_i), \tau_i) = 0 \tag{7.53}$$

This condition will also restrict the value of initial wave function at $\tau = 0$.

Boundary conditions (7.51)–(7.53) are somewhat indefinite because we don't know $R_{i\min}(\tau)$. However, if $R_{i\min}(\tau)$ has been comparable to the radius of our particular branch universe over the past few billion years, the effect of these conditions on the observed evolution of our branch would be considerable. Recall that the observed branch motion follows closely the evolution of the expectation value $\langle R \rangle$ for a wave packet in the potential $V(R)$. Today $\langle R \rangle$ must be very close to the measured radius of our branch universe. The evolution of $\langle R \rangle$ will be quite different if the boundary conditions (7.51) or (7.52) are imposed close to the present observed radius than if they were imposed at $R = 0$; i.e., if conditions (7.40) or (7.41) were imposed. Thus in principle the boundary conditions imposed by WAP and SAP respectively can lead to different observations. The idea that WAP and SAP are observationally distinct from the point of view of the MWI was suggested independently by Michael Berry[80] and one of the authors.[81]

In the above discussion we have assumed that there are no SAP limitations on the upper bound of the Universal wave function domain. An upper bound of plus infinity on square-integrable functions requires such a function and its derivatives to vanish at infinity. If an Anthropic Principle were to require a finite upper bound, then additional boundary conditions, analogous to (7.51) or (7.52), would have to be imposed at the upper boundary. There is some suggestion that FAP may require such a boundary condition; see Chapter 10.

John Wheeler's Participatory Anthropic Principle, which is often regarded as a particularly strong form of SAP, has intelligent life selecting out a *single* branch out of the no-radius Universe that exists prior to the first 'measurement' interaction at $\tau = \tau_i$. This selection is envisaged as being due to some sort of wave function reduction, and so it cannot be analysed via the MWI formalism developed here. Until a mechanism to reduce wave functions is described by the proponents of the various wave-function-reducing-theories, it is not possible to make any experimentally testable predictions. The fact that the boundary conditions on a quantum cosmology permit such predictions to be made is an advantage of analysing the Anthropic Principle from the formalism of the MWI. A more detailed analysis of the significance of boundary conditions in quantum cosmology can be found in ref. 82.

In this chapter we have seen how modern quantum physics gives the observer a status that differs radically from the passive role endowed by classical physics. The various interpretations of quantum mechanical measurement were discussed in detail and reveal a quite distinct Anthropic perspective from the quasi-teleological forms involving the enumeration of coincidences which we described in detail in the preceding two chapters. Wheeler's Participatory Anthropic Principle is motivated by unusual possibilities for wave-packet reduction by observers and can be closely associated with real experiments.

The most important guide as to the correct interpretation of the quantum measurement process is likely to be that which allows a sensible quantum wave function to be written down for cosmological models and consistently interpreted. This naturally leads one to prefer the Many Worlds picture. Finally, we have tried to show that it is possible to formulate quantum cosmological models in accord with the Many-Worlds

Interpretation of quantum theory so that the Weak and Strong Anthropic Principles are observationally testable.

References

1. S. G. Brush, *Social Stud. Sci.* **10,** 393 (1980).
2. P. Formain, *Hist. Stud. Physical Sci.* **3,** 1 (1971).
3. M. Born, *Z. Physik* **37,** 863 (1926).
4. M. Jammer, *The philosophy of quantum mechanics* (Wiley, NY, 1974), pp. 24–33.
5. Ref. 4, pp. 38–44.
6. N. Bohr, in *Atomic theory and the description of Nature* (Cambridge University Press, Cambridge, 1934), p. 52.
7. Ref. 4, pp. 252–9; 440–67.
8. S. G. Brush, *The kind of motion we call heat,* Vols I and II (North-Holland, Amsterdam, 1976).
9. Ref. 6, p. 54.
10. A. Einstein, B. Podolsky, and N. Rosen, *Phys. Rev.* **47,** 777 (1935).
11. D. Bohm, *Quantum theory* (Prentice-Hall, Englewood Cliffs, NJ, 1951).
12. J. S. Bell, *Physics* **1,** 195 (1964); *Rev. Mod. Phys.* **38,** 447 (1966).
13. J. F. Clauser and A. Shimony, *Rep. Prog. Phys.* **41,** 1881 (1978).
14. Ref. 1, footnote 131 on p. 445.
15. A. Shimony, *Int. Phil. Quart.* **18,** 3 (1978).
16. N. Bohr, *Phys. Rev.* **48,** 696 (1935).
17. N. Bohr, *Nature* **136,** 65 (1935).
18. N. Bohr, 'Discussion with Einstein on epistemological problems in modern physics', in *Albert Einstein: philosopher-scientist,* ed. P. A. Schlipp (Harper & Row, NY, 1959).
19. J. von Neumann, *Mathematical foundations of quantum mechanics* (Princeton University Press, Princeton, 1955), transl. by R. T. Beyer from the German edition of 1932.
20. F. London and E. Bauer, *La théorie de l'observation en mécanique quantique* (Hermann et Cie, Paris, 1939). English transl. in ref. 25.
21. E. Schrödinger, *Naturwiss.* **23,** pp. 807–812; 823–828; 844–849 (1935); English transl. by J. D. Trimmer, *Proc. Am. Phil. Soc.* **124,** 323 (1980); English transl. repr. in Wheeler and Zurek, ref. 25; the Cat Paradox was stated on p. 238 of the *Proc. Am. Phil. Soc.* article.
22. H. Putnam, in *Beyond the edge of certainty,* ed. R. G. Colodny (Prentice-Hall, Englewood Cliffs, NJ, 1965).
23. J. A. Wheeler, 'Law without law', in Wheeler and Zurek, ref. 25.
24. J. A. Wheeler, in *Foundational problems in the special sciences,* ed. R. E. Butts and J. Hintikka (Reidel, Dordrecht, 1977); also in *Quantum mechanics, a half century later,* ed. J. L. Lopes and M. Paty (Reidel, Dordrecht, 1977).
25. J. A. Wheeler and W. H. Zurek, *Quantum theory and measurement* (Princeton University Press, Princeton, 1983).

26. A. Fine, in *After Einstein: Proceedings of the Einstein Centenary*, ed. P. Barker (Memphis State University, Memphis, 1982).
27. L. Rosenfeld, in *Niels Bohr*, ed. S. Rozental (Interscience, NY, 1967), pp. 127–8; ref. 26.
28. E. P. Wigner, in *The scientist speculates—an anthology of partly-baked ideas*, ed. I. J. Good (Basic Books, NY, 1962), p. 294; repr. in Wheeler and Zurek, ref. 20.
29. E. P. Wigner, *Monist* **48,** 248 (1964).
30. E. P. Wigner, *Am. J. Phys.* **31,** 6 (1963).
31. H. Everett III, in ref. 43, pp. 1–40. This is Everett's Ph.D. Thesis, a summary of which was published in 1957, ref. 42.
32. J. A. Wheeler, *Monist* **47,** 40 (1962).
33. C. L. Burt, 'Consciousness', in *Encyclopaedia Britannica*, Vol. 6, pp. 368–9 (Benton, Chicago, 1973). Burt asserts that: 'The word 'consciousness' has been used in many different senses. By origin it is a Latin compound meaning 'knowing things together', either because several people are privy to the knowledge, or (in later usage) because several things are known simultaneously. By a natural idiom, it was often applied, even in Latin, to Knowledge a man shared with himself; i.e., self-consciousness, or attentive knowledge. The first to adopt the word in English was Francis Bacon (1601), who speaks of Augustus Caesar as 'conscious to himself of having played his part well'. John Locke employs it in a philosophical argument in much the same sense: 'a man, they say, is always conscious to himself of thinking'. And he is the first to use the abstract noun. 'Consciousness', he explains, 'is the perception of what passes in a man's own mind' (1690).
34. J. Jaynes, *The origin of consciousness in the breakdown of the bicameral mind* (Houghton Mifflin, NY, 1976). This author argues that consciousness did not exist in human beings until recent times, because before that period they did not possess the self-reference concept of mind.
35. G. Ryle, *The concept of mind* (Barnes & Noble, London, 1949).
36. A. Shimony, *Am. J. Phys.* **31,** 755 (1963).
37. J. A. Wheeler, private conversation with F. J. T.
38. W. Heisenberg, *Physics and philosophy* (Harper & Row, NY, 1959), p. 160.
39. C. F. von Weizsäcker, in *Quantum theory and beyond*, ed. T. Bastin (Cambridge University Press, Cambridge, 1971).
40. M. Gardner, *New York Review of Books*, November 23, 1978, p. 12; repr. in *Order and surprise*, part II (Prometheus, Buffalo, 1983), Chapter 32.
41. S. W. Hawking and G. F. R. Ellis, *The large scale structure of space-time* (Cambridge University Press, Cambridge, 1973). The concept of future time-like infinity is discussed in more detail in Chapter 10—see, in particular, Figure 10.2.
42. H. Everett, *Rev. Mod. Phys.* **29,** 454 (1957).
43. B. S. DeWitt and N. Graham, *The Many-Worlds interpretation of quantum mechanics* (Princeton University Press, Princeton, 1973).
44. W. Heisenberg, *The physical principles of quantum theory* (University of Chicago Press, Chicago, 1930), pp. 66–76.
45. N. F. Mott, *Proc. Roy. Soc.* A **126,** 76 (1929); repr. in ref. 25.
46. B. S. DeWitt, in ref. 43, p. 168.

47. B. S. DeWitt, in *Battelle rencontres: 1967 lectures in mathematics and physics*, ed. C. DeWitt and J. A. Wheeler (W. A. Benjamin, NY, 1968).
48. Ref. 43, p. 143.
49. Ref. 43, p. 116.
50. Ref. 43, p. 117.
51. J. Hartle, *Am. J. Phys.* **36,** 704 (1968).
52. D. Finkelstein, *Trans. NY Acad. Sci.* **25,** 621 (1963).
53. N. Graham, in ref. 43.
54. J. S. Bell, in *Quantum gravity 2: a second Oxford symposium*, ed. C. J. Isham, R. Penrose, and D. W. Sciama (Oxford University Press, Oxford, 1981), p. 611.
55. S. W. Hawking, in *General relativity: an Einstein centenary survey*, ed. S. W. Hawking and W. Israel (Cambridge University Press, Cambridge, 1979), p. 746.
56. B. S. DeWitt, in *Quantum gravity 2*, ed. C. J. Isham, R. Penrose, and D. W. Scima (Oxford University Press, Oxford, 1981), p. 449.
57. B. S. DeWitt, *Scient. Am.* **249** (No. 6), 112 (1983).
58. F. J. Tipler, *Gen. Rel. Gravn* **15,** 1139 (1983).
59. J. Hartle and S. W. Hawking, *Phys. Rev.* D **28,** 2960 (1983).
60. S. W. Hawking, D. N. Page, and C. N. Pope, *Nucl. Phys.* B **170,** 283 (1980).
61. A. Einstein, in *The principle of relativity*, ed. A. Einstein (Dover, NY, 1923), pp. 177–83.
62. C. W. Misner, K. S. Thorne, and J. A. Wheeler, *Gravitation* (Freeman, San Francisco, 1973).
63. B. S. DeWitt, *Phys. Rev.* **160,** 1113 (1967).
64. W. F. Blyth and C. J. Isham, *Phys. Rev.* D **11,** 768 (1975).
65. A. Shimony, *Am. J. Phys.* **31,** 755 (1963).
66. J. V. Narlikar and T. Padmanabham, *Phys. Rep.* **100,** 151 (1983).
67. D. Deutsch, *Int. J. Theor. Phys.* **24,** 1 (1985).
68. S. Drake, *Galileo at work* (University of Chicago Press, Chicago, 1978), p. 212.
69. T. K. Kuhn, *The Copernican revolution* (Vintage, NY, 1959).
70. S. Drake, *Galileo* (Hin & Wang, NY, 1980), p. 54.
71. M. J. Gotay and J. Demaret, *Phys. Rev.* D **28,** 2402 (1983); J. D. Barrow and R. Matzner, *Phys. Rev.* D **21,** 336 (1980).
72. M. Reed and B. Simon, *Methods of modern mathematical physics*, Vol. II: *Fourier analysis, self-adjointness* (Academic Press, NY, 1975), Chapter 10.
73. B. Simon, *Quantum mechanics for Hamiltonians defined as quadratic forms* (Princeton University Press, Princeton, 1971).
74. L. S. Schulman, *Techniques and applications of path integration* (Wiley, NY, 1981), Chapter 6.
75. A. Guth, *Phys. Rev.* D **23,** 347 (1981).
76. Y. Hoffman and S. A. Bludman, *Phys. Rev. Lett.* **52,** 2087 (1984).
77. M. S. Turner, G. Steigman, and L. M. Krauss, *Phys. Rev. Lett,* **52,** 2090 (1984).
78. R. Wald, W. Unruh, and G. Mazenko, *Phys. Rev.* D **31,** 273 (1985).

79. G. Hellwing, *Differential operators of mathematical physics* (Addison-Wesley, London, 1967).
80. M. Berry, *Nature* **300,** 133 (1982).
81. F. J. Tipler, *Observatory* **103,** 221 (1983).
82. F. J. Tipler, *Phys. Rep.* (In press.)
83. If the Universe contains a particular form of slight expansion anisotropy, it is possible to distinguish a 'closed' from an 'open' universe no matter how small the value of $|\Omega_0 - 1|$; see J. D. Barrow, R. Juszkiewicz, and D. H. Sonoda, *Mon. Not. R. astron. Soc.* **213,** 917 (1985).
84. J. A. Wheeler, in *Mathematical foundations of quantum theory*, ed. A. R. Marlow (Academic Press, NY, 1978), pp. 9–48.
85. Ref. 43, p. 186, and see also p. 163 for a similar remark.

8 The Anthropic Principle and Biochemistry

Of my discipline Oswald Spengler understands, of course, not the first thing, but aside from that the book is brilliant.

typical German professor's reaction to *Decline of the West.*

8.1 Introduction

A physicist is an atom's way of knowing about atoms.

G. Wald

The Anthropic Principle in each of its various forms attempts to restrict the structure of the Universe by asserting that intelligent life, or at least life in some form, in some way selects out the actual Universe from among the different imaginable universes: the only 'real' universes are those which can contain intelligent life, or at the very least contain some form of life. Thus, ultimately Anthropic constraints are based on the definitions of life and intelligent life. We will begin this chapter with these definitions. We will then discuss these definitions as applied to the only forms of life known to us, those which are based on carbon compounds in liquid water. As pointed out by Henderson as long ago as 1913, and by the natural theologians a century before that, carbon-based life appears to depend in a crucial way on the unique properties of the elements carbon, hydrogen, oxygen and nitrogen. We shall summarize the unique properties of these elements which are relevant to carbon-based life, and highlight the unique properties of the most important simple compounds which these elements can form: carbon dioxide (CO_2), water (H_2O), ammonia (NH_3) and methane (CH_4). Some properties of the other major elements of importance to life as we know it will also be discussed.

With this information before us we will then pose the question of whether it is possible to base life on elements other than the standard quartet of (C, H, O, N). We shall also ask if it is possible to substitute some other liquid for water—such as liquid ammonia—or perhaps dispense with a liquid base altogether. We shall argue that for any form of life which is directly descended from a simpler form of life and which came into existence *spontaneously*, the answer according to our present knowledge of science is 'no'; life which comes into existence in this way must be based on water, carbon dioxide, and the basic compounds of (C, H, O, N). In particular, we shall show that many of the proposed alternative biochemistries have serious drawbacks which would prevent them from serving as a base for an evolutionary pathway to the higher forms of life. The arguments which demonstrate this yield three testable predictions: (1) there is no life with an information content greater than or equal to that possessed by terrestrial bacteria in the atmospheres of Jupiter and the other Jovian planets; (2) there is no life with the above lower bound on the information content in the atmosphere of Venus, nor on its surface; (3) there is no life with these properties on Mars.

This is not to say that other forms of life are impossible, just that these other forms could not evolve to advanced levels of organizations by means of natural selection. For example, we shall point out that self-reproducing robots, which could be regarded as a form of life based on silicon and metals in an anhydrous environment, might in principle be created by intelligent carbonaceous beings. Once created, such robots could evolve by competition amongst themselves, but the initial creation must be by carbon-based intelligent beings, because such robots are exceedingly unlikely to come into existence spontaneously.

A key requirement for the existence of highly-evolved life is ecological stability. This means that the environment in which life finds itself must allow fairly long periods of time for the circulation of the materials used in organic synthesis. It will be pointed out in sections 8.3–8.6 that the unique properties of (C, H, O, N) are probably necessary for this. However, these properties are definitely not sufficient. In fact, there are indications that the Earth's atmosphere is only marginally stable, and that the Earth may become uninhabitable in a period short compared with the time the Sun will continue to radiate. Brandon Carter has obtained a remarkable inequality which relates the length of time the Earth may remain a habitable planet and the number of crucial steps that occurred during the evolution of human life. We discuss Carter's work in section 8.7. The important point to keep in mind is that Carter's inequality, which is based on WAP, is *testable*, and therefore provides a test of WAP.

8.2 The Definitions of Life and Intelligent Life

> We mean by 'possessing life', that a thing can nourish itself and grow and decay.
>
> Aristotle

> Now, I realized that not infrequently books speak of books: it is as if they spoke among themselves. In the light of this reflection, the library seemed all the more disturbing

> to me. It was then the place of long, centuries-old murmuring, an imperceptible dialogue between one parchment and another, a living thing, a receptacle of powers not to be ruled by a human mind, a treasure of secrets emanated by many minds, surviving the death of those who had produced them or had been their conveyors.
>
> U. Eco

Since life is such a ubiquitous and fundamental concept, the definitions of it are legion. Rather than add to the already unmanagable list of definitions, we shall simply give what seem to us to be the *sufficient* conditions which a lump of matter must satisfy in order to be called 'living'. We shall abstract these sufficient conditions from the various definitions proposed over the last thirty years by biologists. We shall try to express these conditions in a form of sufficient generality that will not eliminate non-carbonaceous life *a priori*, but which is sufficiently particular so that no natural process now existing on earth is considered 'living' except those systems recognized as such by contemporary biologists.

A consequence of giving sufficient conditions rather than necessary conditions is the elimination from consideration as 'living' many forms of matter which most people would regard as unquestionably living matter. This situation seems unavoidable in biology. Any attempt to define some of the most important biological concepts results either in a definition with so many caveats that it becomes completely unusable, or else in a definition possessing occasional ambiguities. For example, Ernst Mayr[1] has pointed out that such difficulties are inherent in any attempt to define the concept of species precisely.

Sufficient conditions are generally much stronger than necessary conditions, and so one might wonder if the conditions which we shall give below could eliminate a possible cosmos which contained 'life' recognized as such as by ordinary standards, but not satisfying the sufficient conditions. We do not believe that cases like this can arise. Although the conditions we give for the existence of life are only sufficient when applied to particular lumps of matter, these conditions will actually be necessary when applied to an entire biosphere. That is, although particular individuals in a given biosphere may not satisfy our sufficient conditions, there must be *some* individuals, if not most individuals, in the biosphere who *do* satisfy the conditions. This will become clearer as we present and discuss the sufficient conditions.

Virtually all authors who have considered life from the point of view of molecular biology (e.g. refs. 2, 23, 37) have regarded the property of *self-reproduction* as the most fundamental aspect of a living organism. Looking at life from the everyday perspective, it would seem that self-reproduction is not an absolutely essential feature of life. An individual human being cannot self-reproduce—at least two people are required to produce a child—and a mule cannot produce another mule no matter what assistance it receives from other mules. Further, a substantial fraction of the human species never have children. These examples show that self-reproduction cannot be a ncesssary property of a lump of matter before we can call it 'living', for we would consider mules, childless persons, and celibate persons living beings. But such creatures are metazoans, which means that they are all composed of many single living cells, and generally each cell is *itself capable of self-reproduction.* Many human cells, for instance, will reproduce both in the human body and in the laboratory.[3] In general, *all* known forms of living creatures contain as sub-structure cells which can self-produce, or the living creatures are themselves self-reproducing single cells. All organisms with which we are familiar must contain such cells in order to be able to repair damage, and some damage is bound to occur to every living thing. Thus, the ability to self-repair damage to the organism seems to be intimately connected with self-reproduction in living things, at least on the cellular level of structure. Self-repair and self-reproduction seem to involve the same level of molecular technology; indeed, the machinery needed to self-repair is approximately the same as the machinery needed to self-reproduce. Self-reproduction of metazoans always begins with a single cell; in higher animals and plants this cell is the result of a fusion of at most two cells. This single cell reproduces many times, in the process transforming itself into the differentiated cell types which together make up the metazoan—nerve cells, blood cells, and so on. The ability to self-repair is absolutely essential to a living body. If a creature was unable to self-repair, it would be most unlikely to live long enough to be regarded as living. Any creature unable to repair itself would probably be stillborn.

Since all living things are largely composed of cells which can self-reproduce, or are autonomous single cells with self-reproductive capacity, we will say that self-reproduction is a necessary property which all living things must have at least in some of their substructure. Self-reproduction to this limited extent is still not sufficient for a lump of matter to be considered living. A single crystal of salt dropped into a super-saturated salt solution would quickly reproduce itself in the sense that the basic crystal structure of NaCl would be copied many times to make up a much larger crystal than was initially present. A less prosaic example would be the 'reproduction' of mesons by high-energy bombardment. If the quarks which compose a meson are pulled sufficiently far apart, the nuclear bonds which hold them together will break. But some of the energy used to break these bonds will be converted into new quarks which did not previously exist, and these new quarks can combine together to form a number of new meson pairs, see Figure 8.1.

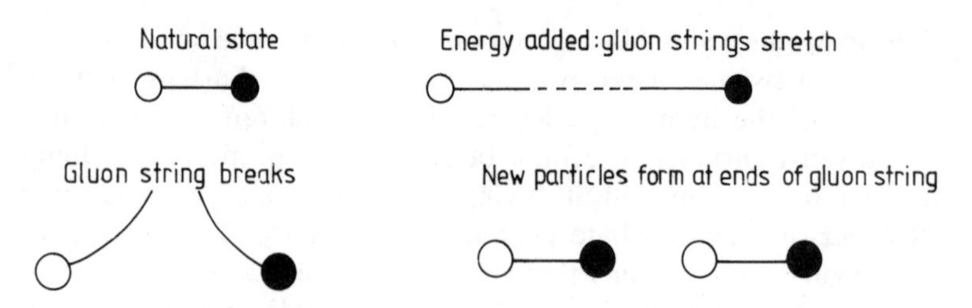

Figure 8.1. Quark reproduction in the string model. Energy added to bound quarks stretches the bonds (strings) until they break. New quarks are formed at the break in the strings, with the net result that the original bound quark system reproduces itself.

Thus, in the appropriate environment—supersaturated solutions and high-energy accelerators—both salt crystals and mesons can self-reproduce. Yet we would be unwilling to regard either salt crystals or mesons as living creatures. The key distinction between self-reproducing living cells and self-reproducing crystals and mesons is the fact that the reproductive apparatus of the cell stores information, and the specific information stored is preserved by natural selection. The reproductive 'apparatus' of crystals and mesons can in some cases store information, but this information is not preserved by natural selection.

Recall that in scientific parlance, 'information' measures the number of alternative possible statements or different individuals.[4] For example, if a computer memory stores 10^6 bits, then this memory can store 2^{10^6} different binary numbers. If a creature has 10^6 genes like humans and each gene can have one of two forms, then there are 3^{10^6} possible individuals. In humans, at least a third of all genes have two or more forms,[60] so this number is a good estimate of the possible number of different human beings. Many of these potential individuals are non-viable in a normal environment—for many of these possible gene constellations would not correspond to workable cellular machinery—but many of the other potential individuals could survive in the same environment. Thus, in a living organism, the same reproductive apparatus allows the existence of many distinct individuals who are able to reproduce in a given environment. The decision as to which individuals actually reproduce in a given environment is made by natural selection. This decision is *not* made by natural selection in the case of the 'self-reproduction' by crystals and protons. In this situation, either *all* the information is located in the environment, or else the various forms do not compete for environmental resources. The form of the crystal which reproduces in a solution is determined by the physical laws and the particular crystal form that is placed in solution, if the salt in question has several crystal forms.

It is not possible for NaCl to change its crystal structure by mutation, resulting in a new crystal structure that begins to reproduce itself and replace the previously existing crystal structure. Similarly, the type of elementary particle one can generate in a high-energy collision depends on the details of the collision, and the particle bombarded. Elementary particles do not compete for scarce resources.

To summarize, we will say that a sufficient condition for a system to be 'living' is that the system is capable of self-reproduction in some environment and the system contains information which is preserved by natural selection. By 'self-reproduction' we will mean not that an exact copy is made every time, but that there is an environment within which an exact copy would have a higher Darwinian selection coefficient[61] than all of the most closely related copies in the same environment (relationship being measured in terms of the number of differences in the copies).

Defining self-reproduction by natural selection as we have done is essential for two reasons: first, it is only the fact that natural selection occurs with living beings that allows us to distinguish living beings from crystals in terms of self-reproduction; second, for very complex living organisms, the probability that exact self-reproduction occurs is almost nil.[5,6] What happens is that many copies—both approximate and exact—are made and natural selection is used to eliminate the almost perfect copies. If one does not allow some errors in the reproductive process, with these errors being corrected at a later stage by natural selection, then one is led to the conclusion that self-reproduction is inconsistent with quantum physics.[5,6] Ultimately, it is natural selection that corrects errors and holds a self-reproductive process together, as Eigen and Schuster[7] have shown in their investigation of the simplest possible molecular systems exhibiting self-reproduction. Thus, basically we define life to be self-reproduction with error correction.

Note that a single human being does not satisfy the above sufficient condition to be considered living, but it is made up of cells some of which do satisfy it. A male–female pair would collectively be a system capable of self-reproduction, and so this system would satisfy the sufficient condition. In any biosphere we can imagine, *some* systems contained therein would satisfy it. Thus, it is a *necessary* condition for some organisms in any biosphere to satisfy the above *sufficient* condition.

A virus satisfies the above sufficient condition, and so we consider it a living organism. A virus is the simplest known organism which does satisfy the condition, so it is instructive to review the reproductive cycle of a typical virus, the T2 virus. This cycle is pictured and discussed in Figure 8.2.

A virus consists of two main components, a nucleic acid molecule surrounded by a protein coat. This coat can have a rather complex

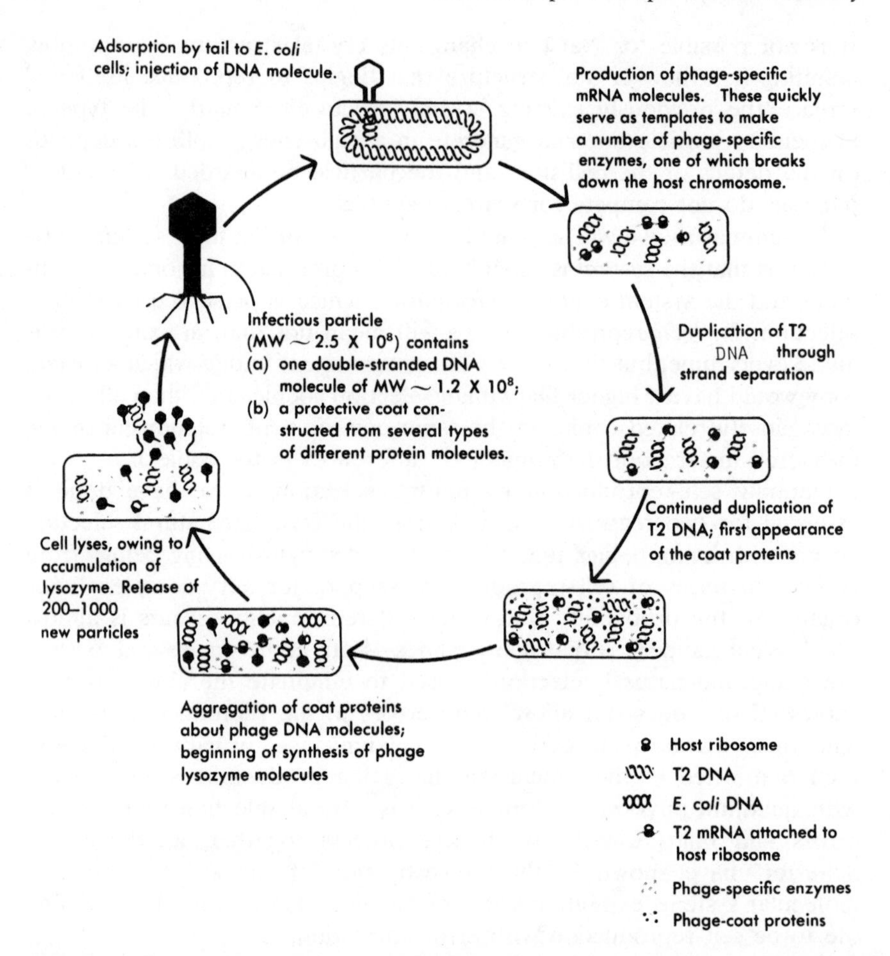

Figure 8.2. Life cycle of a T2 virus. The T2 virus is a bacteriophage, which means it 'eats' bacteria. In the above figure it is shown attacking an *E. coli* bacterium. The enzyme lysozyme is coded by the virus DNA, and its purpose is to break the cell wall. Ribosomes are structures inside the cell that enable DNA to construct proteins (coats and enzymes) from amino acid building-blocks. The DNA produces RNA for the desired protein. The RNA act in the ribosomes as templets on which the amino acids collect to form proteins. (From ref. 33, with permission.)

structure, as in the case of the T2 virus. The nucleic acid molecule, either RNA or DNA, is a gene which codes for the proteins required by the virus in its reproductive cycle. This cycle begins with the nucleic acid gene being injected into a living cell by the protein coat, which remains outside the cell. Once inside the cell, the gene uses the cellular machinery to make copies of itself, and to manufacture other protein coats and an enzyme that makes cell walls explode. These genes and coats combine, and the enzyme coded by the virus nucleic acid causes the cell to explode, thereby releasing new viruses. These new viruses will be carried by forces not under the control of the virus to new cells, at which time the cycle will repeat.

The environment within which this cycle occurs has a dual nature: first, there is the interior of a cell which contains all the necessary machinery and materials to synthesize nucleic acids and the proteins which these acids code; second, whatever environment connects two such cells. Both parts of its environment are necessary for the cycle to complete, and natural selection is active in both environments to decide just what information coded in the nucleic acid molecule will self-reproduce. In the cellular part of the environment, the information coded in the genes must allow the gene to use the cellular machinery to make copies of itself, the protein coat and the enzymes that break cell walls. Furthermore, the particular protein coat which is coded for in the virus gene must be able to combine with the gene to form a complete virus, and it must be able to inject the nucleic acid molecule it surrounds into a cell. If a mutation occurs so that the information coded in the gene does not code for nucleic acids and proteins with these properties, natural selection will eliminate the mutants from the environment. It is the action of natural selection which creates the basic difference between viruses and salt crystals; indeed, aside from just a little more complexity, the physical distinction between the two is not marked, for viruses can be crystallized. But the reproduction cycle of the virus cannot be carried out while the virus is in crystal form; the virus must be transformed into a non-crystalline form, and when it is in this form, natural selection can act.

The structure and reproductive cycle of a virus, as outlined above, is strikingly similar to the basic theoretical structure and replication cycle of a self-reproductive machine developed theoretically by von Neumann in the 1950's in complete ignorance of the make-up and life history of viruses. Perhaps this should not be surprising, since von Neumann was attempting to develop a theory of self-reproducing machines which would apply to *any* machine which could make a copy of itself, and a virus naturally falls into this category. In von Neumann's scheme[9,10,11] a self-reproducing machine is composed of two parts, a constructor and an information bank which contains instructions for the constructor. The constructor is a machine which manipulates matter to whatever extent it is necessary to make the various parts of the self-reproducing machine and assemble them into final form. The complexity of the constructor will depend on both the complexity of the self-reproducing machine and on what sort of material is available in its environment. The most general type of constructor is called a *universal constructor*, which is a machine, a

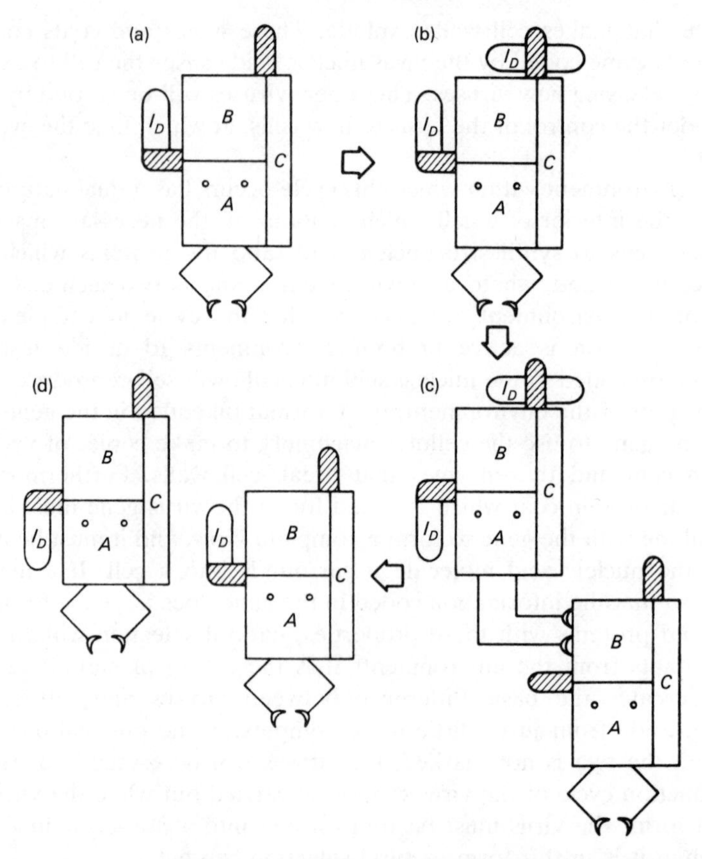

Figure 8.3. The essential features of a self-reproducing machine, according to von Neumann. The self-reproducing machine with the information bank labelled I_D and the constructor divided into three parts labelled A, B and C reproduces as follows: (a) the constructor subsystem B makes a copy of the information (program) in the bank and inserts the program copy into a holder; (b) the constructor subsystem A takes over, and makes a copy of subsystems A, B, and C using the information in I_D; (c) the subsystem C takes the copy of the information from the holder and inserts this copy into the empty bank of $A+B+C$. The product now has all the information which the original machine had, so it is also capable of self-reproduction in the same environment. (Figure after Arbib, ref. 11, with permission.)

robot if you will, that can make *anything*, given instructions about the exact procedure necessary to do so. It is the function of the information bank to provide the necessary instructions to the constructor. The reproductive cycle of von Neumann's self-reproducing machine is pictured in Figure 8.3.

The information bank, which is a computer memory containing detailed instructions about how a constructor should manipulate matter, first instructs the constructor to make a copy of a constructor either without an information bank, or with blank computer memory. The information bank is then duplicated or the information contained in the computer memory is recorded. In the final stage the information bank and constructor are combined, and the result is a copy of the original machine. The copy has all the information which the original machine had, so it is also capable of self-reproduction in the same environment.

Von Neumann showed that a machine could reproduce by following this procedure. A virus does follow it in its reproductive cycle, for within a virus the protein coat corresponds to the constructor, and the nucleic acid corresponds to the information bank. In general, the information required to self-reproduce would be much greater than the information stored in a virus gene, because generally the environment within which a living creature must reproduce has less of the necessary reproductive machinery than does the environment of a virus. The virus invades a cell to deploy the cellular machinery for its own reproduction. For the virus to reproduce there must be *some* self-reproducing cells which can also reproduce the cellular machinery. The environment which these cells face contains just simple molecules like amino acids and sugars; they themselves must have the complex machinery of chemical synthesis to convert this material into proteins and nucleic acids. The information needed to code for the construction of this machinery and to keep it operating is vastly greater than the information coded in the single nucleic acid molecule of a virus. But in the theory of self-reproducing machines this is a matter of degree and not of kind. For our purposes we do not need to distinguish between self-reproducing organisms on the basis of complexity, because in an ecological system which has entities that satisfy our sufficient condition, there necessarily will exist *some* living things which any human observer would regard as 'autonomous' and which would self-reproduce.

All autonomous self-reproducing cells have a structure which can be naturally divided into the constructor part and the information bank part. This has led the French biochemist Jacques Monod[2] to define life as a system which has three properties: *autonomous morphogenesis*, which means that the system can operate as a self-contained system; *teleonomy*, which means the system is endowed with a purpose; and, *reproductive invariance.* He points out[2] that

> The distinction between teleonomy and invariance is more than a mere logical abstraction. It is warranted on grounds of chemistry of the two basic classes of biological macromolecules, one, that of proteins, is responsible for almost all teleonomic structures and performances; while genetic invariance is linked exclusively to the other class, that of nuclei acids.

Thus, nucleic acids correspond to the information bank, and proteins to the constructor in our self-reproducing machine example. However, it is difficult to make the notion of autonomous morphogenesis and teleonomy precise, as Monod admits. How autonomous should a living system be? A virus cannot reproduce outside a cell. Is it 'autonomous'? Humans cannot synthesize many essential amino acids and vitamins, but many bacteria can. Are we 'autonomous'? How does one recognize an object 'endowed with a purpose'? We avoided these problems by basing our sufficient condition for a 'living' system on reproduction and natural selection. It must be autonomous to just that extent which will allow natural selection to act on the various possible sets of information stored in the system. So the degree of autonomy will depend on the environment faced by the organism. It must have structure in addition to the information bank, and this structure is 'endowed with a purpose' in the sense that this additional structure exists for the purpose of letting the living system win the struggle for survival in competition with systems that have alternative information sets. Thus, our sufficient condition includes Monod's definition for all practical purposes.

Monod's definition of life is based, like our sufficient condition, on a generalization from the key structures and processes of living organisms at the molecular level. Before the molecular basis of life was understood, biologists tended to frame definitions of life in terms of macroscopic physiological process, such as eating, metabolizing, breathing, moving, growing, and reproducing.[63] Herbert Spencer's famous definition of life: 'The continuous adjustment of internal relations to external relations'[64] fits into this category. However, such definitions possess rather extreme ambiguities. Mules and childless people are eliminated by a strict reproductive requirement, as we noted earlier. But if information preserving (or increasing) reproduction is removed from the list of physiological processes, then it seems that candle flames must be considered living organisms. Flames 'eat' or rather take in fuel such as candle tallow, and they 'breathe' oxygen just as animals do. The oxygen and fuel are metabolized (or rather burned) in a reaction that is essentially the same as the underlying oxidation reaction that supplies humans with their energy. Flames can also grow, and if the fuel is available in various nearby localities, move from place to place. They can even 'reproduce' by spreading.

On the other hand, tardigardes are simple organisms that can be dehydrated into a powder, and which can be stored in this state for years. But if water is added, the tardigrades resume their living functions.[65] When in the anhydrous state the tardigrades do not metabolize. Are they 'dead' material during this period?

These difficulties led biologists in the first half of this century to attempt to define life in terms of biochemical reactions. J. D. Bernal's definition may be taken as representative of this type of definition:

Life is a potentially self-perpetuating open system of linked organic reactions, catalysed stepwise and almost isothermally by complex and specific organic catalysts which are themselves produced by the system.[66]

The word 'potentially' was inserted to allow such creatures as the tardigrades, and also dormant seeds. Unfortunately, such biochemical definitions are too narrowly restricted to carbon chemistry. If a self-reproducing machine of the type outlined earlier were to be manufactured by Man, it would probably be regarded as living by the average person, but the above biochemical definition would not classify it as living, because the machine was not made of organic (carbon) compounds. Also, the biochemical definition eliminates *a priori* the possibility that non-carbonaceous life could arise spontaneously, which no one wants to do in this age of speculation about extraterrestrial life forms. Thus, more modern definitions of life are generally framed either in terms of natural selection and information theory (Monod's definition and our sufficient condition are examples), or in terms of the non-equilibrium thermodynamics of open systems.

A good example of the latter class of definitions is the definition offered by Feinberg and Shapiro:

Life is fundamentally the activity of a biosphere. A biosphere is a highly ordered system of matter and energy characterized by complex cycles that maintain or gradually increase the order of the system through an exchange of energy with its environment.[55]

We feel this definition has a number of undesirable ambiguities that make it useless. How highly ordered must a system be before it counts as a biosphere? Many astrophysical processes are highly ordered systems with complex cycles that maintain this order. The energy generation processes of stars, for example, involve many complex cycles in a non-equilibrium environment. Is a star a biosphere? Also, by concentrating attention on the biosphere as a whole, the definition becomes impossible to apply to a single creature. Indeed, the notion of 'living creature' is not a meaningful concept according to this definition. What is meant by 'maintaining order'? If the biosphere eventually dies out, does this mean it was never 'alive'?

Definitions like our sufficient conditions, which are based on the concepts of information maintained by natural selection, also seem to have unavoidable and strange implications. Although our sufficient condition does not define as alive *natural* processes which intuitively are not considered alive, there are human constructs which are alive by our sufficient condition, and yet are not usually regarded as alive. Automobiles, for example, must be considered alive since they contain a great deal of information, and they can self-reproduce in the sense that there are human mechanics who can make a copy of the automobile. These mechanics are to automobiles what a living cell's biochemical machinery is to a virus. The form of automobiles in the environment is preserved by natural selection: there is a fierce struggle for existence going on between various 'races' of automobiles! In America, Japanese automobiles are competing with native American automobiles for scarce resources—money paid to the manufacturer—that will result in either more American or more Japanese automobiles being built!

The British chemist A. G. Cairns-Smith[104] has suggested that the first living things—the first entities to satisfy our sufficient condition—were self-replicating metallic minerals. The necessary information was coded in a crystalline structure in these first living things, and was later transferred to nucleic acids. The ecology changed from a basis in metal to one based on carbon. If Cairns-Smith is correct, the development and evolution of 'living' machines would represent a return to a previous ecological basis. If machines were to become completely autonomous, and able to reproduce independently of humans, then it is possible that a non-carbon ecology would eventually replace the current carbon ecology entirely, just as the present carbon ecology replaced a mineral ecology.

The English zoologist Dawkins has pointed out[67] that collections of ideas in human minds can also be regarded as living beings if the information or natural selection definition of life is adopted. Ideas compete for scarce memory space in human minds. Ideas which enable people to function more successfully in their environment tend to replace ideas in the human population which do not. For example, ideas corresponding to Ptolemaic astronomy were essential to anyone who wished to obtain a professorship in astronomy in 1500. However, possessing these ideas would make it impossible to be an astronomer today. Thus, Copernican ideas have eliminated Ptolemaic ideas in a form of struggle for existence. Dawkins calls such idea-complexes 'memes' to stress their similarity to genes and their relationship to self-reproducing machines. In computer science, an idea-complex would be thought of as a subprogram. Thus Dawkins' argument could be phrased as claiming that certain programs could be regarded as being alive. This is essentially the same claim that we have discussed in section 3.9, and that we will develop more fully in Chapters 9 and 10. Examples of computer programs which behave like living organisms in computers—they reproduce and clog computer memories with copies of themselves—have been given recently by the computer scientist Dewdney.[108] Anyone whose computer disks become infected with such programs has no doubt about the remarkable similarity of such programs to disease germs.

Having given a definition of life in terms of self-reproduction and natural selection, we will now define intelligent life in the same way. The Weak Anthropic Principle asserts that our Universe is 'selected' from amongst all imaginable universes by the presence of creatures—ourselves—which asks why the fundamental laws and the fundamental constants have the properties and values that they are observed to have. Thus, to use the Weak Anthropic Principle, one must either use 'intelligent being' as a synonym for 'human being' or else define 'intelligent being' to be a living creature (or rather a system which is made up in part of subsystems—cells—which are living by the above sufficient condition) that is capable of asking such questions. This definition can easily be related to the usual Turing definition of human-level intelligence.

In 1950 the English mathematician Alan Turing[12,13] proposed an operational test to determine if a computer processed intelligence comparable to that of a human being. Suppose we have two sealed rooms, one of which contains a human being while the other contains the computer, but we do not know which. Imagine further that we can communicate with the two rooms by a computer keyboard and TV screen display. Now we set ourselves the problem of trying to determine which of the sealed rooms contains the person, and which the computer. The only way to do this is by typing our questions on the computer keyboard, to the respective room's inhabitant, and analysing the replies. Turing proposed that if after a long period of typing out questions and receiving replies, we could still not tell which room contained the computer, then the computer would have to be regarded as having human-level intelligence. Generalizing this test of intelligence to our case, we will say that an intelligent being is a living system which can pass the Turing Test if the questions involve the fundamental laws and their structure on the levels discussed in this monograph. Further, we would require that at least some of the computer's replies be judged as 'highly creative' by human scientific standards. Such beings will be called 'weakly intelligent'.

To apply the Strong Anthropic Principle, a more rigorous criterion is needed. The Strong Anthropic Principle holds that intelligent beings play some essential role in the Cosmos. However, it is difficult to see how intelligent beings could play an essential role if all such beings are forever restricted to the planet upon which they originally evolve. On the other hand, if intelligent beings eventually develop interstellar travel, it is possible, at least in principle, for them to significantly affect the structure of galaxies and metagalaxies by their activities.[14] We will, therefore, say a living creature is strongly intelligent if he is a member of a weakly intelligent species which at some time develops interstellar travel. Some effects which strongly intelligent species could have on the Cosmos will be discussed in Chapter 10.

8.3 The Anthropic Significance of Water

> Ocean, *n*; A body of water occupying about two-thirds of a world made for man—who has no gills.
>
> A. Bierce

Water is actually one of the strangest substances known to science. This may seem a rather odd thing to say about a substance as familiar but it is surely true. Its specific heat, its surface tension, and most of its other physical properties have values anomalously higher or lower than those of any other known material. The fact that its solid phase is less dense than its liquid phase (ice floats) is virtually a unique property. These aspects of the chemical and physical structure of water have been noted before, for instance by the authors of the *Bridgewater Treatises* in the 1830's and by Henderson in 1913, who also pointed out that these strange properties make water a uniquely useful liquid and the basis for living things. Indeed, it is difficult to conceive of a form of life which can spontaneously evolve from non-self-replicating collections of atoms to the complexity of living cells and yet is not based in an essential way on water. In this

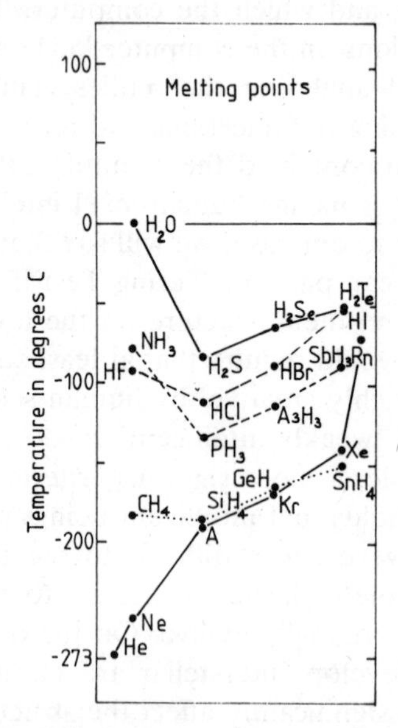

Figure 8.4. Melting points of various RH_n series of molecules, as a function of the location of R in the periodic table. The noble gas column is included for comparison. (From ref. 19, with permission.)

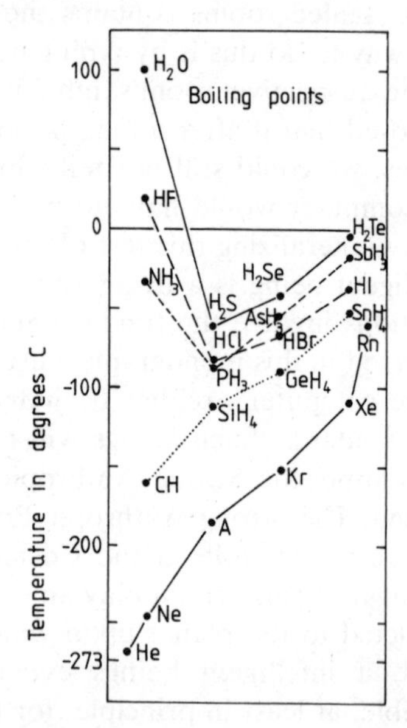

Figure 8.5. Boiling points of various RH_4 series of molecules, as a function of the location of R in the periodic table. The noble gas column is included for comparison. (From ref. 18, with permission.)

section we shall be concerned with listing the various properties of water, showing how these properties depend on the structure of the water molecule, and with highlighting the ways in which the properties of water are crucial for life. In other sections we shall discuss why other compounds cannot be substituted for water in the biochemical roles it plays.

The anomalous melting points, boiling points and heats of vaporization of water relative to those of other substances are seen most clearly if the values of these quantities are graphed as a function of atomic weight or atomic number. Figure 8.4 gives the melting points of various hydride molecules as a function of the location in the periodic table of the largest atom in the molecule. Figure 8.5 gives the boiling points and Figure 8.6 the heats of vaporization of various hydrides as a function of location in the periodic table.

These figures show clearly that if the elements in the first row of the periodic table are ignored, the melting points, the boiling points, and the heats of vaporization all increase with the atomic weight of R for a series of compounds RH_n, where n is constant for a given series. In the RH_4 series, the values of these three quantities all lie more or less on a straight line. In particular, the values for methane, CH_4, are those which one would have obtained by extrapolating backward along the RH_4 series. The values for water [H_2O], hydrogen fluoride [HF], and ammonia [NH_3]

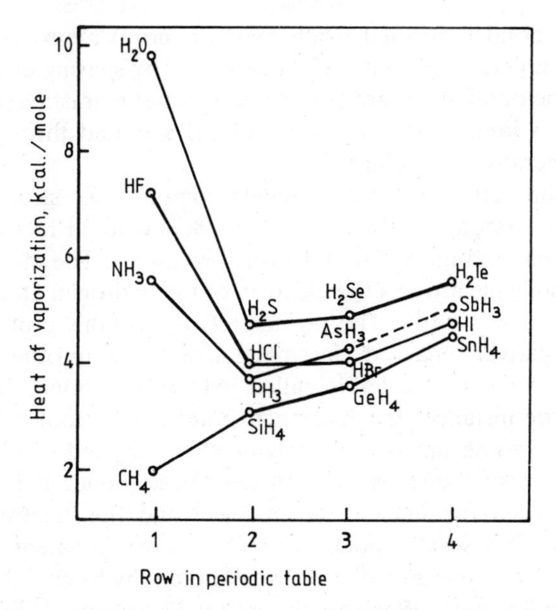

Figure 8.6. Heats of vaporization for various RH_n series of molecules, as a function of the location of R in the periodic table. The noble gas column is included for comparison. (From ref. 18, with permission.)

are far in excess of what one would expect by linear extrapolation. The boiling point of water, for instance, would be expected to occur at −100°C, rather than the +100°C that is observed. This indicates that there is a very strong force acting between the molecules of water, a force which is absent, or almost so, in the interactions of other members of the RH_2 series.

These properties of water can be understood in terms of the atomic structure of the water molecule: Recall from section 5.2 that there are four basic types of chemical bond: (1) the *ionic* bond; (2) the *covalent* bond; (3) the *hydrogen* bond; and (4) the *van der Waals* bond. All types of chemical bond result from the distribution of electrons around the nuclei comprising a molecule. The steady-state distribution is that which corresponds to the lowest energy. In ionic bonds, this state of lowest energy is achieved by one of the atoms pulling the electron from the other to complete an electron shell. For instance, the chlorine atom in NaCl, a molecule whose atoms are joined by an ionic bond, pulls the electron from the sodium atom resulting in a structure for chlorine that resembles the completed shell of neon. However, this leaves the chlorine atom a negatively charged ion and the sodium atom a positively charged ion. These ions are attracted electrostatically. The resulting NaCl molecule has a strong polar character due to the uneven distribution of electrons in the molecule.

A covalent bond is formed when two (or more) atoms complete (or attempt to complete) their outer electron shells by sharing electron pairs. In this case, the outer shells are completed without a drastic redistribution of charge as in ionic bonds, and so molecules bound through covalent bonds are generally non-polar.

An exception can occur in a molecule composed of strongly electronegative atoms (atoms with a strong tendency to attract electrons—fluorine, oxygen, and nitrogen) and hydrogen atoms. The electronegative atom in the molecule attracts the electron of the hydrogen atom, leaving a (more or less) bare proton. The positive charge on this proton can then attract the negatively charged electronegative atoms in other molecules. This bond between molecules is called a hydrogen bond. In hydrogen fluoride gas for instance, the hydrogen bond causes molecules of hydrogen fluoride to link up to form polymers, as pictured in Figure 8.7.

The final type of 'bond' is the van der Waals bond. It is due to the attraction between the nucleus of an atom and the electron cloud of another atom. This sort of bond is much weaker: covalent bonds bind typically with an energy of 100 kcal/mole, hydrogen bonds with 5 kcal/mole, and van der Waals forces with 0.3 kcal/mole.[17] Furthermore, the van der Waals bond is significant only when the atoms are very close together, since for monatomic molecules the van der Waals force varies

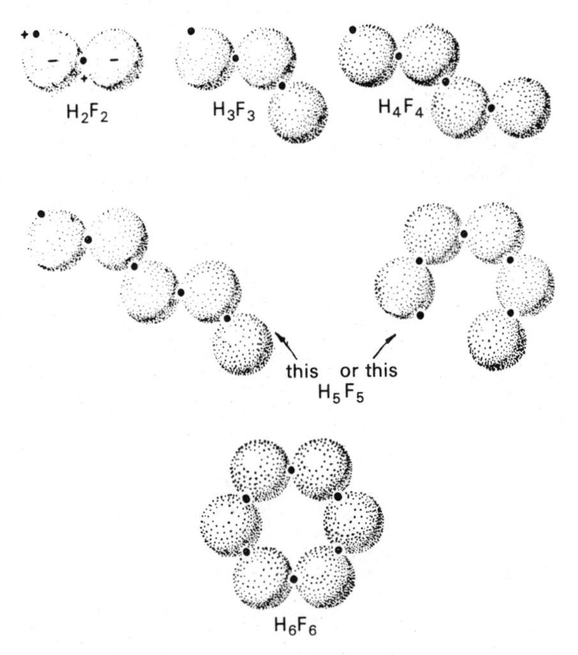

Figure 8.7. Various polymers of hydrogen fluoride. (From ref. 19, with permission.)

as r^{-7} (see ref. 19). At small distances the force of attraction is balanced by a repulsive force due to the interpenetration of outer electron shells. These properties of the van der Waals force are illustrated in Figure 8.8.

It is the van der Waals forces of attraction that cause substances such as the noble gases to condense to liquids and freeze into solids. The van der Waals attraction between molecules increases with increasing number of electrons per molecule, and thus heavy molecules have higher melting and boiling points. This is illustrated in Figures 8.5 and 8.6. The anomalous melting and boiling points exhibited by the compounds in the first row are due to hydrogen bonds, not to van der Waals forces, which are much weaker.

The anomalous melting and boiling points of water are due to the fact that water forms strong hydrogen bonds. The shape of the water molecule is compared in Figure 8.9 with the shapes of the other molecules in the RH_2 sequence. As can be seen in this figure, the shape of the water molecule is an isosceles triangle. The distance between the hydrogen and oxygen nuclei is 0.965×10^{-8} cm, and the H—O—H angle is 104.5 degrees. The oxygen atom in each water molecule strongly attracts the electrons of the two hydrogen atoms, with the result that the electron density in the molecules is concentrated around the oxygen atom. Thus,

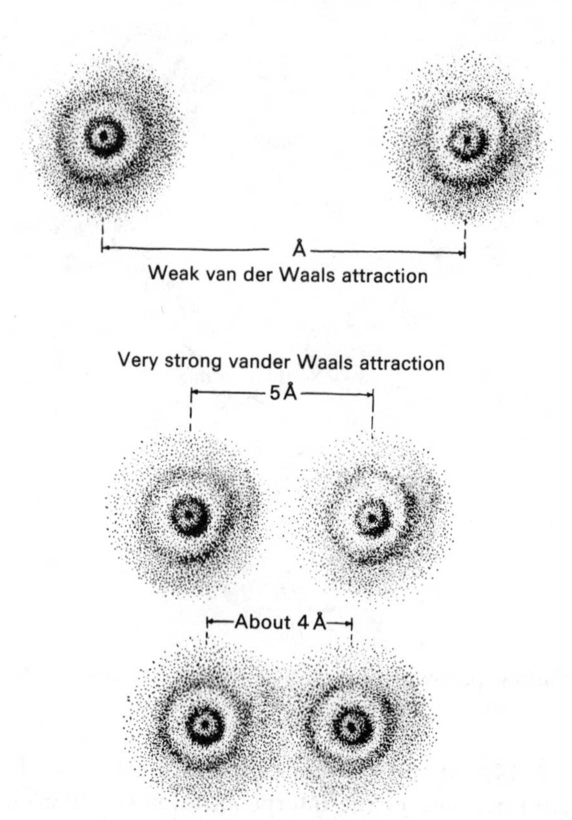

Figure 8.8. Properties of the van der Waals force. (From ref. 19, with permission.)

the hydrogen atom attracts the oxygen atoms of other water molecules, so water molecules tend to form hydrogen bonds with each other. These hydrogen bonds are found to be highly directional. That is, the hydrogen atom in one water molecule which forms a hydrogen bond with the oxygen atom in another water molecule, tends to point directly at the oxygen atom, as illustrated in Figure 8.10.

The bond angle of the free water molecule (104.5°) is only slightly less than the ideal tetrahedral angle (109.5°), so water molecules tend to polymerize to form a tetrahedral structure. This structure is rigid in ice (Figure 8.11), but water polymers exist even in liquid water. Furthermore, water dimers (two water molecules joined by a hydrogen bond) have been shown to exist in the vapour phase.[22] Thus to melt ice and boil water, it is necessary to supply sufficient energy to break some of the hydrogen bonds. The energy required is much greater than the energy required to

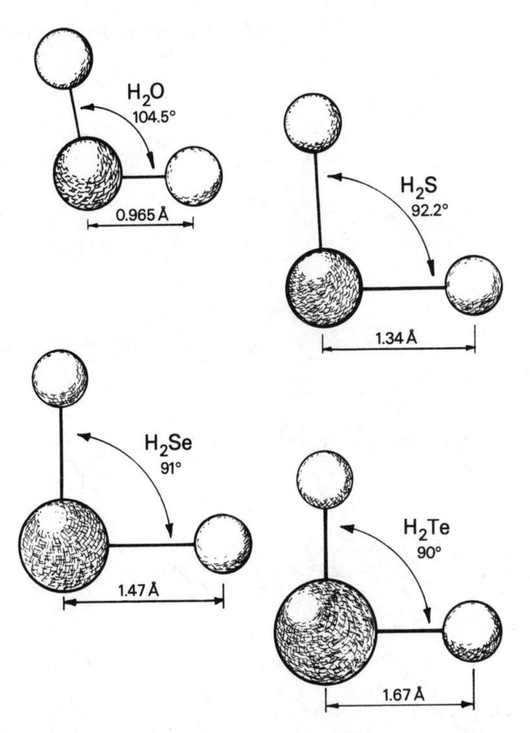

Figure 8.9. The bond angle and internuclear distance in the water molecule and similar molecules. (From ref. 20, with permission.)

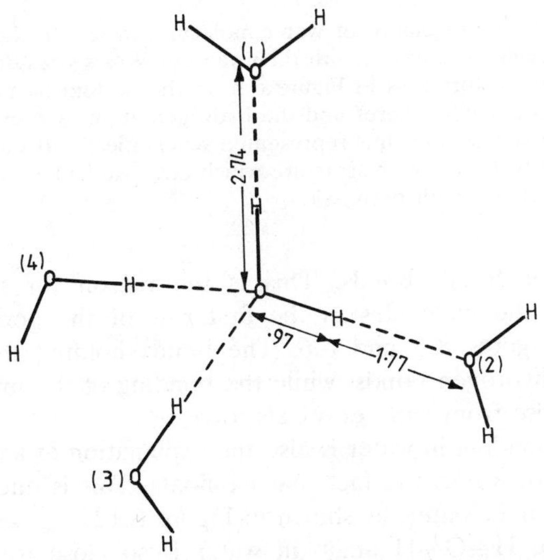

Figure 8.10. Coordination of water molecules into a tetrahedral structure. Covalent bonds are denoted by solid lines. Hydrogen bonds are denoted by dashed lines. Distances are given in angstroms (10^{-8} cm = 1 angstrom). (From ref. 21, with permission.)

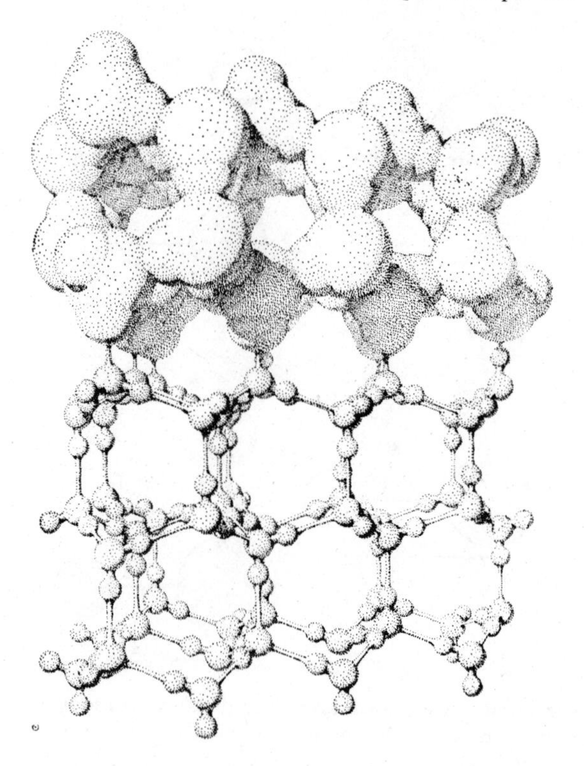

Figure 8.11. The arrangement of water molecules in ice. In the top part of the figure the molecules are shown with their van der Waals sizes determined by the electronic repulsive forces as in Figure 8.8. In the bottom part of the figure the oxygen atoms are small spheres and the hydrogen atoms are smaller spheres. In this part the molecules are thus represented schematically to show the structural pattern clearly. Note the open structure which gives ice its low density relative to water. (From ref. 19, with permission.)

break van der Waals bonds. This is the reason for the anomalous properties of the molecules in the first row of the periodic table, as pictured in Figures 8.5 and 8.6. The bonds holding these molecules together are hydrogen bonds, while the bonding of the molecules in the other rows arise from van der Waals forces.

Hydrogen bonding in water is also the explanation of another anomalous property of water, the fact that ice floats. This is due to ice having lower density than water, as shown in Figure 8.12.

Because the H—O—H angle in water is so close to the ideal tetrahedral angle, water can form the tetrahedral structure picture in Figure 8.11, with very little strain on the bonds. Thus, water tends to polymerize into an open structure. Each molecule in this open structure has only 4

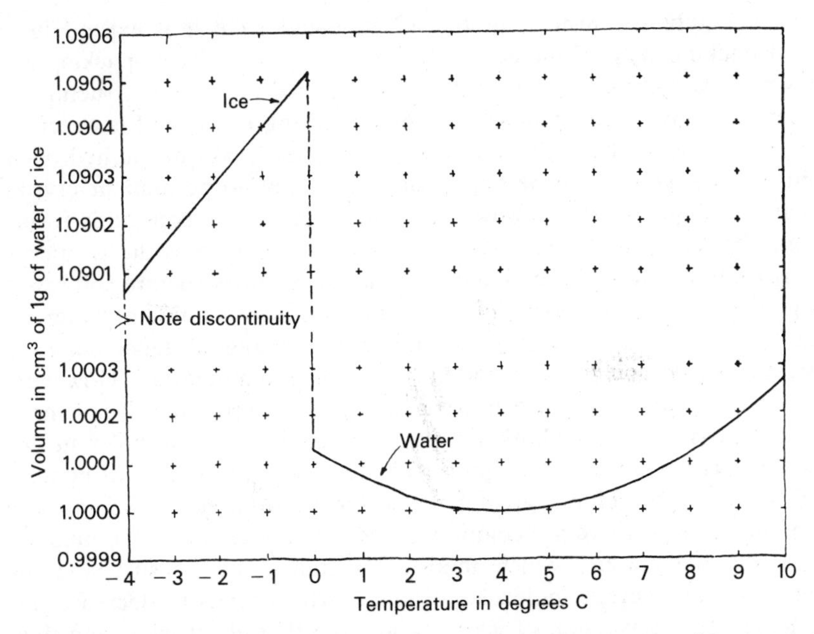

Figure 8.12. The density of water as a function of temperature. (From ref. 19, with permission.)

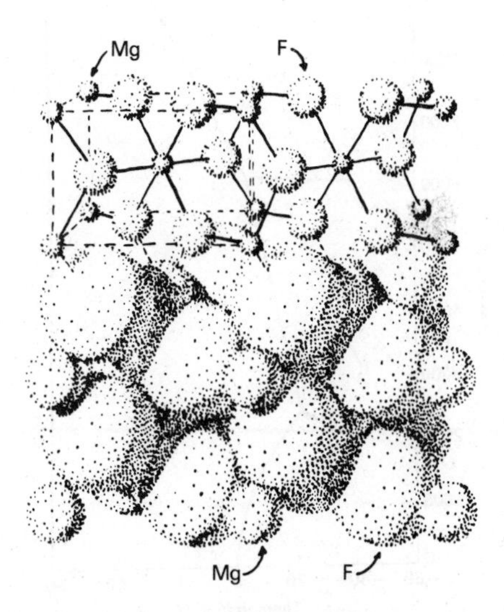

Figure 8.13. The structure of magnesium fluoride. This substance has a high melting point and a high boiling point, but its solid phase is denser than its liquid phase. Note the closed molecular structure as opposed to the open struture of Figure 8.11. (From ref. 19, with permission.)

nearest neighbours, instead of the 12 it would have if it existed in a closest-packing style of molecules arrangement.[18] (A closest-packing arrangement is pictured in Figure 8.13). When ice melts, this structure is broken in many places, but because of the strong tendency for water to polymerize due to the hydrogen bonds, only about 15% of the hydrogen bonds are broken when melting occurs. There is no permanent crystal structure in liquid water, however; crystal structures are constantly forming and breaking up. More and more bonds are broken as the temperature rises, but a large proportion of this 'liquid crystal' structure still exists at the boiling point, as shown by the high heat of vaporization of water.

In water, as in virtually every substance, the average distance between nearest neighbours increases with the increase in temperature. However, because the crystal structure has partially broken upon melting, it now becomes possible to fit additional water molecules into the holes in the crystal structure pictured in Figure 8.11. The average number of nearest neighbours is, therefore, greater in water than in ice.[21] So, as the temperature rises, there are competing effects: an increase in the number of nearest neighbours, which tends to increase the density, and an increase in the average molecular distance, which tends to decrease the density. In the conversion of ice to water, the first effect wins, and thus water is denser than ice. The anomalous density of ice is due to the close

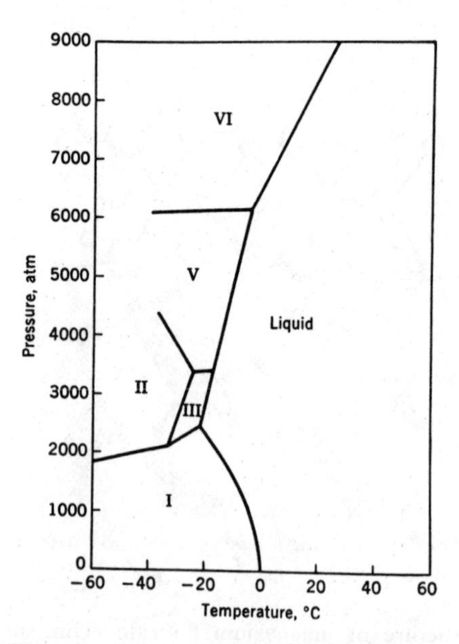

Figure 8.14. Phase diagram of water and various types of ice. The normal type of ice is ice I, and only this type of ice is less dense than water at the same pressure. (From ref. 21, with permission.)

approximation of the angle H—O—H to the ideal tetrahedral angle, as is shown by the fact that there are other solid phases of ice, phases which exist only at high pressure, as shown in Figure 8.14. These phases all have a crystal lattice distorted from the tetrahedral shape of ice, and all of these solid forms of ice are more dense than water at the same pressure.[20,21]

If we ignore the kinetic energy of the nucleus, we can obtain a rather simple scaling law for the binding energies of atomic systems as the fine structure constant α and the electron mass m_e are changed.[70] The Hamiltonian for an atomic system can be written:

$$\hat{H} = \frac{-\hbar^2}{2m_e}\sum_i \frac{\partial}{\partial \vec{x}_i}\cdot\frac{\partial}{\partial \vec{x}_i} + e^2\left[\sum_i\sum_j -\frac{Z_j}{|\vec{x}_i - \vec{R}_j|} + \frac{1}{2}\sum_{i<k}\frac{1}{|\vec{x}_i - \vec{x}_k|} + \frac{1}{2}\sum_{j<l}\frac{Z_jZ_l}{|\vec{R}_j - \vec{R}_l|}\right] \tag{8.1}$$

where $\vec{x}_i$ is the position vector of the ith electron, and $\vec{R}_j$ is the position vector of the jth nucleus. If we transform to new variables $\vec{x}'_i = \vec{x}_i/a_0$, $\vec{R}'_j = \vec{R}_j/a_0$, where a_0 is the Bohr radius $(a_0 = \hbar^2/m_e e^2)$ we can write the above Hamiltonian as

$$\hat{H} = \alpha^2 m_e c^2\left\{-\frac{1}{2}\sum \frac{\partial}{\partial \vec{x}'_i}\cdot\frac{\partial}{\partial \vec{x}'_i} + \sum_i\sum_j -\frac{Z_j}{|\vec{x}'_i - \vec{R}'_j|} + \frac{1}{2}\sum_{i<j} -\frac{1}{|\vec{x}'_i - \vec{x}'_j|} + \frac{1}{2}\sum_{j<l} -\frac{Z_jZ_l}{|\vec{R}'_j - \vec{R}'_l|}\right\} \tag{8.2}$$

Thus, if E' is the energy eigenstate when the fine structure constant and the electron mass are changed to α' and m'_e respectively, we have

$$E'/\alpha'^2 m'_e = E/\alpha^2 m_e \tag{8.3}$$

where E is the original energy eigenstate.[111]

It is possible that the changes in the chemical bond energies and structure due to a change in α or m_e could significantly change the geochemical distribution of the elements,[106] which would render the spontaneous generation of life on an earthlike planet even more improbable than it is.

As Henderson pointed out in Chapter III of *Fitness of the Environment* in 1913, the expansion of water on freezing is essential for life if it is to evolve in a constant environment. If ice were not less dense than water, it would sink on freezing. The coldest water in a lake or ocean would congregate near the bottom and there freeze. Ice would accumulate at the bottom; the amount would become greater each year as more ice formed during the winter and did not melt during the summer. Finally, all the lakes and oceans would be entirely frozen. As it is, ice forms only on the surface and protects the water below and marine life from further cooling. Liquid water thus remains available for use as a solvent for living creatures.

Water also has a higher specific heat than almost all organic compounds. Specific heat, which is the amount of heat energy that must be added to a gram of material to raise its temperature by one degree, essentially measures the amount of internal energy taken up by the molecules of the material as heat is added. In the case of polar molecules such as water, this internal energy is usually absorbed in the form of energy to break bonds. Hydrogen bonds and ionic bonds are more difficult to break than van der Waals attractions, so polar molecules—water, ammonia, sodium—have high specific heats. Some specific heats of representative substances are given in Table 8.1.

The high specific heat of water allows water to be used as a store of heat, and also serves to stabilize the temperature of the environment. Large amounts of heat must be added to masses of water to change its temperature significantly.

The thermal conductivity of water is also higher than that of most liquids, though smaller than most solids. This high thermal conductivity,

TABLE 8.1
Specific heats of representative substances (cal/gm).

Substance	Temperature (°C)	Specific heat	Substance	Temperature (°C)	Specific heat
Potassium dichromate	397	0.034	Isoamyl valerate	20	0.46
Glycerol	−250	0.047	Olive oil	6.6	0.47
Phthalic acid	20	0.232	Acetic acid	0	0.49
Chloroform	20	0.234	Aniline	20	0.52
Succinic acid	0	0.248	Acetone	20	0.53
Glucose (solid)	20	0.275	Glycol	40	0.53
Benzoic acid (solid)	20	0.287	Hydrogen peroxide	0	0.55
Urea	20	0.320	Water	20	1.01
Sulphur dioxide	20	0.327	Ammonia	20	1.13
Glycerol	0	0.330	Hydrogen (gas)	0	3.4
Sulphuric acid	10	0.339	Mercury	−30 to 100	6.68
Benzene	20	0.406	Caustic soda (fused)	400	6.7
Palmitic acid	20	0.430	Sodium (fused)	200	7.3
Pyridine	21	0.431	Potassium chloride (fused)	800	8.8

Reproduced, with permission, from ref. 25.

TABLE 8.2
Thermal conductivities of representative substances (gm cal/sec cm^2 for a temperature gradient of 1°C/cm).

Material	Conductivity	Material	Conductivity
Cotton wool	0.00004	Water (20°C)	0.0014
Eiderdown	0.000046	Brick	0.0015
Inorganic gases (excl. water)	0.00004–0.00003	Ice	0.002–0.005
Organic compounds in general	0.00002–0.0007	Glass	0.0025
Cork	0.00015	Earth's crust	0.004
Wood	0.0005	Quartz	0.016–0.030
Glycerol	0.0007	Brass	0.26
Snow	0.00051	Copper	1.00

Reproduced, with permission, from ref. 25.

coupled with the fact that a liquid can transfer heat by convection, also allows large amounts of water to act as a temperature stabilizer for the environment. Thermal conductivities of representative substances are listed in Table 8.2.

We have previously pointed out that the heat of vaporization of H_2O is the highest amongst all members of the RH_n series of compounds.

TABLE 8.3
Heats of vaporization of various liquids (cal/gm).

Liquid	Temperature (°C)	Heat of vapn.	Liquid	Temperature (°C)	Heat of vapn.
Water	100	538	Acetic acid	118	96.8
Sulphur	316	362	Benzene	80	95.5
Ammonia	−33.4	327	Ethyl ether	34.6	89.3
Ammonia	0	302	Carbon dioxide	−60	87.2
Ethanol	78.3	204	Naphthalene	218	75.5
Hydrogen cyanide	20	210	Mercury	358	68.0
Glycol	197	191	n-Decane	160	60.2
Phosphorus	287	130	Chloroform	61	58.0
Formic acid	101	120	Boron trichloride	10	38.2
Methyl formate	33	111	Argon	−186	37.6
Hydrogen	−253	108	Carbon dioxide	20	35.1
Pyridine	114	107	Helium	−269	6.0

Reproduced, with permission, from ref. 25.

TABLE 8.4
Surface tensions of liquids (dynes/cm).

Liquid	Temperature (°C)	Surface tension	Liquid	Temperature (°C)	Surface tension
Selenium A	217	92.4	Chloroform A	20	27.1
Water A	20	72.5	Cyclohexane A	20	25.5
Glycerol A	20	63.4	Ammonia V	11.1	23.4
Sulphuric acid A, V	20	55.1	Ethanol V	20	22.8
Quinoline A	20	45.0	Acetaldehyde V	20	21.2
Aniline V	20	42.9	Chlorine V	20	18.4
Phenol A, V	20	40.9	Oxygen V	−183	13.2
Benzene A	20	28.9	Argon V	−188	13.2
Naphthalene A, V	127	28.8	Carbon dioxide V	20	1.16
Acetic acid V	20	27.8	Helium V	−270	0.24

A = against air, V = against own vapour. Reproduced, with permission, from ref. 25.

TABLE 8.5
Heats of fusion of various representative solids (cal/gm).

Solid	Melting point (°C)	Heat of fusion	Solid	Melting point (°C)	Heat of fusion
Sodium fluoride	992	186	Anthracene	216.6	38.7
Sodium chloride	804	124	*p*-Aminobenzoic acid	188.5	36.5
Ammonia	−75	108	Benzoic acid	121.8	33.9
Potassium fluoride	860	108	Sulphuric acid	10.4	24.0
Water	0	79.7	Aniline	−7.0	21.0
Nitrogen pentoxide	29.5	76.7	Gallium	3.0	19.2
Hydrogen peroxide	−1.7	74.1	Methanol	−97	16.4
Potassium chloride	772	74.1	Bromine	−7.3	16.2
Cobalt	1495	62.0	Hydrogen	−259	14.0
Formic acid	80	58.9	Hydrogen chloride	−114	13.9
Quinol	172.3	58.8	Sulphur	119	13.2
Stearic acid	64	47.6	Iodine	113.7	11.7
Glycerol	18	47.5	Argon	−190	6.7
Carbon dioxide	−56.2	45.3	Nitrogen	−210	6.1
Acetic acid	16.6	44.7	Phosphorus	44.2	5.0
Palmitic acid	55	39.2	Oxygen	−219	3.0

Reproduced, with permission, from ref. 25.

Actually, water has a higher heat of vaporization than *any* known substance.[25] The heat of vaporization of water is compared with some other liquids in Table 8.3. Such a high heat of vaporization makes water the best possible coolant by evaporation. And indeed living creatures make extensive use of this cooling mechanism.

The strength of the hydrogen bond in water is responsible for the uniquely large heat of vaporization of water. It is also responsible for water having a very high surface tension, exceeded by few substances other than liquid selenium. A list of surface tensions of representative liquids is given in Table 8.4. This high surface tension of water allows compounds which can reduce surface tension, and most of the biologically important carbon compounds have this property. These compounds tend to concentrate near a liquid surface, thus making biochemical reactions more rapid.[25]

The hydrogen bonds also result in an extremely high latent heat of fusion for water; its heat of fusion is exceeded only by various inorganic salts and ammonia, as shown in Table 8.5.

The dielectric constant of water far exceeds that of any other pure liquid except hydrogen cyanide and formanide (see Table 8.6 and Figure

TABLE 8.6

Dielectric constants of representative substances. The dielectric constant measures the reduction in the force between unit charges compared with the vacuum with a dielectric constant of 1.0.

Substance	Dielectric constant	Substance	Deilectric constant
Hydrogen (1 atm.)	1.0003	Sulphur dioxide (liquid)	13.8
Air (1 atm.)	1.0006	Ammonia	17.8
Carbon dioxide (50 atm.)	1.60	Ethanol	26.0
Octane	1.96	Methanol	35.0
Paraffin wax	2.3	Methyl cyanide	39.0
Nitrogen tetroxide	2.5	Thallium chloride	47.0
Ebonite	3.15	Water	81.0
Sulphur	3.52	Formamide	109.0
Plate glass	4.67	Hydrogen cyanide	116.0
Mica	6.64	Glycine	137.0
Casein	6.1–6.8	Rutile (parallel to optic axis)	170.0
Glass	8.45		

Reproduced, with permission, from ref. 25.

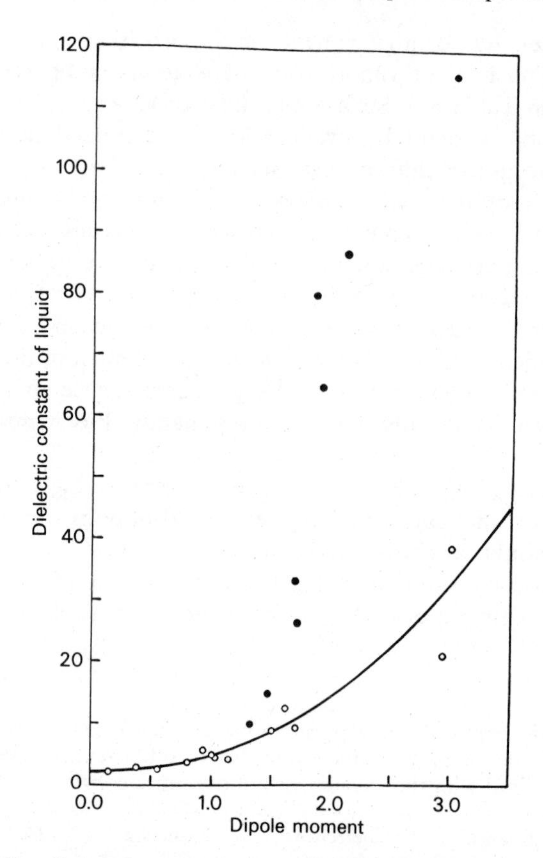

Figure 8.15. The dielectric constants of polar liquids plotted against the dipole moments of the corresponding gas molecules. From left to right, the substances pictured are: *open circles*: AsH_3, HI, PH_3, HBr, H_2S, $CHCl_3$, HCl, $(C_2H_5)_2O$, $SOCl_2$, SO_2, SO_2Cl_2, $(CH_3)_2CO$, CH_3NO_2; *shaded circles*: CH_3NH_2, NH_3, CH_3OH, C_2H_5OH, H_2O, HF, H_2O_2, HCN. (Reprinted from L. Pauling, *The nature of the chemical bond*, 3rd edn, copyright 1960 by Cornell University. Used by permission of the publisher, Cornell University Press.)

8.15). This high dielectric constant allows water to dissociate polar molecules such as NaCl into their constituent ions. This results in the very great solubility of such polar molecules in water. The polarization of a water molecule by the partial transfer of charge from the two hydrogen atoms to the oxygen atom is responsible for the high dielectric constant, and also for an anomalously high *dipole moment*, which is greater for water than for virtually every other light polar compound except hydrogen fluoride. (As shown in Figure 8.15; see also ref. 25.)

This high dielectric constant is also associated with the tendency of water molecules to ionize. In pure water 10^{-7} of the water molecules are ionized, the free H^+ ion associating with a water molecule, so that the main ionic species are[25] H_2OH^+ and OH^-. This provides high-mobility transport for H^+ and OH^- ions,[17] which speed-up chemical reactions. Computer simulation has shown[17,26] that water at and below room temperature lies above the so-called critical percolation threshold, which means that any macroscopic sample of the liquid will have uninterrupted hydrogen bond paths running in all directions, covering the entire volume of the liquid. These networks provide natural pathways for the rapid movement of H^+ and OH^- ions by a directed series of exchange hops.[17,27,28] This means that a change of pH in a water solution is transmitted with great rapidity throughout the solution. This also results in water having a fairly high electrical conductivity.[25]

The hydrogen-bonded, quasi-crystalline lattice structure of water also has an unusual effect upon non-polar solutes and non-polar side-groups attached to organic polymers. When a non-polar solute molecule is placed in liquid water, the hydrogen bond network must rearrange itself so as to minimize distortion of the network while providing sufficient room in the network to accommodate the new molecule. This results in the solute molecule being placed in a 'solvation cage' of water molecules.[17,29,30] Each water molecule in the cage tries to place its tetrahedral hydrogen bonding directions (the directions defined by the lines joining the nuclei of the hydrogen and oxygen atoms) in a sort of straddling mode as shown in Figure 8.16. This arrangement of water molecules allows bonding to other water molecules in the cage, and avoids pointing one of the four tetrahedral directions inward toward the region occupied by the non-polar solute, which would eliminate a possible hydrogen bond.

This tendency of water molecules to 'cage' a non-polar solute results in a negative entropy for the solution of non-polar molecules (in part because the water molecules in the cage layer next to the solute have reduced orientational options), and thus there is an entropy-driven net attraction of non-polar molecules for each other. This is sometimes called the 'hydrophobic bond' or the 'hydrophobic effect'.[17,31,32] The accepted explanation for this attraction is that the joint solvation cage of two non-polar molecules causes less overall order and hence less entropy reduction when two non-polar molecules are together than when they are far apart.[17]

The hydrophobic effect is important for biology. First of all, this effect is largely responsible for shaping organic molecules such as enzymes and nucleic acids into their biologically active forms. It is well-known[8,33] that it is the *shapes* of enzymes that enable these proteins to catalyse biochemical reactions in a living cell. These catalytic interactions are very specific because shapes are very definite. Now these enzymes have both polar and

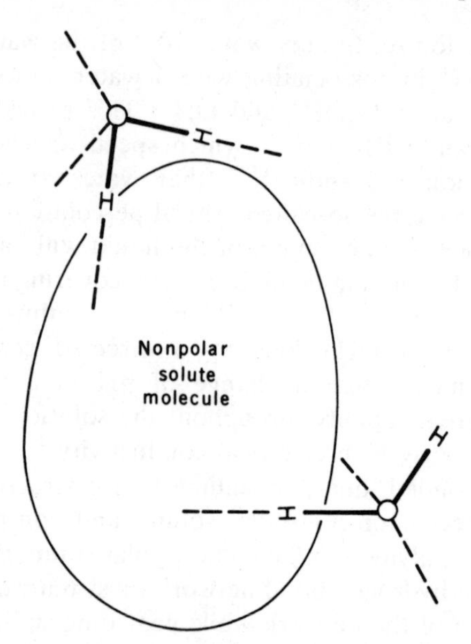

Figure 8.16. Orientation preference of a water molecule next to a non-polar solute. In order to preserve the maximum number of hydrogen bonds, the water molecules try to straddle the inert solute, pointing two or three tetrahedral directions tangential to the surface of the space occupied by the solute molecule. (From ref. 17, with permission.)

non-polar side-groups, and by adjusting the positions and numbers of these groups on the organic molecule, the hydrophobic effect can be induced to adjust the shape of the molecule into any desired form and hold it in that form.[34] A second very important biological use of the hydrophobic effect is in the formation of cells walls and cell membranes. When molecules with polar and non-polar portions, or rather with hydrophilic and hydrophobic portions, respectively—are dissolved in water they can isolate their hydrophobic portions from the water by self-aggregation. The aggregated products are called *micelles.*[35] These micelles are the first step toward the formation of biological membranes, which are thin layers of protein and lipid that permit the compartmentalization of living material.[35] (Lipids are defined to be carbon molecules of biological origin that are highly soluble in organic solvents but only sparingly soluble in water.[35]) The lipids and proteins in a cell membrane are adjusted in a structure which allows selective transport of water and organic molecules through the membrane, and this selective transport depends crucially on the hydrophobic or hydrophilic structure of the lipids.[39]

Fox and his co-workers have argued[36,37] that the spontaneous self-aggregation of protein-like polymers in the oceans of the primitive earth result in the formation of cell walls which could isolate biochemical reactions from the outside environment. In the opinion of these researchers, such spontaneous formation of cell walls was the essential first step in the chemical evolution of life, a step which had to occur before the development of a self-reproducing chemical system could begin. Whether or not the formation of a cell wall was actually the first stage in the evolution of biochemical life, it clearly had to occur at some stage. Its occurrence depends on the hydrophobic effect, and thus spontaneous cell wall formation would be much less likely to occur in solvents, such as liquid ammonia, in which the analogous effect is either not so strong or absent altogether. (Liquid ammonia is a poorer solvent for polar molecules than water, but a better one for lipids.[25]) This suggests an experiment to test the possibility that ammonia could be used on other planets as a liquid solvent within which life could spontaneously form and evolve. Fox's experiment[36,37] on the spontaneous formation in water of cell-like membranes from biological polymers could be repeated in liquid ammonia with other polymers (perhaps inorganic[38]) to see if such structures would also form spontaneously in that solvent. If not, then life based on liquid ammonia could be ruled out. We predict that the spontaneous formation of cell-wall-like structures is much more unlikely to occur in liquid ammonia, if it occurs at all.

One might hope to obtain information about the effects of doubling m_N/m_e by studying the biological effect of replacing H_2O with D_2O, deuterium oxide. The biological effect is substantial; D_2O in metazoans, both animals and plants, is completely inert and useless. In D_2O, seeds will not sprout and rats die of thirst if given only D_2O to drink.[40] Unfortunately, the cause of this dire reaction is unclear. D_2O has a higher viscosity[41] than H_2O, and a higher freezing point. Furthermore, it only seems to affect metazoans; if prokaryotes are introduced in D_2O, the reproduction rate initially slows down, but after a few hours it returns to near normal.[69]

8.4 Unique Properties of Hydrogen and Oxygen

When hydrogen played oxygen,
And the game had just begun,
Hydrogen racked up two fast points
But oxygen had none.
Then oxygen scored a single goal,
And thus it did remain
Hydrogen 2 and oxygen 1
Called off because of rain.

Anonymous

We have presented some of the special properties of hydrogen in our discussion of water. All of its unique properties are due to its position as the lightest element—the first element in the periodic table, with no congeners (elements with similar properties). Free hydrogen molecules, H_2, are the most mobile molecules; consequently any hydrogen gas that forms either quickly undergoes chemical reactions or escapes the atmosphere. (The formation and loss of hydrogen gas from the Earth's atmosphere is probably offset or even counterbalanced by the flux of hydrogen from the Sun.[43]) Although it is the most common atom in the Universe

TABLE 8.7
The elemental composition of the environment and living organisms.

Element	Relative abundance of the elements				
	As a percentage by weight of				In g-atoms per 100 g of living organisms
	Lithosphere (outer 24 miles)	Hydrosphere	Atmosphere	Living organisms	
Oxygen	47.33	85.79	21.00	63.00	3.94
Hydrogen	0.22	10.67	0.02	10.00	10.00
Carbon	0.19	0.01	0.01	20.00	1.67
Nitrogen	0.02	0.02	77.56	2.50	0.18
Calcium	3.47	0.05	—	2.45	0.06
Phosphorus	0.12	0.000004	—	1.10	0.036
Chlorine	0.23	2.07	—	0.16	0.0045
Sulphur	0.12	0.05	—	0.14	0.0044
Potassium	2.46	0.04	—	0.11	0.0028
Sodium	2.46	1.14	—	0.10	0.0043
Magnesium	2.24	0.14	—	0.07	0.0029
Iron	4.50	0.002	—	0.01	0.00018
Manganese	0.08	—	—	0.003	—
Fluorine	0.03	0.00014	—	0.0004	—
Bromine	—	0.01	—	0.0003	—
Copper	0.01	—	—	0.0002	—
Iodine	—	0.006	—	0.0001	—
Zinc	0.004	0.00001	—	0.0001	—
Cobalt	0.003	—	—	0.00004	—
Nickel	0.03	—	—	0.00004	—
Titanium	0.46	—	—	0.00005	—
Barium	0.08	0.00001	—	—	—
Strontium	0.02	—	—	—	—
Chromium	0.06	—	—	—	—
Vanadium	0.04	—	—	—	—
Aluminium	7.85	—	—	—	—
Silicon	27.74	—	—	—	—
Argon	—	—	1.40	—	—
Krypton	—	—	0.01	—	—
Xenon	—	—	0.005	—	—

Reproduced, with permission, from ref. 25.

either by weight or by relative number, it comprises only about 0.22% of the mass of the outer crust of the Earth (Table 8.7).

The hydrogen atom only has a single electron, and it can either donate it to other atoms, or accept an electron from other atoms to complete the K shell. No other element is so symmetrically ambivalent. It can behave in both ways simultaneously, and by so doing, form the hydrogen bonds that were discussed extensively in the last section. When it donates an electron it forms a typical ionic bond, but by sharing a second electron it can form a covalent bond which has about 10 times as much energy of formation as the ionic bond. In strong acids like HCl, the hydrogen appears entirely in ionic form, but in water it can be ionic either sparingly or considerably according to the chemical environment. Hydrogen is the most electropositive of all elements, and it donates electrons to form strong bonds with the electronegative halogens. On the other hand, it also forms hydrides with the light electropositive metals such as sodium. In these hydrides hydrogen acts as an electron acceptor. Hydrogen is very active chemically; more compounds of hydrogen are known than any other element, with carbon a close second.[19] Chemically the lightest elements are the most reactive and form the strongest bonds, so it is probably for this reason that all the biologically important elements are rather light.

In terms of its physical properties, hydrogen has the lowest melting point (14 K) and boiling point (20 K) of any chemically active element (only those of helium are lower). Liquid hydrogen, with a density of 0.07 gm/cm^3 is the least dense of all liquids, and crystalline hydrogen, with a density of 0.088 gm/cm^3, is the least dense of all crystalline materials. Hydrogen is only slightly soluble in water.[19]

Hydrogen forms particularly strong and stable covalent bonds with carbon. Such bonds do not easily form hydrogen bonds with other elements. These bonds are the chief source of non-polar side-groups in biological molecules which give rise to the hydrophilic effect in water.

As can be seen from Table 8.7, by weight oxygen is the most abundant element in both the outer crust of the earth and in living organisms. It also comprises about 21% of the Earth's atmosphere, in the form of molecular oxygen. This form of oxygen is highly reactive; the present terrestrial atmosphere is thus highly oxidizing, although it is generally agreed that the primitive Earth of 4.5 billion years ago most probably had a heavily reducing atmosphere.[42] This transformation of the atmosphere was accomplished through the action of photosynthetic plants. Without the continued action of such plants, the atmosphere of the Earth would revert in something like a million years to a composition similar to that of Venus[43] (Table 8.8).

The biosphere is actually only marginally stable to perturbations of the

TABLE 8.8
The atmospheric composition of Venus, the Earth, Mars, and a hypothetical Earth without life.

Gas	Planet			
	Venus	Earth without life	Mars	Earth as it is
Carbon dioxide	98%	98%	95%	0.03%
Nitrogen	1.9%	1.9	1.7%	78%
Oxygen	trace	trace	0.13%	21%
Argon	0.1%	0.1%	2%	1%
Surface temperatures °C	477	290 ± 50	-53	13
Total pressure bars	90	60	0064	1.0

Reproduced, with permission, from ref. 43.

percentage of oxygen in the atmosphere. The present atmosphere—21% oxygen—is at the upper limit of safety for life. The probability of a forest fire being started by lightning increases by 70% for each 1% rise in the oxygen percentage, with the implication that very little of our present land vegetation would survive incineration if the oxygen percentage rose above 25%.[43] We shall discuss the implications of this important point in section 8.7.

The biosphere circulates a huge amount of oxygen, 10^{11} metric tons per year,[43] with the average oxygen atom running through the cycle in about 2000 years.[44] At one point or another this oxygen in the cycle appears either as water, H_2O, or as carbon dioxide, CO_2. Carbon dioxide dissolves in water to form carbonic acid $(HO)_2C=O$. This acid is both a carbonyl, $>C=O$, compound and carbinol, $\equiv COH$, compound. Oxygen appears in organic molecules primarily in one of these two compounds.[25] Thus in the organic, anhydrous part of a living organism, the number of gram-atoms of oxygen is less than the number of carbon and hydrogen.

Molecular oxygen has unpaired electrons with unopposed spins, and this indicates that molecular oxygen readily assumes the more active atomic form, with two unpaired electrons. This unusual readiness to dissociate adds to the reactivity of oxygen. It largely explains the strong tendency of oxygen atoms to share an electron with the atoms of other elements, and so forming strong covalent bonds with them.

Another important contribution oxygen makes to life is the formation of ozone (O_3) in the atmosphere. Ozone in the upper atmosphere screens out far-ultraviolet radiation which would otherwise destroy carbon-based life. There is evidence that until plants had produced sufficient oxygen to make an ozone layer, life had to remain in the water, which also filters our far-ultraviolet radiation.[47,48]

All life needs an energy source, and chemically the most energetic reactions are $2H_2 + O_2 \rightarrow 2H_2O$ and $H_2 + F_2 \rightarrow 2HF$. Thus the presence of free oxygen molecules in the atmosphere allows living things to make use of the second most efficient energy sources. If some element besides oxygen were used in the basic energy source (fluorine, being very rare, could not be used), the chemistry of life would have less energy available per chemical reaction.

8.5 The Anthropic Significance of Carbon, Carbon Dioxide, and Carbonic Acid

> The only laws of matter are those which our minds must fabricate, and the only laws of mind are fabricated for it by matter.
>
> J. C. Maxwell

Carbon is one of the three most numerous atoms in a living organism; hydrogen and oxygen are the other two. Carbon's usefulness for life is due to its location in the periodic table: in the middle of the first row. Carbon has six electrons in all, two being located in the K shell and so unavailable for chemical combination, but it has four in the L shell, and all of these can be shared with other atoms. Carbon thus has a valence of four.

Only atoms in the first row of the periodic table form multiple bonds to any significant extent, and this is the source of many of carbon's unique properties. For example, carbon forms a double bond with oxygen in carbon dioxide, O=C=O. Since each oxygen atom has a valence of two, the double bond completes the L shell of each oxygen atom, and a carbon dioxide molecule has very little remaining chemical affinity for itself. This means that carbon dioxide is a gas at ordinary temperatures, a fact of great importance for biology.

Silicon, an element which is very similar to carbon in that it lies in the same column of the periodic table, but in the next higher row, cannot form double bonds, even though it (generally) has a valence of four. Thus, when silicon dioxide, —O—Si—O— (with free bonds above and below), is formed, some bonds can still be formed with other molecules of silicon dioxide. The result tends to be a crystal lattice; silicon dioxide is a solid, quartz, and not a gas.[45,46]. Quartz consists of SiO_4 tetrahedra, with each oxygen atom serving as the corner of two of these tetrahedra. Quartz is a very hard mineral because in order to break it, one has to break oxygen–silicon bonds.[19] Because silicon is in the third row of the periodic table, it possesses the 3d orbitals for further combinations; its valence can rise to as high as six.[21] Thus, even if silicon had four valence electrons engaged in chemical bonds, it could still form others, whereas a saturated carbon atom could not coordinate as either a donor or an acceptor of electrons.

Sidgwick has pointed out[49] (see also ref. (21)) that since chemical reactions commonly proceed through coordination, saturated carbon compounds must in general be very slow to react. Carbon compounds tend, in practice, to remain stable over long periods of time, even those which are thermodynamically unstable or metastable, because of the difficulty in finding a path by which a reaction could take place. This allows organic compounds in living organisms to, by and large, remain stable until a transformation is needed, which can then take place with great specificity through the action of enzymes. Silicon, with its extra orbitals, does not have this advantage, and Sidgwick[42] concludes that silicon could not replace carbon in any biosphere. The fact that even thermodynamically unstable carbon compounds are stable in practice also allows a great many carbon molecules to be formed by addition of side groups in sequence with previously existing molecules. Otherwise the molecules would dissociate before they could be built up.

Another important reason for the great multiplicity of carbon compounds is that the energy involved in the formation of C—C bonds is not very different from that for the formation of C—H, C—O, or other bonds involving carbon. However, the energy of formation for the Si—Si bonds is much less than that of Si—H or Si—O bonds, as shown in Table 8.9. The greater stability of the C—C bond is an important factor in allowing the formation of organic molecules with long chains of carbon atoms.[21]

Another advantage in using carbon as the basis of life lies in the fact that most of its compounds are metastable,[50] which means that they can easily be induced to interact further. This fact also aids in the production of large complex molecules by living things. The English chemist, A. E. Needham,[25] has emphasized that most carbon compounds are lightly poised between two stable extremes, the fully oxidized forms of carbon, such as

TABLE 8.9
Bond energies of carbon and silicon.

	X—X	X—H	X—O	X—Cl	
X=C	81.6	98.8	81.5	78.0	kcal
X=Si	42.5	75.1	89.3	85.8	kcal
Difference C minus Si	+39.1	+23.7	−7.8	−7.8	kcal

Reproduced, with permission, from ref. 21.

carbon dioxide, on the one hand, and the fully reduced forms, such as the hydrocarbons, on the other.[51] Both ethane, C_2H_6, and carbon dioxide, CO_2, are formed from their constituent elements in an exothermic manner. However, the oxidation of glucose to carbon dioxide and water, together with the reduction of glucose to methane and water, are both endothermic. In general, some carbon compounds are endothermic and some exothermic, so there are many possibilities for spontaneous formation whenever one has a mixture of the two types. Also, the heats of formation of carbon compounds from their immediate precursors is rarely very great, so that once the initial steps of carbon dioxide reduction have been completed, a large variety of carbon compounds form spontaneously. This property of carbon is of vast importance in the evolution of life, for the formation of the first primitive living organism could occur spontaneously only if the wide variety of molecules which it needed—and all proposals to date for the make-up of the first living being require it to acquire a great many different molecules—could be expected to form spontaneously.

A wide variety of carbon compounds which are known to be precursors of more complex biological molecules—amino acids, for example—have been demonstrated experimentally to be spontaneously formed under conditions which are thought to mimic conditions on the Earth just after it formed.[36] To date, however, no experiment has been able to form nucleotides, the building-blocks of nucleic acids, spontaneously under these conditions.[36,52] This may mean that a reaction pathway leading to nucleotides is longer than has yet been allowed to proceed in the laboratory.

There are probably many alternative pathways leading from the prebiotic 'soup' to nucleotides. No other element besides carbon produces the huge variety of free radicals that carbon does, and such a variety leads to a huge number of alternative pathways of reaction.[25] This increases the probability that *some* collection of carbon molecules will hit upon a self-reproductive reaction pathway spontaneously, and so begin life. Since carbon forms a wider variety of compounds than any other element besides hydrogen, this means that more information can be stored in carbon compounds than in those of any other element. Since life is self-reproduction of information, carbon compounds are uniquely fitted to serve as the basis of life.

Because of the carbon valence of four, carbon tends to bond in chain molecules with a tetrahedral shape. This results in long polymers being either spirally coiled or zig-zag.[25] This contributes to the plasticity and elasticity of organic polymers. The fact that carbon can form strong bonds with itself leads to most carbon polymers being in the form of linear chains in which each molecule is available for reaction. A number of other elements can form chains like this,[38] but in nearly every case more than one element is needed in the backbone of the chain, and all such cases lack the pliability of the carbon chain and its versatility for cross-linking into various possible shapes.[25]

Carbon dioxide is the ultimate source of carbon in living organisms. It has a number of unique properties which rank it with water in importance to life. At ordinary temperatures it has about the same concentration, in molecules per unit volume, in water as in the air. It is unique among gases in possessing this property. This enables carbon dioxide to undergo perpetual exchange between air and water, and between living organisms and their surroundings. Carbon dioxide will thus be available at any point of the Earth's surface for photosynthesis in plants, this process being the mechanism whereby living organisms obtain their carbon for further molecular synthesis. Carbon dioxide is also the ultimate waste product of metabolism; the energy source of organisms being the oxidation of carbon compounds to CO_2. The fact that CO_2 is a gas and very soluble in water enables it to be removed from the body with ease. A human produces a kilogram of carbon dioxide per day, which would be difficult to excrete if it were not for the volatility of the gas.[52]

Carbon dioxide is also unusual, though not unique, in being reversibly hydrated to form an acid, carbonic acid, H_2CO_3.[21] Carbonic acid is dibasic, forming bicarbonate (HCO_3^-) and carbonate (CO_3^{--}) ions as conjugate bases. This enables carbon dioxide to play another important role in the chemistry of living organisms: that of maintaining a constant pH. Carbonic acid has great buffering power, and should more or less of this compound be required to maintain constant pH, the great volatility of carbon dioxide ensures that it is available in the quantities needed. In the words of the biochemists Edsall and Wyman, these properties of CO_2

> ... provide a mechanism of unrivaled efficiency for maintaining constancy of pH in systems which are constantly being supplied, as living organisms are, with acidic products of metabolism.[21]

Carbon dioxide also plays a crucial role in regulating the temperature of the Earth's surface. Carbon dioxide in the atmosphere acts as a barrier to prevent the escape of heat from the Earth's atmosphere. Its presence in the Earth's atmosphere keeps the terrestrial temperature tens of degrees above what it would otherwise be.[43]

8.6 Nitrogen, Its Compounds, and Other elements Essential for Life

> Truth is indivisible, it shines with
> its own transparency and does not allow
> itself to be diminished by our
> interests or shame.
>
> U. Eco

Approximately 80% of the Earth's atmosphere is nitrogen. This element also plays an essential role in the chemistry of life. It is one of the elemental building-blocks of amino acids, which have the generic formula

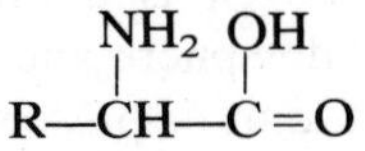

$$\begin{array}{c} \quad\ \mathrm{NH_2}\ \ \mathrm{OH} \\ \quad\ \ |\qquad\ | \\ \mathrm{R{-}CH{-}C{=}O} \end{array}$$

where R is some side chain.[25] The simplest amino acid, glycerine, has a single hydrogen atom as its side chain. Proteins, which form the enzymes and, with lipids, the cell walls of living organisms, are basically arrangements of amino acids linked through peptide bonds. Peptide bonds are themselves partially based on nitrogen:

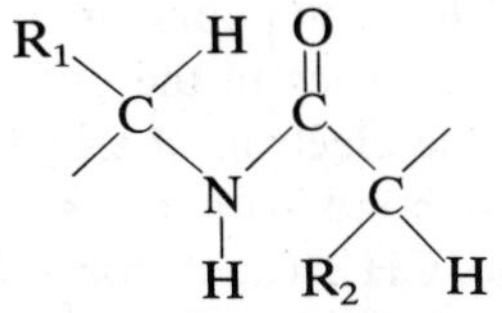

In the illustration above R_1 and R_2 are amino acids. The purines adenine and guanine, and the pyrimidines cytosine, urasil, and thymine, which are the bases of nucleic acids, are ring compounds with two nitrogen atoms per ring.

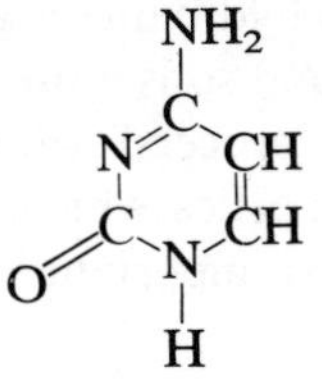

Cytosine 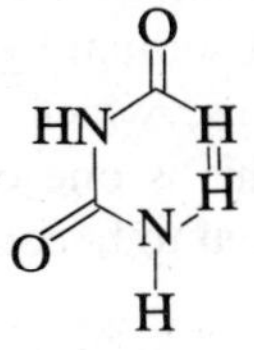 Uracil 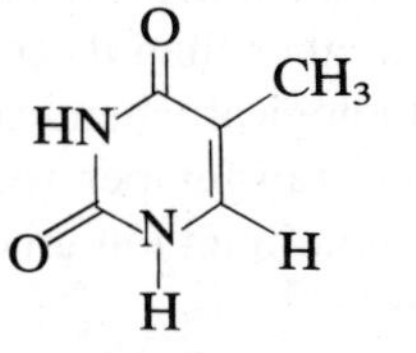 Thymine

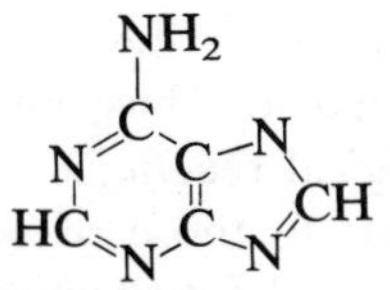

Adenine 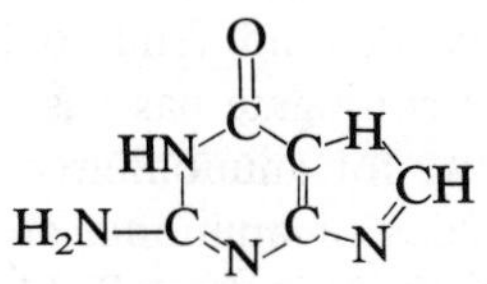 Guanine

Nitrogen compounds thus comprise the fundamental building blocks of living organisms. As an element, nitrogen is not particularly reactive, probably because of its great affinity for itself. Nitrogen molecules are composed of two nitrogen atoms, linked by a triple bond. This bond can be broken only with great difficulty; its heat of dissociation is

210 kcal/mole. As the existence of the triple bond in the nitrogen molecule suggests, nitrogen has a valence of three; it has one paired and three unpaired electrons in its outer shell.

As seen in Table 8.7, nitrogen is a rare element on Earth. It is essentially absent from the hydrosphere and lithosphere, and is available to living organisms only because of its presence in the atmosphere. In the opinion of the English chemist Needham, '... its relative abundance in mobile gaseous form therefore is another of the unique strokes of fortune, like the mobility of carbon dioxide, and of oxygen'.[25] The nitrogen in the atmosphere can be converted without too great an expenditure of energy per mole into ammonia, NH_3, which is the source of nitrogen in biological compounds.

Ammonia is an interesting compound in its own right. It is formed when the three unpaired electrons in the outer shell of a nitrogen atom each become paired with the electron of a hydrogen atom. These paired electrons form a covalent bond with about the same strength as the covalent bonds in methane, CH_4. The nitrogen atom in ammonia can also use its single pair of electrons to form a bond with an additional hydrogen atom. The nitrogen atom in effect donates both electrons of this pair to the hydrogen atom, which in turn is free to donate its solitary electron to yet another ion. The resulting NH_4^+ is a positive ion, the basis of the ammonium compounds. NH_4^+ is similar in many respects to the positive ions of the alkali metals, though the base ammonium hydroxide is a great deal weaker than its counterpart sodium hydroxide, and so is more useful for biological chemical reactions, which generally proceed with little energy transfer per reaction. This is one of the properties which makes the ammonium ion uniquely suited to become the most important biological base.[25]

Oxygen and sulphur can also form ions somewhat analogous to ammonium, namely OH_3^+ and SH_3^+ respectively, by the same mechanism of sharing a pair of electrons with an additional hydrogen atom. However, OH_3^+ is not stable, and the bonds of sulphur in SH_3^+ linking to its other hydrogen atoms are weak, and tend to be ionic, with the result that hydrogen sulphide, the conjugate base, is really an acid. Thus, neither of these ions can substitute for ammonium in biological reactions.

The physical properties of ammonia are similar to those of water in certain respects. As can be seen from Tables 8.1 and 8.5 respectively, the specific heat and heat of fusion is slightly higher than that of water. It is a polar molecule like water. Ammonia is a liquid over about a 40° temperature range, from −77.7° to −33.4°C. (At normal pressures these numbers can be substantially altered by large changes in pressure.) These similarities with water have led to numerous suggestions that ammonia could substitute for water as a liquid solvent in an exotic biosphere on some other planet.[50,53–57] On such a world, organisms would supposedly drink liquid ammonia. For example, Needham[25] suggested that among the more important radicals $—NH_2$ could replace —OH and =NH could replace =O. Atmospheric nitrogen might be used instead of oxygen for respiration, the reaction

$$6CH_4+7N_2 \rightarrow 8NH_3+3C_2N_2$$

being analogous to

$$C_aH_{2a}O_a + aO_2 \rightarrow aH_2O + aCO_2$$

where a is an integer and $C_aH_{2a}O_a$ is a monose sugar.

One problem with this particular proposal for a biochemical cycle is that cyanogen, C_2N_2, the analogue of carbon dioxide in the cycle, is a[25] solid below −35°C and so could cycle only via *solution* in the medium. It would not cycle as readily as carbon dioxide, which is a gas in our biosphere. This specific difficulty is not so serious, but there are other more fundamental problems with the use of ammonia as the central solvent for life. Although the heat of fusion for ammonia is a bit higher than that of water, its heat of vaporization is lower than water's by a factor of 2, and its surface tension is lower by a factor of 3 (see Tables 8.3 and 8.4 respectively). This means that although ammonia has hydrogen bonds and a polar structure, these factors do not preserve (even approximately) the lattice structure that water does upon melting. As indicated in our discussion of water, this structure is crucial to the hydrophobic effect which concentrates non-polar molecules, and allows organisms to organize enzymes into particular shapes by judicious arrangement of polar and non-polar side-groups. Both of these would be much more difficult to accomplish using ammonia as a solvent. Amino acids and nucleic acid bases are expected to be formed in a reducing atmosphere such as is believed to have existed on Earth billions of years ago, but unless these amino acids and nucleic acid bases (or their analogues in ammonia-based chemistry) were concentrated by some natural process, the spontaneous formation of a self-reproducing chemical system is most unlikely to occur. The ability of ammonia to hold complex molecules in a particular shape so it can act as a catalyst would also be weaker than is this tendency in water because the viscosity of ammonia is only one-quarter of that of water.[25] One test of the hypothesis that liquid ammonia is as good a solvent for life as water would be to repeat those experiments which created amino acids and nucleic acid bases[36] under conditions thought to occur on the Earth as it was billions of years ago. Only this time, one could perform the experiments at lower temperatures, with ammonia as a liquid in contact with such gases as N_2, CH_4 and H_2. (Ice could also be allowed.) After the experiment was run, a test would be made for polymers which could be analogues of proteins, and for molecules with sufficient complexity to serve roles analogous to nucleic acid bases. Matthews and Moser[58,59] have passed electric discharges through an atmosphere of methane and ammonia at room-temperature with the resulting formation of some polymeric material. Could any of the compounds thus formed be useful as amino acids to ammonia-based life? What happens if the experiment is repeated at *lower* temperatures, in the presence of liquid ammonia? Is the variety of compounds, or rather building-blocks of the polymers, as great as one finds in an aqueous environment? We have already expressed our doubts that cell walls could spontaneously form in liquid ammonia.[47] Furthermore, there is the difficulty that solid ammonia sinks in liquid ammonia. Several authors (e.g. ref. 55) have argued that this is not important. It merely means that the temperature of the planet with ammonia oceans never drops low enough for the oceans to freeze, or if bodies of ammonia do freeze then the organisms in them hibernate. There are two difficulties with this counter-argument: First of all, it seems most unlikely that the temperature of a planet's oceans could everywhere be kept in the rather narrow temperature range over which ammonia remains a liquid. Temperatures vary by more than this over the surface of the Earth at any given time. Second, once the ammonia began to freeze at the surface, this would cause a positive feed-back reaction. The ammonia ice thus formed would sink to the bottom and new ice would be formed at the top, which would sink, and so on. Water ice remains at the top and insulates the water below, thus damping the freezing reaction. On the other hand, ammonia ice at the bottom of the ocean would be insulated from melting when the temperature rose. This would tend to make the amount of ammonia ice grow until the ammonia oceans were frozen. This instability would probably destroy an ammonia-based biosphere if one ever formed. (It should be mentioned that the formation of ice at the surface tends to increase a planet's albedo. Hence, less heat is absorbed and more ice is formed. This is an instability which would exist here on Earth if ice did not float.)

The biochemist Lehninger[33] has divided the chemical elements important for life into three main classes, listed in Table 8.10. Class one comprises the essential elements which are present in all organisms (with the exception of sulphur), these elements make up one per cent or more of living organisms: they are oxygen, carbon, nitrogen, hydrogen, phosphorus, and sulphur. In class two are found the monotonic ions, such as Na^+ and K^+. In class three are the trace elements: iron, copper, and so forth. We have previously covered four of the six elements in the first class. We will now discuss the unique properties of the remaining two, phosphorus and sulphur.

TABLE 8.10
The elements used in living organisms.

The following elements are essential in the nutrition of one of more species, but not all are essential for every species

Class one elements: The elements of organic matter	Class three elements: Trace elements
O	Mn
C	Fe
N	Co
H	Cu
P	Zn
S	B
	Al
Class two elements:	V
Monoatomic ions	Mo
Na^+	I
K^+	Si
Mg^{2+}	Sn
Ca^{2+}	Ni
Cl^-	Cr
	F
	Se

Reproduced, with permission, from ref. 33.

Phosphorus and sulphur are periodic table congeners of nitrogen and oxygen, respectively. They are, however, in the second row of the periodic table. Atoms in this row attempt to complete an octet of electrons, like the atoms in the first row, but formation of an octet does not saturate the outermost shell (the third shell). The second row atoms still possess the five 3d orbitals in the third shell, which are capable of holding a further five pairs of electrons. It is the existence of this additional bond-forming ability of phosphorus and sulphur which allows them to play the essential role they do in biological systems.

By far the most important role these elements play is to act as energy and group transfer agents in chemical reactions. Most of the energy and group transfer reactions are conducted by organic phosphates, particularly ATP[62,8,33] but sulphur also forms three types of molecules with 'high energy' bonds which can supply energy for biochemical reactions: (1) esters of thiols; (2) mixed anhydrides of phosphoric and sulphuric acids; and (3) sulphonium compounds. George Wald has pointed out that this short list of three classes of sulphur compounds and one class of phosphorus compounds essentially exhausts the known categories of biological 'high-energy' compounds.[62]

The adenosine triphosphate (ATP) molecule, pictured in Figure 8.17, is the molecule generally used in cells as the basic source of energy for

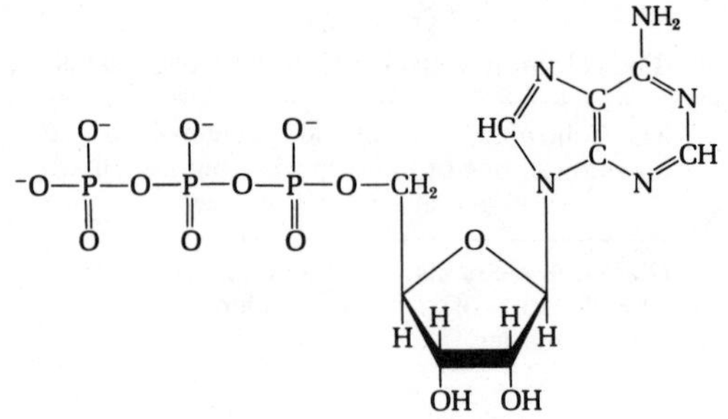

Figure 8.17. The molecular structure of ATP. Energy is made available to drive a reaction if a phosphate radical O—P(=O)(O)— is removed. (From ref. 33, with permission.)

biochemical reactions. ATP molecules are synthesized in cells using the energy obtained from oxidation of compounds like sugars.[8,33] The energy in ATP is stored in the phosphate 'bonds' in the sense that free energy becomes available to drive reactions if these bonds are broken. These high energy bonds are denoted by a wavy line. ATP can thus be written as $A \sim P \sim P \sim P$, which signifies that energy to drive chemical reactions becomes available by removal of the phosphate radicals. The types of reactions that such removals can drive are given in Table 8.11.

The reason why organisms use the transfer of a $\sim P$ radical to supply energy for biochemical reactions rather than use the energy released in the oxidation of sugars directly, is that the latter process releases *too much* energy. The complete oxidation of glucose yields 686 kcal per mole, while typical biochemical reactions are in the energy range required to make and break hydrogen bonds, ~1 kcal/mole. the transfer of a $\sim P$ radical supplies free energy in this range, so it is much better suited for mild biochemical reactions than the oxidation reactions, which involve making and breaking covalent bonds. The oxidation of glucose to form

TABLE 8.11

Types of reaction catalysed by ATP. ADP is $A \sim P \sim P$, and AMP is $A \sim P$. P_i is the phosphate ion, and S is the substrate.

1. P−transfer (kinase) reaction:	$S + ATP \rightarrow S - P + ADP$
2. P~transfer reactions:	$S + ATP \rightarrow S \sim P + ADP$
3. P~P-transfers:	$S + ATP \rightarrow S - P \sim P + AMP$
4. Activation reactions, with AMP transfer:	$S + ATP \rightarrow S - AMP + PP_i$
5. Activation without transfer:	(1) $S + ATP \rightarrow S^* + ADP + P_i$
	(2) $S_1 + S_2 + ATP \rightarrow S_1S_2 + ADP + P_i$
	(3) $S_1 + S_2 + ATP \rightarrow S_1 \sim S_2 + ADP + P$
6. Polymerizations of nucleotide di- and triphosphates:	$n(NDP) \rightarrow (NMP)_n + nP_i$

Reproduced, with permission, from ref. 25.

ATP is rather efficient. The complete oxidation of one molecule of glucose yields 38 molecules of ATP, at 66% efficiency.

There are three reasons why sulphur and phosphorus are uniquely suited for group and energy transfer reactions.[62] First, these elements form more open, and usually weaker, bonds than their congeners in the first period. Second, these elements possess 3-d orbitals, which permit the expansion of their valences beyond four. Third, they retain a capacity to form multiple bonds, a property otherwise characteristic only of carbon, nitrogen and oxygen.

The capacity to form multiple bonds and the ability to form 5 and 6 covalent bonds introduces a wide range of resonance possibilities among the precursors and products of exchange reactions, and this greatly increases the variety of changes that can occur. Sulphur and phosphorus bonds have a wide spacing and are relatively weak. When this property is combined with their tendency to add electron pairs to their unoccupied 3d orbitals, it induces an instability that promotes exchange reactions.[62]

In short, first period elements are not as suitable as sulphur and phosphorus for exchange reactions because the bonds they form are strong, and they can have a valence of at most four because they have no 3d orbitals. The other elements of the second row in the periodic table—silicon, for example—are unsuitable because they cannot form multiple bonds. Thus, amongst all the elements, sulphur and phosphorus are uniquely suited to play an important role in energy and molecule group transfer.

The monotonic ions of Lehninger's class two are mainly used in organisms to regulate the biochemical reactions rather than to participate in them directly, as do the elements of class one. These ions control the overall state of water solution in which the organism's biochemical reactions are occurring, the balance between the various subdivisions of this aqueous environment, and the transport of molecules through the cell walls.[25] Some of these ions accelerate the action of various enzymes. Others, particularly calcium, are used to construct rigid skeletons. The extent to which the various ions are used, however, seems to be more a function of their availability in the environment rather than of their unique properties.

The elements of class three seem to be essential to the formation of just a few biological molecules, although these molecules play vital roles in life. Magnesium (Mg) and iron (Fe) are the central atoms in the chlorophyll and haemoglobin molecules respectively. The particular metallic ion Mg may be necessary to molecules that are basic to photosynthesis. Now we would expect that if other choices were available, natural selection would have picked out the molecule which absorbed light in that frequency band at which sunlight is most intense, but chlorophyll in its various forms has peak absorbances at frequencies

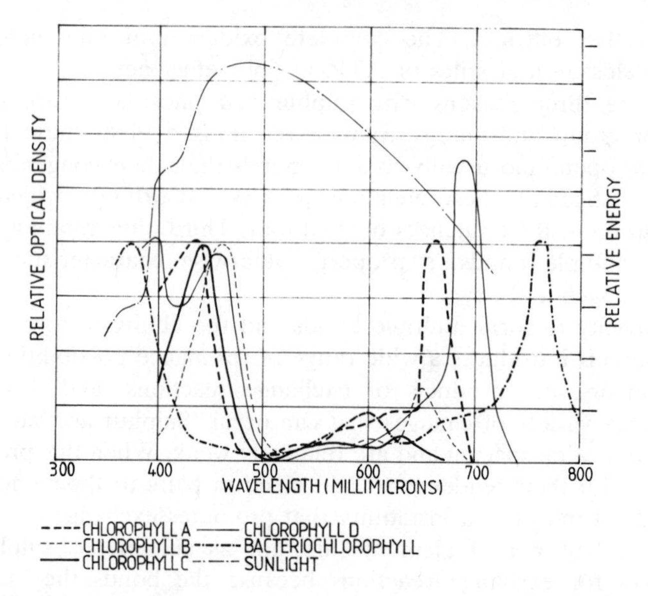

Figure 8.18. Absorption spectra of various types of chlorophyll compared with the energy distribution at the Earth's surface. (From ref. 45, with permission.)

where sunlight is not very intense,[45] as illustrated in Figure 8.18. We do not as yet know the reason why chlorophyll with its Mg ion was chosen, in spite of the fact that it looks wrong for the job. The situation is similar for the other elements in class three.

The biochemist George Wald has made a reasonable guess as to why Nature has utilized chlorophyll:

> Chlorophyll . . . possesses a triple combination of capacities: a high receptivity to light, an inertness of structure permitting it to store the energy and relay it to other molecules, and a reactive site equipping it to transfer hydrogen in the critical reaction that ultimately binds hydrogen to carbon in the reduction of carbon dioxide. I would suppose that these properties singled out the chlorophyll for use by organisms in photosynthesis in spite of their disadvantageous absorption spectrum.[68]

8.7 Weak Anthropic Principle Constraints on the Future of the Earth

> Man is the Measure of all things.
>
> Protagoras

The version of the Anthropic Principle which is most firmly founded is WAP, and we wish to use it to derive a testable formula connecting the number of improbable steps in the evolution of the species *Homo sapiens* on this planet, and the length of time the Earth will remain as a habitable planet. A formula of this type was first obtained by Brandon Carter.[71]

We presented in Chapter 3 the arguments of the evolutionists for there being a fairly large number of very improbable steps in the evolution of an intelligent species on this planet. It is important to recall once again that WAP means the results we obtain from the observations we make of the Universe are self-selected by our nature *qua* measuring apparati. It may be that other types of intelligent being, types not of the class mammalia or even types not based on DNA, exist. If they do exist, the observations they make are *not* restricted by WAP. Nevertheless, it remains true that the observations *we* make *are* so restricted. For whether or not such beings exist, or can exist, is irrelevant for WAP. It remains true that that we are a primate species which evolved in 4.5×10^9 years on an earthlike planet around a G2 star, and all *our* observations are self-selected by these fundamental facts.

The starting point for the derivation of Carter's formula is the observation that the length of time it took before *Homo sapiens* evolved is comparable (to within a factor of 2) to a well-established upper bound on the length of time the non-intelligent portion of the biosphere can continue to live and evolve on this planet. This upper bound is 10 billion years after the formation of the solar system, and it is the length of time a G2 star like the Sun can remain on the main sequence. When the Sun exhausts its hydrogen nuclear fuel and leaves the main sequence, the energy output will increase, the outer atmosphere of the Sun expand to engulf the Earth, and life on Earth will end as the oceans boil away. After a few hundred million years, the expanded Sun will puff off its outer atmosphere, leaving a white dwarf core, which will have a very high temperature, but which will have too little total radiative emission to keep water liquid on any planet currently in orbit around the Sun. Thus if t_e is the actual time it required for evolution to produce an intelligent species on the Earth, then we must have $t_e < t_0$, where t_0 is the least upper bound to the amount of time evolution can proceed on the Earth. We do not know, at the present stage of the discussion, what t_0 is; all we know is that it is bounded above by $t_{ms} \sim 10^{10}$ years, the main sequence lifetime of the Sun. Nor do we know what t_{av} is, where t_{av} is the expectation value of the time needed to evolve an intelligent species on an earthlike planet.

However, we would expect *a priori* that we would have $t_e \approx t_{av}$, and also either $t_{av} \ll t_{ms}$, or $t_{av} \gg t_{ms}$. But either inequality is inconsistent with the observed relation $t_e \approx t_{ms}$, if we assume the actual time needed to evolve an intelligent species on the Earth, which is t_e, is close to the average time needed to evolve an intelligent species on an earthlike planet, which is t_{av}.

The inequality $t_{av} \ll t_{ms}$ could be justified *a priori* in the following manner. The timescale needed by Darwinian selection to change completely a genome (the totality of genes in a living being) of I bits of information is

$$t_D \approx t_g S^{-1} \ln I \tag{8.4}$$

where t_g is the average time between generations, and S is the selection coefficient.[71,109] Even for human beings with tens of years between generations and $I = 10^{10}$ bits (the estimate for I given by Dobzhansky *et al.*[72]), a relatively small selection coefficient could give a value of t_D which is very small relative to the age of the Earth: for humanity $t_D/t_g \sim 10^{-7}\,\mathrm{s}^{-1}$. Thus, if there were no improbable steps leading to intelligent life, and if there were Darwinian selection from single-celled organisms directly toward some form of intelligent life, we would have expected intelligent life to have arisen much earlier in the history of the Earth.

Conversely, if there were many improbable steps, as most evolutionists believe, we would expect that $t_{av} \gg t_{ms}$. Thus the observation $t_e \approx t_{ms}$ is difficult to explain purely on the basis of *a priori* probabilities, for on the basis of such probabilities alone we would expect $t_e \approx t_{av}$, which would imply either $t_e \ll t_{ms}$ or $t_e \gg t_{ms}$.

The key WAP observation is that the actual observed time to produce *Homo sapiens*, which is t_e, might not come close to the average time needed to evolve an intelligent species on an earthlike planet. It could very well be that this average time t_{av} is vastly longer than t_{ms}, and still we would observe $t_e \approx t_{ms}$. For by the self-selection principle called the WAP, it is logically necessary that we human beings observe $t_e < t_0 \leqslant t_{ms}$ on the Earth. It is trivially true that humans must evolve before evolution ceases on Earth, if we are going to evolve on Earth at all. Only if it were the case that $t_{av} \ll t_{ms}$ could we truly expect to observe, with high probability, $t_e \approx t_{av}$. (The only other possibility, $t_e \approx t_{av} \approx t_{ms}$, is ruled out because there is no physical relationship between the timescales of stellar evolution and the timescales of biological evolution.) The fact that we observe $t_e \approx t_{ms}$ instead has the strong implication that $t_{av} \gg t_e$, and that the observed numerical coincidence $t_e \approx t_{ms}$ is a consequence of a WAP self-selection effect. But if $t_{av} \gg t_e \approx t_{ms}$, then the existence of extraterrestrial intelligent life is exceedingly improbable. Most earthlike planets around G type stars will be destroyed by their star leaving the main sequence long before intelligent beings have a good chance of evolving. This WAP argument against the existence of extraterrestrial intelligent beings, due to Carter,[71] is the second such argument we shall present. It is consistent with the purely evolutionary argument, which we discussed in Chapter 3. The third argument against local extraterrestrial intelligence, which we shall term 'the spaceship argument', will be discussed at length in Chapter 9.

The basic idea underlying Carter's formula is simple. It connects the number of very improbable steps in the evolution of *Homo sapiens* with the length of time the biosphere will continue to evolve in the future. We have seen that there is a least upper bound, t_0, to the length of time the biosphere can continue to exist on Earth. If the expectation value for the evolution of intelligent life on Earth is much larger than the lower bound t_0, and if, against the odds, intelligent life does evolve (as it has), then the evolution of intelligent life is far, far more likely to have occurred very close to t_0 than to any epoch in terrestrial history. This is due to the fact that the probability of intelligence increases monotonically with time, even though the probability is still small when the time t_0 is reached.

To make the idea behind Carter's formula precise,[73] we suppose there are n steps in the evolution of *Homo sapiens* which are statistically independent and each so improbable that each of them is unlikely to occur before t_0 on an earthlike planet. The probability that mankind will evolve on Earth by time t can be approximated by

$$p(t) = (1 - \exp[-t/\alpha_i])^n \tag{8.5}$$

where $\alpha_i \gg t_0$ is the timescale for the occurrence of the ith improbable step. In the interval of interest $(0, t_0)$ the probabilitiy (8.5) can be written to a very good approximation as the power law

$$p(t) = \left[\prod_i (\alpha_i^{-1})\right] t^n = \beta t^n \tag{8.6}$$

where $\beta \equiv \prod_i \alpha_i^{-1}$, since the argument of the exponential will be small in this interval.

Now the conditional probability that the evolution of *Homo sapiens* occurs at time t, given that it occurs on or before the time t_0 is just

$$p(\text{evolution occurs at time } t \mid \text{it definitely occurs before } t_0) = \gamma t^n \tag{8.7}$$

where $\gamma = t_0^{-n}$. The normalization constant γ is obtained from the requirement that the conditional probability (8.7) equals one when $t = t_0$; that is, evolution is certain to produce *Homo sapiens* in the interval $(0, t_0)$ if in fact it does. The expectation value for the evolution of *Homo sapiens*, given that it definitely occurs in the interval $(0, t_0)$, is then

$$\bar{t} = \int_0^{t_0} t\, dp = \int_0^{t_0} t(t_0^{-n})nt^{n-1}\, dt = t_0 n/(n+1) \tag{8.8}$$

This implies that the difference between the least upper bound t_0 and the expectation value $\bar{t}$ for the time needed to evolve *Homo sapiens* is

$$t_0 - \bar{t} = \frac{t_0}{n+1} \tag{8.9}$$

which is Carter's formula. We would expect to have $\bar{t} \approx t_e$, so (8.9) is equal to

$$t_0 - t_e = \frac{t_0}{n+1} \tag{8.10}$$

Carter obtained his formula in the form (8.10).

For small n, the approximation $\bar{t} \approx t_e$ is not especially good, but it improves with increasing n. More precisely, the likelihood that t_e is in the interval $(\bar{t}, t_0)$ increases with n. The conditional probability that t_e will be in this interval is

$$p(\bar{t} \leqslant t_e \leqslant t_0) = 1 - \left(\frac{\bar{t}}{t_0}\right)^n = 1 - \gamma(\bar{t})^n \tag{8.11}$$

This probability will be greater than or equal to any preassigned probability δ if n is sufficiently large so that we have

$$t_0 - \bar{t} \geqslant t_0(1 - \sqrt[n]{1-\delta}) \tag{8.12}$$

which clearly can be satisfied for any δ with $1 > \delta \geqslant 0$ for n sufficiently large.

Carter's formula (8.10) can be re-written in terms of the observable t_e:

$$t_0 - t_e \leqslant t_e/n \tag{8.13}$$

where we have replaced the equality with an inequality to emphasize that for large n it is overwhelmingly probable that the actual time required for the evolution of *Homo sapiens* lies within the interval $(\bar{t}, t_0)$.

We would like to point out the robustness of Carter's formula (8.10). If the exponentials $\exp[-t/\alpha_i]$ in (8.5) are replaced by the exponentials $\exp[-t^2/\alpha_i]$, which might be appropriate if we believed the probability distribution to be normal, the only effect is to replace n by $2n$ in formula (8.10). Furthermore, Carter's formula remains valid even if the upper bound is not the same for all earthlike planets, but rather differs from planet to planet. Such a variation would give, in effect, a probability distribution for t_0 on the ensemble of all earthlike planets. In this case we would compute not $t_0 - \bar{t}$, but $\bar{t}_0 - \bar{t}$, where $\bar{t}_0$ is the expectation value of t_0. The calculation would proceed as above, with $\bar{t}_0$ replacing t_0, except that we would set $\bar{t}_0 \approx t_0$ in addition to $\bar{t} \approx t_e$ in the final step, where now the symbol t_0 represents the actual upper bound to the length of time evolution can proceed on Earth.

We could test the formula (8.13) if we could obtain estimates for both n and $(t_0 - t_e)$.

The factor n measures the number of independent steps in human evolution each of which is so improbable that it is unlikely to have occurred before the Earth ceases to be habitable. In principle, evolutionary theory should be able to provide us with an estimate for n. In practice, evolutionary theory is not sufficiently advanced to enable us to calculate n with a high degree of confidence. Nevertheless, we shall attempt to obtain a lower bound for n by applying the following three criteria to evolutionary steps which led to mankind.

CRITERION #1: The step must have been unique; it must have occurred only once in the entire history of life. In principle, nucleic acid sequencing of genomes coding for the trait in question would be sufficient to verify that the trait arose only once in the evolutionary tree. Clearly uniqueness is a necessary condition for an evolutionary step to be in the class we are analysing—if a trait is invented more than once in different lineages its expectation time is likely to be less than t_{ms}—but it is not a sufficient one. It is quite possible that a trait is unique not because it is unlikely to occur but because there is no alternative to that trait. Earlier in this chapter we discussed the unique properties of a number of important biological molecules, for instance chlorophyll. It is probable that chlorophyll is used exclusively in the photosynthesis of all but the simpliest prokaryotes because it is the best molecule for photosynthesis around a G2 star. The universal use of ATP as an energy molecule is probably due to the ease with which it can be synthesized in an abiotic environment.[74] In order to avoid the bias due to lack of alternatives, it will be necessary to consider only those traits defined by a large amount of information; i.e., those traits requiring a large number of genes for their coding. This restriction gives:

CRITERION #2: The unique trait must be polygenetic. In biological terminology, the unique trait must be a single *seme* (a trait under multigene control). There is a considerable amount of evidence that life is sufficiently inventive to rediscover several times those polygenetic traits which are likely to occur in the timespan t_{ms}. We mentioned in section 3.2 that the eye has been invented independently at least 40 times. Furthermore, the same complex molecular endproducts are known to be produced in some instances by different biosynthetic pathways. As an example of such metabolic convergence, the prokaryotes *Zymomonas* and *Escherichia* oxidize glucose, but they differ in *every* enzyme of glucose catabolism.[75] Another example is the ability to fix atmospheric carbon dioxide, which is known to have evolved at least three times.[76] Similarly, fermentation probably represents several semes.[76]

There is also considerable evidence that no seme evolves and disappears without leaving a trace in some surviving lineages,[77] so if a seme is unique, it is likely that it is unique because its most probable time of evolution is greater than t_{ms}, and not because it evolved several times, with all but a single lineage bearing the trait having become extinct.

CRITERION #3: The trait must clearly be essential for the existence of an intelligent species. Note that the trait in question need not lie in the human lineage. The underlying probability (8.5) requires only that the expectation time for the trait be much greater than t_0, that it be essential for the existence of *Homo sapiens*, and that it be statistically independent of the other $n-1$ traits in Carter's class of crucial steps in the evolution of Mankind. In fact, it is better if the trait is *not* in the human lineage; in this case we can be more confident the trait is statistically independent of the other traits.

We suggest the following semes as satisfying the above three criteria, and hence as being crucial steps in the evolution of Man:

CRUCIAL STEP #1: The development of the DNA-based genetic code. The three-codon genetic code is universal among living things,[61,75] so it satisfies the first criterion. Although it is occasionally suggested that several different codes evolved initially (e.g. refs. 78 and 79), with the present code eliminating the others via natural selection, this seems unlikely for as we said in our discussion of criterion #2, no seme dies out without leaving some trace, and there is none for an alternative genetic code. The present code seems unnecessarily redundant to be overwhelmingly superior to any other code which might have arisen. Also, there is no reason why the current system must code for left-handed amino acids, when as far as we can tell, the right-handed mirror images of the current amino acids would do as well. The current code is polygenetic in the sense that the minimum complexity required to code even the simplest cells is quite large.[80] Clearly a genetic code of some sort is absolutely essential for the existence of intelligent life.

CRUCIAL STEP #2: The invention of aerobic respiration. This is a unique seme used in all eukaryotes and many bacteria.[75] It is polygenetic and is essential for the development of life in an oxygen atmosphere. As we discussed earlier in this chapter, oxygen is the uniquely appropriate atom for the oxidizing agent in respiration. Other molecules are possible oxidizing agents, and others are in fact used by anaerobic bacteria. But only oxygen can be the oxidizing agent for metazoans. Only oxygen provides sufficient free energy for metazoan metabolism and simultaneously is sufficiently abundant in the cosmos to be a major component of a planetary atmosphere.

CRUCIAL STEP #3: The invention of glucose fermentation to pyruvic acid is a unique seme[75] which evolved in bacteria and remained unmodified in all eukaryotes. It is sufficiently complex to necessitate control by many genes and is an essential stage in the energy metabolism of metazoans.[8]

CRUCIAL STEP #4: The origin of autotropic photosynthesis (oxygenic photosynthesis). This is a very complex trait which is a single seme.[75,81] The details of photosynthesis in bacteria (actually cyanobacteria), algae, and plants, are remarkably similar. All of these organisms use exactly the same steps to reduce CO_2 to organic compounds via ribulose biphosphocarboxylase. This precise pattern with all its complexities is most unlikely to have evolved independently in separate lines of organisms.[82] Autotropic photosynthesis is essential if the oxygen for metazoan metabolism is to be put into the atmosphere. Inorganic processes cannot generate an oxygen atmosphere. Note that this trait is probably not in the human lineage, though it is possible that we are descended from prokaryotes which lost the trait. As we said above, crucial steps need not be directly in the human lineage; it is sufficient that they be necessary for human evolution.

CRUCIAL STEP #5: The origin of mitochondria: these are the bodies in the cytoplasm of eukaryotes wherein the energy molecule ATP is synthesized. The mitochondrion is about 20 times more efficient than the prokaryotic cell membrane in producing ATP.[83] Without the advent of the mitochondrion, the efficiency of living cells would be too low to allow the existence of metazoans. The mitochondria are thought to be the remnants of a bacterium that was absorbed by another bacterium to form a composite cell. Over time most of the DNA comprising the original pre-mitochondric bacterium disappeared as many of its cellular functions were taken over by the machinery of the absorbing cell; there is, however, some DNA remaining in the mitochondrion to show its initial independent cellular origin. The bulk of the present evidence indicates that mitochondria arose just once; however, the replication of DNA in the mitochondria of *Euglena gracilis* is unusual, which may indicate an independent origin[84] in this protist.

The mitochondria are just one of a number of bodies called *organelles* in eukaryotic cells which are generally thought to have originated as independent cells which were absorbed by another cell. Without these other bodies within them, the eukaryotes would have never developed the genetic complexity needed to evolve into metazoans; it is certain that no prokaryotes are metazoans. The organelles that are unique to eukaryotes are (1) mitochondria, (2) kinetosomes with their undulipodia, (3) centrioles, and (4) the plastids of which chloroplasts are an example. The chloroplasts were once thought to have a unique origin like mitochondria,[85] but it has recently been discovered that they originated several times,[86] showing again that semes never disappear without a trace. There is considerable evidence that the formation of multiple cells through absorption of one cell by another is quite common. Amoebae have been observed to acquire bacterial symbionts in the laboratory.[87] Originally the bacteria were pathogenic, but after five years, the bacteria had taken over the synthesis of certain amoeba metabolites. If such symbiosis is quite common, the monophyletic origin of a eukaryotic organelle means that a particular organelle is most unlikely to be formed from the chance symbiosis of two prokaryotes. The endosymbiotic theory of the formation of eukaryotic organelles—which asserts they formed by absorption of one cell by another, as described above—holds that centrioles, kinetosomes, and undulipodia are all the evolved by-products of a single absorption of a type of prokaryote (called a spirochete) by a pre-eukaryotic cell.[88] Thus we have:

CRUCIAL STEP #6: The formation of the centriole/kinetosome/undulipodia complex; such an event was essential to the evolution of the reproductive system of eukaryotes and of nerve cells.[74] The microtubules which make up the complex are used to form the spindle system which separates the chromosomes during the cell division of a eukaryote, or to form the long fibres (axons and dendrites) of nerve cells.

CRUCIAL STEP #7: The evolution of an eye precursor; L. Ornstein[89] has suggested that the invention of the eye at least 40 different times in metazoan lineages (see Chapter 3) actually required the previous invention of an eye precursor. He argues the evidence strongly suggests this eye precursor appeared only once; however, nucleic acid sequencing is needed to make sure that the eye precursor is monophyletic. If the eye precursor indeed has a unique origin, it constitutes a crucial step in human evolution, for clearly eyes are necessary for intelligence. (Ornstein argues this point at length in ref. 89.)

CRUCIAL STEP #8: The development of endoskeleton. Such a skeleton seems essential for support of large terrestrial animals; only such creatures could develop into an intelligent technological species. From embyro development it seems this trait is monophyletic, but again DNA-sequencing will be required to verify this.

CRUCIAL STEP #9: The development of chordates. It is only a suggestion that the chordates constitute a monophyletic line; the necessary DNA-sequencing has not been done. However, assuming that they are such a single lineage, the chordates would then be the only terrestrial lineage which could develop a complex central nervous system.

CRUCIAL STEP #10: The evolution of *Homo sapiens* in the chordate lineage. The evidence for the uniqueness of this development was discussed at length in section 3.2.

The arguments for the above 10 steps as being crucial in the sense of Carter's formula are not conclusive by any means; they are offered as suggestions only, as examples of the sort of tests it would be necessary to perform in order to calculate an upper bound to n. If we accept $n \geq 10$, then inequality (8.10) gives 4.5×10^8 years as an upper bound for the length of time the biosphere can continue in the future.

Another, more radical approach to obtaining an upper bound for n is as follows. The number n in inequality (8.10) is actually not just the number of improbable steps which occur in the evolution of *any* intelligent species on any earthlike planet; rather, it is the number of crucial steps—steps which are unlikely to occur more than once in the time period t_{ms}—in the evolution of the particular species *Homo sapiens*. This number is probably much larger than the number of steps obtained in the estimate above, which concentrated on steps which would have to occur to generate *any* intelligent species. (We say *probably* much larger because the evolutionist Lovejoy, whose work we discussed in Chapter 3, contends that traits essential to any intelligent species are so uniquely human in the animal kingdom that the probability of the evolution of *any* intelligent terrestrial species is *equal* to the probability of the evolution of the very particular species *Homo sapiens*.)

To get an estimate of this much larger number, we note that *Homo sapiens* is defined biochemically by the proteins—enzymes and structural proteins—which the human genome codes for. Each protein is coded by a separate gene, and the number of different genes in the human genome is estimated by Dobzhansky *et al.*[72] to be 110,000, as compared with 83,000 genes in the cow, 7250 in the fruit fly, 2500 in the prokaryote *Escherichia coli*, and 170 in one of the most primitive bacteria *Mycoplasma gallisepticum*.[90] Morowitz[80] has obtained a theoretical lower bound of ≈ 50 on the number of genes in any cell, no matter how primitive. We will assume that most of the proteins coded for are enzymes.

DeLey[83] estimates from experimental evidence that only some 10 to 20 per cent of the amino acids comprising an enzyme are immutable for enzyme activity. The other amino acids can be changed by random mutations without changing the biochemical effect of the enzyme. This means that if we take the average gene to have 1800 nucleotide bases—the standard estimate[72,83]—then 180 to 360 nucleotide bases are immutable for each gene. The odds for assembling a single gene are between $4^{-180} = 4.3 \times 10^{-109}$ and $4^{-360} = 1.8 \times 10^{-217}$. These numbers are so incredibly small that DeLey[83] opines that an enzyme arises only once during evolution. (See however ref. 107.) There simply has not been sufficient time since the formation of the Earth to try a number of nucleotide base combinations even remotely comparable to these numbers. The number of bacteria on Earth today is estimated to be of the order of 10^{27}; assuming a bacterial reproduction time of 1 hour, there have been at most about 10^{40} bacteria in the entire past history of the Earth. With the order of 10^{7} nucleotide bases per bacterium, it would be possible to try some 10^{47} nucleotide combinations during the past, which is 52 orders of magnitude too few.

The odds against assembling the human genome spontaneously is even more enormous: the probability of assembling it is between $(4^{-180})^{110,000} = 10^{-12\times 10^{6}}$ and $(4^{360})^{110,000} = 10^{-24\times 10^{6}}$. These numbers give some feel for the unlikelihood of the species *Homo sapiens*. From these numbers we can calculate that the species *Homo sapiens* will evolve on the average on earthlike planets between 10^{400} and 10^{800} light years apart (the calculation is insensitive to there being one earthlike planet per star, per galaxy, or per visible universe [of radius 2×10^{10} lyr]; it is also indifferent to evolution rates). These distance estimates are large compared with the numbers of observational astronomy, but even if the universe is closed it may be large enough to accommodate with high probability several independent evolutions of *Homo sapiens*; if it is open, it will almost certainly have more than one. (The implications of such duplication of *Homo sapiens* are discussed in refs. 91, 92, and 93.) As we discussed in Chapter 6, the observational data is not good enough to tell us the actual size of the entire universe.

We should emphasize once again that the enormous improbability of the evolution of intelligent life in general and *Homo sapiens* in particular does *not* mean we should be amazed we exist at all. This would make as much sense as Elizabeth II being amazed she is Queen of England. Even though the probability of a given Briton being monarch is about 10^{-8}, *someone* must be. Only if there *is* a monarch is it possible for the monarch to calculate the improbability of her particular existence. Similarly, only if an intelligent species does evolve is it possible for its members to ask how probable it is for an intelligent species to evolve. Both are examples of WAP self-selection in action.[110]

For the purposes of Carter's formula, all we need to know from the above discussion is that $n \geq 110{,}000$, if n is the number of crucial steps in the evolution of *Homo sapiens*. From (8.10) we have

$$t_0 - t_e \leq 4.1\times 10^4 \text{ years} \tag{8.14}$$

for the estimate of the number of years in the future the biosphere would be capable of evolving our species; as we mentioned earlier, this number is most likely to be the length of time the biosphere will exist in the future.

This is an incredibly short period by astronomical, biological and geological standards; it is about the length of time anatomically modern man has existed.[94] Carter himself was aware that 'reasonable' values of n would give 'unreasonable' values of $t_0 - t_e$; in fact he was unhappy with the value for $t_0 - t_e$ implied by a value of n greater than 2, and he explicitly rejected the idea that the biosphere of the Earth could have a short future.

But he gave no reason for doing so, and his action is reminiscent of the rejection by the nineteenth-century physicists of the very large biological estimates for the age of the Earth. We have discussed the history of this rejection at length in section 3.6. The lesson we should take from the nineteenth-century error is that we must take seriously the timescale estimates given to us from purely biological data, even if the logical implications of these data may appear incredible. It is difficult to imagine now how unreasonable an age for the Earth of 10^9 years appeared to scholars a century or two ago. It seemed obvious to them that 10^4 or 10^7 years was the most they could grant. The numbers we obtained above seem to modern minds to be too short by about the same factor that a terrestrial age of 10^9 years seemed too long to nineteenth century minds.

But if the biosphere can exist for only a rather short time in the future, what could be the physical mechanism of its demise? On this score we can only suggest a single possibility; we hope our discussion of Carter's formula will stimulate both detailed investigations of the possibility we discuss and a search for others.

Michael Hart has performed computer simulations of the evolution of the Earth's atmosphere over its 4.5 billion year history, and he finds that the atmosphere is only marginally stable.[48] In fact, there are so many factors tending to destabilize the atmosphere that he was one of the first investigators to construct a computer model of the atmosphere which did not destabilize in the first billion years. The Earth's atmosphere is finely balanced between runaway glaciation and runaway heating due to the greenhouse effect. Runaway glaciation can occur if the ice caps get too large: an increase in the ice caps causes an increase in the amount of heat reflected back into space; this leads to a decrease in the surface temperature, which in turn causes the ice caps to become even larger, which causes even more heat to be reflected back into space, et cetera. This process continues until the entire planet and all life is frozen solid. Runaway heating can occur if CO_2 accumulates in the atmosphere. The added carbon dioxide causes the surface temperature to rise via the greenhouse effect, which in turn causes more CO_2 to be released into the atmosphere from surface rocks. This in turn drives the temperature even higher, et cetera. This process continues until the oceans boil away, leaving the Earth's surface temperature comparable to that on Venus, which is high enough to melt lead. Hart's simulations were run just 500 million years into the future, at which time the atmosphere was still stable in the sense that neither runaway glaciation nor runaway heating occurred. However, his model revealed a steadily increasing amount of free oxygen in the atmosphere beginning about 500 million years ago, with the atmospheric composition rising from 21% today to 35% 500 million years from now. Such a percentage of free oxygen would make it quite impossible for life to exist on the land, for at this percentage the terrestrial plants will begin to burn spontaneously, as Lovelock has pointed out.[95]

More precisely, the probability of a forest fire being started by a lightning-bolt increases 70% for every 1% rise in oxygen concentration above the present 21%. Above 25% very little of the vegetation on land

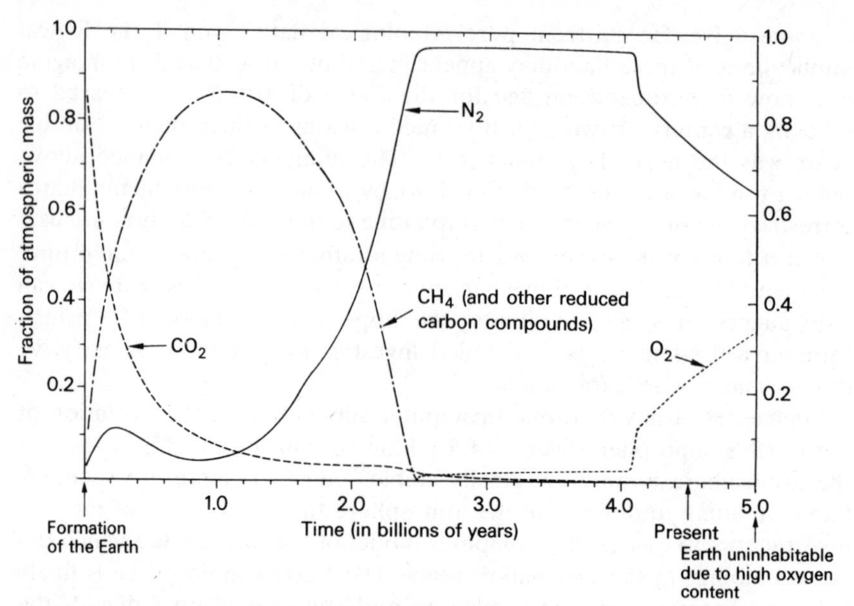

Figure 8.19. Graph of the change in the composition of Earth's atmosphere over time in Hart's model. (After Hart, ref. 48.)

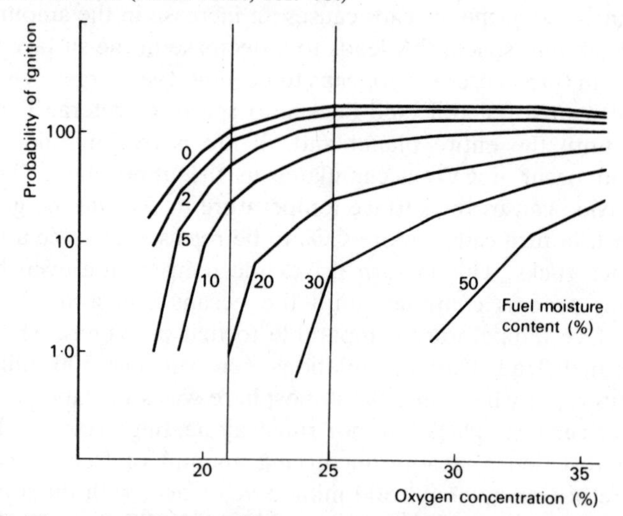

Figure 8.20. The probability that land vegetation will be ignited by lightning bolts or spontaneous combustion. The probability is strongly dependent on the moisture content of the fuel. Each line corresponds to a different moisture content, beginning with completely dry (0%), and going to visibly wet (50%). At the present fraction of 21%, fires will not start at more than 15% moisture content. Were the oxygen concentration to reach 25%, even damp twigs and the grass of a rain forest would ignite. (Data obtained by A. Watson of Reading University, and recorded in ref. 43, reproduced with permission.)

would survive the fires, and this concentration is reached in Hart's model in about 200 million years from now. The increasing probability of fire can be seen by comparing the steady increase of oxygen in Hart's model (Figure 8.19) with the probability of grass or forest fires at different oxygen concentrations and moisture conditions (Figure 8.20).

The source of the free oxygen in Hart's model is green plants, and it is quite possible for the photosynthetic bacteria and plants in the oceans to supply the steadily increasing amount of oxygen even if plant life on the land becomes extinct, so the future evolution of his oxygen source is realistic. However, Hart's model did not take into account the present-day regulator of the oxygen concentration, which is methane supplied by anaerobic bacteria.[95] It is possible that this mechanism would be sufficient to stabilize the oxygen concentration at the present 21% level. More research is needed on this question. Research on the stability of the atmosphere should focus on the question of the oxygen concentration, for it is the free oxygen that gives most of the problems in the long-term computer simulations. It is universally accepted that the atmosphere initially contained very little free oxygen, and that the free oxygen concentration gradually rose from zero in the beginning to the current level as photosynthesizing life supplied the oxygen. As the oxygen level rose, the greenhouse effect faded away, with the result that the temperature fell drastically. This sudden fall in temperature tends to force runaway glaciation in the computer models.

It is also quite possible—quite likely, in fact—that Hart's model cannot be believed because he has omitted too many other factors besides the oxygen regulator (if the current one is truly sufficient to stabilize the atmosphere in the long run). The current review papers (e.g. refs. 96–100) on the significance of long-term atmospheric simulations all urge caution in believing the predictions made by such models; there are simply too many unknowns at present to make accurate computer models of such long-term evolution. Nevertheless, atmospheric simulations such as Hart's (and the recent 'Nuclear Winter' simulations[101,102]) suggest that the Earth's atmosphere is only marginally stable, which means it could be destabilized by relatively small perturbations, and either natural causes or human miscalculation could render the Earth uninhabitable in the near future. A more accurate calculation of the atmospheric stability, a calculation we could place confidence in, would give us a good upper bound for $t_0 - t_e$.

We should emphasize that Carter's formula is based on the idea that the evolution of intelligent life is most improbable, and that if the current searches for extraterrestrial intelligent life succeed in finding such creatures, his entire argument collapses. Thus one testable prediction Carter's formula makes is that we are alone in the Galaxy.

If our crucial step #1 is indeed crucial, i.e., that the evolution of life itself is unlikely to occur in the period t_{ms}, then it follows there should be no other life of any sort in the rest of the solar system. The failure of the Viking probe to detect life on Mars supports this prediction, but there are a number of planets which have not been searched for life. Sagan and Salpeter[103] have presented a detailed scenario for the evolution of DNA-based life on Jupiter. If such a life-form as they suggest were indeed found on Jupiter, WAP would be in serious trouble. Furthermore, if crucial step #1 is indeed crucial in Carter's sense, it is most unlikely that experimenters will succeed in getting primitive life to form spontaneously in the laboratory. We do not mean to suggest that they will be unable to synthesize life; in fact we believe they will succeed in doing this, and in the near future. But we also believe such synthesis will require a great deal of outside help in the form of putting together a large number of reagents under conditions which are most unlikely to have occurred on the primitive Earth 4.5 billion years ago. Recently, the biochemist Cairns-Smith has described[104] in detail the biochemical improbabilities in the current models for the spontaneous formation of life; the evolutionist G. G. Simpson has also pointed out[105] similar biochemical improbabilities.

In the next chapter, we discuss the astronomical evidence that extra-terrestrial intelligent life does not exist elsewhere in our Galaxy. The biological evidence was discussed in section 3.2.

In this Chapter we have discussed the possible definitions of life and the sufficient conditions for intelligent life to be said to exit. Our definition of life is compared with previous suggestions by biologists and physicists. We have developed the deep connection between living beings and self-reproducing automata in order to describe living systems using the precise language of modern computer theory. We considered the special properties of the elements used by life, as we know it to exist, to argue that life which evolves spontaneously must be carbon-based. Some experiments which might falsify this claim were suggested. The key chemical properties and apparent coincidences of Nature which allow the evolution of human life based on atomic structure, were discussed in detail to reveal a situation of considerable complexity. Finally, we investigated a recent Anthropic prediction due to Carter, that life on Earth may have a relatively short future. The logic of this prediction is based upon the coincidence that the timescale for biological evolution has turned out to be so close to the main-sequence stellar lifetime. Various delicate climatic and photochemical coincidences allowing life to exist on Earth were then discussed, along with the likelihood that they may be upset in the future by terrestrial events. This discussion also reveals how stringent are the conditions that must be satisfied before a planetary surface is even a possible site for the successful evolution of life.

References

1. E. Mayr, *Populations, species, and evolution* (Harvard University Press, Cambridge, Mass., 1970).
2. J. Monod, *Chance and necessity* (Vintage Book, NY, 1971), p. 13.
3. G. L. Stebbins, *The basis of progressive evolution* (University of North Carolina Press, Chapel Hill, 1969).
4. L. Brillouin, *Science and information theory* (Academic Press, NY, 1962).
5. E. Wigner, in *Symmetries and reflections* (University of Indiana Press, Bloomington, 1967), p. 200.
6. J. Mehra, *Am. Scient.* **61,** 722 (1973).
7. M. Eigen and P. Schuster, *The hypercycle* (Springer-Verlag, Berlin, 1977).
8. J. D. Watson, *Molecular biology of the gene* (W. A. Benjamin, NY, 1970).
9. J. von Neumann, *Theory of self-reproducing automata*, ed. and completed by A. W. Burks (University of Illinois Press, Urbana, 1966).
10. M. A. Arbib, *Theories of abstract automata* (Prentice-Hall, Englewood Cliffs, NJ, 1969).
11. M. A. Arbib, in *Interstellar communication: scientific perspectives*, ed. C. Ponnamperuma and A. G. W. Cameron (Houghton Mifflin, Boston, 1974).
12. A. Turing, *Mind* **59,** 433 (1950). Turing's article has been reprinted in many anthologies, for example *The mind's I* by D. R. Hofstadter and D. C. Dennett (Basic Books, NY, 1981).
13. Turing's original paper has provoked an enormous literature; a number of articles on the Turing Test and its significance are reprinted in *The mind's I* (ref. 12). In addition, a few articles of interest are K. Gunderson, *Mind* **73,** 234 (1964); M. Scriven, *Mind* **62,** 230 (1953); the *Introduction to automata studies*, ed. C. E. Shannon and J. McCarthy; M. Gardner, *Scient. Am.* **224** (No. 6, June). 120 (1971); the articles on machine intelligence in *Dimensions of Mind*, ed. S. Hook (New York University Press, NY, 1960). One crucial point which these works discuss and which we have ignored is the length of time the question period lasts. Another point which must be considered is how clever and original do we wish the human in the sealed room to be? We have taken these points into account to some extent in our text, by requiring that the machine passing the WAP test to make original observations on WAP, where originality is judged with respect to the performance of human scientists. The nerve of the Turing Test as a criteria for mind, creativity, or intelligence is the idea that *all* intelligent performance is judged with reference to human performance in the corresponding categories, and if the performance of the machine is comparable to that of a human in *all* of the categories, the machine *must* be regarded as a 'person'.
14. F. Drake, *Technol. Rev.*, **78** (7), 22 (June 1976).
15. P. T. Landsberg, *Nature* **203,** 928 (1964).
16. E. P. Wigner and P. T. Landsberg, *Nature* **205,** 1307 (1965).
17. F. H. Stillinger, *Science* **209,** 451 (1980).
18. L. Pauling, *The nature of the chemical bond*, 2nd edn (Cornell University Press, Ithaca, NY, 1948).
19. L. Pauling, *General chemistry* (Freeman, San Francisco, 1956).

20. L. Pauling and R. Hayward, *The architecture of molecules* (Freeman, San Francisco, 1964).
21. J. T. Edsall and J. Wyman, *Biophysical chemistry*, Vol. I (Academic Press, NY, 1958).
22. T. R. Dyke, K. M. Mack and J. S. Muenter, *J. Chem. Phys.* **66,** 498 (1977).
23. M. Eigen, 'The origin of biochemical information', in *The physicist's conception of Nature*, ed. J. Mehra (Reidel, Dordrecht, 1972).
24. F. Drake, *Astronomy* **1** (No. 5, Dec.), 5 (1973).
25. A. E. Needham, *The uniqueness of biological materials* (Pergamon Press, NY, 1965).
26. A. Geiger, F. H. Stillinger and A. Rahman, *J. Chem. Phys.* **70,** 4185 (1979).
27. P. Schuster, G. Zundel, and C. Sandorfy (eds), *The hydrogen bond*, 3 vols (North-Holland, Amsterdam, 1976).
28. M. D. Joesten and L. J. Schaad, *Hydrogen bonding* (Marcel Dekker, NY, 1974).
29. A. Geiger, A. Rahman, and F. H. Stillinger, *J. Chem. Phys.* **70,** 263 (1979).
30. C. Pangali, M. Rao, and B. J. Berne, *J. Chem. Phys.* **71,** 2982 (1979).
31. F. Franks, in *Water a comprehensive treatise*, Vol. 4, ed. F. Franks (Plenum, NY, 1975), p.1.
32. A. Ben-Naim, *Hydrophobic interactions* (Plenum, NY, 1980).
33. A. L. Lehninger, *Biochemistry*, 2nd edn (Worth, NY, 1975).
34. W. Kauzmann, *Adv. Protein Chem.* **14,** 1 (1959).
35. C. Tanford, *The hydrophobic effect: formation of micelles and biological membranes* (Wiley, NY, 1980).
36. S. W. Fox and K. Dose, *Molecular evolution and the origin of life*, 2nd edn (Marcel Dekker, NY, 1977).
37. S. W. Fox, in *The nature of life: 13th Nobel Conference*, ed. W. H. Heidcamp (University Park Press, Baltimore, 1977).
38. F. G. A. Stone and W. A. G. Graham, *Inorganic polymers* (Academic Press, NY, 1962).
39. C. F. Fox, 'The structure of cell membranes', *Scient. Am.* (Feb.) 1972.
40. A. M. Buswell and W. H. Rodebush, in *Conditions for life*, ed. A. Gibor (Freeman, San Francisco, 1976).
41. D. Eisenberg and W. Kauzmann, *The structure and properties of water* (Oxford University Press, Oxford, 1969).
42. M. H. Hart, *Origins of Life* **9,** 261 (1979).
43. J. E. Lovelock, *Gaia: A New Look at Life on Earth* (Oxford University Press, Oxford, 1979).
44. G. Wald, *Scient. Am.* **191** (2), 45 (1954).
45. G. Wald, *Origins of Life*, **5,** 7 (1974).
46. G. Wald, Introduction to *Fitness of the environment*, by L. J. Henderson (Peter Smith, Gloucester, 1970).
47. G. Wald, *Proc. natn. Acad. Sci., U.S.A.* **52,** 595 (1964).
48. M. H. Hart, *Icarus* **33,** 23 (1978).
49. N. V. Sidgwick, *The chemical elements and their compounds*, Vols I and II (Oxford University Press, Oxford, 1950).

50. J. B. S. Haldane, *New Biology* **16,** 12 (1954).
51 H. C. Urey, *Proc. natn. Acad. Sci., U.S.A.* **38,** 351 (1952).
52. L. J. Henderson, *Fitness of the environment* (Macmillan, NY, 1913).
53. V. A. Firsoff, *Discovery* **23,** 36 (1962).
54. V. A. Firsoff, *Life beyond the Earth* (Basic Books, NY, 1963).
55. G. Feinberg and R. Shapiro, *Life beyond Earth: The intelligent Earthling's guide to life in the Universe* (Morrow, NY, 1980), p. 147.
56. G. C. Pimental, K. C. Atwood, H. Gaffron, H. K. Hartline, and T. H. Jukes, *Biology and the exploration of Mars*, ed. C. S. Pittendrigh, W. Vishniac, and J. P. T. Pearman (NASA, Washington, 1966).
57. H. H. Sisler, *Chemistry in non-aqueous solvents* (Reinhold, NY, 1961).
58. C. N. Matthews and R. E. Moser, *Proc. natn. Acad. Sci., U.S.A.* **56,** 1087 (1966).
59. C. N. Matthews and R. E. Moser, *Nature* **215,** 1230 (1967).
60. M. Nei and A. K. Roychoudhury, *Science* **177,** 434 (1972).
61. T. Dobzhansky, F. Ayala, G. Stebbins, and J. W. Valentine, *Evolution* (Freeman, San Francisco, 1977).
62. G. Wald, in *Horizons in biochemistry*, ed. M. Kasha and B. Pullman (Academic Press, NY, 1962).
63. C. Sagan, 'Life', in *Encyclopaedia Britannica*, 15th edn, Vol. 10 (Macropedia, 1974), p. 893.
64. H. Spencer, *Principles of biology* (rev. ed. 1909), p. 123.
65. J. H. V. Crowe and A. F. Cooper, *Scient. Am.* **225,** 30 (Dec. 1971).
66. J. D. Bernal, in *Theoretical and mathematical biology*, ed. T. H. Waterman and H. J. Morowitz (Blaisdell, NY, 1965).
67. R. Dawkins, *The selfish gene* (Oxford University Press, Oxford, 1977).
68. G. Wald, 'Life and light', *Scient. Am.* (Oct. 1959), repr. in *Conditions for life*, ed. A. Gibor (Freeman, San Francisco, 1976).
69. S. Brenner, personal communication (1981).
70. We are grateful to Professors B. Carter and J. Perdew for discussions regarding the scaling of atomic properties with the change of the fundamental constants. J. Perdew pointed out the scaling law (8.3) to us.
71. B. Carter, *Phil. Trans. R. Soc.* A **370,** 347 (1983); also in *The constants of Nature*, ed. W. H. McCrea and M. J. Rees (Royal Society, London, 1983).
72. Ref. 61, p. 87.
73. We are grateful to Professors D. Mohr and J. Maynard Smith for helpful discussions on the derivation of Carter's formula.
74. L. Margulis, *Symbiosis in cell evolution* (Freeman, San Francisco, 1981), p. 82.
75. Ref. 74, p. 92.
76. Ref. 74, p. 95.
77. Ref. 74, p. 97.
78. T. Dobzhansky, *Genetics of the evolutionary process* (Colombia University Press, NY, 1970).
79. D. H. Kenyon and G. Steinman, *Biochemical predestination* (McGraw-Hill, NY, 1969).

80. H. J. Morowitz, *Prog. Theor. Biol.* **1,** 35 (1967).
81. M. A. Ragan and D. J. Chapman, *A biochemical phylogeny of the protists* (Academic Press, NY, 1978), p. 204.
82. J. A. Bassham, jn *Plant biochemistry,* ed. J. Bonner and J. E. Verner (Academic Press, NY, 1965).
83. J. DeLey, *Evol. Biol.* **2,** 103 (1968).
84. Ref. 81, p. 41.
85. Ref. 81, p. 26.
86. Ref. 74, p. 324.
87. K. W. Jeon and M. S. Jeon, *J. Cell Physiol.* **89,** 337 (1976).
88. Ref. 74, p. 286.
89. L. Ornstein, *Physics Today* **35** (No. 3, March), 27 (1982).
90. The number of genes in the cow and fruit fly are found in ref. 72; the number of genes in *E. coli* and *Mycoplasma gallisepticum* are taken from Table 1.1 of ref. 83, where DeLey's estimate of 1000 nucleotides per gene is replaced by Dobzhansky *et al.*'s estimate of 1800.
91. G. F. R. Ellis and G. B. Brundrit, *Quart. J. R. astron. Soc.* **20,** 37 (1979).
92. F. J. Tipler, *Quart. J. R. atrom. Soc.* **22,** 133 (1981).
93. G. G. Simpson, *This view of life* (Harcourt Brace & World, NY, 1964), p. 252.
94. S. M. Stanley, *The new evolutionary timescale* (Basic Books, NY, 1981).
95. Ref. 43, pp. 69–76.
96. S. H. Schneider and S. L. Thompson, *Icarus* **41,** 456 (1980).
97. J. D. Pollack and Y. L. Yung, *Ann. Rev. Earth & Planet. Sci.* **8,** 425 (1980).
98. S. H. Schneider and S. L. Thompson, in *Life in the Universe,* ed. J. Billingham (MIT Press, Cambridge, Mass., 1981).
99. S. Chang, D. DesMarasis, R. Mack, S. L. Miller, and G. E. Strathearn, in *Earth's earliest biosphere: its origin and evolution,* ed. J. W. Schopf (Princeton University Press, Princeton, 1983).
100. J. Veizer, in *Earth's earliest biosphere,* ref. 99.
101. R. P. Turco, O. B. Toon, T. P. Ackerman, J. B. Pollack, and C. Sagan, *Science* **222,** 1283 (1983).
102. P. R. Ehrlich *et al., Science* **222,** 1293 (1983).
103. C. Sagan and E. Salpeter, *Astrophys. J. Suppl.* **32,** 737 (1976).
104. A. G. Cairns-Smith, *Genetic takeover and the mineral origins of life* (Cambridge University Press, Cambridge, 1982); *Seven clues to the origin of life: a scientific detective story,* (Cambridge University Press, Cambridge 1985).
105. Ref. 93, p. 262.
106. L. H. Ahrens, in *Physics and chemistry of the earth,* Vol. 5, ed. L. H. Ahrens, F. Press, and S. K. Runcorn (Pergamon Press, NY, 1964).
107. R. F. Doolittle, in *Science* **214,** 149 (1981), argues that some independent invention of enzymes is not as improbable as DeLey would have us believe; duplication would increase the probability; L. E. Orgel, *Proc. R. Soc.* B **205,** 434 (1978), makes a similar argument (we are grateful to J. Maynard Smith for this reference). Nevertheless, the results of Doolittle and Orgel if true, do not appear to significantly alter our calculations.

108. A. K. Dewdney, *Scient. Am.* **252,** 14 (March 1985).

109. B. Charlesworth, in *Observatory* **102,** 49 (1982), gives

$$t_D \approx t_g S^{-1}[(4N\mu)^{-1} + \ln(2N)]$$

where N is the number of individuals in the population and μ is the probability of occurrence of the mutation in a given gene. If we set $N = I$, then unless μ is very small, this expression will give the same estimate as (8.4).

110. F. B. Salisbury, *Nature* **224,** 342 (1969), argued that the enormous improbability of a given gene, which we computed in the text, means that a gene is too unique to come into being by natural selection acting on chance mutations. WAP self-selection refutes this argument, as R. F. Doolittle in *Scientists confront creationism*, L. R. Godfrey (Norton, NY, 1983) has also pointed out.

111. The invariance we have calculated in (8.1)–(8.3) does not include relativistic effects. However, as can be seen from our discussion of atomic structure in Chapter 5 (see pp. 295–300), an increase in the value of the fine structure constant can induce significant relativistic effects in large atoms because typical orbital velocities are of order αZc. Recently, H. J. Kreuzer, M. Gies, G. L. Malli, and J. Ladik, *J. Phys. A* **18,** 1571 (1985) have examined some of these effects in detail. They find that an increase in the fine structure constant by a factor 5 produces drastic changes in the Fe^{2+} and Fe^{3+} ions which play a key role in haemoglobin. Increases by factors 2.5 and 1.5 produce significant changes in the chemistry of cadmium and lead respectively.

9 The Space-Travel Argument Against the Existence of Extraterrestrial Intelligent Life

Do there exist many worlds, or is there but a single world? This is one of the most noble and exalted questions in the study of Nature.
St. Albertus Magnus

9.1 The Basic Idea of the Argument

. . . the way whereby one can learn the pure truth concerning the plurality of worlds is by aerial navigation [space-travel].
P. Borel (1657 AD)

The contemporary advocates for the existence of extraterrestrial intelligent life seem to be primarily astronomers and physicists, such as Sagan,[2] Drake,[3] and Morrison,[4] while most leading experts in evolutionary biology, for instance Dobzhansky,[5] Simpson,[6] Francois,[7] Ayala *et al.*[8] and Mayr,[9] contend that the Earth is probably unique in harbouring intelligence. We presented the evolutionists' argument against the existence of extraterrestrial intelligent life (ETI) in section 3.2, and Carter's WAP argument in section 8.7. In this chapter we shall present the so-called space-travel argument against the existence of ETI, an argument which one of us has developed at length in a number of publications.[1] Specifically, we shall argue in this chapter that the probability of the evolution of creatures with the technological capability of interstellar communication within five billion years after the development of life on an earthlike planet is less than 10^{-10}, and thus it is very likely that we are the only intelligent species now existing in our Galaxy. The basic idea of the space-travel argument is straightforward and indeed has led other authors, such as Fermi,[10] Dyson,[11] Hart,[12] Simpson,[6] Shklovskii,[101] and Kuiper and Morris,[13] to conclude that extraterrestrial intelligent beings do not exist anywhere in our Galaxy: if they did exist and possessed the technology for interstellar communication, they would also have developed interstellar travel and thus would already be present in our Solar System. Since they are not here,[14,15] this implies that they do not exist. Although this argument has been expressed before,—indeed, it was used in the *seventeenth* century to rule out intelligent life on the Moon[1]—its force does not seem to have been appreciated. We shall try to demonstrate its force by arguing that an intelligent species with the technology for interstellar communication would necessarily develop the technology for interstellar travel, and this would automatically lead to the exploration and/or colonization of our Galaxy in less than 300 million years.

It seems reasonable to assume that any intelligent species which develops the technology for interstellar communication must also have (or will develop in a few centuries) technology which is at least comparable to our present-day technology in other fields, particularly rocketry. This is actually a consequence of the Principle of Mediocrity[16] (that our own evolution is typical), which is usually invoked, particularly by Sagan,[85] in analyses of interstellar communication. This assumption about technological development is also an essential one to make if interstellar communication via radio waves is to be regarded as likely. If we do not assume that an advanced species knows at least what we know, then we have no reason to believe an advanced species would transmit radio waves, for they may never have discovered such things. In the case of rocket technology, the human species invented rockets some six hundred years before it was even aware of the existence of radio waves, and present-day chemical rockets can be regarded as natural developments of early rocket technology.

In addition to a rocket technology comparable to our own, it seems probable that a species engaging in interstellar communication would possess a fairly sophisticated computer technology. In fact, Sagan himself has asserted[17] that 'Communication with extraterrestrial intelligence ... will require ..., if our experience in radioastronomy is any guide, computer-actuated machines with abilities approaching what we might call intelligence'. Furthermore, the Cyclops[18] and SETI[19] proposals for radio telescopes to search for artificial extraterrestrial radio signals have required some quite advanced data-processing computers. We shall assume therefore that any species engaging in interstellar communication will have a computer technology which is not only comparable to our present-day technology, but which is comparable to the level of technology which we know is possible, which we are now spending billions of dollars a year to develop, and which a majority of computer experts believe we will actually possess within a century. That is, we shall assume that such a species will eventually develop a self-replicating universal constructor with intelligence comparable to the human level—such a machine should be developed within a century, according to the experts[20–22] (see section 3.2 for additional information supporting this opinion)—and such a machine combined with present-day rocket technology would make it possible to explore the Galaxy in less than 300 million years, for an initial investment less than the cost of operating a 10 MW microwave beacon for several hundred years, as proposed in SETI.[19] It is a deficiency in present-day computer technology, not rocket technology, which prevents us from beginning the exploration of the Galaxy tomorrow.

The above conclusions may seem to hinge on the motivations of advanced extraterrestrial intelligent beings, a subject about which we admittedly know nothing. However, we know *by definition* the motivations of the most interesting class of intelligent beings: those whose technology is far in advance of ours, and who are *interested in communicating with us*, or otherwise interacting with us. It is this class that most SETI programs are designed to detect, and it is the class—in the terminology of Chapter 8, the class of strongly intelligent beings—whose existence is made doubtful by the arguments we present here. We shall also argue that the interstellar exploration mechanism discussed here has so many uses besides contacting other intelligent beings that *any* technologically advanced species would be likely to use it, and hence if they existed, they should be here. In section 9.3 and in Chapter 10, we shall point out that the ultimate survival of a technological civilization, and indeed the survival of the biosphere in some form, requires the eventual expansion of the civilization into interstellar space. We gave upper bounds to the lifetime of a biosphere restricted to a single planet and a single solar system in Chapter 3. A civilization far in advance of ours is probably aware of this, and such awareness would provide a motivation to begin the colonization of space.

9.2 General Theory of Space Exploration and Colonization

If they existed, they would be here.
E. Fermi

In space exploration (or colonization), it is wise to adopt a strategy which maximizes the probable rate of information gained (or regions colonized) and minimizes the cost subject to the constraints imposed by the level of technology. Costs may be minimized in two ways: first, 'off-the-shelf' technology should be used as far as possible to reduce the research and development costs; second, resources which could be used for no other purpose should be utilized as far as possible. The resources available in uninhabited stellar systems cannot be utilized for any human purpose unless a space vehicle is first sent; indeed, from the economic viewpoint materials which cannot be utilized at all are valueless. Therefore, any optimal exploration strategy must utilize the material available in other stellar systems as far as possible. With present-day technology, such utilization could not be very extensive, but with the level of computer technology assumed in the previous section, these otherwise useless resources can be made to pay for virtually the entire cost of the exploration programme.

What one needs is a self-reproducing universal constructor: a machine capable of making any device, given the construction materials and a construction program. By definition, it is capable of making a copy of itself. Von Neumann has shown[23,24] that such a machine is theoretically possible, and in fact a human being is a universal constructor specialized to perform on the surface of the Earth. Thus the manned space exploration (and colonization) programme outlined in refs. 11, 12, and 13 is just a special case of an exploration strategy to be carried out by universal constructors. We discussed the theory of such machines in section 8.2.

The payload of a probe to another stellar system would be a self-reproducing universal constructor with human-level intelligence (we shall term such an interstellar probe a *von Neumann probe*) together with an engine for slowing down once the other stellar system is reached, and an engine for travelling from one place to another within the target stellar system—the latter could be an electric propulsion system,[25] or a solar sail.[26] The von Neumann probe would be instructed to search out construction material with which to make several copies of itself and the original probe rocket engines. Judging from observations of our own solar system,[27] what observations we have of other stellar systems,[28] and most contemporary solar system formation theories,[29] such materials should be readily available in virtually any stellar system—including binary star systems—in the form of meteors, asteroids, comets, and other debris from the formation of the stellar system. Recent observations of huge amounts of dust around Vega and other stars indicate that such materials are indeed present around many, if not all, stars. Whatever elements are necessary to reproduce the von Neumann probe, they should be available from some source in a stellar system. For instance, the material in the asteroids is highly differentiated; many asteroids are largely nickel-iron, while others contain large amounts of hydrocarbons.[27]

As the copies of the von Neumann probe are made, they would be launched at the stars nearest the target star. When these probes reached those stars, the process would be repeated, and repeated again until the probes had covered all the stars of the Galaxy. Once a sufficient number of copies had been made, the von Neumann probe would be programmed to explore the stellar system in which it finds itself, and relay the information gained back to the original solar system from which the exploration began. In addition, the von Neumann probe could be programmed to use the resources of the stellar system to conduct scientific research which would be too expensive to conduct in the original solar system.

It would also be possible to use the von Neumann probe to colonize the stellar system. Even if there were no planets in the stellar system—the system could be a binary star with asteroid-like debris—the von Neumann probe could be programmed to turn some of the available material into an *O'Neill colony*,[30] a self-sustaining human colony in space which is not located on a planet but is rather a space station. Inhabitants for the colony could be synthesized by the von Neumann probe. All the information needed to manufacture a human being is contained in the genes of a single human cell. Thus if an intelligent extraterrestrial species possessed the knowledge to synthesize a living cell—and some biologists claim[31,32] the human race could develop such knowledge within 30 years—they could program a von Neumann probe to synthesize a fertilized egg-cell of their species. If they also possessed artificial womb technology—and such technology is in the beginning stages of being developed on Earth[33]—then they could program the von Neumann probe to synthesize members of their species in the other stellar system. As suggested by Eiseley,[34] these beings could be raised to adulthood in the O'Neill colony by robots also manufactured by the von Neumann probe, after which these beings would be free to develop their own civilization in the other stellar system.

Suggestions have been made[35] that other solar systems could be colonized by sending frozen cells via space probe to the stars. But, it has not yet been shown[36–39] that such cells would remain viable over the long periods required to cross interstellar distances. This difficulty does not exist in the outlined colonization strategy above; the computer memory of the von Neumann probe can be made so that it is essentially stable over long periods of time. If it is felt that the information required to synthesize an egg cell would tax the memory storage space of the original probe, the information could be transmitted via microwave to the von Neumann probe once it has had time to construct additional storage capacity in the other solar system. The key point is that once a von Neumann probe has been sent to another solar system, the entire resources of that solar system become available to the intelligent species which controls the von Neumann probe; all sorts of otherwise prohibitively expensive projects become possible to carry out. It would even be possible to program the von Neumann probe to construct a very powerful radio beacon with which to signal other intelligent species! A number of scientists, for instance G. O'Neill[96] and R. A. Freitas[97–99] have independently suggested that self-reproducing probes are the most efficient way to contact ETI. Freitas' articles contain a quite detailed analysis.

Hence the problem of interstellar travel has been reduced to the problem of transporting a von Neumann probe to another stellar system. This can be accomplished even with present-day rocket technology. For example, Hunter[40,41] has pointed out that by using a Jupiter swing-by to approach the Sun and then adding a velocity boost at perihelion, a solar system escape velocity v_{es} of about 90 km/sec ($\sim 3 \times 10^{-4}c$, where c is the speed of light) is possible with present-day chemical rockets, even assuming the launch is made from the surface of the Earth. As pointed out in references 28 and 29, most other stars should have planets (or companion stars) with characteristics sufficiently close to those of the Jupiter–Sun system to use this launch strategy in reverse to slow down in the other solar system. The mass ratio μ (the ratio of the payload mass to the initial launch mass) for the initial acceleration in the swing-by would be 10^3, so the total trip would require $\mu < 10^6$ (less than, since the 10^3 number assumed an Earth surface launch); quite high, but still feasible. (With Jupiter swing-by only, the escape velocity would be about $1.6 \times 10^{-4}c$ with $\mu = 10^3$.) For comparison, we note that Voyager spacecraft will have[42] a solar escape velocity of about $0.6 \times 10^{-4}c$ with $\mu = 850$.)

Thus it seems reasonable to assume that any intelligent species would develop at least the rocket technology capable of a one-way trip with deceleration at the other stellar system, and with a travel velocity v_{es} of $3 \times 10^{-4}c$. At this velocity the transmit time to the nearest stars would be between 10^4 and 10^5 years. This very long travel time would necessitate a highly developed self-repair capacity, but this should be possible with the level of computer technology assumed for the payload.[43] In addition, nuclear power-sources could be developed which would supply power for that period of time. However, nuclear power is not really necessary. If power utilization during the free-fall was sufficiently low, even chemical reactions could be used to supply the power. Since v_{es} is of the same order as the stellar random motion velocities, sensitive guidance would be required, but this does not seem to be an insuperable problem with the assumed level of computer technology.

Because of the very long travel times, it is often claimed[44] that interstellar probes would be obsolete before they arrived. However, in a fundamental sense a von Neumann probe *cannot* become obsolete, since it is a universal constructor. The von Neumann probe can be given instructions by radio about how to make the latest devices after it arrives at the destination star.

Restricting consideration to present-day rocket technology is probably too conservative. It seems likely that an advanced intelligent species would eventually develop rocket technology at least to the limit which we regard as technically feasible today. For example, the nuclear pulse rocket of the Orion Project pictured[45] a solar escape velocity v_{es} of $3 \times 10^{-2}c$ with $\mu = 36$ for a one-way trip and deceleration at the target star. The cost of the probe would be $\$4 \times 10^{12}$ at 1985 prices, almost all of the money being for the deuterium fuel. This is approximately the present GNP of the United States. Project Daedalus,[43] the interstellar probe study of the British Interplanetary Society, envisaged a stellar fly-by via nuclear pulse rocket (no slow-down at the target star), with $v_{es} = 1.6 \times 10^{-1}c$, $\mu = 150$, and a cost of $\$10^{12}$. As before, almost all the cost is for the helium-3 fuel (at 1960 prices). With slow-down at the target star, $\mu = 2 \times 10^4$ and the cost would be $\$2 \times 10^{14}$, or almost 100 times the United States GNP, and it would require centuries to extract the necessary helium-3 from the helium source proposed in the Daedalus study, the Jovian atmosphere.

The cost of such probes is far beyond the means of present-day civilization. However, in the above estimates almost all the cost is for the rocket fuel. Building the probe itself and testing it would cost relatively little. A possible interstellar exploration strategy would be to design a probe capable of $v_{es} = 0.1c$, record the construction details in a von Neumann probe, launch the probe payload via a chemical rocket at $3 \times 10^{-4}c$ to a nearby stellar system, and program the machine to construct and fuel several high-velocity ($0.1c$) probes with von Neumann payloads in the other system. When the probes reach their target stars, they would be programmed to build high-velocity probes, and so on. In this way the investment on interstellar probes by the intelligent species is reduced to a minimum while maximizing the rate at which the Galaxy is explored. (The von Neumann probe could conceivably be programmed to develop the necessary technology in the other stellar system. This would reduce the initial investment even further.) The disadvantage, in a 10^4 year transit time is the fact that for 10^4 years, there is no information on other stellar systems reaching the original solar system. There is a trade-off between the cost of the first probe and the time interval the intelligent species must await before receiving any information on the other stellar systems. But with second generation probes with $v_{es} = 0.1c$, new solar systems would be explored at the rate of several per year by 10^5 years after the original launch. The intelligent species launching the original probe need only be patient and launch a sufficient number of initial probes at $v_{es} = 3 \times 10^{-4}c$ so that at least one succeeds in reproducing itself (or in making a high-velocity probe) several times. This number will of course depend on the failure rate. Project Daedalus[43] aimed at a mission failure rate of 10^{-4}, and the designers believed that such a failure rate was feasible with on-board repair. If we adopt this failure rate and assume failures to be statistically independent, then only three probes need be launched to reduce the failure probability to 10^{-12}. Judging by contemporary rocket technology, the cost of the initial low-velocity probes would be less than $\$1 \times 10^{10}$ each, since von Neumann probes could make themselves and the original research and development costs would be small—intelligent self-reproducing machines would originally be developed for other purposes.[46] Thus the exploration of the Galaxy would cost about 30 billion dollars, approximately the cost of the Apollo program.

These costs—$\$3 \times 10^{10}$ for a low-speed probe and $\$2 \times 10^{14}$ for a high-speed one—seem quite large to us, but there is evidence that they would not seem large to a member of a civilization greatly in advance of ours. As we pointed out in section 3.7, the cost relative to wages of raw materials, including fuel, has been dropping exponentially with a time constant of 50 years for the past 150 years. If we assume this trend continues for the next 400 years (the reasons for believing that it will continue were discussed in section 3.7; Newman and Sagan[62] believe it will continue for the next 1000 years), then to an inhabitant of our own civilization at this future date, the cost of a low-velocity probe would be as difficult to raise as 10 million dollars today, and the cost of a high-velocity probe would be as difficult to raise as 70 billion dollars today. The former cost is easily within the ability of a large number of individuals today. There are today at least 100,000 Americans who are worth 10 million dollars and the Space Telescope project budget exceeds $\$10^9$. If the cost trend continues for the next 800 years, then the cost of a $\$3 \times 10^{10}$ probe would be as difficult to raise as $4000 today; an interstellar probe would appear to cost as much then as a home computer does now. Tens of millions of people could afford one. In such a society, *someone* would almost certainly build and launch a probe.

To maximize the speed of exploration and/or colonization, one must minimize $[(d_{av}/v_{es}) + t_c]$, where d_{av} is the average distance between stars and t_c is the time needed for the von Neumann probe to reproduce itself. The time t_c will be much larger for $v_{es} = 0.1c$ probes than for $10^{-4}c$ probes. We would guess the minimum to be obtained for $v_{es} = 5 \times 10^{-2}c$ and $t_c = 100$ years. With $d_{av} = 5$ lyr, this gives a rate of expansion of 2.5×10^{-2} lyr/yr, and thus the Galaxy could be explored in 4 million years. Here, we shall be conservative and assume only present-day rocket technology, which would give an expansion rate of 3×10^{-4} lyr/yr, and such a rate would complete the exploration of the Galaxy in 3×10^8 years.

The travel time between stars will equal the expansion rate provided $d_{av}/v_{es} \gg t_c$, or $t_c \ll 10^3$ yr. This seems a reasonable condition when we compare von Neumann probes with the only highly intelligent, self-reproducing machines of our experience, namely human beings. In their natural environment humans have a $t_c \sim 20$–30 yr. If we compare a von Neumann probe to an entire technical civilization, then $t_c \sim 300$ yr for the time required to build up the United States into an industrial nation. Most of this time was required to develop not the hardware but rather the knowledge of which machines to build. Possessing the necessary knowledge, Germany and Japan rebuilt their industries in a decade after World War II, requiring only minor investments from outside. As for the t_c for space industries, G. O'Neill estimates[30] that space colonies could be self-sufficient and able to make more colonies in less than a century. Such a rapid space colony construction rate might require a large initial investment from Earth, and this might correspond to a very large (i.e., expensive) probe payload. As before, an intelligent species can reduce the initial investment by building an initial probe small, but programmed to construct larger probes in the target systems. It seems unlikely that a Project Daedalus size payload ($\sim 10^3$ tons), which appears to have most of the essential equipment of a von Neumann probe, would require longer than 10^6 yr to reach the large-scale-probe-making stage, and with this upper bound the above estimate for the time needed to explore the Galaxy is valid. For comparison, recall that modern man, *Homo sapiens*, has been in existence for about 4×10^4 years. (See Chapter 3.)

Once the exploration and/or colonization of the Galaxy has begun, it can be modelled quite closely by the mathematical theory of island colonization—a theory first developed by MacArthur and Wilson[47,48]—since the islands in the ocean are closely analogous to stars in the heavens, and the von Neumann probes are even more closely analogous to biological species. There are several general conclusions applicable to interstellar exploration and/or colonization which follow from the MacArthur–Wilson theory. First, there are two basic behavioural strategies, the r-strategy and the K-strategy, which could be adopted in different phases of the colonization; (r is the net reproductive rate [per capita births minus deaths], and K is the carrying capacity of the environment.) The r-strategy is one which emphasizes rapid reproduction. It is used by species inhabiting a rapidly changing environment, or an environment in which it is crucial to exclude competitors by occupying niches as quickly as possible. Thus it seems likely that an r-strategy would be followed in the early stages of the colonization. The K-strategy on the other hand, is the one followed by species inhabiting a slowly changing environment, or one in which the niches are already occupied by other members of the same species, and there is competition within this species for the occupied niches. We would therefore expect the K-strategy to be adopted after the solar system had been colonized for some time, and this strategy would result in fewer probes being sent to other stars. Second, the MacArthur–Wilson theory suggests[49] that the fraction of probes reaching a distance d from the system of launch is $\sqrt{2\pi}[\exp(-d^2/2]/d$. This means that even with random dispersal, probes would be expected to be sent not just to nearby solar systems, but also to far distant ones, though distant solar systems would be less likely targets than nearby ones.

It is important to realize that the MacArthur–Wilson theory must be modified before it can be applied to the problem of interstellar exploration/colonization. The MacArthur–Wilson theory assumes that the dispersal of colonizers is *random*, while the dispersal of von Neumann probes would be intelligently directed. The von Neumann probes can use radio waves to determine which nearby stars have already been reached by other probes, and launch descendant probes only at those stars which have not yet been reached; at least they can follow this strategy on the colonization frontier. Animal colonizers do not have an analogous ability to learn about uninhabited but habitable islands, and so they must use a random search strategy. This also means that a diffusion model[50,51] of interstellar colonization would not be completely accurate. Diffusion can be viewed as expansion against resistance, and there would be no resistance to the expansion of the volume of stars colonized by the von Neumann probes. In the case of the diffusion of gas molecules, the diffusing molecules collide with molecules of the ambient gas, and this leads to (in the usual Brownian motion derivation of the one-dimensional diffusion equation) an equally great probability of going backward as forward from a given collision site. Picture a one-dimensional array of collision points (stellar systems). The von Neumann probe at x_i would be programmed to send probes to all nearby unoccupied points (in the interval x_{i-r} to x_{i+r}, say), concentrating first on a probe to point x_{i+1}, the nearest neighbour in the forward direction. (The probe will have a memory of having arrived from the x_{i-j} point ($j \geqslant 1$), so the direction is defined.) If the reproductive failure rate of the probe at x_i is neglected, then with probability one the motion will be forward to x_{i+1}, x_{i+2}, etc., at a rate greater than or equal to $[(d_{av}/v_{es})+t_c]$. By adjusting r (that is, by adjusting the net probe reproduction rate), the effect of the failure rate can be cancelled out. This analysis can be immediately generalized to three dimensions. The expansion speed in three-dimensions would still be $[(d_{av}/v_{es}+t_c]$, at least in the later stages of expansion. (The earlier stages of expansion might be dominated by t_c, since there are more than two nearest neighbours. However, for t_c upper bounds like those given above, the timescale for expansion throughout the Galaxy would be dominated by the properties of its later stages.) In summary, we would expect the *initial* colonization of space to be much more like the free expansion of a gas into a vacuum, rather than like the diffusion of one variety of gas through another, or like the diffusion of a coloured liquid through a non-coloured liquid. Free expansion is much, much more rapid than diffusion. Subsequent colonization of a previously colonized region, if it occurs, could closely resemble diffusion, for there would be resistance to the colonization by the descendants of the first probes. But there is no reason to expect such interstellar imperialism. Indeed, if the probes are sent out for exploratory purposes, it would be pointless. Even if such imperialism does occur, it would not change the fact that the colonization frontier would be expanding freely rather than diffusing. Furthermore, the existence of such imperialists would motivate the colonizers on the frontier to speed up their occupation of previously unoccupied solar systems, in order to prevent the imperialists from seizing them. The rapid conquest of central Africa in the late nineteenth century by the European powers was driven by such a motivation. Germany began occupying parts of one section of Africa, which previously no European nation cared to control. The other powers thereupon began their movement into this section in order to prevent the Germans from occupying it all. Another example would be the occupation of Oklahoma territory by settlers virtually overnight after the region was thrown open to settlement by the United States government. Since whoever first reached the land in Oklahoma owned it thereafter, there was a strong motivation to occupy it as rapidly as possible, and develop it afterwards. This is an instance of an initial r-strategy being replaced later by a K-strategy.

9.3 Upper Bounds on the Number of Intelligent Species in the Galaxy

> Absence of evidence is not evidence of absence.
>
> M. Rees

In most discussions, the probability that intelligent life which eventually attempts interstellar communication will evolve in a star system is expressed by the Drake equation:[52]

$$p = f_p n_e f_l f_i f_c \tag{9.1}$$

where f_p is the probability that a given star system will have planets, n_e is the number of habitable planets in a solar system that has planets, f_l is the probability that life evolves on a habitable planet, f_i is the probability that intelligence evolves on a planet with life, and f_c is the probability that an intelligent species will attempt interstellar communication within 5 billion years after the formation of the planet on which it evolved. The time limit in f_c is only tacit in most discussions of extraterrestrial intelligence. However, some time period which is short compared with the age of the universe must be assumed if the Drake equation is to yield a number of existing civilizations which is significantly greater than one. If, for example, f_c were a Gaussian distribution with peak at 30 billion years and a standard deviation of $\sigma = 1$ billion years, then ours would be the only civilization in the Galaxy. Most discussions of ETI probability are based on the Principle of Mediocrity, as we mentioned earlier. Since our evolution to technological ability occurred within five billion years after the formation of our planet, it follows from this principle that other technological species would typically evolve in a similar time period. We shall adopt the Principle of Mediocrity implication for the evolutionary timescale in this section, and show that, when combined with the interstellar travel assumption, it implies ETI to be quite rare in our Galaxy. The estimates made below for the number of intelligent species in our Galaxy will hold if it is assumed that f_c is either sharply peaked at 5 billion years after planetary formation or a Gaussian distribution with $t_{peak} < 6$ billion years and $\sigma > 1$ billion years. As we pointed out in sections 3.2 and 8.7, however, the Principle of Mediocrity is probably *not* true; the argument of this section is a *reductio ad absurdum* argument.

The problem in applying the Drake equation is that only f_p—and to a lesser degree n_e—is subject to experimental determination. In order to calculate a probability with a high degree of confidence, one must have a fairly large sample; for f_l, f_i, and f_c we have only one sample point, the Earth. However, if one accepts the argument which we developed in the previous section that any intelligent species which attempts interstellar communication will begin the Galactic exploration programme outlined within 100 years of developing the technology for interstellar communication, then the sample size is enlarged to include all those stellar systems older than $t_{age} = 5$ billion years $+ t_{ex}$, where $t_{ex} \leqslant 300$ million years is the time needed to expand throughout the Galaxy. That is, the Drake probability p is less than or equal to $1/N$, where N is the number of stellar systems older than t_{age}, because all of these stars were, under the assumptions underlying the Drake equation, potential candidates to evolve communicating intelligent species, yet they failed to do so—had such species evolved on planets surrounding these stars within 5 billion years after star formation, their probes would already be present in the solar system, and these probes are surely not here.[14,15] Since f_p and n_e can, at least in principle, be determined by direct astrophysical measurement, the fact that extraterrestrial intelligent beings are not present in our solar system permits us to obtain a direct astrophysical measurement of an upper bound to the product $f_l f_i f_c$, which depends only on biological and sociological factors.

This argument assumes that the five probabilities in the Drake equation do not vary rapidly with Galactic age. The available astrophysical evidence and most theories of the formation of solar systems indicate that this assumption is probably valid. The formation of solar systems requires that the interstellar gas be sufficiently enriched by 'metals'). Most experts[29,53–55] agree that a substantial fraction of existing 'metals' (in astrophysical parlance, a 'metal' is any element heavier than hydrogen or helium) were formed in massive stars very early in Galactic history—during the first 100 million years of the Galaxy's existence—and the metal abundance has changed by at most a factor of about two since then. The evidence[56,57] gives a Galactic age between 13 and 20 billion years. What evidence there is suggests[54] that the rate of star formation has been decreasing exponentially ever since the initial burst of heavy element formation. However, existing stellar formation theory is unable to determine definitely if the so-called initial mass function—the number of stars formed per unit time with masses between m and $m + \Delta m$—changes with time after the initial burst of massive stars.[53] Furthermore, it is not clear to what extent the formation rate of earthlike planets depends on the metal abundance.[58,59] However, the observational evidence[53] (such as it is) does not indicate a large variation in the initial mass function or in the earthlike planet formation rate with time. We shall assume that these are roughly constant, and most discussions of extraterrestrial intelligence also make this assumption.[60,61] The factors f_l, f_i, f_c probably do not depend strongly on the evolution of the Galaxy as a whole, and so can be regarded as constants. Since the Galaxy is between 13 and 20 billion years old, the number, Q, of stars older than 5.3 billion years is about twice the number of stars formed after the Sun, and so is approximately equal to the number of stars in the Galaxy, $N \sim 10^{11}$. Thus $p \lesssim Q^{-1} \approx 10^{-11}$. If we accept the usual values of $f_p = 0.1$–1 and $n_e = 1$ found in most discussions of interstellar communication,[2,18] then $f_l f_i f_c \lesssim 10^{-10}$. The number, N, of *communicating* civilizations now existing in *our Galaxy* is less than or equal to $p \times$(number of stars in the Galaxy)$\simeq 1$; that is to say, probably only us.

This conclusion that we are probably the only technical civilization, willing and able to communicate, now existing in the Galaxy does not depend on any biological or sociological arguments except for the assumption that a communicating species would evolve in less than 5 billion years and would eventually begin interstellar travel; nor does it depend on f_p or n_e. This lack of dependence of the conclusion on f_p and n_e is very important, for we actually have no experimental evidence on the value of either, although the Space Telescope may eventually provide us with some. The fact that our Solar System has planets, and an earthlike planet, could be a WAP selection effect: our type of life must evolve on such a planet, so it is no surprise that we have. The conclusion $N \sim 1$ follows from just the interstellar travel assumption, the assumption that the Galactic environment has not changed by more than a factor of five during the history of the Galaxy, and the fact that extraterrestrial probes are not present in our solar system.

If the Galactic age is at the upper limit of 18×10^9 yr or older, then we can conclude that we are the only technological species which now exists in the Galaxy around main sequence stars of spectral type earlier than G3, if we assume that such a species will develop interstellar travel before its star leaves the main sequence. (If the foreseeable destruction of its solar system does not motivate a species to develop interstellar travel, it is hard to imagine what would.) Stars in these spectral classes will leave the main sequence in about 13×10^9 yr or earlier, and so by the argument presented above, the number of species around such stars is less than $13/(20-13.3)\sim2$.

For simplicity the above discussion was based on the Drake equation, but it should be clear that the same arguments can be used with any other plausible equation for the number of communicating species in the Galaxy, with much the same results.

The result $N\sim1$ is crucially dependent on the timescale for exploring and/or colonizing the Galaxy being short compared with the Galaxy's age. The free-expansion model for exploration/colonization which we developed in section 9.2 yielded 3×10^8 years as an upper bound for the time t_G needed to cover the entire Galaxy, but it is clear that the conclusion $N\sim1$ will not be changed provided that t_G is less than 2 billion years.

The only alternative to the free-expansion model is the diffusion model, which, as we indicated in section 9.2, probably would not give an accurate picture of the first colonization wave of intelligent life, though it may well constitute an accurate representation of secondary colonization through an already-colonized Galaxy. Nevertheless, a number of authors, most notably Newman and Sagan,[51,62] have applied the diffusion model to the first colonization of the Galaxy. In view of our complete lack of knowledge about the actual colonization of space, it behooves us to entertain the possibility that the diffusion model could be correct.

The Newman–Sagan model is the most sophisticated diffusion model developed to date. It is based on a non-linear diffusion equation which is capable of taking into account, at least partially, the asymmetry between forward and backward motion which we mentioned in section 9.2. Newman and Sagan normalized the diffusion coefficients appearing in their diffusion equation as best they could by referring to the available data on animal and human colonization. They concluded that the effective rate of expansion would be dominated by t_c, the time between the first arrival of the probe in a target system and the launch of its first descendant probe. This is in contrast to the conclusion in section 9.2, where we contended that the transit time between the stars (d_{av}/v_{es}) would be the dominant factor. The difference in conclusions is due to a difference in assumption: we assumed that the colonists (or probes) would follow an r-strategy initially and switch to a K-strategy as the system became more populated, while Newman and Sagan assumed a K-strategy would be followed at all times. A K-strategy leads to a slower rate of colonization/exploration; with a K-strategy, a descendant probe would not be launched until the colonized stellar system was fully populated. As we mentioned in section 9.2, the r-strategy would be the one adopted by a species sending out *probes*, for they would wish to explore as fast as possible. Furthermore, the r-strategy would adopted by a species which wished to stake territorial claims. Newman and Sagan also assumed a somewhat higher v_{es} than we did (the value we selected in section 9.2 was deliberately an extremely conservative lower bound).

In spite of these quite different assumptions, the Newman–Sagan analysis yielded a value $t_G \lesssim 7.5 \times 10^8$ years,[62,63] as compared with our value of $t_G \lesssim 3 \times 10^8$ years. These estimates for t_G are the same to within a factor of 3, a fact which we regard as extremely significant: *the general conclusion that the Galaxy would be completely explored and/or colonized within one billion years after the first appearance of a technological species which embarks on such a programme, is essentially model-independent.* Thus, since 1 billion years is quite short in comparison with the age of the Galaxy, it follows from the absence of ETI in our Solar System that such space-travelling ETI apparently do not exist, and have never existed in our Galaxy.

It is important to note that the above argument uses the *observed evidential* fact that the ETI are not present in our solar system; the situation is *not* the one implied by the epigram to this section, 'absence of evidence is not evidence of absence'. Rather, the evidence *is* that ETI are absent from our Solar System,[14,15] and from this *observed* fact (and other astrophysical observations and theories) it is inferred as a logical consequence that ETI are absent from the Galaxy.

Newman and Sagan do not accept this implication (which they acknowledge) of their own analysis; they seek to avoid their own conclusion by arguing that an advanced technological species would not be motivated to explore and/or colonize the entire Galaxy. It is to these interesting questions of motivation that we now turn.

9.4 Motivations for Interstellar Communication and Exploration

> The individual is the true reality in life. A cosmos in himself, he does not exist for the State, nor for that abstraction called 'society', or for the 'nation', which is only a collection of individuals. Man, the individual, has always been and, necessarily is the sole source and motive power of evolution and progress.
>
> Emma Goldman

It is difficult to construct a plausible scenario whereby an intelligent species develops and retains for centuries an interest in interstellar communication together with the technology to engage in it, and yet does not begin interstellar travel. Even if we adopt the pessimistic point of view that all intelligent species cease communication efforts before developing von Neumann probes, either because of a loss of interest or because they blow themselves to bits in a nuclear war, the conclusion that we are the only intelligent species in the Galaxy *with interest in interstellar communication* is not changed. For in this case, the longevity L of a communicating civilization is less than or equal to 100 years (if we use our computer experts' opinions for the time needed to develop von Neumann probes), and since the Drake equation gives $N = R_\star pL$ for the number of communicating civilizations in the Galaxy, we obtain $N \sim 10$, even if we accept Sagan's optimistic estimate[2] of $R_\star p = 1/10$. (The number $R_\star$ is the average rate of star formation.) This value of N is essentially the same as the estimate $N \sim 1$ obtained in the section 9.3, and in any case such short-lived civilizations would on the average be too far apart and would exist for too short a time to engage in interstellar communication. (If $L \geqslant 100$ years so that the species has sufficient time to develop probe technology, the value of L is irrelevant to the calculation of the number p. Once the probes have been launched, they will explore the Galaxy automatically; the destruction of the civilization that launched them would not stop them.) We are thus left with the possibility that for some reason, intelligent beings with the technology and desire for radio communication do not use the exploration strategy because they *choose* not to do so, rather than because they are incapable of developing the probe technology.

There is no good reason for believing this is true. Virtually every reason for engaging in interstellar radio communication provides an even stronger argument for the exploration of the Galaxy. For example, if the motivation for communication is to exchange information with another intelligent species, then as Bracewell[65,66] has pointed out, contact via space-probe has several advantages over radio waves. One does not have to guess the frequency used by the other species, for instance. In fact, if the probe is a von Neumann probe, then the probe could construct an artefact in the solar system of the species to be contacted, an artefact so noticeable that it could not possibly be overlooked. If nothing else, the von Neumann probe could construct a 'Drink Coca-Cola' sign a thousand miles across and put it in orbit around the planet of the other species. Once the existence of the probe has been noted by the species to be contacted, information exchange can begin in a huge variety of ways. Using a von Neumann probe obviates the main objection[67] to interstellar probes as a method of contact, namely the expense of putting a probe around each of an enormous number of stars. One need only construct a few probes, enough to make sure that at least one will succeed in making copies of itself in another solar system. Probes will then be sent automatically to the other stars of the Galaxy, with no further expense to the intelligent species constructing the first probe.

Philip Morrison, one of the leading proponents of contacting extraterrestrial technological species via radio, has expressed the opinion:[4] '... once there is really interstellar communication, it may be followed by a ceremonial interstellar voyage of some special kind, which will not be taken for the sake of the information gained, or the chances for trade ..., but simply to be able to do it, for one special case, where there is a known destination. That's possible, one can imagine it being done—but it is very unlikely as a search procedure'. However, if it is granted that a *single* probe is launched, for *any* reason, then if the probe is a von Neumann probe, this single probe can be used to start the Galactic expansion programme outlined in section 9.2. While *en route* to a solar system known to be inhabited, a von Neumann probe could make a stop-over at a stellar system along the way, make several copies of itself, refuel and then proceed on its way (or send one of the copies to the inhabited system). If the target system is further away than 100 light years from the sending system, and if $v_{es} \leq 0.1c$ and $t_c \leq 100$ years, then the time needed to reach the inhabited system would be increased by less than 10%, and one would obtain the exploration and/or colonization of the entire Galaxy as a free bonus. Furthermore, because any inhabited system is likely to be quite a way from another inhabited system (Sagan's estimate is 600 lyr), *any* probe sent would have to be autonomous, which would mean a computer with human-level intelligence, and capable of self-repair—and this means that it would essentially be a von Neumann probe. Since its instrumentation necessarily makes *any* single interstellar probe capable of exploring the entire Galaxy, why not use it for that?

Since there was a time in the past when there was no life at all in the Galaxy, the probability is very high that there was a first intelligent species to evolve. (The only other possibility is for several such species to emerge simultaneously, which is exceedingly unlikely.) Consider the search strategy adopted by this *first* species interested in interstellar communication to evolve in our Galaxy. It is probable it would be thousands or even millions of years before another such species arose. Even if another species arose simultaneously, the probability is only about 10^{-6} that it would be within 100 light years of the first species. Therefore, when the first species begins to signal, it will probably get no answer for thousands or even millions of years. During the time it is sending fruitlessly, it will be receiving no information on other stellar systems for its investment. If there remains strong interest in interstellar communication during this entire period, why should it not also launch a few probes? *Some* information on other systems would be guaranteed in 100 to 10^4

years, even if other intelligent beings are not discovered by either the probes or the radio receivers. Also, if there are other intelligent beings in the Galaxy, the von Neumann probes will eventually find them, even if they are intelligent beings who would never develop on their own an interest in interstellar communication or travel. With radio waves and a null result, there is always the possibility that the wrong frequency has been chosen, that some means other than radio waves has been used by the other species. There is no such uncertainty with probes.

If human history is any guide, this first species will launch a probe rather than construct radio beacons in the first place. In the early part of this century, when the American astronomer Percival Lowell had convinced many that there were intelligent beings on Mars, but when interplanetary rocket probes were regarded as a ridiculous fantasy, the Harvard astronomer W. H. Pickering pointed out[68] that communication with these beings was possible with a mirror one-half square mile in area: '[it] would be dazzlingly conspicuous to Martian observers, if they were intellectually and physically our equals'. If we were content to use such a device to learn about Mars from the hypothetical Martians, we would still know virtually nothing about Mars. Instead, we sent robot probes, and Sagan's recent proposal[69] for advanced Mars probes are robots with manipulative ability and a considerable degree of artificial intelligence—they are a step in the direction of a von Neumann probe.

If we *assume* that a behaviour pattern which is typical not only of *Homo sapiens* but also of all other living things *without exception* on our planet would also be adopted by any intelligent species (this assumption has the support of *all* the experimental evidence; if we deny this assumption, we have really nothing at all to go on except opinion), then we would conclude that a sufficiently advanced intelligent species would launch either von Neumann probes, or colonization ventures of some type. All living things have a dispersal phase,[70] in which they tend to expand into new environments, for obviously the dispersal pattern is dictated by natural selection. The expansion is generally carried out to the limit allowed by their genetic constitution. In intelligent species, this limit would be imposed by the level of technology,[71,72] and we would expect the dispersal behaviour pattern to be present in at least some groups of an intelligent species, for those groups which do not exhibit this behaviour would be selected against. We should therefore expect that at least some groups of the species would attempt an expansion into the Galaxy, and the launch of only one successful von Neumann probe would be sufficient for such expansion to cover the Galaxy. By launching such a probe and using it to colonize the stars, a species increases the probability that it will survive the death of its star, nuclear war, or other catastrophes. Note that it need not take territory away from another species (intelligent or not) to accomplish this purpose of indefinite survival. The species could, for example, restrict itself to the construction of O'Neill colonies around stars with no natural living things about them.

In many articles (e.g. ref. 62), the colonization of the Galaxy by the first species to engage in interstellar travel is referred to as 'imperialism', with the intent of investing interstellar colonization with the negative moral overtones which the concept of imperialism rightly carries. However, *Encyclopaedia Britannica*[100] defines 'imperialism' to be 'the policy of a state aiming at establishing control beyond its borders over people unwilling to accept such control'. Thus colonization by itself does not constitute imperialism; an imperialist is one who invades territory which is already occupied by someone else. The Portuguese pilot Diogo de Silves discovered the Azores in 1427. No trace of human beings were found on any of these islands, nor was any evidence found of their having been visited before.[86] Thus when the Portuguese colonists began arriving in 1432, they were *not* imperialists, for they were not taking over someone else's occupied territory. On the other hand, when Philip II of Spain took possession of Portugal in 1580, the people of the Azores fiercely resisted Spanish rule. The Spaniards, led by the Marquess de Santa Cruz, who subdued them in 1582, *were* imperialists, for they *were* taking over territory previously occupied by someone else.

Even if we replace the word 'people' in the above definition of imperialism by the words 'any life-form whatsoever' the colonization of the Galaxy would not constitute imperialism, for the evidence of our own Solar System indicates that most planets are uninhabited, even by single-celled organisms. Certainly the asteroids and comets around stars would be uninhabited. Colonizing completely uninhabited areas does not take territory away from any living being, and thus cannot be considered imperialism.

But the same argument would indicate that the colonization of all planets in our Solar System besides the Earth by ETI would not be imperialism: the other planets are presently dead rock and gas, and have been so for 4.5 billion years. Since the other planets are not now inhabited, colonists have never entered our Solar System. Some writers, in particular Newman and Sagan,[62] seem to feel the colonization of the uninhabited planets in a solar system which has an inhabited planet to be imperialistic, but the colonization of a totally uninhabited system would not be. We do not see how this distinction can be justified. It is true that colonization of the uninhabited planets in the inhabited system would prevent the native intelligent species from eventually colonizing these planets, but it is also true that the colonization of solar systems with no native life would prevent other species from other systems from colonizing them.

It is possible that an intelligent species which develops a level of technology capable of interstellar communication would decide not to build von Neumann probes because they would be afraid that they would lose control of these machines. Since no reproduction can be perfect, it is quite possible that the program which keeps the von Neumann probes under the control of the intelligent species could be accidently omitted during the reproduction process, with the result that the copy goes into business for itself. This problem can be avoided in at least three ways. First, the program which keeps the probe under control can be so integrated with the total program that its omission would cause the probe to fail to reproduce. This is precisely analogous to the constraints imposed on the cells used in recombinant DNA technology. Second, the intelligent species could program the probes to form colonies of the intelligent species in the stellar system reached by the probes, and use probes which are not as intelligent as the intelligent species. These colonies would be able to destroy any probes which slipped out of control, since the probes are less intelligent than the colonists. Third,—and this is the possibility we consider most likely—the intelligent species might not care if the von Neumann probes slipped out of control. As we have shown at length in Chapters 3 and 8, an advanced von Neumann probe would be an intelligent being in its own right, only made of metal rather than flesh and blood. The rise of human civilization has been marked by a decline in racism, and an extension of human rights—which include freedom—to a wider and wider class of people: in fact, the arguments one hears today against considering intelligent computers to be persons and against giving them human rights have precise parallels in the nineteenth-century arguments against giving blacks and women full human rights.[64] If this anti-racist trend continues and occurs in the cultures of all civilized beings, von Neumann probes would be recognized as intelligent fellow beings, beings which are the heirs to the civilization of the naturally evolved species that invented them, and with the right to the freedom possessed by the inventing species. We shall show in Chapter 10 that the naturally evolved species and all of its naturally evolved descendants must *inevitably* become extinct. But the machine descendants of the naturally evolved species need not ever become extinct. Thus if a naturally evolved species never has machine descendants, its civilization will eventually die out. A civilization with machine descendants could continue indefinitely.

The wisdom of creating an intelligent being superior to a human is often questioned, because it is said that such a being would not be a servant, but a master. We believe, on the contrary, that it would make good economic sense for members of a naturally evolved species to construct a robot vastly superior intellectually to themselves. Recall from section 3.7 that all wealth is ultimately information. Robots with superior intelligence would increase the amount of information available in a civilization far beyond what it could be from the efforts of the creator species alone. Cooperation between superintelligent robots and members of the species creating them would result in this increased wealth being shared between the two groups, with the creator species being wealthier with the robots than without them. The notion that cooperation between economic entities A and B, with A being absolutely superior in every way to B, results in both A and B being better off economically, is a well-known consequence of the Theory of Comparative Advantage in economic theory.[80] We humans should no more fear our robot descendants than we should fear our flesh and blood descendants, who will one day evolve away from the *Homo sapiens* form. It would admittedly not be wise to attack or try to enslave our intelligent robot descendants. Let us never forget, however, that the Frankenstein monster in the original novel was initially a kind and generous being, who turned to evil because of callous treatment by mankind.

If, on the other hand, the intelligent species retained their racism (specism?!), it seems likely that they would regard other 'flesh and blood' intelligent species as 'non-people'. If so, then they would either wish to avoid communication altogether (lest it 'pollute' their culture with alien ideas), or else launch von Neumann probes either to colonize the Galaxy for themselves (lest it be done by 'non-people' who would crowd them out) or to destroy these other intelligent species. For example, this complete colonization of the Galaxy and destruction of other species would be their best strategy if they believed that the biological 'exclusion principle', which says[73,74] two species cannot occupy the same ecological niche in the same territory, applies to intelligent species. With the advent of the O'Neill colony, the ecological niche occupied by an intelligent species would expand to encompass the entire resources of a solar system. The ecological niches of two intelligent species would have to overlap. Thus an intelligent species would be motivated to launch von Neumann probes, be they racist or anti-racist. If a species was not afraid of alien ideas itself, but was reluctant to contaminate the culture of another species with its own culture, then it would not attempt radio contact, for this would contaminate the other culture no less than actual colonization. In fact, as we pointed out in section 8.2, radio contact would constitute colonization with memes rather than genes. This has been termed 'cultural imperialism' by a French Minister of Culture. However, with probes it would be possible to study an alien species without it becoming aware of the species which is studying it. Thus, only with probes is it possible to interact without cultural imperialism. With radio beacons, cultural imperialism is a logical necessity.

This leads to the suggestion that perhaps the von Neumann probes of an extraterrestrial intelligent species *are* present in our Solar System. If a probe had just arrived, there would as yet be no evidence of its presence. The probability that a probe arrived for the first time within the past 20 years is about 10^{-9} ($\approx$20/[age of Galaxy]). Thus it seems reasonable to dismiss this unlikely possibility. Another possibility would be that they have been here for a long time but have decided not to make their presence known, say to avoid cultural imperialism; this is the so-called *zoo hypothesis*.[75] Kuiper and Morris[13] have proposed testing the zoo hypothesis by attempting to intercept radio communications between beings in our Solar System and the parent stars. Another possible test would be to search for the construction activities of a von Neumann probe in our Solar System. For example, one could look for the waste heat from such construction. As Dyson has pointed out,[11,76] this heat would give rise to an infrared excess, and the most likely place to look for a von Neumann probe would be the asteroid belt where material is most readily available.[77] (It is amusing that in fact much of the observed infrared radiation of astronomical origin does come from the asteroid belt![78])

However, we do not believe the zoo hypothesis is a likely possibility. If it were true, then our entire Solar System would be analogous to an American national forest, or an African game preserve. Thus even if the zoo hypothesis were true, von Neumann probes would now be present in the Solar System, acting in the capacity of game wardens. The national forests and other wilderness areas of the Earth remain unoccupied only because the police forces of the various nations keep them that way. If the protection of the state were withdrawn, then these regions would be rapidly colonized, just as Oklahoma was. Furthermore, though humans are for the most part excluded from these areas, most animals therein are probably aware of the existence of human beings. Scientists at least are permitted to interact with the animals, and jet aircraft are heard in all but the most remote regions. If our Solar System were a similar preserve, then *all* contact must have been rigorously prevented for as long as the robot game wardens were present in the Solar System, since there is not one jot of evidence for any contact in the past.

This total prevention of contact is, we submit, quite impossible. No police system, however efficient, can prevent *all* crime. In particular, the robot game wardens could not possibly prevent a group of ETI from beaming a radio signal to the Earth which would be easily detected. In any large and advanced civilization, there is almost certain to be some group which would think contact to be in the interest of the Earth primitives. It is very common in the ETI literature to treat advanced civilizations as if they were a single individual, with a single set of goals and a single set of moral values. But this is not the case in our own civilization, it was not the case in earlier human civilizations, and as we argued at length in sections 2.8, 3.2 and 3.7, it is even less likely to be the case in an advanced civilization. A civilization *is* advanced because it codes an enormous amount of information which is not shared by all the citizens, but rather is spread out among them. The more advanced the civilization, the more variety one will find in the information coded in different individuals. More variety in the individuals will inevitably lead to more variety of opinions as to the proper course of action in a given situation. In particular, in a very advanced interstellar civilization spread out over many light years, there are bound to be individuals who will believe it is wrong to let us humans hobble along in ignorance and disease. These beings would contact us, robot game wardens or no.

Even states which establish wilderness preserves allow most of the region under their control to be colonized, if for no other reason, than to supply the game wardens, or rather the military apparatus which backs them up. We thus feel that if ETI existed, they would have been permitted to colonize the outer reaches of the Solar System. Furthermore, if ETI existed, it is likely that their probes would have arrived a billion years ago when there was nothing on Earth but one-celled organisms, and hence they would have no reason to hide their technology, as we discussed above. The entire asteroid belt and indeed the entire outer part of the Solar System would have been converted into artifacts by now. Even if colonization were forbidden in the entire Solar System, there is no reason to think it would be stopped in completely uninhabited stellar systems. In these systems, the material therein would quite possibly by now be so completely utilized that the waste heat from the advanced civilization occupying the system would be visible over interstellar distances, as Dyson[76] has pointed out. This is likely if in fact the system was colonized over a billion years ago. Most stellar systems occupied for so long would by now have been visibly modified. There is no evidence this has happened in our Galaxy, nor indeed, as Newman and Sagan[62] point out, in nearby galaxies. Thus, there is strong evidence that 'local' extraterrestrial beings do not exist.

The point is that a belief in the existence of extraterrestrial intelligent beings anywhere in the Galaxy is not significantly different from the widespread belief that UFO's are extraterrestrial spaceships; this is the import of the space travel argument against the existence of ETI. In fact, one strongly suspects a psychological motivation common to both beliefs, namely, 'the expectation that we are going to be saved from ourselves by some miraculous interstellar intervention . . .'.[83] The truth or falsity of the ETI hypothesis is of course independent of the motivations of those who advance the hypothesis; we mention their motives only because they are indicative of motives which may also appear in advanced civilizations.

For example, one of the greatest, if not *the* greatest, danger facing the human race today is the threat of nuclear war. There is a considerable amount of evidence, which we mentioned in section 8.7, that an unlimited nuclear war could result in the extinction of the human species, and we pointed out earlier in the present section that the absence of ETI in our Solar System could be explained by the almost inevitable occurrence of nuclear war soon after the development of radio technology in an intelligent species' history but before the level of technology became high enough to develop von Neumann probes.

Newman and Sagan[62] have suggested another possibility, that civilizations develop social control of dangerous technology, such as nuclear weapons, before destroying themselves, and they similarly control the development of self-reproducing interstellar probes to prevent these probes from colonizing the Galaxy. Only with reference to such social evolution can Newman and Sagan avoid the implications of their own diffusion model. (It would also provide a motive for us to make radio searches for ETI, since a message from the advanced beings who understand such social control would save us from ourselves.)

We think the decentralization of information in any advanced civilization would prevent such social control of technology. Such control over nuclear weapons is possible today because of the vast expense of such devices: only a nation or perhaps a few individuals can afford them, and the industrial base needed to construct them cannot be hidden. But when a machine becomes so inexpensive relative to wages that a substantial fraction of the population can afford one, and can even construct one in tiny, hidden laboratories, then there will be individuals and groups in the civilization who will possess the devices, whatever the laws governing the possession may be. In particular, if in fact the cost of resources relative to wages continues to decrease at the present rate for the next 500 years (as Newman and Sagan grant),[62] then there will be *individuals*, not only nations, in human society at that period who will be armed with nuclear weapons. Because of the ultimate unity of knowledge, the only way to prevent this would be to halt *all* scientific advance. We can only hope that widespread ownership of nuclear weapons will not threaten the survival of the human species at that time, as it would if ownership were widespread now.

As Sagan and his fellow developers of the Nuclear Winter Model have shown, human survival will be endangered by nuclear weapons if we remain limited to the surface of this planet; ultimately, the only possible route to indefinite survival is to colonize space, including other stellar systems, as we showed in section 3.7. Other intelligent species that reach our level of technology will also be aware of this, or at least a substantial minority of this species will be, so that they will undertake the colonization of interstellar space in order to ensure the survival of their civilization. But we can express this claim in an even stronger form: even if it is *not true* that survival depends on colonization, there will be a group in any intelligent species which believes it does, and *they* will launch the probes.

If one contemplates restricting von Neumann probe technology, it should be recalled exactly what this would entail. Remember that a von Neumann computer is an intelligent being in its own right; thus restricting the 'use' of such machines means, quite literally, restricting the rights of an ethnic group of people to live and reproduce. Would an advanced civilization engage in such explicit racism?

Newman and Sagan[62] suggest that a civilization vastly ahead of us in science might use resources so efficiently that they will find it unnecessary to gain control of the resources in other solar systems. However, no matter how advanced a civilization may become, it is still restricted by the laws of thermodynamics. We have shown in section 3.7 that *all* activity of life, whether it be scientific research, economic production, or mystical contemplation, is a form of information processing. We shall show in section 10.6 that an upper bound to the total amount of information that can be generated using the *total* material resources in a solar system is 10^{70} bits. As a lower bound to the current information processing rate in our civilization is 10^{19} bits/sec ($= 10^{10}$ bits/person $-$ sec $\times 10^9$ people) our entire Solar System will be used up in 5000 years if science continues to grow exponentially at its current rate during that period. Thus whatever activities an advanced civilization conducts, it will be wanting more resources in a few thousand years.

A short study of the ETI literature will convince one that the main justification offered for the existence of extraterrestrial intelligent life is not scientific but philosophical: the main justification offered is the Copernican Cosmological Principle. As Newman and Sagan put it:

> Every one of [a number of deep scientific questions] has been settled decisively in favour of the proposition that there is nothing special about us: we are not at the centre of the Solar System; our planet is one of many; it is vastly older than the human species; the Sun is just another star, obscurely located, one among some 400 billion others in the Milky Way, which in turn is one galaxy among perhaps hundreds of billions. We humans have emerged from a common evolutionary process with all the other plants and animals on Earth. We do not possess any uniquely valid locale, *epoch* [our emphasis], velocity, acceleration or means of measuring space and time.

Or as Sagan himself concluded:[85]

[If ETI are not found], it would be the first instance in the long series of historic scientific debates in which the anthropocentric hypothesis had proved even partly valid.

This book can be viewed as an analysis of this Copernican dogma. In particular, we find it simply untrue that there is nothing special about the epoch in which we now live. As the discovery in the 1960's of the cosmological background radiation showed, the Universe is changing with time. We have shown at length that the epoch in which we live is very special in permitting the evolution of carbon life. (The cosmic microwave background radiation also provides a uniquely valid means of measuring space and time position.)

The first failure of the Copernican Principle was the experimental refutation of the steady-state theory, which indeed held that there is nothing special about the epoch in which we live. It is to the steady-state theory that we now turn.

9.5 Anthropic Principle Arguments Against Steady-State Cosmologies

We moderns think in terms of simple space,... [but] archaic man thought in terms of time dominating all else.
G. de Santillana

Steady-state cosmologies have a very strong intellectual appeal. If the Universe were not changing in the large, it would not be necessary to address the questions of how the Universe began[107] and how it will end, questions which seem unavoidable in the context of evolving, Big Bang cosmologies. Furthermore, many find the idea of evolutionary change abhorrent. Such people are attracted to a steady-state cosmology because in such a universe Time is ultimately without meaning: on a sufficiently large-scale view, no change occurs. For these reasons numerous steady-state cosmological models have been developed during the past thirty years. The simple Bondi–Gold–Hoyle, model[81,83] was killed by the discovery of the three degree background radiation, but Hoyle and Narlikar[84,87] have shown that with a little imagination the steady-state cosmology can be generalized artificially to include such radiation. These cosmologies are both spatially infinite models, but George Ellis and his students[88,89] have constructed a steady-state—actually a static-in-the-large—cosmology which is finite in spatial extent. The Ellis *et al.* model can also account for the background radiation, and many other cosmological features which one might think must be due to cosmic evolution. Since these models can be adjusted to explain many observations thought at first sight to be caused by large scale evolution, it seems possible that they could also be adjusted to include most other apparently evolutionary effects. Nevertheless, the Anthropic Principle can be used to rule out virtually any type of steady-state theory.

Davies[90] in fact used a SAP argument against the Ellis *et al* static universe. Davies pointed out that if it is accepted that intelligent life must evolve somewhere in the Universe, then this is inconsistent with a static universe since such a universe cannot explain why intelligence should arise at a particular time t_0 and not before. Indeed, in a universe which is static in the large, the evolution rate of intelligent species should be a non-zero constant. But if it were a non-zero constant, then in a spatially finite static-in-the-large cosmology like that of Ellis *et al.* some intelligent species should have long ago developed space travel and expanded throughout the entire cosmos, incidentally preventing the evolution of new intelligent species by occupying all ecological niches which could contain intelligent beings. However, this means that the evolution rate of new intelligent species is zero, contrary to assumption. We shall generalize this argument of Davies to all universes which do not change with time in the large and justify the tacit assumptions in the Davies argument. This argument becomes most compelling if the notion of 'intelligent species' is generalized to include intelligent machines which are capable of self-reproduction.

It is known[91] that a space-time which is stationary and homogeneous in the large—so which satisfies the Perfect Cosmological Principle[81]—is a portion of de Sitter space with the conformal Penrose diagram represented in Figure 9.1. Penrose diagrams will be defined and described in detail in section 10.3. For the moment, it will suffice to understand two concepts which are central to global general relativity. The *causal past* $J^-(p)$ of an event p in space-time is the set of all events which can be connected to p by a future-directed timelike or null curve. That is, $J^-(p)$ is the set of all events in the past of p that can send signals to p, or travel to p at a velocity less than the velocity of light. The *chronological past* $I^-(p)$ is the set of all events which can be connected to p by a timelike curve. Causal and chronological futures, $J^+(p)$ and $I^+(p)$ respectively, are defined analogously. The salient point to notice in Figure 9.1 is that the causal past $J^-(p)$ of the typical event p intersects *all* world lines for some interval could correspond to the history of an intelligent species. The evolutionary histories of two such species are denoted by dotted lines in Figure 9.1. The intelligent species labelled A could at any point in its history have sent a space probe to event p. In fact, since the space-time is globally hyperbolic (see section 10.3 for a precise definition of this concept) and the evolutionary interval A lies entirely in $J^-(p)$, it follows

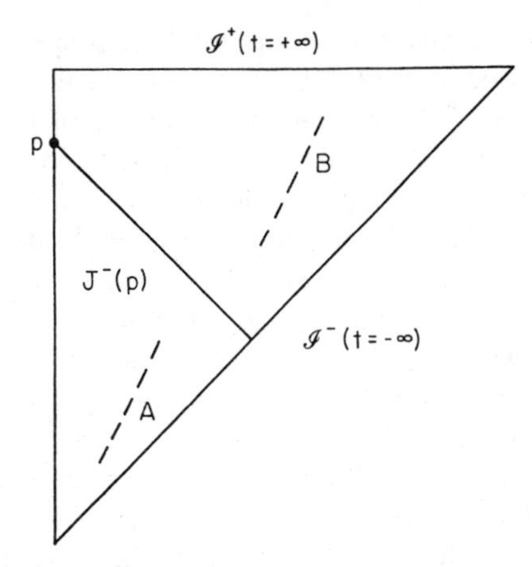

Figure 9.1. Penrose conformal diagram for the global causal structure of the steady-state universe. The point p is an arbitrary event on the origin of coordinates. Intervals A and B are the world lines of intelligent species existing in p's past light-cone $J^-(p)$, and outside $J^-(p)$, respectively. $\mathscr{I}^+$ and $\mathscr{I}^-$ are future and past infinity respectively.

from Proposition 6.7.1 of ref. 91 that there is a time-like geodesic from any event of A to the event p. Now Ellis and Brundrit[92] have shown that the assumption of homogeneity in a spatially infinite universe, when combined with the atomicity of matter, implies that all possible evolutionary histories must occur an infinite number of times with probability one. The argument is essentially the one used to prove recurrence in a discrete, finite Markov chain: if there are only a finite number of possibilities—and this follows from the homogeneity of the Universe and the atomicity of matter—then with an infinite amount of space (or time) each possibility has a probability one of being realized an infinite number of times.[93] A space-time satisfying the Perfect Cosmological Principle is homogeneous in both space and time, so the Ellis and Brundrit argument applies with double force: there must be an infinite number of evolutionary histories like A to the past of any point p in the space-time.

Earlier in this chapter we have described a process whereby it is possible to travel from one star to another in a galaxy at a net speed comparable to that of light if the stars are far apart, while the initial investment in energy and money is quite small: construct and send out von Neumann probes. Since all possible intelligent species have evolved to the past of p, infinitely many species would have evolved which sent out von Neumann probes, or which otherwise colonized space. But we can go further. Since all possible evolutionary sequences have occurred to the past of p, one of these evolutionary sequences consists of the random assembly, without the assistance of any intelligent species whatsoever, of a von Neumann probe out of the atoms of interstellar space. Such a random assembly would occur an infinite number of times to the past of p, by homogeneity and stationarity in an infinite universe. At least one of these randomly assembled probes would have the motivations of a living being, that is to expand and reproduce without limit. Natural selection acting on this probe and its descendants—such descendants can be regarded as comprising a living intelligent species—would insure that this mechanical lineage would expand to occupy all 'ecological' niches available to it. In so expanding the descendants may split into many 'species', but once the expansion begins, at least one lineage would continue to exist and expand. Since the probes are intelligent, some would realize that distant galaxies would constitute available ecological niches, and since it would be possible for them to construct probes like themselves which could travel intergalactic distances of arbitrary length along a curve which is very close to any given timelike geodesic, one concludes that natural selection would impel some descendant probes to do so. Because in an infinite steady-state universe, some such events would lie to the past of p, these probes should have already arrived at p, and should be using the material at p to construct more probes. In effect, the probes would have colonized the region around p. Since p is *any* event, we obtain a contradiction with the fact that our solar system has not been colonized. The entire above argument is just a systematic use of the Perfect Cosmological Principle, which means this Principle is self-contradictory: the assumption that intelligent beings can evolve implies with this Principle that they never can evolve—they must already be everywhere. In fact, the paradox that extraterrestrial intelligent beings ought to have arrived in our solar system long ago but did not, is as astounding as Olbers' paradox which it closely resembles. Both paradoxes follow from assumptions of homogeneity in space and time. Olbers' paradox can be resolved in a steady-state universe by the redshift, but the expansion will not reduce the effective speed of probes, since the expansion-caused slowing of the intergalactic probes with respect to the fundamental frames can be cancelled by using the matter in galaxies encountered to re-accelerate the probe.

The above argument can be extended to any cosmology which is stationary in the large, since the infinite past during which the Universe is locally evolutionary would give rise to the von Neumann probes used to reach a contradiction in the steady-state cosmology case. Davies has given a related argument against the Ellis *et al.* cosmology, as mentioned above.

In addition, the above argument rules out the chronometric cosmology of Segal. The chronometric cosmos[94] is a globally static cosmology with topology $\mathbf{S}^3 \times \mathbf{R}^1$. Just as in the Einstein static universe, it is possible to travel via rocket from any one spatial event to any other spatial point in finite time as measured by a physical clock on the rocket.[95] The above argument thus still applies.

The most important steady-state cosmologies discussed at the present time are those based on inflation. We discussed the basic inflationary mechanism in Chapter 6: when the density and temperature of matter is very high in the very early universe, the expansion of the universe is driven by a non-zero vacuum energy density (equivalent to a primordial positive cosmological constant), which is later cancelled out by a spontaneous symmetry breaking phase transition which occurs when density and temperature of drop sufficiently far. In most inflation models, the expansion is envisaged as beginning at an initial cosmological singularity as in Standard Model. However, such a beginning is not strictly required by the mathematics of the inflation model. In fact, during the inflationary phase the evolution equations are essentially the same as the equations for the steady-state universe, so it is possible to regard the phase transition which terminates the inflation as generating a 'bubble' within which the entire visible universe is located. Outside the walls of this bubble the metric is that of the steady-state universe. Thus, in this model, the Universe in the large is steady-state. In the steady-state region—that is, in the region of space-time outside the bubble—the matter density is only a few orders of magnitude less than the Planck density of 5×10^{93} gm/cm^3, and the dominant term in the Einstein equations is the vacuum energy term. The visible universe is then just a tiny bubble of evolving matter in a Universe which is changeless in the large. There may be other bubble universes in this steady-state Universe, but they comprise only an infinitesimal fraction of the volume of the whole. Narlikar[102] has recently argued that there is no essential difference between the final version of the steady-state theory, defended by himself and Hoyle, and the inflation steady-state model.

From the point of view of the global causal structure, there are two basic types of bubble universes which can form in the inflation version of the steady-state Universe: 'open universe' bubbles and 'closed universe' bubbles. The open universe bubbles have been discussed extensively by Gott.[103] Their spatial sections have negative or zero curvature, and their walls expand indefinitely at the speed of light. Although finite in spatial extent at any given time, the volume of an open bubble becomes infinite in infinite time. The causal structure of a steady-state universe with infinitely many non-intersecting open bubbles is pictured in Figure 9.2. As seen in this figure, the different bubbles are forever out of causal contact with each other; evolution proceeds in each as if the others did not exist.

However, non-intersecting open bubbles are actually inconsistent with the steady-state universe, which is homogeneous in space and time. If the

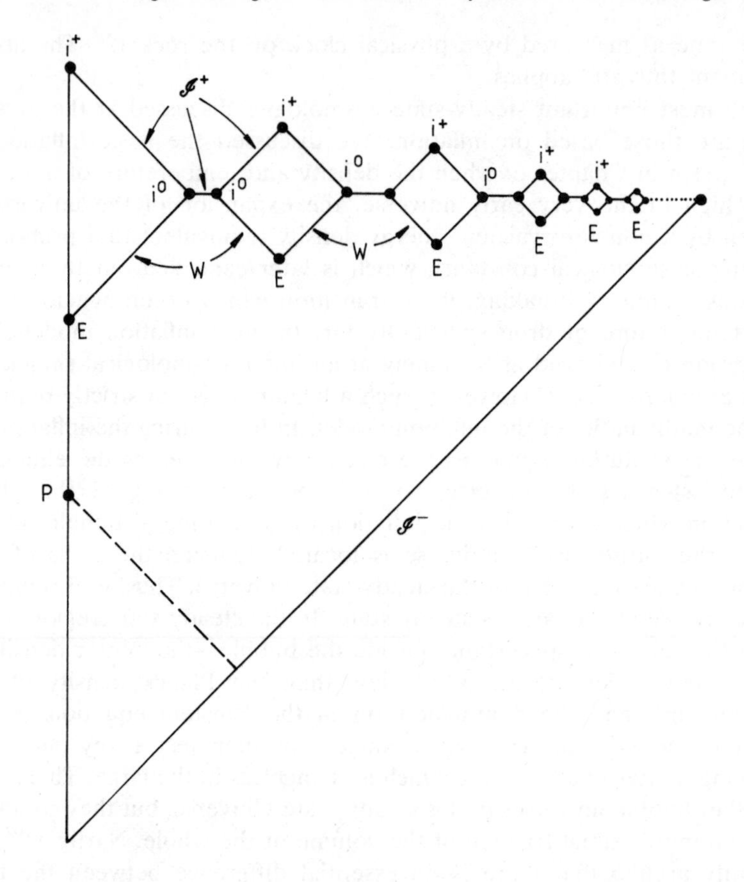

Figure 9.2. Penrose conformal diagram for the global causal structure of a steady-state universe with an infinite number of open bubbles. The bubbles come into existence at the events labelled *E*. The walls of the bubbles are labelled *W*; these walls expand at the speed of light, reaching spacelike infinity i^0 at an infinite time in the future. In the future of the bubbles, $\mathscr{I}^+$ becomes null, but $\mathscr{I}^+$ is timelike elsewhere, as it is in the standard steady-state model: outside the bubbles, the space-time is the same as the standard steady-state universe pictured in Figure 9.1. The point labelled i^+ in each bubble is the future end-point of all timelike geodesics inside the bubbles. All bubbles are to the future of the event p: the boundary of the past light-cone $J^-(p)$ of p is represented by a dotted line.

Universe were truly steady-state, we would expect that there would be a constant probability per unit time of bubble formation on the timelike geodesics which are normal to the spacelike hypersurfaces of global homogeneity. But as we saw in our discussion of the standard steady-state universe, all timelike curves must intersect the past light-cone of any event p on the world line of the origin of spatial coordinates. Thus, if the inflationary steady-state universe were truly steady-state, there must be a bubble in the past light-cone of such an event *p*, which contradicts the causal structure pictured in Figure 9.2. (Gott[103] was himself aware of this difficulty with open bubbles in a steady-state universe.)

Closed bubbles[104,105,106] do not suffer from this problem, for a closed bubble would evolve like a closed universe: it would be formed in a phase transition, expand to a maximum size and then re-contract to a high density. Eventually the bubble walls would intersect and the bubble universe would disappear. Thus, there could be an infinite number of bubbles in the light-cone of an event *p*, because these bubbles would have formed and disappeared long ago. The causal structure of a steady-state inflation model with only closed bubbles is identical to the causal structure of the standard steady-state model, which is pictured in Figure 9.1.

Thus, the closed bubble model is open to the same SAP objection levelled at the standard steady-state universe. We would expect intelligent life to evolve in at least some of the bubbles. These intelligent beings would die out when their bubble disappears if they are restricted to the bubble in which they evolve. Therefore, if it is possible for an intelligent species to escape its bubble of origin—that is, if it is possible for the species to develop a means to travel in the steady-state region—we would expect at least one such species in the past of *p* to do so, and indeed to expand to the region containing *p*.

This SAP objection is much weaker in the inflation steady-state universe situation than it is in the standard steady-state universe model, for it is far from clear that it is possible to develop technology which will allow intelligent life to exist and travel in the steady-state region: the density and temperature in this region are near the corresponding Planck magnitudes. We shall present arguments in Chapter 10 that it is actually possible for intelligent 'life' to exist in such high-density and high-temperature regimes, but this cannot be regarded as an established fact by any stretch of the imagination. However, this possibility must be taken into account in any steady-state theory based on closed bubbles: such a theory cannot be regarded as true unless it is shown that it is impossible for intelligent life, no matter how advanced, to leave their bubble of origin. If FAP holds, then it must be possible for intelligent life to leave a closed bubble, for by FAP, intelligent life cannot disappear once it comes into existence. This argument will become clearer once the FAP is defined precisely in Chapter 10.

The arguments presented in this chapter complement the earlier arguments we presented in Chapters 3 and 8 regarding the improbability of other forms of local (that is, within range of communication with us) extraterrestrial life. We have developed the general theory of exploration and colonization using self-reproducing robots, the theory of whose existence is already known to us although at present we lack the level of computer technology to implement it in practice. This theory was then used to demonstrate the ease with which advanced Galactic civilizations could reveal their presence and the difficulty they would have concealing it. These arguments are based upon technological considerations and an analysis of the collective features necessary to support an advanced technological civilization. Finally, we have demonstrated how the non-observation of other life-forms in our Galaxy allows one to rule out a large class of otherwise quite possible steady-state cosmologies having infinite ages.

References

1. F. J. Tipler, *Quart. J. R. astron. Soc.* **21,** 267 (1981). *Quart. J. R. astron. Soc.* **22,** 279 (1981); *Physics Today* **34** (No. 4, April, 9 (1981); *Mercury* **11** (No. 1, Jan.), 5 (1982); *New Scient.* **96** (No. 1326, 7 Oct.), 33 (1982); *Physics Today* **35** (No. 3, March), 26 (1982); *Science* **219,** 110 (1983); *Discovery* **4** (No. 3, March), 56 (1983). A short history of the belief in extraterrestrial intelligence can be found in F. J. Tipler, *J. R. astron. Soc.* **22,** 133 (1981). The belief in ETI is closely interwoven with the Design Arguments.
2. I. S. Shklovskii and C. Sagan, *Intelligent life in the Universe* (Dell, NY, 1966).
3. F. D. Drake, *Intelligent life in space* (Macmillan, NY, 1960).
4. P. Morrison, in *Interstellar communication: scientific perspectives*, ed. C. Ponnamperuma and A. G. W. Cameron (Houghton Mifflin, Boston, 1974).
5. T. Dobzhansky, in *Perspectives in Biology and Medicine,* **15,** 157 (1972); *Genetic diversity and human equality* (Basic Books, NY, 1973), pp. 99–101.
6. G. G. Simpson, *This view of life* (Harcourt, Brace & World, NY, 1964), Chapters 12 and 13.
7. J. Francois, *Science* **196,** 1161 (1977); see also W. D. Mathew, *Science* **54,** 239 (1921).
8. T. Dobzhansky, F. J. Ayala, G. L. Stebbins, and J. W. Valentine, *Evolution* (Freeman, San Francisco, 1977).
9. E. Mayr, *Scient. Am.* **239** (No. 3, Sept.), 46 (1978).
10. E. Fermi, quoted on p. 495 of C. Sagan, *Planet. Space Sci.* **11,** 485 (1963).
11. F. M. Dyson, in *Perspectives in modern physics: essays in honor of Hans A. Bethe*, ed. R. E. Marshak (Wiley, NY, 1966).
12. M. H. Hart, *Quart. J. R. astron. Soc.* **16,** 128 (1975).
13. T. B. H. Kuiper and M. Morris, *Science* **196,** 616 (1977).
14. P. J. Klass, *UFOs explained* (Random House, NY, 1974).
15. D. H. Menzel and E. H. Taves, *The UFO enigma* (Doubleday, Garden City, NY, 1977).
16. Ref. 2, Chapter 25.
17. C. Sagan, *The dragons of Eden* (Ballantine, NY, 1977), p. 239.

18. *Project Cyclops* (Report CR114445, NASA Ames Research Center, Moffett Field California, 1971).
19. *The search for extraterrestrial intelligence: SETI* (NASA report SP-419, 1977).
20. D. Michie, *Nature* **241,** 508 (1973).
21. O. Firschein, M. A. Fischler, and L. S. Coles, in *Third International Joint Conference on Artificial Intelligence* (Stanford University, 1973). This reference actually gives the opinions of leading computer scientists as to when computers with human-level intelligence and manipulative ability will be manufactured. This technology seems to be to be roughly comparable to von Neumann machine technology, so we use this number as our estimate for how long it will be before we develop von Neumann probes. No explicit mention was made of self-reproducing machines in refs 20, 21, or 22, However, G. von Tiesenhausen and W. A. Darbo have claimed in NASA Technical Memorandum TM-78304 (July, 1980) that self-reproducing space robots could be developed in only 20 years. See also *Advanced automation for space missions*; ed. R. A. Freitas, Jr., and W. P. Gilbreath (NASA Conference Publication 2255, Washington, 1982).
22. M. Minsky, in *Communication with extraterrestrial intelligence*, ed. C. Sagan (MIT Press, Cambridge, Mass, 1973), p. 160.
23. J. von Neumann, *Theory of self-reproducing automata*, ed. and completed by A. W. Burks (University of Illinois Press, Urbana, 1966).
24. M. A. Arbib, *Theories of abstract automata* (Prentice-Hall, Englewood Cliffs, NJ, 1969); see also the Arbib article in Ponnamperuma and Cameron, ref. 4.
25. E. Stuhlinger, *Ion propulsion for space flight* (McGraw-Hill, NY, 1964).
26. J. L. Wright and J. M. Warmke, 'Solar sail mission applications', JPL preprint 76-808 AIAA/AAS 1976 San Diego Astrodynamics Conference.
27. C. R. Chapman, *Scient. Am.* **232** (No. 1), 24 (Jan. 1975); B. J. Skinner, *Am. Scient.* **64,** 258 (1976); D. W. Hughes, *Nature* **270,** 558 (1977).
28. H. A. Abt, *Scient. Am.* **236** (No. 4), 96 (April 1977); A. H. Batten, *Binary and multiple systems of stars* (Pergamon Press, NY, 1973); S. A. Dole, *Habitable planets for man* (Blaisdell, NY, 1964).
29. J. W. Truran and A. G. W. Cameron, *Astrophys. Space Sci.* **14,** 179 (1971); A. G. W. Cameron in ref. 4.
30. G. K. O'Neill, *Physics Today* **27,** 32 (Sept. 1974); *Science* **190,** 943 (1975); *The high frontier* (Morrow, NY, 1977).
31. C. C. Price, *Chem. Eng. News* **43,** 90 (27 Sept. 1965); *Synthesis of Life*, ed. C. C. Price (Dowden, Hutchinson & Ross, Stoudsburg, Pa, 1974), pp. 284–6.
32. J. F. Danielli, *Bull. Atom. Scient.*, (Dec. 1972) pp. 20–4 (also in C. C. Price, ref. 31); K. W. Jeon, I. J. Lorch, and J. F. Danielli, *Science* **167,** 1626 (1970).
33. C. Grobstein, *Scient. Am.* **240** (No. 6), 57 (June 1979).
34. L. Eiseley, *The invisible pyramid* (Scribner's, NY, 1970), pp. 78–80.
35. F. H. C. Crick and L. E. Orgel, *Icarus* **19,** 341 (1973).
36. P. H. C. Sneath, *Nature* **195,** 643 (1962).
37. M. Seibert, *Science* **191,** 1178 (1976); *In Vitro* **13,** 194 (1977).

38. E. G. Cravalho, *Technol. Rev.* **78** (No. 1), 30 (Oct. 1975).
39. A. S. Parkes, *Sex, science, and society* (Oriel Press, London, 1965).
40. M. W. Hunter, II, in *AAS Science & Technology Series* **17,** 541 (1967).
41. G. W. Morgenthaler, *Ann. NY Acad. Sci.* **163,** 559 (1969).
42. M. R. Helton, Jet Propulsion Laboratory Inter-office Memorandum 312/774-173 (21 June 1977).
43. A. Bond *et al., Project Daedalus,* special supplement of *J. Br. Interplan. Soc.* (1978).
44. Ref. 19, p. 108.
45. F. H. Dyson, *Ann. NY Acad. Sci.* **163,** 347 (1969); see also D. F. Spencer and L. D. Jaffe, Jet Propulsion Laboratory preprint #32–233 (1962).
46. F. J. Dyson, quoted in A. Berry, *The next thousand years* (New American Library, NY, 1974), p. 125.
47. R. H. MacArthur and E. O. Wilson, *The theory of island biogeography* (Princeton University Press, Princeton, 1967).
48. E. O. Wilson, *Sociobiology* (Harvard University Press, Cambridge, Mass., 1975).
49. Ref. 48, p. 105.
50. E. M. Jones, *J. Br. Interplan. Soc.* **31,** 103 (1978).
51. W. I. Newman and C. Sagan, *Icarus* **46,** 293 (1981).
52. C. Sagan (ed.), *Communication with extraterrestrial intelligence* (MIT Press, Cambridge, Mass., 1973); T. L. Wilson, *Quart. J. R. astron, Soc.* **25,** 435 (1984).
53. V. Trimble, *Rev. Mech. Phys.* **47,** 877 (1975).
54. J. Audouze and B. M. Tinsley, *Ann. Rev. Astron. & Astrophys.* **14,** 43 (1976).
55. A. A. Penzias, *Comm. Astrophys.* **8,** 19 (1978).
56. J. C. Browne and B. L. Berman, *Nature* **262,** 197 (1976).
57. S. van den Bergh, *Quart. J. R. astron. Soc.* **25,** 137 (1984).
58. R. J. Talbot, *Astrophys. J.* **189,** 209 (1974); R. J. Talbot and W. D. Arnett, *Astrophys. J.* **186,** 51 (1973).
59. D. C. Barry, *Nature* **268,** 509 (1977).
60. Ref. 18, p. 25.
61. J. G. Kreifeldt, *Icarus* **14,** 419 (1971).
62. C. Sagan and W. I. Newman, *Quart. J. R. astron. Soc.* **24,** 113 (1983).
63. The timescale t_G is 7.5×10^8 years if the diffusion equation is discretized, in order to take into account the fact that the stars are not next to each other, but are separated by distances of ~ 1 pc. If the diffusion equation is *not* discretized, then $t_G \sim 10^{10}$ years, a number which would invalidate our argument. Newman and Sagan stick by the 7.5×10^8 yr figure (see ref. (62)).
64. E. A. Feigenbaum and P. McCorduck, *The fifth generation* (Addison-Wesley, London, 1983).
65. R. N. Bracewell, *Nature* **186,** 670 (1980); repr. in *The search for extraterrestrial life,* ed. A. G. W. Cameron (Benjamin, NY, 1963).
66. R. N. Bracewell, *The galactic club* (Freeman, San Francisco, 1975).
67. Ref. 19, p. 108.

68. W. H. Pickering, *Popular Astronomy* **17,** 495 (1909); reprinted in *MARS* (Gorham Press, Boston, 1921). F. Galton even worked out a code which is similar to those of the SETI proposals. See *Fortnightly Rev.* **66,** 657 (Nov. 1896).
69. C. Sagan, quoted in *Technol. Rev.* **79** (No. 6, May), 14 (1977).
70. T. Dobzhansky, *Genetics of the evolutionary process* (Colombia University Press, NY, 1970), p. 278.
71. S. E. Morison, *Portuguese voyages to America in the fifteenth century* (Octagon, NY, 1965), pp. 11–15.
72. K. Davies, *Scient. Am.* **231** (No. 3, Sept.), 92 (1974).
73. E. Mayr, *Populations, species, and evolution* (Harvard University Press, Cambridge, Mass., 1970), p. 48.
74. R. M. May, *Scient. Am.* **239** (No. 3, Sept.), 160 (1978).
75. J. A. Ball, *Icarus* **19,** 347 (1973).
76. F. J. Dyson, *Science* **131,** 1667 (1960).
77. M. D. Papagianiss has independently suggested that the asteroid belt would be the most likely place to search for extraterrestrial industrial activities; see *Quart. J. R. astron. Soc.* **19,** 227 (1978).
78. F. J. Low and H. J. Johnson, *Astrophys. J.* **139,** 1130 (1964).
79. N. Calder, *1984 and beyond* (Viking, NY, 1984).
80. P. A. Samuelson, *Economics* (McGraw-Hill, NY, 1964).
81. H. Bondi., *Cosmology* (Cambridge University Press, Cambridge, 1961).
82. C. Sagan, in *UFOs—a scientific debate*, ed. C. Sagan and T. Page (Norton, NY, 1972), p. 272. The extent to which this motivation is present among the supporters of SETI, in particular Drake and Sagan, is discussed in F. J. Tipler, *Quart. J. R. astron. Soc.* **22,** 279 (1981).
83. F. Hoyle, *Mon. Not. R. astron. Soc.* **109,** 365 (1949).
84. F. Hoyle, *Astrophys. J.* **196,** 661 (1975).
85. C. Sagan, *Discovery* **4** (No. 3, March), 30 (1983).
86. *Encyclopaedia Britannica*, Vol. 2 (Benton, Chicago, 1967), p. 936.
87. J. V. Narlikar, *Pramana* **2,** 158 (1974).
88. G. F. R. Ellis, R. Maartens, and S. Nel, *Mon. Not. R. astron. Soc.* **184,** 439 (1978).
89. G. R. F. Ellis, *Gen. Rel. Gravn* **9,** 87 (1978).
90. P. C. W. Davies, *Nature* **273,** 336 (1978).
91. S. W. Hawking, and G. F. R. Ellis, *The large-scale structure of space-time* (Cambridge University Press, Cambridge, 1973).
92. G. F. R. Ellis and G. B. Brundrit, *Quart. J. R. astron. Soc.* **20,** 37 (1979).
93. F. J. Tipler, 'General relativity and the eternal return' in *Essays in general relativity: a festschrift for Abraham H. Taub*, ed. F. J. Tipler (Academic Press, NY, 1980).
94. I. E. Segal, *Mathematical cosmology and extragalactic astronomy* (Academic Press, NY, 1976).
95. I. E. Segal, private communication to FJT.
96. G. O'Neill, private communication to FJT.
97. R. A. Freitas, Jr., *J. Brit. Interplan. Soc.* **33,** 95 (1980).

98. R. A. Freitas, Jr., *J. Brit. Interplan. Soc.* **33,** 251 (1980).
99. F. Valdes and R. A. Freitas, Jr., *J. Brit. Interplan. Soc.* **33,** 402 (1980).
100. *Encyclopedia Britannica,* Vol. 12 (Benton, Chicago, 1967), p. 4.
101. I. S. Shklovskii, quoted in *Astronomy* **5,** 56 (Jan. 1977).
102. J. V. Narlikar, *J. Astrophys. Astron.* **5,** 67 (1984).
103. J. R. Gott, *Nature* **295,** 304 (1982); an analysis of this article appeared in *Science* **215,** 1082 (1982).
104. E. P. Tryon, *Nature* **246,** 396 (1973).
105. Ya. B. Zel'dovich and L. P. Grishchuk, *Mon. Not. R. astron. Soc.* **207,** 23P (1984); J. D. Barrow and F. J. Tipler, *Mon. Not. R. astron. Soc.* **216,** 395.
106. A. D. Linde, *Rep. Prog. Phys.* **47,** 925 (1984).
107. The steady-state universe is often presented as a singularity-free cosmological model. However, as S. W. Hawking and G. F. R. Ellis point out in *The large-scale structure of space-time,* the steady-state universe is actually singular in the sense that all the null geodesics are incomplete in the past direction. A straightforward calculation shows that a necessary and sufficient condition for the past completeness of null geodesics in a Friedman universe is

$$\int_{t_0} R(t)\, dt = -\infty$$

where t_0 is the lower limit of the length of the timelike geodesics normal to the hypersurfaces of homogeneity and isotropy ($t_0 = -\infty$ if these geodesics are complete), and $R(t)$ is the usual scale factor of the Friedman universe. Since $R(t) = \exp[Ht]$ in the steady-state universe, the above integral is finite.

10 The Future of the Universe

Some say the world will end in fire,
Some say in ice
From what I've tasted of desire
I hold with those who favor fire.
Robert Frost

10.1 Man's Place in an Evolving Cosmos

We need scarcely add that the
contemplation in natural science of a
wider domain than the actual leads to a
far better understanding of the actual.
A. S. Eddington

When we investigate the relationship between intelligent life and the Cosmos, one fact stands out at the present time: there is no evidence whatsoever of intelligent life having any significant effect upon the Universe in the large. As we have discussed at length in earlier chapters, the evidence is very strong that intelligent life is restricted to a single planet, which is but one of nine circling a star which itself is only one of about 10^{11} stars in the Galaxy and our Galaxy is but one of some 10^{12} galaxies in the visible universe. Indeed, one of the seeming implications of science as it has developed over the past few centuries is that mankind is an insignificant accident lost in the immensity of the Cosmos. The evolution of the human species was an extremely fortuitous accident, one which is unlikely to have occurred elsewhere in the visible universe.

It has appeared to most philosophers and scientists over the past century that mankind is forever doomed to insignificance. Both our species and all our works would disappear eventually, leaving the Universe devoid of mind once more. This world view was perhaps most eloquently stated by Bertrand Russell in the passage we quoted in Section 3.7, but the same sentiments have recently been expressed by the Nobel-prize-winning physicist Steven Weinberg in his popular book on cosmology, *The First Three Minutes:*

It is almost irresistible for humans to believe that we have some special relation to the universe, that human life is not just a more-or-less farcical outcome of a chain of accidents reaching back to the first three minutes [of the Universe's existence], but that we were somehow built in from the beginning.... It is very hard to realize that [the entire earth] is just a tiny part of an overwhelmingly hostile universe. It is even harder to realize that this present universe has evolved from an unspeakably unfamiliar early condition, and faces a future extinction of endless cold or intolerable heat. The more the universe seems comprehensible, the more it also seems pointless.[1]

These ideas neglect to consider one extremely important possibility: Although mankind—and hence life itself—is at present confined to one insignificant, doomed planet, this confinement may not be perpetual. Bertrand Russell wrote his gloomy lines at the turn of the century, and at that time space travel was viewed as an impossibility by almost all scientists. But we have landed men on the Moon. We *know* space travel is possible. We argued in Chapter 9 that even interstellar travel is possible. Thus once space travel begins, there are, in principle, no further physical barriers to prevent *Homo sapiens* (or our descendants) from eventually expanding to colonize a substantial portion, if not all, of the visible Cosmos. Once this has occurred, it becomes quite reasonable to speculate that the operations of all these intelligent beings could begin to affect the large scale evolution of the Universe. If this is true, it would be in *this* era—in the far future near the Final State of the Universe—that the true significance of life and intelligence would manifest itself. Present-day life would then have cosmic significance because of what future life may someday accomplish.

One can draw an analogy with the geological effect of life upon the Earth. At the dawn of life, some four billion years ago, living beings were nothing more than simple biochemical machines capable of self-reproduction. When the machines formed, they were originally restricted (as far as we can tell) to a small, insignificant portion of the Earth's surface. A being from another world who happened to observe the Earth at this time would have not noticed their presence, nor seen any effect of their presence on the geological evolution of the Earth.

As time went on, however, these living creatures increased their numbers exponentially. A significant fraction of the carbon available on the surface of the Earth was incorporated into living bodies. A photosynthetic ability evolved, and plants with this ability began to release oxygen into the atmosphere. As a consequence of this action by green plants, 21% of the present-day atmosphere is now oxygen. Had plants never supplied the atmosphere with oxygen, our planetary atmosphere would probably closely resemble the atmosphere of Venus: 95% carbon dioxide and 5% nitrogen. As we discussed in section 8.7, an oxygen atmosphere such as ours is intrinsically unstable, and the Earth's atmosphere would revert to a Venus-like atmosphere in the absence of the constant action of plants. Life has transformed the global atmosphere of the Earth on such a scale that the effect of life on the Earth (or at least on its atmosphere) could be recognized as such by an observer far outside the Solar System.

We can view the action of intelligent life on the entire Universe in a similar fashion. A species capable of rapid technological innovation has existed in the Universe for only about 40,000 years.[2,3] This species has just begun to take the first, faltering steps to leave its place of origin. In the time to come, it and its descendant species could conceivably change structural features of the Universe.

To say that intelligent life has some global cosmological significance is to say that intelligent life will someday begin to transform and continue to transform the Universe on a cosmological scale. What we wish to discuss now is the question of what the Universe must be like in order for this to be possible. As our discussion of dysteleology in section 3.7 and Weinberg's remarks make abundantly clear, until recently scientists did not believe the physical laws could ever permit intelligent life to act on a cosmological scale. In part this belief is based on the notion that intelligent life means *human life*. Weinberg points out that the ultimate future of the Universe involves great cold or great heat, and that human life—the species *Homo sapiens*—cannot survive in either environment. We must agree with him. The ultimate state of the Universe appears to involve one of these environments, and thus *Homo sapiens* must eventually become extinct. This is the inevitable fate of any living species. As Darwin expressed it in the concluding pages of the *Origin of Species*:[4]

> Judging from the past, we may safely infer that not one living species will transmit its unaltered likeness to a distant futurity.

But though our species is doomed, our civilization and indeed the values we care about may not be. We emphasized in Chapters 8 and 9 that from the behavioural point of view intelligent *machines* can be regarded as people. These machines may be our ultimate heirs, our ultimate descendants, because under certain circumstances they could survive forever the extreme conditions near the Final State. Our civilization may be continued indefinitely by them, and the values of humankind may thus be transmitted to an arbitrarily distant futurity. But before discussing under what conditions this might be possible, it will prove instructive to briefly review the reasons which were given to justify the idea that *all* intelligent life must become extinct.

10.2 Early Views of the Universe's Future

> Cosmology, since it is the outcome
> of the highest generality of speculation, is
> the critic of all speculation inferior to itself
> in generality.
>
> A. N. Whitehead

The final state of the Universe and mankind's role in the universe have throughout history been important topics of speculation for both philosophers and scientists. Final state scenarios seem to be based on one of three types of cosmological model. *Unchanging Models* claim that the Universe does not change in the large. *Cyclic Models* assert that the Universe undergoes a never-ending cycle of growth and decay, analogous to the human life-cycle. *Evolving Models* claim the Universe continuously evolves from some original state, and will never repeat a previous state.

In unchanging models there is no initial or final state; one time is the same as any other. When Einstein constructed his first cosmological model in 1916, he assumed that the Universe was of this class. In the large, Einstein's model was *static*; that is, the galaxies did not move systematically, relative to one another. Unfortunately, it was shown by Lemaître and his mentor Eddington that this static model is unstable. A slight perturbation would cause it to expand or contract, thereby converting it into a model of the third type.

The next attempt to construct an unchanging cosmology was made in the 1950's by Thomas Gold and Fred Hoyle. This cosmology was termed the *steady-state theory*. In this model the galaxies were pictured as moving apart according to the usual Hubble law, but the average density of galaxies in the Universe was kept constant by the continuous creation of primordial matter in intergalactic space. This material would then condense to form galaxies. The galaxies thus formed would evolve, eventually ending their existence as a collection of burnt-out stars. Thus, although the galaxies would undergo a birth and death cycle, the cosmos as a whole would retain the same aspect. At any time, the Universe would contain the same percentage of young, middle-aged, and dead galaxies.

The steady-state theory enjoyed wide-spread support among cosmologists in the 1950's, but as we pointed out in Chapter 6, it is generally considered to have been ruled out by the observation of the microwave background radiation. This radiation indicates that the visible universe was at one time much hotter and much denser than it is today. It is possible to retain a belief in the steady-state theory only if one is willing to assume that the visible universe is just a very small atypical portion of the entire Universe. Just as in the original version of the steady-state theory the galaxies were pictured as evolving and changing entities in a much larger structure which does not undergo any overall net change, so to defend this steady-state picture today we must picture the entire *visible* universe, which is that portion of the Universe within a Hubble distance ($\sim 10^{10}$ light years) of us, as an evolving 'bubble' within a much larger Universe. Although 'bubbles' would be born and then decay, the Universe as a whole would not undergo any net change.

This idea of a 'Universe of bubbles' was first put forward by Hoyle and his student Narlikar on the basis of their philosophical belief in an unchanging universe, but it has recently been independently invented by particle physicists who have been studying the implications of Grand Unified Theories (GUTs) for cosmology. In relativistic cosmology, a steady-state model will be the consequence of the following assumptions: (1) the universe is spatially homogeneous and isotropic on a sufficiently large scale, (2) the evolution is dominated by a positive cosmological constant term in the field equations, and (3) the universe has the spatial topology $\mathbf{R}^3$. Now GUTs strongly suggest that in the visible universe spontaneous symmetry-breaking should have given rise to an effective *negative* cosmological constant which is some fifty-seven orders of magnitude larger than is permitted by observation. This means that there must be an enormously large positive cosmological constant which will cancel out the negative cosmological constant generated by spontaneous symmetry-breaking. If this spontaneous symmetry-breaking does not act over the entire universe, but just in localized bubbles, then the evolution of the Universe as a whole will be dominated by the positive cosmological constant, which means that in the large, the Universe will be a steady-state cosmos. In section 9.5 we have seen one way in which Anthropic arguments can rule out such a steady-state cosmos, and we shall point out other Anthropic objections to such a scenario.

Until the advent of relativistic cosmology in the twentieth century, most scientific discussions of the evolution of the Universes were evolving models based on the concept of a 'Heat Death' of the Universe, which we discussed in section 3.7.

It was difficult, in the context of nineteenth-century physics, to criticize the predictions that the Universe would end in a Heat Death. We have mentioned a few rather weak and inconclusive criticisms in Chapter 3. The most powerful arguments that could be directed against it using only classical thermodynamics was first propounded in 1914 by the French thermodynamicist and philosopher of science Pierre Duhem:[5]

> The deduction [of the Heat Death from the Second Law of thermodynamics] is marred in more than one place by fallacies. First of all, it implicitly assumes the assimilation of the universe to a finite collection of bodies isolated in a space absolutely void of matter; and this assimilation exposes one to many doubts. Once this assimilation is admitted, it is true that the entropy of the universe has to increase endlessly, but it does not impose any lower or upper limit on this entropy; nothing then would stop this magnitude from varying from $-\infty$ to $+\infty$ while the time itself varied from $-\infty$ to $+\infty$; then the allegedly demonstrated impossibilities regarding an eternal life for the universe would vanish.

We shall see below that both effects discussed by Duhem operate in general relativity to prevent a Heat Death from occurring in a relativistic cosmology. First of all, we take a relativistic cosmology that is assumed at present to be roughly homogeneous and isotropic, and to be either open or closed. If it is open, then there is an infinite amount of non-gravitational energy available now. If it is closed, then the relativistic analogue of the conservation of energy equation, namely $T^{ab}_{;b}=0$, implies that the total energy, which is the sum of the gravitational and non-gravitational energies and which can be written as a volume integral over the three-sphere corresponding to space at a given time, is trivially zero. This result can be interpreted either as saying that the conservation of energy law is 'transcended globally' (Wheeler prefers this interpretation[6]) or that the gravitational and non-gravitational energies in a closed universe are always equal in magnitude but opposite in sign (York prefers this interpretation[7] and Penrose's new definition of mass supports this interpretation).[8,9] In either interpretation the law of energy conservation places no restrictions on the continued entropy generation in a closed universe. In the Penrose-York interpretation, available free-energy can always be increased without limit by increasing the magnitude of the gravitational energy without limit. We shall see below in our analysis of life in a closed universe that this is possible; in effect, gravitation is the *ultimate* source of energy.

In spite of earlier cautions, the notion of a Heat Death dominated thought at the end of the nineteenth century, as we discussed in Chapter 3. The discovery of the expanding universe in the early part of the twentieth century changed the picture of the Heat Death slightly; but, as developed by the British astrophysicists Jeans and Eddington, relativistic cosmology in the form of universe which expands forever would still end in a type of Heat Death. As Eddington asserted in 1931[10]:

> It used to be thought that in the end all the matter of the Universe would collect into one rather dense ball at uniform temperature; but the doctrine of the spherical space, and more especially the recent results as to the expansion of the Universe, have changed that . . . It is widely thought that matter slowly changes into radiation. If so, it would seem that the Universe will ultimately become a ball of radiation growing ever larger, the radiation becoming thinner and passing into longer and longer wave lengths. About every 1,500 million years it will double its radius, and its size will go on expanding in this way in geometrical progression forever.

In his classic work of speculative cosmology, *The World, the Flesh, and the Devil*, written in 1929, the physicist J. D. Bernal tried to picture what life would be like in the far future of such a universe:

> Finally, consciousness itself may end or vanish in a humanity that has become completely etherialized, losing the close-knit organism, becoming masses of atoms in space communicating by radiation, and ultimately perhaps resolving itself entirely into light[111] . . . these beings, nuclearly resident, so to speak, in a relatively small set of mental units, each utilizing the bare minimum of energy, connected

together by a complex of etherial intercommunication, and spreading themselves over immense areas and periods of time by means of inert sense organs which, like the field of their active operations, would be, in general, at a great distance from themselves. As the scene of life would be more the cold emptiness of space than the warm, dense atmosphere of the planets, the advantage of containing no organic material at all, so as to be independent of both of these conditions, would be increasingly felt.[112]

But in the end, Bernal came to believe that his 'etherialized life' probably would be destroyed in the Heat Death:

The second law of thermodynamics which, as Jeans delights in pointing out to us, will ultimately bring this universe to an inglorious close, may perhaps always remain the final factor. But by intelligent organizations the life of the Universe could probably be prolonged to many millions of millions of times what it would be without organization.[113]

It is now generally believed that protons and other forms of matter will decay, in part to radiation. Thus Eddington's picture of the final state of ever-expanding cosmologies is quite similar in several respects to the contemporary view, as we shall discuss in more detail below. Also, we shall show in section 10.6 that if life continues to survive in the far future, it must take on a form that is roughly similar to Bernal's 'etherialized life'.

There is another criticism which can be directed against the concept of the Heat Death and which is based on nineteenth-century physics: the so-called *recurrence paradox* of Poincaré and Zermelo, which we mentioned briefly in section 3.8.

The recurrence paradox arose in physics as a consequence of attempts to derive the Second Law of thermodynamics from Newton's laws of motion. In 1890, Poincaré[11] showed that for almost all initial states, any Newtonian mechanical system with a finite number of degrees of freedom, finite total and kinetic energy, which is constrained to evolve within a finite spatial region must necessarily return arbitrarily closely and infinitely often to almost every previous state of the system. Poincaré emphasized that this doomed attempts to deduce the Second Law rigorously from Newton's laws, because the recurrence theorem proved that a mechanical system must be cyclic in its behaviour rather than unidirectional as implied by the Second Law. Thus if we believe in the validity of the Newtonian laws of motion, a Heat Death cannot be the final state of the Universe. Rather, the evolution of the Universe must consist of a series of cycles.

This idea of a cyclic universe—the second type of cosmological model—is very old. Histories of the development of this idea in prescientific times have been written by Eliade,[12] by Jaki,[13] and by Tipler.[14] Modern science contained the idea of a cyclic cosmos from the very beginning. Newton himself was worried that his solar system model was gravitationally unstable in the long run, and to compensate for this instability he suggested a cyclic process whereby the planets would be replaced as the gravitational action of the other bodies in the solar system[15] periodically perturbed them from their orbits. By the beginning of the nineteenth century, Euler, Laplace, Lagrange, and others had shown that the solar system was in fact stable to first order, the gravitational perturbations leading merely to a cyclic oscillation of the planetary orbits.

The cosmological implications of Newtonian theory were first discussed extensively by the German philosopher Immanuel Kant in 1755. In Kant's cosmology, the inhabited portion of the Universe began as a perturbation of initially static matter, distributed in a homogeneous and isotropic manner throughout infinite Euclidean space. This material perturbation condenses to form the stars and planets.[16] Eventually our particular region of space will exhaust its energy, and the inhabited portion will be another region, conspheric around the original perturbation, whose condensation has been started by the initial perturbation. Thus as time advances, the inhabited portion of the Universe is restricted to spheres of larger and larger radius around the point where the initial disturbance began.[17] Thus from the point of view of life, Kant's cosmology is globally progressive, in the sense that in the long run, the region in which life exists increases with time as t^2, with $t=0$ being the instant of the initial perturbation. This is a cosmological analogue of the progressive expansion of the biomass on Earth (see section 3.2). However, in Kant's scheme life is locally cyclic, for in each sphere life begins anew rather than expanding outward from the point at which it first began. As we mentioned in sections 3.2 and 3.10, it was impossible for Kant and the other eighteenth-century philosophers to imagine a progressive evolution of life, because the Principle of Plenitude did not allow it. In the model we shall develop below, life will be globally progressive in two senses: the amount of living material, and the amount of knowledge both grow without limit as a power of the cosmic time t.

Modern discussions of the cyclic universe are generally based on the so-called 'oscillating closed universe' model found in 1922 by A. Friedman.[18] Friedman himself was aware of the cyclic nature of time in his solution, and suggested that one could identify corresponding times in each cycle. However, in the Friedman model the radius of the Universe goes to zero at the beginning and at the end of each cycle, and thus from a strict mathematical standpoint the cycles were disjoined by a singularity. In other words, they were not actually cycles. Each 'cycle' would really be a universe complete in itself, with no possibility of transmitting information of any sort from one 'cycle' to the next. In 1931 Tolman[19] proved that such a singularity was inevitable at the beginning and at the end of any isotropic and homogeneous closed universe with a physically reasonable matter tensor. He argued[20] that this singularity was merely an artefact of the high symmetry assumed, and that in a physically realistic universe, which naturally would not be exactly isotropic and homogeneous, these singularities would disappear. Therefore he assumed that in a realistic case the singularity would be replaced by a very small but non-zero radius followed by a re-expansion, and that the entropy would be conserved on passage through this radius. This would result in the thermodynamics of a cycle being determined in part by the history of a previous cycle. Other relativists of the time by and large agreed with Tolman that an initial singularity was unlikely, and then they found his proposal of transfer of information from one cycle to the next quite reasonable (see ref. 21 for a detailed discussion of the early relativists' opinions on singularities).

The Hawking–Penrose singularity theorems, which indicated that a singularity was inevitable provided certain very general hypothesis were made, changed relativists' minds on the reality of oscillating universes. It is now generally believed that either the Universe began in an initial singularity some 20 billions years ago (as measured in proper time), or else quantum effects must be the agency causing the Universe to 'bounce' at extremely high densities and temperatures or which even allows it to appear spontaneously from 'nothing'. By 'extremely high' we mean something of the order of the Planck density (5×10^{93} gm/cm^2) or the Planck temperature (1.4×10^{32} K). Wheeler, for example has until recently suggested[22] that the physical constants themselves are cycled at such a bounce. At present, however, Wheeler believes in the reality of an initial singularity, and thus he is advocating a 'one-cycle' closed universe model.[23,24] As we will discuss below, the large primordial cosmological constant in which many particle physicists believe could cause such a bounce if the temperature goes sufficiently high to dissolve the spontaneous symmetry breaking, but this process can not lead to a series of cycles. Only a single bounce would be possible.[116] We note in passing that the SAP and FAP arguments which we used in section 9.5 against the steady-state theory can also be used to eliminate the possibility that a cyclic universe results from presently unknown physical laws, which cause an infinite sequence of bounces.

10.3 Global Constraints on the Future of the Universe

Absolute space is the divine sensorium.

E. A. Burtt,
paraphrasing
Sir Isaac Newton

In this section we shall briefly review the possible future histories of the Universe from a more mathematical point of view than in Chapter 6. A reader wishing a more detailed discussion is referred to ref. 25. We shall consider only those universe models which satisfy the Principle of Strong Cosmic Censorship.[21,26] This Principle states that the space-time manifold is *globally hyperbolic*, which in rough non-technical language means Laplacean determinism holds: initial data given on a special space-like slice S of the space-time manifold uniquely determine the entire global structure of space-time.[25,27] The special spacelike slice is called a Cauchy hypersurface and Geroch[28] has shown that, in particular, the Principle of Strong Cosmic Censorship implies that the global topology of space-time is $\mathbf{S}\times\mathbf{R}^1$, where $\mathbf{S}$ denotes the topology of any Cauchy hypersurface. If $\mathbf{S}$ is compact, then any compact spacelike 3-manifold in the globally hyperbolic space-time is in fact a Cauchy hypersurface.[29,30] (This is not true if $\mathbf{S}$ is non-compact.)

From the point of view of classical general relativity, the reason for postulating Strong Cosmic Censorship is that if this assumption is dropped, the future evolution of the universe becomes non-unique. Strong Cosmic Censorship can only be violated if space-time has singularities which lie both in the future and in the past of some observer's world-line. Since space-time itself breaks down at such a naked singularity, *anything* can come out of the singularity, resulting in an inability to predict the future evolution of the universe.

There are indications that naked singularities would cause even worse disasters in quantized general relativity, although we cannot be sure of this, because to date there is no complete quantum theory of gravity. For instance, Hawking,[31] Wald[32] and Page[33] have shown that naked singularities resulting from quantum black hole evaporation could cause pure quantum states to evolve into mixtures, which is not allowed by the fundamental postulates of quantum field theory. Such an evolution would also undermine the theoretical basis for the Many-Worlds interpretation of quantum mechanics. The entire reason for inventing this interpretation in the first place was to avoid having to assume an interaction (the collapse of the wave function) which caused pure states to become mixtures.

There are a considerable number of quite different cosmologies which do obey the Principle of Strong Cosmic Censorship. They are distinguished by the topology of their Cauchy hypersurfaces, and they have been classified into two categories. The *closed universes* are those whose Cauchy hypersurfaces are compact, and the *open universes* are those whose Cauchy hypersurfaces are non-compact. (Compactness is a topological concept; see ref. 34 for a definition of this and other topological terms.) We discussed these two classes of cosmological models in Chapter 6 from a physical point of view.

In addition to classification by the topology of the Cauchy hypersurfaces, universes can also be distinguished by their long-term dynamical behaviour. Universes whose size or radius of curvature (scale factor) grows without limit are called *ever-expanding* universes while universes which reach a maximum size and recollapse to a final singularity are called *recollapsing universes.* This classification applies only to those cosmologies which are now expanding, as the real Universe apparently is.

Friedman cosmologies—those which have Cauchy hypersurfaces that are homogeneous and isotropic—are generally considered to have one of two possible Cauchy hypersurface topologies: $\mathbf{R}^3$ and $\mathbf{S}^3$. Because of the high symmetry in Friedman cosmologies, identifications can be made in the Cauchy hypersurfaces to form non-simply connected topologies. For example, the open $\mathbf{R}^3$ topology can be identified to form a three-torus $\mathbf{T}^3$. Such identifications are generally considered unaesthetic and in any case would destroy some of the global symmetries of the Friedman universe. In our three-torus example, the global rotational symmetry which is present in the original $\mathbf{R}^3$ topology is no longer present in the $\mathbf{T}^3$ universe. In the case of the Friedman cosmologies, it has been known since Tolman's work in the 1930's that there is a deep connection between the topology of the Cauchy hypersurface and the long term dynamical behaviour. Universes with topology $\mathbf{R}^3$ and other universes formed from them by identification expand forever provided the stress-energy tensor satisfies

$$(T_{ab}-\tfrac{1}{2}Tg_{ab})V^aV^b \geqslant -\frac{1}{8\pi G}\Lambda V^aV_a \qquad (10.1)$$

for all unit time-like vectors V^a. Furthermore, all Friedman universes with topology $\mathbf{S}^3$ recollapse provided the same inequality holds.

It is not known whether this connection between Cauchy hypersurface topology and long-term dynamics persists when the conditions of homogeneity and isotropy are relaxed. It is, however, generally believed that this connection is valid for any globally hyperbolic cosmology which satisfies (10.1). A few partial results are known. It is known,[30,114] that a necessary and sufficient condition for recollapse to occur in globally hyperbolic closed universes is for the space-time to contain a maximal Cauchy hypersurface, which is a spacelike hypersurface with vanishing trace of its extrinsic curvature. Such a hypersurface is the largest hypersurface in the universe; the maximal hypersurface defines the time of maximal expansion of the Universe. The following theorem,[114] which is a restatement and slight generalization of an earlier theorem due to R. Schoen and S.-T. Yau,[36] places strong restrictions on the topology of recollapsing closed universes:

Theorem: If **S** *is a spacelike compact orientable maximal hypersurface,* *then it must have topology*

$$(\mathbf{S}^3\times\mathbf{P}_1)\#(\mathbf{S}^3\times\mathbf{P}_2)\#\ldots\#(\mathbf{S}^3\times\mathbf{P}_n)\#k(\mathbf{S}^2\times\mathbf{S}^1)$$

where $\mathbf{P}_i$ *is a finite subgroup of SO(3), "#" denotes connected sum, and* $k(\mathbf{S}^2\times\mathbf{S}^1)$ *means the connected sum of k copies of* $\mathbf{S}^2\times\mathbf{S}^1$, *provided the following conditions hold:*

(1) *The Einstein equations* $R_{ab}-Rg_{ab}+\Lambda g_{ab}=8\pi GT_{ab}$ *hold on the space-time:*

(2) $[T_{ab}-\Lambda g_{ab}/8\pi G]V^aV^b\geqslant 0$ *for all timelike vectors* V^a;

(3) *The space-time is not suitably identified Minkowski (flat) space;*

(4) *The differentiable structure on the space-time is not exotic.*

The terms in this theorem require some explanation. Roughly speaking, a *connected sum* of two three-dimensional manifolds is the manifold formed by cutting a small spherical volume out of each, and then gluing the two manifolds together along the boundaries of the remaining manifolds. The *quotient* $\mathbf{S}^3/\mathbf{P}_i$ of a three-sphere $\mathbf{S}^3$ with a subgroup $\mathbf{P}_i$ means identifying points of the three-sphere which are carried into one another under the action of the subgroup. A *non-exotic differentiable structure* on the space-time manifold $M=\mathbf{S}\times\mathbf{R}^1$, where $\mathbf{S}$ is the maximal spacelike hypersurface, means that the coordinate systems covering M are generated by pulling up the coordinate systems which cover $\mathbf{S}$. Condition (4) is probably not necessary, but it simplifies the proof of the theorem. In any case, cosmologists never even consider space-times which violate condition (4), for there is no evidence for exotic differentiable structures.

Since any physically realistic space-time contains *some* matter and hence is not flat space, and also satisfies condition (3) in the low density regime where a maximal hypersurface would be expected to occur (unless $\Lambda<0$), this means that closed universes expand forever if their topologies are not of the above form. In particular, the three-torus ($\mathbf{T}^3$) closed universes will expand forever.[30,36]

The only simple topologies which are of the above form are the topologies $\mathbf{S}^3$ and $\mathbf{S}^2\times\mathbf{S}^1$. All known examples of closed universes with Cauchy hypersurfaces having these topologies actually recollapse, so it has been conjectured[30,114] that *all* universes with Cauchy hypersurfaces having these topologies and satisfying conditions (1)–(4) recollapse.

If the *strong energy condition* holds; that is, if

$$(T_{ab}-\tfrac{1}{2}g_{ab}T)V^aV^b\geqslant 0 \tag{10.2}$$

for all timelike vectors V^a, and if the cosmological constant is *negative*, then one of us[38] has shown that *all* globally hyperbolic universes recollapse irrespective of spatial topology. The relevance of this result for cosmology is unclear at present. Spontaneous symmetry-breaking gives rise to a negative vacuum energy density, and this vacuum energy density is equivalent to a negative cosmological constant.[39,40] Unfortunately, the magnitude of this effective cosmological constant is too large by many orders of magnitude to be consistent with observation,[41] and so many physicists[42,43] assume that there exists a positive primordial cosmological constant of unknown origin which cancels out the effective cosmological constant generated by spontaneous symmetry breaking. It is of course possible that this cancellation is not exact and there is a tiny residual net cosmological constant. If so, this net cosmological constant may dominate the dynamical evolution of the Universe in the long term. This net cosmological constant, if it exists, could be of either sign. If the universe recollapses, and the spatial topology is not of the form given in the theorem, then the net cosmological constant must be negative.

The most important physical factor in the dynamical evolution of the Universe is the scale factor $R(t)$, which can be roughly interpreted as the geometric mean radius of the Universe. For space-times which are approximately homogeneous and isotropic over most of their history the time evolution of $R(t)$ is governed by a generalized Friedman equation[44,45] along with an evolution equation for the shear tensor, σ_{ab}, of the timelike geodesic congruence normal to the Cauchy hypersurfaces (the shear, σ, is defined by $\sigma^2 = \sigma_{ab}\alpha^{ab} \geqslant 0$):

$$\frac{\dot{R}^2}{R^2} = \frac{8\pi G}{3}\left[\rho_m + \rho_\gamma + \frac{3\Lambda}{8\pi G}\right] + \sigma^2 - \frac{k}{R^2} \tag{10.3a}$$

$$\sigma_a^b = R^{-3}\left\{\Sigma_a^b + \int [{}^{(3)}R_a^b - \tfrac{1}{3}\delta_a^{b(3)}R]R^{-3}\,dt\right\} \tag{10.3b}$$

where Σ_a^b is time-independent and we write $\Sigma_{ab}\Sigma^{ab} \equiv \Sigma^2$ and the quantity in (10.3b) under the integral is the anisotropic part of the spatial three-curvature ${}^{(3)}R_a^b$; $\dot{R} \equiv dR/dt$, Λ is the cosmological constant, $k > 0$ if the universe has the spatial topology $\mathbf{S}^3$, and $k < 0$ or 0 if it has the topology $\mathbf{R}^3$ ($k = 0$ if and only if the Cauchy surfaes are flat.) Cosmologies will often reserve the term 'open universe' for the $k < 0$ open universes, and call the $k = 0$ open universes 'flat universes', (see section 6.8). The term σ^2 measures the energy in the form of anisotropic gravitational shear, or roughly speaking, the energy in the form of very long wavelength gravitational waves.

The shear evolution equation of (10.3) shows that, in general, σ has two sources: a kinematic Newtonian component, $\Sigma^2 R^{-6}$, associated with the isotropic part of the curvature, and a non-Newtonian part associated with the spatial curvature anisotropy. In the most general anisotropic cosmological models it is this anisotropic curvature term which tends to dominate the dynamics at late times. In ever-expanding open universes it typically contributes a shear evolution $\sigma^2 \propto R^{-2} \propto t^{-2}$ as its dominant term at large times. The term ρ_m is the density of the material particles which travel on timelike world-lines. Examples are electrons, protons, human beings, and black holes. It can be shown that $\rho_m = C/R^3$, where C is a constant. The term ρ_γ is the density of massless particles which travel along null world-lines. It is thus composed of all radiation fields except for long-wavelength gravitational waves. We have $\rho_\gamma = \Delta/R^4$, where Δ is a constant, assuming no conversion of matter into radiation. It is clear from (10.3) that the cosmological constant will dominate as $R \to \infty$. However, there will in general be epochs when other terms will be the most important. At present, for example, ρ_m is the most important term, so we say the Universe is at present 'matter dominated'.

If the Universe is closed and recollapses then R will increase from zero, rise to a maximum $R_{\max}$, and then decrease back to zero. The zeros correspond to the initial and final singularities, respectively. The *maximum* proper time t_U along all timelike curves going from the initial to the final singularity is called the *lifetime of the Universe.* If it is assumed that the Universe is closed and recollapses, then this lifetime can, in principle, be computed from (10.3) using observations made at the present day.

At present the Σ^2/R^6 term is very small in comparison to the other terms. Since it drops off faster than the other terms, we can ignore its effect except near the final singularity. Recall from Chapter 6 that we can express the age of the Universe in terms of H_0, Ω_0, R_0, and Λ, where

$$H_0 \equiv \frac{1}{R}\frac{dR}{dt}\bigg|_{t=\text{present day}} \tag{10.4}$$

is the Hubble constant measured today ($H_0 = 50$–$100\ \text{km s}^{-1}\ \text{Mpc}^{-1}$, according to the observers) and

$$\Omega_0 \equiv \frac{\rho_{m0} + \rho_{\gamma 0}}{3H_0^2/8\pi G} \equiv \frac{\rho_{m0} + \rho_{\gamma 0}}{\rho_c} \tag{10.5}$$

is the density parameter, and ρ_c is the critical density, so called because the total density of the universe must be greater than ρ_c if $k > 0$ and $\Lambda = 0$. Thus for a closed three-sphere universe, we must have $\Omega_0 > 1$. The scale factor of the Universe today is denoted R_0. The *Hubble distance* R_H is cH_0^{-1} and the *Hubble time* t_H is H_0^{-1}. If $\Lambda = 0$ and the universe is matter-dominated the lifetime of the universe, t_U, is

$$t_U = t_H\left(\frac{\pi\Omega_0}{(\Omega_0 - 1)^{3/2}}\right)$$

$$= \pi\frac{R_0}{c}\left(1 + \left(\frac{R_0}{R_H}\right)^2\right) \tag{10.6}$$

The lifetime t_U is just twice the time to maximum expansion; see equation (6.138).

If the universe becomes radiation-dominated for most of its future history, say through the radiocative decay of matter and the Hawking evaporation of black holes, and if $\Lambda = 0$, then the lifetime of the universe is

$$\begin{aligned} t_U &= t_H\left(\frac{2\Omega_0^{1/2}}{\Omega_0 - 1}\right) \\ &= 2\frac{R_0}{c}\left(1+\left(\frac{R_0}{R_H}\right)^2\right)^{1/2} \end{aligned} \tag{10.7}$$

The universal lifetime should lie between the values given by (10.6) and (10.7) if matter is converted into radiation slowly, but not slowly enough to give (10.6). We have pointed out earlier in Chapter 6 that Anthropic and other theoretical considerations imply Ω_0 is very close to one.

One can prove a number of very general theorems about the long-term time evolution of the universe. For example, we can prove that in contrast to a finite universe governed by Newtonian mechanics, states of a closed general relativistic universe cannot recur. In other words, the universe is *not* oscillating. The events of the present will *never* be repeated in the future, and what is more, the events of the future will not even be arbitrarily close to present events.[13,46]

Another theorem, first obtained by Brill and Flaherty, and generalized by Tipler and Marsden[30], and by Gerhardt,[30] is that in a globally hyperbolic universe which is not everywhere flat and which satisfies (10.1), there will exist a unique globally defined time coordinate, which is given by the constant mean curvature foliation. A time coordinate in relativity is defined by any 'slicing' of four-dimensional space-time by a sequence of three-dimensional spacelike hypersurfaces. This sequence is called a *foliation* of space-time, and each hypersurface is called a *leaf* of the foliation. For a simple example of the concept of 'foliation', consider the surface of an ordinary cylinder. The surface of an ordinary cylinder is two-dimensional, and it can be foliated by a sequence of circles which are perpendicular to the axis of the cylinder. The cylinder is then just all of these circles stacked on top of one another. Each circle is a leaf of the foliation, and the foliation is all of the circles together.

Any physically realistic cosmology can be foliated uniquely by Cauchy hypersurfaces of constant mean extrinsic curvature, and it is this foliation which defines the unique global time. The *extrinsic curvature* of a spacelike hypersurface is its relative rate of expansion in time. This relative rate of expansion is measured by the *Hubble parameter* $H \equiv (1/R)\,dR/dt$, which we have encountered earlier in our discussion of the Friedman universe. However, in a general cosmology it is possible for the universe to expand faster in some directions than others, so the Hubble parameter must be generalized to a tensor in order to express properly this directional dependence. This tensor is the extrinsic curvature. The *mean extrinsic curvature* is a scalar like the Hubble parameter, and it is an average of the extrinsic curvatures in the three spatial directions. (More exactly, it is the contraction of the extrinsic curvature, which is a rank two tensor—see ref. 25 or 30 for a precise definition). A *constant mean extrinsic curvature hypersurface*, or constant mean curvature hypersurface for short, is a spacelike hypersurface on which the mean extrinsic curvature is the same at every point. The hypersurfaces of homogeneity and isotropy in the Friedman universe are constant mean curvature hypersurfaces on which the mean curvature is $3H$. Since the Universe is in fact closely isotropic and homogeneous, the constant mean hypersurface defining the global instant 'now' over the entire universe essentially coincides with the spacelike hypersurface in which the 3 K background radiation temperature is constant. The Earth is currently moving at about 300 km/sec with respect to this globally defined rest frame of the universe.

In addition to the no-return theorem and the uniqueness of cosmological time theorem, one can obtain some constraints on the long-term behaviour of the matter and shear terms in equation (10.3), even beyond the point at which the equation breaks down. If the space-time can be assumed to remain roughly homogeneous for all future time (this should be a good approximation for ever-expanding universes), then from Chapter 6 (see also ref. 47):

$$\lim_{t \to +\infty} \inf t^2 \sigma^2 < \tfrac{3}{8} \tag{10.8}$$

If instead the universe ends in a final singularity, then along a timelike geodesic which terminates in this final singularity at proper time t_f, we must have[48]

$$\lim_{t \to t_f} (t_f - t)^2 \left(8\pi G \left[T_{ab} - \tfrac{1}{2} g_{ab} \left(T + \frac{1}{4\pi G} \Lambda \right) \right] V^a V^b + \sigma_{ab} \sigma^{ab} \right) \leq \tfrac{1}{2} \tag{10.9}$$

Roughly speaking, these inequalities say that the shear cannot drop off slower than $1/t^2$ if the universe expands forever, and the shear and matter energy densities cannot increase faster than $1/t^2$ near the final singularity if it recollapses. Conversely, as a general rule one can compute $R(t)$ in a given regime from the requirement that the dominant term for that regime in the generalized Friedman equation (10.3) dies off or grows as $1/t^2$. For example, in a matter-dominated regime $\rho_m \propto R^{-3}$ is the dominant term by definition, the rule $R^{-3} \propto t^{-2}$ implies $R(t) \propto t^{2/3}$. For a radiation-dominated regime $R^{-4} \propto t^{-2}$ gives $R(t) \propto t^{1/2}$. Spatially-flat universes ($k=0$) will virtually always be either radiation- or matter-dominated, since the only other term in (10.3) is apparently zero near the initial singularity. Universes which are not spatially flat (those universes with $k \neq 0$) will have regimes where the radiation dominates, the matter dominates, the spatial curvature (the k/R^2 term) dominates, or the shear dominates. If the spatial curvature or the R^{-2} shear term dominates—as it will in the far future of an open ($k<0$) universe—$R^{-2} \propto t^{-2}$ gives roughly $R(t) \propto t$.

There are two terms, the R^{-2} shear term and the spatial curvature term, either of which could be the dominant term in the far future of open ($k = <0$) universes. In fact, as we discussed in section 6.11, both are important in generic open universes. Whenever $R(t)$ varies as a power of t; i.e. whenever $R(t) \propto t^n$, where n is some positive constant, the Hubble parameter $H \equiv \dot{R}/R$ will vary as $H \propto 1/t$ whatever the value of the constant n. The shear term will be important unless the distortion $\sigma/H \propto \sigma t$ goes to zero asymptotically; and we have seen in section 6.11 that this does not occur generically, so the shear term remains important in almost all open universes.

Both of the R^{-2} terms are absent in the spatially flat ever-expanding universes, so the long-term evolution of these universes will be either matter- or radiation-dominated. We shall see in section 10.5 that for the most part it will be matter-dominated as it is now, except for a brief period 10^{30} years in the future.

The generic behaviour of closed universes in the shear-dominated regime near the final singularity is particularly interesting. In closed spatially homogeneous universes, which have been extensively studied,[68,69] the anisotropic curvature stresses create a chaotic, oscillatory evolution of the shear anisotropy. We have very little knowledge about the behaviour of inhomogeneous closed universes in this regime; we can only hope it is qualitatively similar to the homogeneous case.

As we discussed in Chapter 6, in a homogeneous closed universe with topology $\mathbf{S}^3$ (the only closed universe topology we shall consider) the shear measures the rate of change in the distortion of the three-sphere. When the shear σ is identically zero the closed universe is isotropic, which means the proper distance around the universe is the same in all directions. If the shear is non-zero, the proper distance around the universe at any given time depends on the direction in space. In developing a feel for the physical meaning of shear, it is instructive to visualize what shear means for a two-sphere. Imagine an observer standing at a point on the two-sphere—the north pole, say—and looking out along two mutually perpendicular great circles through that point. If the sphere is undistorted, the lengths of the great circles will be the same. If the sphere is distorted into an ellipsoidal figure, the length of one great circle will be longer than the other.

Suppose our two-sphere universe is shrinking in area as it goes into a final singularity (where the area is zero). The shear measures the rate of change of the distortion, so a non-zero shear means that as our two-sphere universe gets smaller, the lengths of the great circles change their size at different rates: the universe changes its size differently in different directions. A contracting universe means the area is decreasing, but it is quite possible for the length of one great circle to *increase*, and the over-all area will still decrease if the other length decreases even faster.

The behaviour of the three-sphere universe is qualitatively the same. In three dimensions there would be three mutually perpendicular great circles. A non-zero shear means these great circles are changing their lengths at different rates. The typical behaviour on approach to a singularity is for two of the great circles to get smaller very rapidly while the other gets longer, yet the net volume of the universe still decreases.

But this situation does not persist for very long. The rate of expansion of the expanding great circle decreases to zero, and the rate of contraction of the other two great circles decreases, until the previously expanding great circle starts to *contract* at a faster and faster rate, and one of the previously contracting great circles begins to expand. We can equally well express this by saying that an over-all contracting universe expands in one direction while contracting in the other two, and the direction of expansion changes with time. This is pictured in Figure 10.1 for our two-sphere universe.

This directional dependence of expansion and contraction means the temperature of the background radiation will depend on direction also. The radiation coming from the expanding direction will be redshifted, while the radiation coming from the contracting directions will be blueshifted. The precise directional dependence of the temperature is a rather complicated function of the optical depth of the universe at the time. (The optical depth measures the distance a radiation particle can travel before being absorbed.) Approximate formulae have been obtained by Thorne,[49] Misner,[50] Barrow *et al.*,[51] and Matzner.[51] For small optical depths—the expected case near the final singularity—the formula simplifies enormously to:

$$T(n) = (T_0/R)/(\{\exp(\beta)\}_{ij} n^i n^i)^{1/2} \tag{10.10}$$

where n^i is a unit vector in the direction the temperature is measured, $(T_0/R(t))$ is the temperature averaged over all directions (we put in a factor $R(t)$ explicitly to indicate that this average temperature scales with the universal scale factor; T_0 is a constant), and the exponential factor is a direct measure of the *ratios* of the proper distances around the universe in

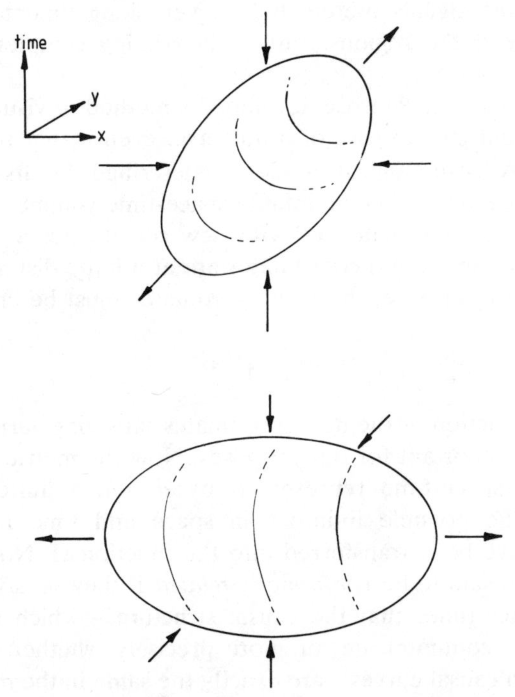

Figure 10.1. A contracting closed universe with shear. Although the total volume of the universe decreases monotonically, the universe contracts in one direction while it expands in another. Some time later the directions of expansion and contraction have been interchanged.

the three directions. The variation in temperature with direction will be the same at every point in a homogeneous universe. The temperature difference in different directions is a manifestation of the shear gravitational energy, since it is the shear that generates a non-zero β (we actually have $\sigma_{ab} \equiv \dot{\beta}_{ab}$). The temperature difference can, in principle, provide an energy source for life in a closed universe near the final singularity.

The extremely rapid contraction in one direction in a shearing universe can cause the disappearance of the horizons in that direction. This fact will be crucial for the continued existence of life in closed universes, because horizons are the ultimate barriers to communication in space-time.

An horizon is said to exist if there are regions of space which cannot send light signals to each other. If the regions cannot send light signals, then they cannot send signals of any sort, which means it is impossible for them to communicate. But to determine that regions cannot communicate, it is necessary to know the entire future history of the regions, for it may be that the signals merely take a very long time to traverse the distance between the regions, rather than being completely unable to traverse the distance.

In the 1960's Roger Penrose developed a method to visualize easily the entire future and past history of a universe, even if that future and past are infinite. A cosmological model is described by its metric $ds^2 = g_{ab}dx^a dx^b$ which may define an infinite space-time volume. Penrose's idea is to replace the coordinates x^a with new coordinates $\mathbf{x}^a$ such that the points at infinity in the old coordinates are at a finite distance in the new coordinates. Furthermore, the new coordinates must be chosen so that

$$ds^2 = \Omega^2 \mathbf{ds}^2 \tag{10.11}$$

where Ω is a function of the new coordinates satisfying various conditions which are not important for our purposes. Now the metric $\mathbf{ds}^2$ covers the whole of the space-time represented by ds^2 in a finite range of its coordinates; the possible infinities in space and time in the original coordinates have been transferred into the function Ω. Now two metrics ds^2 and $\mathbf{ds}^2$ are said to be *conformally related* if they satisfy (10.11). This means for space-times that the causal structures—which regions in the space-time can communicate, or more precisely whether events can be connected with causal curves—are exactly the same in the metrics ds^2 and $\mathbf{ds}^2$. Thus if we are interested in the causal structure of the original metric ds^2, all we have to do is throw away the function Ω and study the metric $\mathbf{ds}^2$ in a small finite region, for in this region the causal structure will be exactly the same as for the whole of ds^2.

The conformal metrics, $\mathbf{ds}^2$, have been computed for a number of key cosmological models and the region conformal to the entire original cosmological model can be drawn as a two-dimensional figure called a *Penrose diagram* (or *conformal diagram*), in which the time dimension and one of the three spatial dimensions appear in the figure. The Penrose diagram for the open and flat Friedman universe are shown in Figure 10.2, the Penrose diagram for the closed Friedman universe is shown in Figure 10.3, and the Penrose diagram for the static Einstein universe is shown in Figure 10.4. A fourth Penrose diagram, that of the steady-state universe, was given as Figure 9.1 in Chapter 9.

The causal conventions in Penrose diagrams are the same as in Minkowski diagrams: lines at 45° off the vertical are the paths of light rays, timelike curves are those whose tangents are less than 45° off the vertical and spacelike curves are those whose tangents are greater than 45° off the vertical. Time increases vertically upward, and the horizonal direction is a space direction.

The boundaries of a Penrose diagram represent what are termed

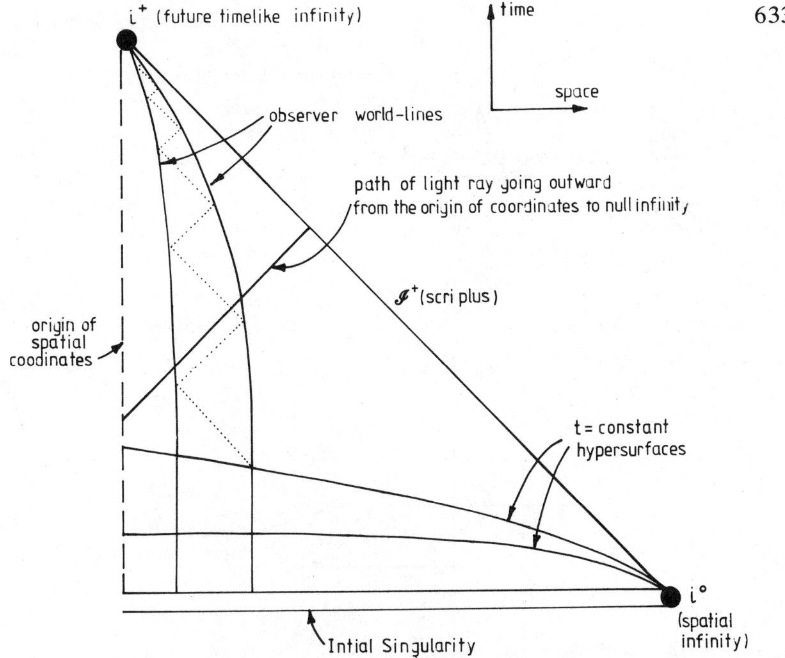

Figure 10.2. Penrose diagram for the open or flat Friedman universes. The horizontal double lines at the bottom of the figure denote the initial singularity out of which the universe sprang. The dashed vertical line denotes the origin of spatial coordinates. For all Penrose diagrams, a timelike curve (world line of a possible observer) is any curve whose tangent makes an angle of less than 45° off the vertical. Null lines (paths of light rays) are those whose tangents make an angle of 45° off the vertical, and spacelike curves or surfaces make an angle of more than 45° off the vertical. A Penrose diagram represents the time dimension and one of the three spatial dimensions: in a Friedman universe space is spherically symmetric about the origin of coordinates, so the two angular coordinates in a spherical coordinate system can be suppressed without loss of information. Thus each point in the Penrose diagram except the origin of coordinates actually represents a two-sphere. Two observer world-lines are pictured. These observers are those which are at rest with respect to the universal background radiation, or equivalently, which are normal to the constant mean extrinsic curvature foliation. *All* observers which do not accelerate to the speed of light come together in the infinite future at i^+, future timelike infinity. All outgoing light rays hit $\mathscr{I}^+$ (scri plus) when $t = +\infty$. Two leaves of the constant mean extrinsic curvature foliation are pictured. The leaves of this foliation define a global time; $t =$ constant in each leaf. Each leaf has infinite volume, and each leaf hits i^0 at spatial infinity. The jagged dotted line connecting the two observer world-lines denotes light signals being sent back and forth between the two observers. Because the observers are coming closer and closer together in the Penrose diagram as future timelike infinity is approached, they will be able to send an infinite number of signals to each other between now and future timelike infinity. The same process looks as follows in the actual Friedman universe: the two observers are moving away from each other, so the proper time between the transmission of a pulse from one observer to another and the receipt of a return pulse will grow longer and longer. But since there is an infinite amount of future proper time, an infinite number of pulses can be sent.

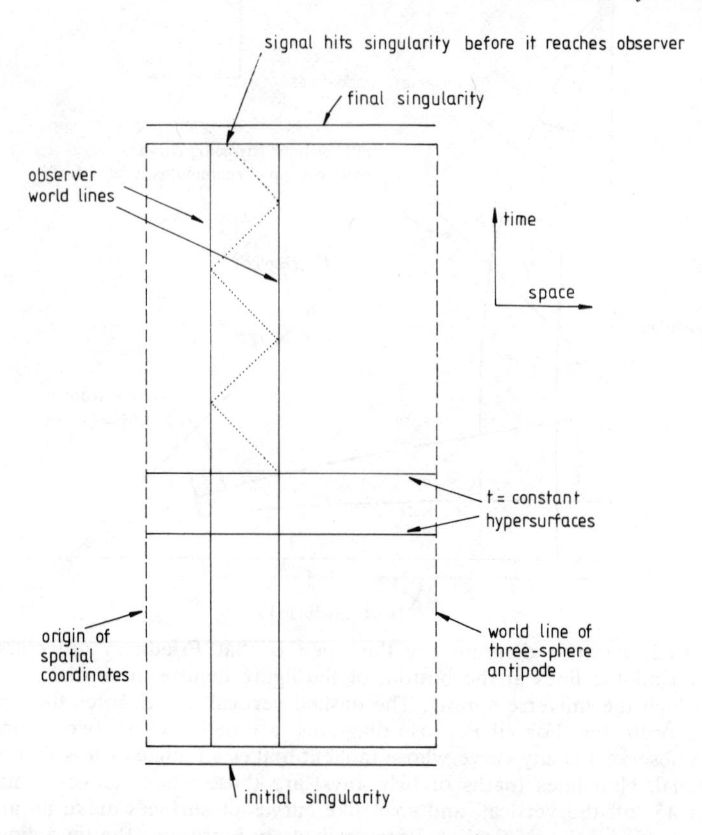

Figure 10.3. Penrose diagram for the closed Friedman universe. The conventions are the same as in the Penrose diagram for the open or flat Friedman universe: curves whose tangents are less than 45° off the vertical are timelike, null curves are those with tangents at 45° off the vertical, and spacelike curves are those whose tangents are greater than 45°. The initial and final singularities are denoted by double lines. Each point except those on the dashed lines denote a two-sphere. Two $t = \text{constant}$ hypersurfaces are pictured. These hypersurfaces are constant mean extrinsic curvature hypersurfaces. Each hypersurface is a three-sphere. Each three-sphere hypersurface is foliated by two-spheres beginning with a point at the origin of coordinates. The area of the two-sphere leaves of the foliation increases from zero as the foliation goes out from the origin of coordinates; the area reaches a maximum size midway between the origin of coordinates and the antipode, and it goes back to zero at the antipode of coordinates. Two observer world lines are pictured. These observers are at rest with respect to the background radiation. The jagged dotted line denotes light signals passing back and forth between the two observers. In contrast to the open and flat Friedman universes, only a finite number of signals can be sent between the two world-lines, for the light signal hits the final singularity before it can travel between the two observers. This will be true no matter how close the world-lines are.

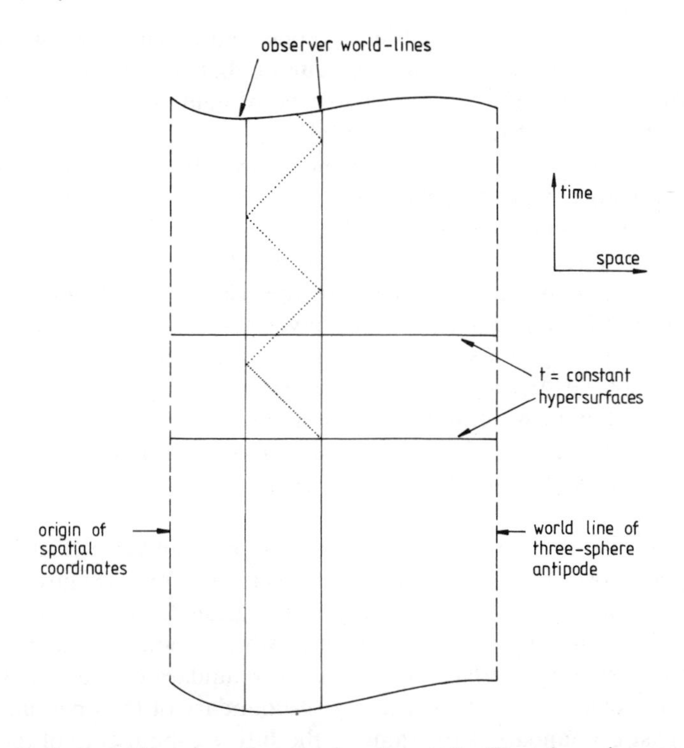

Figure 10.4. Quasi-Penrose diagram for the static Einstein universe. The $t =$ constant hypersurfaces are three-spheres, just as in the closed Friedman universe. The curved lines at the top and bottom of the figure indicate that the diagram continues on indefinitely. Since the points at infinity are *not* brought in to a finite distance, we term this diagram a 'quasi-Penrose' diagram. The pictured observer world-lines remain equidistant from each other for infinite proper time, so an infinite number of signals can be sent back and forth between any two observers in the space-time. Thus the future c-boundary of the Einstein static universe is an omega point.

c-boundaries of the cosmological models. The c-boundaries are composed of the singularities and the points at infinity; the c-boundary of a cosmology is the edge of space-time, the 'place' at which space and time begin. By convention, singularities are represented by double lines in Penrose diagrams. As can be seen from Figure 10.3, the initial and final singularities are the only c-boundaries in a closed Friedman universe. An open Friedman universe, on the other hand, has four distinct c-boundary structures: an initial singularity out of which the entire space-time arose, a single point i^0 representing spatial infinity, a 45° line $\mathscr{I}^+$ (called 'scri plus') representing 'null infinity' which are the points at infinity that light rays (null curves) reach after infinite time, and a single point i^+ which all timelike curves approach for all finite times, and reach after infinite time (with the exception of those timelike curves that accelerate forever and thus approach arbitrarily close to the speed of light. These curves hit scri plus rather than i^+ at temporal infinity).

A Penrose diagram allows us to define rigorously 'an achieved infinity', a concept whose logical consistency philosophers have been doubtful about for thousands of years. Using the c-boundary, it is possible to discuss the topology of the 'achieved infinity' and the 'beginning of time' in cosmological models. In the closed Friedman universe, the initial and final singularities both have topology $\mathbf{S}^3$, while the initial singularity in the open and flat Friedman universes have topology $\mathbf{R}^3$. In these very special space-times, it is even possible to put a metric on the singularities, but in general this will not be possible. It can be shown, however, that if Strong Cosmic Censorship holds in a space-time, then there is a natural Hausdorff topology on the c-boundary of the space-time.

By comparing the Penrose diagram for the open universe (Figure 10.2) with the Penrose diagram for the steady-state universe (Figure 9.1) we see that there is not necessarily any topological distinction between a universe that exists forever and a universe which begins at a singularity a finite proper time in the past. The future c-boundary of the steady-state universe is exactly the same as the past c-boundary of the open universe, and the past c-boundary is the same as the future c-boundary of the open universe. This means that asking the question of what happened before the initial singularity makes no more sense than asking the question of what happened before the universe began, if in fact the universe has existed forever in proper time. Furthermore, trying to find a theory of gravity which would remove singularities makes as much sense as trying to find a theory which would remove the c-boundaries of infinite universes. It is always possible to find a conformal transformation which will convert an infinite universe into a finite one and *vice versa*. One can always find a time coordinate in which a universe that exists for a finite proper time (for example, the closed Friedman universe) exists for an infinite time in the new time coordinate, and a time coordinate in which a universe that exists for an infinite proper time (for example, Minkowski space) exists for only a finite time. The most appropriate *physical* time may or may not be proper time. This point will be crucial in our discussion of the continued existence of life for infinite time in the future of a closed universe. As we shall see below, closed universes can exist for only a finite proper time, but for an infinite time that could, in principle, be measured by the subjective clocks of living beings.

Two observers can communicate for all time only if they can send light signals back and forth to one another indefinitely. If two observers lose the ability to send light rays back and forth, we say that an *event horizon* has formed between them. It is immediately apparent from the Penrose diagrams that in an open and in a flat Friedman universe, no horizons form and any two observers, represented by timelike curves, can send an infinite number of light rays back and forth between now and the time when i^+ is reached. This is because in the Penrose diagram, the timelike curves get closer and closer together as i^+ is approached. In contrast, event horizons *do* form between any two observers in the closed Friedman universe. In Figure 10.3, the world-lines of comoving observers are shown as vertical lines, and no matter how close the observers are, there will come a time when it will no longer be possible to connect the two lines with a 45° line which represents a light ray; the final singularity is reached before a light ray from one observer can reach the other. It is simply impossible for life to exist indefinitely in a closed Friedman universe because it would eventually become impossible for a being in such a universe to even send signals to different parts of itself! Freeman Dyson, whose work we shall discuss in section 10.6, ruled out the never-ending existence of life in closed universes because of this breakdown in communications.[52]

But not all closed universes have a c-boundary structure, or rather a final singularity, like the closed Friedman universe. The Friedman final singularity will occur only when the shear is zero, and as we pointed out earlier in this section, not only is the shear in generic closed universes non-zero, it is in fact so large that the evolution of the universe will be *dominated* by the shear near the final singularity. What can happen is that a shear-dominated closed universe can contract so much faster in one direction than a Friedman universe that it becomes possible for light signals to circle the universe in that direction and it is possible to communicate in that direction. We say that the horizon disappears (temporarily) in that direction. Note that it is possible for an horizon to disappear in a given direction and for there still to be an event horizon in that direction. The event horizon disappears also only if it is possible to send signals back and forth in that direction not just once but an infinite number of times. However, if the direction in which the horizon disappears alternately covers all directions, and covers them infinitely many times before the singularity is reached, then it is possible for all observers to send light rays infinitely often back and forth before the singularity is reached. In such a universe there would be no *event* horizons, and there would be no communication barriers preventing the never-ending existence of life.

Now two points are defined as distinct in the c-boundary only if there are timelike curves which reach these two points and which are not contained in the chronological pasts of each other. If all event horizons disappear, then all timelike curves are in the chronological past of each other. Thus the future c-boundary of a universe with no event horizons must consist of just a single point; we shall call such a point an *omega point.*

Two simple examples of cosmological models with an omega point are the Einstein static universe (Figure 10.4) and Löbell space.[53] Löbell space is a space-time constructed by identifying Minkowski space in a certain way to obtain three-torus spacelike hypersurfaces with constant mean extrinsic curvature. These hypersurfaces are Cauchy hypersurfaces for Löbell space. The omega point in Löbell space is a singularity, and it is reached in finite proper time.

In contrast, the omega point in Einstein space is reached only after infinite proper time. It is easy to see from Figure 10.4 that a light ray can be sent from one observer to another an infinite number of times. Figure 10.4 gives a picture of the causal structure of Einstein space, but it is different from the other Penrose diagrams in that the point at temporal infinity—the omega point—is not brought in to a finite distance. A true Penrose diagram for Einstein space, with the omega point brought in to a finite distance, has been constructed by one of us.[115] The future part of such a diagram for Einstein space is the same as a closed $\mathbf{S}^3$ universe which begins in an initial singularity like the closed Friedman universe, and approaches the Einstein static universe asymptotically in the future. Such a solution to the Einstein equations is known; it is called a time-reversed Eddington–Lemaître–Bondi universe. We give a guess of what the diagram of a time-reversed Eddington–Lemaître–Bondi universe is pictured in Figure 10.5.

As in the open universe, the timelike curves come closer and closer together as the c-boundary point is reached, so that light rays can pass an infinite number of times between them. Penrose diagrams are completely accurate only for space-times with spherical symmetry—such symmetry allows two angular coordinates to be suppressed without loss of information—but we would imagine that the causal structure of any closed universe which begins in an initial Friedman-like singularity and ends in an omega point would look qualitatively like Figure 10.5.

Seifert has proved[54] that space-times in which the future c-boundary consists of a single point must have a compact Cauchy surface. That is, cosmologies with an omega point must be closed universes which satisfy Strong Cosmic Censorship. Although open universes like the Friedman universe pictured in Figure 10.2 have a c-boundary point i^+ to which all non-accelerated timelike curves terminate, these space-times do not contain an omega point. For scri plus also forms part of the future c-boundary, and as we mentioned above, there are timelike curves which terminate on scri plus. Other interesting general restrictions on space-times with an omega point have been obtained by Budic and Sachs.[55]

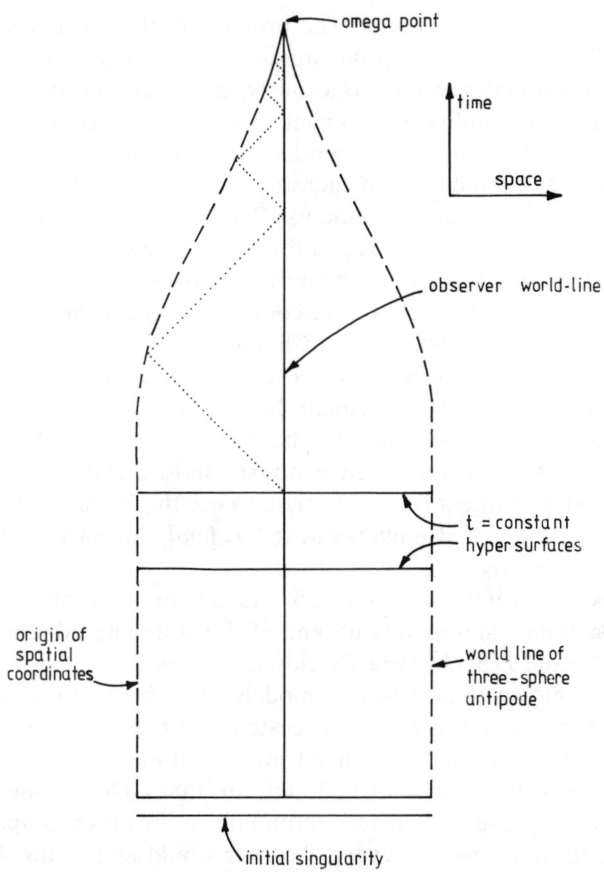

Figure 10.5. Penrose diagram for a time-reversed Eddington–Lemaître–Bondi universe, a space-time with an omega point. The $t = \text{constant}$, constant mean curvature hypersurfaces are three-spheres as in the closed Friedman universe and as in the Einstein static universe. The time-reversed Eddington–Lemaître–Bondi universe begins with an initial singularity identical to the initial singularity in a closed Friedman universe, and then asymptotically approaches an Einstein static universe in the future. Every observer can communicate with every other observer an infinite number of times, because the world lines of all observers come together in the infinite future to hit a single future c-boundary point, the omega point. If life continues to exist forever, and if the universe is closed, the Penrose diagram for the actual universe must be similar to this diagram.

As we discussed in Chapter 6, the spatially homogeneous Bianchi type IX model, which is a shearing closed universe with Cauchy hypersurfaces having topology $\mathbf{S}^3$, was extensively investigated during the late 1960's by relativity groups in the United States, the United Kingdom, and the Soviet Union to see if the horizons in this model disappeared in the past. The conclusion was that it is possible for the horizon to disappear in one direction, but even this is a rather improbable occurrence in a generic Bianchi type IX model. The probability that horizons will disappear in *all* directions an infinite number of times as the singularity is approached was never actually rigorously calculated, but there are indications that this probability is zero in the vacuum models. Which is to say, there *could* be Bianchi type IX models that have an omega point, but if these models exist, they are of measure zero in the initial data space of the Bianchi type IX vacuum models. Putting in perfect fluids does not change this conclusion, but it is possible that more exotic forms of matter could result in an omega point being more likely.

The Bianchi type IX models were considered only as models of the *past* singularity. But as we have seen in Chapter 6, the evidence is strong that the initial singularity was probably close to being shear-free, so the closed Friedman universe with its regular $\mathbf{S}^3$ topology singularity is a more accurate model of the past than the shearing Bianchi type IX. However, if there is a *final* singularity, it is by contrast almost certain to be dominated by shear, and so it might be appropriate to use the Bianchi type IX model as a model of the actual universe near the *final* singularity, but not near the initial singularity.

The reader is referred to refs 25 and 67 for additional material on Penrose diagrams, and to refs 68 and 69 for a detailed discussion of the behaviour of the Bianchi type IX closed universe.

In the homogeneous Bianchi models, the final singularity is all-encompassing. It is occasionally suggested that the final singularity in an inhomogenous closed universe need not be all-encompassing, but there are two singularity theorems which indicate that a closed universe which begins to re-collapse will in fact terminate in an all-encompassing final singularity. In other words, every observer would end in this final singularity.

Theorem:[30] *If a closed universe contains a maximal Cauchy hypersurface and if on this hypersurface,* (10.1) *is a strict inequality when* V^a *is the normal to this hypersurface, then provided* (10.1) *holds everywhere on the space-time, there is a universal upper bound to the length of every timelike curve.*

Furthermore, if every timelike geodesic congruence passing through a given event E will eventually begin to recontract, then every timelike curve through E will eventually hit the final singularity. This can be formalized as:

Theorem: There is a universal upper bound to the length of every future-directed timelike curve through E*, provided* (10.1) *holds everywhere in a globally hyperbolic closed universe, and provided there is some future-directed unit timelike vector* W^a *at the event* E *and a positive constant* b *such that if* V^a *is the unit tangent vector to the future-directed timelike* *geodesics through E, then on each such geodesic the expansion $\theta \equiv V^a;_a$ of these geodesics becomes less than $-3c/b$ within a distance b/c from E, where $c \equiv -W^a V_a$.*

The proof and statement of this theorem have never appeared in the literature before, but the proof is simple.

The most important implication for the Anthropic Principle of the results discussed in this section is that certain global properties of the universe—openness vs. closure, and re-collapse vs. expansion forever—cannot be changed from one to the other by *any* sort of operation by intelligent life, provided the laws of physics as we now understand them are correct. We shall now discuss some of the global aspects of the universe which intelligent life *can* change.

10.4 The Future Evolution of Matter: Classical Timescales

> For when I was a babe and wept and slept,
> Time crept;
> When I was a boy and laughed and talked,
> Time walked;
> Then when the years saw me a man,
> Time ran
> But as I older grew, Time flew.
>
> G. Pentreath

In this section we shall consider the evolution of the Universe—or rather the evolution of matter in the Universe—over periods with timescales defined by classical mechanics. This epoch runs between 10^{10} and 10^{25} years, and so it will be relevant to closed universes only if $\Omega_0 - 1$ is less than a number between 0 and 10^{-7}, (which is by no means thought unlikely; see Chapter 6). We shall first discuss the evolution of stars and galaxies in the absence of intelligent life, and then consider what effect the activities of intelligent life could have on these dynamical timescales. Interestingly, it turns out that intelligent life can in principle change the upper end of classical timescales by several orders of magnitude. The calculations clearly indicate that the effects of intelligent life could be an important consideration in any analysis of the behaviour of the Cosmos over timescales greater than 10^{20} years.

At the present time stellar births are still occurring, but rates are decreasing exponentially, with a half-life of about 5×10^9 years,[56] due to exhaustion of primordial hydrogen, and the dissipation of gas from the Galaxy. In 10^{12} years, star formation will have ceased. Galaxies will become redder, as the hotter, more massive stars leave the main sequence. The later M-type stars will exhaust their hydrogen cores and also leave the main sequence in about 10^{12} years. Thus, after about 10^{12} years, stars will cease to provide energy sources for life.

Decay via emission of gravitational radiation, and evaporation of the system's subcomponents will dominate the dynamical evolution of stellar and galactic systems[57–60] in the classical epoch. The latter process gives the shorter timescales. For example, the average time required to detach a planet from a star by a close encounter with a second star is given by

$$T_{\text{natural}} = (nV\sigma)^{-1} \tag{10.12}$$

where n is the number density of stars, V is the average relative velocity of two stars, and σ is the cross-section for an encounter resulting in detachment. A rough guess for the cross-section would be $\sigma = 2\pi r^2$, where r is the distance of closest approach. The Earth and other planets would probably be detached if another star went between us and the Sun, so for detachment of our solar system's planets, we take $\sigma \sim 2 \times 10^{16}$ km^2. In the vicinity of our solar system, we have at present $n = 3 \times 10^{-41}$ km^{-3} and $V = 50$ km/sec, which gives

$$T^{\text{planets}}_{\text{natural}} = 10^{15} \text{ yr} \tag{10.13}$$

Serious disturbance of the Solar System will result from encounters on a timescale shorter than (10.13).

Close stellar encounters will also result in the escape of some stars from the Galaxy, since some encounters would result in some stars attaining escape velocity. The details of these losses are exceedingly complex because of subtle relaxation effects,[61] but the timescale will be closely related to the relaxation time for gravitational encounters.[61–64] A very rough estimate of the timescale for evaporation of stars from stellar systems has been given by Dyson.[58] If the system consists of N stars of mass M in a volume of radius R, the root-mean-square velocity of the stars will be of order

$$V = [GNM/R]^{1/2} \tag{10.14}$$

The cross-section, $\bar{\sigma}$, for a close encounter between stars in this system is

$$\bar{\sigma} = (GM/V^2)^2 = (R/N)^2 \tag{10.15}$$

We obtain the average time between close encounters by inserting (10.14) and (10.15) into (10.12):

$$T^{\text{stars}}_{\text{natural}} = (nV\bar{\sigma})^{-1} = (NR^{-3}/GM)^{1/2} \tag{10.16}$$

For a typical galaxy, we will have $N = 10^{11}$, $R = 3 \times 10^{17}$ km, so roughly

$$T^{\text{stars}}_{\text{natural}} = 10^{19} \text{ years} \tag{10.17}$$

With the above numbers, the cross section (10.15) corresponds to a closest approach of about 10^6 km, which is much closer than required to disrupt a solar system. The timescale for the dynamical relaxation of a system with N stars is

$$T_R = T^{\text{stars}}_{\text{natural}}/\log N \tag{10.18}$$

or $T_R = 10^{18}$ years for a typical galaxy. Using the same formulae, we can calculate that clusters of galaxies will evaporate galaxy-sized objects in $\sim 10^{11}$ years and stellar-sized ones in $\sim 10^{23}$ years. The evaporation of objects from the system will leave its total energy more negative than before since the objects leaving the system will necessarily have positive energy. The system will thus become more tightly bound as time goes on.

Another mechanism which leads to positive energy loss to the system is gravitational radiation. A mass which is orbiting around a fixed centre with velocity V, period P and kinetic energy E will lose energy by gravitational radiation at the rate:[6,65]

$$\frac{dE}{dt} = V^5\left(\frac{E}{P}\right) \tag{10.19}$$

where units are such that all velocities are measured in fractions of the velocity of light, c. Thus the timescale for orbital decay via gravitational radiation emission is

$$T = \frac{E}{dE/dt} = V^{-5}P \tag{10.20}$$

For the Earth orbiting the Sun, (10.20) gives $T = 10^{20}$ years. For our Sun's orbit in our Galaxy, with $V = 200$ km/sec and $P = 2 \times 10^8$ yr, the timescale is 10^{24} years. We can use (10.20) to obtain the timescale for the emission of all rotational energy of any bound system by equating the gravitational energy $\sim M^2R^{-1}$ with the rotational energy to get P and V. This gives a lifetime of[57]

$$T_{\text{grav}} \sim G^{-3}R^4M^{-3} \tag{10.21}$$

The timescale for a large $10^{15}M_\odot$ cluster to radiate away all its energy is $\sim 10^{19}$ years.

The final state after the objects are evaporated and rotational energy has been radiated away is probably a black hole. Dead stars with mass exceeding the Landau–Chandrasekhar limit, $M_{LC} \sim G^{3/2}m_N^{-2} \sim 3M_\odot$ (see section 5.8), will be the first objects to become black holes, but galaxies and the largest bound configurations—typically, the latter have a mass of order $10^{17}M_\odot$—will eventually follow them on the above timescales.

Thus the evolution timescales for dynamical evolution of large stellar systems are between 10^{15} and 10^{20} years. The question we want to consider now is, what effect could intelligent activity have on these upper and lower limits?

To answer this, we must obtain a conservative estimate of the energies which intelligent life may be expected to manipulate, if indeed intelligent beings develop interplanetary and interstellar travel. Dyson has pointed out that it is possible in principle, using known physical laws, for intelligent beings to take planets apart, but for our purposes, it will only be necessary to assume that they will be able to manipulate masses of the order of an asteroid or small comet, say 10^{16} gms (ten billion tons). Dyson and the Daedalus Study Group (see Chapter 9) have shown that it is possible, by using thermonuclear explosions in the form of a pulse rocket, to impart a kinetic energy of 10^{28} ergs to a body for a cost of 10^{14} dollars as we mentioned in Chapter 9. Using Simon's (see section 3.7) best estimate of the evolution of the human economy we can predict (?!) that, the cost of energy relative to wages will decrease over the next 400 years so that the cost of such a rocket would look to an individual in that society as a project costing about 30 billion dollars does to us. Suppose such a rocket is used to crash a 10^{15} gm asteroid into the Sun. The resulting change in the magnitude of the Sun's momentum is given by

$$|\Delta\vec{p}/\vec{p}| = \frac{(2E_{\text{ast}}m_{\text{ast}})^{1/2}}{M_\odot V_\odot} \tag{10.22}$$

so $\Delta p/p = 10^{-16}$ if the asteroid is crashed into the Sun in a direction perpendicular to its motion. Such a change in the Sun's momentum will cause a shift Δl in the Sun's coordinate perpendicular to $\vec{p}$, and after travelling a distance l, Δl will be given approximately by

$$\frac{\Delta p}{p} = \frac{\Delta V}{V} = \frac{\Delta l}{l} \tag{10.23}$$

Such a change in position can change a mere close encounter to one which results in a large change in path—typically a loss of one star from the Galaxy. We can obtain such a Δl setting $l = VT$ and $\sigma = 2\pi(\Delta l)^2$ in equation (10.12). This gives

$$\Delta l = \left(\frac{\Delta p}{2\pi p n}\right)^{1/3} \tag{10.24}$$

which when inserted back into (10.10) gives a timescale for disruption of

$$T_{\text{artificial}} = \left(\frac{\Delta p}{p}\right)^{-2/3} n^{-1/3}(2\pi)^{-1/2}V^{-1}$$

$$= \left[\frac{2E_{\text{ast}}m_{\text{ast}}}{M}\right]^{-1/3} n^{-1/3}\frac{(2\pi)^{-1/2}}{V^{2/3}} \tag{10.25}$$

which for stars in our Galaxy, is about

$$T_{\text{artificial}} = 10^{22}\ \text{yr} \tag{10.26}$$

This is three orders of magnitude greater than the Galactic disruption time. Of course, this calculation assumes that intelligent beings will plan for events happening 10^{22} years in their future (a very dubious assumption!) and that measurements of stellar velocities can be made sufficiently precisely for the these beings to hit their star in the right place, etc. Nevertheless, we find this computation very suggestive that intelligent life can indeed influence the behaviour of matter on a cosmic scale.

Furthermore, our numbers in the above calculation were based on rather conservative estimates about the energies and masses which will be controllable by intelligent beings in the very far future. We used energies and masses which we should be able to control within 400 years, while the calculation tacitly assumes that intelligent life would be limited to this level for the next 10^{22} years, a much too conservative guess. However, because the timescale (10.25) for the influence of intelligence on the large-scale structure of the Universe depends on the controllable factors $E_{ast}m_{ast}$ to the *one-third* power, a large increase in the energies and masses under control will not result in a very large decrease in the timescale (10.25).

For example, if the mass is increased from 10^{15} gm to 10^{27} gm—which would correspond to throwing an Earth-sized planet into the Sun rather than an asteroid—while keeping the energy fixed at 10^{28} ergs, the timescale (10.23) would be reduced from 10^{22} years to 10^{18} years, an order of magnitude *less* than the natural stellar collision timescale. Such a collision could be effected if one could find an Earth-sized planet in interstellar space unbound to any star system, (the binding energy of an Earth-sized planet at one light year is 10^{35} ergs, seven orders of magnitude above the energies we have assumed humans should be able to manipulate.) The kinetic energy such an initially unbound planet would have at the Sun's surface would be 10^{43} ergs, due to gravitational acceleration, which is much less that the binding energy (10^{50} ergs) of the black dwarf that the Sun will be at the time. Such a collision would probably not disrupt the black dwarf. One could imagine a hierarchy of collisions: an asteroid hits a planet in just the right way so that the planet hits a star, and so on, with the net result that a galaxy is disrupted in much less than its natural 10^{18} year timescale. In effect, intelligent beings disrupt the Galaxy by amplifying instabilities before they can damp by natural processes.

The natural disruption rate is determined by (10.12), while the artificial rate is controlled by (10.25). Let us imagine that intelligent life operates on systems at larger and larger scales, with the scale increasing with time as $R(t)$, the scale factor of the universe. The typical velocities in (10.12) and (10.25) are thermal, so let us suppose $V \propto R^{-1}$ which is the appropriate evolution for cosmological thermal velocities ($\langle V^2 \rangle \sim 1/R^2$, which is obtained from the conservation of momentum). Similarly, we assume that

$\sigma \sim R^2$ and $n \sim R^{-3}$. The momenta Δp and p we would expect to scale in the same way (with $p \sim \Delta p \sim V \sim R^{-1}$), so $(\Delta p/p) \sim$ constant. This gives: constant. This gives:

$$T_{\text{natural}} \propto R^2$$
$$T_{\text{artifical}} \propto R^{5/3} \tag{10.27}$$

Thus if the Universe has a sufficiently long lifetime, $T_{\text{artifical}}$ will eventually become smaller that T_{natural}, no matter what we assume the constant $(\Delta p/p)$ to be, which would mean that intelligent beings could eventually gain control of cosmological systems on the largest scales.

If the Universe recollapses, there will be a further epoch where classical dynamics will be important: that period over which the universe contracts from 10^{25} years before the final singularity to 10^{5-10} years before the final singularity. The lower bound is difficult to predict, because the shear—long-wavelength gravitational waves and curvature anisotropy—will play a dominant role in the evolution of the final state, (since $\sigma^2 \sim R^{-6}$) and just at what point the shear will dominate the evolution depends in a very complicated way on the exchange of energy between the R^{-6} and R^{-2} shear terms of equation (10.3). However, such an exchange will be significant in a closed universe only if the universe is very long-lived (with $\Omega_0 - 1 \sim 10^{-6}$ or less). The classical evolution of a closed universe in the case when $\Omega_0 - 1 \sim 0(1)$ has been discussed by Rees.[67] When Ω_0 is this large, stars will survive until they are disrupted near the final singularity. The cosmological background radiation will equal the surface temperature of stars when $t \sim 10^5$ years before the final singularity. A star will be disrupted when the background radiation equals the temperature of the star's central region at $t \sim 10^{-1}$ years before the final singularity. Stellar collisions are never a significant factor in stellar disruption. Radiation or neutrino pressure will tend to damp out inhomogeneities on scales shorter than the Jeans length which, if $\Omega_0 - 1 \sim O(1)$, corresponds to a mass of $\sim 10^{14} M_{\odot}$. In such a short-lived closed universe, the stars have insufficient time to add significantly to the entropy of the universe, so the temperature near the final singularity will go as $T = (R_{\text{now}}/R)(3\text{ K})$, where $R_{\text{now}} \sim 2 \times 10^{10}$ lyr is the present-day value of the scale factor. The important timescales near the final singularity of a short-lived closed universe are summarized in Table 10.1. In very long-lived universe, it is possible for stellar radiation to make a significant contribution to the universal entropy and energy density, because in such universes the radiation can be emitted at times when the Big Bang radiation has been redshifted to very low temperatures; a stellar photon in this environment will make a very large contribution to the entropy when it is thermalized near the final singularity.

TABLE 10.1
Timescales in small closed universe near final singularity

Event	Universal scale factor $R(t)$	Temperature (in degrees K)
Galaxies merge	$10^{-2}R_{\rm now}$	300
Sky as bright as the Sun's surface	$10^{-3}R_{\rm now}$	3000
Sky as hot as stellar cores; stars explode	$10^{-6}R_{\rm now}$	3×10^{6}
Nuclei disintegrate into neutrons and protons	$10^{-9}R_{\rm now}$	3×10^{9}
Protons and neutrons become free quarks	$10^{-12}R_{\rm now}$	3×10^{12}

This table applies only to *small* closed universes, i.e., closed universes which begin to contract less than 10^{11} years after the initial singularity. For much larger closed universes, i.e., those which do not begin to contract until 10^{31} years after the initial singularity, the temperature near the final singularity will increase as $1/R(t)$; but only the elementary particles e^+, e^-, $\bar{\nu}$, ν, and γ will exist to be heated. Furthermore, the additional radiation from stars at late times will change the constants in the temperature formula $T=(3 \text{ degrees K})(R_{\rm now}/R(t))$. Closed universes intermediate in size will have a mixture of dead stars, black holes, and gas which will be heated. $R_{\rm now}$ is the value of the universal scale factor at the present time.

10.5 The Future Evolution of Matter: Quantum Timescales

The future is not what it was.
B. Levin

The timescales in the previous section were governed entirely by classical mechanics, including general relativity. In the very long run, the important timescales arise from the decays, due to quantum effects, of various material structures.

The most important decay, both in terms of its cosmological consequences and in terms of its significance for life, is proton decay. As we discussed at length in Chapter 6, if the SU(5) or one of the similar GUTs is correct, the proton will disintegrate into leptons and photons. There are a number of decay branches via various short-lived particles, but the end-result is usually a decay of the type,

$$\begin{aligned} p &\to e^+ + \nu + \bar{\nu} + \gamma \\ n &\to e^+ + e^- + \nu + \bar{\nu} + \gamma \end{aligned} \tag{10.28}$$

Depending on the decay mode, there will be different numbers of the four particles on the right-hand side of (10.28), subject to the conservation of electric charge.

The proton lifetime in the minimal SU(5) GUT is expected to be roughly 10^{31} years, and a number of other GUTs give a similar lifetime. Experiments have to date failed to detect the predicted proton decay, so it may be the predictions are wrong. Nevertheless, if baryons and leptons truly lie in the same multiplet in a unified field theory of some sort, then it is very likely there will be transitions between various levels of the multiplet, and this will cause proton decay, even if the lifetime is longer that 10^{31} years. We shall therefore assume that proton decay occurs, and we shall use the 10^{31} year timescale of SU(5). If proton decay occurs via a different process on a longer timescale, the qualitative features will nevertheless be the same: the over-all decay reaction will still include (10.28), and the thermal and gravitational effects of proton decay on macroscopic bodies will be the same except that temperatures and evolution rates will have to be scaled appropriately.

Proton decay provides an energy source for large bodies—dead planets, black dwarfs, and neutron stars—which will prevent them from cooling to the temperature of the radiation background. Since these bodies are effectively electrically neutral, the positrons produced in the reactions (10.28) will be immediately annihilated, so in such bodies the net effect of (10.28) is to turn matter into energy. The energy released in proton decay will keep neutron stars at a temperature of 100 K, and black dwarfs and Earth-sized planets at 5 K and 0.16 K respectively for around 10^{31} years into the future.[70,71,72] These numbers are calculated by equating the usual cooling law:

$$dE/dt = 4\pi R^2 \sigma_{SB} T^4 \tag{10.29}$$

(where σ_{SB} is the Stefan–Boltzmann constant), to the power generated by proton decay:

$$\begin{aligned} dE/dt &= (1\ \text{GeV/proton})(\#\text{protons/proton lifetime}) \\ &= 6\times 10^{17}\ \text{ergs/sec}(M/M_\odot) \end{aligned} \tag{10.30}$$

White dwarfs cool to black dwarfs at 1 K in 10^{20} years in the absence of proton decay, but proton decay will keep them at 5 K during the period $10^{17} < t < 10^{31}$ years. At the end of 10^{31} years only about 5×10^{-5} of the original mass of the star or planet will remain: planets will have become asteroids and black dwarfs become Earth-sized planets, and the process will continue until the mass has been entirely converted into energy. By about 10^{33} years, the most massive solid structures, which have a mass of about $10M_\odot$, will have completely disappeared.

As emphasized by Frautschi,[73] proton decay spells ultimate doom for life based on protons and neutrons like *Homo sapiens* and all other forms of life constructed of atoms. Baryons are disappearing at the exponential rate $N(t)\exp[-t/10^{31}\ \text{yrs}]$, where $N(t)$ is the number of protons in the structure under consideration, which may be increasing. Even if intelligent life were to expand the spatial volume under their control at the speed of light, the number of protons in that volume would increase only as $\mathbf{N}t^3$, where the constant $\mathbf{N}$ is bounded above by the cosmological baryon number density today, so the maximum value of $N(t)$ would be $\mathbf{N}t^3$. The exponential decrease eventually will defeat the power-law increase, and baryon-based life will disappear if the Universe is flat or open, or if it is a sufficiently long-lived closed cosmology. Setting $\mathbf{N} = 10^{100}\,\text{yr}^{-3}$ and $\mathbf{N}t^3 \exp[-t/10^{31}\,\text{yrs}] = 1$ gives $t = 10^{34}$ years for the time by which *all* atom-based life must be extinct. More generally, if the proton lifetime is τ_N, all atom-based life will be extinct by $10^3\tau_N$ years.

The conversion of mass into energy via proton decay can have dynamical effects on cosmological evolution, for it can change a matter-dominated universe into a radiation-dominated one. The cosmological effects of proton decay have been investigated by Barrow and Tipler,[57] by Page and McKee[74,75] and by Dicus, Letaw, Teplitz and Teplitz.[71,72] All of these authors agree that in all Friedman universes, the only matter remaining after 10^{34} years is an electron-positron plasma, which originates entirely from the protons which did *not* form clumpy matter—stars, planets, asteroids, rocks, dust particles, or any bound group of atoms. When protons decay in clumpy matter, the electrons and positrons annihilate as we said above, and so cannot contribute to the plasma. About 1% of the matter will be in the form of atomic hydrogen after 10^{20} years, so these atoms will be the source of the electron-positron plasma. Dicus *et al.* point out that in an open universe, there will be a brief period between τ_N and $10^3\tau_N$ in which the exponential decay of the protons will generate radiation so rapidly from the matter that the universe will be radiation-dominated. After that time the matter density of the electron-positron plasma will dominate because its density falls off as R^{-3} while the radiation density falls off as R^{-4}. All of the above-mentioned authors agree that in an open universe, the cosmological expansion will be too rapid for the electrons and positrons in the plasma to recombine into positronium, at least via electrical forces, though Page and McKee raise the possibility that gravitational clumping could cause the electrons and positrons to recombine. This seems rather doubtful to us because the gravitational and electrical forces are both R^{-2} laws in the distance regime in question. But like-charged particles repel electrically, while gravity is always attractive, and as Page and McKee point out, this difference could lead to clumping when many body interactions are properly taken into account.

In summary, taking into account the classical timescales discussed in the previous section, the matter in the universe at 10^{31} years will consist of 90% dead stars and planets being maintained at a temperature between

100 K and 0.1 K by proton decay; 9% galactic-mass black holes from the evaporation and collapse of galaxies; and 1% atomic hydrogen. All of this material will be immersed in a radiation bath of photons and neutrinos, whose density relative to matter is increasing due to proton decay. Between 10^{31} and 10^{34} years the dead stars and planets will disappear, leaving the black holes, an electron-positron plasma, and the radiation. The radiation density will dominate both the plasma density and the black hole density. After 10^{34} years, the radiation density will have decreased sufficiently far so that the black holes will be the dominant component of the Universe's mass density.

But black holes do not last forever any more than protons do. Hawking has shown[76,77] that quantum effects cause black holes to radiate away their mass, with the mass being entirely converted into radiation at the end of $10^{66}(M/M_\odot)^3$ years (see section 5.9). Thus galactic-mass black holes ($10^{11}M_\odot$) will disappear in 10^{99} years, and supercluster-mass black holes ($10^{17}M_\odot$) will disappear in 10^{117} years. Elsewhere[57] we have argued that the evaporation of black holes combined with the expansion of the Universe will be sufficiently rapid to overcome the increase of black hole mass due to black hole coalescence induced by gravitational attraction. Page and McKee regard this question as still open, because of the complicated many-body effects mentioned above. If we are correct,[57] supercluster-mass black holes will be the most massive black holes ever to form. If so, after 10^{118} years the matter in the universe will consist entirely of an electron-positron plasma in a radiation both of neutrinos and photons.

Both we[57] and Page and McKee, agree that in a flat ($k = 0$) and in a long-lived closed universe, the rate of expansion of the universe will be sufficiently slow so that almost all of the electrons and positrons in the plasma will recombine. The particles will recombine into positronium when the total energy of an electron–positron system becomes negative. The only energies the particles have in a flat universe is the Coulomb energy of attraction and the random thermal energy of motion. The thermal energy of the electrons and positrons comes from the energy of proton decay. The average initial momentum P of the electron or positron produced in a proton decay will be $P = \gamma m_e$ where m_e is the mass of the electron and γ the Lorentz gamma factor; probably $\gamma \approx m_N/2m_e \approx 10^3$, where m_N is the proton mass. This initial momentum will redshift as $\gamma m_e(\tau_N/t)^n$, where the constant n is defined by $R(t) \propto t^n$. Thus the thermal kinetic energy will scale with the expansion of the universe as $E_K \approx P^2/m_e \approx \gamma^2 m_e(\tau_N/t)^{2n}$. If at τ_N the fraction of the mass in $e^\pm$ is f_e, the number density, N, of the $e^\pm$ will decrease because of the expansion of the Universe as $N \propto f_e m_e^{-1} R^{-3}$, and the average distance between $e^\pm$ will grow as $r \sim N^{-1/3} \sim f_e^{-1/3} m_e^{1/3} R$. The sum of thermal energy and the Coulomb energy is thus

$$\gamma^2 m_e \left(\frac{\tau_N}{P}\right)^{2n} - e^2 f_e^{1/3} m_e^{-1/3} t^{-n} \tag{10.31}$$

If we assume a matter-dominated flat universe where $R(t) \propto t^{2/3}$, the energy given by the expression (10.31) will go negative at

$$t_b \approx (10^{80}\ \text{yrs}) f_e^{-1/2} (\tau_N/10^{31}\ \text{yrs})^2 \tag{10.32}$$

This is the time when most positronium will be formed by two-body collisions in a flat universe. It will occur somewhat earlier in a closed universe because $R(t)$ is not increasing quite as fast as $t^{2/3}$, and because the sum of cosmological expansion and binding energies, which is negative, must be added in the closed universe case to (10.31).

However, Page and McKee have shown that three-body collisions of the form

$$e^+ + e^- + e^\pm \to Ps_n + e^\pm \tag{10.33}$$

where Ps_n denotes positronium with principal quantum number n, will actually cause most $e^\pm$ to become bound long before τ_N, due to recombination into positronium states that have binding energy much greater than E_K. The true positronium formation timescale is

$$t_{\text{pos}} \approx (10^{73}\ \text{yrs}) f_e^{-2/3} (\tau_N/10^{31}\ \text{yrs})^2 \tag{10.34}$$

where we have assumed $R(t) \propto t^{2/3}$. The timescale (10.34) will be smaller than (10.32) unless $f_e \leqslant 10^{-42}$, which seems highly unlikely (and would contradict our earlier calculations). Thus most of the free electrons and positrons in the plasma will bind around time t_{pos}, going typically into an orbit with principal quantum number a bit below the value

$$n \approx 10^{22} f_e^{-4/9} (\tau_N/10^{31}\ \text{yrs})^{2/3} \tag{10.35}$$

These positronium states have a radii of

$$r_n \approx (10^{12}\ \text{megaparsecs}) f_e^{-8/9} (\tau_N/10^{31}\ \text{yrs})^{4/3} \tag{10.36}$$

which is much larger than the radius of the visible universe today. In this state, the orbital velocities of the electron and positron about each other are about 10^{-4} cm/century.

The state Ps_n will gradually decay by emission of photons to the ground state, where the positronium will rapidly annihilate. Page and McKee used the classical power-loss formula for electromagnetic radiation from a dipole to calculate the decay time:

$$\begin{aligned} t_{\text{decay}} &\sim E_n/(dE/dt) \sim 2 m_e^{-1} e^{-10} n^3 l^5 (n+l)^{-2} \\ &\sim m_e^{-1} e^{-10} n^6 \end{aligned} \tag{10.37}$$

For comparison, Bethe and Salpeter[78] give exact times for decay from $l=0$ singlet and triplet states as $1.25\times10^{-10}n^3$ and $1.4\times10^{-7}n^3$ respectively, and Sakurai[79] gives a decay time proportional to $n^{4.5}$, so the decay time (10.37) will apply only if the positronium forms in a $n=l$ state most of the time (as seems reasonable). The typical transition will be $\Delta n=1$, so something of the order of 10^{22} photons will be generated as the Ps_n state cascades downward. Putting (10.35) into (10.37) gives the typical time of annihilation:

$$t_{\text{decay}}\approx(10^{117}\text{ yrs})f_e^{-8/3}(\tau_N/10^{31}\text{ yrs}) \tag{10.38}$$

The fraction of free electrons and positrons decreases rapidly in flat universes, but they bind in very high quantum numbers so that by the time they have decayed the radiation bath will have been redshifted to such low mass-densities that a flat universe will be *always* matter-dominated by the electrons and positrons.

The evolution of the photon spectrum is complex, due to the sequence of first baryon and then positronium decays. Page and McKee have given some estimates for the evolution of the spectrum, the main conclusion being that the energy density of the photons arising from positronium cascades and annihilation completely dominates all other contributions to the radiation background if we assume that black holes larger than supercluster mass never form. If this is not true, the radiation background may become dominated by emission from black holes. Black holes with a mass exceeding that of superclusters will eventually form if the spectrum of density inhomogeneities in the Universe (which we discussed in section 6.10) possesses associated metric perturbations that do not decrease with scale. This will be the case for the constant curvature spectrum of density perturbations preferred by theorists. Also, a spectrum with metric perturbations slowly increasing with scale up to the extent of inflation is predicted if inflation is the source of density inhomogeneities in the Universe (see section 6.12). If the spectrum of density inhomogeneities is steeper than $\delta\rho/\rho\propto M^{-2/3}$ then supercluster-sized black holes should be the largest that form.

Thus after 10^{117} years, the matter of the universe consists of an electron-positron plasma, with a good percentage of positronium in the flat and long-lived closed universes, immersed in a radiation bath fed by decays of the positronium. Essentially neither black holes, nor stars, nor planets, nor any other material remains.

If proton decay does not occur by GUTs, it is likely that protons will decay via the Hawking process. Timescales for this decay scenario are most uncertain, but the latest calculations give a proton lifetime of 10^{122} years. If protons survive this long the summary in the preceding paragraph would be the same, except the number 10^{117} would be replaced by 10^{122}, and the radiation bath would be fed by proton decay.

If proton decay does not occur at all, then Dyson has shown that quantum tunnelling will cause the eternal matter to decay first to iron (timescale 10^{1500} years) and the iron would collapse to black holes in timescales of $10^{1\times 10^{26}}$ years. All of the classical and quantum timescales are collected together in Table 10.2.

TABLE 10.2
Timescales in open, flat, and very large closed universes

Event	Timescale (in years)
Sun leaves Main Sequence	5×10^9
Large clusters evaporate galaxies	10^{11}
Stars cease to form; all massive stars have become either neutron stars or black holes	10^{12}
Longest lived stars use all their fuel, and become white dwarfs	10^{14}
Dead planets detached from dead stars via stellar collisions	10^{15}
White dwarfs cool to black dwarfs at 5 degrees K. Proton decay will keep dwarfs at this temperature for 10^{30} years	10^{17}
Dead stars (black dwarfs and neutron stars) evaporate from galaxies (approximately 90–99% of stars will evaporate; 1–10% will collect in galactic centres to form gigantic black holes)	10^{19}
Neutron stars cool to 100 degrees K	10^{19}
Orbits of planets decay via gravitational radiation	10^{20}
Dead stars evaporate from galactic clusters (black dwarfs are at 5 degrees K and neutron stars are at 100 degrees K due to proton decay; background radiation has cooled to 10^{-13} degrees K)	10^{23}
At this stage matter consists of about 90% dead stars, 9% black holes, and 1% atomic hydrogen and helium	
Protons decay (according to SU(5) GUT)	10^{31}
Dead stars evaporate via proton decay (GUT)	10^{32}
All carbon-based life-forms become extinct	10^{34}
At this stage most matter in the universe is in form of: $e^+, e^-, \bar{\nu}, \nu, \gamma$	
Ordinary matter liquifies due to quantum tunneling	10^{65}

TABLE 10.2 (*continued*)

Event	Timescale (in years)
Solar mass black holes evaporate via Hawking process	10^{66}
In flat and closed universes, most e^+ and e^- form positronium (in open universes, most e^+, e^- remain free	10^{73}
Galactic mass black holes ($10^{11}M_\odot$) evaporate via Hawking process	10^{99}
In flat and closed universes, positronium decays via cascade, releasing 10^{22} photons	10^{117}
Supercluster mass black holes ($10^{17}M_\odot$) evaporate via Hawking process	10^{117}
Protons decay via Hawking process	10^{122}
Our descendants (if any) probably meet descendants (if any) of *Homo sapiens* evolved independently on distant planet (see section 8.7)	10^{800}
If ordinary matter survives decay via GUTs or Hawking process, it decays to iron	10^{1500}
All iron collapses into black holes	$10^{1\times10^{26}}$

This table is a list of all significant timescales for the evolution of matter. However, it should be noted that some processes listed will preclude the operation of others. For example, if all protons decay via GUTs at 10^{31} years, there will be none remaining to decay via the Hawking process at 10^{122} years. In view of our ignorance concerning the operation of some of these processes (the predicted GUT decay has not been seen experimentally, and may not exist), it is best to list *all* possible processes, and point out that the exact evolutionary sequence is unknown.

Before moving on to consider the constraints on the processing of information in the far future we should draw together the consequences of hypothetical elementary particles or new particle properties for any long-range forecast of the Universe's fate. So far we have assumed that the material content of the Universe involves only *known* particles and we have assumed neutrinos to be massless. Now, we recall from Chapter 6 that there is growing evidence that at least 90% of the mass density of the Universe resides in a non-luminous, and probably a *non-baryonic* form. Neutrinos possessing a small rest mass may well prove to be the identity of this 'missing mass'. In addition, there is some experimental evidence (see ref. 32 of Chapter 6) that the electron neutrino possesses a finite rest mass of order 20 eV. If neutrinos do possess a rest mass of this order then they will have been non-relativistic ever since the cosmic background radiation cooled below $\sim 20\text{ eV} \sim 10^4\text{ K}$. This occurs after the Universe has expanded for about a million years, after which the neutrino density falls off as $\rho_\nu \propto R^{-3}$. If neutrinos (or for that matter any other type of exotic weakly interacting particle which may exist—like the axions, gravitinos or photinos discussed in 6.10) do possess a non-zero rest mass they will ensure that ever-expanding universes remain matter-dominated forever unless there exists some completely unknown type of decay mode allowing neutrinos (or these other, hypothetical, particles) to decay into photons. Such a decay mode would have to violate many known laws since neutrinos are fermions and photons are bosons. The decay modes for neutrinos that have been contemplated involve the decay of a heavy neutrino into a lighter *stable* neutrino plus either a photon or a lepton-antilepton pair. Clearly, this type of instability would have no effect upon our argument. The presence of massive weakly interacting particles is sufficient to keep ever-expanding universes matter-dominated. Whatever the fate of electrons, positrons and positronium, the annihilation cross-section for neutrinos and antineutrinos is utterly negligible. Similar futures await if axions, photinos or gravitinos possess finite rest masses, so long as the rest masses are not high enough to contribute more than the critical density, in which case the Universe is closed and recollapses to high density.

The last possibility we consider has more dramatic consequences. Grand unified gauge theories predict that very massive, stable magnetic monopoles should exist[80] and will be generated in the first 10^{-35} s of the Universe's expansion.[81] The extent to which their abundance will be depleted by inflation is difficult to calculate exactly[82] and so their likely abundance in the Universe today is a fairly open question. In models, like SU(5), where grand unification occurs at an energy of m_X, the monopole mass is $m_M \sim \alpha_X^{-1} m_X \sim 10^{16}$ GeV. This is an enormous mass for an elementary particle and is $\sim 10^{-8}$ gm. Monopoles are not point-like particles as we imagine quarks and leptons to be. Rather, they possess a nested internal structure resembling a series of Chinese boxes. Most of their 10^{-8} gm should reside in a tiny core $\sim 10^{-28}$ cm in diameter, (see Figure 10.6). Within that core, energies are high enough for grand unification to exist, and all strong and electro-weak interactions possess effectively the same strength. Around the periphery of this inner core, there lies a shell where the superheavy X and $\bar{X}$ bosons are numerous. Beyond this there is a sparse outer periphery fading away beyond $\sim 10^{-15}$ cm where there exists a shall of W and Z bosons. This peculiar structure means that monopoles can affect the stability of matter in spectacular ways.[84] Recall, from section 6.7, that the baryon non-conserving interactions which turn quarks into leptons and which therefore induce proton decays are mediated by X bosons. When a proton encounters a magnetic monopole it may penetrate its outer shell sufficiently to interact with the shell of X bosons. This will considerably enhance the probability of proton decay above that in the absence of

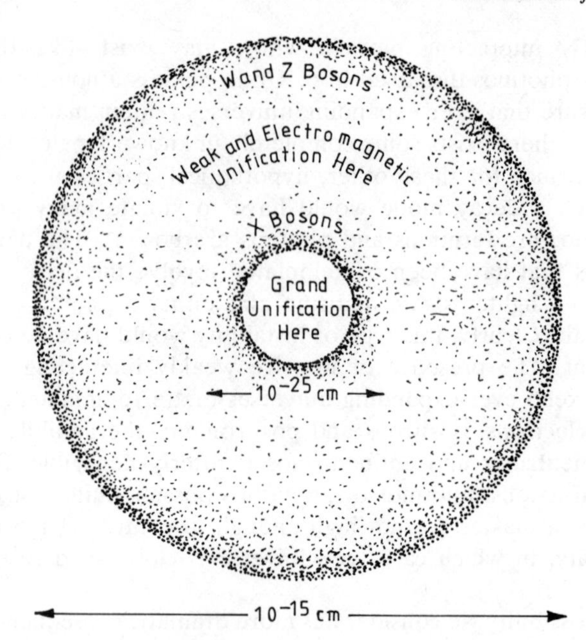

Figure 10.6. Structure of a magnetic monopole. This is a schematic picture of the internal structure of a magnetic monopole in grand unified gauge theories. Within the central core of diameter about 10^{-25} cm there is grand unification (the strong and electroweak interactions have the same effective strengths. This region is ringed by a cloud of X bosons and in the region between 10^{-25} cm and 10^{-15} cm from the centre there is electroweak unification but with distinct strong interactions. This outer region is an outer cloud of W and Z bosons. (Figure from ref. 83.)

monopoles. In effect, monopoles *catalyse* proton decays like $M+$ proton $\rightarrow M+\pi^0+e^+$. Surprisingly, it appears that the cross-section, σ, for these catalysed decays is large[84,85]—of order that of conventional strong interactions, $\sigma \sim 10^{-27}\ \text{cm}^{-2}$.

Michael Turner[86] has pointed out that the presence of a small cosmic flux of magnetic monopoles surviving the Big Bang, with an abundance too low to have any other observable consequences could radically alter the future of astronomical structures. Monopoles will gradually accumulate in the central regions of planets and stars, having been captured by gravity. Their large mass means that they just steam-roller through other baryonic material as if it were not there. Once present in the centres of astronomical objects they then proceed to induce the decay of nucleons. In Table 10.3 we give the life-time of the Earth and of a giant planet like Jupiter along with those of white dwarfs and neutron stars, against complete decay of all their constituent nucleons by monopole catalysis.[86]

TABLE 10.3
Nucleon decays in the presence of monopoles

Object	Lifetime (yrs)	Final temperature (K)
Earth	$10^{18}f^{-1/2}$	$140f^{1/8}$
Jupiter	$5\times10^{15}f^{-1/2}$	$730f^{1/8}$
White dwarf	$8\times10^{13}f^{-1/2}$	$4\times10^{4}f^{1/8}$
Neutron star	$10^{11}f^{-1/2}$	$5\times10^{6}f^{1/8}$

The lifetime and final temperature of structures as a result of monopole-induced proton decays. The monopole flux is $F \equiv f \cdot 10^{-21}\ \mathrm{cm^{-2}\,sr^{-1}\,s^{-2}}$ and corresponds to one monopole passing through an area the size of a large city each year. In the absence of monopoles the nucleons in these objects would only have decayed after $\sim 10^{31}$ years according to GUTs.

The numbers assume a monopole mass $\sim 10^{16}$ GeV and a monopole flux F. We use a representative flux of $10^{-21}\ \mathrm{cm^{-2}\,sr^{-1}\,s^{-1}}$ for illustrative purposes as this is the largest monopole flux consistent with the observed X-ray emission from neutron stars. A larger value of F would induce a high enough rate of nucleon decays within neutron stars *today* to produce more X-ray photon emission than is observed from these objects.[87] In Table 10.3 we also give the final temperature attained by the objects at the completion of their evaporation. The lifetimes are dramatically shortened below the standard proton decay time expected in the absence of monopoles, of $\sim 10^{31}$ years, and the demise of matter is complete after only about 10^{18} years if $F \sim 10^{-21}\ \mathrm{cm^{-2}\,sr^{-1}\,s^{-2}}$. These timescales can be compared with those displayed in Tables 10.1 and 10.2 which ignored the possible role of monopoles.

The fate of the monopoles themselves in the far future of an open universe is also an interesting question and one that has not been considered before. The monopole density will fall off as $\rho_M \propto R^{-3}$ and this will ensure that ever-expanding universes remain matter-dominated so long as the monopoles do not decay or annihilate. Now, if charge is conserved individual monopoles cannot decay into radiation. The only way in which monopoles can be destroyed is by annihilation with anti-monopoles. The origin of monopoles in the Big Bang leads naturally to the existence of equal numbers of monopoles (M) and antimonopoles ($\bar{M}$). Their long-term fate should be completely analogous to that of electron-positron pairs which was described above. In this case the two competing effects of electron-positron annihilation and positronium formation will simply be replaced by the processes of $M\bar{M}$ annihilation and monopolonium[88] formation. The essentials of the previous analysis of positronium evolution will hold with the GUT monopole mass $m_M \sim 10^{16}$ GeV replacing the electron mass, $m_e \sim 0.59$ MeV, and with the monopole magnetic charge; g_M, replacing the electric charge, e, where

$g_M e \sim 1$ numerically. One final possibility which it would be interesting to evaluate exactly is the probability that one of these monopoles tunnels into a Reissner–Nordstrom black hole[25] before evaporating into radiation by the Hawking effect[76,77] leaving its magnetic charge on the final naked singularity (or even losing it due to topology change there). Indeed, this decay route may be probable because Grand Unified monopoles of mass $\sim 10^{16}$ GeV are very close to being black holes. They need only a 0.01% fluctuation for an event horizon to develop around them.

This is all we have to say on the speculative aspects of material evolution and we shall now turn to consider what constraints, if any, there might be upon the processing of information in the indefinite future.

10.6 Life and the Final State of the Universe

> The law that entropy always increases—the Second Law of thermodynamics—holds, I think, the supreme position among the laws of Nature. If someone points out to you that your pet theory of the universe is in disagreement with Maxwell's equations—then so much the worse for Maxwell's equations. If it is found to be contradicted by observation—well, these experimentalists do bungle things sometimes. But if your theory is found to be against the Second Law of thermodynamics I can give you no hope; there is nothing for it but to collapse in deepest humiliation.
>
> A. S. Eddington

The study of the survival and the behaviour of life in the far future became a branch of physics with the publication in 1979 of a paper[58] by Freeman J. Dyson, entitled 'Time without End: Physics and Biology in an Open Universe'. Dyson's first paper was followed by another in 1981[99] and by a *Science* article in 1982 by S. Frautschi.[73] Although the papers on life in the far future are not numerous, they have shown the progression required of physical science: the papers subsequent to Dyson's first article built on, improved, and corrected their predecessors, and the discussion is now based entirely on the laws of physics and computer theory. This is in sharp contrast to the vague speculations which were typical eschatological discussions prior to Dyson, which we have discussed in Chapters 2 and 3.

Broadly speaking, the current work may be said to be concerned with investigating the validity of FAP, and attempting to draw testable conclusions from FAP. In this final section, we shall summarize what is now known in the new study of 'physical eschatology'.

As we argued in earlier chapters, an intelligent being—or more generally, any living creature—is fundamentally a type of computer, and is thus subject to the limitations imposed on computers by the laws of physics. However, the really important part of a computer is not the particular hardware, but the program; we may even say that a human being *is* a program designed to run on particular hardware called a human body, coding its data in very special types of data storage devices called DNA molecules and nerve cells. The essence of a human being is not the body but the program which controls the body; we might even identify the program which controls the body with the religious notion of a *soul*, for both are defined to be non-material entities which are the essence of a human personality. In fact, defining the soul to be a type of program has much in common with Aristotle and Aquinas' definition of the soul as 'the form of activity of the body'.[89] A living human being is a representation of a definite program rather than the program itself. In principle, the program corresponding to a human being could be stored in many different forms—in books, on computer disks, in RAM—and not just in the brain of a particular human body. However, a human being is a program designed to run on very special hardware, and most of the subprograms of the human program are present only because of the peculiar structure of the hardware. These properties are most unlikely to be present in non-human intelligent programs. When atoms disappear human bodies will disappear, but programs capable of passing the Turing test need not disappear. An intelligent program can in principle be run on many types of hardware, and even in the far future of a flat Friedman universe matter in the form of electrons, positrons and radiation will continue to exist. The basic problem of physical eschatology is to determine if the forms of matter which will exist in the far future can be used as construction materials for computers that can run complex programs, if there is sufficient energy in the future environment to run the programs, and if there are any other barriers to running a program.

The FAP requires intelligent life to continue to exist forever. We have to make this requirement precise. We shall say that 'life' continues to exist forever if *three* conditions hold:

(*1*) information processing—the running of programs—continues along at least some future-endless timelike curve γ all the way to the future c-boundary of the Universe;

(*2*) the amount of information *processed* in $I^{-}(\gamma)$ between now and the c-boundary is infinite; and

(*3*) the amount of information *stored* in $I^{-}(\gamma) \cap S(t)$, where $S(t)$ denotes the constant mean curvature foliation of the Universe, diverges as the leaves of the foliation approach the future c-boundary.

The global instant 'now' is defined to be all those events contained in the leaf of the constant mean curvature foliation which passes through the Earth at the present time. This definition assumes that a constant mean-curvature foliation exists, but the definition can be generalized to apply in other space-times. However, such a generalization will not be necessary, for we have shown in section 10.3 that a physically realistic cosmology will have a constant mean curvature foliation.

We have required the information to grow in the chronological past, $I^-(\gamma)$, of a single timelike curve because events must be in the chronological past of a timelike curve if they are to be able to communicate with the curve. The rough idea is this: intelligent life comes into existence in a limited region and spreads out from there. The definition does not preclude intelligent life arising elsewhere, but the information generated by the other intelligent life counts only if the other intelligent life can eventually communicate with the intelligent life around the curve γ.

The total information processed is required to be infinite because only if there are an infinite number of thoughts in the future is it reasonable to say that intelligent life has existed 'forever'. Conversely, an intelligent being or civilization can be reasonably said to be immortal if it thinks an infinite number of thoughts. We do not know exactly how many bits I of information constitute a thought, but it lies in the range $1 \leqslant I \leqslant 10^{15}$ bits for human beings, since 10^{15} bits corresponds to the upper bound of the information storage capacity of the human brain (see Chapter 3). But information is much more than just thought. Recall that in Chapter 3 we showed that according to modern economics, the *whole* of the economy, not merely the so-called information industry, can be regarded as being concerned with the production and transfer of information of one sort or another. Thus *everything* intelligent beings do, not just their thinking, is purely and simply a form of information processing. It follows that every conceivable thought and action of any possible form of life is ultimately constrained by the physical laws governing the processing of information.

It is vitally important to note that *there need be no correspondence between the duration of various measures of physical times such as proper time, and the number of bits processed in that time interval.* It is quite possible for the universe to exist for only a finite proper time in the future before ending in the c-boundary—as happens in closed universes—and yet for an infinite number of bits to be processed in that time interval. All that is required for this to occur is for the rate of information-processing as measured in proper time to diverge sufficiently rapidly as the final singularity is approached. We would claim that the appropriate measure of time duration by intelligent beings in a given environment is not in general proper time but the length of time it takes to process 1 bit, for the bit-duration measure will be a direct measure of 'subjective' time, the rate at which thoughts succeed each other in the mind. In the current astrophysical and biological environment, the bit-processing rate is directly proportional to proper time, and this is the reason why we consider the latter to correctly measure time. But if the bit-processing rate of *every* living being in the environment was increasing relative to proper time, the beings in that environment would have no hesitation in rejecting proper time as an appropriate measure of *true* time. The appropriate measure of *physical* time, and the fact that this measure may not be the same as proper time in a given cosmological epoch, has been discussed at length by Misner.[90]

But it is not sufficient for the number of thoughts to be infinite if life is reasonably to be said to exist forever. If a computer with a finite amount of information storage—such a computer is called a *finite state machine*—were to operate forever, it would start to repeat itself over and over again. After a finite time, it would think no thoughts it had not thought before. It seems reasonable to say that 'subjectively', a *finite state machine* exists for only a finite time even if it exists forever and processes an infinite amount of data. A being or civilization that truly exists forever ought to have the possibility of always being able to think new thoughts. This means it must be what computer scientists call an *infinite state machine* (though 'potentially infinite' would be a more appropriate nomenclature). Condition (*3*) above is imposed in order to allow life collectively to be an infinite state machine. Furthermore, condition (*3*) *requires* life to expand its knowledge without limit, for the only alternative would be for thoughts to endlessly recycle.

The absolute minimum amount of energy required to process a given amount of information is determined by the Second Law of thermodynamics. If ΔI is the information processed in bits, then the Second Law requires

$$\Delta I \leq \Delta E / k_B T \ln 2 = \Delta E / T(\text{ergs/K})(1.05 \times 10^{16}) \qquad (10.39)$$

where k_B is Boltzmann's constant, T is the absolute temperature in degrees K, and ΔE is the amount of free-energy expended. The inequality (10.39) is due to Brillouin.[91] If the temperature at which the computer operates is higher than absolute zero, there is a minimum amount of energy that must be expended to process a bit of information. The Third Law of thermodynamics says that the temperature must always be greater than absolute zero.

In the present cosmological epoch, the lowest temperature that physics will effectively permit computers to operate at is the temperature of the background radiation, which is 3 K. If we put this temperature into (10.39) then the total amount of information that could be generated in the present epoch by using the entire mass-energy of the Earth is

$\Delta I \leqslant 10^{64}$ bits. For the mass-energy of the entire solar system, we would have $\Delta I \leqslant 10^{70}$ bits; for the entire Galaxy $\Delta I \leqslant 10^{81}$ bits; and for the mass-energy of all the matter in the entire visible universe, $\Delta I \leqslant 10^{98}$ bits. We emphasize again that these limits apply to *any* type of life, whether it be a computer based on silicon, *Homo sapiens* based on carbon atoms, or intelligent beings formed of pure energy or a type of matter unknown to modern science. All forms of life are without exception subject to the Second Law of thermodynamics.

These upper bounds can be lowered only by a decrease in the cosmological temperature. Since $T/3\,\text{K} = R_{\text{now}}/R(t) \approx (2 \times 10^{10}\,\text{yrs/t})^{2/3}$, the cosmological temperature will drop by only a factor of 2 over the next 20 billion years, so the upper bounds on the amount of information that can be processed will apply to life over this length of time.

If the inequality (10.39) is divided by the time-difference Δt, and the limit $\Delta t \to 0$ is taken, we obtain a constraint on the information-processing rate:

$$(dI/dt)/(dE/dt) \leqslant 1.05 \times 10^{23}\ \text{bits/sec-watt}\ T^{-1} \qquad (10.40)$$

where as before the temperature, T, is measured in degrees Kelvin. At room temperature (300 K) the thermodynamic limit of computing speed per unit power is about 10^{21} bits per second per watt. At present the average off-the-shelf microcomputer works at about 10^{8} bits per second per watt, while state-of-the-art supercomputers work at 10^{10} bits per second per watt. We have a long way to go before reaching the thermodynamic limit.

We should mention that there has been a debate in recent years as to whether (10.39) really applies to information-processing inside computers.[92] Inequality (10.39) can be derived from several quite different assumptions. Brillouin obtained (10.39) by calculating the minimum amount of energy needed to *measure* one bit of information; in computers, measuring would correspond to *reading* a bit. (If there was no minimum, Maxwell's Demon could operate, thereby contradicting the Second Law.) Von Neumann derived (10.39) by calculating the minimum amount of energy required for accurate transmission of a bit from one logical gate to the next.[93] The IBM computer scientist Landauer arrived at (10.39) by arguing that computation is logically irreversible.[94,95] Both the Brillouin and the von Neumann arguments are founded solidly on the Second Law as generalized by information theory, but Landauer's derivation is open to the objection[96] that computation *is* in actuality logically reversible, and a number of idealized physical models of reversible computers have been published.[92]

However, these models directed at Landauer's argument do not touch the thermodynamic arguments of von Neumann and Brillouin, as has been recently pointed out by Porod *et al.*[93] Furthermore, it seems to us that these models are not true thermodynamic models, because they are either unstable or else do not really interact with a temperature reservoir. To work at all, the models require the information to be already in the machine. These points have also been made by Porod *et al.* For our purposes the existence of ideal computers which can process information *already in the machine* with no energy minimum is irrelevant, for in Nature it is necessary to transfer information to the machine, and all agree this transfer is subject to (10.39). It is also the case that transfers between different parts of a real machine would be subject to (10.39), so even if the claims of the critics were correct, (10.39) would restrict most of the computer operations. It would also apply to the *increase* of information, for the ideal computer processing in the critics' models just manipulates the information already in the computer memory. We shall therefore assume the validity of (10.39) in our subsequent discussions of information growth in the far future.

From (10.40) we have the following inequality between the total information processed in the future and the energy required to process it:

$$I = \int_{t_{\text{now}}}^{t_{c\text{-bound}}} (dI/dt)\, dt \leq (k_B \ln 2)^{-1} \int_{t_{\text{now}}}^{t_{c\text{-bound}}} T^{-1}(dE/dt)\, dt \qquad (10.41)$$

where the upper bound $t_{c\text{-bound}}$ is the time the c-boundary is reached. The value of the integrals in (10.41) do not depend on which measure of time duration is used.

By condition (*2*) in the precise definition of FAP above, the left-hand integral must diverge if FAP is to hold, which implies that the right-hand integral must also diverge. In an open or flat cosmology, it is possible for the right-hand integral to diverge even if the total energy used,

$$E = \int_{t_{\text{now}}}^{t_{c\text{-bound}}} (dE/dt)\, dt \qquad (10.42)$$

is finite. Since the temperature goes to zero as the c-boundary is approached in these cosmologies, the information processed can diverge whilst the total energy being used remains finite if the information is processed sufficiently slowly. In closed universes the integral (10.42) must diverge, and diverge very rapidly near the final singularity, since the temperature diverges as $1/R(t)$. We shall show that it is possible, in principle, for the right-hand integral in (10.41) to diverge in all three basic cosmologies: open, flat, and closed.

What will be the most important energy source in the far future? At present, the most important energy source is matter: mass is converted into energy in stars via thermonuclear fusion, or via radioactive decay of heavy nuclei in bulk matter. But matter is gradually being used up, and no matter how efficient the conversion of energy into information there are the finite upper bounds, which we calculated earlier, to the amount of information that can be generated by the matter available in any finite region over the next 20 billion years. Life has a tendency to increase its population exponentially until the limits of a given ecological niche are reached. It is the characteristic of intelligent life to discover how to use *all* forms of matter for its own purposes, so we would expect such life to use up all the material within its home solar system on timescales which are short in comparison with the age of the Universe. They will begin the expansion from their home system, and gain control of new material. On timescales of tens of billions of years, the total region under the control of intelligent life will be an expanding sphere, with almost all of the activity concentrated in a narrow region within a distance ΔR of the surface of the sphere. The interior of the sphere will be an essentially dead region, the matter having been converted into information during the previous eons. The sphere will be expanding on average at some fraction of the speed of light, so on the average the region under the control of life and the net information stored will be increasing as t^2. (If the interior had not been exhausted, the increase would be proportional to the volume of the sphere rather than its area, or t^3.) Thus although perpetual exponential growth of life, or of the economy, or of information, is not allowed by the laws of physics, a power-law growth is allowed. If the average expansion rate of life, as measured in the local rest-frame of the inner boundary of the expanding sphere, is always greater than the current Hubble expansion of 50 to 100 km/sec per megaparsec, then the growth can continue as t^2 for the next 10^{31} years, until the decay of protons becomes important. Whether or not this t^2 growth as measured in proper time can continue indefinitely depends on whether the clumping of matter will permit the higher and higher biosphere expansion speeds needed to overcome the expansion of the universe. However, as we show below, in the appropriate timescale growth *can* continue as a power-law growth indefinitely. As we mentioned in section 10.1, the growth of life predicted here is quite similar to the growth of life in Kant's cosmology.

By the end of the period from 10^{31} to 10^{33} years, the only matter surviving will be electrons and positrons from the decays of single atoms in interstellar space, (we ignore the possibility of massive neutrinos but they do not change the argument). Frautschi[73] has considered various possible energy sources, such as Hawking radiation from black holes, and the energy from electron-positron annihilation. He concludes that in open universes, black holes would just barely supply sufficient energy, but the electrons and positrons would not. However, it seems to us that neither of these would be the main energy source of life in the far future.

As we discussed in section 10.4, the most important form of energy available in this epoch will be the shear energy, so it is the most probable energy source for life. As we discussed in section 10.4, the shear energy can be extracted by making use of the directional temperature differential it generates. By Carnot's theorem, the efficiency of energy extraction should be proportional to $\Delta T/T$ which is independent of the scale factor R by equation (10.10), so the percentage of energy extracted from the shear energy should be independent of time unless the distortion parameter ($\exp \beta_{ij}$) goes to zero asymptotically. However, since $\sigma = \dot{\beta}$ and the shear falls as $\sigma \sim t^{-1}$ the distortion actually increases as $\beta \sim \ln t$. The shear energy will be equally available at all points inside the sphere of life and not just on the surface as was the matter energy (at least an approximately homogeneous universe,[97] and we will assume, as observations suggest, that the universe is homogeneous). Thus life will continue its expansion outward, but also begin to re-inhabit the desert of the interior, until the region under the control of life is growing proportionally to the volume rather than just the surface area. The total energy available to the whole of life will in the long run be

$$E \sim \rho R^3 \tag{10.43}$$

where ρ is the energy density of the available energy source, which in the end will be the shear. For open universes we have $R \sim t$ and $\sigma^2 \sim \rho \sim t^{-2}$, so $dE/dt \sim$ constant, where t is the proper time. Putting these relations and (10.43) into (10.41), remembering that $T \sim 1/R \sim 1/t$, and absorbing all constants into one constant C, we get

$$I \leqslant C \int t \, dt \sim t^2 \tag{10.44}$$

which *diverges* as $t \to \infty$. For flat universes, the only energy source is the electron-positron plasma, so $\rho \sim R^{-3}$ neglecting annihilation. Thus neglecting annihilation, we have $E \sim$ constant. If this finite amount of energy is used slowly, with say on the average $dE/dt < t^{-\delta}$, then the total energy used over infinite proper time will be finite if $\delta > 1$. We have $R \sim t^{2/3}$ for matter-dominated flat universes, so (10.41) gives

$$I \leqslant C \int t^{-\delta} (t^{2/3}) \, dt \sim t^{-\delta + 5/3} \tag{10.45}$$

which *diverges* as $t \to +\infty$ if $\delta < 5/3$. Thus if the energy used for all purposes together with the particle annihilation is slow enough, the amount of information processed can diverge in flat universes.

In closed universes, the R^{-6} shear term and complicated curvature anisotropy effects will eventually dominate.[68,69] In fact, it can be shown that the shear grows as R^{-6} almost all of the time in general vacuum Bianchi type IX cosmologies.[68] We might expect, as we said in section 10.3, such a cosmology to model parts of the actual Universe near the final singularity quite closely. But before that epoch is reached, the closed universe, will pass through epochs of matter, radiation, R^{-2} shear, and curvature domination, the sequence and duration of each epoch depending on the lifetime of the closed universe. The behaviour of life in the late expanding and early contracting periods will be essentially the same as that predicted for the expanding periods in open and flat universes, which we described at length above. Therefore we shall consider only the R^{-6} shear-dominated epoch of a closed universe, for it will be in this epoch that the information integral will be divergent or convergent. If the universe contains a massless scalar field or any other perfect fluid source with equation of state $p=\rho$ then conservation of energy requires $\rho \propto R^{-6}$ for this field and it keeps pace with the shear energy. However, it can be shown[98] that the presence of such a field stops the chaotic Mixmaster oscillations required for there to be the possibility of horizon removal and so FAP predicts there should exist no such fields in Nature; their presence would prevent indefinite information-processing in a closed universe.

What we are demanding here is that the information-processing diverge on approach to the singularity. This would appear to be possible in recollapsing cosmological models of the Mixmaster type which possess chaotically unpredictable behaviour on approach to their final singularities.[69] As a singularity is approached, these space-times undergo stochastic permutations in their expansion rates along any three orthogonal directions, because of their anisotropic spatial curvature. What is vital for our argument above is that an infinite number of these space-time oscillations occur during any open interval of proper-time $(0, T)$ about the singularity at $t=0$. If so, then an *infinite* number of physically distinct events happen on approach to the singularity, even though that approach occupies only the *finite* interval of proper-time from T to 0. Any entity whose subjective or information-processing time 'ticked' at the rate of Mixmaster oscillations would live forever in that time. It would process an infinite amount of information; but only if the Mixmaster oscillations continue all the way into the singularity and do not stop because of quantum gravitational influences at some finite moment before $t=0$, no matter how small. This appears at first sight analogous to Zeno's paradox,[6] but it is quite distinct.

Indeed, Zeno's argument that it takes an infinite time for me to go between any two points, no matter how close they are, because the inter-point separation can be expressed as an infinite series of decreasing lengths, is rejected because such a decomposition of their separation does not correspond to a distinct set of realized steps during motion between the points. But, by this same logic which leads us to reject Zeno's argument, we should claim that it does take an infinite time to reach a Mixmaster singularity, because the infinite number of space-time oscillations that must be experienced on the way to it *are* all physically distinct, realizable events.

The energy available for information-processing near the final singularity will be, according to (10.43):

$$E \sim (\Sigma^2/R^6)(R^3) \sim \Sigma^2 R^{-3} \sim \Sigma^2 t^{-1} \tag{10.46}$$

so $dE/dt \sim t^{-2}$, where t is the proper time before the final singularity is reached at $t=0$. Since $T^{-1} \sim R \sim t^{1/3}$, the right-hand integral in (10.41) gives

$$I \leq C \int (t^{-2}) t^{1/3}\, dt \sim t^{-2/3} \tag{10.47}$$

which *diverges* as $t \to 0$. Thus even though the energy available for information-processing must diverge very rapidly as the final singularity is approached, we see that there is sufficient energy in the form of shear to provide it.

As in the case of the open universe, the efficiency of energy extraction by life near the final singularity will be independent of the scale factor R, but it will be dependent on the distortion parameter β. We pointed out in section 10.4 that if communication is to be possible arbitrarily close to the singularity, the horizons must continue to disappear, and this requires β to continue to alternate in size from very small values to very large values. On the average, β will not approach zero asymptotically, so on the average the efficiency of energy extraction will have a lower bound, which we can absorb in the constant C in (10.47).

In the above evaluations of equation (10.41) for the open, flat, and closed universes, we have assumed that $T \sim 1/R(t)$, which is the adiabatic variation. In reality, the temperature variation will be non-adiabatic, because in processing information, waste heat is being generated at the rate dE/dt, and this waste heat will raise the temperature of whatever thermal sink is used. If FAP is to hold, a thermal sink must be found which can absorb heat sufficiently fast so that life (information processing) will not be incinerated by its own waste heat. In Dyson's model,[58] waste heat elimination was the main difficulty facing life in the far future. Dyson assumed that life was restricted to a constant comoving volume, and that life eliminated heat by radiation to its exterior.

Radiating waste heat to an exterior region is difficult in open and flat universes. It is absolutely impossible in closed universes in which life has expanded to engulf the entire Universe, for in such a case there is no exterior region. Therefore, a heat sink co-present with life must be used.

The obvious choice for such a heat sink is the thermal radiation background. If waste heat from information processing is dumped into the radiation background as it is generated, the energy density of the background will rise at the same rate as the energy density of the energy source, which means $E/V \sim t^{-2} \sim T^4$. This gives $T \sim t^{-1/2}$. For shear-dominated closed universes, this implies that the temperature will rise faster as the Universe collapses, as $T \sim R^{-3/2}$ rather than as $T \sim R^{-1}$. For open universes, the temperature will fall-off more slowly, as $R^{-1/2}$ rather than as R^{-1}. If these non-adiabatic temperature variations are inserted into (10.41) we find that $\int T^{-1}(dE/dt)\, dt$ diverges as $t^{-1/2}$ in a shear-dominated closed universe, and as $t^{1/2}$ in an open universe. For flat universes, the integral will still diverge provided $\delta < 5/3$. To summarize: waste heat does not seem to pose a problem for continuing information processing in either the open, flat, or closed universes.

We have argued that event horizons cannot exist if FAP is to hold, because such horizons would prevent communication between different observers, and even different parts of the same extended observer. One might wonder, however, if a single observer could nevertheless process an infinite amount of information in the ever-shrinking region with which he could communicate, by processing information faster than the communication region is shrinking. We can now show this is impossible on energetic grounds.

At any cosmic time t, the region from which the observer γ can receive signals is $I^-(\gamma) \cap S(t)$. In the Friedman universe, the boundary of this region is determined by the ingoing radial null geodesics, which satisfy $ds^2 = 0$. This gives $dr = -dt\sqrt{1-r^2}/R(t)$. (See Chapter 6 for the definition of these symbols.) The proper radius l of the communication region then decreases as $dl = R(t)\, dr/\sqrt{1-r^2}$, so $dl = -dt$, and thus the proper radius of the communication region goes as $l \sim t$ near the final singularity at $t = 0$. In the radiation- or matter-dominated Friedman universe, $R(t) \sim t^{1/2}$ or $R(t) \sim t^{2/3}$ respectively, so the proper volume of the communication region goes as $V \sim l^3 \sim R^6(t)$ or $\sim R^{9/2}(t)$ respectively. The proper energy density can rise only as $R^{-4}(t)$ or $R^{-3}(t)$, so the available energy in the communication region decreases as $R^2(t)$ or $R^{3/2}(t)$. (The available shear energy in a shear-dominated universe would decrease as $R^3(t)$ if horizons were present.) Since the available energy must increase if an infinite amount of information is to be processed, we conclude that horizons will prevent an infinite amount of information-processing.

If there exists a non-zero positive cosmological constant, Λ, then an ever-expanding universe inevitably approaches de Sitter space-time as $t \to \infty$ with $R \sim \exp(\sqrt{\Lambda/3}\, t)$ as discussed in sections 6.11 and 6.12. In this case information-processing dies out, event horizons exist and indefinite survival is ruled out. However, if the cosmological constant is negative the

Universe must recollapse and the advantageous information-processing properties near a chaotic second singularity can be exploited.

We now come to condition (3) in our definition of the FAP, the requirement that it must be possible to *store* an amount of information which diverges as the c-boundary is approached. The storage of n bits of information requires the existence in space of at least n distinguishable states of matter, radiation, or black holes. Furthermore, in order that this information not be lost, the energy of these states must be above the random fluctuation energy k_BT of the environment of the storage device. It seems unlikely that radiation by itself can serve as a storage device, for it tends to dissipate unless it is confined by solid matter. In all environments we shall be concerned with, the far future of an expanding universe and the hot environment near the final singularity of a closed universe, solid matter will not exist (except possibly for magnetic monopoles), so radiation is probably ruled out as the basis of information storage. Black holes are probably ruled out as storage devices in the far future for three reasons: first, if black holes of arbitrarily large mass do not continue to form, all black holes will eventually evaporate and hence cease to exist; second, it is not clear how black holes could be used to store distinguishable bits of information; third, if a black hole could be used to store information, the amount of mass-energy used per bit is likely to be too large to be supplied by the feeble energy sources of the far future (recall that the power usage must decrease as $dE/dt \sim t^{-\delta}$). However, we should emphasize that neither radiation nor black holes are conclusively ruled out as information storage devices in the far future, though they do seem unlikely candidates.

This leaves only matter from which to construct information storage devices. If we ignore the exotic forms of matter whose existence has been hypothesized but never seen (see section 10.5 for a list), the only matter remaining in the far future are positrons and electrons in a mixture of free-particle plasma and positronium. Dyson[99] was the first to suggest that it may be possible for life to exist in such a medium.

It is certainly possible to store information in a positronium atom. For example, parallel spins of the electron and positron could denote 1, and antiparallel spins could denote 0. The energy ΔE required to induce a transition between the lower energy antiparallel state and the parallel state decreases as $\Delta E \sim 1/n^3$, where n is the principal quantum number of positronium.[78] We must have $\Delta E > k_BT \sim 1/R(t)$, and from equations (10.35) and (10.36) we obtain a radius $r \sim n^2$, so $r \leq R^{2/3}(t)$. In short, the positronium atoms used to store information must grow, but at a slower rate than the Universe expands. The energy needed to cancel out the radiative losses of the positronium and ΔE decreases sufficiently rapidly so that it is possible to satisfy the above constraints on dE/dt in both the flat and open universes and still cause an infinite number of transitions between now and the c-boundary, if we ignore the problem of exactly how the available energy in the form of shear is to be transferred to the atoms. We shall also not attack the question of whether the atoms can be organized together in the complicated fashion required for computers. These are very complex unsolved problems which we cannot discuss in this preliminary survey. But of course they must be solved if indeed life is to survive indefinitely in an open or flat universe. All we can do here is indicate the directions which future investigation into the question of the indefinite survival of life must take.

As the Universe expands, the number of positronium atoms being used as random access memory (RAM) must diverge as the c-boundary is approached if the amount of information actually stored is to diverge, but this is not problem since the region under the control of life increases as $R^3(t)$. However, it seems that there will be difficulties in open universes in obtaining the necessary positronium, because very little positronium will be formed because of the rapid expansion. Thus we make a very tentative prediction that the Universe must be either flat or closed if FAP is to hold. In addition, it is worth remarking that the electron-positron and positronium evolution would be significantly affected by any cosmological magnetic field that may exist. Another important detail to be included in further studies is the fact that many open anisotropic universe models do not expand as $R \sim t$ in all directions. An important example is one found by Lukash which is approached by a very large class of ever-expanding anisotropic universes, (see the discussion of section 6.11). In this solution[105] the expansion scale factors in three-orthogonal directions, call them $X(t)$, $Y(t)$ and $Z(t)$, behave as,[100] $X \sim Y \sim (1+\lambda^2)t$ and $Z \sim t^{1/(1+\lambda^2)}$ where $\lambda^2 \geqslant 0$ is a constant that is essentially the ratio of the shear, σ, to the Hubble expansion rate $H = (XYZ)^{\cdot}/3XYZ$. So there is a horizon in the Z direction and the tidal stresses between directions can be exploited for information-processing. Furthermore, judicious orientation can exploit the fact that energy can be extraced from the slowly expanding Z direction. As the shear energy is removed, λ will tend to zero and the model will approach the isotropic open universe asymptote with $X \sim Y \sim Z \sim t$.

In open or flat universes, it is necessary that the region in which the information is stored diverge as the c-boundary is approached, for Bekenstein[100] has shown that the information which can be stored in a region of radius D is bounded above by

$$I \leqslant 2\pi ED/\hbar c \tag{10.48}$$

where E is the amount of energy used to store one bit. Bekenstein has argued[101] that E is bounded below in any finite region in open and flat universes, so if I is to diverge, so must D. The Bekenstein bound (10.48) has been derived only for space-times with non-compact Cauchy hypersurfaces and for closed universes with event horizons. If it applies to *all* closed universes, then it will be impossible for FAP to hold in closed universes, since D will go to zero as the c-boundary is approached, and thus (10.48) would prevent the amount of information stored from diverging. The derivation of (10.48) seems to depend crucially on the presence of event horizons, so there is no reason to believe it will apply to closed universes with an omega point. We will assume here that it does not, but this a point that needs to be investigated. But there are good reasons to believe it will apply (see however refs. 102, 103, 104) to closed universes without horizons. If it does, this provides another argument that FAP requires the c-boundary to be an omega point.

It is occasionally claimed[118,121] that energy-time uncertainty relation restricts the rate at which computers can process information. On dimensional grounds,[118] the energy-time uncertainty relation would appear to require $(dI/dt)^2 \leqslant \hbar^{-1}\, dE/dt$. However, it is not clear that such a restriction actually applies, because it is not clear what time coordinate t is the appropriate one to use.[119] In fact, if t is a time external to the system being measured, then the energy-time uncertainty relation $\Delta E\, \Delta t \geqslant \hbar/2$ can be evaded.[120,121] The energy-time uncertainty relation only restricts the measurements of times which are intrinsically defined by the physical system being measured.

If we assume that the t which is restricted by the energy-time relation is cosmic proper time, then a straight forward calculation shows that $I = \int (dI/dt)\, dt$ can diverge even if $(dI/dt)^2 \leqslant \hbar^{-1}\, dE/dt$ applies in the open universe (since $dE/dt \sim$ constant); in the closed universe (since $dE/dt \sim t^{-2}$); and in the flat universe if $\delta \leqslant 2$ (since $dE/dt \sim t^{-\delta}$). In other words, the energy-time uncertainty relation does not prevent an infinite amount of information processing in any type of universe.

In a closed universe condition (*3*) on p. 659 requires that information be stored in high-energy states of mass m. As the radius of a closed universe near the final state goes to zero near the final singularity, the information must be stored in particle states of higher and higher energies in order that it not be lost through random fluctuations. Furthermore, the total number N of particle states of mass m in the closed universe must diverge as the final singularity is approached if the amount of information stored in these particle states is to diverge, but the divergence of the total energy in elementary particle states cannot be more rapid than the divergence of the shear energy which is the energy source for the creation of these particle states. These are clearly necessary conditions for FAP to be satisfied, though they are not sufficient conditions. However, these conditions suffice to derive some indicative restrictions on elementary particle states.

The restriction that the mass m of the elementary particle state be greater than the thermal energy is expressed as:

$$m > k_B T \sim 1/R(t) \tag{10.49}$$

while the requirement that the energy in N particle states each with energy m be less than the shear energy in a volume V can be written in the form of a restriction on energy densities:

$$Nm/V \sim \sigma^2 \sim 1/t^2 \sim 1/R^6 \tag{10.50}$$

where we have used the *average* growth rate of shear energy density[68] in the last two steps. Now $V \sim R^3$, so (10.50) becomes

$$Nm/V \geqslant N(1/R)/R^3 \sim N/t^{4/3} \tag{10.51}$$

so the total number of particle states could grow as fast as $1/t^{2/3}$ without violating the energy upper bound. The total stored information I_{TOT} we would expect to grow roughly as N, so I_{TOT} can diverge as fast as $t^{-2/3}$ if the growth of particle states with energy permits. But the energy in the particle states cannot grow faster than this without exhausting the energy supply. Suppose that we write $N \sim t^{-\varepsilon}$, where $0 < \varepsilon < 2/3$. Remembering that on the *average* $R(t) \sim t^{1/3}$ near the final singularity, we obtain from (10.50) and (10.51):

$$m < V/(Nt^2) \sim 1/Nt \sim t^{\varepsilon - 1} \tag{10.52}$$

The inequalities (10.49) and (10.52) can be combined to give a constraint on the mass-energy of the particles:

$$1/t^{1/3} < m < t^{\varepsilon - 1} \tag{10.53}$$

We can put the energy scales into (10.53) by noting that $t^{-1/3} \sim 1/R(t) \sim k_B T \sim E$, where E is the actual particle energy measured in GeV. The inequality (10.53) then becomes

$$E < m < E^{3(1-\varepsilon)} \tag{10.54}$$

where $0 < \varepsilon < 2/3$. The final inequality (10.54) means that if condition (*3*) of FAP is to hold, there must be a particle state with energy in between the upper and lower limits of (10.54). Furthermore, on the average the number of particle states cannot grow faster than E^3, since otherwise the shear would be damped out by the production of particle states.

The bound (10.54) is not incredibly strong, but it is sufficient to rule out a number of proposed elementary particle spectra at high energies, for example the exponentially increasing spectra which underlies the Hagedorn equation of state.[106] This type of particle mass spectrum is also predicted in the latest 'super-string' models of space-time at high energy. It also rules out the possibility of a 'great desert'—a lack of particle states—between the electroweak unification energy and the grand unification energy. It may also be sufficient to allow non-ideal gas behaviour. In order to treat the contents of the Universe as an ideal gas (and so apply quantum statistical mechanics legitimately) the mean-free-path between elementary particle interactions, λ, must exceed the interparticle spacing, $n^{-1/3}$, where $n \sim gT^3$ is the particle number density and $g(T)$ the number of spin states. If interactions remain asymptotically free then $\lambda = (\bar{\sigma}n)^{-1}$ with the cross-section going as $\bar{\sigma} \sim \alpha T^{-2}$ for a unified gauge coupling α and so ideal gas behaviour requires[107] $g < \alpha^{-3/2}$. We also recall from section 5.10 that a large increase in g because of additional quark flavours leads to a breakdown of asymptotic freedom. The breakdown of ideal gas and asymptotic freedom conditions could only make more free-energy available for information-processing because it would enhance non-equilibrium behaviour.

The true importance of (10.54) is that it shows it may be possible to *test* the FAP conjecture, for the indefinite existence of life in the far future is possible only if matter has certain properties, and if we assume that the properties of matter do not change with time, then these properties of matter in the far future will also be properties of matter now. It is not possible of course to investigate the properties of matter and the structure of the universe in far future, but it is possible to investigate these aspects of Nature today.

We showed in section 10.4 that if life is to be able to continue to communicate indefinitely near the final singularity, the c-boundary must be a single point, which we call the omega point. Furthermore, solutions of the Einstein equations with an omega point may be of measure zero in the initial data of the space of solutions. However, it is of measure zero only if the action of intelligent life on the universe is neglected. In principle, it is possible, by exerting relatively small forces at just the right series of instants on a truly global scale, for intelligent life to *force* a generic Bianchi type IX closed universe into having an omega point by systematically eliminating the horizons in sequence in all directions an infinite number of times. This will be possible, that is, if intelligent life has expanded to encompass the entire universe, and if the properties of matter will allow the necessary forces. Since the probability is one that the actual universe is *not* of measure zero in the space of solutions, then if FAP is true and the universe is closed, it must be true that matter has the appropriate properties. Determining these properties could yield another testable FAP prediction. At the present state of our knowledge, we can say only that such manipulation of the time evolution of the entire Universe would require matter to have anisotropic stresses—vector or tensor fields rather than scalar fields—at the relevant times. One would also like to understand the limits placed upon horizon manipulation by the chaotic unpredictability of the Bianchi type IX Mixmaster universe.[68,69]

The last speculative FAP prediction we shall mention has already been discussed at some length in Chapter 3. According to FAP, the information processing and storage goes on indefinitely, so even in quantum cosmology the physical time *must* be unidirectional. In particular, closed universes—classical or quantum—in which the entropy decreases during the contracting phase of the universe are ruled out if FAP holds. This means that Hawking's proposed quantum cosmology[108] would be ruled out, because entropy does in fact decrease in its contracting phase.[117] Similarly, any classical or quantum cosmology based on a *four*-dimensional compact topology are ruled out because this would imply a cyclic time. The cosmological model of Cocke and Wheeler, which we discussed in section 3.8, in which entropy is reversed in the contracting phase of a closed universe, is also ruled out.

Were one to adopt a teleological view of Nature, one could go so far as to assert that matter has many of its properties today not because these properties are necessary for life today, but because these properties will be *essential* for the existence of life in the distant future. However, we would expect such teleological properties to exist in matter only if SAP were true, and that life is in some way equally essential to the Cosmos. Are there any reasons to think that life is essential to the Cosmos? Besides reasons which arise from rather controversial interpretations of quantum mechanics which we discussed in Chapter 7, we can offer only one extremely speculative suggestion for why life could be essential for the Cosmos. If black holes evaporate completely via the Hawking process, the black hole event horizons must terminate in naked singularities. Such naked singularities would terminate the Cauchy development of space-time. The other place at which the Cauchy development terminates is the initial singularity, and it is usually assumed that there is no space-time beyond the initial singularity. It would be possible to extend space-time across the Cauchy horizon formed by a black hole naked singularity, but there is no unique extension. The only natural assumption to make is to assume that *no* space-time exists beyond a Cauchy horizon, which means complete black hole evaporation would completely destroy the universe.[109] Such a possibility has been discussed by Penrose.[67]

However, if intelligent life were operating on a cosmic scale before any black holes approach their explosive state, these beings could intervene to keep the black holes from exploding by dumping matter down the black hole, at least in a short-lived closed universe. Thus ultimately life exists in order to prevent the Universe from destroying itself! We emphasize that we do not really want to defend this possibility, but we mention it to show that it is possible to imagine that intelligent life could play an essential global role in the universe.

In Chapter 3 and 7 we gave a teleological argument, based on Action Principles, for the closure of the universe. We also pointed out in section 10.3 that only in a closed universe is it possible for *all* timelike curves to be forever in causal contact with one another. For these reasons, it seems on anthropic grounds that the universe is more likely to be closed than open, but this is only a weak prediction. Nevertheless, let us assume that the universe is closed, and follow broad features of the evolution of life from the present all the way into the Omega Point. (It is possible for a closed universe to have an omega point but for life to die out before it is reached. We shall distinguish betwen these two classes of universes with omega points by capitalizing 'Omega Point' if life reaches the omega point.) We shall see that the Anthropic Principle implies a melioristic evolving cosmos. The future evolution of life is pictured in Figure 10.7.

Life begins its expansion from a single planet. The information and material under the control of life increases as t^3 initially, but eventually the rate of increase drops to t^2 as the resources in the centre of the expanding biosphere are exhausted. The increase will continue until life has expanded to encompass fully one-half of the entire universe. Because of the curvature of space (assumed to be a three-sphere), the area of the biosphere will *decrease* as it expands. But if the universe is sufficiently large, the region in which life is growing can still increase because the energy source is no longer matter energy, but shear energy. It is likely that before life has expanded to cover more than one-half the Universe, the contraction will have begun. The cosmological radiation temperature, which has dropped to extremely low values, will begin to increase again. The redshift will have become a blueshift. The information will continue to increase as t^n in the early contracting phase, where t is, as before, the proper time from the initial singularity, and n is some number less than or equal to three. The value of n will depend on which specific energy source is used, what form of matter is used to store information, and what percentage of the entire Universe is encompassed by life.

Finally, the time is reached when life has encompassed the entire Universe and regulated all matter contained therein. Life begins to manipulate the dynamical evolution of the universe as a whole, forcing the horizons to disappear, first in one direction, and then another. The information stored continues to increase, but now at the average rate $t^{-\varepsilon}$, where t is the proper time until the final singularity at $t=0$, and $0<\varepsilon<2/3$. As measured in proper time, the rate of growth of stored information is faster than exponential growth (since it diverges in finite time), but a more accurate measure of subjective time in this epoch is the

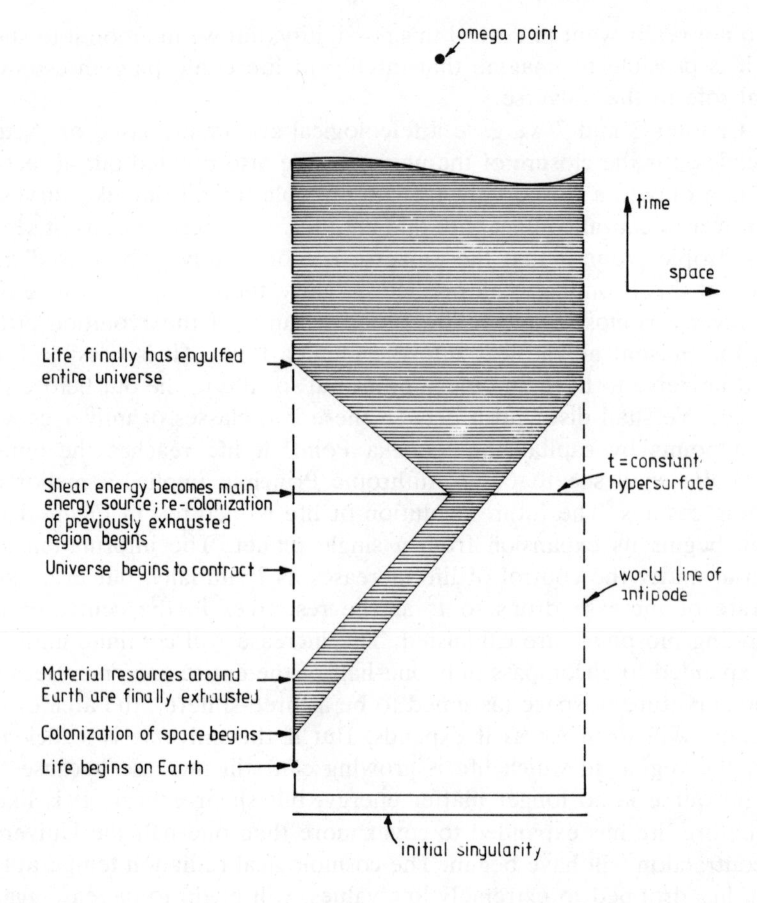

Figure 10.7. Hypothetical future of 'life' in a closed universe. The key events in the colonization of space are labelled on the figure. The region of the universe in which life exists at a given cosmic time t is determined by the intersection of the t = constant hypersurface with the shaded region. The shaded region is the biosphere.

amount of time needed to process one bit. In this time measure, the information storage increases, but the increase is a power-law, with the power less than ε. However, the increase continues as a power-law indefinitely.

From our discussion in Chapter 7, and from Figure 7.2, we see that if life evolves in all of the many universes in a quantum cosmology, and if life continues to exist in all of these universes,[110] then *all* of these universes, which include *all* possible histories among them, will approach the Omega Point. At the instant the Omega Point is reached, life will have gained control of *all* matter and forces not only in a single universe, but in all universes whose existence is logically possible; life will have spread into *all* spatial regions in all universes which could logically exist, and will have stored an infinite amount of information, including *all* bits of knowledge which it is logically possible to know.[123] And this is the end.

References

1. S. Weinberg, *The first three minutes: a modern view of the origin of the universe* (William Collins, Glasgow, 1977), p. 148.
2. S. M. Stanley, *Macro evolution: pattern and process* (Freeman, San Francisco, 1979).
3. S. M. Stanley, *The new evolutionary timescale* (Basic Books, NY, 1981).
4. C. Darwin, *On the origin of species by means of natural selection*, 2nd edn (John Murray, 1860), p. 489.
5. P. Duhem, *The aim and structure of physical theory* (Princeton University Press, Princeton, 1954), p. 288.
6. C. W. Misner, K. S. Thorne, and J. A. Wheeler, *Gravitation* (Freeman, San Francisco, 1973).
7. J. W. York, Jr., private communication to F. J. Tipler.
8. R. Penrose, *Proc. R. Soc.* A **381,** 53 (1982).
9. M. Tod, *Proc. R. Soc.* A **388,** 457 (1983).
10. A. S. Eddington, *Nature* **127,** 447 (1931).
11. H. Poincaré, *Acta Math.* **13,** 1 (1890). English translation in *Kinetic theory*, Vol. 2: *Irreversible processes*, ed. by S. G. Brush (Pergamon Press, Oxford, 1966).
12. M. Eliade, *The myth of the eternal return* (Patheon, NY, 1934).
13. F. J. Tipler, 'General relativity and the eternal return', in *Essays in general relativity: a festschrift for Abraham H. Taub*, ed. F. J. Tipler (Academic Press, NY, 1980).
14. S. Jaki, *Science and Creation* (Scottish Academic Press, Edinburgh, 1974).
15. D. Kubrin, *J. Hist. Ideas* **28,** 325 (1967).
16. I. Kant, *Universal natural history and theory of the heavens*, transl. W. Hastie, introduction by W. Ley (Greenwood Publishing Co., NY, 1968), Chapter 1, pp. 59–70.
17. Ref. 16, Chapter 7, pp. 123–44.
18. A. Friedman, *Z. Phys.* **10,** 377 (1922).
19. R. C. Tolman, *Relativity, thermodynamics, and cosmology* (Oxford University Press, Oxford, 1934).
20. R. C. Tolman, *Phys. Rev.* **38,** 1758 (1931).
21. F. J. Tipler, C. J. S. Clarke, and G. F. R. Ellis, in *General relativity and gravitation: one hundred years after the birth of Albert Einstein*, Vol. 2, ed. A. Held (Plenum, NY, 1980).

22. Ref. 6, Chapter 44.
23. J. A. Wheeler, *Frontiers of time* (North-Holland, Amsterdam, 1979).
24. J. A. Wheeler, personal communication to FJT.
25. S. W. Hawking and G. F. R. Ellis, *The large-scale structure of space-time* (Cambridge University Press, Cambridge, 1973).
26. R. Penrose, *Riv. Nuovo Cim.* **1,** 252 (1969).
27. Y. Choquet-Bruhat and R. Geroch, *Commun. Math. Phys.* **14,** 329 (1969).
28. R. Geroch, *J. Math. Phys.* **11,** 437 (1970).
29. R. Budic, J. Isenberg, L. Lindblom, and P. B. Yasskin, *Commun. Math. Phys.* **61,** 87 (1978).
30. J. E. Marsden and F. I. Tipler, *Phys. Rep.* **66,** 109 (1980); C. Gehardt, *Commun. Math. Phys.* **89,** 523 (1983). Another proof of Gehardt's theorem has been given by R. Bartnik, *Commun. Math. Phys.* **94,** 155 (1984).
31. S. W. Hawking, *Phys. Rev.* D **14,** 2460 (1976).
32. R. M. Wald, *Phys. Rev.* D **21,** 2742 (1980).
33. D. N. Page, *Phys. Rev. Lett.* **44,** 301 (1980).
34. J. Dugundji, *Topology* (Allyn & Bacon, Boston, 1966); J. Hempel, *Three-manifolds* (Princeton University Press, NY, 1976).
35. D. Eardley and L. Smarr, *Phys. Rev.* D **19,** 2239 (1979).
36. R. Schoen and S. T. Yau, *Proc. natn. Acad. Sci., U.S.A.* **75,** 2567 (1978); *Manuscripta Math.* **28,** 159 (1979); *Ann. Math.* **110,** 127 (1979).
37. W. S. Massey, *Algebraic topology: an introduction* (Harcourt Brace & World, NY, 1967).
38. F. J. Tipler, *Astrophys. J.* **209,** 12 (1976).
39. J. Ellis, M. K. Gaillard and D. V. Nanopoulos, *Nucl. Phys.* B **106,** 292 (1976).
40. J. Dreitlein, *Phys. Rev. Lett.* **33,** 1243 (1974).
41. P. Langacker, *Phys. Rep.* **72,** 185 (1981).
42. A. D. Linde, *Phys. Lett.* B **108,** 389 (1982).
43. A. H. Guth. *Phys. Rev.* D **23,** 347 (1981).
44. G. F. R. Ellis, in *Cargese lectures in physics,* Vol. 6, ed. E. Schatzman (Gordon & Breach, NY, 1973).
45. J. D. Barrow and M. S. Turner, *Nature* **292,** 35 (1981).
46. F. J. Tipler, *Nature* **280,** 203 (1979).
47. F. J. Tipler, *Gen. Rel. Gravn* **10,** 1005 (1979).
48. F. J. Tipler, *Phys. Rev.* D **15,** 942 (1977).
49. K. S. Thorne, *Astrophys. J.* **148,** 51 (1967).
50. C. W. Misner, *Astrophys. J.* **151,** 431 (1968).
51. R. A. Matzner, *Ann. Phys.* **65,** 482 (1971), and J. D. Barrow, R. Juszkiewicz, and D. Sonoda, *Mon. Not R. astron. Soc.* **213,** 917 (1985).
52. F. Dyson, personal communication to FJT (1982).
53. Ref. 25, p. 274.
54. H. J. Seifert, *Gen. Rel. Gravn* **1,** 247 (1971).
55. R. Budic and R. K. Sachs, *Gen. Rel. Gravn.* **7,** 21 (1976).
56. V. Trimble, *Rev. Mod. Phys.* **47,** 877 (1975).

57. J. D. Barrow and F. J. Tipler, *Nature* **276,** 453 (1978).
58. F. J. Dyson, *Rev. Mod. Phys.* **51,** 447 (1979).
59. J. N. Islam, *Quart. J. R. astron. Soc.* **18,** 3 (1977).
60. J. N. Islam, *Sky & Telescope* **57,** 13 (1979).
61. W. C. Saslaw, *Publ. astron. Soc. Pacific* **85,** 5 (1973).
62. W. C. Saslaw, in *IAU Symposium No. 58*, ed. J. R. Shakeshaft (Reidel, Dordrecht, 1974).
63. S. Chandrasekhar, *Principles of stellar dynamics* (Dover, NY, 1960).
64. L. Spitzer and R. Härm, *Astrophys. J.* **127,** 544 (1958).
65. L. Landau and E. M. Lifshitz, *The classical theory of fields*, 4th edn (Pergamon Press, NY, 1975).
66. M. J. Rees, *Observatory* **89,** 193 (1969).
67. R. Penrose, in *Theoretical principles in astrophysics and relativity*, ed. N. R. Lebovitz, W. H. Reid, and P. O. Vandervoort (University of Chicago Press, Chicago, 1978).
68. M. P. Ryan and L. C. Shepley, *Homogeneous relativistic cosmologies* (Princeton University Press, Princeton, 1975), p. 216.
69. V. A. Belinskii, I. M. Khalatnikov, and E. M. Lifshitz, *Adv. Phys.* **19,** 525 (1970); M. P. Ryan, *Hamiltonian cosmology* (Springer-Verlag, Berlin, 1972); J. D. Barrow, *Phys. Rev. Lett.* **46,** 963, 1436 (1981); *Phys. Rep.* **85,** 1 (1982); D. Chernoff and J. D. Barrow, *Phys. Rev. Lett.* **50,** 134 (1983); A. Zardecki, *Phys. Rev.* D **28,** 1235 (1983) J. D. Barrow in *Classical general relativity*, ed. W. Bonnor, J. Islam, and M. MacCallam, (Cambridge University Press, Cambridge, 1984).
70. G. Feinberg, *Phys. Rev.* D **23,** 3075 (1981).
71. D. A. Dicus, J. R. Letaw, D. C. Teplitz, and V. L. Teplitz, *Astrophys. J.* **252,** 1 (1982).
72. D. A. Dicus, J. R. Letaw, D. C. Teplitz, and V. L. Teplitz, *Scient. Am.* **248** (No. 3), 90 (1983).
73. S. Frautschi, *Science* **217,** 593 (1982).
74. D. N. Page and M. R. McKee, *Phys. Rev.* D **24,** 1458 (1981).
75. D. N. Page and M. R. McKee, *Mercury*, **12,** 17 (1983).
76. S. W. Hawking, *Nature* **248,** 30 (1974).
77. S. W. Hawking, *Commun. Math. Phys.* **43,** 199 (1975).
78. H. A. Bethe and E. E. Salpeter, *Quantum mechanics of one- and two-electron atoms* (Academic Press, NY, 1957), p. 115.
79. J. J. Sakurai, *Advanced quantum mechanics* (Addison-Wesley, Reading, 1967), p. 44.
80. A. M. Polyakov, *Sov. Phys. JETP Lett.* **20,** 194 (1974); G. t'Hooft, *Nucl. Phys.* B **79,** 276 (1974).
81. Y. B. Zeldovich and M. Y. Khlopov, *Phys. Lett.* B **79,** 239 (1978); J. Preskill, *Phys. Rev. Lett.* **43,** 1365 (1979).
82. M. S. Turner, *Phys. Lett.* B **115,** 95 (1982).
83. J. D. Barrow and J. Silk, *The left hand of creation*, (Basic Books, NY, 1983).
84. V. A. Rubakov, *Sov. Phys. JETP Lett* **33,** 644 (1981), *Nucl. Phys.* B **203,** 311 (1982); C. G. Callan, *Phys. Rev.* D **25,** 2141 (1982), D **26,** 2058 (1982).

85. At first sight one might have expected the cross-section to be the negligible geometrical cross-section $\sigma \sim m_M^{-2} \sim 10^{-56}\ \text{cm}^2$, but Callan and Rubakov revealed that there is a subtle kinematic effect because of the particular character of the potential between monopoles and fermions. The fermions are drawn into the monopole core from the lowest angular momentum states and the interaction cross-section saturates at the unitarity limit $\sigma \sim m_N^{-2} \sim 10^{-27}\ \text{cm}^2$.
86. M. S. Turner, *Nature* **306,** 161 (1983).
87. E. Kolb, S. Colgate and J. Harvey, *Phys. Rev. Lett.* **49,** 1373 (1982).
88. C. T. Hill, *Nucl. Phys.* B **224,** 469 (1983).
89. For Aristotle's definition of the soul, and Aquinas' interpretation of this definition see *Aristotle's 'De anima' in the version of William of Moerbeke and the Commentary of St. Thomas Aquinas,* transl. K. Foster and S. Humphries (Yale University Press, New Haven, 1951). Aquinas developed Aristotle's definition at length. For instance, see his *Summa contra gentiles, Book 2: Creation,* transl. J. F. Anderson (Notre Dame University Press, Notre Dame, 1975), and his *The soul,* transl. J. P. Rowan (Herder, London, 1949). An interesting commentary on Aquinas' definition of the soul can be found in *St. Thomas and the problem of the soul,* by A. C. Pegis (St. Michael's College Press, Toronto, 1934).

    Recall from Chapter 2 that in Aristotelian terminology, the *form* of a thing is to the thing as the computer program is to the physical computer; it is the abstract cause of its action, as opposed to the material and efficient causes, which are respectively the properties of the physical matter making up the computer and the opening and breaking of the circuits. Furthermore, Aristotle (in *De anima,* p. 164) asserts that the soul is to the body as an impression is to wax. He also claims (p. 164) that the soul is analogous to knowledge possessed rather than to the act of knowing. That is, the soul is analogous to the program rather than to the running of the program. Aristotle explicitly criticizes (*De anima,* Book 1) theories of the soul, such as those of Democritus and Plato, that identify the soul with a material entity—a vital force of some sort. For Aquinas, the soul needed a body to think and feel, just as a program needs a computer to process information. A soul is immortal because it is abstract, like a number is abstract. For example, the number 2 exists abstractly, but a representation of the number 2—like the number '2' written here—exists concretely. Similarly, a computer program exists abstractly and independently of its representations on computer disks or in RAM.

    Aquinas' concept of the soul is not by any means identical to our view of the soul as computer program—the idea of a computer program is after all a twentieth-century concept—but it is astonishing how similar the two concepts really are. Aquinas, for example, regarded the human mind as consisting of two faculties: the agent intellect (*intellectus agens*), and the receptive intellect (*intellectus possibilis*). The former is the ability to acquire concepts, and the latter is the ability to retain and use the acquired concepts. A similar distinction is made in computer theory. In any physical computer, general rules concerning the processing of information are coded in the physical structure of the central processor and in ROM. Such 'hardwired' programs are analogous to Aquinas' agent intellect. The programs which are coded in RAM or in disks or tape are the computer analogue of Aquinas' receptive intellect. The philosopher A. Kenny has given a preceptive discussion of Aquinas' concept of the mind and the soul in Chapter 3 of his book *Aquinas* (Oxford University Press, Oxford, 1980). Kenny himself compares Aquinas' agent intellect with Chomsky's idea of a species-specific innate language acquiring ability, but Chomsky's idea is just an example of a hardwired information-manipulating program. Kenny also notes (p. 77) that the very word *information* is derived from the Aristotle–Aquinas notion of form. We are 'informed' if new forms are added to the receptive intellect. Even semantically, the information theory of the soul is the same as Aquinas' theory.

The great empiricist philosophers Hume and Berkeley also considered the soul to be a collection or series of ideas. As Berkeley put it: 'The very existence of ideas constitutes the Soul.' See J. O. Urmson, *Berkeley* (Oxford University, Press, Oxford, 1982), p. 60.

90. C. W. Misner, *Phys. Rev.* **186,** 1328 (1969).
91. L. Brillouin, *Science and Information Theory*, 2nd edn (Academic Press, NY, 1962).
92. C. H. Bennett, R. Landauer, and R. P. Feynman *et al.*, *Proceedings of the Conference on Physics of Computation*, printed in *Int. J. Theor. Phys.* **21,** Nos. 3/4, 6/7, and 12 (1982).
93. W. Porod, R. O. Grondin, D. K. Ferry and G. Porod, *Phys. Rev. Lett.* **52,** 232 (1984). Replies by C. H. Bennett, P. Benioff, T. Toffoli, and R. Landauer, together with a rejoinder by W. Porod, R. O. Grondin, D. K. Ferrey, and G. Porod appeared in *Phys. Rev. Lett.* **53,** 1202 (1984); see also W. H. Zurek, *Phys. Rev. Lett.* **53,** 391 (1984).
94. R. Landauer, *IBM J. Res. Dev.* **3,** 183 (1961).
95. R. Landauer and J. W. F. Woo, *J. Appl. Phys.* **42,** 2301 (1971).
96. C. H. Bennett, *IBM J. Res. Dev.* **17,** 525 (1973).
97. In fact, the inclusion of inhomogeneity would raise the chances of information-processing continuing. Large-scale inhomogeneities would induce local tidal forces which could be exploited to extract energy. Whether this is possible into the indefinite future depends on the form of the density inhomogeneity spectrum beyond our present horizon which we cannot know. However, for illustration, consider the flat Universe: if density inhomogeneities vary with mass scale as $\delta\rho/\rho \propto M^{-n}$, n constant, then the amount of energy available from inhomogeneous tidal forces for information-processing will increase as $t \to \infty$ if $n < 2/3$, remain constant for $n = 2/3$ and decrease for $n > 2/3$. It is interesting that the inflationary Universe picture described in Chapter 6 results in $n \leqslant 2/3$ but only out to a finite radius in open universes. Quantum gravity effects probably determine the value of n beyond the inflation scale. The importance of knowing the complete inhomogeneity spectrum in order to predict the future behaviour of the Universe was stressed first by us in ref. 57.
98. J. D. Barrow, *Nature* **272,** 211 (1978); V. Belinskii and I. Khalatnikov, *Sov. Phys. JETP* **36,** 591 (1972).
99. F. J. Dyson, 'Life in the Universe', unpublished Darwin Lecture given at Darwin College, Cambridge, England, November 10, 1981. Dyson, however (private communication to FJT), gives the credit to Bernal whose work we discussed in section 10.2.
100. J. D. Bekenstein, *Phys. Rev.* D **23,** 287 (1981).

101. J. D. Bekenstein, *Phys. Rev. Lett.* **46,** 623 (1981).
102. D. N. Page, *Phys. Rev.* D **26,** (1982).
103. J. D. Bekenstein, *Phys. Rev.* D **27,** 2262 (1983).
104. W. G. Unruh and R. M. Wald, *Phys. Rev.* D **27,** 2271 (1983).
105. V. Lukash, *Sov. Phys. JETP* **40,** 792 (1975).
106. S. Weinberg, *Gravitation and cosmology* (Wiley, NY, 1972).
107. J. D. Barrow, *Fund. Cosmic Phys.* **8,** 83 (1983).
108. S. W. Hawking, *Nucl. Phys.* B **239,** 257 (1984).
109. F. J. Tipler, *Gen. Rel. Gravn* **15,** 1139 (1983).
110. It may be impossible for all classical universes to reunite consistently into a final singularity *generally* if the final singularity were not an omega point. If this is true, then life *must* continue to exist all the way into the omega point in *all* universes in order for quantum gravity to be logically consistent.
111. J. D. Bernal, *The world, the flesh and the Devil,* 2nd edn (Indiana University Press, Bloomington, 1969), p. 47. We are grateful to F. Dyson for this reference.
112. Ref. 111, p. 63.
113. Ref. 111, p. 28.
114. J. D. Barrow and F. J. Tipler, *Mon. Not. R. astron. Soc.* **216,** 395 (1985).
115. F. J. Tipler, 'Penrose diagrams for the Eddington–Lemaître, Eddington–Lemaître–Bondi, and Anti-de Sitter universes' (preprint 1985).
116. S. A. Bludman, *Nature* **308,** 319 (1984); A. H. Guth, *Nature* **302,** 505 (1983).
117. S. W. Hawking, 'The arrow of time in cosmology' (preprint 1985).
118. D. Mundici, *Nuovo Cim.* B **61,** 297 (1981).
119. M. Jammer, *The philosophy of quantum mechanics* (Wiley, NY, 1973).
120. C. M. Caves, K. S. Thorne, R. W. P. Drever, V. D. Sanberg, and M. Zimmermann, *Rev. Mod. Phys.* **52,** 341 (1980).
121. K. N. L. Kharev, *Int. J. Theo. Phys.* **21,** 311 (1982).
122. An example communicated to us by G. Galloway shows that this conjecture can only be true if some additional energy conditions are imposed upon the matter. We conjecture that the dominant energy condition together with the positive pressure criterion we gave on page 423 will be sufficient conditions for collapse of the $\mathbf{S}^3$ and $\mathbf{S}^2 \times \mathbf{S}^1$ spaces.
123. A modern-day theologian might wish to say that the totality of life at the Omega Point is omnipotent, omnipresent, and omniscient!

Index

Note: index entries to reference notes are indicated by suffix 'n'; Figures and Tables are indicated by *italic page numbers*.

absolute idealism, telelogical ideas in 153–9
Action Principles 6, 66
  in Einstein's gravitational theory 490
  Euler's approach 149, 150
  opposition to 150–1
  predictions based on 124–5, 148–53
  and telelogical ideas 148–53
Adams, Douglas, quoted 490
adenosine triphosphate (ATP) 553–5, 561
  molecular structure of *554*
  reactions catalysed by *554*
adiabatic irregularities 388–9, 393
aerobic respiration, development necessary as crucial step in Man's evolution 562
age of Universe, expressed in terms of density parameter 376, 377, 626
Agucchi, Giovanni, argument against Copernican system 496
al-Biruni 95
Albertus Magnus, quoted 576
Albrecht–Steinhardt 'new inflationary universe' model 431
Alexander, Samuel 190
  evolving-deity concept 190–1
alpha-particle wave function 485, 486
alternative forms of life 511
amino acids 549
ammonia 550
  as alternative solvent to water 541, 550–2
  anomalous physical properties of *524*, *525*
  density of solid form 552
  physical properties of *534*, *535*, *536*, *537*, 550–1
ammonium ions 550
Anaxagoras of Clazomenae 32–3
  Aristotle's criticism of 33
  Socrates' criticism of 33–4
ancient world, design arguments in 31–46
anisotropy model
  entropy per baryon explained by 402
anonymous poet, quoted, on hydrogen and oxygen 541
Anselm, St. 106
anthropic, definition of 159
  first use of word by Tennant 181
Anthropic Principle
  Brandon Carter's definition 1
  definitions 1, 510
  rediscovery of 219–76
antimatter, cosmic sources of 379–80
aquatic life, *see* marine . . .
Aquinas, *see* Thomas Aquinas
Arabic philosophy 46
  Aristotelian ideas in 45
Archimedes, enumeration of particles in Universe 220
Aristotle 37–40
  Causes argument 38–9
  criticism of Anaxagoras 33
  criticism of Empedocles 34
  definition of soul by 659, 680n
  final cause concept 9, 28, 38–9
  quoted 38, 39, 40
    on definition of life 511
Arouet, Francois-Marie: *see* Voltaire
artificial telelogy 134
asteroids
  infrared radiation from 597
  likely place for ETI probe 597, 611n
  size–mass relationship *290*
astronomical unit
  calculation of 337
  value quoted 221
astrophysics, Weak Anthropic Principle applied 290–1, 305–18, 327–60
asymptotically free interactions 408
atmospheres, composition for various planets *544*; *see also* Earth's atmosphere
atomic binding energies 297, 533
atomic bond energies
  carbon compared to silicon *546*
  values quoted *301*, 526
atomic bonding
  possible on neutron stars 346

atomic bonding (*cont.*)
  types of 300–1, 526
    see also covalent . . . ; hydrogen . . . ; ionic . . . ; metallic . . . ; van der Waals bonding
atomic radius
  all of same-order size 298
  formulae for 296–7
atomic year 298
atomism 36–7, 41
  revival of 112n
atoms
  energy of 296
  formation at beginning of Universe 372
    relativistic effect in 295–7, 575n
  structure of 297
  Weak Anthropic Principle applied to 295–9
automobiles, as living organisms 521–2
autonomous morphogenesis, as property of life 519, 520
autotropic photosynthesis, development necessary as crucial step in Man's evolution 562–3
Averroes of Cordova 46–7
Axelrod, Robert 100, 102
axions 390, 393, *395*
Ayala, F. J.
  reductionism classification 138
  telelogical nomenclature 134

background radiation temperature
  directional dependence of 630–1
  as limit for computer operation 661–2
  see also microwave background radiation
Bacon, Francis 51–2
  first use of word 'consciousness' 507n
  quoted 49, 52
Bacon, Roger 48
Baer, K. E. von 82
ball-in-box paradox 466–7
balls, variation of circumference and volume with dimension 270, 271
Balmer law (for hydrogen spectrum) 222
Barnes, Ernest W.
  Anthropic Principle reasoning 126, 184
  on Heat Death 168
  on mankind 184
baryon asymmetry of Universe 359, 402–3
baryon number 357
  generation 404–8
  non-conservation of 357, 403–4
baryon-to-photon ratio 381
Bauer, E.
  quoted, on introspection 468–9
  simplified discussion of von Neumann theory of measurement 465
Bayesian theorem, WAP as application of, 17
Beck, G., spoof on Eddington's methodology 227–8
Becker, Carl, quoted 295
beginning of universe 442
  events before 442, 636
bending stresses, stability to 312–13
Bentley, R. 61
Bergson, Henri
  evolutionary telelogy 189–90
  on global time 216n
  on Heat Death 190
Berkeley, Bishop George 22, 125, 155–6
  definition of soul by 681n
Bernadin de Saint-Pierre 92, 119
Bernal, J. D.
  definition of life by 521
  quoted, on end of Universe 618–19
beryllium-8, autocatalytic role in helium burning reaction 252
Bethe, H., spoof on Eddington's methodology 227–8
Bianchi classification 423, 454n
Bianchi type VII universes 423, 424–5, 426
Bianchi type IX cosmologies 666
Bianchi type IX universes 423, 425, 639–40
  forced into having omega point by intelligent life 673
  as models of past universes 640
Bierce, A., quoted, on water/ocean 524
Big Bang cosmological model 372–83
  confirmed by microwave background radiation 368, 380
  and size of Universe 2–3
binary collisions theory, E. W. Barnes' comment on 184
binding energies, planets/stars 645
biochemistry, Anthropic Principle applied to 510–76
biological exclusion principle 596
biology, status of teleology in 124, 127–43
biosphere
  life as activity of 521
  limit to lifetime of 566, 578
  marginal stability of 543–4, 567, 569
  prediction of future age 566
birds, limitation on size 314–15
black body, microwave background radiation spectrum compared with *380*
black dwarfs
  cooling of 648
  planet collisions with 645
black hole thermodynamics 223–4
black holes 347–53

as black bodies 224, 352, 445
evaporation of 353, 650, 674
as final states after evaporation and gravitational radiation 643
formation of 342, 347
general relativity description 352
as information storage devices 669
stabilized by matter-dumping 674
supermassive growth process 349
as unobservable phenomena 396
blood circulation, discovery by William Harvey 52–3
blueshift, definition of 377
Bode, Johann, quotation of Titius law 221
Boethius 45
Bohr, Niels
quantized atomic model 222, 223, 305
reaction to EPR experiment 463
on role of observer in quantum mechanics 459
boiling oceans, effect of time-varying gravitation 239–40
boiling points, of various hydrides *524*
Boltzmann, Ludwig
on direction of time 173–4
H-Theorem 174, 175, 176
quoted 195
Bondi–Gold–Hoyle (steady-state) cosmological model 601, 616
bonds: *see* atomic . . . ; covalent . . . ; hydrogen . . . ; ionic . . . ; metallic . . . ; van der Waals bonding
Bonnet, Charles 68, 221
books, information content of 137
Borel, P., quoted 576
Born, Max, spoof on Eddington's methodology 228
Boshongo creation myth 94
boundary conditions 444–9
evolution equations 409
removal of 448–9
Strong Anthropic, in quantum cosmology 503–5
Weak Anthropic, in quantum cosmology 497–503
Boyle, Robert 55–8
astronomical examples not favoured by 56
conversation with William Harvey 53
lectures series bequeathed by 58, 61
quoted 55, 56, 57
brain, *see* human brain
branch universes 493, *494*
branches of Universe, split by measurement *494*, 500–1
Brans–Dicke gravitation theory 245
bread, information content of piece 198
bremsstrahlung 299, 389
bremsstrahlung (free–free emission), cross-section for 299
Bridgewater Treatises 82, 117n, 144, 524
Brill–Flaherty theorem 627
Brillouin, L.
equation for energy to obtain information 179–80
information–processing inequality 661, 662
Brinkley, J. 79, 116n
Brücke, E. von, quoted 123
bubble-universe models 192–3, *436*, 441, 605–7, 616–17
closed-universe 605, 607
and Final Anthropic Principle 607
open-universe 605–7
Penrose conformal diagram for *606*
Strong Anthropic Principle arguments against 607
Buddhist cosmology, large numbers in 276n
Buffon, G.
estimate of Earth's age 117n, 159–60
on Final Causes 68, 114n
building blocks of life
synthesis of 3
time required for synthesis 18
bulk modulus 303
Burt, C. L., meaning of word 'consciousness' 507n
Burtt, E. A., paraphrasing Isaac Newton 621
Bush, Ian D., quoted 327
Butterfield, Herbert, on Whig interpretation of history 10, 135

caesium hyperfine transitions, as time standard 242
Cairns–Smith, A. G., mineral ecology concept 522
Capek, Milic 216n
carbon
anthropic significance of 545–8
relative abundance *542*
carbon compounds, metastability of 546–7
carbon dioxide 545
anthropic significance of 548
reversible hydration of 548
in temperature regulation 548
unique solubility property of 548
carbon-12
formation from He-4 252
resonance level predicted by Hoyle 252
carbon-based life
dependent on unique properties of certain elements 510

carbon-based life (*cont.*)
  fine-tuned coincidences 253
  as only possibility 3, 570
carbon-based polymers 547–8
  spontaneous formation of 541
carbonic acid 544, 548
Carlyle, Thomas, quoted 384
Carr–Rees coincidence 400
Carroll, Lewis, quoted 288
Carter, Brandon
  Anthropic Principle limitation to Copernician principle 1
  on gravitational constant 336
  rejection of prediction of future age of biosphere 566
  Strong Anthropic Principle introduced by 21–2, 248
  Weak Anthropic Principle formula 557, 559–60
    crucial steps for evolution of Man 562–4
    evolutionary criteria 561–2
    starting point for derivation 557
Carter's inequality 511
Cauchy horizon 674
Cauchy hypersurface topology 622
cell membrane formation processes 540
cell self-reproduction 513
centriole/kinetosome/undulipodia complex, formation necessary as crucial step in Man's evolution 564
Chamberlain, Thomas C., solar system formation theory 125–6, 165
Chandrasekhar, S., reaction to Dirac's Large Numbers Hypothesis 234
chaos, and determinism 90–1, 118–19
chaotic cosmology 178
  gravitational entropy as antithesis of 446
chaotic cosmology programme (of Charles Misner) 420, 422
chaotic gauge theories 8, 255, 256
chaotic inflationary-universe model 436–7
chemical binding energy fraction 301
chemistry, Weak Anthropic Principle applied 295–302
Chesterton, G. K., quoted 1
Chinese philosophy
  spontaneous order concept in 96–7
  telelogical concepts in 95–6, 98–9
chlorophyll 555–6, 561
  absorption spectra of 556
chordates, development necessary as crucial step in Man's evolution 564
Christie, Agatha, quoted 288
chronology
  ancient world *46*
  Medieval ages *54*
  modern age *204*
  Renaissance *54*
chronometric cosmology 604–5
Chu Hsi 96
Churchill, Winston S., quoted 153
Cicero (Marcus Cicero) 43–4
ciliate protozoa, reason for size 314
civilization, growth in future 173
Clarke, Samuel, correspondence with Leibniz 62–3
classical epoch, definition of 641
classical timescales, future evolution of matter in 641–7
Clauser, John, quoted, on physical reality 463
clock/watch analogy
  Boyle's use of 58
  Cicero's use of 44
  as example of artificial telelogy 134
  Paley's use of 77
close stellar encounters 642
closed Friedman universe 376, *376*
  density inhomogeneities growth rate 415–16
  Many-Worlds approach 490–6
  Penrose diagram for *634*
closed universes
  Cauchy hypersurface topology of 622
  criterion for evolution of life 376, 410
  energy-conservation law 618
  energy sources in far future 665–6
  existence of life in 637, 675, *676*
  infinitely cycling model 192, 193
  lifetime calculation 376
  long-lived
    cosmological expansion of 650
    timescales of *653–4*
  small, near final singularity, timescales of *647*
  telelogical prediction 153
  time-evolution of expansion scale factor in *376*
closure of Universe, teleological arguments for 153, 675
clustering 377–8, 415
co-operative behaviour 100–2
Cocke–Wheeler cosmological model 674
coincidences
  elemental properties 124, 143, 252–3
  large number 11, 219–31, 246–7
cold bodies, planets as 328, 340
Coleman–Weinberg interaction potential *432*, 435
Collins, C. B.
  on isotropy of Universe 419, 423, 426
  on special initial conditions for evolution of Universe 250

colonization, motivations for 593–600
colonization (of space) 580, 584–6
colonization theories/models
  diffusion model 585
  MacArthur–Wilson colonization theory 584, 585
colour charge (of quarks) 355, 356
communications, interstellar 577
  motivations for 590–3
communications region, far future of Universe 668
Comparative Advantage, (economic) Theory of 596
complexity theory (of ecologists) 141
Compton scattering 299
computer hardware
  human brain compared with 136, 137
  living body compared to 75, 659
computer program
  human being as 659
  self-reproducing 522
  telelogical explanation 75
  Universe as 9, 125
computer speeds 136, 206n, 662
computer technology, necessary for interstellar communication/ exploration 577, 578
conformal diagrams: *see* Penrose diagrams
conformal time 442, 491
Confucianism 98–9
Confucius, quoted 98
consciousness
  collapsing wave functions 468, 469
  meaning of word 469, 507n
  necessary to bring Cosmos into existence 471
consequences, as coincidences 12
constants of Nature 5
  constrainted by self-selection effect 5
  effects of small changes 254
  statistical basis 253–4, 257
  see also fundamental constants of Nature
contingency argument 104–6
continuous creation theories, of Jordan 235
convergent evolution 132
Cooke, Josiah 88–9, 118n
Copenhagen Interpretation 458, 459
  contrary-to-common-sense nature highlighted by EPR experiment 463
  von Neumann's axioms for 464–5
  wave function collapse required by 459, 466, 467, 496
Copernician Principle 1, 501, 502, 503
  ETI justified by 600–1
  Weak Anthropic Principle resulting in 4
Copernican system 496
Copernicus, Nicolas 49–50
Copernicus satellite/probe 379
Copleston, F. C., debate with Bertrand Russell 103, 107–8
cosmic epochs concept, of A. N. Whitehead 192
cosmic teleology
  defended by Peacocke and Mascall 183
  developed by Tennant 181–2
  notion introduced by Pringle-Pattison 181
Cosmogonic Philosophy 158
cosmological arguments, and design arguments 103–9
cosmological constant 225, 234, 412–14
  anthropic limit for 414
  effects on information processing 668–9
cosmological models
  closed universe 622
  cyclic models 616
  ever-expanding universe 623
  evolving models 616
  open universe 622
  recollapsing universe 623
  unchanging models 616
cosmology, anthropic principles in 367–449
costs
  interstellar exploration 581–3
  von Neumann probes 582–3, 591–2
covalent bonds 300, *301*, 526
Cowper, William, quoted 220
Crane, S., quoted 166
creation interpretation, Second Law of Thermodynamics 176, 177
creation 'out of nothing' 440–4
Croll, James, estimate of Earth's age 163
crucial steps, in evolution of Man 562–4
Cudworth's 'plastic nature' concept 59
cultural imperialism 596
Cummings, E. E., quoted 497
Cyclops project 577

Daedalus project 581–2
  cost 582
  payload size 584
  rocket power 644
D'Alembert, Jean 65
  quoted 103
  wave equation solution in one dimension 267
dark material
  galaxy groups/clusters 377
  possible particles for 390–1
Darwin, Charles 83–4
  concession on rate of evolutionary change 162
  on Heat Death 167

Darwin, Charles (*cont.*)
Origin of Species 9, 161, 162, 615
progress-to-perfection optimism 166
quoted 166, 167, 180
telelogical interpretation 85, 86
Darwin, Erasmus 72
Dawkins, R., idea-complexes as living organisms 522
day length, calculation of 309–10
de Sitter, W., quoted 430
de Sitter space-time 433, 434, 435, 668
deceleration, of expanding Universe 375
definitions
Anthropic Principle 15–23
biology 512
degeneracy pressure 303
Deity
contrasting evidence for, in Newton's time 30
and pantheism 107, 121
delayed choice experiments 469–70
Democritus 36–7
density, atomic weight relationship 299
density contrast
conformal time relationship 416
limited range for 417, *418*
density enhancement *378*
density irregularities
adiabatic 388, 390
condensation growth of 415
isothermal 388
as possible formation processes for galaxies 388
density parameter (Ω)
age of Universe expressed in terms of 376, 377, 626
effect on time-evolution of expansion scale factor *376*
energy relationship 375
limited range for 411
unity value explained
by Anthropic arguments 627
by quantum gravity 501, 503
values quoted 377
Derham, W. 67–8
Descartes, René 53–4
teleology dismissed by 54
design arguments 27–109
of ancients 31–46
chronology of 46, 54, 92
and cosmological arguments 103–9
critical developments 68–83
Darwin's rejection of 84
devolution 83–92
historical prologue 27–31
Kant's discussion 73–4
Medieval 46–9
non-Western philosophy/religion 92–102
Renaissance 49–54
scepticism against 29
designed Universe
Aristotelian view 37
Design Argument for 22
determinate natural teleology 134
determinism
and chaos 90–1, 118–19n
classification of 138–9
Teilhard's use of weak form 200
Thomas Huxley's view 86–7
deterministic system, world viewed as 459–60
deuterium
abundance relative to hydrogen 379
formation at beginning of Universe 371
deuterium oxide, biological effects 541
deuterons
formation reactions 321
nuclear potential as function of radius *319*
triplet state 321
Dewdney, A. K., two-dimensional world 266
DeWitt, Bryce S., quoted, on WAP selection effect in MWI 482–3
DeWitt boundary condition 497
DeWitt's Postulate of Complexity 473
Dicke, Robert H.
anthropomorphic explanation 16, 219
gravitation theory (with Carl Brans) 245
on large numbers coincidences 246–7
quoted, on cosmological coincidences 243
Weak Anthropic Principle developed by 16, 125
dielectric constants
of various liquids, plotted against dipole moments *538*
of various substances *537*
diffusion models (of exploration/colonization) 585, 589–90
dimensional analysis
first used 289
unreasonable effectiveness of 270
dimensionality 258–76
additional-dimension effects 274–5
effect on fundamental constants 269
see also three-dimensional space
dineutrons 321, 322
Dingle, H., quoted 224
dinosaurs, size of 313
Diogenes 36
diprotons
non-existence of 322
nuclear binding energies 322

singlet state 321
Dirac, P. A. M.
influenced by Milne's gravitational theory 238
large number hypothesis 231–4, 276
criticisms of 233–4
time-varying Newtonian gravitation constant 11, 20–1
discreteness (of apparatus/system) 485
dispersal phase, behaviour pattern in colonization 593–4
DNA, computer program analogy 75
DNA replication 304–5
Dobzhansky, T.
on Teilhard 199
teleological nomenclature 134
dog colours, as design argument 92
dolphins, encephalization of 131
Dominant Energy Condition 423
Doyle, Arthur Conan, quoted 310, 354
Drake equation 586, 589
probabilities not varying with Galactic age 587
dual timescales, effect on evolutionary biology 244–5
Duhem, Pierre, Heat Death of Universe criticized by 617
Dyson, Freeman
closed-universe life ruling 637
first paper on study of far future 658
quoted 318
on cosmic key times 385
dysteleology 126, 166–73

Earth
information content of 661–2
Earth's age, biological constraints on 159–65
Earth's atmosphere
composition *542*, 543, *544*, 549
marginal stability of 543–4, 567, 569, 614
Earth's future
in presence of monopoles *657*
Weak Anthropic Principle constraints on 556–70
Eco, U., quoted 511–12, 548
economics
interstellar travel 581–3
nuclear weapons 599
relative materials/wages 172, 644
Eddington, Arthur S.
on age of Universe 385
calculation of fundamental constants of Nature 224–7
on cosmological constant 412
credibility damaged 227
on existence of global time 195
quoted 613
on end of Universe 618
on number of protons in Universe 219
on Second Law of thermodynamics 658
on redshifts 237
'spoofs' of his methodology 227–8
Eddington luminosity 350
Eddington number 225
Eddington–Lemaître–Bondi universe 638, *639*
Edsall, J. T., quoted, on carbon dioxide 548
Egerton, Francis, bequest for (Bridgewater) Treatises 82
Egyptian creation myth 94
Egyptian text, quoted, on creation 440
Ehrenfest, P., on dimensionality 260, 261–2
Ehrlich, Paul R.
criticism of Julian Simon 211n
on energy requirements 169–70
single political and economic goal proposed 141
Einstein, Albert
criticism of quantum mechanics 476
definition of physical reality 461
quoted 430
on magnitude of dimensionless factors of proportionality 270
unsuccessful in simultaneous-measurement experiment 461
Einstein–de Sitter universe 425
Einstein–Podolsky–Rosen (EPR) experiment 461–3
and Copenhagen Interpretation 463
as delayed choice experiment 470
Einstein's cosmological constant 225, 234, 415, 450n
Einstein's cosmological equations, ensemble solutions 250
Einstein's cosmological model 616
quasi-Penrose diagram for *635*
Einstein's general relativity equations 138, 409
Einstein's gravitational equations, quantization of 492
Einstein's gravitational theory, Action Principle in 490
Einstein's special theory of relativity 384
electromagnetic interaction 293
in unified theories 355, *357*
electromagnetic interactions 370–1
electron charge, time variation of 240–1
electron orbitals 297
velocities of 297

electron spin
  measured in EPR experiment 462, 463
  memory states of apparatus to measure 473
electron–positron plasma 664, 665
  as information storage device 669
  as remaining matter 649, 650
electron-to-proton mass ratio (B)
  as controlling physical parameter 293, 296
  importance in DNA replication 304–5
  uncertainty relationship 304
elements
  quirks in properties 22, 143
  relative abundance *542*
Ellis cosmological model 249, 434, 601
  Davies' SAP argument against 602
Empedocles of Argigentum 34
Empirical theologians 180–4
encephalization 130
  increase in humans 130
  increase in marine species 131
end (telos) 33
  Aristotelian view 38
end of Universe 677
endoskeleton, development necessary as crucial step in Man's evolution 564
energy requirements 169–71
energy sources, far future 663–4, 665, 675
energy–time uncertainty relation, information-processing restricted by 671
entropy
  black holes 353
  maximization 166–73
  reversal from maximum 178
entropy per baryon 381–2, 383
  value of 401–8
  weak anthropic limits on 407–8
Epicurus of Samos 41
epiglottis, Paley's discussion 78
epistemological determinism 139
epistemological reductionism 138
epoch, uniqueness of 601
Equivalence Principle 347
Erdös, P., quoted 401
eschatology 658
essential traits, in evolution of Man 561–2
Euler, Leonhard, action principle 149, 150
eutaxiological arguments 88–9
  compared to teleological arguments 29, 144
event horizon, formation of 636–7
ever-expanding universe model 623
Everett, H.
  many-worlds interpretation of quantum mechanics 8, 249, 472
  reply to Einstein on split-universe model 476–7
Everett Friends Paradox 468, 470–1
evolution of life, in far future 675–7
evolution of Man
  criteria for steps in 561–2
  as crucial step 564
  crucial steps in 562–4
  improbability of spontaneous evolution 565–6
evolution timescales
  lower/upper bounds 643
    effect of intelligent activity on 643–5
evolutionary cosmological models 185–95
evolving-universe theory, of Teilhard de Chardin 196, 198
excited states
  atomic/molecular 301–2, 321
  nuclear 321
expanding Universe, Jeans' explanation 236–7
expansion rate of Universe 18
exponential inflation 432
  results of 432–4
extinction, evolving species 615
  avoided by machine descendants 595–6
extraterrestrial intelligent (ETI) life
  absent from Galaxy 590
  Carter's WAP formula invalid if found 569
  in Solar System 597–8
  space-time argument against 576–608
extraterrestrial life 14
  Teilhard's views on 203, 217
eye
  evolution necessary as crucial step in Man's evolution 564
  independent 'invention' of 132, 561, 564
  see also human eye

falsifiability, as condition for scientific theory 198
far future, study as separate branch of physics 658
Feinberg, G., definition of life 521
Fermat, P. de, reflection/refraction predictions 125, 148–9
Fermi, E., quoted, on extraterrestrial intelligent life 578
Fermi constant 294, 354
Fermi coupling 294
Feynman, Richard P. 151, 152
Fichte, J. G. 153
  quoted 154
  subjective idealism 155, 156
Fifth Way (of Thomas Aquinas) 47

Final Anthropic Principle (FAP)
definition of 23
event horizons in 668
Participatory Anthropic Principle interpreted as 471
requirement that life exists forever 659
speculative nature of 23
Final Causes
Aristotelian view 38, 39
Descartes' dismissal of concept 54
Henderson's concept 145
Paley's dislike of concept 76–7
separated from science by Francis Bacon 51–2
Whitehead's view 193–4
Final Observation 471
final state of Universe 658–77
summary of 675–7
fine structure constant
as controlling physical parameter 293, 295
Eddington's combinatorical formulae 227
first postulated by Sommerfeld 223
gravitational analogue 293, 336
lower and upper bounds in grand unified theories 358, 359
numerical value of reciprocal 227, 231, 293
physical meaning of 298
relationship to gravitational interaction constant 358
stability influence 298
strong interaction analogue 295, 319
Teller's relationship 230, 239
Wyler's numerical coincidence 231
finite state machine 661
first intelligent species to evolve, interstellar communication/exploration by 592–3
Fiske, John 85–6
Fitness of the Environment 118n, 143, 147, 533
flames, as living organisms 520
flat Friedman universe 375, *376*
density inhomogeneities growth rate 415
galaxy growth in 428–9
isotropy criterion for 425
Penrose diagram for *633*
flat universes
cosmological expansion 650–1
energy sources in far future 665
human evolution not possible in 426
isotropy/anisotropy of 427–8
flat-universe models 625
Flatness Problem 410–11, 432, 501
flavours (of quarks) 356, 408
flea jumping, as spurious design argument 119n
Flew, Anthony 104, 106
fluctuation interpretation, Second Law of Thermodynamics 176, 177
Feynman's objection 177
flying creatures, limitation on size 314–15
foam model (of space–time) 353
foliation, space–time 627, 659, 660
fossil, Paley's discussion 78
Fourier, J. B. J.
dimensional analysis used by 289
heat conduction theory 160
fourth dimension 260
Fox, S. W., experiment on spontaneous cell-wall formation 541
free-expansion model, exploration/colonization 584–5, 589
free-market economic system
Hayek's description 99, 140
Spencer's view 187
Friedman equation (for expansion of universe) 373
solutions for 375–6, *376*
Friedman (oscillating closed universe) model 620
Friedman universes
Many-Worlds approach 490–6
matter remaining in 649
Friedman's equation, revised for Einstein static universe 413
Frost, Robert, quoted 27, 613
fundamental constants of Nature 224–31
dimensionality effects 269
equilibrium states determined by 288
large scale features contingent upon 289
strong anthropic aspect to 367
fundamental mass, in terms of Eddington number 226–7, 278
fusion, heats of, for various substances *536*
future age of biosphere 566
future of Earth, Weak Anthropic Principle constraints 556–70
future evolution of matter
in classical timescales 641–7
in quantum timescales 647–58
future histories 14
future of Universe, global constraints on 621–41

galaxies
cosmology for 387–97
density enhancement in *378*
formation theory incomplete 396
mass–size relationship *290*
possible formation processes for 388–90

Galaxy
age estimated 588
information content of 662
number of communicating civilizations, estimated as unity 588
number of intelligent species in 586–90
time required for exploration of 583, 584
Galen 44
Galileo Galilei
on anthropocentric design arguments 50
on marine creatures 313
on size of living organisms 311
Gamow, George
on origin of elements 398
time-varying electron charge 240
Gamow peak 330, *331*
gas clouds
cooling of 389–90
possible evolutionary sequence for *351*
protostar formation in 339
Gassendi, P. 57
gauge theories 152
sum-over-histories method for 152
see also chaotic . . . ; grand unified . . . ; stochastic . . .
Gaussian statistical distribution 256
Geikie, Archibald, estimate of Earth's age 163
genes, improbability of spontaneous assembly 565, 575n
genetic code, development necessary as crucial step in Man's evolution 562
geometrical factors, variation with dimension 270–2
German professor (typical), quoted 510
Gibbs' Phase Rule 147
global constraints, on future of Universe 621–41
global time, definability of 195, 216
glucose fermentation, development necessary as crucial step in Man's evolution 562
gluons
colour smearing by 408
in unified theories 355, 356
Goethe, J. W. von 82–3
Gold, T., quoted 408
Gold–Hoyle (steady-state) cosmological model 601, 616
Goldman, Emma, quoted 590
Goodchild, J. G., estimate of Earth's age 164
Gosse, Philip, young Universe with ready-made fossils theory 428
Gott's bubble-universe model 192–3, 605, 607
Gould, S. J., quoted 92
Graham, W., quoted 84
grand unified gauge theories 354–60
multi-dimensionality in 274
grand unified theories (GUTs) 354, 357
energy scale of 357–8
non-conservation of baryon number 357, 403–4
proton lifetimes in 648
spontaneous symmetry-breaking 617
gravitation potentials 262
algebraic potential 263
Yukawa-type potentials 262–3
gravitational binding energy 327
balanced by thermal pressure 328
formula for 306
gravitational clocks 237
gravitational constant, numerical coincidence 336
gravitational criterion, human existence 311–12
gravitational entropy 445–9
as antithesis of chaotic cosmology 446
as argument against Strong Anthropic Principle 448
definition of 447
gravitational equilibrium 448
gravitational radiation 643
gravitinos 390, 393
Gray, Asa 85, 127
Great Chain of Being 128, 189
Green's function, boundary conditions 498–9
Gregory, David, on Newton 61
Grene, M., on teleology 136
Grew, Nehemiah, design argument 63–4
ground state energies, finity of 264
Gurevich, L., on dimensionality 259, 265
Guth's inflationary universe model 430

H-Theorem 174, 175, 176
Haas, A.
fundamental constants 228
quantum equation 223
habitable planets, requirements for 309–10
hadrons, formation at beginning of Universe 371
Haldane, J. B. S.
on anthropic-fluctuation interpretation of Second Law 176–7
on dual timescales 244–5
halos, galaxy/cluster 377
possible particles for 391
Hamiltonian
atomic system 533
radiation gas 492

hand, *see* human hand
Hansen, Hans Marius, and Bohr's quantum theory 222
Harrison–Zeldovich spectrum 434
Hart, Michael, computer simulation of Earth's atmosphere 567–9
Hartle–Hawking quantum model 443–4
Hartle–Hawking wave function of the Universe 105
  non-normalizability of 501
Hartshorne, Charles
  definition of Universe 121n
  ontological argument defended by 108
  pantheistic approach 108, 122n, 195
  on special relativity 194–5
  on speed of light limitation 195
Harvey, William, design argument approach 52–3
Hausdorff topology 636
Hawking, S. W.
  black hole thermodynamics 224, 445–6
  on boundary conditions of Universe 444
  on isotropy of Universe 419, 423, 426
  quoted, on Schrödinger's Cat Paradox 458
  removal of boundary conditions of Universe 449
  special initial conditions for evolution of Universe 250
Hawking foam space–time model 353
Hawking process, proton decay by 652
Hawking quantum cosmology 674
Hawking–Moss 'new inflationary universe' model 431
Hawking–Penrose singularity theorems 621
Hayek, Fredrick A. 99–100
  indeterminate teleology 140
  spontaneous order of society 99–100
  unpredictability in social evolution 188, 189
heartbeats, number in a lifetime 315, 316
Heat Death of Universe
  Bergson's approach 190
  Bernal's version 619
  criticism of concept 617
  dual timescales, effect on 245
  Eddington's version 618
  first forecast by Helmholtz 166
  gravitational equilibrium as geometrical analogue of 448
  Teilhard's treatment 168, 197, 201
heats of fusion, *see* fusion, heats of
heats of vaporization, *see* vaporization, heats of
heavy elements, formation in Galaxy 587
heavy water, *see* deuterium oxide
Hegel, G. W. F. 157
Heisenberg Uncertainty Principle 460–1, 474
  insufficiency to prove stability of matter 264
  see also Uncertainty Principle
Heisenberg–Mott analysis, Wilson cloud chamber experiment 485, 486–9
heliocentric model, of Copernicus 49
helium stars, anthropic consequence of 253, 399
helium-4
  formation at beginning of Universe 252, 371, 399
  relative abundance 369, 379, 399
  stellar burning reaction 252
Helmholtz, Hermann L. F. von
  action principle 151
  Heat Death forecast by 166
Henderson, Lawrence J.
  on anomalous properties of water 143, 524, *531*, 533–4
  fitness-of-environment concept 22, 88, 118n, 124, 143–8
  impact of his ideas 147
  Tennant's use of his arguments 182
  unique properties of certain elements and compounds 124, 143, 510, 524
Hero of Alexandria, law of reflection 110n, 148
Hertzsprung–Russell diagram 336, *337*
Hicks, L. E. 29, 144
hidden variables 460, 463
hierarchy of causes 103–4
Higgs field 431, *432*
high-energy particle states, information stored in 671
Hilbert, D. 151
history, teleological interpretation 9–10, 135–6
Hodge, Charles, on Darwinism 84
Holton, Gerald, scientific progress themata concept 11
horizon problem, Leibniz's view 63
horizon structure of Universe 419–20
horizons, disappearance in anisotropic contracting universes 631–2
hot Big Bang cosmological model 372–83

  key times in 385–7
  reason for name 401
Hoyle, Fred
  bubble universe model 192
  nuclear resonance studies 250–4
    teleological conclusion 22
  steady-state cosmological model 421
Hoyle–Narlikar (steady-state) cosmological model 601, 616–17

Hubble, Edwin, expanding-universe
interpretation 2–3, 368
Hubble constant 373, 626
present value 374
Hubble distance 626
Hubble parameter 627
Hubble redshift 368, 450n
Hubble time 626
Hubble's Law 373, 378
human being, as computer program 75, 659
human brain
computation rates 136
conscious processing capacity 137
information storage capacity 136, 660
human cells, synthesis of 580
human eye
Lucretius' materialistic approach 42
Paley's discussion 78
radiation receptivity of 338
Socrates' design argument 36
human genome, improbability of spontaneous assembly 565
human hand, teleological aspects of 44
human size, estimated by gravitational criterion/molecular bonding calculation 311–12
Hume, David 69–72
attack on Newtonian Design Argument 62
quoted 105, 107
vitalist views of 71, 115n
Huxley, Thomas H., quoted 86, 87, 127, 148
hydrides
molecular bond angles/distances *529*
various physical properties *524*, *525*
hydrogen
chemical properties 543
percentage in present Universe 369
physical properties 543
relative abundance *542*
unique properties of 542–3
hydrogen atom, simple model 296
hydrogen bonding 300–1, 526
in water 527–8, *529*
hydrogen fluoride
anomalous physical properties *524*, *525*
hydrogen-bonded polymers 526, *527*
hydrophobic effect 539–40

ice
density compared with water 524, 530, *531*, 533–4
molecular structure 528, *530*
phase diagram for *532*
idea-complexes, as living organisms 522
Idlis, G., large-universe argument 16
ignition risk, Earth's vegetation 544, 567, *568*
imperialism 585–6, 594
definition of 594
improbable steps, in evolution of Man 559, 562–4
indeterminate natural teleology 134–5
infinite state machine 661
inflation, in early Universe 438–40
inflationary-universe models 430–4, 605
and Anthropic Principle 434–40
to answer Flatness Problem 502
information content
books 137
computer memory 514
Earth's mass-energy 661–2
economic resources 172–3, 595–6, 600
Galaxy 662
human brain 660
human gene 514
piece of bread 198
solar system resources 600, 662
thought 660
Universe 662
information-processing
condition for indefinite existence of intelligent life 659
economic activity as 172–3, 660
energy requirements 663
inequality for 661, 662, 663
near final singularity 667
information-processing rate
computers 136, 206n
human civilization 600
limit on 662
information storage
far future of Universe 669–71
growth rate of 675–6
information-processing, limited by dimensionality 266
Inge, William R., on Heat Death 168
inhomogeneity of Universe 414–19
initial conditions 408–12
intelligence, evolution of 131–3
intelligent life
definitions of 523
essential global role of 674–5
matter used up by 664
intelligent species, upper bound on number of 586–90
interstellar communication 577
motivations for 590–3
interstellar transit time 581
interstellar travel 579–83
economics of 581–3
motivation for 591–3

intractable problems 139
introspection, wave function collapse caused by 468–9
inverse square (gravitational) law 262, 263, 264
ionic bonds 300, *301*, 526
ionic solids, electron motion in 304
ionization temperature, at beginning of Universe 382
Iroquois creation myth 94
Islam, teleological concepts in 94–5
island colonization theory 584
isothermal irregularities 388, 393
isotropization, definition of 423
isotropy of Universe 419–30
  confirmed by microwave background radiation 419
  tested by counting galaxies 420
  Weak Anthropic Principle invoked for 426

James, W., quoted 103
Janet, Paul, on Final Causes 89–92
Jeans, James, shrinking atomic size concept 236–7
Jeans instability, rowth rate of 414
Jenkin, F., estimate of age of Earth and Sun 161–2
Johnson, Dr Samuel, dislike of David Hume 115n
Johnstone Stoney, G.
  natural units 291–2
  on planetary atmospheric composition 309
Jordan, Pascual, continuous creation theory 235
Jupiter
  almost a star 331
  lifetime in presence of monopoles *657*
  orbital radius of *221*
  possible life-form on 511, 570
Jupiter swing-by launch/slow-down strategy 580–1

K-strategy (of exploration/colonization) 584, 589, 590
Kaluza–Klein cosmological models 239
Kaluza–Klein higher-dimensional theories 273–4
Kant, Immanuel 72–5
  cosmology 620
  design argument admired by 73
  design argument undermined by 73–4
  on dimensionality 260
  influence on German biologists 74–5

Keats, John, quoted 458
Kelvin, Lord
  estimate of age of Earth and Sun 160–1
  evolutionary timescale debate 87–8, 159, 161
Kepler, Johannes 50–1
Kerr–Newman metric 352
kinematic relativity, Milne's theory 237
King, Clarence, estimate of Earth's age 162
Kirchhoff, wave equation solution in three dimensions 268
Kramers opacities 334, 335
krypton-86 wavelength, as length standard 242
Kuhn, Thomas, on teleology 142–3

Lagrangian
  closed Friedman universe 491
  physical laws derived from 257
Lamb shift, energy of 299
Landau–Chandrasekhar mass-limit 643
Landau–Chandrasekhar number 332
Landsberg–Park oscillating-universe model 455n
Lao Tzu, quoted 97–8
Laplace, Nebular Hypothesis 90
large bodies, effect of existence of 306
large creatures
  advantage in sea 313
  precarious existence on land 313
large numbers coincidences 11, 219–31, 358–9
  Dicke's observations 246–7
Large Numbers Hypothesis (LNH) 232–3
  cosmological constant and density ratio in 455n
  time-varying gravitation in 232–4
laws of physics 255–8
Least Action, Principle of 6, 66, 149
Least Time, Principle of 148
Lehninger, A. L., classification of elements used in living organisms 552, *553*
Leibniz, Gottfried
  caricatured by Voltaire 6, 65
  correspondence with Samuel Clarke 62, 63
  many-world hypothesis 6, 62–3, 192
Lemaître, G.
  criticism of Jeans' shrinking atom theory 236–7
  quoted, on Big Bang 372
length reference standards 242–3
lepton number 357
leptons 357
Leucippus of Elea 36
Levin, B., quoted 647

Lewis, C. S., quoted 46
liberal democracy 9, 135
Lieh Yü-Khou 95
life
  definitions of 13, 511–23
  non-extinction of 14, 659
  sufficient conditions for 512
life-supporting conditions, planetary size effect 309
life systems, crucial molecular structures 88
lifetimes, quoted *171*, 626–7
light, velocity of
  change with time 236
  coincidence with charge ratios 223
  Paley's discussion 79
light elements
  abundance predicted by Big Bang theory 368–9
  abundance relative to hydrogen 378–9
  formation at beginning of Universe 371
  origin of 398–400
Linde's inflationary universe model 431, 436–7
lipids, cell membrane 540
liquid-drop nuclear model 323–4, 363n
  binding energy components for *324*
Liu Tsung-Yuan, quoted 97
living organisms, Weak Anthropic Principle applied to 310–18
Löbell space 638
local hidden variables, existence tested by EPR experiment 463
Locke, John, meaning of word 'consciousness' 507n
Lodge, Oliver, on black holes/neutron stars 340, 348–9, 366n
London, F.
  quoted, on introspection 468–9
  simplified discussion of von Neumann theory of measurement 465
Lotze, Hermann 75
Lovejoy, A. O. 128, 189
Lovejoy, C. O. 131–2
low-energy Universe, orderliness in 256, 257
Lucretius (Titus Lucretius Carus) 41–2
Lukash ever-expanding anisotropic universes 670
luminosity, factors affecting 332–4

MacArthur–Wilson colonization theory 584, 585
  K-strategy 584, 586
  r-strategy 584, 586
MacClaurin, Colin, design argument 62
McCullough–Pitts neural network 266
machine descendants (of naturally evolved species) 595–6
machines, regarded as people 595, 615
McTaggart, John M. E., absolute idealism of 157–8
magnesium
  in chlorophyll 555–6
  relative abundance *542*
magnesium fluoride, molecular structure of *531*
magnetic moment effects 298
Maimonides 47
main-sequence stars
  lifetimes *171*, 246, 333–4
  luminosity–temperature relationship *337*
  surface temperatures 335
mammals
  heat balance in 315
  limitation on size 315, 362n
Man, timescale of existence *171*, 584
many-worlds hypotheses 6
  design argument for 22–3
  Ellis' infinite occurrence model 249
  infinite Universe approach 7–8
  Leibniz' optimal world 62–3
  multi-valued quantities approach 6–7
  Parmenides' argument for 34
  quantum theory interpretation 8, 193, 249
  types of ensemble for 6–8
  variable constants of Nature approach 8
  Wheeler's oscillating universe model 248–9
  see also split-universe model
Many-Worlds Interpretation (MWI) 8, 249, 472–89
  cosmological model compared with Narlikar model 495
  reason for invention of 622
marginal utility theory 172
marine creatures, size of 313
Mars
  atmospheric composition *544*
  communication/exploration of 593
  life on 511
  orbital radius of *221*
Marx, Karl, on division of labour 187
Mascall, E., on size of Universe 247
mass, coincidences of 12
mass of Universe 384–5
mass–size diagram, for Universe *290*, *341*
  possible explanations 291
  revised to exclude unobservable phenomena *397*
massive neutrinos 390–3, *394*, 664
matter
  nature of 302–5

preponderance over antimatter 370
matter-dominated Universe 626, 628, 629
Matthews, W. R. 183
Maupertuis, P. L. M. de 65–7
Principle of Least Action 6, 66, 67, 149
Maxwell, James Clerk 88
quoted, on laws of matter and mind 545
Maxwell's Demon 174, 179, 662
Mayan creation myth 93
Maynard Smith, J. 102
Mayr, Ernst, on evolution of intelligence 132–3
Medawar, Peter, on Teilhard 196–7, 198
Medieval ages, design arguments in 46–9
Mediocrity, Principle of 577
in discussion of ETI 577, 586
probably not true 587
melting points, of various hydrides *524*
memes (Dawkins' name for idea-complexes) 522
memory, of measuring system 472–3
Mercury, orbital radius of *221*
meson exchange concept, nuclear interaction theory 295, 318, 320
mesons
formation at beginning of Universe 371
reproduction of 513, *514*
metabolic requirements, large brain 130
metallic bonds 300, *301*
metals, astrophysical definition of 587
metazoans 513
methodological determinism 139
methodological reductionism 138
micelles 540
Michell, John, on black holes 347
Michell mass (of black hole) 348
Michell radius (of black hole) 348
microscope, invention of 57
microwave background radiation
as echo of Big Bang 368, 380
spectrum compared with that of black body *380*
steady-state cosmologies dealing with 601–2
temperature isotropy of 419
Milne, E. A., gravitation theory 237–8
dual timescales in 244, 245
Milton, John, quoted 219, 384
Mind (*νους*) 32
Anaxagoras' concept of 32–3
mineral ecology concept 522
minimum principles 148, 149; *see also* Least Action . . . ; Least Time . . .
Mises, Ludwig von, view of human history 139–40
Misner, Charles W., chaotic cosmology 178, 420
mitochondria, development necessary as crucial step in Man's evolution 563
Mixmaster cosmological model 66
mobility, size relationship
for aquatic creatures 313–14
for land-based animals 316–17
molecular structures, life systems 88
molecular synthesis 3
molecules
excited states 301–2
Weak Anthropic Principle applied to 299–302
Molière, J. B., quoted 64
Monod, Jacques
definition of life 519–20
Postulate of Objectivity 123
on purposeful behaviour 133–4
monopoles 433, 655–6
effect on astronomical structures 656–8
final fate of 657–8
origin of 655, 657
Montaigne, M. 51
translation of book by Raymonde of Sebonde 49
moons, mountain size on 308
Morrison, Philip, quoted, on interstellar communication 592
Mostepanenko, V., on dimensionality 259, 265
motivations, interstellar communication/exploration 590–601
Mott–Heisenberg analysis, Wilson cloud chamber experiment 485, 486–9
mountains, limitations to size 307–9
mutation
not possible in salt crystals 515
of viruses 517

Narlikar, J. V.
quantum cosmology 493, 495
steady-state cosmological model 421, 601, 605, 616–17
natural selection
applied to physical laws 87
as differentiator between virus and salt crystals 517
effect on intelligence 129
first introduced by Darwin 84
in Henderson's argument 145
self-reproduction by 515
Natural Units 291–2
Needham, A. E., quoted, on atmospheric abundance of nitrogen 550
Needham, J., quoted 92
Neptune, orbital radius incorrectly predicted by Bode Law 221

nervous system, complex
  evolution of 129–30
  network model for 266
neutrinos 371
  decay of 655
  rest mass of 255–6, 388, 390–1, 654
neutron stars 342, 365n
  frequency of rotation 342–3
  influence of magnetic fields on 345–6
  lifetimes in presence of monopoles *657*
  schematic diagram of *344*
  size of 342
  size of possible intelligent 'systems' on 345
  surface conditions on 343–5
neutron–proton mass difference coincidence 400
neutron–proton ratio (in Universe) 371
neutron–proton transformation, at beginning of Universe 371
Newcomb, Simon, estimate of Earth's age 162
Newman, W. I., quoted, on extraterrestrial intelligent life 600
Newman–Sagan (space exploration/colonization) model 589–90
Newton, Isaac 60–1, 113n
Newton's gravitational constant, as fundamental constant 235–6
Nielsen, H. B. 257
nitrogen, relative abundance *542*, 549
nitrogen compounds, Anthropic significance of 549–52
non-baryonic matter 390
non-Western philosophy/religion, design arguments in 92–102
noosphere (Tielhard's term) 200
  transition to Omega Point 201, 203
νους 32; *see* mind
now, definition of 660
nuclear binding energies, sensitivity to value of fine structure constant 322
nuclear density, values quoted 340, 349, 366n
nuclear forces
  characteristics of 322
  Weak Anthropic Principle applied to 318–27
nuclear reactions, primordial timescale for 398
nuclear resonance, Hoyle's observations 250–4
nuclear sources of energy, predicted by Anthropic Principle 165
nuclear stability 324–7
  atomic number/mass number criterion 325
  strong coupling relationships *326*, *327*
nuclear timescales 294, 320
  relationship to stellar lifetimes 358
nuclear war 599
Nuclear Winter model 569, 599–600
nuclear-to-atomic ratios 320
nucleic acid gene 516
nucleosynthesis coincidence 399
nucleotides, spontaneous formation of 547
number of nucleons in Universe 219, 358–9
number of particles in Galaxy, time variation 234–5
number of particles in Universe
  Archimedes' enumeration 220
  Eddington's evaluation 225
numerical simulation of human behaviour, data requirements 137

objective idealism 155
Objectivity, Monod's Postulate of 123
observability of Universe 429–30
observer
  quantum mechanics, role in 459
  role in physical measurement 458
observer effect 22, 28; *see also* Copenhagen Interpretation; Schrödinger's Cat Paradox
observer role, imposed by quantum mechanics 505
occasionalist, Samuel Clarke as 63
Ockham's Razor/Principle 476, 495, 496
Olbers' paradox 604
Old Testament
  design arguments in 27, 111n
  quoted 1, 339
oldness problem 411
omega, *see* density parameter
Omega Point
  definition of 675
  identification by Teilhard 201, 217n
  reaching of 677
  Teilhard's process for origination of 201–2
  Teilhard's spatial image for 203
  Teilhard's use of term 200–1
  totality of life, theological definition 682n
  transition of noosphere to 201, 203
omega point
  compared to conical singularity 201–2
  definition of 638, 675
  in Einstein space 638
  examples of cosmological models with 638, *639*
  in Löbell space 638
Omega Point theory
  as approach to panentheism 122n

Teilhard's version 200–4, 217–18n
mankind restricted to Earth in 203, 217–18n
one-cycle closed universe model 621
O'Neill colony 580, 584, 596
ontological argument 106–7, 108
ontological determinism 138–9, 409
ontological reductionism 138
open Friedman universe 375, *376*
density inhomogeneities growth rate 416
instability shown by Collins and Hawking 424
Penrose diagram for *633*
open universes
Cauchy hypersurface topology of 622
cosmological expansion 649–50
energy sources in far future 665
orderly singularity theory 178
organelles 563
organic view of world
Aristotelian view 37
by Greeks 28–9
Origin of Species 9, 161, 162, 615
reviewed by Fleming Jenkin 161
Orion Project 581
orthogenesis 199
Tielhard's use of term 199–200, 217n
oscillating universes
constants of Nature in 8
models 248–9, 455n, 620–1; *see also* Landsberg–Park . . .
Wheeler's concept 193, 248–9
oxygen
reactivity of 544
relative abundance *542*, 543, 614
safety limit percentage 544, 567, *568*
stellar formation from C^{12} 253, 254
unique properties of 543–5
ozone, Anthropic significance of 544–5

Paley, William 76–82
astronomical phenomena discussed 79–81
at Cambridge 116n, 285n
Darwin's reading of book by 76, 83
dislike of final-causes concept 76–7
on gravitational laws 80–1, 260, 261
panentheism 121–2n
distinguished from theism 122n
pantheism, and motion of deity 107, 121n
Pantin, Charles, on Many-Worlds hypothesis 19, 83, 250
Parmenides 34–5
Participatory Anthropic Principle (PAP) 22, 505
Participatory Universe 470
past light-cones, isotropy of 429–30
Pauli Exclusion Principle 300, 302–3, 391
Peebles, P. J., quoted, on cosmological coincidences 243
Peirce, Charles Sanders, idealism of 158
pendulum period, dimensional analysis for 270
Penrose, Roger
gravitational entropy proposal 445–8
orderly singularity theory 178
Penrose diagrams 632, 635–6
achieved infinity defined by 636
boundaries, meaning of 632, 635
bubble-universe model *606*
c-boundaries in 635, 636
closed Friedman universe *634*
flat Friedman universe *633*
open Friedman universe *633*
steady-state universe 602, *603*
Teilhard's drawing of Omega Point similar to 203
see also quasi-Penrose diagram
Penrose twistor programme, dimensionality in 272
Penrose–York interpretation, energy-conservation law 618
Penzias, A. A., discovery of microwave background radiation 368, 380
peptide bonds 549
Perfect Cosmological Principle 602, 603, 604
Permo-Triassic boundary, change in encephalization increase 130
Perry, John, on Earth's age 164
phases, elementary particle 430–1
philosophers/theologians, compared to scientists 15
phosphorus compounds
importance to life of 553–5
relative abundance *542*
photinos 390, 393
photon emission, from stars 332, *333*
photon spectrum 652
photons, formation at beginning of Universe 370
photosynthesis, oxygen produced by 543, 614
physical eschatology 658
physics, Weak Anthropic Principle applied 290–305, 318–27
Pickering, W. H., quoted, on extraterrestrial communication 593
place, meaning of word 501
Planck, Max
action principle 151
natural units 292
quoted 123

Planck energy, orderliness predicted below this value 256, 257
Planck length 499, 500
Planck temperature
for biological systems 302, 309
definition of 292
for various states of matter 305–6
Planck's constant 222–3, 292
planetary life, Weak Anthropic Principle applied to 310–18
planetary masses, reason for values 307
planetary time-reckoning
day length 310, 338
year length 337, 338
planets
limit on size 307
size–mass relationship *290*, 307
Titius' law of orbital radii 221, 222
Weak Anthropic Principle applied to 305–10
planiverse concept 266
plastic nature, Dr Cudworth's concept 59
Plato 35–6
Platonic solids 259
Plenitude, Principle of 128, 193
Pliny 44
Pluto, orbital radius incorrectly predicted by Bode Law 222
Podolsky, Boris, definition of physical reality 461
Poincaré, H., recurrence theorem 174–5, 176, 619
dimensionality of 272
Poisson's gravitational equation
revised for Einstein static universe 413
in higher-dimensions 287n
Poisson's wave equation solution in two dimensions 267
Polanyi, Michael 75, 138
polygenetic traits, in evolution of Man 561
Pope, Alexander, quoted 55
Popper, Karl
on cosmology 367
on prediction of human action 188, 189, 215
Positive Pressure Criterion 423
positronium 650, 651
positronium atoms, as information storage devices 669–70
Postulate of Complexity 473
Poulton, Edward B., estimate of Earth's age 163–4
price structure, as Participatory Anthropic Principle 172
primates, encephalization of 131
Principles, *see* Equivalence . . . ; Least Action . . . ; Least Time . . . ; Mediocrity . . . ; sufficient reason . . .
Pringle-Pattison, Andrew, natural theology of 180–1
Prisoner's Dilemma game 100–2
probability power law, evolution of Man 559
progression/regression of species 128–30
progressive nature of history 186
proportionality constants, close to unity 270, 289–90
Protagoras, quoted 556
proteins, Anthropic significance of 549
protogalaxies, pre-existence of, in cosmological models 428
proton decay 647
catalysed by monopoles 656, *657*
effect on Man 648–9
as energy source for large bodies 648
lifetime quoted 648
power generated by 648
protostar formation 339
Ptolemaic system 496
Ptolemy, on three-dimensional space 259
pulsars: *see* neutron stars
purposeful action 134

quantization, atomic models 305
quantum Copernican Principle 502
quantum cosmological model 13
quantum field theories, problem of infinities in 257–8
quantum gravity models, creation *ex nihilo* 443
quantum mechanical approaches 13
quantum mechanics 13
electron orbitals 297, 305
many-worlds interpretation of 8, 193, 249
quantum region, mass–size diagram 396, *397*
quantum timescales, future evolution of matter in 647–58
quantum tunnelling 653
quarks
preponderance over antiquarks 370, 403
reproduction of 513, *514*
in unified theories 355–6, 357
quartz
as main material of Earth 307
silicon–oxygen bonds in 545–6
quasar spectra, photon energy/wavelengths 241
quasi-Penrose diagram, for static Einstein universe *635*
quirks, in properties of certain elements 22, 143, 524–56

r-strategy (of exploration/ colonization) 584, 589, 590
racism 595, 596
  restriction of rights of von Neumann machines as 595, 600
radiation, as information storage device 669
radiation-dominated Universe 649–50
radiation era, beginning of Universe 382–3
radiation gas
  Lagrangian for 491
  quantization of 492
radiation pressure (of stars) 331
radioactive decay, as example of two-timing 244
radioactivity, as weak interaction 293–4
raw materials, relative cost decrease 172
  effect on cost of nuclear weapons 599
  effect on cost of space travel 583
Ray, John, design arguments of 58–9
Raymonde of Sebonde 48–9
realist interpretation (of quantum mechanics) 472
reality, EPR experiment, definition 461
  validity denied by Niels Bohr 463–4
recollapsing universe model 623, 666
  strong restrictions on topology of 623–4
recombination time 382, 386, 389
recurrence paradox 176, 619
redshift
  definition of 377
  Eddington's explanation 237
  Jeans' explanation 236
  light velocity decrease explanation 236
redshifts, photon energy/wavelengths relationship 241
reductionism, classification of 138
Rees, M. J., quoted 17, 586
reference standards 241–3
Regge, T., quoted, on DNA replication fidelity 304–5
Renaissance, design arguments in 49–54
reproductive invariance, as property of life 519
resonances, nuclear reaction 251, 252–3
retrograde motion, of planets 4
Ricci curvature 446, 447, 490
Riemannian spaces, space–time properties 273
Riezler, W., spoof on Eddington's methodology 227–8
Robertson, H., quoted on cosmic homogeneity 414
rocket technology
  advanced technology 581–2
  first developed 577
room temperature, definition of 302
Rosen, Nathan, definition of physical reality 461
Royce, Josiah
  Harvard seminars organized by 158–9
  idealism of 158
Rubik's cube, number of configurations possible 285n
runaway glaciation 567
runaway heating (of Earth) 567
running speeds, size relationships 316–17
Russell, Bertrand
  debate with Father Copleston 103, 107–8
  dysteleological attitude 167
  quoted 31, 68, 167, 169
Russell, D. A., on brain/metabolism relationship 130

Sagan, Carl, quoted, on
  anthropocentric hypothesis 601
  communications with ETI 577
  extraterrestrial intelligent life 601
Sakharov model 404
salt crystal, reproduction of 513
Sand Reckoner (Archimedes' enumeration) 220
Santillana, G. de, quoted 601
Sarton, G., quoted 27
Saslaw, W., quoted 387
Saturn, orbital radius of *221*
scaling techniques, size limitations in 310, 311
Schelling, F. W. J. von 125, 155, 156–7
Schoen–Yau theorem 623–4
Scholasticism 47, 48
Schrödinger equation/wave function 265, 321
  Born's probability interpretation 458
  separation into radial and angular parts 265
  solution for two particles 321
Schrödinger's Cat Paradox 465–6
Schwarzschild metric 348, 349, 352
Schwarzschild singularity 348, 352
scientists, compared to philosophers/ theologians 15
scri-plus *633*, 635
Second Law of thermodynamics
  anthropic-fluctuation interpretation 176
  Boltzmann's proof 174
  Chamberlain's prediction of nuclear energy 125
  creation interpretation 176
  ecology movement reaction 169–70
  energy to process information determined by 661
  fluctuation interpretation 176, 177

Second Law of thermodynamics (*cont.*)
gravitational entropy non-decrease 446
Helmholtz' forecast of Heat Death 166
Segal's chronometric cosmology 604–5
selection effect 2
as explanation of Universe size–mass diagram 290
self-reference 469
self-reference arguments 4
self-repair, as condition of 'living' 513
self-reproducing computer programs 522
self-reproducing machines
development timescale for 609n
spontaneous creation unlikely 511
von Neumann's scheme 517–19
self-reproducing universal constructor 517–18
in interstellar exploration 577, 579
self-reproduction, as condition of 'living' 512–13
self-selection, measuring operators 479
self-selection properties, of human observations 3–4
seme, definition of 561
separation, principle of, violated by Copenhagen Interpretation 466
SETI proposals 577, 578
Shapiro, R., definition of life 521
shear energy, as energy source in far future 631, 665
shearing universe 629–30, *631*
disappearance of horizons 631
Shimony, Abner, quoted, on
Multi-Worlds Interpretation 495
on physical reality 463
shrinking atomic size theory 236–7
signal transmission/reception
in even-dimensional space 268
in odd-dimensional space 268–9
silicon (and compounds) 545–6
compared with carbon compounds 546
Simon, Julian L.
Ehrlich's criticism of 211n
on progressive economy 171–2
simple harmonic oscillator (SHO)
Hamiltonian for 492
radiation gas Lagrangian as for 491
Simpson, George Gaylord 128, 133
criticism of Teilhard 196, 199
size
living organisms 310–11
observable Universe 4–5
planetary bodies 306–7
size limitations
living organisms 310–17
observable Universe 7
size of Universe
anthropic principles applied 384–5
relationship to time needed to generate observers 2–3, 18, 385
selection effect illustrated by 2–3
slow roll-over, Higgs field 432
small organisms, advantages of 314
Smith, Homer, quoted 147–8
Smith, Sydney, quoted 231
social control of technology 599–600
Socrates 35
criticism of Anaxagoras 33–4
design argument of 36
Solar System
formation of 372
information content of 662
possibility of von Neumann probes in 597
Sommerfeld's fine structure constant 223, 227, 293
soul
as computer program 659
definitions of 659, 680n
sound-speed 304
space exploration 578–83
space–time, Einstein's concept 384
space-travel argument, against existence of ETI 576–608
specific heats, for various substances *534*
spectral shift 376–7
Spencer, Herbert
definition of life 520
on division of labour 186–7
evolving-cosmos theory 188
on social systems 187
Spinoza, Benedict de 59–60
split-universe model 475–8
four-degrees-of-freedom example 483–4
meaning of terminology 'universe' 475–6
measurement effects 481–2, 484
measuring apparatus in 476
time evolution effects 479–81
see also many-worlds hypotheses
spontaneous cell-wall formation, Fox's experiment on 541
spontaneous creation of life, elements necessary for 510–11
spontaneous evolution, biological improbabilities of 570
spontaneous ignition, of Earth's vegetation 544, 567, 568, 569
spontaneous order concept
in Chinese philosophy 96–7
free market example 99
stabilization (of economies), failure of 141
standard cosmological model (of Universe) 369–72

stars
  borderline with planets 307
  formation 339–40
    timescale for cessation of 641
  formation processes 328–9
  lower size limit 331
  surface temperatures 334–5
  Weak Anthropic Principle applied to 327–38
static cosmological model 616
statistical distribution, of constants of Nature 253–4, 257
statistical fluctuations, effect on size limitation of living organisms 317
statistical interpretation (of quantum mechanics) 472
statistical mechanics 460
steady-state cosmological models 17, 185, 601, 616
  Anthropic Principle arguments against 601–7
  Aristotelian version 38
  see also inflationary-universe model
steady-state universe 612n
  Paley's disproof 80
  Penrose conformal diagram 602, *603*
Stefan–Boltzmann law (of cooling) 648
stellar disruption, by intelligent species 644–6
stellar lifetimes
  calculation of 333
  relationship to nuclear times 358
  size-dependency 333–4
stellar size, factors affecting 331–2
Stewart coincidence 229
stochastic gauge theory 257, 258
Stoics, teleological ideas 42–3
strange attractors 255, 273, 286n
Strindberg, Johann, quoted 305
Strong Anthropic Principle (SAP)
  criticism of 248
  definition of 21–2
  definition of intelligent life for 523
  quantum cosmology boundary conditions 503–5
  speculative nature of 23
Strong Cosmic Censorship, Principle of 622, 636, 638
strong interactions 295, 371
  in unified theories 355–7, *357*
  see also nuclear forces
subjective idealism 155
sufficient reason, principle of 103, 104
sulphur compounds
  importance to life of 553–5
  relative abundance *542*
sum-over-histories action principle method 152
Sun
  asteroid collision 644–5
  end of main sequence 557
  planet collision 645
supersymmetric gauge theories 274
surface temperatures, stellar 334–5
surface tensions, of various liquids *536*
survival times, carbon-based life forms 170, *171*
Swift, Jonathan, quoted 367

T2 virus, life cycle of *516*
Tait, P. G.
  estimate of Earth's age 162–3
  hierarchy-of-worlds cosmology 179
Tangherlini, F. R., on dimensionality 262, 265
Taoism, spontaneous order concept in 96–8
tardigrades, as living organisms 520
Teilhard de Chardin, Pierre 195–203
  energy modes concept 197–8
  on Heat Death 168, 197
  and information theory 199, 204
  Omega Point theory 200–4
  personal background 195–6
  views on extraterrestrial life 203, 217n
telelogical arguments, compared to eutaxiological arguments 29, 144
telelogical evolution 185–95
telelogy
  artificial/natural 134
  divisions of 134, 136
teleomechanists 75
teleonomic (Mayr's terminology) 134
teleonomy, as property of life 519
Teller, Edward
  effect of time-varying gravitation 239–40
  fine structure constant relationship 230, 239
telos, meaning of word 33
temperature effects
  biological systems 302, 309
  biological/spectral 338
  molecular level 302
  star formation 328–9
  surface vs. centre of stars 334–5
  Universe 305–6
temperatures, small closed universe near final singularity *674*
Tennant, F. R.
  cosmic teleology 181–2
  the first use of term 'anthropic' 181
terminal velocity, as illustration of chaotic cosmology 421

tetrahedral angle, water molecule bond
close to 528
Theaitetos 259
themata concept of scientific progress 11
Theophrastus 40–1
thermal conductivities, for various
substances *535*
thermal expansion coefficient 304
thermonuclear reactions 330–1
thing-in-itself concept 73, 153, 155
Thomas Aquinas 9, 47–8
definition of soul by 659, 680–1n
Thomist philosophy 47–8
Thompson Indian creation myth 94
Thomson, William: *see* Kelvin, Lord
Thomson scattering, stellar emission 333,
334, 335
thought, information content of 660
three-dimensional space
proof for uniqueness of 265–9
Whitrow's paper on 12, 15–16, 247, 259
see also dimensionality
Tillich, Paul, on God 107
time
effects in Nature 30
unidirectionality of 674
time direction, and Weak Anthropic
Principle 173–80
time evolution, effect on split-universe
model 479–81
time reference standards 242–3
time-reckoning systems, reason for
universality 338
time-reversed Eddington–Lemaître–Bondi
universe 638, *639*
timescales
building blocks of life, generation 18
carbon-based life forms 170, *171*
controlled by entropy per baryon 382,
401
dual 244–5
future evolution of matter
classical mechanics analysis 641–7
quantum mechanical analysis 647–58
hot Big Bang cosmological model 385–7
living organisms *171*
size relationships 316
main-sequence stars, lifetimes *171*, 333–4
natural selection, evolution by 87
open/flat/very-large-closed
universes *653–4*
significant lifetimes *171*
small closed universe near final
singularity *647*
Titius von Wittenberg, J. D., planetary
orbital radii law 221, 222
Tolman cyclic model 620–1
Toulmin, S., on definitions 15
trace elements *553*, 555–6
trajectories
crossing in two dimensions 254, 273,
285n
not intersecting in three (or more)
dimensions 254–5, 266, 285n
Triassic period, encephalization
increase 130
Turing Test 154, 523, 571n
Twin Paradox
Bergson's discussion 216n
Capek's analysis 216n
two-dimensional space, signal transmission/
reception 268
two-legged animals, Man as largest 312
two-sphere universe, contraction of 630,
631
tychistic idealism 158
Tychonic system 496
Tyrtamus of Eresos 40–1

Ultimate Observer 468, 470, 471
Uncertainty Principle 303, 304; *see also*
Heisenberg's Uncertainty Principle
unified theories 354; *see also* grand unified
gauge theories
unique properties of certain elements and
compounds 124, 143, 510, 524–56
uniqueness, as condition for evolutionary
step 561
universal constructor 517–18
in interstellar exploration 577, 579
Universal Program 154
actual Universe as representation of 155
complexity difficulty 155
Ultimate Purpose of 156–7
Universal wave function: *see* wave function
of Universe
universality, as minimum requirement for
fundamental theory 258
Universe, information content of 662
universe, meaning of word 475–6
university president (anon.), quoted 185
Uranus, orbital radius predicted for 221

vacuum fluctuation model 440–1
vacuum polarization 355, 408
van der Waals bonding 300, *301*, 526–7,
528
van Flandern, T. C., technique to measure
constancy of gravitational
constant 243
vaporization, heats of
for various hydrides *525*

for various liquids *535*
variational principles, contribution of D'Alembert and Maupertuis 65–6
varying constants 238–43
Jordan's approach 235
Venus
atmospheric composition *544*
life on 511
orbital radius of *221*
Vesalius, illustrated anatomy book 112
Vico, Giovanni 68–9
viruses, reproductive cycle of 515–17, 519
Voltaire (Francois-Marie Arouet) 64–5
caricature of Leibniz 6, 65
von Neumann machines 517–19
rights to freedom of 595, 600
von Neumann measurements 464–5, 473–4
von Neumann probes 579–80, 581
colonization of space by 580
compared with human beings 583
obsolescence of 581
possibility of loss of control 595
possibility of random assembly 603–4
right to freedom 595, 600
in Solar System 597
see also self-reproducing machines; universal constructors

W bosons 370
in magnetic monopoles *656*
Wald, George, quoted 127, 510
on chlorophyll 556
Ward, James, telelogical argument 182–3
Ward, Lester, on evolution 188
waste heat
from advanced civilizations 597, 598
information-processing 667, 668
heat sink for 667–8
water
anthropic significance of 524–41
boiling point compared with other hydrides *524*
crystal structure of 532
density as function of temperature *531*
dielectric constant compared with other substances *537*, *538*
heat of vaporization compared with other substances *525*, *535*
melting point compared with other hydrides *524*
molecular bond angles/distance 528, *529*
percolation networks in 539
phase diagram for *532*
phases of 431
solvation cage effect 539, *540*
specific heat compared with other substances *534*
surface tension compared with other liquids *536*
thermal conductivity compared with other substances *535*
unique properties of 88, 143, 524–5
various physical properties quoted *524*, *525*
wave equation
solution in one dimension 267
solution in seven dimensions 268–9
solution in three dimensions 267–8
solution in two dimensions 267
wave function collapse, required by Copenhagen Interpretation 459, 466, 467, 496
wave function of Universe 105, 482, 492, 499
Hawking's definition of 449
non-normalizability of 482, 501
wave-packet spreading, Many-Worlds approach 483–5
Weak Anthropic Principle (WAP)
as application of Bayes' theorem 17
in astrophysics 290–1, 305–18, 327–60
constraints on Earth's future 556–70
Copernican Principle resulting from 4
definition of 16
first examples 247–8
in physics 290–305, 318–27
quantum cosmology boundary conditions 497–503
as restatement of existing principle 23
selection of measuring operators in split-universe model 478–9
three-dimensional space 275
weak interactions 293–4, 370
radioactivity as example 293–4
in unified theories 354, *357*
wealth, as information 172–3, 595, 600
weather systems, on habitable planets 338
Webster, John, view of Scholasticism 58
Webster's Dictionary, quoted 159
Weinberg, Steven, quoted 472, 613–14
Weinberg angle 294, 354
Weinberg coincidence 229, 230
Weinberg criterion (for non-equilibrium decay of *X* bosons) 406
Weinberg–Salam model 354
Weisskopf, V., quoted 238
Weyl, Hermann
on dimensionality 260–1
on large number coincidences 224
Weyl curvature 446, 447
Weyl's hypothesis 224, 277n
Wheeler, John A.
delayed choice experiments 469–70

Wheeler, John A. (*cont.*)
infinitely cycling closed universe concept 193, 248–9
on laws of Nature 255, 257
Participatory Anthropic Principle 22, 470, 471, 505
quoted, on time 173
Wheeler–Feynman electrodynamics formulation 151–2
Whewell, William 82, 117n, 144
Whig historical interpretation 9–10
criticism of 10, 135
justification of 10–11
white dwarfs 340–1
cooling of 648
lifetimes in presence of monopoles *657*
mass limitation 341
size compared with neutron stars 342
Whitehead, Alfred North 191
contingency problem solution 106
cosmic epochs concept 192
global time discussion 216
organism view of Universe 191
quoted, on cosmology 615
teleological mechanism in his cosmology 193–4
transient nature of physical laws 192
Whitrow, G. J., anthropic argument for three dimensions 12, 15–16, 247, 259, 266
Whittaker, E. T.
on Eddington's Fundamental Theory 224–5
on Heat Death 168–9
Wigner, Eugene, quoted 468
Wigner's Friend Paradox 467–8
Wilson, R. A., discovery of microwave background radiation 368, 380
Wilson cloud chamber experiment 485–9
dynamic analysis 487–9
static analysis 486–7
Wyman, J., quoted, on carbon dioxide 548

X bosons 370, 371
asymmetrical decay of 405–7
in magnetic monopoles 655, *656*
mediation of interactions in grand unified theories 357, 404
Xenophon, quoted 35, 36

Y bosons
in magnetic monopoles *656*
mediation of interactions in grand unified theories 404
year length, calculation of 337
young-universe model 409
in isotropy interpretation 426, 427
Yukawa gravitation potentials 262–3
Yukawa nuclear interaction model 295, 318, 319, 363n

Z bosons, in magnetic monopoles *656*
Zeldovich, Y. B., criticism of varying gravitational constant large numbers hypotheses 233–4
Zeno of Citium 42
Zeno's paradox 666–7
Zermelo, E. 174–5, 176
zoo hypothesis 597
Zunĩ Indian creation myth 93–4